BIG IDEAS MATH®

A Common Core Curriculum

TEACHING EDITION

GREEN

Ron Larson
Laurie Boswell

Erie, Pennsylvania
BigIdeasLearning.com

Big Ideas Learning, LLC
1762 Norcross Road
Erie, PA 16510-3838
USA

For product information and customer support, contact Big Ideas Learning
at **1-877-552-7766** or visit us at ***BigIdeasLearning.com***.

Copyright © 2012 by Big Ideas Learning, LLC. All rights reserved.

No part of this work may be reproduced or transmitted in any form or by any means, electronic or mechanical, including, but not limited to, photocopying and recording, or by any information storage or retrieval system, without prior written permission of Big Ideas Learning, LLC unless such copying is expressly permitted by copyright law. Address inquiries to Permissions, Big Ideas Learning, LLC, 1762 Norcross Road, Erie, PA 16510.

Big Ideas Learning and *Big Ideas Math* are registered trademarks of Larson Texts, Inc.

Common Core State Standards: © Copyright 2010. National Governors Association Center for Best Practices and Council of Chief State School Officers. All rights reserved.

Printed in the U.S.A.

ISBN 13: 978-1-60840-229-8
ISBN 10: 1-60840-229-0

4 5 6 7 8 9 10 WEB 15 14 13 12

AUTHORS

Ron Larson is a professor of mathematics at Penn State Erie, The Behrend College, where he has taught since receiving his Ph.D. in mathematics from the University of Colorado in 1970. Dr. Larson is well known as the lead author of a comprehensive program for mathematics that spans middle school, high school, and college courses. His high school and Advanced Placement books are published by Holt McDougal. Ron's numerous professional activities keep him in constant touch with the needs of students, teachers, and supervisors. Ron and Laurie Boswell began writing together in 1992. Since that time, they have authored over two dozen textbooks. In their collaboration, Ron is primarily responsible for the pupil edition and Laurie is primarily responsible for the teaching edition of the text.

Laurie Boswell is the Head of School and a mathematics teacher at the Riverside School in Lyndonville, Vermont. Dr. Boswell received her Ed.D. from the University of Vermont in 2010. She is a recipient of the Presidential Award for Excellence in Mathematics Teaching. Laurie has taught math to students at all levels, elementary through college. In addition, Laurie was a Tandy Technology Scholar, and served on the NCTM Board of Directors from 2002 to 2005. She currently serves on the board of NCSM, and is a popular national speaker. Along with Ron, Laurie has co-authored numerous math programs.

ABOUT THE BOOK

The traditional mile-wide and inch-deep programs that have been followed for years have clearly not worked. The Common Core State Standards for Mathematical Practice and Content are the foundation of the Big Ideas Math program. The program has been systematically developed using learning and instructional theory to ensure the quality of instruction. Big Ideas Math provides middle school students a well-articulated curriculum consisting of fewer and more focused standards, conceptual understanding of key ideas, and a continual building on what has been previously taught.

- **DEEPER** Each section is designed for 2–3 day coverage.
- **DYNAMIC** Each section begins with a full class period of active learning.
- **DOABLE** Each section is accompanied by full student and teacher support.
- **DAZZLING** How else can we say this? This book puts the dazzle back in math!

Ron Larson *Laurie Boswell*

TEACHER REVIEWERS

- Aaron Eisberg
 Napa Valley Unified School District
 Napa, CA

- Gail Englert
 Norfolk Public Schools
 Norfolk, VA

- Alexis Kaplan
 Lindenwold Public Schools
 Lindenwold, NJ

- Lou Kwiatkowski
 Millcreek Township School District
 Erie, PA

- Marcela Mansur
 Broward County Public Schools
 Fort Lauderdale, FL

- Bonnie Pendergast
 Tolleson Union High School District
 Tolleson, AZ

- Tammy Rush
 Hillsborough County Public Schools
 Tampa, FL

- Patricia D. Seger
 Polk County Public Schools
 Bartow, FL

- Denise Walston
 Norfolk Public Schools
 Norfolk, VA

STUDENT REVIEWERS

- Ashley Benovic
- Vanessa Bowser
- Sara Chinsky
- Kaitlyn Grimm
- Lakota Noble
- Norhan Omar
- Jack Puckett
- Abby Quinn
- Victoria Royal
- Madeline Su
- Lance Williams

CONSULTANTS

- **Patsy Davis**
 Educational Consultant
 Knoxville, Tennessee

- **Bob Fulenwider**
 Mathematics Consultant
 Bakersfield, California

- **Deb Johnson**
 Differentiated Instruction Consultant
 Missoula, Montana

- **Mark Johnson**
 Mathematics Assessment Consultant
 Raymond, New Hampshire

- **Ryan Keating**
 Special Education Advisor
 Gilbert, Arizona

- **Michael McDowell**
 Project-Based Instruction Specialist
 Tahoe City, California

- **Sean McKeighan**
 Interdisciplinary Advisor
 Norman, Oklahoma

- **Bonnie Spence**
 Differentiated Instruction Consultant
 Missoula, Montana

Expressions and Number Properties

	What You Learned Before	1
Section 1.1	**Evaluating Algebraic Expressions**	
	Activity	2
	Lesson	4
Section 1.2	**Writing Expressions**	
	Activity	8
	Lesson	10
Section 1.3	**Properties of Addition and Multiplication**	
	Activity	14
	Lesson	16
	Study Help	20
	1.1–1.3 Quiz	21
Section 1.4	**The Distributive Property**	
	Activity	22
	Lesson	24
Section 1.5	**Using Formulas to Solve Problems**	
	Activity	28
	Lesson	30
	1.4–1.5 Quiz	34
	Chapter Review	35
	Chapter Test	38
	Standardized Test Practice	39

"I love my math book. It has so many interesting examples and homework problems. I have always liked math, but I didn't know how it could be used. Now I have lots of ideas."

Multiplying and Dividing Fractions

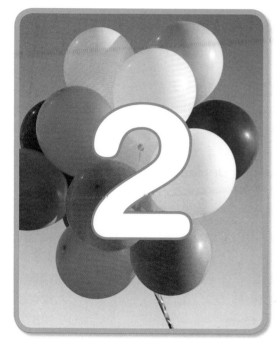

	What You Learned Before	43
Section 2.1	**Fractions and Estimation**	
	Activity	44
	Lesson	46
Section 2.2	**Multiplying Fractions and Whole Numbers**	
	Activity	50
	Lesson	52
Section 2.3	**Multiplying Fractions**	
	Activity	56
	Lesson	58
Section 2.4	**Multiplying Mixed Numbers**	
	Activity	62
	Lesson	64
	Study Help	68
	2.1–2.4 Quiz	69
Section 2.5	**Dividing Fractions**	
	Activity	70
	Lesson	72
Section 2.6	**Dividing Mixed Numbers**	
	Activity	78
	Lesson	80
Section 2.7	**Writing Decimals as Fractions**	
	Activity	84
	Lesson	86
Section 2.8	**Writing Fractions as Decimals**	
	Activity	90
	Lesson	92
	2.5–2.8 Quiz	96
	Chapter Review	97
	Chapter Test	100
	Standardized Test Practice	101

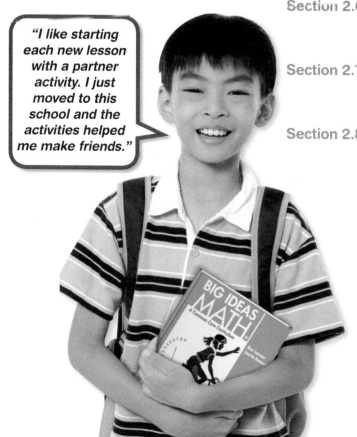

"I like starting each new lesson with a partner activity. I just moved to this school and the activities helped me make friends."

Multiplying and Dividing Decimals

	What You Learned Before	105
Section 3.1	**Decimals and Estimation**	
	Activity	106
	Lesson	108
Section 3.2	**Multiplying Decimals and Whole Numbers**	
	Activity	112
	Lesson	114
Section 3.3	**Multiplying Decimals**	
	Activity	118
	Lesson	120
	Study Help	124
	3.1–3.3 Quiz	125
Section 3.4	**Dividing Decimals by Whole Numbers**	
	Activity	126
	Lesson	128
Section 3.5	**Dividing Decimals**	
	Activity	132
	Lesson	134
	3.4–3.5 Quiz	138
	Chapter Review	139
	Chapter Test	142
	Standardized Test Practice	143

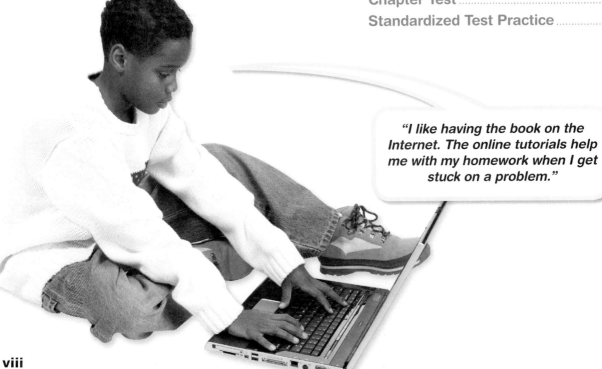

"I like having the book on the Internet. The online tutorials help me with my homework when I get stuck on a problem."

Fractions, Decimals, and Percents

	What You Learned Before	147
Section 4.1	**Percents and Fractions**	
	Activity	148
	Lesson	150
Section 4.2	**Percents and Decimals**	
	Activity	154
	Lesson	156
Section 4.3	**Comparing and Ordering Fractions, Decimals, and Percents**	
	Activity	160
	Lesson	162
	Study Help	166
	4.1–4.3 Quiz	167
Section 4.4	**Finding the Percent of a Number**	
	Activity	168
	Lesson	170
Section 4.5	**Percents and Estimation**	
	Activity	174
	Lesson	176
	4.4–4.5 Quiz	180
	Chapter Review	181
	Chapter Test	184
	Standardized Test Practice	185

"I love the cartoons. They are funny and they help me remember the math. I want to be a cartoonist some day."

Ratios, Rates, and Data Analysis

	What You Learned Before	189
Section 5.1	**Ratios**	
	Activity	190
	Lesson	192
Section 5.2	**Rates**	
	Activity	196
	Lesson	198
Section 5.3	**Solving Rate Problems**	
	Activity	202
	Lesson	204
	Study Help	208
	5.1–5.3 Quiz	209
Section 5.4	**Mean**	
	Activity	210
	Lesson	212
Section 5.5	**Median, Mode, and Range**	
	Activity	216
	Lesson	218
Section 5.6	**Analyzing Data Sets**	
	Activity	222
	Lesson	224
5.6b	**Data Displays**	227A
	5.4–5.6 Quiz	228
	Chapter Review	229
	Chapter Test	232
	Standardized Test Practice	233

"I like how I can click on the words in the book that is online and hear them read to me. I like to pronounce words correctly, but sometimes I don't know how to do that by just reading the words."

Circles and Area

	What You Learned Before	237
Section 6.1	**Circles and Circumference**	
	Activity	238
	Lesson	240
Section 6.2	**Perimeters of Composite Figures**	
	Activity	246
	Lesson	248
	Study Help	252
	6.1–6.2 Quiz	253
Section 6.3	**Areas of Circles**	
	Activity	254
	Lesson	256
Section 6.4	**Areas of Composite Figures**	
	Activity	260
	Lesson	262
	6.3–6.4 Quiz	266
	Chapter Review	267
	Chapter Test	270
	Standardized Test Practice	271

"I really liked the projects at the end of the book. The history project on ancient Egypt was my favorite. Someday I would like to visit Egypt and go to the pyramids."

Equations

	What You Learned Before	275
Section 7.1	**Writing Equations in One Variable**	
	Activity	276
	Lesson	278
Section 7.2	**Solving Equations Using Addition or Subtraction**	
	Activity	282
	Lesson	284
Section 7.3	**Solving Equations Using Multiplication or Division**	
	Activity	290
	Lesson	292
Section 7.4	**Solving Two-Step Equations**	
	Activity	296
	Lesson	298
	Study Help	304
	7.1–7.4 Quiz	305
Section 7.5	**Finding Dimensions of Plane Figures**	
	Activity	306
	Lesson	308
Section 7.6	**Finding Dimensions of Prisms**	
	Activity	312
	Lesson	314
7.6b	**Volume and Surface Area**	317A
	7.5–7.6 Quiz	318
	Chapter Review	319
	Chapter Test	322
	Standardized Test Practice	323

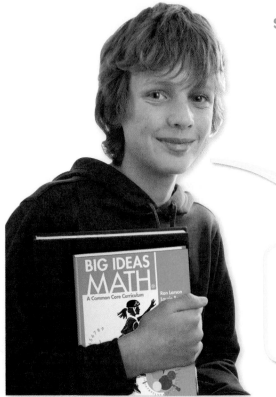

"I like how the glossary in the book is part of the index. When I couldn't remember how a vocabulary word was defined, I could go to the index and find where the word was defined in the book."

Inequalities

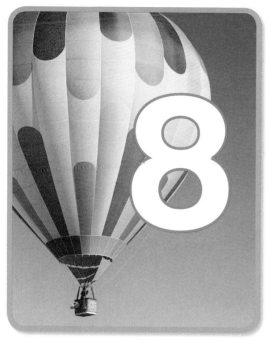

	What You Learned Before	327
Section 8.1	**Writing and Graphing Inequalities**	
	Activity	328
	Lesson	330
Section 8.2	**Solving Inequalities Using Addition or Subtraction**	
	Activity	336
	Lesson	338
	Study Help	342
	8.1–8.2 Quiz	343
Section 8.3	**Solving Inequalities Using Multiplication or Division**	
	Activity	344
	Lesson	346
Section 8.4	**Solving Two-Step Inequalities**	
	Activity	350
	Lesson	352
	8.3–8.4 Quiz	356
	Chapter Review	357
	Chapter Test	360
	Standardized Test Practice	361

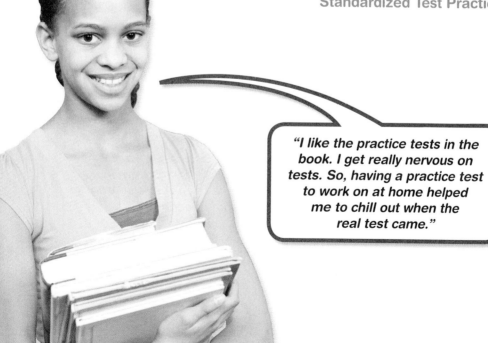

"I like the practice tests in the book. I get really nervous on tests. So, having a practice test to work on at home helped me to chill out when the real test came."

9 Tables, Graphs, and Functions

What You Learned Before 365

Section 9.1 Mapping Diagrams
Activity 366
Lesson 368

Section 9.2 Functions as Words and Equations
Activity 372
Lesson 374

Section 9.3 Input-Output Tables
Activity 378
Lesson 380
Study Help 384
9.1–9.3 Quiz 385

Section 9.4 Graphs
Activity 386
Lesson 388

Section 9.5 Analyzing Graphs
Activity 392
Lesson 394
9.4–9.5 Quiz 398
Chapter Review 399
Chapter Test 402
Standardized Test Practice 403

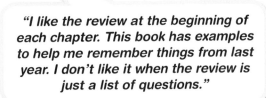

"I like the review at the beginning of each chapter. This book has examples to help me remember things from last year. I don't like it when the review is just a list of questions."

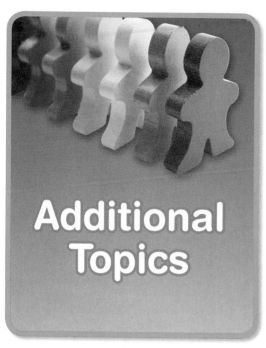

Additional Topics

	What You Learned Before	407
Topic 1	The Number Line	408
Topic 2	Comparing Numbers	410
Topic 3	Absolute Value	412
Topic 4	The Coordinate Plane	414

Appendix A: My Big Ideas Projects

Section A.1	Literature Project	A2
Section A.2	History Project	A4
Section A.3	Art Project	A6
Section A.4	Science Project	A8

Selected Answers A10
Key Vocabulary Index A39
Student Index A40
Additional Answers A49
Mathematics Reference Sheet B1

"I like the workbook (Record and Practice Journal). It saved me a lot of work to not have to copy all the questions and graphs."

PROGRAM OVERVIEW
Print
Available in print, online, and in digital format

- Pupil Edition
- Teaching Edition
- Record and Practice Journal
- Assessment Book
 - Pre-Course Test
 - Quizzes
 - Chapter Tests
 - Standardized Test Practice
 - Alternative Assessment
 - End-of-Course Tests
- Resources by Chapter
 - Start Thinking! and Warm Up
 - Family and Community Involvement: English and Spanish
 - School-to-Work
 - Graphic Organizers/Study Help
 - Financial Literacy
 - Technology Connection
 - Life Connections
 - Stories in History
 - Extra Practice
 - Enrichment and Extension
 - Puzzle Time
 - Projects with Rubrics
 - Cumulative Practice

- Differentiating the Lesson
- Skills Review Handbook
- Basic Skills Handbook
- Worked-Out Solutions
- Lesson Plans
- Math Tool Paper

Skills Review Handbook

Basic Skills Handbook

Technology

- **Dynamic Assessment Resources**
 - ExamView® Assessment Suite
 - LessonView® Dynamic Planning Tool
 - PuzzleView® Vocabulary Puzzle Builder
 - MindPoint® QuizShow

- **Dynamic Teaching Resources**
 - Answer Presentation Tool
 - Dynamic Classroom
 - Interactive Whiteboard Lessons
 - Support for Mathematical Practices
 - Editable Ancillaries
 - Teaching Edition

- **Dynamic Student Edition**
 - Interactive Pupil Edition
 - Tutorials
 - Interactive Multi-Language Glossary: English and Spanish
 - Record and Practice Journal
 - Basic Skills Handbook
 - Skills Review Handbook

- *BigIdeasMath.com*
 - Dynamic Student Materials
 - Dynamic Teaching Materials

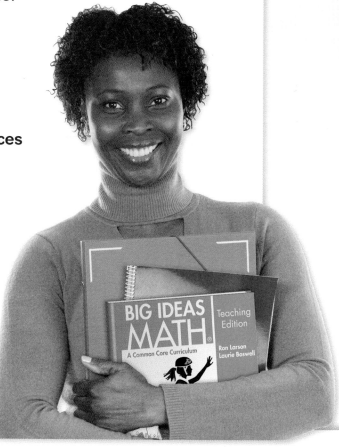

COMMON CORE STATE STANDARDS TO BOOK CORRELATION

After a standard is introduced, it is revisited many times in subsequent activities, lessons, and exercises.

Domain: Ratios and Proportional Relationships

Standards

6.RP.1 Understand the concept of a ratio and use ratio language to describe a ratio relationship between two quantities.
- **Section 4.2** Percents and Decimals
- **Section 5.1** Ratios
- **Section 5.2** Rates

6.RP.2 Understand the concept of a unit rate a/b associated with a ratio $a : b$ with $b \neq 0$, and use rate language in the context of a ratio relationship.
- **Section 5.2** Rates
- **Section 5.3** Solving Rate Problems

6.RP.3 Use ratio and rate reasoning to solve real-world and mathematical problems.
 a. Make tables of equivalent ratios relating quantities with whole-number measurements, find missing values in the tables, and plot the pairs of values on the coordinate plane. Use tables to compare ratios.
 - **Section 5.3** Solving Rate Problems

 b. Solve unit rate problems including those involving unit pricing and constant speed.
 - **Section 5.2** Rates
 - **Section 5.3** Solving Rate Problems

 c. Find a percent of a quantity as a rate per 100; solve problems involving finding the whole, given a part and the percent.
 - **Section 4.1** Percents and Fractions
 - **Section 4.2** Percents and Decimals
 - **Section 4.4** Finding the Percent of a Number
 - **Section 4.5** Percents and Estimation

 d. Use ratio reasoning to convert measurement units; manipulate and transform units appropriately when multiplying or dividing quantities.
 - **Section 5.2** Rates
 - **Section 5.3** Solving Rate Problems

Domain: The Number System

Standards

6.NS.1 Interpret and compute quotients of fractions, and solve word problems involving division of fractions by fractions.
- **Section 2.1** Fractions and Estimation
- **Section 2.2** Multiplying Fractions and Whole Numbers
- **Section 2.3** Multiplying Fractions
- **Section 2.4** Multiplying Mixed Numbers
- **Section 2.5** Dividing Fractions
- **Section 2.6** Dividing Mixed Numbers
- **Section 2.7** Writing Decimals as Fractions

6.NS.2 Fluently divide multi-digit numbers using the standard algorithm.
- **Section 2.8** Writing Fractions as Decimals
- **Section 3.4** Dividing Decimals by Whole Numbers
- **Section 3.5** Dividing Decimals

6.NS.3 Fluently add, subtract, multiply, and divide multi-digit decimals using the standard algorithm for each operation.
- **Section 3.1** Decimals and Estimation
- **Section 3.2** Multiplying Decimals and Whole Numbers
- **Section 3.3** Multiplying Decimals
- **Section 3.4** Dividing Decimals by Whole Numbers
- **Section 3.5** Dividing Decimals

6.NS.4 Find the greatest common factor of two whole numbers less than or equal to 100 and the least common multiple of two whole numbers less than or equal to 12. Use the distributive property to express a sum of two whole numbers 1–100 with a common factor as a multiple of a sum of two whole numbers with no common factor.
- **Section 1.4** The Distributive Property

6.NS.5 Understand that positive and negative numbers are used together to describe quantities having opposite directions or values; use positive and negative numbers to represent quantities in real-world contexts, explaining the meaning of 0 in each situation.
- **Topic 1** The Number Line
- **Topic 2** Comparing Numbers
- **Topic 3** Absolute Value
- **Topic 4** The Coordinate Plane

6.NS.6 Understand a rational number as a point on the number line. Extend number line diagrams and coordinate axes familiar from previous grades to represent points on the line and in the plane with negative number coordinates.

 a. Recognize opposite signs of numbers as indicating locations on opposite sides of 0 on the number line; recognize that the opposite of the opposite of a number is the number itself, and that 0 is its own opposite.
- **Topic 1** The Number Line

 b. Understand signs of numbers in ordered pairs as indicating locations in quadrants of the coordinate plane; recognize that when two ordered pairs differ only by signs, the locations of the points are related by reflections across one or both axes.
- **Topic 4** The Coordinate Plane

 c. Find and position integers and other rational numbers on a horizontal or vertical number line diagram; find and position pairs of integers and other rational numbers on a coordinate plane.
- **Topic 1** The Number Line
- **Topic 2** Comparing Numbers
- **Topic 3** Absolute Value
- **Topic 4** The Coordinate Plane

6.NS.7 Understand ordering and absolute value of rational numbers.

 a. Interpret statements of inequality as statements about the relative position of two numbers on a number line diagram.
- **Section 4.3** Comparing and Ordering Fractions, Decimals, and Percents
- **Topic 2** Comparing Numbers
- **Topic 3** Absolute Value

 b. Write, interpret, and explain statements of order for rational numbers in real-world contexts.
- **Section 4.3** Comparing and Ordering Fractions, Decimals, and Percents
- **Topic 2** Comparing Numbers
- **Topic 3** Absolute Value

 c. Understand the absolute value of a rational number as its distance from 0 on the number line; interpret absolute value as magnitude for a positive or negative quantity in a real-world situation.
- **Topic 3** Absolute Value

 d. Distinguish comparisons of absolute value from statements about order.
- **Topic 3** Absolute Value

6.NS.8 Solve real-world and mathematical problems by graphing points in all four quadrants of the coordinate plane. Include use of coordinates and absolute value to find distances between points with the same first coordinate or the same second coordinate.
- **Topic 4** The Coordinate Plane

Domain: Expressions and Equations

Standards

6.EE.1 Write and evaluate numerical expressions involving whole-number exponents.
- **Section 1.1** Evaluating Algebraic Expressions

6.EE.2 Write, read, and evaluate expressions in which letters stand for numbers.
 a. Write expressions that record operations with numbers and with letters standing for numbers.
 - **Section 1.2** Writing Expressions

 b. Identify parts of an expression using mathematical terms (sum, term, product, factor, quotient, coefficient); view one or more parts of an expression as a single entity.
 - **Section 1.2** Writing Expressions
 - **Section 1.3** Properties of Addition and Multiplication
 - **Section 1.4** The Distributive Property
 - **Section 7.1** Writing Equations in One Variable
 - **Section 7.4** Solving Two-Step Equations

 c. Evaluate expressions at specific values of their variables. Include expressions that arise from formulas used in real-world problems. Perform arithmetic operations, including those involving whole-number exponents, in the conventional order when there are no parentheses to specify a particular order (Order of Operations).
 - **Section 1.1** Evaluating Algebraic Expressions
 - **Section 1.5** Using Formulas to Solve Problems
 - **Section 6.1** Circles and Circumference
 - **Section 6.2** Perimeters of Composite Figures
 - **Section 6.3** Areas of Circles
 - **Section 6.4** Areas of Composite Figures

6.EE.3 Apply the properties of operations to generate equivalent expressions.
- **Section 1.3** Properties of Addition and Multiplication
- **Section 1.4** The Distributive Property

6.EE.4 Identify when two expressions are equivalent.
- **Section 1.3** Properties of Addition and Multiplication
- **Section 1.4** The Distributive Property

6.EE.5 Understand solving an equation or inequality as a process of answering a question: which values from a specified set, if any, make the equation or inequality true? Use substitution to determine whether a given number in a specified set makes an equation or inequality true.
- **Section 7.2** Solving Equations Using Addition or Subtraction
- **Section 7.3** Solving Equations Using Multiplication or Division
- **Section 7.4** Solving Two-Step Equations
- **Section 8.1** Writing and Graphing Inequalities
- **Section 8.2** Solving Inequalities Using Addition or Subtraction
- **Section 8.3** Solving Inequalities Using Multiplication or Division
- **Section 8.4** Solving Two-Step Inequalities

6.EE.6 Use variables to represent numbers and write expressions when solving a real-world or mathematical problem; understand that a variable can represent an unknown number, or, depending on the purpose at hand, any number in a specified set.
- **Section 1.2** Writing Expressions
- **Section 1.3** Properties of Addition and Multiplication
- **Section 1.4** The Distributive Property

6.EE.7 Solve real-world and mathematical problems by writing and solving equations of the form $x + p = q$ and $px = q$ for cases in which p, q and x are all nonnegative rational numbers.
- **Section 7.1** Writing Equations in One Variable
- **Section 7.2** Solving Equations Using Addition or Subtraction
- **Section 7.3** Solving Equations Using Multiplication or Division
- **Section 7.4** Solving Two-Step Equations
- **Section 7.5** Finding Dimensions of Plane Figures

6.EE.8 Write an inequality of the form $x > c$ or $x < c$ to represent a constraint or condition in a real-world or mathematical problem. Recognize that inequalities of the form $x > c$ or $x < c$ have infinitely many solutions; represent solutions of such inequalities on number line diagrams.
- **Section 8.1** Writing and Graphing Inequalities
- **Section 8.2** Solving Inequalities Using Addition or Subtraction
- **Section 8.3** Solving Inequalities Using Multiplication or Division
- **Section 8.4** Solving Two-Step Inequalities

6.EE.9 Use variables to represent two quantities in a real-world problem that change in relationship to one another; write an equation to express one quantity, thought of as the dependent variable, in terms of the other quantity, thought of as the independent variable. Analyze the relationship between the dependent and independent variables using graphs and tables, and relate these to the equation.
- **Section 9.1** Mapping Diagrams
- **Section 9.2** Functions as Words and Equations
- **Section 9.3** Input-Output Tables
- **Section 9.4** Graphs
- **Section 9.5** Analyzing Graphs

Domain: Geometry

Standards

6.G.1 Find the area of right triangles, other triangles, special quadrilaterals, and polygons by composing into rectangles or decomposing into triangles and other shapes; apply these techniques in the context of solving real-world and mathematical problems.
- **Section 6.4** Areas of Composite Figures
- **Section 7.5** Finding Dimensions of Plane Figures

6.G.2 Find the volume of a right rectangular prism with fractional edge lengths by packing it with unit cubes of the appropriate unit fraction edge lengths, and show that the volume is the same as would be found by multiplying the edge lengths of the prism. Apply the formulas $V = \ell wh$ and $V = bh$ to find volumes of right rectangular prisms with fractional edge lengths in the context of solving real-world and mathematical problems.
- **Section 7.6** Finding Dimensions of Prisms
- **Lesson 7.6b** Volume and Surface Area

6.G.3 Draw polygons in the coordinate plane given coordinates for the vertices; use coordinates to find the length of a side joining points with the same first coordinate or the same second coordinate. Apply these techniques in the context of solving real-world and mathematical problems.
- **Topic 4** The Coordinate Plane

6.G.4 Represent three-dimensional figures using nets made up of rectangles and triangles, and use the nets to find the surface area of these figures. Apply these techniques in the context of solving real-world and mathematical problems.
- **Lesson 7.6b** Volume and Surface Area

Domain: Statistics and Probability

Standards

6.SP.1 Recognize a statistical question as one that anticipates variability in the data related to the question and accounts for it in the answers.
- **Section 5.4** Mean

6.SP.2 Understand that a set of data collected to answer a statistical question has a distribution which can be described by its center, spread, and overall shape.
- **Section 5.4** Mean
- **Section 5.5** Median, Mode, and Range
- **Section 5.6** Analyzing Data Sets
- **Lesson 5.6b** Data Displays

6.SP.3 Recognize that a measure of center for a numerical data set summarizes all of its values with a single number, while a measure of variation describes how its values vary with a single number.
- **Section 5.4** Mean
- **Section 5.5** Median, Mode, and Range
- **Lesson 5.6b** Data Displays

6.SP.4 Display numerical data in plots on a number line, including dot plots, histograms, and box plots.
- **Lesson 5.6b** Data Displays

6.SP.5 Summarize numerical data sets in relation to their context, such as by:
 a. Reporting the number of observations.
 - **Section 5.4** Mean
 - **Section 5.5** Median, Mode, and Range
 - **Section 5.6** Analyzing Data Sets

 b. Describing the nature of the attribute under investigation, including how it was measured and its units of measurement.
 - **Section 5.4** Mean
 - **Section 5.5** Median, Mode, and Range
 - **Section 5.6** Analyzing Data Sets

 c. Giving quantitative measures of center (median and/or mean) and variability (interquartile range and/or mean absolute deviation), as well as describing any overall pattern and any striking deviations from the overall pattern with reference to the context in which the data were gathered.
 - **Section 5.4** Mean
 - **Section 5.5** Median, Mode, and Range
 - **Section 5.6** Analyzing Data Sets
 - **Lesson 5.6b** Data Displays

 d. Relating the choice of measures of center and variability to the shape of the data distribution and the context in which the data were gathered.
 - **Section 5.5** Median, Mode, and Range
 - **Section 5.6** Analyzing Data Sets
 - **Lesson 5.6b** Data Displays

BOOK TO COMMON CORE STATE STANDARDS CORRELATION

Chapter 1 Expressions and Number Properties

1.1 Evaluating Algebraic Expressions
- **6.EE.1** Write and evaluate numerical expressions involving whole-number exponents.
- **6.EE.2c** Evaluate expressions at specific values of their variables. Include expressions that arise from formulas used in real-world problems. Perform arithmetic operations, including those involving whole-number exponents, in the conventional order when there are no parentheses to specify a particular order (Order of Operations).

1.2 Writing Expressions
- **6.EE.2a** Write expressions that record operations with numbers and with letters standing for numbers.
- **6.EE.2b** Identify parts of an expression using mathematical terms (sum, term, product, factor, quotient, coefficient); view one or more parts of an expression as a single entity.
- **6.EE.6** Use variables to represent numbers and write expressions when solving a real-world or mathematical problem; understand that a variable can represent an unknown number, or, depending on the purpose at hand, any number in a specified set.

1.3 Properties of Addition and Multiplication
- **6.EE.2b** Identify parts of an expression using mathematical terms (sum, term, product, factor, quotient, coefficient); view one or more parts of an expression as a single entity.
- **6.EE.3** Apply the properties of operations to generate equivalent expressions.
- **6.EE.4** Identify when two expressions are equivalent.
- **6.EE.6** Use variables to represent numbers and write expressions when solving a real-world or mathematical problem; understand that a variable can represent an unknown number, or, depending on the purpose at hand, any number in a specified set.

1.4 The Distributive Property
- **6.NS.4** Find the greatest common factor of two whole numbers less than or equal to 100 and the least common multiple of two whole numbers less than or equal to 12. Use the distributive property to express a sum of two whole numbers 1–100 with a common factor as a multiple of a sum of two whole numbers with no common factor.
- **6.EE.2b** Identify parts of an expression using mathematical terms (sum, term, product, factor, quotient, coefficient); view one or more parts of an expression as a single entity.
- **6.EE.3** Apply the properties of operations to generate equivalent expressions.

1.4 **The Distributive Property** *(continued)*
- **6.EE.4** Identify when two expressions are equivalent.
- **6.EE.6** Use variables to represent numbers and write expressions when solving a real-world or mathematical problem; understand that a variable can represent an unknown number, or, depending on the purpose at hand, any number in a specified set.

1.5 **Using Formulas to Solve Problems**
- **6.EE.2c** Evaluate expressions at specific values of their variables. Include expressions that arise from formulas used in real-world problems. Perform arithmetic operations, including those involving whole-number exponents, in the conventional order when there are no parentheses to specify a particular order (Order of Operations).

Chapter 2 Multiplying and Dividing Fractions

2.1 **Fractions and Estimation**
- **6.NS.1** Interpret and compute quotients of fractions, and solve word problems involving division of fractions by fractions.

2.2 **Multiplying Fractions and Whole Numbers**
- **6.NS.1** Interpret and compute quotients of fractions, and solve word problems involving division of fractions by fractions.

2.3 **Multiplying Fractions**
- **6.NS.1** Interpret and compute quotients of fractions, and solve word problems involving division of fractions by fractions.

2.4 **Multiplying Mixed Numbers**
- **6.NS.1** Interpret and compute quotients of fractions, and solve word problems involving division of fractions by fractions.

2.5 **Dividing Fractions**
- **6.NS.1** Interpret and compute quotients of fractions, and solve word problems involving division of fractions by fractions.

2.6 **Dividing Mixed Numbers**
- **6.NS.1** Interpret and compute quotients of fractions, and solve word problems involving division of fractions by fractions.

2.7 **Writing Decimals as Fractions**
- **6.NS.1** Interpret and compute quotients of fractions, and solve word problems involving division of fractions by fractions.

2.8 **Writing Fractions as Decimals**
- **6.NS.2** Fluently divide multi-digit numbers using the standard algorithm.

Chapter 3 Multiplying and Dividing Decimals

3.1 **Decimals and Estimation**
- **6.NS.3** Fluently add, subtract, multiply, and divide multi-digit decimals using the standard algorithm for each operation.

3.2 **Multiplying Decimals and Whole Numbers**
- **6.NS.3** Fluently add, subtract, multiply, and divide multi-digit decimals using the standard algorithm for each operation.

3.3 **Multiplying Decimals**
- **6.NS.3** Fluently add, subtract, multiply, and divide multi-digit decimals using the standard algorithm for each operation.

3.4 **Dividing Decimals by Whole Numbers**
- **6.NS.2** Fluently divide multi-digit numbers using the standard algorithm.
- **6.NS.3** Fluently add, subtract, multiply, and divide multi-digit decimals using the standard algorithm for each operation.

3.5 **Dividing Decimals**
- **6.NS.2** Fluently divide multi-digit numbers using the standard algorithm.
- **6.NS.3** Fluently add, subtract, multiply, and divide multi-digit decimals using the standard algorithm for each operation.

Chapter 4 Fractions, Decimals, and Percents

4.1 **Percents and Fractions**
- **6.RP.3c** Find a percent of a quantity as a rate per 100; solve problems involving finding the whole, given a part and the percent.

4.2 **Percents and Decimals**
- **6.RP.1** Understand the concept of a ratio and use ratio language to describe a ratio relationship between two quantities.
- **6.RP.3c** Find a percent of a quantity as a rate per 100; solve problems involving finding the whole, given a part and the percent.

4.3 **Comparing and Ordering Fractions, Decimals, and Percents**
- **6.NS.7a** Interpret statements of inequality as statements about the relative position of two numbers on a number line diagram.
- **6.NS.7b** Write, interpret, and explain statements of order for rational numbers in real-world contexts.

4.4 **Finding the Percent of a Number**
- **6.RP.3c** Find a percent of a quantity as a rate per 100; solve problems involving finding the whole, given a part and the percent.

4.5 **Percents and Estimation**
- **6.RP.3c** Find a percent of a quantity as a rate per 100; solve problems involving finding the whole, given a part and the percent.

Chapter 5 Ratios, Rates, and Data Analysis

5.1 **Ratios**
- **6.RP.1** Understand the concept of a ratio and use ratio language to describe a ratio relationship between two quantities.

5.2 **Rates**
- **6.RP.1** Understand the concept of a ratio and use ratio language to describe a ratio relationship between two quantities.
- **6.RP.2** Understand the concept of a unit rate a/b associated with a ratio $a:b$ with $b \neq 0$, and use rate language in the context of a ratio relationship.
- **6.RP.3b** Solve unit rate problems including those involving unit pricing and constant speed.
- **6.RP.3d** Use ratio reasoning to convert measurement units; manipulate and transform units appropriately when multiplying or dividing quantities.

5.3 Solving Rate Problems
- **6.RP.2** Understand the concept of a unit rate a/b associated with a ratio $a:b$ with $b \neq 0$, and use rate language in the context of a ratio relationship.
- **6.RP.3a** Make tables of equivalent ratios relating quantities with whole-number measurements, find missing values in the tables, and plot the pairs of values on the coordinate plane. Use tables to compare ratios.
- **6.RP.3b** Solve unit rate problems including those involving unit pricing and constant speed.
- **6.RP.3d** Use ratio reasoning to convert measurement units; manipulate and transform units appropriately when multiplying or dividing quantities.

5.4 Mean
- **6.SP.1** Recognize a statistical question as one that anticipates variability in the data related to the question and accounts for it in the answers.
- **6.SP.2** Understand that a set of data collected to answer a statistical question has a distribution which can be described by its center, spread, and overall shape.
- **6.SP.3** Recognize that a measure of center for a numerical data set summarizes all of its values with a single number, while a measure of variation describes how its values vary with a single number.
- **6.SP.5a** Summarize numerical data sets in relation to their context by reporting the number of observations.
- **6.SP.5b** Summarize numerical data sets in relation to their context by describing the nature of the attribute under investigation, including how it was measured and its units of measurement.
- **6.SP.5c** Summarize numerical data sets in relation to their context by giving quantitative measures of center (median and/or mean) and variability (interquartile range and/or mean absolute deviation), as well as describing any overall pattern and any striking deviations from the overall pattern with reference to the context in which the data were gathered.

5.5 Median, Mode, and Range
- **6.SP.2** Understand that a set of data collected to answer a statistical question has a distribution which can be described by its center, spread, and overall shape.
- **6.SP.3** Recognize that a measure of center for a numerical data set summarizes all of its values with a single number, while a measure of variation describes how its values vary with a single number.
- **6.SP.5a** Summarize numerical data sets in relation to their context by reporting the number of observations.
- **6.SP.5b** Summarize numerical data sets in relation to their context by describing the nature of the attribute under investigation, including how it was measured and its units of measurement.
- **6.SP.5c** Summarize numerical data sets in relation to their context by giving quantitative measures of center (median and/or mean) and variability (interquartile range and/or mean absolute deviation), as well as describing any overall pattern and any striking deviations from the overall pattern with reference to the context in which the data were gathered.
- **6.SP.5d** Summarize numerical data sets in relation to their context by relating the choice of measures of center and variability to the shape of the data distribution and the context in which the data were gathered.

5.6 Analyzing Data Sets
- **6.SP.2** Understand that a set of data collected to answer a statistical question has a distribution which can be described by its center, spread, and overall shape.
- **6.SP.5a** Summarize numerical data sets in relation to their context by reporting the number of observations.
- **6.SP.5b** Summarize numerical data sets in relation to their context by describing the nature of the attribute under investigation, including how it was measured and its units of measurement.
- **6.SP.5c** Summarize numerical data sets in relation to their context by giving quantitative measures of center (median and/or mean) and variability (interquartile range and/or mean absolute deviation), as well as describing any overall pattern and any striking deviations from the overall pattern with reference to the context in which the data were gathered.
- **6.SP.5d** Summarize numerical data sets in relation to their context by relating the choice of measures of center and variability to the shape of the data distribution and the context in which the data were gathered.

5.6b Data Displays
- **6.SP.2** Understand that a set of data collected to answer a statistical question has a distribution which can be described by its center, spread, and overall shape.
- **6.SP.3** Recognize that a measure of center for a numerical data set summarizes all of its values with a single number, while a measure of variation describes how its values vary with a single number.
- **6.SP.4** Display numerical data in plots on a number line, including dot plots, histograms, and box plots.
- **6.SP.5c** Summarize numerical data sets in relation to their context by giving quantitative measures of center (median and/or mean) and variability (interquartile range and/or mean absolute deviation), as well as describing any overall pattern and any striking deviations from the overall pattern with reference to the context in which the data were gathered.
- **6.SP.5d** Summarize numerical data sets in relation to their context by relating the choice of measures of center and variability to the shape of the data distribution and the context in which the data were gathered.

Chapter 6 Circles and Area

6.1 Circles and Circumference
- **6.EE.2c** Evaluate expressions at specific values of their variables. Include expressions that arise from formulas used in real-world problems. Perform arithmetic operations, including those involving whole-number exponents, in the conventional order when there are no parentheses to specify a particular order (Order of Operations).

6.2 Perimeters of Composite Figures
- **6.EE.2c** Evaluate expressions at specific values of their variables. Include expressions that arise from formulas used in real-world problems. Perform arithmetic operations, including those involving whole-number exponents, in the conventional order when there are no parentheses to specify a particular order (Order of Operations).

6.3 **Areas of Circles**
- **6.EE.2c** Evaluate expressions at specific values of their variables. Include expressions that arise from formulas used in real-world problems. Perform arithmetic operations, including those involving whole-number exponents, in the conventional order when there are no parentheses to specify a particular order (Order of Operations).

6.4 **Areas of Composite Figures**
- **6.EE.2c** Evaluate expressions at specific values of their variables. Include expressions that arise from formulas used in real-world problems. Perform arithmetic operations, including those involving whole-number exponents, in the conventional order when there are no parentheses to specify a particular order (Order of Operations).
- **6.G.1** Find the area of right triangles, other triangles, special quadrilaterals, and polygons by composing into rectangles or decomposing into triangles and other shapes; apply these techniques in the context of solving real-world and mathematical problems.

Chapter 7 Equations

7.1 **Writing Equations in One Variable**
- **6.EE.2b** Identify parts of an expression using mathematical terms (sum, term, product, factor, quotient, coefficient); view one or more parts of an expression as a single entity.
- **6.EE.7** Solve real-world and mathematical problems by writing and solving equations of the form $x + p = q$ and $px = q$ for cases in which p, q and x are all nonnegative rational numbers.

7.2 **Solving Equations Using Addition or Subtraction**
- **6.EE.5** Understand solving an equation or inequality as a process of answering a question: which values from a specified set, if any, make the equation or inequality true? Use substitution to determine whether a given number in a specified set makes an equation or inequality true.
- **6.EE.7** Solve real-world and mathematical problems by writing and solving equations of the form $x + p = q$ and $px = q$ for cases in which p, q and x are all nonnegative rational numbers.

7.3 **Solving Equations Using Multiplication or Division**
- **6.EE.5** Understand solving an equation or inequality as a process of answering a question: which values from a specified set, if any, make the equation or inequality true? Use substitution to determine whether a given number in a specified set makes an equation or inequality true.
- **6.EE.7** Solve real-world and mathematical problems by writing and solving equations of the form $x + p = q$ and $px = q$ for cases in which p, q and x are all nonnegative rational numbers.

7.4 **Solving Two-Step Equations**
- **6.EE.2b** Identify parts of an expression using mathematical terms (sum, term, product, factor, quotient, coefficient); view one or more parts of an expression as a single entity.
- **6.EE.5** Understand solving an equation or inequality as a process of answering a question: which values from a specified set, if any, make the equation or inequality true? Use substitution to determine whether a given number in a specified set makes an equation or inequality true.

- **6.EE.7** Solve real-world and mathematical problems by writing and solving equations of the form $x + p = q$ and $px = q$ for cases in which p, q, and x are all nonnegative rational numbers.

7.5 Finding Dimensions of Plane Figures
- **6.EE.7** Solve real-world and mathematical problems by writing and solving equations of the form $x + p = q$ and $px = q$ for cases in which p, q, and x are all nonnegative rational numbers.
- **6.G.1** Find the area of right triangles, other triangles, special quadrilaterals, and polygons by composing into rectangles or decomposing into triangles and other shapes; apply these techniques in the context of solving real-world and mathematical problems

7.6 Finding Dimensions of Prisms
- **6.G.2** Find the volume of a right rectangular prism with fractional edge lengths by packing it with unit cubes of the appropriate unit fraction edge lengths, and show that the volume is the same as would be found by multiplying the edge lengths of the prism. Apply the formulas $V = \ell wh$ and $V = bh$ to find volumes of right rectangular prisms with fractional edge lengths in the context of solving real-world and mathematical problems.

7.6b Volume and Surface Area
- **6.G.2** Find the volume of a right rectangular prism with fractional edge lengths by packing it with unit cubes of the appropriate unit fraction edge lengths, and show that the volume is the same as would be found by multiplying the edge lengths of the prism. Apply the formulas $V = \ell wh$ and $V = bh$ to find volumes of right rectangular prisms with fractional edge lengths in the context of solving real-world and mathematical problems.
- **6.G.4** Represent three-dimensional figures using nets made up of rectangles and triangles, and use the nets to find the surface area of these figures. Apply these techniques in the context of solving real-world and mathematical problems.

Chapter 8 Inequalities

8.1 Writing and Graphing Inequalities
- **6.EE.5** Understand solving an equation or inequality as a process of answering a question: which values from a specified set, if any, make the equation or inequality true? Use substitution to determine whether a given number in a specified set makes an equation or inequality true.
- **6.EE.8** Write an inequality of the form $x > c$ or $x < c$ to represent a constraint or condition in a real-world or mathematical problem. Recognize that inequalities of the form $x > c$ or $x < c$ have infinitely many solutions; represent solutions of such inequalities on number line diagrams.

8.2 Solving Inequalities Using Addition or Subtraction
- **6.EE.5** Understand solving an equation or inequality as a process of answering a question: which values from a specified set, if any, make the equation or inequality true? Use substitution to determine whether a given number in a specified set makes an equation or inequality true.
- **6.EE.8** Write an inequality of the form $x > c$ or $x < c$ to represent a constraint or condition in a real-world or mathematical problem. Recognize that inequalities of the form $x > c$ or $x < c$ have infinitely many solutions; represent solutions of such inequalities on number line diagrams.

8.3 Solving Inequalities Using Multiplication or Division
- **6.EE.5** Understand solving an equation or inequality as a process of answering a question: which values from a specified set, if any, make the equation or inequality true? Use substitution to determine whether a given number in a specified set makes an equation or inequality true.
- **6.EE.8** Write an inequality of the form $x > c$ or $x < c$ to represent a constraint or condition in a real-world or mathematical problem. Recognize that inequalities of the form $x > c$ or $x < c$ have infinitely many solutions; represent solutions of such inequalities on number line diagrams.

8.4 Solving Two-Step Inequalities
- **6.EE.5** Understand solving an equation or inequality as a process of answering a question: which values from a specified set, if any, make the equation or inequality true? Use substitution to determine whether a given number in a specified set makes an equation or inequality true.
- **6.EE.8** Write an inequality of the form $x > c$ or $x < c$ to represent a constraint or condition in a real-world or mathematical problem. Recognize that inequalities of the form $x > c$ or $x < c$ have infinitely many solutions; represent solutions of such inequalities on number line diagrams.

Chapter 9 Tables, Graphs, and Functions

9.1 Mapping Diagrams
- **6.EE.9** Use variables to represent two quantities in a real-world problem that change in relationship to one another; write an equation to express one quantity, thought of as the dependent variable, in terms of the other quantity, thought of as the independent variable. Analyze the relationship between the dependent and independent variables using graphs and tables, and relate these to the equation.

9.2 Functions as Words and Equations
- **6.EE.9** Use variables to represent two quantities in a real-world problem that change in relationship to one another; write an equation to express one quantity, thought of as the dependent variable, in terms of the other quantity, thought of as the independent variable. Analyze the relationship between the dependent and independent variables using graphs and tables, and relate these to the equation.

9.3 Input-Output Tables
- **6.EE.9** Use variables to represent two quantities in a real-world problem that change in relationship to one another; write an equation to express one quantity, thought of as the dependent variable, in terms of the other quantity, thought of as the independent variable. Analyze the relationship between the dependent and independent variables using graphs and tables, and relate these to the equation.

9.4 Graphs
- **6.EE.9** Use variables to represent two quantities in a real-world problem that change in relationship to one another; write an equation to express one quantity, thought of as the dependent variable, in terms of the other quantity, thought of as the independent variable. Analyze the relationship between the dependent and independent variables using graphs and tables, and relate these to the equation.

9.5 Analyzing Graphs
- **6.EE.9** Use variables to represent two quantities in a real-world problem that change in relationship to one another; write an equation to express one quantity, thought of as the dependent variable, in terms of the other quantity, thought of as the independent variable. Analyze the relationship between the dependent and independent variables using graphs and tables, and relate these to the equation.

Additional Topics

Topic 1 The Number Line
- **6.NS.5** Understand that positive and negative numbers are used together to describe quantities having opposite directions or values; use positive and negative numbers to represent quantities in real-world contexts, explaining the meaning of 0 in each situation.
- **6.NS.6a** Recognize opposite signs of numbers as indicating locations on opposite sides of 0 on the number line; recognize that the opposite of the opposite of a number is the number itself, and that 0 is its own opposite.
- **6.NS.6c** Find and position integers and other rational numbers on a horizontal or vertical number line diagram; find and position pairs of integers and other rational numbers on a coordinate plane.

Topic 2 Comparing Numbers
- **6.NS.5** Understand that positive and negative numbers are used together to describe quantities having opposite directions or values; use positive and negative numbers to represent quantities in real-world contexts, explaining the meaning of 0 in each situation.
- **6.NS.6c** Find and position integers and other rational numbers on a horizontal or vertical number line diagram; find and position pairs of integers and other rational numbers on a coordinate plane.
- **6.NS.7a** Interpret statements of inequality as statements about the relative position of two numbers on a number line diagram.
- **6.NS.7b** Write, interpret, and explain statements of order for rational numbers in real-world contexts.

Topic 3 Absolute Value
- **6.NS.5** Understand that positive and negative numbers are used together to describe quantities having opposite directions or values; use positive and negative numbers to represent quantities in real-world contexts, explaining the meaning of 0 in each situation.
- **6.NS.6c** Find and position integers and other rational numbers on a horizontal or vertical number line diagram; find and position pairs of integers and other rational numbers on a coordinate plane.
- **6.NS.7a** Interpret statements of inequality as statements about the relative position of two numbers on a number line diagram.
- **6.NS.7b** Write, interpret, and explain statements of order for rational numbers in real-world contexts.
- **6.NS.7c** Understand the absolute value of a rational number as its distance from 0 on the number line; interpret absolute value as magnitude for a positive or negative quantity in a real-world situation.
- **6.NS.7d** Distinguish comparisons of absolute value from statements about order.

Topic 4 The Coordinate Plane

- **6.NS.5** Understand that positive and negative numbers are used together to describe quantities having opposite directions or values; use positive and negative numbers to represent quantities in real-world contexts, explaining the meaning of 0 in each situation.
- **6.NS.6b** Understand signs of numbers in ordered pairs as indicating locations in quadrants of the coordinate plane; recognize that when two ordered pairs differ only by signs, the locations of the points are related by reflections across one or both axes.
- **6.NS.6c** Find and position integers and other rational numbers on a horizontal or vertical number line diagram; find and position pairs of integers and other rational numbers on a coordinate plane.
- **6.NS.8** Solve real-world and mathematical problems by graphing points in all four quadrants of the coordinate plane. Include use of coordinates and absolute value to find distances between points with the same first coordinate or the same second coordinate.
- **6.G.3** Draw polygons in the coordinate plane given coordinates for the vertices; use coordinates to find the length of a side joining points with the same first coordinate or the same second coordinate. Apply these techniques in the context of solving real-world and mathematical problems.

NARROWER AND DEEPER

The writers of the Common Core State Standards ask, "What does mathematical understanding look like?" One indication of understanding is the ability for students to justify why a mathematical statement is true or where a mathematical rule comes from.

Big Ideas Math's *Narrower and Deeper* program is a revolutionary combination of the discovery and direct instruction approaches. Students gain a deeper understanding of the math concepts by narrowing their focus to fewer topics and by delving into concepts through inductive reasoning opportunities, engaging activities, concise examples, and rich, thought-provoking exercises.

A BALANCED APPROACH

DISCOVERY

Each section begins with a 2-page *Activity* that is introduced by an Essential Question.

- *Deeper*
- *Dynamic*
- *Doable*
- *Dazzling*

TO INSTRUCTION

DIRECT INSTRUCTION

After the concept has been introduced with a full-class period **Activity**, it is extended the following day through the **Lesson**.

- **Key Ideas**
- **Examples**
- **"On Your Own" Questions**

ENGAGING PUPIL BOOKS

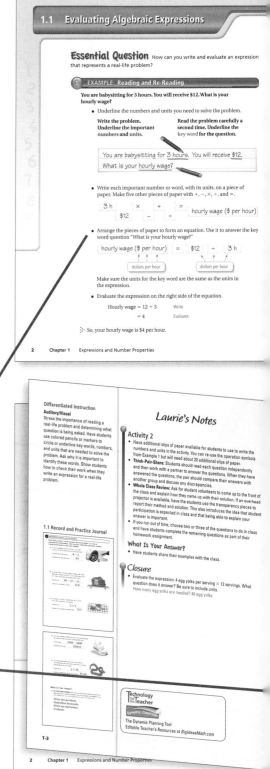

ACTIVITY

Each section begins with a 2-page *Activity* that is introduced by an Essential Question.
- **Deeper**
- **Dynamic**
- **Doable**
- **Dazzling**

Students gain a deeper understanding of topics through inductive reasoning and exploration.

Students develop communication and problem-solving skills by answering Essential Questions.

INSPIRING TEACHER BOOKS

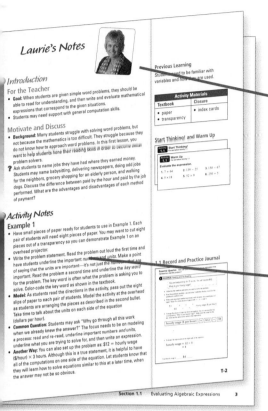

Teachers have the benefit of **Laurie Boswell's** 20-plus years of classroom experience reflected in her lively and informative notes.

Teachers are offered suggestions for questioning that help guide students toward better understanding.

"Laurie's Notes"
COMPLETE ACTIVITY AND TIME MANAGEMENT SUPPORT FROM A MASTER CLASSROOM TEACHER

CLEAR PUPIL BOOKS

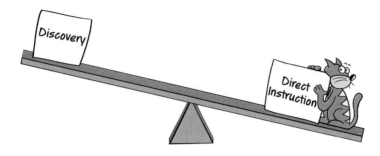

LESSON

After the concept has been introduced with a full-class period **Activity**, it is extended the following day through the **Lesson**.

- **Key Ideas**
- **Examples**
- **"On Your Own" Questions**

Each concept is accompanied by clear, stepped-out examples.

Each example in the pupil edition is accompanied by teaching suggestions from Laurie.

xl

INSIGHTFUL TEACHER BOOKS

Student-friendly and teacher-tested motivation activities start each lesson.

Laurie shares insights she has gained through years of teaching experience.

"Laurie's Notes"
COMPREHENSIVE TEACHING SUPPORT FROM A MASTER CLASSROOM TEACHER

DIFFERENTIATED INSTRUCTION

Opening Doors to Learning

Two primary concerns while developing the Big Ideas Math program were the diversity of the student population and their different learning profiles. The authors developed a curriculum that helps teachers create classrooms that concentrate on learner needs by using the Universal Design for Learning model (UDL). The curriculum is designed to incorporate a wide variety of options to achieve its goals; offering materials, methods, and assessments so that the curriculum in its entirety is flexible and accommodating of individual student needs. By using Differentiated Instruction, teachers open doors to learning that students are unable to open themselves.

English Language Learners

The Big Ideas Math program recognizes that English Language Learners are a highly heterogeneous and complex group of students with diverse gifts, educational needs, backgrounds, languages, and goals. The writers used researched-based recommendations while developing the program that were designed to specifically assist English Language Learners (ELL). In addition to global support, such as curriculum that is organized around Essential Questions that involve both reading and writing, the program includes at-point-of-use ELL notes for the teacher; Home and Community Letters; ebooks with audio (English and Spanish); and a visual glossary.

Differentiating the Lesson Ancillary

Differentiating the Lesson is an online ancillary available at *BigIdeasMath.com*. This ancillary provides complete teaching notes and worksheets that address the needs of the diverse learners in the classroom. The lessons engage students in activities that often incorporate visual learning and kinesthetic learning. Some lessons present an alternative approach to teaching the content while other lessons extend the concepts of the text in a challenging way for advanced students. Each chapter of the *Differentiating the Lesson* ancillary begins with an overview that outlines the differentiated lessons in the chapter, and describes the students who would most benefit from the approach used in each lesson.

RESPONSE TO INTERVENTION

Through print and digital resources, the Big Ideas Math program completely supports the 3-tier RTI model. Opportunities for daily assessment help identify areas of needs and easy-to-use resources are provided to support the education of all students.

Tier 1: Daily Intervention

The Big Ideas Math program uses research-based instructional strategies to ensure quality instruction. Vocabulary support, cooperative learning opportunities, and graphic organizers are included in the Pupil Edition. Additional strategies can be found throughout the program. Daily student reviews and assessment guarantee that every student is making regular progress. Complete support helps teachers personalize instruction for every student.

- On Your Own
- Mini-Assessment
- Differentiated Instruction
- Skills Review Handbook

Tier 2: Strategic Intervention

The Big Ideas Math program facilitates increased time and focus on instruction for students who are not responding effectively to Tier 1 intervention. Additional support to assist teachers with these struggling learners can be found in the ancillary materials. Extra Examples, Fair Game Reviews, Graphic Organizers, Study Tips, and Real-Life Applications have been specifically written to enhance learning and to engage the diverse students within today's math classrooms. Using the classroom and online resources provided, teachers can reach, challenge, and motivate each student with germane, high-quality instruction targeted to their individual needs.

- Differentiating the Lesson
- Lesson Tutorials
- Basic Skills Handbook
- Record and Practice Journal
- Chapter Resource Books

Tier 3: Customized Learning Intervention

Support for students working below grade level is also available.

Skills Review Handbook Basic Skills Handbook

PACING GUIDE

Each page in the book is in the *Pacing Guide*.
Chapters 1–9: 156 days
(Including Additional Topics)

Chapter 1 (15 Days)

Chapter Opener	1 Day
Activity 1.1	1 Day
Lesson 1.1	1 Day
Activity 1.2	1 Day
Lesson 1.2	1 Day
Activity 1.3	1 Day
Lesson 1.3	1 Day
Study Help/Quiz	1 Day
Activity 1.4	1 Day
Lesson 1.4	1 Day
Activity 1.5	1 Day
Lesson 1.5	1 Day
Quiz/Chapter Review	1 Day
Chapter Test	1 Day
Standardized Test Practice	1 Day

Chapter 2 (22 Days)

Chapter Opener	1 Day
Activity 2.1	1 Day
Lesson 2.1	1 Day
Activity 2.2	1 Day
Lesson 2.2	1 Day
Activity 2.3	1 Day
Lesson 2.3	1 Day
Activity 2.4	1 Day
Lesson 2.4	1 Day
Study Help/Quiz	1 Day
Activity 2.5	1 Day
Lesson 2.5	2 Days
Activity 2.6	1 Day
Lesson 2.6	1 Day
Activity 2.7	1 Day
Lesson 2.7	1 Day
Activity 2.8	1 Day
Lesson 2.8	1 Day
Quiz/Chapter Review	1 Day
Chapter Test	1 Day
Standardized Test Practice	1 Day

Chapter 3 (15 Days)

Chapter Opener	1 Day
Activity 3.1	1 Day
Lesson 3.1	1 Day
Activity 3.2	1 Day
Lesson 3.2	1 Day
Activity 3.3	1 Day
Lesson 3.3	1 Day
Study Help/Quiz	1 Day
Activity 3.4	1 Day
Lesson 3.4	1 Day
Activity 3.5	1 Day
Lesson 3.5	1 Day
Quiz/Chapter Review	1 Day
Chapter Test	1 Day
Standardized Test Practice	1 Day

Chapter 4 (15 Days)

Chapter Opener	1 Day
Activity 4.1	1 Day
Lesson 4.1	1 Day
Activity 4.2	1 Day
Lesson 4.2	1 Day
Activity 4.3	1 Day
Lesson 4.3	1 Day
Study Help/Quiz	1 Day
Activity 4.4	1 Day
Lesson 4.4	1 Day
Activity 4.5	1 Day
Lesson 4.5	1 Day
Quiz/Chapter Review	1 Day
Chapter Test	1 Day
Standardized Test Practice	1 Day

Chapter 5 (20 Days)

Chapter Opener	1 Day
Activity 5.1	1 Day
Lesson 5.1	1 Day
Activity 5.2	1 Day
Lesson 5.2	1 Day
Activity 5.3	1 Day
Lesson 5.3	1 Day
Study Help/Quiz	1 Day

Activity 5.4	1 Day
Lesson 5.4	1 Day
Activity 5.5	1 Day
Lesson 5.5	1 Day
Activity 5.6	2 Days
Lesson 5.6	1 Day
Lesson 5.6b	2 Days
Quiz/Chapter Review	**1 Day**
Chapter Test	**1 Day**
Standardized Test Practice	**1 Day**

Chapter 6 (14 Days)

Chapter Opener	1 Day
Activity 6.1	1 Day
Lesson 6.1	1 Day
Activity 6.2	2 Days
Lesson 6.2	1 Day
Study Help/Quiz	**1 Day**
Activity 6.3	1 Day
Lesson 6.3	1 Day
Activity 6.4	1 Day
Lesson 6.4	1 Day
Quiz/Chapter Review	**1 Day**
Chapter Test	**1 Day**
Standardized Test Practice	**1 Day**

Chapter 7 (21 Days)

Chapter Opener	1 Day
Activity 7.1	1 Day
Lesson 7.1	1 Day
Activity 7.2	1 Day
Lesson 7.2	2 Days
Activity 7.3	1 Day
Lesson 7.3	1 Day
Study Help/Quiz	**1 Day**
Activity 7.4	1 Day
Lesson 7.4	2 Days
Activity 7.5	1 Day
Lesson 7.5	1 Day
Activity 7.6	1 Day
Lesson 7.6	1 Day
Lesson 7.6b	2 Days
Quiz/Chapter Review	**1 Day**
Chapter Test	**1 Day**
Standardized Test Practice	**1 Day**

Chapter 8 (14 Days)

Chapter Opener	1 Day
Activity 8.1	1 Day
Lesson 8.1	2 Days
Activity 8.2	1 Day
Lesson 8.2	1 Day
Study Help/Quiz	**1 Day**
Activity 8.3	1 Day
Lesson 8.3	1 Day
Activity 8.4	1 Day
Lesson 8.4	1 Day
Quiz/Chapter Review	**1 Day**
Chapter Test	**1 Day**
Standardized Test Practice	**1 Day**

Chapter 9 (15 Days)

Chapter Opener	1 Day
Activity 9.1	1 Day
Lesson 9.1	1 Day
Activity 9.2	1 Day
Lesson 9.2	1 Day
Activity 9.3	1 Day
Lesson 9.3	1 Day
Study Help/Quiz	**1 Day**
Activity 9.4	1 Day
Lesson 9.4	1 Day
Activity 9.5	1 Day
Lesson 9.5	1 Day
Quiz/Chapter Review	**1 Day**
Chapter Test	**1 Day**
Standardized Test Practice	**1 Day**

Additional Topics (5 Days)

Opener	**1 Day**
Topic 1	1 Day
Topic 2	1 Day
Topic 3	1 Day
Topic 4	1 Day

PROFESSIONAL DEVELOPMENT

Big Ideas Learning, LLC is a professional development and publishing company founded by Dr. Ron Larson. We are dedicated to providing 21st century teaching and learning in the area of mathematics. We work with middle schools across the country as they implement world-class standards for mathematics.

As teachers and school districts move forward in implementing new Common Core State Standards, the fundamental ideas of rigor and relevance and big ideas that are deep and focused take on new meaning. Big Ideas Learning provides a rich, hands-on experience that allows for astute understanding and practice, not only of the challenging world-class standards, but of the underlying mathematics pedagogy as well.

WORKSHOPS

- Creating Highly Motivating Classrooms
- Implementing the Common Core State Standards

Activities for the Mathematics Classroom:

- Teaching More with Less
- Best Practices in the Mathematics Classroom
- Questioning in the Mathematics Classroom
- The Three R's: Rigor, Relevance and Reality
- Reading in the Content Areas
- Games for Numerical Fluency
- Engaging the Tech Natives

Our professional staff of experienced instructors can also assist you in creating customized training sessions tailored to achieving your desired outcome.

Common Core State Standards for Mathematical Practice

Make sense of problems and persevere in solving them.
- Multiple representations are presented to help students move from concrete to representative and into abstract thinking
- *Essential Questions* help students focus and analyze
- *In Your Own Words* provide opportunities for students to look for meaning and entry points to a problem

Reason abstractly and quantitatively.
- Visual problem solving models help students create a coherent representation of the problem
- Opportunities for students to decontextualize and contextualize problems are presented in every lesson

Construct viable arguments and critique the reasoning of others.
- *Error Analysis*; *Different Words, Same Question*; and *Which One Doesn't Belong* features provide students the opportunity to construct arguments and critique the reasoning of others
- *Inductive Reasoning* activities help students make conjectures and build a logical progression of statements to explore their conjecture

Model with mathematics.
- Real-life situations are translated into diagrams, tables, equations, and graphs to help students analyze relations and to draw conclusions
- Real-life problems are provided to help students learn to apply the mathematics that they are learning to everyday life

Use appropriate tools strategically.
- *Graphic Organizers* support the thought process of what, when, and how to solve problems
- A variety of tool papers, such as graph paper, number lines, and manipulatives, are available as students consider how to approach a problem
- Opportunities to use the web, graphing calculators, and spreadsheets support student learning

Attend to precision.
- *On Your Own* questions encourage students to formulate consistent and appropriate reasoning
- Cooperative learning opportunities support precise communication

Look for and make use of structure.
- *Inductive Reasoning* activities provide students the opportunity to see patterns and structure in mathematics
- Real-world problems help students use the structure of mathematics to break down and solve more difficult problems

Look for and express regularity in repeated reasoning.
- Opportunities are provided to help students make generalizations
- Students are continually encouraged to check for reasonableness in their solutions

Go to *BigIdeasMath.com* for more information on the Common Core State Standards for Mathematical Practice.

Common Core State Standards for Mathematical Content for Grade 6

Chapter Coverage for Standards

1 – 2 – 3 – 4 – 5 – 6 – 7 – 8 – 9 – AT

Domain: Ratios and Proportional Relationships
- Understand ratio concepts and use ratio reasoning to solve problems.

1 – 2 – 3 – 4 – 5 – 6 – 7 – 8 – 9 – AT

Domain: The Number System
- Apply and extend previous understandings of multiplication and division to divide fractions by fractions.
- Compute fluently with multi-digit numbers and find common factors and multiples.
- Apply and extend previous understandings of numbers to the system of rational numbers.

1 – 2 – 3 – 4 – 5 – 6 – 7 – 8 – 9 – AT

Domain: Expressions and Equations
- Apply and extend previous understandings of arithmetic to algebraic expressions.
- Reason about and solve one-variable equations and inequalities.
- Represent and analyze quantitative relationships between dependent and independent variables.

1 – 2 – 3 – 4 – 5 – 6 – 7 – 8 – 9 – AT

Domain: Geometry
- Solve real-world and mathematical problems involving area, surface area, and volume.

1 – 2 – 3 – 4 – 5 – 6 – 7 – 8 – 9 – AT

Domain: Statistics and Probability
- Develop understanding of statistical variability.
- Summarize and describe distributions.

Go to *BigIdeasMath.com* for more information on the Common Core State Standards for Mathematical Content.

How to Use Your Math Book

- Read the **Essential Question** in the activity.

 Work with a partner to decide **What Is Your Answer?**

 Now you are ready to do the Practice problems.

- Find the Key Vocabulary words, **highlighted in yellow**.

 Read their definitions. Study the concepts in each Key Idea.
 If you forget a definition, you can look it up online in the

 Multi-Language Glossary at BigIdeasMath.com.

- After you study each **EXAMPLE**, do the exercises in the On Your Own.

 Now You're Ready to do the exercises that correspond to the example.

 As you study, look for a Study Tip or a Common Error.

- The exercises are divided into 3 parts.

 Vocabulary and Concept Check

 Practice and Problem Solving

 Fair Game Review

 If an exercise has a ① next to it, look back at Example 1 for help with that exercise.

 More help is available at .

- To help study for your test, use the following.

 Quiz Study Help

 Chapter Review Chapter Test

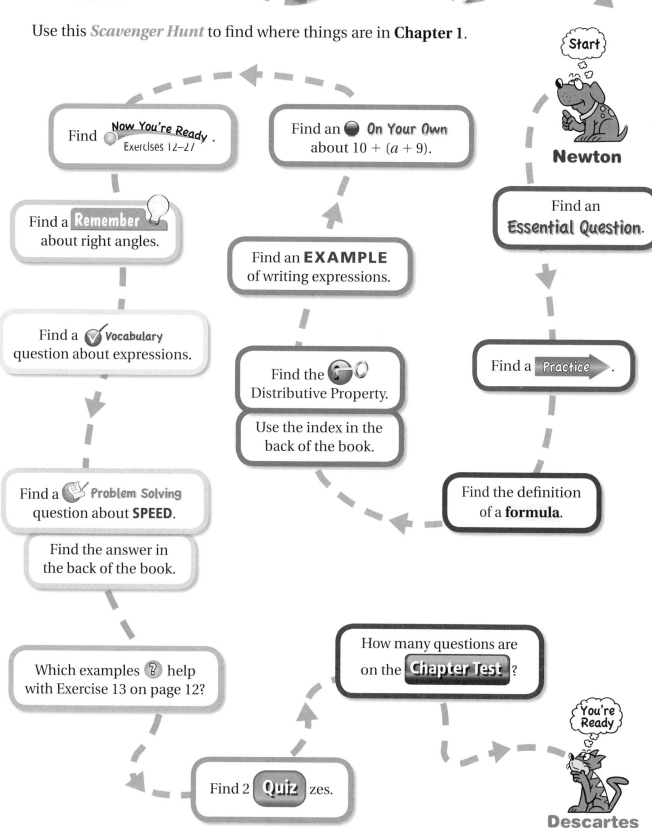

1 Expressions and Number Properties

1.1 **Evaluating Algebraic Expressions**
1.2 **Writing Expressions**
1.3 **Properties of Addition and Multiplication**
1.4 **The Distributive Property**
1.5 **Using Formulas to Solve Problems**

Connections to Previous Learning

- Recognize and use variables.
- Recognize and write algebraic expressions for functions with two operations.
- Describe mathematical relationships using expressions and visual representations.

- Use order of operations including exponents and parentheses.
- Divide multi-digit whole numbers efficiently, including real-world problems

- Write and evaluate mathematical expressions in real-world situations.
- Solve problems using a formula.
- Apply the Commutative, Associative, and Distributive Properties to show expressions are equivalent.

Math in History

The formalization of various properties of addition and multiplication tended to come late in the history of mathematics.

★ The first known naming of the Distributive Property was by a French man named Francois Joseph Servois (1768–1847) in a memoir published in 1814 in *Annales de Gergonne*. He introduced the name as follows, "Because $a(x + y) = ax + ay$, the factor "a" is said to distribute."

★ Ancient Egyptians used the Distributive Property to multiply whole numbers. Here is how they multiplied 57 by 27.

How it Looked		The Reason it Worked
1	57	$57(27) = 57(1 + 2 + 8 + 16)$
2	114	$= 57(1) + 57(2) + 57(8) + 57(16)$
8	456	$= 57 + 114 + 456 + 912$
16	912	$= 1539$
27	1539	

Pacing Guide for Chapter 1

Chapter Opener	1 Day
Section 1	
Activity	1 Day
Lesson	1 Day
Section 2	
Activity	1 Day
Lesson	1 Day
Section 3	
Activity	1 Day
Lesson	1 Day
Study Help / Quiz	1 Day
Section 4	
Activity	1 Day
Lesson	1 Day
Section 5	
Activity	1 Day
Lesson	1 Day
Quiz / Chapter Review	1 Day
Chapter Test	1 Day
Standardized Test Practice	1 Day
Total Chapter 1	15 Days
Year-to-Date	15 Days

Check Your Resources

- Record and Practice Journal
- Resources by Chapter
- Skills Review Handbook
- Assessment Book
- Worked-Out Solutions

Technology For the Teacher

The Dynamic Planning Tool
Editable Teacher's Resources at
BigIdeasMath.com

Common Core State Standards

6.G.1 Find the area of right triangles, other triangles, special quadrilaterals, and polygons by composing into rectangles or decomposing into triangles and other shapes

5.OA.1 Use parentheses, brackets, or braces in numerical expressions, and evaluate expressions with these symbols.

Additional Topics for Review
- Line, Bar, and Coordinate Graphs

Try It Yourself
1. 18 cm^2
2. 42 in.2
3. 21 mm^2
4. 19
5. 5
6. 6
7. 38
8. 16
9. 9

Record and Practice Journal
1. 9 yd^2
2. 40 m^2
3. 42 ft^2
4. 60 in.2
5. 66 cm^2
6. 18 cm^2
7. 120 in.2
8. 7
9. 31
10. 14
11. 22
12. 42
13. 2
14. 16
15. 1
16. $6.00

Math Background Notes

Vocabulary Review
- Parallelogram
- Trapezoid
- Exponent

Finding Area
- Students should be familiar with area formulas for rectangles, trapezoids, parallelograms, and triangles.
- Remind students that the area formula in each example is derived from the area formula for a rectangle.
- You may want to review units with students. A complete answer will include a numeric solution and the correct units. Remind students that area is always measured in square units.
- The base and height of a rectangle are interchangeable. Some students may think that the base of the rectangle must always be the side that sits on the ground. Turn the rectangle vertically to demonstrate that the base and height are relative to interpretation.
- **Common Error:** Be sure that students use the correct order of operations to compute the area of a trapezoid. Remind students to work inside the parentheses first and then move on to multiplication and division from left to right (see order of operations below).

Using Order of Operations
- Students should know the order of operations.
- You may want to review the correct order of operations with students. Many students probably learned the pneumonic device *Please Excuse My Dear Aunt Sally*. Ask a volunteer to explain why this phrase is helpful.
- You may choose to review exponents with students. Remind students that the exponent tells you how many times the base acts as a factor.
- **Common Error:** Students may misinterpret the exponent in Example 5. Watch for students who incorrectly compute $4^2 = 4 \cdot 2 = 8$.

Reteaching and Enrichment Strategies

If students need help...	If students got it...
Record and Practice Journal • Fair Game Review Skills Review Handbook Lesson Tutorials	Game Closet at *BigIdeasMath.com* Start the next section

What You Learned Before

"Great! You're up to x = 2. Let's keep going."

Finding Area (4.OA.3)

Find the area of each figure.

Example 1 Rectangle

$A = bh$
$= 4 \cdot 2$
$= 8 \text{ cm}^2$

Example 2 Trapezoid

$A = h(b + B) \div 2$
$= 6(3 + 5) \div 2$
$= 48 \div 2$
$= 24 \text{ in.}^2$

Example 3 Parallelogram

$A = bh$
$= 11 \cdot 4$
$= 44 \text{ mm}^2$

Example 4 Triangle

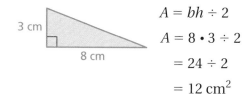

$A = bh \div 2$
$A = 8 \cdot 3 \div 2$
$= 24 \div 2$
$= 12 \text{ cm}^2$

Try It Yourself
Find the area.

1.

2.

3.

Using Order of Operations (5.OA.1)

Example 5 Simplify $4^2 \div 2 + 3(9 - 5)$.

First: Parentheses
Second: Exponents
Third: Multiplication and Division (from left to right)
Fourth: Addition and Subtraction (from left to right)

$4^2 \div 2 + 3(9 - 5) = 4^2 \div 2 + 3 \cdot 4$
$= 16 \div 2 + 3 \cdot 4$
$= 8 + 12$
$= 20$

Try It Yourself
Simplify the expression.

4. $3^2 + 5(4 - 2)$
5. $3 + 4 \div 2$
6. $10 \div 5 \cdot 3$
7. $4(3^3 - 8) \div 2$
8. $3 \cdot 6 - 4 \div 2$
9. $12 + 7 \cdot 3 - 24$

1.1 Evaluating Algebraic Expressions

COMMON CORE STATE STANDARDS
6.EE.1
6.EE.2c

Essential Question How can you write and evaluate an expression that represents a real-life problem?

1 EXAMPLE: Reading and Re-Reading

You are babysitting for 3 hours. You will receive $12. What is your hourly wage?

- Underline the numbers and units you need to solve the problem.

Write the problem. Underline the important numbers and units.

Read the problem carefully a second time. Underline the key word for the question.

> You are babysitting for 3 hours. You will receive $12.
> What is your hourly wage?

- Write each important number or word, with its units, on a piece of paper. Make five other pieces of paper with +, −, ×, ÷, and =.

 | 3 h | × | + | = | hourly wage ($ per hour) |
 | $12 | − | ÷ |

- Arrange the pieces of paper to form an equation. Use it to answer the key word question "What is your hourly wage?"

 hourly wage ($ per hour) = $12 ÷ 3 h
 (dollars per hour) (dollars per hour)

 Make sure the units for the key word are the same as the units in the expression.

- Evaluate the expression on the right side of the equation.

 Hourly wage = 12 ÷ 3 Write.
 = 4 Evaluate.

So, your hourly wage is $4 per hour.

2 Chapter 1 Expressions and Number Properties

Laurie's Notes

Introduction

For the Teacher
- **Goal:** When students are given simple word problems, they should be able to read for understanding, and then write and evaluate mathematical expressions that correspond to the given situations.
- Students may need support with general computation skills.

Motivate and Discuss
- **Background:** Many students struggle with solving word problems, but not because the mathematics is too difficult. They struggle because they do not know how to approach word problems. In this first lesson, you want to help students hone their reading skills in order to become better problem solvers.
- Ask students to name jobs they have had where they earned money. Students may name babysitting, delivering newspapers, doing odd jobs for the neighbors, grocery shopping for an elderly person, and walking dogs. Discuss the difference between paid by the hour and paid by the job performed. What are the advantages and disadvantages of each method of payment?

Activity Notes

Example 1
- Have small pieces of paper ready for students to use in Example 1. Each pair of students will need eight pieces of paper. You may want to cut eight pieces out of a transparency so you can demonstrate Example 1 on an overhead projector.
- Write the problem statement. Read the problem out loud the first time and have students underline the important *numbers* and *units*. Make a point of saying that the units are important—it's not just the numbers that are important. Read the problem a second time and underline the *key word* for the problem. The key word is often what the problem is asking you to solve. Color-code the key word as shown in the textbook.
- **Model:** As students read the directions in the activity, pass out the eight slips of paper to each pair of students. Model the activity at the overhead as students are arranging the pieces as described in the second bullet. Take time to talk about the units on each side of the equation (dollars per hour).
- **Common Question:** Students may ask "Why go through all this work when we already knew the answer?" The focus needs to be on modeling a process: read and re-read, underline important numbers *and* units, underline what you are trying to solve for, and then write an expression.
- **Another Way:** You can also set up the problem as: $12 = hourly wage ($/hour) × 3 hours. Although this is a true statement, it is helpful to have all of the computations on one side of the equation. Let students know that they will learn how to solve equations similar to this at a later time, when the answer may not be so obvious.

Common Core State Standards

6.EE.1 Write and evaluate numerical expressions involving whole-number exponents.

6.EE.2c Evaluate expressions at specific values of their variables. Include expressions that arise from formulas used in real-world problems. Perform arithmetic operations, including those involving whole-number exponents, in the conventional order when there are no parentheses to specify a particular order (Order of Operations).

Previous Learning
Students need to be familiar with variables and how they are used.

Activity Materials	
Textbook	Closure
• paper • transparency	• index cards

Start Thinking! and Warm Up

1.1 Record and Practice Journal

T-2

Differentiated Instruction

Auditory/Visual

Stress the importance of reading a real-life problem and determining what question is being asked. Have students use colored pencils or markers to circle or underline key words, numbers, and units that are needed to solve the problem. Ask why it is important to identify these words. Show students how to check their work when they write an expression for a real-life problem.

1.1 Record and Practice Journal

Laurie's Notes

Activity 2

- Have additional slips of paper available for students to use to write the numbers and units in the activity. You can re-use the operation symbols from Example 1 but will need about 20 additional slips of paper.
- **Think-Pair-Share:** Students should read each question independently and then work with a partner to answer the questions. When they have answered the questions, the pair should compare their answers with another group and discuss any discrepancies.
- **Whole Class Review:** Ask for student volunteers to come up to the front of the class and explain how they came up with their solution. If an overhead projector is available, have the students use the transparency pieces to report their method and solution. This also introduces the idea that student participation is expected in class and that being able to explain your answer is important.
- If you run out of time, choose two or three of the questions to do in class and have students complete the remaining questions as part of their homework assignment.

What Is Your Answer?

- Have students share their examples with the class.

Closure

- Evaluate the expression: 4 egg yolks per serving × 12 servings. What question does it answer? Be sure to include units.
 How many egg yolks are needed? 48 egg yolks

The Dynamic Planning Tool
Editable Teacher's Resources at *BigIdeasMath.com*

2 ACTIVITY: Reading and Re-Reading

Work with a partner. Use the strategy shown in Example 1 to write an expression for each problem. After you have written the expression, evaluate it using mental math or some other method.

a. You are washing cars for 2 hours. You will receive $6. How much will you earn per hour?

b. You have $60. You buy a pair of jeans and a shirt. The pair of jeans costs $27. You come home with $15. How much did you spend on the shirt?

c. For lunch, you buy 5 sandwiches that cost $3 each. How much do you spend?

d. You are running a 4500-foot race. How much farther do you have left to go after running 2000 feet?

e. A young rattlesnake grows at a rate of about 20 centimeters per year. How much does a young rattlesnake grow in 2 years?

What Is Your Answer?

3. **IN YOUR OWN WORDS** How can you write and evaluate an expression that represents a real-life problem? Give one example with addition, one with subtraction, one with multiplication, and one with division.

Use what you learned about evaluating expressions to complete Exercises 4–7 on page 6.

Section 1.1 Evaluating Algebraic Expressions **3**

1.1 Lesson

Check It Out
Lesson Tutorials
BigIdeasMath ✓com

Key Vocabulary 🔊
numerical expression, p. 4
algebraic expression, p. 4
evaluate, p. 4

A **numerical expression** contains only numbers and operations. An **algebraic expression** may contain numbers, operations, and one or more variables. Here are some examples.

Numerical Expression *Algebraic Expression*
 $15 + 9 \cdot 3$ $45 \div p - q$

To **evaluate** an algebraic expression, substitute a number for each variable. Then use the order of operations to find the value of the numerical expression.

EXAMPLE 1 Evaluating Algebraic Expressions

a. Evaluate $k + 10$ when $k = 25$.

$k + 10 = 25 + 10$ Substitute 25 for k.

$\qquad = 35$ Add 25 and 10.

b. Evaluate $4 \cdot n$ when $n = 12$.

$4 \cdot n = 4 \cdot 12$ Substitute 12 for n.

$\qquad = 48$ Multiply 4 and 12.

Study Tip
You can write the product of 4 and n in several ways.
$4 \cdot n$
$4n$
$4(n)$

On Your Own

1. Evaluate $24 + c$ when $c = 9$. **2.** Evaluate $d - 17$ when $d = 30$.

EXAMPLE 2 Evaluating an Expression with Two Variables

Evaluate $a \div b$ when $a = 27$ and $b = 3$.

Substitute 27 for a.
$a \div b = 27 \div 3$
Substitute 3 for b.

$\qquad = 9$ Divide 27 by 3.

On Your Own

Now You're Ready
Exercises 8–19

Evaluate the expression when $p = 24$ and $q = 8$.

3. $p \div q$ **4.** $q + p$ **5.** $p - q$ **6.** pq

4 Chapter 1 Expressions and Number Properties 🔊 Multi-Language Glossary at BigIdeasMath✓com.

Laurie's Notes

Introduction

Connect
- **Yesterday:** Students explored how to write and evaluate an expression written in words, focusing on units.
- **Today:** The lesson explains how to evaluate an algebraic expression containing a variable.

Motivate
- **Whole Class Activity:** Ask five students to stand at the front of the room. Give each student an index card with a number written on the card (possible numbers: 1, 3, 4, 5, 8). Write simple expressions on the board, drawing a box where the variable would be. Students should walk in front of the expression, place their index card in the box, and announce the value of the expression for their number. Examples:

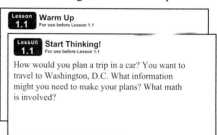

- You can vary the expressions and numbers to fit the level of your class. The goal is for students to recognize that the value changes (or varies) for each number substituted.
- **?** After several examples, have the students sit down. Erase the boxes and write variables to replace them. "What do the letters mean? Does anyone know what the expressions are called?" The variables have the same meaning as the boxes (an unknown value that represents a number). The expressions are called algebraic expressions.

Words of Wisdom
- Students can be intimidated by the phrases "numerical expression" and "algebraic expression." Reassure students that they have been working with each of these types of expressions since they began studying math. An early **numerical expression** might have been $1 + 2$. An early **algebraic expression** might have been ____ $+ 2$, where they picked a number to put in the blank. The concepts are the same but the language is simply more formal, preparing students for future courses.

Lesson Notes

Example 1
- Work through the written examples in part (a) and part (b) as a class. Use color-coding to help all students recognize that substitution has occurred.

Example 2
- Encourage students to use color-coding to differentiate between the two values to be substituted.

On Your Own
- Have students work together on the questions. They should share their answers orally or on the board. Check students' work.

Goal Today's lesson is evaluating algebraic expressions.

Lesson Materials
Introduction
• index cards

Start Thinking! and Warm Up

Lesson 1.1 Warm Up For use before Lesson 1.1

Lesson 1.1 Start Thinking! For use before Lesson 1.1
How would you plan a trip in a car? You want to travel to Washington, D.C. What information might you need to make your plans? What math is involved?

English Language Learners
Vocabulary
The word *variable* is often used in algebra and is represented by a letter. The letter stands for a number that changes, or *varies*. Students find it helpful when the letter is meaningful to the problem. For instance, use the letter t to represent a unit of time or use the letter d to represent distance.

Extra Example 1
a. Evaluate $15 - y$ when $y = 3$. 12
b. Evaluate $24 \div w$ when $w = 4$. 6

On Your Own
1. 33
2. 13

Extra Example 2
Evaluate $f \cdot g$ when $f = 7$ and $g = 11$. 77

On Your Own
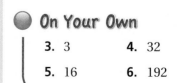
3. 3
4. 32
5. 16
6. 192

T-4

Extra Example 3

a. Evaluate $\frac{t}{5} + 6$ when $t = 45$. 15

b. Evaluate $v^2 - 4$ when $v = 9$. 77

On Your Own

7. 31 8. 26
9. 29 10. 37.5

Extra Example 4

Your uncle gives you $10 and you save $2 a week. The expression $10 + 2w$ gives the amount you save after w weeks.

a. How much money will you have saved after 2 weeks, 5 weeks, and 10 weeks? $14, $20, $30

b. After 10 weeks, do you have enough money to buy a video game that costs $25? Explain. Yes, after 10 weeks you have $30 and the video game costs $25.

On Your Own

11. Yes.
 Because $45 + 4(20) = 125$, you will be able to buy the skateboard.

Laurie's Notes

Example 3
- Students are asked to evaluate expressions involving more than one operation. They will need to remember the order of operations.
- Write the expression from the example and ask how many operations will be performed.
- ❓ "What operation should you do first?" Part a. multiplication Part b. exponent
- This question usually elicits a comment about the order of operations, at which time you can probe for understanding.

On Your Own
- Students should work the questions and then share their answers either orally, on individual white boards, or at the board. Check students' work looking for errors with the order of operations.

Example 4
- Work through the example. Model the strategy from yesterday's class. Money saved ($) = amount at the beginning ($) + $3 per week × # of weeks
- This example is good practice for students to set up a table to help to organize their work.

On Your Own
- ❓ "Describe what changes in the example to produce the expression $45 + 4w$." saving $4 per week instead of $3

Words of Wisdom
- If students are having difficulty, determine if the problem is a matter of computation, reading comprehension, order of operations, or some other difficulty.

Closure
- **Exit Ticket:** In Question 11, you found that by saving $4 per week for 20 weeks you would have enough to purchase a skateboard. Will you have enough after 40 weeks to purchase two skateboards? Justify your answer. No. Initially you had $45 at the beginning, and you were adding 4 dollars a week to it. Unless you had another $45 to add in, you would not be able to afford 2 skateboards.

Technology For the Teacher

The Dynamic Planning Tool
Editable Teacher's Resources at *BigIdeasMath.com*

EXAMPLE 3 — Evaluating Expressions with Two Operations

a. Evaluate $3x - 14$ when $x = 5$.

$3x - 14 = 3(5) - 14$ Substitute 5 for x.

$ = 15 - 14$ Using order of operations, multiply 3 and 5.

$ = 1$ Subtract 14 from 15.

b. Evaluate $z^2 + 8.5$ when $z = 2$.

$z^2 + 8.5 = (2)^2 + 8.5$ Substitute 2 for z.

$ = 4 + 8.5$ Using order of operations, evaluate 2^2.

$ = 12.5$ Add 4 and 8.5.

On Your Own

Exercises 25–33

Evaluate the expression when $y = 6$.

7. $5y + 1$ 8. $30 - 24 \div y$ 9. $y^2 - 7$ 10. $1.5 + y^2$

EXAMPLE 4 — Real-Life Application

You are saving for a skateboard. Your aunt gives you $45 to start and you save $3 each week. The expression $45 + 3w$ gives the amount of money you save after w weeks.

a. How much will you have after 4 weeks, 10 weeks, and 20 weeks?

b. After 20 weeks, can you buy the skateboard? Explain.

a. *Substitute the given number of weeks for w.*

Number of Weeks, w	45 + 3w	Amount Saved
4	45 + 3(4)	45 + 12 = $57
10	45 + 3(10)	45 + 30 = $75
20	45 + 3(20)	45 + 60 = $105

b. After 20 weeks, you have $105. So, you cannot buy the $125 skateboard.

On Your Own

11. **WHAT IF?** In Example 4, the expression for how much money you have after w weeks is $45 + 4w$. Can you buy the skateboard after 20 weeks? Explain.

1.1 Exercises

Vocabulary and Concept Check

1. **VOCABULARY** Copy and complete the table.

Algebraic Expression	Numbers	Variables	Operations	
$x - 8$	8	x	Subtraction	*Sample*
$3w + 9$				
$6y - 12$				

2. **NUMBER SENSE** Which step in the order of operations is first? second? third? fourth?

 Add and subtract from left to right. Multiply and divide from left to right.

 Evaluate powers. Evaluate expressions inside grouping symbols.

3. **NUMBER SENSE** Will the value of the expression $20 - x$ increase, *decrease*, or *stay the same* as x increases? Explain.

Practice and Problem Solving

Write and evaluate an expression for the problem.

4. You receive $8 for raking leaves for 2 hours. What is your hourly wage?

5. Music lessons cost $20 per week. How much do 6 weeks of lessons cost?

6. The scores on your first two history tests were 82 and 95. By how many points did you improve on your second test?

7. You buy a hat for $12 and give the cashier a $20 bill. How much change do you receive?

ALGEBRA Evaluate the expression when $a = 3$, $b = 2$, and $c = 12$.

8. $6 + a$ 9. $b \cdot 5$ 10. $c - 1$ 11. $27 \div a$

12. $12 - b$ 13. $c + 5$ 14. $2a$ 15. $c \div 6$

16. $a + b$ 17. $c - a$ 18. $\dfrac{c}{a}$ 19. $b \cdot c$

20. **ERROR ANALYSIS** Describe and correct the error in evaluating the expression when $m = 8$.

 $5m + 3 = 5 \cdot 8 + 3$
 $= 5 \cdot 11$
 $= 55$

21. **LAWNS** You earn $15n$ dollars for mowing n lawns. How much do you earn for mowing one lawn? seven lawns?

Assignment Guide and Homework Check

Level	Day 1 Activity Assignment	Day 2 Lesson Assignment	Homework Check
Basic	4–7, 37–41	1–3, 9–19 odd, 20–22, 23–29 odd	17, 20, 22, 27, 29
Average	4–7, 37–41	1–3, 13–19 odd, 20–34 even	17, 20, 22, 30, 34
Advanced	4–7, 37–41	1–3, 18–20, 26–32 even, 33–36	18, 20, 33, 35

Common Errors

- **Exercises 8–19** Students may substitute the wrong value for the variable. Tell students to write out the expression, and then write the value of the variable underneath the variable before substituting the value into the expression.
- **Exercises 25–33** Students may forget about the order of operations or they may substitute the wrong value for the variable. Review the order of operations. Before substituting, have students write or identify what operations are involved in the problem and which will be evaluated first.
- **Exercise 34** Students may focus on the variable and forget to double the given length. Ask students what they are finding (perimeter) and ask the students to think about how to find the perimeter of a rectangle.

Technology For the Teacher
Answer Presentation Tool
Quiz*Show*

1.1 Record and Practice Journal

Evaluate the expression when $a = 4$, $b = 5$, and $c = 10$.

1. $a + 7$ — 11
2. $b - 3$ — 2
3. $9c$ — 90
4. $25 + b$ — 5
5. $a \cdot c$ — 40
6. $b - a$ — 1
7. $a + b + c$ — 19
8. $\frac{c}{b}$ — 2
9. $4a - 7$ — 9
10. $b^2 + 2.5$ — 27.5
11. $\frac{c}{5} + 6a$ — 26
12. $bc - 13$ — 37

13. After m minutes, an adult blinks about $10m + 3$ times. About how many times does an adult blink in three minutes? in two hours?
 33; 1203

14. You need $2b$ cups of flour for making b loaves of bread. You have 8 cups of flour. Do you have enough for 5 loaves of bread? Explain.
 No. Five loaves of bread requires 10 cups of flour. You only have 8 cups of flour.

15. The expression $9a + 6s$ is the cost for a adults and s students to see a musical performance.
 a. Find the total cost for three adults and five students.
 $57
 b. The number of adults and students in a group both double. Does the cost double? Explain your answer using an example.
 Yes
 c. The number of students doubles, but the number of adults is cut in half. Is the cost the same? Explain your answer using an example.
 No

Vocabulary and Concept Check

1. See Additional Answers.
2. First: Evaluate expressions inside grouping symbols. Second: Evaluate powers. Third: Multiply and divide from left to right. Fourth: Add and subtract from left to right.
3. decrease; When you subtract greater and greater values from 20, you will have less and less left.

Practice and Problem Solving

4. $4 per hour
5. $120
6. 13 points
7. $8
8. 9
9. 10
10. 11
11. 9
12. 10
13. 17
14. 6
15. 2
16. 5
17. 9
18. 4
19. 24
20. Multiplication should be done first, then addition.
 $5m + 3 = 5 \cdot 8 + 3$
 $= 40 + 3$
 $= 43$
21. $15; $105
22. 34 mm; 118 mm
23.

x	3	6	9
x · 8	24	48	72

24.

x	2	4	8
64 ÷ x	32	16	8

25. 23
26. 8.2
27. 6
28. 11
29. 22
30. 82
31. 46
32. 48.3
33. 24

T-6

34. $2(w + 72)$

35. See *Taking Math Deeper*.

36. a. $392

b. yes; *Sample answer:* For 8 adults and 48 students, the cost is $20(8) + 13(48) = \$784$. Because $2(392) = 784$, the cost doubles when the number of adults and students in the group both double.

c. no; *Sample answer:* For 8 adults and 12 students, the cost is $20(8) + 13(12) = \$316$. The cost is not the same.

Fair Game Review

37–40. See Additional Answers.

41. C

Mini-Assessment

Evaluate the expression when $x = 2$, $y = 5$, and $z = 10$.

1. $5 + x$ 7
2. $3y$ 15
3. $z - y$ 5
4. $x \cdot y$ 10
5. Your friend earns $7x$ dollars for working x hours. How much does your friend earn for working 16 hours? 40 hours? $112; $280

Taking Math Deeper

Exercise 35

Remind students that they don't have to see how to start the solution before they put their pencil to paper.
- Doodle some notes.
- Draw a diagram.
- Jot down some facts.
- Rewrite the problem.
- Underline the important words.

These things kick students' brains into gear. Soon they see a way to the solution.

 How can an area be represented by $128 + s^2$?
128 could be a rectangle. Draw a couple.

 This one looks like a good deck.

 Here is a straightforward solution.
- You have the 128. How about the s^2?
- It could be the area of a square. Add this to your diagram.

 Here are more creative solutions.

Project

A builder needs to present four different decks to a customer using the information in Exercise 35. Design four different decks for him to use in his sales presentation. Explain how they meet the requirements of Exercise 35.

Reteaching and Enrichment Strategies

If students need help...	If students got it...
Resources by Chapter • Practice A and Practice B • Puzzle Time Record and Practice Journal Practice Differentiating the Lesson Lesson Tutorials Skills Review Handbook	Resources by Chapter • Enrichment and Extension Start the next section

22. FINGERNAIL After m months, the length of a fingernail is $10 + 3m$ millimeters. How long is the fingernail after eight months? three years?

Copy and complete the table.

23.
x	3	6	9
x · 8			

24.
x	2	4	8
64 ÷ x			

ALGEBRA Evaluate the expression when $a = 10$, $b = 9$, and $c = 4$.

25. $2a + 3$

26. $4c - 7.8$

27. $\dfrac{a}{2} + 1$

28. $\dfrac{27}{b} + 8$

29. $c^2 + 6$

30. $a^2 - 18$

31. $a + 9c$

32. $bc + 12.3$

33. $3a + 2b - 6c$

34. QUILT You are bordering a quilt with a blue ribbon. Which expression tells you how much ribbon you will need? Draw a diagram to help explain your reasoning.

$2(w + 72)$ $2w + 72$

35. DECK Your deck has an area of 128 square feet. After adding a section, the area will be $s^2 + 128$ square feet. Draw a diagram of how this can happen.

36. SCIENCE CENTER The expression $20a + 13c$ is the cost for a adults and c students to enter the Science Center.

 a. Find the total cost for 4 adults and 24 students.

 b. You find the cost for a group. Then the numbers of adults and students in the group both double. Does the cost double? Explain your answer using an example.

 c. **Reasoning** In part (a), the number of adults doubles, but the number of students is cut in half. Is the cost the same? Explain your answer using an example.

Fair Game Review What you learned in previous grades & lessons

Plot the points in a coordinate plane. *(Skills Review Handbook)*

37. (3, 2) 38. (2, 3) 39. (5, 1) 40. (1, 8)

41. MULTIPLE CHOICE Which numbers have a least common multiple of 24? *(Skills Review Handbook)*

 Ⓐ 4, 6 Ⓑ 2, 22 Ⓒ 3, 8 Ⓓ 6, 12

1.2 Writing Expressions

COMMON CORE STATE STANDARDS
6.EE.2a
6.EE.2b

Essential Question Which words correspond to the four operations of addition, subtraction, multiplication, and division?

1 ACTIVITY: Words That Imply Addition or Subtraction

Work with a partner.

a. Copy and complete the table.

Variable	Phrase	Expression
n	4 **more than** a number	$n + 4$ or $4 + n$
m	The **difference** of a number and 3	$m - 3$
x	The **sum** of a number and 8	
p	10 **less than** a number	
n	7 units **farther** away	$n + 7$ or $7 + n$
t	8 minutes **sooner**	
w	12 minutes **later**	
y	A number **increased** by 9	

Here is a word problem that uses one of the expressions in the table.

You arrive at the pizza shop 8 minutes sooner than your friend. Your friend arrives at 6:42 P.M. When did you arrive?

b. Which expression from the table can you use to solve the problem?

c. Write a problem that uses a different expression from the table.

2 ACTIVITY: Words That Imply Multiplication or Division

Work with a partner. Match each phrase with an expression.

The product of a number and 3	$n \div 3$
The quotient of 3 and a number	$4p$
4 times a number	$n \cdot 3$
A number divided by 3	$2m$
Twice a number	$3 \div n$

8 Chapter 1 Expressions and Number Properties

Laurie's Notes

Introduction

For the Teacher
- **Goal:** Students should be able to write algebraic expressions to represent phrases that include words corresponding to the operations of addition, subtraction, multiplication, and division.
- Students not working at the independent level may need support with general computation skills. English language learners may need support with vocabulary and identifying the meaning of words.

Motivate
- A homonym is one of two or more words that have the same sound and often the same spelling, but differ in meaning. Example: bank (embankment) and bank (where money is kept), and their, there, and they're.
- As a fun warm up, have groups of students make lists of homonyms (both with the same spelling and different spellings). When students are done, share each group's list with the class.

Discuss
- In math class you often hear words or phrases that suggest a particular operation. In fact, there are several words that suggest the same operation. Today's lesson is about translating phrases into algebraic expressions.

Activity Notes

Activity 1
- **Class Activity:** Introduce the activity and work through the first two phrases as a class. Students should now work with a partner to finish parts (a) through (c).
- **Common Error:** Subtraction is not commutative, and therefore students must be careful with this operation. Make a point of explaining that the difference of a number and 3, $m - 3$, is not the same as the difference of 3 and a number, $3 - m$.
- **Struggling Students:** Some students may find Activity 2 easier than Activity 1. You may want to start with Activity 2 or do Activity 1 together as a class.

Activity 2
- **Common Error:** Give students the same reminder about division as was given about subtraction. Explain that the quotient of a number and 3, $n \div 3$, is not the same as the quotient of 3 and a number, $3 \div n$.

Words of Wisdom
- Before moving to the third activity, review the key words or phrases that suggest each of the four operations. Students may suggest phrases different from those mentioned in the student book.

Common Core State Standards

6.EE.2a Write expressions that record operations with numbers and with letters standing for numbers.
6.EE.2b Identify parts of an expression using mathematical terms (sum, term, product, factor, quotient, coefficient); view one or more parts of an expression as a single entity.

Previous Learning
Students need to be familiar with variables and how they are used.

Activity Materials

Activity Materials
Closure
• index cards

Start Thinking! and Warm Up

1.2 Record and Practice Journal

English Language Learners
Organization
Create a table so that students can practice translating variable expressions. Add blank rows for students to add additional expressions.

Variable Expression	Meaning	Operation
$x + 14$	x plus 14	Addition
$y - 75$	y minus 75	Subtraction
$3 \div z, \dfrac{3}{z}$	3 divided by z	Division
$6n, 6 \cdot n, (6)(n)$	6 times n	Multiplication

Use the table at the beginning of the lesson to review the key words and phrases.

1.2 Record and Practice Journal

Laurie's Notes

Activity 3
- In this activity, students need to think about what the ID numbers for each ship might mean. For example, $PQ + 4T - 3$ contains the following expressions: $Q + 4$ (4 more than a number), $4T$ (4 times a number), and $T - 3$ (a number decreased by 3).
- When students record or state their answers, it may be easiest for them to refer to the ships by the color of the flag at the top of the ship.
- **Management Tip:** If you think your time is going to run short, omit part (b) of this activity.
- **Extension:** For students who finish early, have them determine if there is more than one gate the non-intruder ships could go through.

What Is Your Answer?
- Pay special attention to Questions 5(b) and 5(c). Check that the order is correct for Question 5(b): *the difference of 4 and a number*. Question 5(c) involves two operations, multiplication and addition. A student might say "1 more than the product of 3 and a number."
- ? "What are some additional phrases for each algebraic expression in Question 5?"

Closure

Classroom Activity: This is a nice summary activity that can be done with the whole class.
- Write words or phrases on index cards for each of the four operations so that each student can have an index card. Distribute the cards. Some of the phrases can be the same.
- Designate a corner of the room for each of the four operations. Have students move to the corner of the room associated with the operation suggested by the word or phrase on their index card.
- Each of the four clusters of students should verify that all of the group's index cards match the correct operation.
- Have each cluster of students read the words on their index cards aloud to the class.
- Collect the cards to use again.

The Dynamic Planning Tool
Editable Teacher's Resources at *BigIdeasMath.com*

3 ACTIVITY: Find the Intruder

Work with a partner.

A ship can pass through a gate only if the gate expression can be found in its ID number.

For instance, the ID number JK−5+NM contains the expression K − 5. So the ship is allowed through the gate marked "5 less than a number".

a. Which ships are allowed in the lake? Which are intruders?

b. Make up your own "Find the Intruder" game. Trade games with your partner. Solve the game and talk about your solution.

What Is Your Answer?

4. Which words correspond to each operation? In each box, write any words that imply the operation.

+	−	×	÷

5. Write a phrase for each expression.

 a. $n + 6$ b. $4 - x$ c. $3x + 1$ d. $2x$

Practice — Use what you learned about writing expressions to complete Exercises 9–12 on page 12.

Section 1.2 Writing Expressions 9

1.2 Lesson

Some words imply math operations.

Operation	Addition	Subtraction	Multiplication	Division
Key Words and Phrases	added to plus sum of more than increased by total of and	subtracted from minus difference of less than decreased by fewer than take away	multiplied by times product of twice of	divided by quotient of

EXAMPLE 1 Writing Numerical Expressions

Write the phrase as an expression.

a. 8 fewer than 21

 21 − 8 The phrase "fewer than" means subtraction.

b. The product of 30 and 9

 30 × 9, or 30 • 9 The phrase "product of" means multiplication.

EXAMPLE 2 Writing Algebraic Expressions

Write the phrase as an expression.

a. 14 more than a number x

 $x + 14$ The phrase "more than" means addition.

b. A number y minus 75

 $y - 75$ The phrase "minus" means subtraction.

c. The quotient of 3 and a number z

 $3 \div z$, or $\frac{3}{z}$ The phrase "quotient of" means division.

Common Error

When writing expressions involving subtraction or division, order is important. For example, the quotient of a number x and 2 means
$x \div 2$, not $2 \div x$.

On Your Own

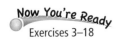
Now You're Ready
Exercises 3–18

Write the phrase as an expression.

1. The sum of 18 and 35
2. 6 times 50
3. 25 less than a number b
4. A number x divided by 4
5. The total of a number t and 11
6. 100 decreased by a number k

Laurie's Notes

Introduction

Connect
- **Yesterday:** Students explored how to write an algebraic expression that represented a verbal phrase.
- **Today:** Today's lesson presents the content more formally.

Motivate
- Write "price + $4" on the board and ask how this phrase might be expressed in words. For example: "the price increased by $4" or "$4 added to the price." Students should be able to think of several more interpretations.
- Below "price + $4," write "$p + 4$." Explain that you can use the variable p to represent the price, instead of writing it out.
- Additional examples:

Age $-$ 2	$2.50 \times number	Quarts \div 4
$a - 2$	$2.5n$	$q \div 4$ or $\frac{q}{4}$

- **Whole Class Activity:** Recall the sorting activity that closed the class yesterday. Make a chart on the board similar to the start of Lesson 1.2 and fill it in as a class. Students should *not* have their books open. Use the sorted index cards from yesterday. Ask for 4 volunteers to simply read any versions not already listed.

Lesson Notes

Example 1
- Note the use of color to connect the words and the operations. If possible, use different colors to write problems on the board or at the overhead projector.

Example 2
- Work through the example. Again note that subtraction and division are not commutative. For example in 2(c), note that the solution is *not* $z \div 3$.

Words of Wisdom
- No example is shown for multiplication; however, you may wish to include one in order to point out the conventional way in which multiplication with a variable is represented.
- ❓ "Are $n3$ and $3n$ equivalent? Why?" Multiplication is commutative, so they are equivalent.

On Your Own
- **Neighbor Check:** Have students work independently and then have their neighbor check their work. Have students discuss any discrepancies.

Goal
Today's lesson is writing mathematical expressions that correspond to given situations.

Start Thinking! and Warm Up

> **Lesson 1.2 Warm Up**
> For use before Lesson 1.2

> **Lesson 1.2 Start Thinking!**
> For use before Lesson 1.2
>
> What does it mean when you hear the sentence "To get to school, Susie walks 5 blocks less than Bill does."?
>
> Can the sentence be written mathematically?
>
> If you knew how many blocks Bill walked to school, how would you figure out how many blocks Susie walked?

Differentiated Instruction

Auditory

On the board, write a few expressions containing a variable and one operation. Have students practice reading them aloud in various ways. For example, $x - 4$ can be read as "x minus four," "the difference of x and four," or "4 less than x."

Extra Example 1

Write the phrase as an expression.
a. 34 divided by 2 $34 \div 2$
b. 35 increased by 43 $35 + 43$

Extra Example 2

Write the phrase as an expression.
a. the difference of a number t and 95 $t - 95$
b. twice a number w $2w$

On Your Own

1. $18 + 35$ or $35 + 18$
2. $6 \cdot 50$ or $50 \cdot 6$
3. $b - 25$ 4. $x \div 4$
5. $t + 11$ or $11 + t$
6. $100 - k$

Extra Example 3

The number of students in the science club is 4 more than twice the number of students in the math club. Let m be the number of students in the math club. Write an expression to find the number of students in the science club.
$2m + 4$

Extra Example 4

A bamboo shoot is 4 inches tall. During its growing season it grows 3 inches a day.

a. Make a table and graph that show the height of the bamboo for 4 days. Then write an expression for the height after d days. $4 + 3d$

Day, d	Height (inches)
0	4
1	$4 + 3(1) = 7$
2	$4 + 3(2) = 10$
3	$4 + 3(3) = 13$
4	$4 + 3(4) = 16$

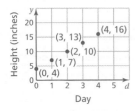

b. What is the height after 14 days?
46 inches

 On Your Own

7. $2t + 5$ **8.** 250 in.

Laurie's Notes

Example 3
- Because this expression involves more than one operation, it can be confusing to students. Order of operations must be considered when reading the problem.
- When reading the problem aloud, pause after the phrase "3 less than" to bring focus to the phrase "4 times the number from Alabama." 3 is subtracted from this product.
- **Common Error:** Choice C is a very common *wrong* answer. Ask students to think about what this expression would equal when $a = 1$. (Something negative is the common response!)

Example 4
- Remind students that an ordered pair is a pair of numbers that describes the location of a point in the coordinate plane.
- Say, "We want to organize some information about this problem, so let's set up a table to keep track of the height each year."
- Make a table with two columns (year, height). It is not natural to start with the year 0, so be sure to discuss this. Each year students should add 15 to the last number.
- "Can you find the height after 9 years without computing all of the years between 4 and 9?" Write an expression.
- Work with students to develop the general expression $10 + 15t$ and verify that it indeed works for years 0 through 4.

On Your Own
- Note that some students might write $5 + 2t$ and others might write $2t + 5$. Addition is commutative, so this is fine. You may want to caution students that subtraction is not commutative, so there is only one way to correctly write "five less than twice as many tokens."

Words of Wisdom
- English Language Learners will likely need additional vocabulary support.

Closure
- **Matching:**

 $3x - 2$ A. The difference of 2 and 3 times a number
 $2 - 3x$ B. Three less than twice a number
 $2x - 3$ C. The difference of three times a number and 2

 Answers in order: C, A, B

The Dynamic Planning Tool
Editable Teacher's Resources at *BigIdeasMath.com*

EXAMPLE 3 — Standardized Test Practice

In 2009, the number of members from Florida in the House of Representatives was 3 less than 4 times the number from Alabama. Let a be the number of members from Alabama. Which expression could be used to find the number of members from Florida?

Ⓐ $4a - 3$ **Ⓑ** $4a + 3$ **Ⓒ** $3 - 4a$ **Ⓓ** $4 - 3a$

> The phrase "times" means multiplication. So, multiply 4 and a.

> The phrase "less than" means subtraction. So, subtract 3 from $4a$.

$$4a - 3$$

∴ The correct answer is **Ⓐ**.

EXAMPLE 4 — Real-Life Application

You plant a cypress tree that is 10 inches tall. Each year, its height increases by 15 inches.

a. Make a table and graph that show the height of the tree for 4 years. Then write an expression for the height after t years.

b. What is the height after 9 years?

a. The height is *increasing*, so *add* $15t$ as shown in the table. From the table, you can then graph the ordered pairs.

Year, t	Height (inches)
0	10
1	$10 + 15(1) = 25$
2	$10 + 15(2) = 40$
3	$10 + 15(3) = 55$
4	$10 + 15(4) = 70$

> When t is 0, the height is 10 inches.

> You can see that an expression is $10 + 15t$.

Cypress Tree Height

∴ So, the height after year t is $10 + 15t$.

b. Evaluate $10 + 15t$ when $t = 9$.

$$10 + 15t = 10 + 15(9) = 145$$

∴ After 9 years, the height of the tree is 145 inches.

On Your Own

7. Your friend has 5 more than twice as many game tokens as your sister. Let t be the number of game tokens your sister has. Write an expression for the number of game tokens your friend has.

8. WHAT IF? In Example 4, what is the height of the cypress tree after 16 years?

1.2 Exercises

Vocabulary and Concept Check

1. **DIFFERENT WORDS, SAME QUESTION** Which is different? Write "both" expressions.

 | 12 more than x | x increased by 12 | x take away 12 | the sum of x and 12 |

2. **OPEN-ENDED** Write a real-world problem represented by the expression $x \div 3$.

Practice and Problem Solving

Write the phrase as an expression.

3. 5 less than 8
4. the product of 3 and 12
5. 28 divided by 7
6. the total of 6 and 10
7. 3 fewer than 18
8. 17 added to 15
9. 13 subtracted from a number x
10. 5 times a number d
11. the quotient of 18 and a number a
12. the difference of a number s and 6
13. 7 increased by a number w
14. a number b multiplied by 9
15. the sum of a number y and 4
16. the difference of 12 and a number x
17. twice a number z
18. a number t take away 8

ERROR ANALYSIS Describe and correct the error in writing the phrase as an expression.

19. the quotient of 8 and a number y

20. 16 decreased by a number x

21. **DINNER** Five friends share the cost of a dinner equally.

 a. Write an expression for the cost per person.

 b. Make up a total cost and test your expression. Is the result reasonable?

22. **TV SHOW** A television show has 19 episodes per season.

 a. Copy and complete the table.
 b. Write an expression for the number of episodes in n seasons.

Seasons	1	2	3	4	5
Episodes					

Give two ways to write the expression as a phrase.

23. $n + 6$
24. $4w$
25. $15 - b$
26. $14 - 3z$

12 Chapter 1 Expressions and Number Properties

Assignment Guide and Homework Check

Level	Day 1 Activity Assignment	Day 2 Lesson Assignment	Homework Check
Basic	9–12, 36–40	1, 2, 3, 5, 7, 13–21 odd, 22, 27, 28	13, 17, 22, 27
Average	9–12, 36–40	1, 2, 13–19 odd, 20–26 even, 27–33 odd	17, 26, 27, 33
Advanced	9–12, 36–40	1, 2, 19, 20, 24–32 even, 33–35	26, 30, 33, 34

For Your Information
- **Exercise 1** *Different Words, Same Question* is a new type of exercise. Three of the four choices pose the same question using different words. The remaining choice poses a different question. So there are two answers.

Common Errors
- **Exercises 3–18** Students may write the subtraction and/or division problems in the wrong order. For example, they may write "5 less than 8" as $5 - 8$ instead of $8 - 5$.
- **Exercises 3–18** For subtraction problems, have students look for the key words "from" and "than" to know that the order of the numbers must be switched from the way it is written. For example, "9 subtracted *from* 14" means $14 - 9$; "9 fewer *than* 14" means $14 - 9$.
- **Exercises 27–30** Students may write the problem in the wrong order such as $\frac{y-3}{4}$ instead of $\frac{y}{4} - 3$. Tell students to write out the phrase and put parentheses around the different parts of the expression. Then evaluate the expression using order of operations. For example, "3 less than (the quotient of a number *y* and 4)."

1.2 Record and Practice Journal

Technology For the Teacher
Answer Presentation Tool
QuizShow

Vocabulary and Concept Check

1. x take away 12; $x - 12$; $x + 12$
2. *Sample answer:* A pizza is divided among three friends. How many slices will each person receive?

Practice and Problem Solving

3. $8 - 5$
4. $3 \cdot 12$ or $12 \cdot 3$
5. $28 \div 7$
6. $6 + 10$ or $10 + 6$
7. $18 - 3$
8. $15 + 17$ or $17 + 15$
9. $x - 13$
10. $5 \cdot d$ or $d \cdot 5$
11. $18 \div a$ 12. $s - 6$
13. $7 + w$ or $w + 7$
14. $b \cdot 9$ or $9 \cdot b$
15. $y + 4$ or $4 + y$
16. $12 - x$
17. $2 \cdot z$ or $z \cdot 2$
18. $t - 8$
19. The expression is not written in the correct order; $\frac{8}{y}$
20. The expression is not written in the correct order; $16 - x$
21. a. $x \div 5$
 b. *Sample answer:* If the total cost is $30, then the cost per person is $x \div 5 = 30 \div 5 = \$6$. The result is reasonable.
22. a.

Seasons	1	2	3	4	5
Episodes	19	38	57	76	95

b. $19n$

T-12

Practice and Problem Solving

23–26. Sample answers are given.

23. The sum of n and 6; 6 more than a number n

24. 4 times a number w; The product of 4 and a number w

25. A number b less than 15; 15 take away a number b

26. 3 times a number z less than 14; 14 minus 3 times a number z

27. $\frac{y}{4} - 3$; 2 28. $\frac{x+4}{3}$; 3

29. $8x + 6$; 46 30. $\frac{40}{y-16}$; 10

31–32. See Additional Answers.

33. See *Taking Math Deeper*.

34. $a + 8$

35. $\frac{x}{4}$

Fair Game Review

36. 45 37. 59
38. 390 39. 140
40. D

Mini-Assessment

Write the phrase as an expression.

1. the product of 5 and 7 $5 \cdot 7$
2. the sum of a number b and 3 $b + 3$
3. the quotient of 9 and a number y $\frac{9}{y}$
4. 8 less than a number x $x - 8$
5. It costs $25 per day to rent a surfboard and $15 to rent a wetsuit. Write an expression for the cost of renting the surfboard for y days.
 $25y + 15$

T-13

Taking Math Deeper

Exercise 33

Tables are wonderful ways to organize information. This problem doesn't ask students to make a table. That is something they can do *just because it helps*.

 Make a table. Find a pattern.

Round	0	1	2	3	4	5
Singers	140	125	110	95	80	65

15 fewer for each round

 a. Use the pattern to write an expression.

Number of Singers in Round n = $140 - 15n$

Test the expression.
When $n = 1$, $140 - 15(1) = 125$. ✓
When $n = 2$, $140 - 15(2) = 110$. ✓

 b. Use the expression to evaluate.
When $n = 8$, $140 - 15(8) = 20$.

There are 20 people left after round 8. So, there are 20 people in round 9.

Check by extending the table.

Round	5	6	7	8
Singers	65	50	35	20

Subtract 15 for each round.

Solve it two ways.

Project

Investigate and write a report on singing competitions for middle school choruses.

Reteaching and Enrichment Strategies

If students need help...	If students got it...
Resources by Chapter • Practice A and Practice B • Puzzle Time Record and Practice Journal Practice Differentiating the Lesson Lesson Tutorials Skills Review Handbook	Resources by Chapter • Enrichment and Extension Start the next section

Write the phrase as an expression. Then evaluate when $x = 5$ and $y = 20$.

27. 3 less than the quotient of a number y and 4

28. the sum of a number x and 4, all divided by 3

29. 6 more than the product of 8 and a number x

30. the quotient of 40 and the difference of a number y and 16

31. BOWLING It costs $3 to bowl a game and $2 for shoe rental.

 a. Make a table for the cost of up to 5 games.

 b. Draw a graph of the data in your table.

 c. Write an expression for the cost of g games.

 d. Use your expression to find the cost of 8 games.

32. PUZZLE Florida has 8 less than 5 times the number of counties in Arizona. Georgia has 25 more than twice the number of counties in Florida.

 a. Write an expression for the number of counties in Florida.

 b. Write an expression for the number of counties in Georgia.

 c. Arizona has 15 counties. How many do Florida and Georgia have?

33. COMPETITION There are 140 people in a singing competition. The graph shows the results for the first five rounds.

 a. Write an expression for the number of people after each round.

 b. How many people are in the ninth round? Explain your reasoning.

34. STRUCTURE The difference between two numbers is 8. The lesser number is a. Write an expression for the greater number.

35. *Reasoning* One number is four times another. The greater number is x. Write an expression for the lesser number.

 What you learned in previous grades & lessons

Evaluate the expression. *(Skills Review Handbook)*

36. $8 + (22 + 15)$ **37.** $(13 + 9) + 37$ **38.** $(13 \times 6) \times 5$ **39.** $4 \times (7 \times 5)$

40. MULTIPLE CHOICE A grocery store is making fruit baskets using 144 apples, 108 oranges, and 90 pears. Each basket will be identical. What is the greatest number of fruit baskets the store can make using all the fruit? *(Skills Review Handbook)*

 Ⓐ 6 Ⓑ 9 Ⓒ 16 Ⓓ 18

1.3 Properties of Addition and Multiplication

COMMON CORE STATE STANDARDS

6.EE.3
6.EE.6

Essential Question Does the order in which you perform an operation matter?

1 ACTIVITY: Does Order Matter?

Work with a partner. Place each statement in the correct oval.

a. Fasten 5 shirt buttons.
b. Put on a shirt and tie.
c. Fill and seal an envelope.
d. Floss your teeth.
e. Put on your shoes.
f. Chew and swallow.

Order Matters **Order Doesn't Matter**

Think of some math problems using the four operations where order matters and some where order doesn't matter.

The Meaning of a Word • Commute

When you **commute** the positions of two stuffed animals on a shelf, you switch their positions.

2 ACTIVITY: Commutative Properties

Work with a partner.

a. Which of the following are true?

$3 + 5 \stackrel{?}{=} 5 + 3$ $3 - 5 \stackrel{?}{=} 5 - 3$

$9 \times 3 \stackrel{?}{=} 3 \times 9$ $9 \div 3 \stackrel{?}{=} 3 \div 9$

b. The true equations show the Commutative Properties of Addition and Multiplication. Why are they called "commutative?" Write the properties.

14 Chapter 1 Expressions and Number Properties

Laurie's Notes

Introduction

For the Teacher
- **Goal:** The goal of this activity is for students to gain an understanding of what is meant by *order* and *grouping* as they apply to number operations.
- Students should understand the idea of *order* as it relates to performing everyday tasks, such as putting on shoes and socks (order matters) versus brushing teeth and hair (order doesn't matter).

Motivate
- **Act It Out:** If there is a bit of an actor in you, start the class by pretending that you're brushing your teeth (about 10 seconds) and then take the tube of toothpaste out and apply some to the toothbrush. This will clearly evoke a few comments about the order in which you performed the two tasks.
- ? "Oh, does the order matter? Hmmm…"
- If brushing your teeth doesn't work for you, wear a sock over a shoe, or push the number on the intercom phone and then pick up the receiver (anything obviously in the wrong order). Do something that students will remember. Catch their attention!

Discuss
? "What are some common things we do where order matters?"

Activity Notes

Activity 1
- Introduce the activity and have students work with a partner to complete it. They'll have fun thinking of a few situations on their own where order matters. Allow some sharing of ideas, and then steer the conversation to math. Students have seen the Commutative Property before, so they may know which operations are commutative and which are not.

Meaning of the Word
- Discuss the meaning of the word *commute*. An example is shown. Talk about *commuters* who go to work and back home, switching the direction.

Activity 2
- This activity is *not* asking students to evaluate each side of the equation, only which statements are true.
- Point out the question mark above the equal sign. Explain that it is used when you are checking to see that the two sides are equal—when you are *questioning* whether the two sides are equal.
- Students should do this activity with a partner.
- Ask for volunteers to share their answers with the class.
- ? "Can anyone write the property using variables (like *a* and *b*)?"

Common Core State Standards

6.EE.3 Apply the properties of operations to generate equivalent expressions.
6.EE.6 Use variables to represent numbers and write expressions when solving a real-world or mathematical problem; understand that a variable can represent an unknown number, or, depending on the purpose at hand, any number in a specified set.

Previous Learning

Students need to be familiar with basic computation facts. Applying mental math strategies will be helpful.

Start Thinking! and Warm Up

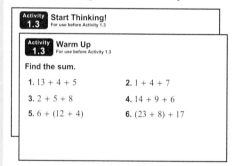

1.3 Record and Practice Journal

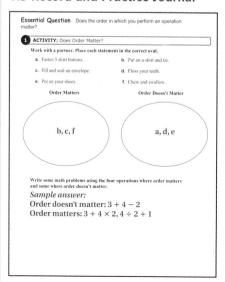

T-14

English Language Learners
Vocabulary
Write the word *association* on the board. Emphasize the meaning as *a group of people*. Demonstrate that the Associative Properties of Addition and Multiplication state that you can *group numbers so they are associated in different ways* and the answer will be the same.

1.3 Record and Practice Journal

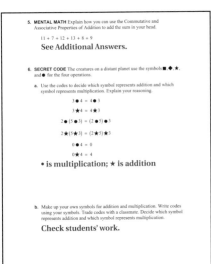

Laurie's Notes

The Meaning of a Word
- Discuss the meaning of the word *associate*, putting two things together. You *associate* dogs with barking, Arizona with the Grand Canyon, and sports cars with speed.
- For each of these examples, it is possible to *associate* a third word, like Arizona, Grand Canyon, and national parks. Numbers are different because you can only operate on two numbers at a time, even though there might be three numbers. So, you use parentheses to group two numbers at a time.
- Remind students that grouping symbols, such as parentheses, tell us to perform the operations inside them first.

Activity 3
- This activity is asking students to decide if the statement is true. In other words, does each side of the equation have the same value?
- Discuss class answers to part (a). Ask for a volunteer to share the answer to part (b). Again, encourage the use of variables.

What Is Your Answer?
- For Question 5, students should recognize the pairs that sum to 20, although some students might think there are easier combinations.
- For Question 6, encourage students to share their reasoning. The last two problems are key. Have students share their own codes, if time permits.

Closure
- **Share This Story:** Years ago it was important for shopkeepers to add quickly and efficiently in their heads. Or, if there were many items being purchased, they would write the prices down on the brown paper used to wrap the items (or a brown paper bag) and quickly add the prices together. There were no adding machines or cash registers. Here is a list of items being purchased.
- ❓ "How would you add them if you weren't allowed to use a calculator, adding machine, or cash register?"

 $4.40
 $1.22
 $3.30
 $1.08
 $2.30

The Dynamic Planning Tool
Editable Teacher's Resources at *BigIdeasMath.com*

The Meaning of a Word • Associate

You have two best friends. Sometimes you **associate** with one of them.

And sometimes you **associate** with the other.

③ ACTIVITY: Associative Properties

Work with a partner.

a. Which of the following are true?

$8 + (3 + 1) \stackrel{?}{=} (8 + 3) + 1$ \qquad $8 - (3 - 1) \stackrel{?}{=} (8 - 3) - 1$

$12 \times (6 \times 2) \stackrel{?}{=} (12 \times 6) \times 2$ \qquad $12 \div (6 \div 2) \stackrel{?}{=} (12 \div 6) \div 2$

b. The true equations show the Associative Properties of Addition and Multiplication. Why are they called "associative?" Write the properties.

What Is Your Answer?

4. **IN YOUR OWN WORDS** Does the order in which you perform an operation matter?

5. **MENTAL MATH** Explain how you can use the Commutative and Associative Properties of Addition to add the sum in your head.

$$11 + 7 + 12 + 13 + 8 + 9$$

6. **SECRET CODE** The creatures on a distant planet use the symbols ■, ◆, ★, and ● for the four operations.

a. Use the codes to decide which symbol represents addition and which symbol represents multiplication. Explain your reasoning.

$3 ● 4 = 4 ● 3$

$3 ★ 4 = 4 ★ 3$

$2 ● (5 ● 3) = (2 ● 5) ● 3$

$2 ★ (5 ★ 3) = (2 ★ 5) ★ 3$

$0 ● 4 = 0$

$0 ★ 4 = 4$

b. Make up your own symbols for addition and multiplication. Write codes using your symbols. Trade codes with a classmate. Decide which symbol represents addition and which symbol represents multiplication.

Practice

Use what you learned about the properties of addition and multiplication to complete Exercises 5–8 on page 18.

1.3 Lesson

Key Vocabulary
equivalent
 expressions, *p. 16*

Expressions with the same value, like 12 + 7 and 7 + 12, are **equivalent expressions**. The commutative and associative properties can be used to write equivalent expressions.

 Key Ideas

Commutative Properties

Words Changing the order of addends or factors does not change the sum or product.

Numbers $5 + 8 = 8 + 5$ **Algebra** $a + b = b + a$
$5 \cdot 8 = 8 \cdot 5$ $a \cdot b = b \cdot a$

Associative Properties

Words Changing the grouping of addends or factors does not change the sum or product.

Numbers $(7 + 4) + 2 = 7 + (4 + 2)$
$(7 \cdot 4) \cdot 2 = 7 \cdot (4 \cdot 2)$

Algebra $(a + b) + c = a + (b + c)$
$(a \cdot b) \cdot c = a \cdot (b \cdot c)$

EXAMPLE 1 **Using Properties to Simplify Expressions**

a. Simplify the expression $7 + (12 + x)$.

$7 + (12 + x) = (7 + 12) + x$ Associative Property of Addition
$= 19 + x$ Add 7 and 12.

b. Simplify the expression $(6.1 + x) + 8.4$.

$(6.1 + x) + 8.4 = (x + 6.1) + 8.4$ Commutative Property of Addition
$= x + (6.1 + 8.4)$ Associative Property of Addition
$= x + 14.5$ Add 6.1 and 8.4.

c. Simplify the expression $5(11y)$.

$5(11y) = (5 \cdot 11)y$ Associative Property of Multiplication
$= 55y$ Multiply 5 and 11.

On Your Own

Now You're Ready
Exercises 5–8

Simplify the expression. Explain each step.

1. $10 + (a + 9)$ 2. $(c + 25.3) + 17.9$ 3. $5(4n)$

16 Chapter 1 Expressions and Number Properties Multi-Language Glossary at BigIdeasMath✓com.

Laurie's Notes

Introduction

Connect
- **Yesterday:** Students developed an understanding of the Commutative (order) and Associative (grouping) Properties.
- **Today:** Students will use those two properties, and two other properties, to show that expressions are equivalent.

Discuss
- Begin with a discussion of what the word *equivalent* means. Examples: a dozen and 12 items, cell phone and mobile phone, etc.
- In mathematics you have equivalent fractions $\left(\text{like } \frac{1}{2} \text{ and } \frac{2}{4}\right)$ and equivalent measures (like 12 inches and 1 foot).
- Define **equivalent expressions.** Relate this to the *equivalent* things previously discussed.

Lesson Notes

Key Ideas
- Write the Commutative and Associative Properties. The key words to remember are *order* for the Commutative Property and *grouping* for the Associative Property.
- For the Commutative Property, multiplication can be represented in different ways: $ab = ba$ or $a \cdot b = b \cdot a$. This is also true for the Associative Property: $(ab)c = a(bc)$ versus $(a \cdot b) \cdot c = a \cdot (b \cdot c)$.

Example 1
- Work through the three parts of the example. The challenges students have with these problems are that the solutions are obvious! Students will say that it seems okay to simply combine the numbers and then tag on the variable. Remind students that it is okay to *just combine numbers* for only the operations of addition and multiplication. If the first problem was $7 - (12 - x)$, combining the numbers would result in the wrong answer.
- **Connection:** Because part (b) involves decimals, it is a good place to check on previous skills. What if the problem had been $(6 + x) + 8.4$? Ask students how they would simplify the problem. You are trying to determine if students remember that when you add a whole number and a decimal, you must locate the decimal point to the right of the whole number and line up the decimal points. Check for understanding of adding like place values.

On Your Own
- **Neighbor Check:** Have students work independently and then have their neighbor check their work. Have students discuss any discrepancies.
- Note how Question 3 is represented without the multiplication dot.

Goal
Today's lesson is using properties to show that two expressions are equivalent.

Start Thinking! and Warm Up

Differentiated Instruction

Visual
Write numbers, addition operators, and multiplication operators on index cards. Model expressions and let students rearrange the cards to model the Commutative and Associative Properties. Create expressions so students see the benefits of using the properties.

Sum to multiples of 10:
$$17 + 6 + 3 + 4 = (17 + 3) + (6 + 4)$$

Multiply to get a product of 100:
$$25 \times 15 \times 4 = (25 \times 4) \times 15$$
$$= 100 \times 15$$
$$= 1500$$

Extra Example 1
a. Simplify the expression $(r + 3) + 6$.
 $r + 9$
b. Simplify the expression $4(3p)$.
 $12p$
c. Simplify the expression $(6 \cdot b) \cdot 7$.
 $42b$

On Your Own

1. $10 + (a + 9) =$
 $10 + (9 + a)$
 Comm. Prop. of Add.
 $= (10 + 9) + a$
 Assoc. Prop. of Add.
 $= 19 + a$
 Add 10 and 9.

On Your Own

2. $(c + 25.3) + 17.9 =$
 $c + (25.3 + 17.9)$
 Assoc. Prop. of Add.
 $= c + 43.2$
 Add 25.3 and 17.9.

3. $5(4n) = (5 \cdot 4)n$
 Assoc. Prop. of Mult.
 $= 20n$
 Multiply 5 and 4.

Extra Example 2

a. Simplify the expression $0 + (s + 4)$.
 $s + 4$

b. Simplify the expression $(h \cdot 1) \cdot 8$.
 $8h$

Extra Example 3

You hand out 425 programs on the first night of your school's variety show, p programs on the second night, and 520 programs on the third night. Write an expression for the total number of programs you handed out. Then simplify the expression. $p + 945$

On Your Own

4. $12 \cdot b \cdot 0 = 12 \cdot (b \cdot 0)$
 Assoc. Prop. of Mult.
 $= 12 \cdot 0$
 Mult. Prop. of Zero
 $= 0$
 Mult. Prop. of Zero

5. $1 \cdot m \cdot 24 = (1 \cdot m) \cdot 24$
 Assoc. Prop. of Mult.
 $= m \cdot 24$
 Mult. Prop. of One
 $= 24m$
 Comm. Prop. of Mult.

6. $(t + 15) + 0 = t + (15 + 0)$
 Assoc. Prop. of Add.
 $= t + 15$
 Add. Prop. of Zero

7. $7x + 154.75$

Laurie's Notes

Key Idea

- Write and discuss the three new properties. Students may say (or at least think) that these properties are obvious.
- Adding zero, in any form, to a quantity does not change the value of the quantity. Remind students that zero can be represented in different ways: 0, (4 − 4), or (−7) + 7. This is also called the Additive Identity.
- Multiplying by zero, in any form, produces a product of 0.
- Multiplying by 1, in any form, does not change the value of the quantity.
 Remind students that 1 can be represented in different ways: 1, $\left(\frac{1}{2} + \frac{1}{2}\right)$, or $\frac{3}{3}$. This is also called the Multiplicative Identity.

Example 2

- Work through the two parts and point out the properties being used. Mention to students that the first step in both parts involves the Associative Property. In part (a) you would get the same result if $(0 \cdot p)$ had been grouped together first.

Example 3

- Ask a student to read the problem aloud. Ask how many T-shirts were purchased (note the common error in the margin next to the example). Explain that the total amount spent on T-shirts was 7 times the cost of one shirt.
- **Connection:** Here is an example of a whole number being added to a decimal. Did your students compute correctly?
- **? Extension:** "What if the sponsor bought 3 extra T-shirts, for a total of 10, and the shirts were $12.45 each? What would be the cost of the T-shirts?" $124.50
- This is a quick review of how to multiply by 10.

On Your Own

- **Think-Pair-Share:** Students should read each question independently and then work with a partner to answer the questions. When they have answered the questions, the pair should compare their answers with another group and discuss any discrepancies.

Closure

- On the board in your class, ask student volunteers to share examples of:
 Commutative Property using a variable and a number
 Commutative Property using two variables
 Associative Property using 3 numbers

The Dynamic Planning Tool
Editable Teacher's Resources at *BigIdeasMath.com*

 Key Ideas

Addition Property of Zero

Words The sum of any number and 0 is that number.

Numbers $7 + 0 = 7$ **Algebra** $a + 0 = a$

Multiplication Properties of Zero and One

Words The product of any number and 0 is 0.

The product of any number and 1 is that number.

Numbers $9 \times 0 = 0$ **Algebra** $a \cdot 0 = 0$
$4 \times 1 = 4$ $a \cdot 1 = a$

EXAMPLE 2 **Using Properties to Simplify Expressions**

a. Simplify the expression $9 \cdot 0 \cdot p$.

$9 \cdot 0 \cdot p = (9 \cdot 0) \cdot p$ Associative Property of Multiplication

$= 0 \cdot p = 0$ Multiplication Property of Zero

b. Simplify the expression $4.5 \cdot r \cdot 1$.

$4.5 \cdot r \cdot 1 = 4.5 \cdot (r \cdot 1)$ Associative Property of Multiplication

$= 4.5 \cdot r$ Multiplication Property of One

$= 4.5r$

EXAMPLE 3 **Real-Life Application**

You and six friends play on a basketball team. A sponsor paid $100 for the league fee, x dollars for each player's T-shirt, and $68.25 for trophies. Write an expression for the total amount paid by the sponsor.

Add the entry fee, the cost of the T-shirts, and the cost of the trophies.

$100 + 7x + 68.25 = 7x + 100 + 68.25$ Commutative Property of Addition

$= 7x + 168.25$ Add 100 and 68.25.

∴ An expression for the total amount is $7x + 168.25$.

Common Error

You **and** six friends are on the team, so use the expression $7x$, not $6x$, to represent the cost of the T-shirts.

On Your Own

Exercises 9–23

Simplify the expression. Explain each step.

4. $12 \cdot b \cdot 0$ 5. $1 \cdot m \cdot 24$ 6. $(t + 15) + 0$

7. **WHAT IF?** In Example 3, your sponsor paid $54.75 for trophies. Write an expression for the total amount paid by the sponsor.

1.3 Exercises

Check It Out
Help with Homework
BigIdeasMath.com

Vocabulary and Concept Check

1. **NUMBER SENSE** Write an example of a sum of fractions. Show that the Commutative Property of Addition is true for the sum.

2. **OPEN-ENDED** Write an algebraic expression that can be simplified using the Associative Property of Addition.

3. **OPEN-ENDED** Write an algebraic expression that can be simplified using the Associative Property of Multiplication and the Multiplication Property of One.

4. **WHICH ONE DOESN'T BELONG?** Which statement does *not* belong with the other three? Explain your reasoning.

 $7 + (x + 4) = 7 + (4 + x)$ $(3 + b) + 2 = (b + 3) + 2$

 $9 + (7 + w) = (9 + 7) + w$ $(4 + n) + 6 = (n + 4) + 6$

Practice and Problem Solving

Tell which property is illustrated by the statement.

5. $5 \cdot p = p \cdot 5$
6. $2 + (12 + r) = (2 + 12) + r$
7. $4 \cdot (x \cdot 10) = (4 \cdot x) \cdot 10$
8. $x + 7.5 = 7.5 + x$
9. $(c + 2) + 0 = c + 2$
10. $a \cdot 1 = a$

11. **ERROR ANALYSIS** Describe and correct the error in stating the property illustrated by the statement.

$(7 + x) + 3 = (x + 7) + 3$
Associative Property of Addition

Simplify the expression. Explain each step.

12. $6 + (5 + x)$
13. $(14 + y) + 3$
14. $6(2b)$
15. $7(9w)$
16. $3.2 + (x + 5.1)$
17. $(0 + a) + 8$
18. $9 \cdot c \cdot 4$
19. $(18.6 \cdot d) \cdot 1$
20. $(3k + 4.2) + 8.6$
21. $(2.4 + 4n) + 9$
22. $(3s) \cdot 8$
23. $z \cdot 0 \cdot 12$

24. **GEOMETRY** The expression $12 + x + 4$ represents the perimeter of a triangle. Simplify the expression.

25. **SCOUT COOKIES** A case of Scout cookies has 10 cartons. A carton has 12 boxes. The amount you make on a whole case is $10(12x)$ dollars.

 a. What does x represent?
 b. Simplify the expression.

18 Chapter 1 Expressions and Number Properties

Assignment Guide and Homework Check

Level	Day 1 Activity Assignment	Day 2 Lesson Assignment	Homework Check
Basic	5–8, 35–43	1–4, 5–23 odd, 24, 25, 27	7, 17, 21, 24, 27
Average	5–8, 35–43	1–4, 11–23 odd, 24–32 even	17, 21, 24, 26, 30
Advanced	5–8, 35–43	1–4, 11, 12–22 even, 28–34 even	16, 22, 32, 34

Common Errors

- **Exercises 12–23** Students are often confused by the differences between the Commutative and Associative Properties and will incorrectly label steps. Show students that the Associative Property moves the parentheses but does not change the position of the numbers. The Commutative Property changes the positions of the numbers.
- **Exercises 29–33** Some students may still be struggling with the differences between the two properties. Ask students to think about how they would complete the statement without identifying a property. Once they know how to simplify it, ask them to look at the definitions of the properties and identify which property they used.
- **Exercise 34** Students may think that there is only one correct expression. Ask students to identify which property leads to the other expression.

1.3 Record and Practice Journal

Technology For the Teacher
Answer Presentation Tool
QuizShow

 Vocabulary and Concept Check

1. *Sample answer:*
$$\frac{1}{5} + \frac{3}{5} = \frac{3}{5} + \frac{1}{5}$$
$$\frac{4}{5} = \frac{4}{5}$$

2. *Sample answer:*
$(x + 8) + 4 = x + (8 + 4)$
$ = x + 12$

3. *Sample answer:*
$(5 \cdot x) \cdot 1 = 5 \cdot (x \cdot 1)$
$ = 5x$

4. $9 + (7 + w) = (9 + 7) + w$ does not belong because it demonstrates the Associative Property of Addition whereas the other statements demonstrate the Commutative Property of Addition.

 Practice and Problem Solving

5. Comm. Prop. of Mult.
6. Assoc. Prop. of Add.
7. Assoc. Prop. of Mult.
8. Comm. Prop. of Add.
9. Add. Prop. of Zero
10. Mult. Prop. of One
11. The grouping of the numbers did not change. The statement illustrates the Commutative Property of Addition because the order of the addends changed.
12. $6 + (5 + x) = (6 + 5) + x$
Assoc. Prop. of Add.
$= 11 + x$
Add 6 and 5.
13. $(14 + y) + 3 = (y + 14) + 3$
Comm. Prop. of Add.
$= y + (14 + 3)$
Assoc. Prop. of Add.
$= y + 17$
Add 14 and 3.

T-18

Practice and Problem Solving

14. $6(2b) = (6 \cdot 2)b$
Assoc. Prop. of Mult.
$= 12b$
Multiply 6 and 2.

15. $7(9w) = (7 \cdot 9)w$
Assoc. Prop. of Mult.
$= 63w$
Multiply 7 and 9.

16. $3.2 + (x + 5.1) =$
$3.2 + (5.1 + x)$
Comm. Prop. of Add.
$= (3.2 + 5.1) + x$
Assoc. Prop. of Add.
$= 8.3 + x$
Add 3.2 and 5.1.

17–25. See Additional Answers.

26. See *Taking Math Deeper*.

27–34. See Additional Answers.

Fair Game Review

35. 98 **36.** 108
37. 90 **38.** 200
39. 37 is already prime.
40. $2^4 \times 3^2$
41. 3×7^2 **42.** 5×41
43. B

Mini-Assessment

Simplify the expression. Explain each step.

1. $7 + (3 + y)$ $10 + y$; Assoc. Prop. of Add.

2. $4(3x)$ $12x$; Assoc. Prop. of Mult.

3. $6 + (b + 4)$ $10 + b$; Comm. Prop. of Add.; Assoc. Prop. of Add.

4. $(7 \cdot d) \cdot 1$ $7d$; Assoc. Prop. of Mult.; Mult. Prop. of One

5. The expression $5 + x + 5$ represents the perimeter of a triangle. Simplify the expression. $10 + x$

Taking Math Deeper

Exercise 26

This problem emphasizes the power of visualization. In the drawing, *x* appears to be about half of the length, or about 6. Without redrawing, it is difficult for most people to "see" a different shape. So, encourage students to draw the prism for each of the values of *x*: 0.25, 12.5, and 144.

① 1 by 12.5 by 0.25

Looks like a ruler.

② 1 by 12.5 by 12.5

Looks like a floor tile.

③ 1 by 12.5 by 144

Looks like siding.

Project

Think of three different objects you could use for the items in Exercise 26. Draw the objects and give reasonable values for *x*.

Reteaching and Enrichment Strategies

If students need help...	If students got it...
Resources by Chapter • Practice A and Practice B • Puzzle Time Record and Practice Journal Practice Differentiating the Lesson Lesson Tutorials Skills Review Handbook	Resources by Chapter • Enrichment and Extension • School-to-Work Start the next section

26. **GEOMETRY** The volume of the rectangular prism is 12.5 • x • 1.

 a. Simplify the expression.
 b. Match x = 0.25, 12.5, and 144 with the object.

 A. siding for a house B. ruler C. square floor tile

Write the phrase as an expression. Then, simplify the expression.

27. 7 plus the sum of a number x and 5

28. the product of 8 and a number y multiplied by 9

Copy and complete the statement using the specified property.

	Property	Statement
29.	Associative Property of Multiplication	7(2y) =
30.	Commutative Property of Multiplication	13.2 • (x • 1) =
31.	Associative Property of Addition	17 + (6 + 2x) =
32.	Addition Property of Zero	2 + (c + 0) =
33.	Multiplication Property of One	1 • w • 16 =

34. **HATS** You and a friend sell hats at a fair booth. You sell 16 hats on the first shift and 21 hats on the third shift. Your friend sells x hats on the second shift.

 a. Write an expression for the number of hats sold.
 b. The expression 37(14) + 10x represents the amount made for both of you. How can you tell that your friend was selling the hats for a discounted price?
 c. **Reasoning** You took in more money than your friend. What can you say about the value of x?

Fair Game Review What you learned in previous grades & lessons

Evaluate the expression. *(Skills Review Handbook)*

35. 7(10 + 4) 36. 12(10 − 1) 37. 6(5 + 10) 38. 8(30 − 5)

Find the prime factorization of the number. *(Skills Review Handbook)*

39. 37 40. 144 41. 147 42. 205

43. **MULTIPLE CHOICE** A bag has 16 blue, 20 red, and 24 green marbles. What fraction of the marbles in the bag are blue? *(Skills Review Handbook)*

 Ⓐ $\frac{1}{5}$ Ⓑ $\frac{4}{15}$ Ⓒ $\frac{4}{11}$ Ⓓ $\frac{11}{15}$

1 Study Help

You can use an **information frame** to help you organize and remember concepts. Here is an example of an information frame for the Commutative Property of Addition.

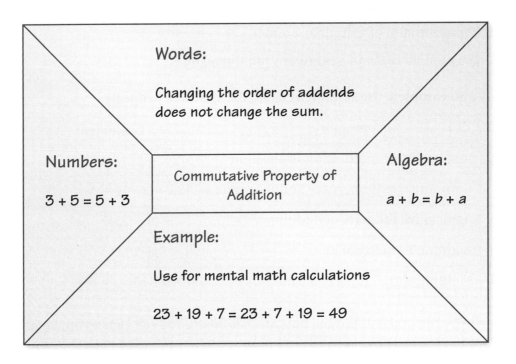

On Your Own

Make an information frame to help you study these topics.

1. Commutative Property of Multiplication
2. Associative Property of Addition
3. Associative Property of Multiplication
4. Addition Property of Zero
5. Multiplication Property of Zero
6. Multiplication Property of One
7. Expressions

After you complete this chapter, make information frames for the following topics.

8. Distributive Property
9. Formulas

"Dear Mom, I am sending you an information frame card for Mother's Day!"

Sample Answers

1.

2.

3.

4.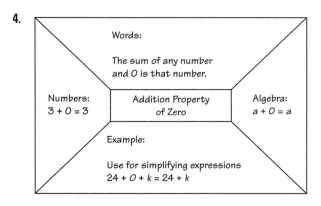

5–7. Available at *BigIdeasMath.com*.

List of Organizers
Available at *BigIdeasMath.com*

Comparison Chart
Concept Circle
Definition (Idea) and Example Chart
Example and Non-Example Chart
Formula Triangle
Four Square
Information Frame
Information Wheel
Notetaking Organizer
Process Diagram
Summary Triangle
Word Magnet
Y Chart

About this Organizer

An **Information Frame** can be used to help students organize and remember concepts. Students write the topic in the middle rectangle. Then students write related concepts in the spaces around the rectangle. Related concepts can include *Words, Numbers, Algebra,* and *Example* as shown in the sample answers, but other related concepts can include *Definition, Non-Example, Visual, Procedure, Details,* and *Vocabulary.* Students can place their information frames on note cards to use as a quick study reference.

Technology For the Teacher
Vocabulary Puzzle Builder

Answers

1. 8
2. 6
3. 6

4.
x	x + 6
1	7
2	8
3	9

5.
x	3x − 5
3	4
6	13
9	22

6. $28 + 35$ or $35 + 28$

7. $x \div 2$

8. $m \cdot 23$ or $23m$

9. $a - 10$

10. Comm. Prop. of Mult.

11. Assoc. Prop. of Add.

12. $3.2 + (b + 5.7) =$
 $3.2 + (5.7 + b)$
 Comm. Prop. of Add.
 $= (3.2 + 5.7) + b$
 Assoc. Prop. of Add.
 $= 8.9 + b$
 Add 3.2 and 5.7.

13–16. See Additional Answers.

Assessment Book

Alternative Quiz Ideas

100% Quiz	Math Log
Error Notebook	Notebook Quiz
Group Quiz	**Partner Quiz**
Homework Quiz	Pass the Paper

Partner Quiz

- Partner quizzes are to be completed by students working in pairs. Student pairs can be selected by the teacher, by students, through a random process, or any way that works for your class.
- Students are permitted to use their notebooks and other appropriate materials.
- Each pair submits a draft of the quiz for teacher feedback. Then they revise their work, and turn it in for a grade.
- When the pair is finished they can submit one paper, or each can submit their own.
- Teachers can give feedback in a variety of ways. It is important that the teacher does not reteach or provide the solution. The teacher can tell students which questions they have answered correctly, if they are on the right track, or if they need to rethink a problem.

Reteaching and Enrichment Strategies

If students need help...	If students got it...
Resources by Chapter • Study Help • Practice A and Practice B • Puzzle Time Lesson Tutorials BigIdeasMath.com Practice Quiz Practice from the Test Generator	Resources by Chapter • Enrichment and Extension • School-to-Work Game Closet at BigIdeasMath.com Start the next section

Technology For the Teacher

Answer Presentation Tool
Big Ideas Test Generator

1.1–1.3 Quiz

Evaluate the expression when $a = 3$, $b = 2$, and $c = 12$. *(Section 1.1)*

1. $a + 5$
2. ab
3. $c \div b$

Copy and complete the table. *(Section 1.1)*

4.

x	x + 6
1	
2	
3	

5.

x	3x – 5
3	
6	
9	

Write the phrase as an expression. *(Section 1.2)*

6. The sum of 28 and 35
7. A number x divided by 2
8. The product of a number m and 23
9. 10 less than a number a

Tell which property is illustrated by the statement. *(Section 1.3)*

10. $3.5 \cdot z = z \cdot 3.5$
11. $14 + (35 + w) = (14 + 35) + w$

Simplify the expression. Explain each step. *(Section 1.3)*

12. $3.2 + (b + 5.7)$
13. $6 \cdot (10 \cdot k)$

14. **COUPON** The expression $p - 15$ is the amount you pay after using the coupon on a purchase of p dollars. How much do you pay for a purchase of $83? *(Section 1.1)*

15. **MOVING TRUCK** To rent a moving truck for the day, it costs $33 plus $0.90 for each mile driven. *(Section 1.2)*

 a. Write an expression for the cost to rent the truck.
 b. You drive the truck 300 miles. How much do you pay?

16. **GEOMETRY** The expression $18 + 7 + (18 + 2x) + 7$ represents the perimeter of the trapezoid. Simplify the expression. *(Section 1.3)*

1.4 The Distributive Property

COMMON CORE STATE STANDARDS
6.EE.3
6.EE.6

Essential Question
How do you multiply two 2-digit numbers using mental math?

1 ACTIVITY: Finding Products Involving Multiples of 10

Working with a partner, take turns using mental math to find the product.

Read the expression to your partner. Then ask your partner to write the answer.

a. 10×20 b. 10×30 c. 10×13 d. 24×10
e. 20×25 f. 30×12 g. 13×40 h. 30×70

In Activity 1, you used mental math to find simple products. You can use the *Distributive Property* and mental math to find more complicated products.

The Meaning of a Word Distribute

When you **distribute** something to each person in a group,

you give that thing to each person in the group.

2 ACTIVITY: Using Mental Math

Work with a partner. Use the Distributive Property and mental math to find the product.

a. **Sample:** 6×23

$$6 \times 23 = 6 \times (20 + 3) \quad \text{Write 23 as the sum of 20 and 3.}$$
$$= (6 \times 20) + (6 \times 3) \quad \text{Distribute the 6 over the sum.}$$
$$= 120 + 18 \quad \text{Find the products.}$$
$$= 138 \quad \text{Add.}$$

∴ So, $6 \times 23 = 138$.

b. 4×17 c. 8×26 d. 7×33 e. 9×47

Chapter 1 Expressions and Number Properties

Laurie's Notes

Introduction

For the Teacher
- **Goal:** The goal of this activity is for students to gain an understanding of how the Distributive Property can be used to perform multiplication problems using mental math.
- Students should understand how a 2-digit number can be written as the sum or difference of its tens and ones digits. This form of the number can make it easier to multiply by a 1-digit number using mental math.

Motivate
- Play *Equal Corners*. Write four problems in the corners of a rectangle on the board. Two of the four problems should have the same product. Students then circle the two equal products. Try several rounds. *Sample:*

20×8	4×40
60×10	20×40

Discuss
- **?** "How did you determine which products were equal?"
- **Listen:** Pay attention to how students explain multiplying by multiples of 10. Students should multiply the nonzero numbers, then insert zeros equal to the total number in the two factors.

Activity Notes

Activity 1
- Introduce the activity and have students work with a partner to complete it. Students could write answers on small white boards or at the board.

Activity 2
- Discuss what the word *distribute* means. When you distribute homework, each student in the class receives a paper.
- Students should recognize that when you multiply a 2-digit number and a 1-digit number, two multiplications are performed.
- After reminding students that $23 = 20 + 3$, draw the following:

$$\begin{array}{r} 23 \\ \times\ 6 \\ \hline 18 \\ +\ 120 \\ \hline 138 \end{array}$$

$6\ \boxed{\begin{array}{c|c} 20 & 3 \\ \hline 120 & 18 \end{array}} = 138$

- **?** "Can you hold one answer in your head while doing a second multiplication problem? Now add the answers together."
- Model how you use mental math to multiply 6×23.
- Ask for volunteers to share the process they used.
- **Connection:** In each problem, the 1-digit number had to be multiplied by (*distributed over*) each digit of the 2-digit number.

Common Core State Standards
6.EE.3 Apply the properties of operations to generate equivalent expressions.
6.EE.6 Use variables to represent numbers and write expressions when solving a real-world or mathematical problem; understand that a variable can represent an unknown number, or, depending on the purpose at hand, any number in a specified set.

Previous Learning
Students need to be familiar with basic computation facts. Applying mental math strategies will be helpful.

Start Thinking! and Warm Up

1.4 Record and Practice Journal

Differentiated Instruction

Auditory

Some students may prefer to learn a verbal statement of the Distributive Property: "*a* times the quantity *b* plus *c* equals *a* times *b* plus *a* times *c*" or "*a* times the quantity *b* minus *c* equals *a* times *b* minus *a* times *c*." Remind students that the quantity *b* plus *c* means that *b* and *c* are grouped together.

1.4 Record and Practice Journal

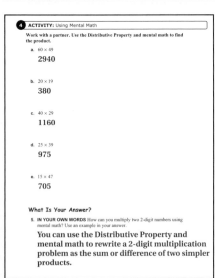

Laurie's Notes

Activity 3

- When two 2-digit numbers are multiplied, there are four products that get computed. For instance, 63×28 involves four different products.
- Work through each method. Be sure to point out the connection between the methods.
- You might also want to try the rectangular method:

	60	3	
20	1200	60	1260
8	480	24	+ 504
			1764

- Ask volunteers to share their processes and answers at the board. Ask if there is anyone who can do the problems using mental math. Some students can, and they usually keep a *running sum* as they go instead of adding the terms at the end.
- **Connection:** In each problem, each digit of the 2-digit number had to be multiplied by (*distributed over*) each digit of the second 2-digit number.

Activity 4

- This activity should help students recognize that any 2-digit number can be represented as a sum or difference. Example: $49 = 40 + 9 = 50 - 1$
- The last two parts involve *nice numbers*. Ask students how they did these problems.
 Example: $25 \times 39 = 25 \times (40 - 1) = (25 \times 40) - (25 \times 1) = 1000 - 25 = 975$
- If time permits, discuss how order of operations is involved in this problem. The parentheses are not needed around the two products (25×40) and (25×1).

What Is Your Answer?

- For Question 5, you are listening for the idea that a 2-digit number can be written as a sum or a difference.
- Have students share their answers.

Closure

- Discuss with students that even though 8×36 is represented with one multiplication symbol, two products are found. Multiplying 8×36 involves distributing the 8 over the 36 by multiplying 8×30 and 8×6.

The Dynamic Planning Tool
Editable Teacher's Resources at *BigIdeasMath.com*

3 ACTIVITY: Two Ways to Multiply

Work with a partner. Find the product two different ways. Compare the two methods.

a. **Sample:** 63 × 28

Method 1

```
    63
  × 28
   504
 +1260
  1764
```

Method 2

28 is 20 + 8. 63 is 60 + 3.

```
    63
  × 28
  1200    Multiply 20 and 60.
    60    Multiply 20 and 3.
   480    Multiply 8 and 60.
 +  24    Multiply 8 and 3.
  1764    Add.
```

b. 32 × 45 c. 37 × 61
d. 28 × 57 e. 17 × 43

4 ACTIVITY: Using Mental Math

Work with a partner. Use the Distributive Property and mental math to find the product.

a. **Sample:** 60 × 49

$60 \times 49 = 60 \times (50 - 1)$ Write 49 as the difference of 50 and 1.

$ = (60 \times 50) - (60 \times 1)$ Distribute the 60 over the difference.

$ = 3000 - 60$ Find the products.

$ = 2940$ Subtract.

∴ So, 60 × 49 = 2940.

b. 20 × 19 c. 40 × 29
d. 25 × 39 e. 15 × 47

What Is Your Answer?

5. **IN YOUR OWN WORDS** How can you multiply two 2-digit numbers using mental math? Use an example in your answer.

Practice

Use what you learned about the Distributive Property to complete Exercises 4–7 on page 26.

Section 1.4 The Distributive Property

1.4 Lesson

Key Idea

Distributive Property

Words To multiply a sum or difference by a number, multiply each number in the sum or difference by the number outside the parentheses. Then evaluate.

Numbers $3(7 + 2) = 3 \times 7 + 3 \times 2$

$3(7 - 2) = 3 \times 7 - 3 \times 2$

Algebra $a(b + c) = ab + ac$

$a(b - c) = ab - ac$

EXAMPLE 1 Using Mental Math

Use the Distributive Property and mental math to find 8×53.

$8 \times 53 = 8(50 + 3)$ Write 53 as $50 + 3$.

$= 8(50) + 8(3)$ Distributive Property

$= 400 + 24$ Multiply.

$= 424$ Add.

On Your Own

Use the Distributive Property and mental math to find the product.

1. 5×41 **2.** 12×32 **3.** 9×19 **4.** $6(37)$

EXAMPLE 2 Simplifying Algebraic Expressions

Use the Distributive Property to simplify the expression.

a. $4(n + 5)$

$4(n + 5) = 4(n) + 4(5)$ Distributive Property

$= 4n + 20$ Multiply.

b. $12(y - 3)$

$12(y - 3) = 12(y) - 12(3)$ Distributive Property

$= 12y - 36$ Multiply.

On Your Own

Use the Distributive Property to simplify the expression.

5. $7(a + 2)$ **6.** $10(9 + b)$ **7.** $6(b - 7)$ **8.** $3(d - 11)$

24 Chapter 1 Expressions and Number Properties

Laurie's Notes

Introduction

Connect
- **Yesterday:** Students developed an understanding that multiplying numbers with more than one digit involves partial products: the one digit is *distributed over* the digits of the 2-digit number.
- **Today:** Students will represent this process with algebra and refer to it as the Distributive Property.

Discuss
- Begin with a discussion of how multiplication can be represented
 $8 \times 53 \qquad 8 \cdot 53 \qquad 8(53)$
- There is also a fourth representation for variables.
 $8 \times n \qquad 8 \cdot n \qquad 8(n) \qquad 8n$

Lesson Notes

Key Idea
- **Representation:** Draw attention to how the multiplication is being shown.
- Students may ask why you would write 3×9 as $3(7 + 2)$ and 3×5 as $3(7 - 2)$. The answer is that you wouldn't! Simple values are being used to demonstrate the property instead of using a problem such as 8×67.
- Spend more time discussing the algebraic representation. In the problem $a(b + c)$, ask what *operations* students see. Not all students will see two. Some may only see addition.
- Explain that the blue arrows are used to remind us that we are distributing one number over another. The arrows can also be written above the problem.

Example 1
- Work through the problem. Draw a connection to what students did yesterday. The only thing that has changed is how multiplication is being represented, which is easier for math teachers to see than for students!

On Your Own
- **Neighbor Check:** Have students work independently and then have their neighbor check their work. Have students discuss any discrepancies.
- Note how Question 4 is represented without the multiplication symbol.
- Probe to see if 9×19 was done as $9(10 + 9)$ or as $9(20 - 1)$.
- Point out that 5×41 is the same as 41×5 because of the Commutative Property. So, the Distributive Property can also be written as $(b + c)a$.

Example 2
- Slowly work through the two parts. Introducing a variable expression is *not* a small change! The use of arrows to show that the 4 is being distributed over the n and the 5 is helpful.

Goal Today's lesson is applying the Distributive Property to show that two expressions are equivalent.

Start Thinking! and Warm Up

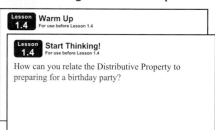

Extra Example 1
Use the Distributive Property and mental math to find 6×82. 492

 On Your Own
1. 205
2. 384
3. 171
4. 222

Extra Example 2
Use the Distributive Property to simplify the expression.
a. $11(x + 7)$ $11x + 77$
b. $8(c - 4)$ $8c - 32$

On Your Own
5. $7a + 14$
6. $90 + 10b$
7. $6b - 42$
8. $3d - 33$

English Language Learners

Visual

You may want to use an area model to show how the Distributive Property relates multiplication and addition.

From the model, students should see that the total area can be represented by $6 \times 23 = 6(20 + 3)$ or as the sum of two smaller areas $6(20) + 6(3)$. So, $6(20 + 3) = 6(20) + 6(3)$.

Extra Example 3

Use the Distributive Property to simplify $3(4 + r + 3)$. $3r + 21$

On Your Own

9. $2n + 34$
10. $15y + 150$
11. $5z + 95$
12. $3x + 45$
13. $56 + 7d$
14. $360 + 8w$

Extra Example 4

Maria is x years old. Her father is 25 years older than Maria. Maria's grandfather is two times older than Maria's father. Write and simplify an expression that represents Maria's grandfather's age in years.
$2(x + 25) = 2x + 50$

On Your Own

15. $5(x + 7) = 5x + 35$

Laurie's Notes

Words of Wisdom

- The Distributive Property is used frequently in mental math problems, even when we don't think about it. These examples may not be appropriate for students yet, but keep them in mind when students do encounter them.

 Examples: $\dfrac{2}{3} \times 3\dfrac{1}{2} = \dfrac{2}{3} \times \left(3 + \dfrac{1}{2}\right)$

 15% of \$42 = 10% of \$42 + 5% of \$42

 3 pizzas at \$12.50 each is $3 \times \$12 + 3 \times \0.50

Example 3

- You may want to have students work through the steps of the problem without showing them the solution. Based on the answer choices they make, you can tell the types of errors students are making.
 - Students who choose answer choice A added $6 + 2$ to get 8 and didn't distribute the 9.
 - Students who choose answer choice B multiplied 9 by 6, but didn't multiply 9 by 2.
 - Students who choose answer choice C multiplied 9 by 6 and added 2, not the product of 9 and 2.

On Your Own

- **Think-Pair-Share:** Students should read each question independently and then work with a partner to answer the questions. When they have answered the questions, the pair should compare their answers with another group and discuss any discrepancies.
- Some students may recognize that $2(n + 5 + 12)$ is equivalent to $2(n + 17)$.

Example 4

- Write the table representing each person's age in words and algebraically. Connect this problem back to Section 1.1.
- **? Extension:** "If José is 12, find each person's age." Felipe is 14 and Maria is 42.

On Your Own

- **Think-Pair-Share:** Students should read the question independently and then work with a partner to answer the question. When they have answered the question, the pair should compare their answer with another group and discuss any discrepancies.

Closure

- **Writing:** Ask students to explain the Distributive Property in words. As a writing activity, have students write a paragraph explaining how to do the problem $5(n + 8)$.

The Dynamic Planning Tool
Editable Teacher's Resources at *BigIdeasMath.com*

EXAMPLE 3 **Standardized Test Practice**

Which expression is equivalent to $9(6 + x + 2)$?

(A) $9x + 8$ (B) $9x + 54$ (C) $9x + 56$ (D) $9x + 72$

$9(6 + x + 2) = 9(6) + 9(x) + 9(2)$ Distributive Property
$ = 54 + 9x + 18$ Multiply.
$ = 9x + 54 + 18$ Commutative Property of Addition
$ = 9x + 72$ Add 54 and 18.

∴ The correct answer is (D).

On Your Own

Now You're Ready
Exercises 12–27

Use the Distributive Property to simplify the expression.

 9. $2(n + 5 + 12)$ 10. $15(y + 3 + 7)$ 11. $5(10 + z + 9)$
12. $3(x + 11 + 4)$ 13. $7(2 + 6 + d)$ 14. $8(20 + 25 + w)$

EXAMPLE 4 **Real-Life Application**

José is x years old. His brother, Felipe, is 2 years older than José. Their aunt, Maria, is three times as old as Felipe. Write and simplify an expression that represents Maria's age in years.

Name	Description	Expression
José	He is x years old.	x
Felipe	He is 2 years *older* than José. So, *add* 2 to x.	$x + 2$
Maria	She is three *times* as old as Felipe. So, *multiply* 3 and $(x + 2)$.	$3(x + 2)$

$3(x + 2) = 3(x) + 3(2)$ Distributive Property
$ = 3x + 6$ Multiply.

∴ Maria's age in years is represented by the expression $3x + 6$.

On Your Own

15. Alexis is x years old. Her sister, Gloria, is 7 years older than Alexis. Their grandfather is five times as old as Gloria. Write and simplify an expression that represents their grandfather's age in years.

1.4 Exercises

Vocabulary and Concept Check

1. **WRITING** One meaning of the word *distribute* is to give something to each member of a group. How can this help you remember the Distributive Property?

2. **OPEN-ENDED** Write an algebraic expression in which you use the Distributive Property and then the Associative Property of Addition to simplify.

3. **WHICH ONE DOESN'T BELONG?** Which expression does *not* belong with the other three? Explain your reasoning.

$2(x + 2)$ $5(x - 8)$ $4 + (x \cdot 4)$ $8(9 - x)$

Practice and Problem Solving

Use the Distributive Property and mental math to find the product.

4. 3×21
5. 9×76
6. $12(43)$
7. $5(88)$
8. 18×52
9. 8×27
10. $8(63)$
11. $7(28)$

Use the Distributive Property to simplify the expression.

12. $3(x + 4)$
13. $10(b - 6)$
14. $6(s - 9)$
15. $7(8 + y)$
16. $8(12 + a)$
17. $9(n + 1)$
18. $12(6 - k)$
19. $18(w + 5)$
20. $9(3 + c + 4)$
21. $7(8 + x + 2)$
22. $8(g + 5 + 2)$
23. $6(10 + z + 3)$
24. $6(x + 4) + 3$
25. $5 + 8(3 + x)$
26. $9 + 8(x + 2)$
27. $5(8 + x) + 12$

28. **ERROR ANALYSIS** Describe and correct the error in rewriting the expression.

✗ $6(y + 8) = 6y + 8$

29. **ART MUSEUM** A class of 30 students visits an art museum and a special exhibit while there.

 a. Use the Distributive Property to write and simplify an expression for the cost.
 b. Estimate a reasonable value for x. Explain.
 c. Use your estimate for x to evaluate both expressions in part (a). Are the values the same?

 PRICES
	Museum	Exhibit
Child (under 5)	Free	Free
Student	$8	$x
Regular	$12	$4
Senior	$10	$3

30. **FITNESS** Each day, you run on a treadmill for r minutes and lift weights for 15 minutes. Which expressions can you use to find how many minutes of exercise you do in 5 days? Explain your reasoning.

$5(r + 15)$ $5r + 5 \cdot 15$ $5r + 15$ $r(5 + 15)$

Assignment Guide and Homework Check

Level	Day 1 Activity Assignment	Day 2 Lesson Assignment	Homework Check
Basic	4–7, 45–53	1–3, 13–27 odd, 28–30, 33	13, 21, 27, 29, 30
Average	4–7, 45–53	1–3, 13–27 odd, 28, 31, 32, 35, 39	13, 21, 27, 31, 35
Advanced	4–7, 45–53	1–3, 20–28 even, 31, 34, 40–44	22, 26, 31, 42, 43

Common Errors

- **Exercises 12–27** Students might forget to distribute to each part of the expression especially when there are more than two parts within the parentheses. Tell students to write the expression on their paper and draw arrows from the number being distributed to each part.
- **Exercise 27** Students might try to use the Commutative and Associative Properties before distributing. This is a misunderstanding of the order of operations. Remind students to multiply first before they add.
- **Exercises 35–37** Students might only use *Guess, Check, and Revise* to solve. Encourage students to use the Distributive Property and then compare the expressions to find the value of *x*.

Technology For the Teacher
Answer Presentation Tool
QuizShow

1.4 Record and Practice Journal

Use the Distributive Property and mental math to find the product.
1. 4×31 2. 7×49 3. 13×102
 124 343 1326
4. $6(19)$ 5. $23(42)$ 6. $16(38)$
 114 966 608

Use the Distributive Property to simplify the expression.
7. $8(5 + w)$ 8. $11(9 + d)$ 9. $15(p - 4)$
 $40 + 8w$ $99 + 11d$ $15p - 60$
10. $4(21 - n)$ 11. $20(y + 10)$ 12. $9(12 - k)$
 $84 - 4n$ $20y + 200$ $108 - 9k$

13. A practice field is 75 yards long and 29 yards wide. Use the Distributive Property and mental math to find the area of the practice field.
 2175 yd^2

14. You have a rectangular fort that is 7 feet long and 6 feet wide. You decide to add *x* feet to the length of the fort.
 a. Find the area of the fort before the addition.
 42 ft^2
 b. Use the Distributive Property to write and simplify an expression for the area of the fort with the addition.
 $6(7 + x) = 42 + 6x$
 c. You have 90 square feet of space available for your fort. What is the maximum value of *x* feet that you can add to the length?
 8 ft

Vocabulary and Concept Check

1. *Sample answer:* You must distribute or give the number outside the parentheses to the numbers inside the parentheses.

2. *Sample answer:*
$6 + [2(2 + x)]$
$= 6 + [2(2) + 2(x)]$
$= 6 + (4 + 2x)$
$= (6 + 4) + 2x$
$= 10 + 2x$

3. $4 + (x \cdot 4)$ does not belong because it doesn't represent the Distributive Property.

Practice and Problem Solving

4. 63 5. 684
6. 516 7. 440
8. 936 9. 216
10. 504 11. 196
12. $3x + 12$ 13. $10b - 60$
14. $6s - 54$ 15. $56 + 7y$
16. $96 + 8a$ 17. $9n + 9$
18. $72 - 12k$ 19. $18w + 90$
20. $63 + 9c$ 21. $70 + 7x$
22. $8g + 56$ 23. $78 + 6z$
24. $6x + 27$ 25. $29 + 8x$
26. $8x + 25$ 27. $5x + 52$

28. The 6 was not distributed to the 8 inside the parentheses.
$6(y + 8) = 6y + 48$

29. a. $30(8 + x) = 240 + 30x$
 b. *Sample answer:* $2; It is less than the regular price to the exhibit.
 c. *Sample answer:* $300; yes

30. $5(r + 15)$ and $5r + 5 \cdot 15$, which are equal expressions.

T-26

Practice and Problem Solving

31. $10(103 - x) = 1030 - 10x$

32. $8(x + 8) = 8x + 64$

33. $13(7 - x) = 91 - 13x$

34. $9(7 + x + 5) = 108 + 9x$

35. $x = 8$

36. $x = 15$ **37.** $x = 3$

38. $4(x + 9)$ **39.** $2(3 + x)$

40. $5(15 - x)$

41. $7(1 + 2x + 3)$

42. a. $4x + 24$ **b.** $4x + 24$

 c. The results of parts (a) and (b) are the same. Multiplying $(x + 6)$ by 4 and adding 4 groups of $(x + 6)$ are equivalent operations.

43. See *Taking Math Deeper*.

44. See Additional Answers.

Fair Game Review

45. 14 **46.** 8

47. 120 **48.** 19

49. no; $\frac{2}{3}$ **50.** yes

51. no; $\frac{19}{31}$

52. yes **53.** C

Mini-Assessment

Use the Distributive Property to simplify the expression.

1. $6(x + 4)$ $6x + 24$

2. $14(g - 2)$ $14g - 28$

3. $3(a + 10)$ $3a + 30$

4. $8(n + 3 + 4)$ $8n + 56$

5. Mike is y years old. His cousin, Jeremy, is 3 years older than Mike. Their friend, Ryan, is two times as old as Jeremy. Write and simplify an expression that represents Ryan's age in years. $2(y + 3); 2y + 6$

T-27

Taking Math Deeper

Exercise 43

The key to this problem is to read all of the information *carefully!* There are two types of dollar amounts in the problem. One is *cost* (what a business pays). The other is *price* (what a customer pays). The difference between price and cost is the profit for the company.

 Organize information.

 Write expressions for the profit.

 a. Total Profit $= 42(10 - x) + 56(5 - y)$ Add expressions.
 $= 420 - 42x + 280 - 56y$ Distributive Property
 $= 700 - 42x - 56y$ Simplify.

 Find the profit when $x = \$5$ and $y = \$3$.

 b. Total Profit $= 700 - 42x - 56y$
 $= 700 - 42(5) - 56(3)$
 $= 700 - 210 - 168$
 $= \$322$

Project

Make a poster to market and sell your candles. Include mathematical information in your poster. Do you think that advertising will help your sales?

Reteaching and Enrichment Strategies

If students need help...	If students got it...
Resources by Chapter • Practice A and Practice B • Puzzle Time Record and Practice Journal Practice Differentiating the Lesson Lesson Tutorials Skills Review Handbook	Resources by Chapter • Enrichment and Extension • School-to-Work Start the next section

31. **SPEED** A cheetah can run 103 feet per second. A zebra can run *x* feet per second. Use the Distributive Property to write and simplify an expression for how much farther the cheetah can run in 10 seconds.

GEOMETRY Use the Distributive Property to write and simplify an expression for the area of the rectangle.

32.

33.

34.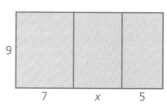

ALGEBRA Find the value of *x* that makes the expressions equivalent.

35. $4(x - 5)$; $32 - 20$
36. $2(x + 9)$; $30 + 18$
37. $7(8 - x)$; $56 - 21$

Use the Distributive Property to rewrite the expression as a product.

38. $4x + 36$
39. $6 + 2x$
40. $75 - 5x$
41. $7 + 14x + 21$

42. **STRUCTURE** Simplify the expressions and compare.

 a. $4(x + 6)$
 b. $(x + 6) + (x + 6) + (x + 6) + (x + 6)$
 c. Compare the expressions and results in parts (a) and (b). What do you notice? Explain.

43. **FUNDRAISER** An art club sells 42 large candles and 56 small candles.

 a. Use the Distributive Property to write and simplify an expression for the profit.
 b. A large candle costs $5 and a small candle costs $3. What is the club's profit?

Price: $10
Cost: $x

Price: $5
Cost: $y

Profit = Price − Cost

44. **Puzzle** Add one set of parentheses to the expression
$7 \cdot x + 3 + 8 \cdot x + 3 \cdot x + 8 - 9$ so that it is equivalent to $2(9x + 10)$.

Fair Game Review *What you learned in previous grades & lessons*

Evaluate the expression when $x = 5$ and $y = 24$. *(Section 1.1)*

45. $x + 9$
46. $y \div 3$
47. xy
48. $y - x$

Tell whether the fraction is in simplest form. If not, simplify it.
(Skills Review Handbook)

49. $\dfrac{8}{12}$
50. $\dfrac{6}{35}$
51. $\dfrac{38}{62}$
52. $\dfrac{16}{27}$

53. **MULTIPLE CHOICE** What is the surface area of a cube that has a side length of 8 feet? *(Skills Review Handbook)*

 Ⓐ 64 ft^2
 Ⓑ 192 ft^2
 Ⓒ 384 ft^2
 Ⓓ 512 ft^2

1.5 Using Formulas to Solve Problems

COMMON CORE STATE STANDARDS
6.EE.2c

Essential Question How can you use formulas to find the area of an object with an unusual shape?

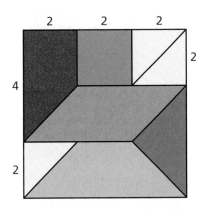

1 ACTIVITY: Using an Area Formula

Work with a partner. Copy and complete the table.

Polygon	Name	Area Formula	Area
square with sides s	Square	$A = s^2$	$s = 2$ $A = 2^2$ $= 4$ square units
trapezoid with h, b, B	Trapezoid	$A = h(b + B) \div 2$	
parallelogram with h, b			
triangle with h, b			
trapezoid with b, h, B			
right triangle with h, b			

28 Chapter 1 Expressions and Number Properties

Laurie's Notes

Introduction

For the Teacher
- **Goal:** The goal of this activity is for students to find the area of composite figures by dividing the figure into common geometric shapes.
- Students will need to see how dimensions of a geometric figure can be read from a diagram.

Motivate
- If you have tangram pieces, or a computer with virtual manipulatives, make a square using tangram pieces. If the area of the tangram square is 16 square units, find the area of each individual piece. This doesn't require any formulas, only reasoning. For an additional aid, draw the diagram on 1-inch grid paper. The dimensions of the square are 4 by 4.

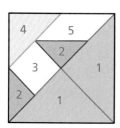

Area of Triangle 1: 4 sq. units
$\left(\frac{1}{4} \text{ of the whole square}\right)$
Area of Triangle 4: 2 sq. units
$\left(\frac{1}{2} \text{ the area of Triangle 1}\right)$
Area of Triangle 2: 1 sq. unit
$\left(\frac{1}{2} \text{ the area of Triangle 4}\right)$
Area of Square 3: 2 sq. units
(2 times the area of Triangle 2)
Area of Parallelogram 5: 2 sq. units
(2 times the area of Triangle 2)

- **Teaching Tip:** Students may overlay the tangram pieces to find the answers. This reinforces the concept that area is related to covering.

Activity Notes

Activity 1
- Note that the tangram from the Motivate section is different than the rongram in the activity.
- Ask questions (i.e. What is a square?) to review vocabulary.
- ❓ Some dimensions are given. Ask how they can find the dimensions when the numbers are not directly on the figure. For instance, how do they know the shorter base of the red trapezoid is 2 units?
- You may want to review the formula for each shape with students. In particular, be sure to discuss that height is a perpendicular distance from the base. In triangles, the height may (in the yellow triangle) or may not (in the blue triangle) be one of the sides.
- ❓ "What is the sum of the areas of the 8 pieces?" 36 sq. units
- You may want to ask students if the area will change when the pieces are rearranged to form another shape. You want to be sure that students understand that area doesn't change when a shape is moved.

Common Core State Standards

6.EE.2c Evaluate expressions at specific values of their variables. Include expressions that arise from formulas used in real-world problems.

Previous Learning

Students need to be familiar with order of operations and evaluating expressions by substituting values for the variables. Students should know the area formulas for triangles, squares, trapezoids, and parallelograms.

Activity Materials	
Introduction	Textbook
• tangram pieces	• rongram pieces

Start Thinking! and Warm Up

1.5 Record and Practice Journal

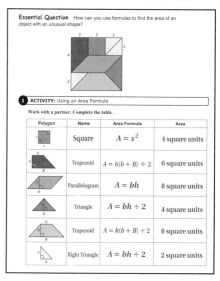

T-28

English Language Learners
Vocabulary
Have students start a *Formula* page in their notebooks with the formulas used in this section. Each formula should be accompanied by a description and a diagram clearly labeled. In the case of area formulas, units of measure should be included with the description (e.g. unit and square unit). As students progress throughout the year, additional formulas can be added to their *Formula* notebook page.

1.5 Record and Practice Journal

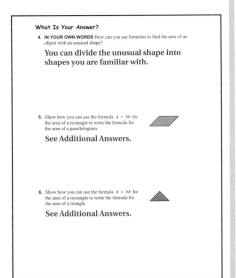

Laurie's Notes

Activity 2
- Ask students how they found the area of the sailboat. Many will probably add the area of each of the pieces. You can also find the area by using the sum of the areas you found in Activity 1 and subtracting the area of the red trapezoid. Ask students which method they think is easier and why.
- **Discuss:** How does the area of the sailboat compare to the area of the large square? Because the red trapezoid is missing, the area of the sailboat is 6 sq. units less.

Activity 3
- Tell students to use some or all of the pieces available to make the shapes. Explain that a green square and a yellow triangle won't make the house shape in part (a) for two reasons: their combined area is not 20 square units, and the hypotenuse of the yellow triangle is longer than a side of the square.
- **Classroom Review:** Students could share their results at the overhead by placing the pieces to correctly form the shapes shown. This activity helps to develop spatial skills. It's not obvious to all students how to manipulate the pieces to form the shapes shown.
- **Differentiating Instruction:** The rabbit is the easiest figure to make because you can see the outline of the shapes. You may want to give students a choice as to which shape(s) they try to make.

What Is Your Answer?
- Question 4 really refers to Activity 1. Students may not have used the formulas after finishing the first activity. Hopefully, they would have recognized that in order to find the area of an unusual shape, they should first divide it into common shapes. Common shapes will help because they know how to find those areas.
- If students have difficulty with Questions 5 and 6, superimpose a rectangle on the diagram as shown.

Closure
- Have students summarize their understanding of the activities. First, they used formulas to find the areas of simple geometric shapes. Then they rearranged the pieces to form unusual shapes.
- Because the areas of the individual pieces did not change when moved, the area of an unusual shape is the sum of the parts that form it.

Technology For the Teacher

Dynamic Classroom

The Dynamic Planning Tool
Editable Teacher's Resources at *BigIdeasMath.com*

2 ACTIVITY: Finding an Area

Work with a partner. Use the shapes from Activity 1 to find the area of the sailboat. Explain your reasoning.

3 ACTIVITY: Finding an Area

Work with a partner. Use the shapes from Activity 1 to create the picture.

a. house b. rabbit c. bird

20 square units 36 square units 32 square units

What Is Your Answer?

4. **IN YOUR OWN WORDS** How can you use formulas to find the area of an object with an unusual shape?

5. Show how you can use the formula $A = bh$ for the area of a rectangle to write the formula for the area of a parallelogram.

6. Show how you can use the formula $A = bh$ for the area of a rectangle to write the formula for the area of a triangle.

Practice

Use what you learned about using formulas to complete Exercises 3–5 on page 32.

Section 1.5 Using Formulas to Solve Problems

1.5 Lesson

Key Vocabulary
formula, *p. 30*
solve a formula, *p. 30*

A **formula** is an equation that tells you how one variable is related to one or more other variables. To **solve a formula**, find the value of one variable by substituting numbers for the other variables.

EXAMPLE 1 — Using a Simple Formula

The formula $M = 220 - a$ gives a person's maximum heart rate M, where a is the person's age in years. Malcolm is 12 years old. His uncle is 40 years old. What is the difference between their maximum heart rates?

Malcolm	His Uncle	
$M = 220 - a$	$M = 220 - a$	Write the formula.
$= 220 - 12$	$= 220 - 40$	Substitute their ages for a.
$= 208$	$= 180$	Subtract.

∴ The difference between their maximum heart rates is $208 - 180$, or 28 beats per minute.

On Your Own

1. What is the difference between the maximum heart rates of Malcolm and his grandmother, who is 85 years old?

EXAMPLE 2 — Using an Area Formula

Find the area of the rectangular jumping surface of the trampoline.

7 ft
14 ft

Use the formula for the area of a rectangle.

$A = bh$	Write the formula.
$= 14 \times 7$	Substitute 14 for b and 7 for h.
$= 98$	Multiply.

∴ The area of the jumping surface is 98 square feet.

On Your Own

2. Find the area of a rectangular trampoline that measures 12 feet by 6 feet.

Laurie's Notes

Introduction

Connect
- **Yesterday:** Students reviewed area formulas and decomposed composite shapes into simple geometric shapes.
- **Today:** Students will use formulas, primarily for area, to solve problems.

Discuss
- Begin with a discussion of what a formula is. What formulas do students know, besides area formulas? They may give examples from science or sports.
- From the examples given, ask students to give possible values of the variables and to solve the formula for these values.
- **Key Vocabulary:** Write the vocabulary, *formula* and *solve a formula*, on the board. List the students' examples under the vocabulary word.

Lesson Notes

Example 1
- Work through the problem.
- When finished, ask students to complete the statement: "As age increases, the maximum heart rate _____." decreases
- If time is available, try several additional values of *a*. Record the data in a table so that students have experience with this method of organizing data. Go back to the statement "As age increases, the maximum heart rate decreases," to verify the response.

On Your Own
- Students should complete this problem independently.

Example 2
- Work through the example.
- To model good problem solving, write all three steps of the solution: write the formula needed, substitute the value(s) of the variable(s), and perform the operation(s). Finally, label the answer with the appropriate units.
- **Extension:** If time permits, estimate the area of the classroom floor, the white board, or a bulletin board.

On Your Own
- Students should complete this problem independently.

Goal Today's lesson is using **formulas** to solve problems.

Start Thinking! and Warm Up

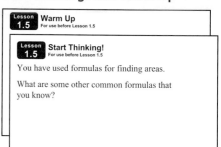

Extra Example 1
Nilyne is 13 years old. Her aunt is 43 years old. What is the difference between their maximum heart rates? 30 beats per minute

On Your Own
1. 73 beats per minute

Extra Example 2
Find the area of a chalkboard with a width of 12 feet and a height of 4 feet. 48 ft^2

On Your Own
2. 72 ft^2

T-30

Laurie's Notes

Extra Example 3
a. Use the formula $A = bh \div 2$ to find the area of a triangle with a base of 8 centimeters and a height of 7 centimeters. 28 cm^2
b. A parallelogram has an area of 30 square centimeters. Is the area of the triangle greater than or less than the area of the parallelogram? less than

Example 3
- **Representation:** Be sure to point out the corner mark in the diagram. Ask about the approximate measures of the remaining angles to review vocabulary (acute and obtuse).
- Write the formula for the area of a trapezoid. Discuss what each variable represents. The two bases are interchangeable, meaning either base could be b or B, although it is common to think of b for the shorter base and B for the longer base.
- ? "What operations are involved in this formula, and in what order will they be performed?" addition, multiplication, division
- Remind students that the h next to the parentheses is multiplication. Off to the side, you might want to write:
 $A = h \times (b + B) \div 2$.
- Work through the steps in this problem to model good problem solving strategies.
- **Homework Note:** It would be a good idea to remind students to include the following when completing their homework: the formula they use, the substitution, and the final answer. You want to see the work, not just the answer (to make sure that the students correctly understand the concepts).

Words of Wisdom
- Some students may need to see the formulas written with words versus variables.
 Example: Area of trapezoid = height \times (shorter base + longer base) \div 2

On Your Own
- **Think-Pair-Share:** Students should read the question independently and then work with a partner to answer the question. When they have answered the question, the pair should compare their answer with another group and discuss any discrepancies.

On Your Own
3. about $56{,}750 \text{ mi}^2$

Differentiated Instruction
Visual
Have visual learners model the areas of the polygons on grid paper. Then compare their answers with the answers found using the area formulas.

Closure
- Give students the following:
 "Here's a formula for how I compute semester grades:
 $A = 0.4H + 0.3T + 0.2Q + 0.1D$, where H = homework average, T = test average, Q = quiz average, D = department average, and A = overall average."
- ? Ask your students to explain in words how grades are computed for the students of the class. The overall average is the sum of 40% of the homework average, 30% of the test average, 20% of the quiz average, and 10% of the department average.

Technology For the Teacher

The Dynamic Planning Tool
Editable Teacher's Resources at *BigIdeasMath.com*

EXAMPLE 3 **Using an Area Formula**

A trapezoid can be used to approximate the shape of Arkansas, as shown on the map.

a. Use the formula $A = h(b + B) \div 2$ to find the area.

b. Mississippi has an area of about 46,907 square miles. Is the area of Arkansas greater than or less than the area of Mississippi?

> **Remember**
> The corner mark ⌐ in a figure means that the angle formed by the sides is a right angle.

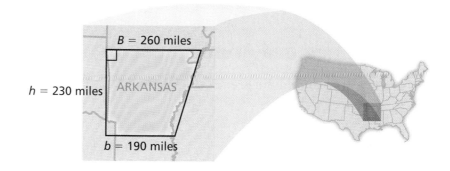

a. $A = h(b + B) \div 2$ Write the formula.

 $= 230(190 + 260) \div 2$ Substitute 230 for h, 190 for b, and 260 for B.

 $= 230(450) \div 2$ Add inside parentheses.

 $= 103{,}500 \div 2$ Multiply 230 and 450.

 $= 51{,}750$ Divide.

∴ The area of Arkansas is about 51,750 square miles.

b. Because 51,750 is greater than 46,907, the area of Arkansas is greater than the area of Mississippi.

On Your Own

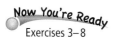
Exercises 3–8

3. A trapezoid can be used to approximate the shape of Nevada, as shown on the map. How much larger is the area of Nevada than the area of Arkansas?

1.5 Exercises

Vocabulary and Concept Check

1. **WRITING** How is using a formula similar to evaluating an expression?
2. **REASONING** The cost C (in dollars) to make x dartboards is $C = 50 + 10x$. What do you need to know to solve this formula? Explain.

Practice and Problem Solving

Use a formula to find the area of the figure.

3.

4.

5.

6.

7.

8.

9. **PARKING SPACE** A parking space is shaped like a parallelogram with a base of 26 feet and a height of 9 feet.

 a. What is the area of the parking space?
 b. Draw a diagram of what the parking space might look like.
 c. Use your diagram to estimate the length of the longest car that will fit in the space. Explain your reasoning.

10. **LIGHTNING** You can estimate how far you are from lightning.
 - When you see the lightning, count the number of seconds ("One one-thousand, two one-thousand, . . .") until you hear the thunder.
 - Divide the number of seconds by 5.
 - This is how many miles you are from lightning.

 You see lightning. After about a second, you hear a crack of thunder and your friend says "Wow, that was close!" Was your friend correct? How close was the lightning?

11. **VOLUME** The formula $V = \ell wh$ represents the volume of a rectangular prism with length ℓ, width w, and height h.

 a. What is the volume of the cereal box?
 b. The volume of a bowl is about 15 cubic inches. How many bowls of cereal does the box hold?

32 Chapter 1 Expressions and Number Properties

Assignment Guide and Homework Check

Level	Day 1 Activity Assignment	Day 2 Lesson Assignment	Homework Check
Basic	3–5, 20–24	1, 2, 6–8, 9–13 odd	7, 9, 11, 13
Average	3–5, 20–24	1, 2, 7–12, 14, 17	7, 9, 12, 17
Advanced	3–5, 20–24	1, 2, 10–16 even, 17–19	12, 14, 17, 18

Common Errors

- **Exercises 3–8** Students may use the wrong formula or forget to do all the operations (especially with the area of a trapezoid in Exercise 4). Tell the students to write out the formula they need to use, then identify the value of each variable before substituting.
- **Exercises 13–15** Students may forget to subtract the area of the white square when finding the area of the shaded region. Give students a visual aid. Draw the rectangle on a piece of paper and cut out the part that the students must subtract to find the area.
- **Exercises 13–15** Some students may try to break the shaded region into smaller rectangles and find the areas of these smaller regions. This is much more difficult. Show students the visual model.

Technology For the Teacher
Answer Presentation Tool
QuizShow

1.5 Record and Practice Journal

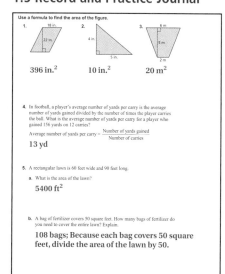

Vocabulary and Concept Check

1. *Sample answer:* You substitute value(s) for the variable(s) to find the value of the formula.
2. You need to know how many dart boards are going to be made. Then you can substitute that value for x and find the cost.

Practice and Problem Solving

3. 48 in.2
4. 49 ft^2
5. 108 in.2
6. 16 m^2
7. 30 ft^2
8. 224 ft^2
9. a. 234 ft^2
 b.
 c. 26 ft; The base of the parking space is related to the length of the car.
10. Yes, it is $\frac{1}{5}$ of a mile away.
11. a. 192 in.3
 b. almost 13 bowls

Practice and Problem Solving

12. 3
13. $4x - 9$
14. $6x + 24 - x^2$
15. $32x - 40 - x^2$
16. Divide the area by the base to find the height of the parallelogram.
17. 24 karats; If you let $k = 24$, then $P = 100$.
18. See *Taking Math Deeper*.
19. Area of black = 252 in.²
 Area of yellow = 244 in.²
 Area of each blue stripe = 328 in.²

Fair Game Review

20. 2
21. $\dfrac{1}{2}$
22. 0
23. 1
24. C

Mini-Assessment

Find the area of each figure.

1. 6 in.²
2. 16 m²
3. 48 ft²
4. A parking space is shaped like a square with a side length of 15 feet. What is the area of the parking space? 225 ft²

Taking Math Deeper

Exercise 18

There are a lot of words in the first sentence. Help your students see that it is possible to break the problem into doable parts.

 Translate words into math.
To find the temperature T in degrees Fahrenheit, take the number c of chirps per minute of a snowy tree cricket and subtract 40. Then, divide by 4. Then add 50.

$\boxed{c - 40}$ $\boxed{\div 4}$ $\boxed{+ 50}$

 Write a formula.
a. $T = \dfrac{c - 40}{4} + 50$

 Evaluate the formula.
b. When $c = 56$,
$T = \dfrac{56 - 40}{4} + 50$
$= 54°F$

c. When $c = 168$,
$T = \dfrac{168 - 40}{4} + 50$
$= 82°F$

Project
Investigate and write a report on the snowy tree cricket. Be sure to include mathematical information in your report.

Reteaching and Enrichment Strategies

If students need help...	If students got it...
Resources by Chapter • Practice A and Practice B • Puzzle Time Record and Practice Journal Practice Differentiating the Lesson Lesson Tutorials Skills Review Handbook	Resources by Chapter • Enrichment and Extension • School-to-Work • Financial Literacy Start the next section

12. **BASEBALL** A pitcher's earned run average is the average number of earned runs given up per nine innings. What is the earned run average of a pitcher who gave up 75 earned runs in 225 innings?

$$\text{Earned Run Average} = \frac{9R}{I}$$

- Number of earned runs
- Number of innings

Write a formula for the area of the shaded region in terms of x.

13.

14.

15.

16. **REASONING** You know a parallelogram's area and base. Explain how you can find its height.

17. **GOLD** The purity of gold is measured in karats or in percent. What number of karats represents 100% pure gold? Explain your reasoning.

$$P = (25 \cdot k) \div 6$$

- Percent
- Karats

18. **SNOWY TREE CRICKET** To find the temperature T in degrees Fahrenheit, take the number c of chirps per minute of a snowy tree cricket and subtract 40. Then, divide by 4. Then, add 50.

 a. Write a formula for the verbal description.
 b. In the morning, a cricket chirps 56 times in one minute. What is the temperature?
 c. Later in the afternoon, a cricket chirps 168 times in one minute. What is the temperature now?

19. **Geometry** Find the area of each region in the flag of the Bahamas.

Fair Game Review *What you learned in previous grades & lessons*

Estimate the sum or difference using benchmarks. *(Skills Review Handbook)*

20. $\dfrac{7}{8} + \dfrac{9}{10}$

21. $\dfrac{1}{6} + \dfrac{2}{5}$

22. $\dfrac{4}{7} - \dfrac{7}{12}$

23. $\dfrac{4}{5} - \dfrac{1}{9}$

24. **MULTIPLE CHOICE** Which expression represents "8 more than x"? *(Section 1.2)*

 Ⓐ $8 - x$ Ⓑ $8x$ Ⓒ $x + 8$ Ⓓ $\dfrac{8}{x}$

1.4–1.5 Quiz

Use the Distributive Property and mental math to find the product. *(Section 1.4)*

1. 6 × 49
2. 7 × 86
3. 35(47)
4. 8(97)

Use the Distributive Property to simplify the expression. *(Section 1.4)*

5. $5(x - 8)$
6. $7(y + 3)$

Use the Distributive Property to write and simplify an expression for the area of the rectangle. *(Section 1.4)*

7.
8.

Use a formula to find the area of the figure. *(Section 1.5)*

9.
10.

11.
12.

13. **MOVIES** You and four of your friends go to a movie. *(Section 1.4)*

 a. Use the Distributive Property to write an expression for the total cost to buy movie tickets and popcorn. Simplify the expression.

 Movie Tickets — Student $8, Adult $10
 Snacks — Candy $3, Popcorn $x

 b. Choose a reasonable value for x. Evaluate the expression.

14. **CALORIES** A formula for the number of calories c in s servings of a breakfast cereal is $c = 160s$. How many calories are there in 5 servings? *(Section 1.5)*

15. **FISH** An expression for the number of angelfish that you can keep in an aquarium is $g \div 3$, where the aquarium holds g gallons of water. How many angelfish can you keep in a 75-gallon aquarium? *(Section 1.5)*

Alternative Assessment Options

Math Chat
Structured Interview
Student Reflective Focus Question
Writing Prompt

Math Chat
- Have students work in pairs. One student describes the Distributive Property, giving examples. The other student probes for more information. Students then switch roles and repeat the process for how to use formulas to find the area of a figure.
- The teacher should walk around the classroom listening to the pairs and asking questions to ensure understanding.

Study Help Sample Answers

Remind students to complete Graphic Organizers for the rest of the chapter.

8.
9.

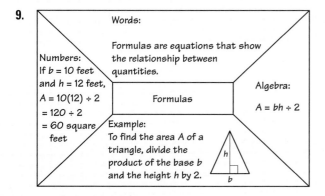

Reteaching and Enrichment Strategies

If students need help...	If students got it...
Resources by Chapter • Study Help • Practice A and Practice B • Puzzle Time Lesson Tutorials BigIdeasMath.com Practice Quiz Practice from the Test Generator	Resources by Chapter • Enrichment and Extension • School-to-Work Game Closet at *BigIdeasMath.com* Start the Chapter Review

Answers

1. 294
2. 602
3. 1645
4. 776
5. $5x - 40$
6. $7y + 21$
7. $5(x + 7) = 5x + 35$
8. $14(6 - x) = 84 - 14x$
9. 84 ft^2
10. 84 ft^2
11. 60 ft^2
12. 45 in.^2
13. a. $5(x + 8) = 5x + 40$
 b. *Sample answer:* $x = 4$; $60
14. 800 calories
15. 25 angelfish

Technology For the Teacher
Answer Presentation Tool

Assessment Book

For the Teacher
Additional Review Options
- Quiz*Show*
- Big Ideas Test Generator
- Game Closet at *BigIdeasMath.com*
- Vocabulary Puzzle Builder
- Resources by Chapter
 Puzzle Time
 Study Help

Answers
1. 4
2. 9
3. 100
4. 32
5. 16
6. 24
7. 80
8. 5
9. 12
10. 305
11. $b - 11$
12. $d \cdot 32$ or $32d$
13. $n + 18$ or $18 + n$
14. $t - 17$
15. **a.** $2p - 4$
 b. 44 points

Review of Common Errors

Exercises 1–10
- Students may substitute the wrong value for the variable.
- Tell students to write out the expression and then write the value of the variable needed before substituting the value into the expression.

Exercises 11–15
- Students may write the subtraction and/or division problems in the wrong order.
- For subtraction problems, have students look for the key words "from" and "than" to know that the order of the numbers must be switched from the way it is written.

Exercises 16–21
- Students often get confused about the difference between the Commutative and Associative Properties and will incorrectly label steps.
- Show students that the Associative Property moves the parentheses (but does not change the position of the numbers) so that there are different numbers or variables inside, while the Commutative Property changes the positions.

Exercises 22–28
- Students may forget to distribute to each part of the expression, especially when there are more than two parts within the parentheses.
- Tell students to write the expression on their paper and draw arrows to each part that needs to be distributed to, then distribute.

Exercises 29–32
- Students may use the wrong formula or forget to do all the operations in a formula.
- Tell students to write out the formula they need to use, then identify the value of each variable before substituting.

1 Chapter Review

Review Key Vocabulary

numerical expression, *p. 4*
algebraic expression, *p. 4*
evaluate, *p. 4*

equivalent expressions, *p. 16*
formula, *p. 30*
solve a formula, *p. 30*

Review Examples and Exercises

1.1 Evaluating Algebraic Expressions (pp. 2–7)

a. Evaluate $a \div b$ when $a = 48$ and $b = 8$.

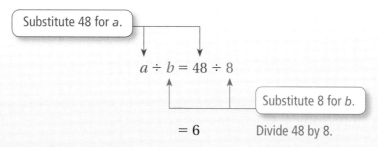

$$= 6 \qquad \text{Divide 48 by 8.}$$

b. Evaluate $2x + 21$ when $x = 6$.

$2x + 21 = 2(6) + 21$ Substitute 6 for x.
$\qquad\qquad = 12 + 21$ Using order of operations, multiply 2 and 6.
$\qquad\qquad = 33$ Add 12 and 21.

c. Evaluate $y^2 - 14$ when $y = 5$.

$y^2 - 14 = (5)^2 - 14$ Substitute 5 for y.
$\qquad\qquad = 25 - 14$ Using order of operations, evaluate 5^2.
$\qquad\qquad = 11$ Subtract 14 from 25.

Exercises

Evaluate the expression when $x = 20$ and $y = 4$.

1. $x \div 5$
2. $36 \div y$
3. $5x$
4. $8 \cdot y$
5. $x - y$
6. $y + x$
7. $x \cdot y$
8. $\dfrac{x}{y}$
9. $8y - x$

10. **GAMING** In a video game, you score p game points and b triple bonus points. An expression for your score is $p + 3b$. What is your score when you earn 245 game points and 20 triple bonus points?

1.2 Writing Expressions (pp. 8–13)

Write the phrase as an expression.

a. a number z decreased by 18

$z - 18$ The phrase "decreased by" means subtraction.

b. the sum of 7 and the product of a number x and 12

$7 + 12x$ The phrase "sum of" means addition. The phrase "product of" means multiplication.

Exercises

Write the phrase as an expression.

11. 11 fewer than a number b
12. the product of a number d and 32
13. 18 added to a number n
14. a number t decreased by 17

15. **BASKETBALL** Your basketball team scored 4 fewer than twice as many points as the other team.

 a. Write an expression for the number of points your team scored.
 b. The other team scored 24 points. How many points did your team score?

1.3 Properties of Addition and Multiplication (pp. 14–19)

a. Simplify the expression $(x + 18) + 4$.

$(x + 18) + 4 = x + (18 + 4)$ Associative Property of Addition
$= x + 22$ Add 18 and 4.

b. Simplify the expression $(5.2 + a) + 0$.

$(5.2 + a) + 0 = 5.2 + (a + 0)$ Associative Property of Addition
$= 5.2 + a$ Addition Property of Zero

c. Simplify the expression $36 \cdot r \cdot 1$.

$36 \cdot r \cdot 1 = 36 \cdot (r \cdot 1)$ Associative Property of Multiplication
$= 36 \cdot r$ Multiplication Property of One
$= 36r$

Exercises

Simplify the expression. Explain each step.

16. $10 + (2 + y)$
17. $(21 + b) + 1$
18. $3(7x)$
19. $1(3.2w)$
20. $5.3 + (w + 1.2)$
21. $(0 + t) + 9$

Review Game
Evaluating Expressions

Materials
- 1 number cube
- 1 pathway
- 2 chips

Players: 2

Big Ideas Game Closet

Directions
1. Players take turns rolling the number cube and substituting for *x* in the expression.
2. Players evaluate the expression and use that number to move their chip along the pathway. A positive answer results in a forward movement and a negative answer results in a backward movement.

Who Wins?
The player who reaches FINISH first wins.

For the Student Additional Practice
- Lesson Tutorials
- Study Help (textbook)
- Student Website
 Multi-Language Glossary
 Practice Assessments

Answers

16. $10 + (2 + y) =$
 $(10 + 2) + y$
 Assoc. Prop. of Add.
 $= 12 + y$
 Add 10 and 2.

17. $(21 + b) + 1 =$
 $(b + 21) + 1$
 Comm. Prop. of Add.
 $= b + (21 + 1)$
 Assoc. Prop. of Add.
 $= b + 22$
 Add 21 and 1.

18. $3(7x) =$
 $(3 \cdot 7)x$
 Assoc. Prop. of Mult.
 $= 21x$
 Multiply 3 and 7.

19. $1(3.2w) =$
 $(1 \cdot 3.2)w$
 Assoc. Prop. of Mult.
 $= 3.2w$
 Mult. Prop. of One

20–21. See Additional Answers.

22. $2x + 24$

23. $11b - 33$ 24. $8s - 8$

25. $36 + 6y$ 26. $25z - 100$

27. $35w - 70$ 28. $52

29. 160 ft^2 30. 45 ft^2

31. 120 in.^2 32. 200 ft^2

T-36

My Thoughts on the Chapter

What worked...

What did not work...

What I would do differently...

Teacher Tip

Not allowed to write in your teaching edition? Use sticky notes to record your thoughts.

1.4 The Distributive Property (pp. 22–27)

Use the Distributive Property to simplify $3(n + 9)$.

$3(n + 9) = 3(n) + 3(9)$ Distributive Property

$\qquad\quad = 3n + 27$ Multiply.

Exercises

Use the Distributive Property to simplify the expression.

22. $2(x + 12)$
23. $11(b - 3)$
24. $8(s - 1)$
25. $6(6 + y)$
26. $25(z - 4)$
27. $35(w - 2)$

28. **HAIRCUT** A family of four goes to a salon for haircuts. The cost of each haircut is $13. Use the Distributive Property and mental math to find the product 4×13 for the total cost.

1.5 Using Formulas to Solve Problems (pp. 28–33)

Find the area of the triangular sail.

The sail is a triangle, so use the formula for the area of a triangle.

$A = \dfrac{b \cdot h}{2}$ Write the formula.

$ = \dfrac{10 \cdot 35}{2}$ Substitute 10 for b and 35 for h.

$ = 175$ Simplify.

∴ The area of the sail is 175 square feet.

Exercises

Use a formula to find the area of the figure.

29.

30.

31.

32.

1 Chapter Test

Evaluate the expression when $a = 6$ and $b = 8$.

1. $4 + a$
2. $b + 1$
3. $a - 6$
4. $15 - b$

Write the phrase as an expression.

5. A number y decreased by 14
6. Twice a number z
7. 25 more than 50
8. 40 divided by 5

Simplify the expression. Explain each step.

9. $3.1 + (8.6 + m)$
10. $(10 \cdot n) \cdot 7$
11. $3(15w)$
12. $40 + (x + 12)$

Use the Distributive Property to simplify the expression.

13. $4(x + 8)$
14. $12(y - 5)$

Use a formula to find the area of the figure.

15.
7 m, 7 m

16.
8 in., 20 in.

17. **SOCCER GAME** Playing time is added at the end of a soccer game to make up for stoppages. An expression for the length of a 90-minute soccer game with x minutes of stoppage time is $90 + x$. How long is a game with 4 minutes of stoppage time?

18. **GEOMETRY** The expression $15 \cdot x \cdot 6$ represents the volume of a rectangular prism with a length of 15, a width of x, and a height of 6. Simplify the volume expression.

19. **VERTICAL JUMP** You can reach 71 inches high while standing. An expression for your vertical jump is $h - 71$, where h is the height you reach with a jump. What is your vertical jump when you reach 79 inches?

79 in.

20. **PARTY FAVORS** You make party favors for an event. You tie 9 inches of ribbon around each party favor. Write an expression for the amount of ribbon you need for n party favors. The ribbon costs $3 for each yard. Write an expression for the total cost of the ribbon.

Test Item References

Chapter Test Questions	Section to Review	Common Core State Standards
1–4, 17, 19	1.1	6.EE.1, 6.EE.2c
5–8, 20	1.2	6.EE.2a, 6.EE.2b
9–12	1.3	6.EE.3, 6.EE.6
13, 14	1.4	6.EE.3, 6.EE.6
15, 16, 18	1.5	6.EE.2c

Test-Taking Strategies

Remind students to quickly look over the entire test before they start so that they can budget their time. They should not spend too much time on any single problem. Urge students to try to work on a part of each problem because partial credit is better than no credit. Teach students to use the Stop and Think strategy before answering. **Stop** and carefully read the question, and **Think** about what the answer should look like.

For Your Information

- **Exercise 17** Students may not know the meaning of the word *stoppage*. Stoppage is discretionary time that is added to make up for time lost due to major fouls or injuries.

Common Assessment Errors

- **Exercises 1–4** Students may not replace the variable correctly. Sometimes it helps to have students rewrite the expression with parentheses and put the value of the variable in the parentheses.
- **Exercise 5** Many students miss this exercise because they read the problem incorrectly. The problem states "a number y decreased by 14," and many will write it as $14 - y$. Remind them that subtraction is not commutative.
- **Exercise 9** Students miss this exercise because of the decimal addition. Remind them to line up the decimal points before adding.
- **Exercises 13 and 14** The Distributive Property could be the reason why students miss these exercises. Have them review the property.

Reteaching and Enrichment Strategies

If students need help. . .	If students got it. . .
Resources by Chapter • Practice A and Practice B • Puzzle Time Record and Practice Journal Practice Differentiating the Lesson Lesson Tutorials Practice from the Test Generator Skills Review Handbook	Resources by Chapter • Enrichment and Extension • School-to-Work • Financial Literacy Game Closet at *BigIdeasMath.com* Start Standardized Test Practice

Answers

1. 10
2. 9
3. 0
4. 7
5. $y - 14$
6. $2z$ or $z \cdot 2$
7. $25 + 50$ or $50 + 25$
8. $40 \div 5$
9. $3.1 + (8.6 + m) =$
 $(3.1 + 8.6) + m$
 Assoc. Prop. of Add.
 $= 11.7 + m$
 Add 3.1 and 8.6.
10. $(10 \cdot n) \cdot 7 =$
 $(n \cdot 10) \cdot 7$
 Comm. Prop. of Mult.
 $= n \cdot (10 \cdot 7)$
 Assoc. Prop. of Mult.
 $= n \cdot 70$
 Multiply 10 and 7.
 $= 70n$
 Comm. Prop. of Mult.

11–12. See Additional Answers.

13. $4x + 32$
14. $12y - 60$
15. 49 m^2
16. 80 in.2
17. 94 min
18. $90x$
19. 8 in.
20. $9n; 3\left(\dfrac{1}{4}n\right)$

Assessment Book

Test-Taking Strategies
Available at *BigIdeasMath.com*

After Answering Easy Questions, Relax
Answer Easy Questions First
Estimate the Answer
Read All Choices before Answering
Read Question before Answering
Solve Directly or Eliminate Choices
Solve Problem before Looking at Choices
Use Intelligent Guessing
Work Backwards

About this Strategy

When taking a multiple choice test, be sure to read each question carefully and thoroughly. Before answering a question, determine exactly what is being asked, then eliminate the wrong answers and select the best choice.

Answers

1. C
2. F
3. $26
4. B

Item Analysis

1. **A.** The student does not consider that there is a cost for *each* game. The student adds $2.00 and $0.25 and does not use a variable to represent the number of games played.
 B. The student multiplies the number of games played by the sum of the admission price and the games price. This expression multiplies the admission price by the number of games played, not just the price for each game.
 C. Correct answer
 D. The student uses *x* to represent the number of admissions, not the number of games played. This expression assumes that there is a flat rate for the games, rather than for the admission.

2. **F.** Correct answer
 G. The student misidentifies the property being used as the Identity Property of Multiplication.
 H. The student misidentifies the property being used as the Associative Property of Multiplication.
 I. The student misidentifies the property being used as the Commutative Property of Multiplication.

3. **Gridded Response:** Correct answer: $26

 Common Error: The student multiplies the cost of a hardcover book by the number of paperback books and the cost of a paperback book by the number of hardcover books to get an answer of $24.

4. **A.** The student does not recognize that this equation is a correct example of the Commutative Property of Multiplication.
 B. Correct answer
 C. The student does not recognize that this equation is a correct example of the Commutative Property of Addition.
 D. The student does not recognize that this equation is a correct example of the Identity Property of Addition.

Standardized Test Practice Icons

 Gridded Response

 Short Response (2-point rubric)

 Extended Response (4-point rubric)

Technology For the Teacher
Big Ideas Test Generator

1 Standardized Test Practice

1. The student council is organizing a school fair. Council members are making signs to show the prices for admission and for each game a person can play.

 SCHOOL FAIR
 Admission $2.00
 Price per Game $0.25

 Let x represent the number of games. Which expression can be used to determine the total amount, in dollars, a person pays for admission and playing x games? *(6.EE.6)*

 A. 2.25

 B. 2.25x

 C. $2 + 0.25x$

 D. $2x + 0.25$

 Test-Taking Strategy
 Solve Directly or Eliminate Choices

 How many hyenas are $5 - 2^2 - 1$ hyenas?
 Ⓐ 0 Ⓑ 2 Ⓒ 8 Ⓓ 10

 Survival strategy: A

 "Which strategy would you use on this one: solve directly or eliminate choices?"

2. Which property does the equation below represent? *(6.EE.3)*

 $$17 \cdot 44 + 17 \cdot 56 = 17 \cdot 100$$

 F. Distributive Property

 G. Identity Property of Multiplication

 H. Associative Property of Multiplication

 I. Commutative Property of Multiplication

3. At a used book store, two types of books can be purchased.

 The expression $3h + 2p$ can be used to find the total cost for h hardcover books and p paperback books. What is the total cost, in dollars, for 6 hardcover books and 4 paperback books? *(6.EE.2c)*

 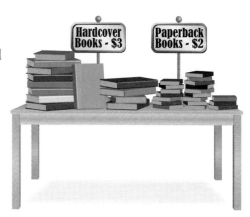

4. Which equation is *not* true for all numbers a and b? *(6.EE.4)*

 A. $a \times b = b \times a$

 B. $a - b = b - a$

 C. $a + b = b + a$

 D. $a + b + 0 = a + b$

Standardized Test Practice 39

5. You are selling items for a fundraiser. You are selling school folders for 75 cents each and school pencils for 20 cents each. Let *f* represent the number of folders and *p* represent the number of pencils. Which expression can be used to determine the total amount of money, in cents, you earn? *(6.EE.6)*

 F. $75f \cdot 20p$ H. $95(f \cdot p)$

 G. $75f + 20p$ I. $95(f + p)$

6. Properties of addition and multiplication are used to simplify an expression.

 $$36 \cdot 23 + 33 \cdot 64 = 36 \cdot 23 + 64 \cdot 33$$
 $$= 36 \cdot 23 + 64 \cdot (23 + 10)$$
 $$= 36 \cdot 23 + 64 \cdot 23 + 64 \cdot 10$$
 $$= x \cdot 23 + 64 \cdot 10$$
 $$= 2300 + 640$$
 $$= 2940$$

 What number belongs in place of the *x*? *(6.EE.3)*

7. Which property was used to simplify the expression? *(6.EE.3)*

 $$(47 \times 125) \times 8 = 47 \times (125 \times 8)$$
 $$= 47 \times 1000$$
 $$= 47{,}000$$

 A. Distributive Property

 B. Identity Property of Multiplication

 C. Associative Property of Multiplication

 D. Commutative Property of Multiplication

8. What is the value of the expression below when $a = 5$, $b = 7$, and $c = 6$? *(6.EE.2c)*

 $$9b - 4a + 2c$$

 F. 29 H. 55

 G. 31 I. 78

Item Analysis (continued)

5. **F.** The student multiplies the earnings from the folders by the earnings from the pencils.

 G. Correct answer

 H. The student adds the prices of a folder and a pencil and then multiplies this sum by the product of the number of folders and the number of pencils.

 I. The student adds the prices of a folder and a pencil and then multiplies this sum by the sum of the number of folders and the number of pencils.

6. **Gridded Response:** Correct answer: 100

 Common Error: The student does not recognize the application of the Distributive Property and instead tries to work backwards from 2300. The student makes a place value error while dividing 2300 by 23 to get 1000.

7. **A.** The student misidentifies the property being used as the Distributive Property.

 B. The student misidentifies the property being used as the Identity Property of Multiplication.

 C. Correct answer

 D. The student misidentifies the property being used as the Commutative Property of Multiplication.

8. **F.** The student substitutes 5 for b and 7 for a.

 G. The student makes an order of operations error, adding $4a$ to $2c$ and then subtracting this sum from $9b$.

 H. Correct answer

 I. The student makes two-digit numbers out of the coefficient and the variable, such as 97 instead of 9×7.

9. **A.** The student does not distribute the multiplication by a to the second addend, c.

 B. Correct answer

 C. The student incorrectly distributes the addition of a instead of multiplication by a.

 D. The student incorrectly distributes the addition of a instead of multiplication by a and then multiplies the resulting sums.

10. **F.** Correct answer

 G. The student multiplies the length of a red-eye tetra by the number of banded rainbows, and the length of a banded rainbow by the number of red-eye tetras.

 H. The student multiplies the lengths of the two fish and then multiplies this product by the sum of the numbers of the two fish.

 I. The student adds the lengths of the two fish and then multiplies this sum by the sum of the numbers of the two fish.

Answers

5. G
6. 100
7. C
8. H

Answers

9. B
10. F
11. 1800
12. D
13. H

Answers for Extra Examples

1. **A.** The student misapplies the Zero Property of Multiplication, thinking that any number multiplied by 0 equals itself.
 B. The student misapplies the Identity Property of Addition, thinking that adding 1 to any number equals the number itself.
 C. Correct answer
 D. The student misapplies the Identity Property of Multiplication, thinking that any number multiplied by itself equals itself.

2. **F.** The student misinterprets "more than" as meaning multiplication.
 G. Correct answer
 H. The student misinterprets the inverse relationship in the problem, thinking that the town received more rain on Tuesday than on Wednesday.
 I. The student subtracts in the wrong order and writes an expression that represents 17 millimeters less rain than the difference of the rain amounts on Tuesday and Wednesday.

Item Analysis (continued)

11. **2 points** The student demonstrates a thorough understanding of the Associative and Commutative Properties of Multiplication. For example, the student explains that the Commutative Property can be used to rewrite the expression as $(18 \times 25) \times 4$. The student then explains that the Associative Property can be used to rewrite the expression as $18 \times (25 \times 4)$. The student then uses compatible numbers to get 18×100 and a final answer of 1800.

 1 point The student demonstrates a partial understanding of the Associative and Commutative Properties of Multiplication. For example, the student shows correct, efficient steps but provides no explanation.

 0 points The student demonstrates insufficient understanding of the Associative and Commutative Properties of Multiplication. The student either simplifies the expression inefficiently (such as left-to-right) or simplifies the expression incorrectly.

12. **A.** The student makes two-digit numbers out of the coefficient and the variable, such as 36 instead of 3×6.
 B. The student substitutes 9 for x and 6 for y.
 C. The student uses the values of the variables as coefficients.
 D. Correct answer

13. **F.** The student subtracts the two terms in the wrong order.
 G. The student confuses the concept of "less than" with the concept of "is less than."
 H. Correct answer
 I. The student confuses the concept of "less than" with the concept of "is less than."

Extra Examples for Standardized Test Practice

1. Which equation is true for all values of n?

 A. $n \times 0 = n$ **C.** $n - n = 0$
 B. $n + 1 = n$ **D.** $n \times n = n$

2. On Wednesday, the town of Mims received 17 millimeters of rain. This was x millimeters more rain than the town received on Tuesday. Which expression represents the amount of rain, in millimeters, the town received on Tuesday?

 F. $17x$ **H.** $x + 17$
 G. $17 - x$ **I.** $x - 17$

9. Which equation correctly demonstrates the Distributive Property? *(6.EE.4)*

 A. $a(b + c) = ab + c$

 B. $a(b + c) = ab + ac$

 C. $a + (b + c) = (a + b) + (a + c)$

 D. $a + (b + c) = (a + b) \cdot (a + c)$

10. A common rule for deciding how many tropical fish to put into an aquarium is

 1 inch of fish for every 1 gallon of water.

 You are considering putting two types of fish into your new 60-gallon aquarium.

 Let t represent the number of red-eye tetras and b represent the number of banded rainbows. Which equation can be used to determine how many of each type of fish you can put into your aquarium? *(6.EE.6)*

 Red-eye Tetra

 |— 3 inches —|

 Banded Rainbow

 |—— 5 inches ——|

 F. $3t + 5b = 60$

 G. $3b + 5t = 60$

 H. $15(t + b) = 60$

 I. $8(b + t) = 60$

11. Use the properties of multiplication to simplify the expression in an efficient way.

 $$(25 \times 18) \times 4$$

 Show your work and explain how you used the properties of multiplication. *(6.EE.3)*

12. You evaluated an expression using $x = 6$ and $y = 9$. You correctly got an answer of 105. Which expression did you evaluate? *(6.EE.2c)*

 A. $3x + 6y$

 B. $5x + 10y$

 C. $6x + 9y$

 D. $10x + 5y$

13. Camilla and Debbie went to the beach and collected seashells. The number of seashells Debbie collected was 3 less than twice the number of shells Camilla collected. Let c represent the number of seashells Camilla collected. Which of the following represents the number of seashells Debbie collected? *(6.EE.6)*

 F. $3 - 2c$

 G. $3 < 2c$

 H. $2c - 3$

 I. $2c < 3$

2 Multiplying and Dividing Fractions

- 2.1 Fractions and Estimation
- 2.2 Multiplying Fractions and Whole Numbers
- 2.3 Multiplying Fractions
- 2.4 Multiplying Mixed Numbers
- 2.5 Dividing Fractions
- 2.6 Dividing Mixed Numbers
- 2.7 Writing Decimals as Fractions
- 2.8 Writing Fractions as Decimals

"Dear Sir: You say that MOST humans use only a fraction of their brain power."

"But, $\frac{3}{2}$ is a fraction. So, does that mean that SOME humans use one and a half of their brain power?"

"One of my homework problems is 'How many halves are in five halves?'"

Connections to Previous Learning

- Simplify fractions.
- Write common fractions as decimals and decimals as fractions.
- Estimate fractions in real-world problems.
- Write equivalent fractions.

- Represent addition and subtraction of fractions using models with like and unlike denominators.
- Add and subtract fractions efficiently.

- Explain and justify procedures for multiplication and division of fractions.
- Multiply and divide fractions efficiently, including real-world problems.
- Estimate the results of computations with fractions, and verify reasonableness.

Math in History

The convention of writing a fraction as numerator over denominator is relatively recent. Early cultures used a variety of methods to write fractions.

★ Here are the symbols for Egyptian numbers.

1	10	100	1,000	10,000	100,000	10^6

Egyptians only wrote fractions whose numerators are one. To write a fraction, they wrote the denominator under an "eye."

For example, the fraction $\frac{1}{5}$ is shown at the right.

★ Al-Hassar (12th century), a mathematician from Maghreb (North Africa), developed the modern symbolic mathematical notation for fractions, where the numerator and denominator are separated by a horizontal bar. This same fractional notation appears soon after in the work of Fibonacci in the 13th century.

Pacing Guide for Chapter 2

Chapter Opener	1 Day
Section 1	
Activity	1 Day
Lesson	1 Day
Section 2	
Activity	1 Day
Lesson	1 Day
Section 3	
Activity	1 Day
Lesson	1 Day
Section 4	
Activity	1 Day
Lesson	1 Day
Study Help / Quiz	1 Day
Section 5	
Activity	1 Day
Lesson	2 Days
Section 6	
Activity	1 Day
Lesson	1 Day
Section 7	
Activity	1 Day
Lesson	1 Day
Section 8	
Activity	1 Day
Lesson	1 Day
Quiz / Chapter Review	1 Day
Chapter Test	1 Day
Standardized Test Practice	1 Day
Total Chapter 2	22 Days
Year-to-Date	37 Days

Check Your Resources for

- Record and Practice Journal
- Resources by Chapter
- Skills Review Handbook
- Assessment Book
- Worked-Out Solutions

The Dynamic Planning Tool
Editable Teacher's Resources at
BigIdeasMath.com

Common Core State Standards

5.NF.1 Add and subtract fractions with unlike denominators . . . by replacing given fractions with equivalent fractions in such a way as to produce an equivalent sum or difference of fractions with like denominators.

4.NF.3c Add and subtract mixed numbers with like denominators, e.g., by replacing each mixed number with an equivalent fraction

Additional Topics for Review
- Least Common Multiple
- Greatest Common Factor
- Division Algorithm

Try It Yourself

1. $\frac{3}{16}$
2. $\frac{23}{45}$
3. $\frac{3}{25}$
4. $\frac{13}{28}$
5. $\frac{7}{2}$
6. $\frac{82}{11}$
7. $\frac{29}{5}$
8. $\frac{32}{3}$
9. $1\frac{3}{25}$
10. $33\frac{1}{3}$
11. $1\frac{2}{15}$
12. $5\frac{7}{10}$

Record and Practice Journal

1. $\frac{5}{6}$
2. $\frac{29}{24}$
3. $\frac{61}{45}$
4. $\frac{1}{2}$
5. $\frac{1}{14}$
6. $\frac{76}{165}$
7. $\frac{13}{24}$ inch
8. $\frac{7}{3}$
9. $\frac{23}{5}$
10. $\frac{99}{10}$
11. $\frac{47}{8}$
12. $1\frac{5}{7}$
13. $2\frac{8}{11}$
14. $8\frac{1}{4}$
15. $4\frac{3}{20}$
16. $\frac{11}{4}$; 11 one-fourth cups

Math Background Notes

Vocabulary Review
- Numerator
- Mixed Number
- Denominator
- Multiple
- Improper Fraction
- Factor

Adding and Subtracting Fractions
- Students should know how to add and subtract fractions. In this chapter, students will learn to multiply and divide fractions.
- Remind students that adding and subtracting fractions always requires a common denominator.
- If two fractions already share a common denominator, simply add or subtract their numerators and keep the common denominator. If two fractions do not have a common denominator, students will need to find one.
- You may want to review finding the Least Common Multiple between two numbers. Remind students that finding the Least Common Multiple among the denominators will produce the least common denominator used to add or subtract fractions.
- You may want to review how to rename fractions using the least common denominator.

Writing Mixed Numbers and Improper Fractions
- Students should be familiar with mixed numbers and improper fractions, although these vocabulary terms might be new to them.
- Remind students that mixed numbers and improper fractions represent the same number in two different ways, much like the relationship between fractions and decimals.
- Both mixed numbers and improper fractions represent parts greater than one whole. Review adding and subtracting fractions (see above) to help students rewrite mixed numbers as improper fractions.
- Review the division algorithm (long division) to help students prepare to rewrite improper fractions as mixed numbers.

Reteaching and Enrichment Strategies

If students need help. . .	If students got it. . .
Record and Practice Journal • Fair Game Review Skills Review Handbook Lesson Tutorials	Game Closet at *BigIdeasMath.com* Start the next section

What You Learned Before

"On a scale from 1 to 10, how do you like my painting?"

● **Adding and Subtracting Fractions** (5.NF.1)

Example 1 Find $\frac{3}{4} + \frac{1}{8}$.

$\frac{3}{4} + \frac{1}{8} = \frac{3 \cdot 2}{4 \cdot 2} + \frac{1}{8}$ ← Use the Least Common Multiple to find a common denominator.

$= \frac{6}{8} + \frac{1}{8}$

$= \frac{7}{8}$

Example 2 Find $\frac{5}{6} - \frac{1}{5}$.

$\frac{5}{6} - \frac{1}{5} = \frac{5 \cdot 5}{6 \cdot 5} - \frac{1 \cdot 6}{5 \cdot 6}$

$= \frac{25}{30} - \frac{6}{30}$

$= \frac{19}{30}$

Try It Yourself

Find the sum or difference.

1. $\frac{1}{16} + \frac{1}{8}$
2. $\frac{1}{15} + \frac{4}{9}$
3. $\frac{1}{5} - \frac{2}{25}$
4. $\frac{3}{4} - \frac{2}{7}$

● **Writing Mixed Numbers and Improper Fractions** (4.NF.3c)

Example 3
Write $2\frac{3}{4}$ as an improper fraction.

$2\frac{3}{4} = 2 + \frac{3}{4}$

$= \frac{8}{4} + \frac{3}{4}$

$= \frac{11}{4}$

Example 4
Write $\frac{23}{2}$ as a mixed number.

$\begin{array}{r} 11 \text{ R}1 \\ 2\overline{)23} \end{array}$

$\frac{23}{2} = 11\frac{1}{2}$

Try It Yourself

Write the mixed number as an improper fraction.

5. $3\frac{1}{2}$
6. $7\frac{5}{11}$
7. $5\frac{4}{5}$
8. $10\frac{2}{3}$

Write the improper fraction as a mixed number.

9. $\frac{28}{25}$
10. $\frac{100}{3}$
11. $\frac{17}{15}$
12. $\frac{57}{10}$

2.1 Fractions and Estimation

Essential Question How can you use estimation to check that your answer is reasonable?

COMMON CORE STATE STANDARDS
6.NS.1

1 ACTIVITY: Using Models for Fractions

Work with a partner. Use the model for the whole to draw a model for the given fractions.

Whole	Model for the Whole	Fractions	Model for Fraction
a. Sample: Circle	○	$\dfrac{5}{8}$	
b. Circle	○	$\dfrac{3}{4}, \dfrac{5}{12}, \dfrac{4}{6}$	
c. Rectangle	▭	$\dfrac{3}{5}, \dfrac{4}{5}, \dfrac{7}{10}$	
d. Counters	○○○○ ○○○○	$\dfrac{1}{2}, \dfrac{3}{8}, \dfrac{3}{4}$	
e. Piece of paper	▭	$\dfrac{7}{8}, \dfrac{1}{8}, \dfrac{1}{4}$	

2 ACTIVITY: Estimating Sums and Differences

Work with a partner. Add or subtract. Then check your answer by using one of the models in Activity 1 to estimate the sum or difference.

a. Sample:

$$\dfrac{1}{6} + \dfrac{1}{4} = \dfrac{2}{12} + \dfrac{3}{12}$$ Write with common denominator.

$$= \dfrac{2+3}{12}$$ Add numerators.

$$= \dfrac{5}{12}$$ Simplify.

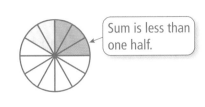

Sum is less than one half.

b. $\dfrac{1}{3} + \dfrac{1}{4}$ **c.** $\dfrac{5}{8} + \dfrac{1}{3}$ **d.** $\dfrac{7}{8} - \dfrac{1}{3}$ **e.** $\dfrac{2}{3} - \dfrac{4}{9}$

Chapter 2 Multiplying and Dividing Fractions

Laurie's Notes

Introduction

For the Teacher
- **Goal:** The goal of this activity is for students to use a variety of models to estimate the answers to fraction computations.

Motivate
- Have students watch you fold a piece of paper in half three times. Without unfolding the paper, ask "What will the paper look like when it is opened?" It will show eight equal rectangles. (Unfold the paper to verify the answer.)
- "If a fourth fold is done, how many equal parts will there be when the paper is unfolded?" 16
- Fold the paper a fourth time and confirm that it is folded into 16 equal parts.
- How would you represent $\frac{5}{16}$ on this paper? Shade 5 of the 16 sections.

Activity Notes

Activity 1
- Discuss two of the common models for fractions, the part-to-whole model and the set model.
- **Set Model:** In a set model, a collection of objects (e.g., students in your class) is the whole and subsets of the whole (the number of girls) make up the fractional parts.
- **Part-to-Whole Model:** The part-to-whole model, also known as the area model, was demonstrated in the paper folding activity. The whole (the sheet of paper) was divided into parts of equal area. Students sometimes forget that the parts must be equal in area.
- Show these models to the students. Ask "Which of these are models of $\frac{1}{4}$?"

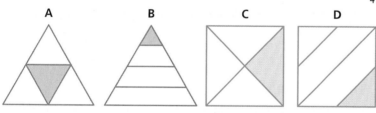

Models A and C represent $\frac{1}{4}$. The other two models do not have four sections of equal area, so they can't represent $\frac{1}{4}$.

- Students should do the activity with a partner. Students might recognize the challenge of making equal parts in a circle. The rectangle is easier.

Activity 2
- Students should be familiar with adding and subtracting fractions. The sample will remind students of the need for a common denominator.
- **Common Error:** Watch for students who add denominators: $\frac{1}{3} + \frac{1}{4} = \frac{2}{7}$.

Common Core State Standards
6.NS.1 Interpret and compute quotients of fractions, and solve word problems involving division of fractions by fractions.

Previous Learning
Students should know how to add and subtract fractions with like and unlike denominators, verify the reasonableness of results, and make reasonable estimates.

Activity Materials
Introduction
• paper

Start Thinking! and Warm Up

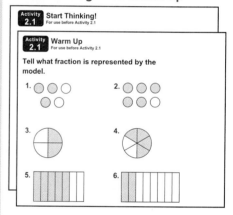

2.1 Record and Practice Journal

T-44

Differentiated Instruction

Auditory

Have students verbally describe the process for adding, subtracting, multiplying, and dividing fractions, as well as for rewriting fractions and simplifying fractions. Throughout the chapter, encourage students to use the words *numerator, denominator, least common denominator, add, subtract, multiply, divide, common factor, improper fraction, reciprocal,* and *simplify*.

2.1 Record and Practice Journal

Laurie's Notes

Activity 3

- ❓ "How would you describe what 3×4 means?" Listen for students to mention repeated addition (3 groups of 4 = 12), or count by 3s, four times (3, 6, 9, 12).

- ❓ "When you multiply two numbers less than 1, such as $\frac{2}{3} \times \frac{7}{8}$, is the answer *greater than* or *less than* 1?" less than 1; help students think of this as taking two-thirds of $\frac{7}{8}$, or as taking a part of $\frac{7}{8}$.

- **Understanding the Model:** The rectangle is a whole, or 1. Make sure that students see that seven of the eight equal parts are shaded blue to represent $\frac{7}{8}$. The blue portion $\left(\frac{7}{8}\right)$ is then divided into three equal parts or thirds. Two of those thirds represent the product of $\frac{2}{3}$ and $\frac{7}{8}$.

- After students have worked parts (b)–(d), have them share their estimates and explain the models they used.

Activity 4

- **Discuss:** Begin with a discussion of what division means. In the sample, a fraction is being divided into two equal parts and you want to know the size of the part. This is different than thinking "how many groups of two are there in five-ninths?"

- **Understanding the Model:** The rectangle is a whole, or 1, and the blue section is $\frac{5}{9}$. Divide $\frac{5}{9}$ into two equal parts. The quotient is one of the two parts.

- Parts (b)–(d) may be challenging for students. Instead of following a rule, they are being asked to think visually about what it means to divide a simple fraction into a whole number of parts. For part (c), think of $\frac{1}{2}$ of a pan of brownies that you divide into 8 equal pieces.

- ❓ Remind students that you name the part by comparing it to the whole—in this case, the whole pan of brownies. Ask, "how many brownies were in the pan before any were taken?" 16

- ❓ How much does one brownie represent? One brownie represents 1 out of 16, or $\frac{1}{16}$.

What Is Your Answer?

- You may want to have students think of contexts for their examples, too.

Closure

- **Exit Ticket:** Estimate the answer and draw a model.

 a. $\frac{1}{2} + \frac{1}{3}$ b. $\frac{1}{2} - \frac{1}{3}$ c. $\frac{1}{2} \times \frac{1}{3}$ d. $\frac{1}{2} \div 3$

Technology For the Teacher

The Dynamic Planning Tool
Editable Teacher's Resources at *BigIdeasMath.com*

3 ACTIVITY: Estimating Products

Work with a partner. Use a fraction model to choose $0, \frac{1}{4}, \frac{1}{2}, \frac{3}{4},$ or 1 as the best estimate of the product.

a. Sample: $\frac{2}{3} \times \frac{7}{8}$

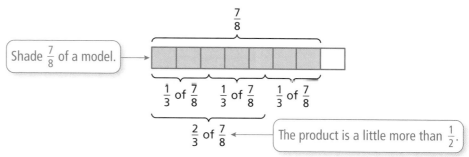

So, the best estimate is $\frac{1}{2}$.

b. $\frac{1}{5} \times \frac{3}{10}$ c. $\frac{3}{4} \times \frac{5}{7}$ d. $\frac{7}{8} \times \frac{7}{8}$

4 ACTIVITY: Estimating Quotients

Work with a partner. Use a fraction model to choose $0, \frac{1}{4}, \frac{1}{2}, \frac{3}{4},$ or 1 as the best estimate of the quotient.

a. Sample: $\frac{5}{9} \div 2$

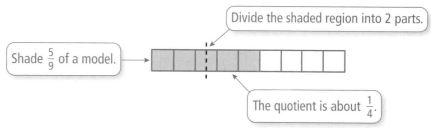

So, the best estimate is $\frac{1}{4}$.

b. $\frac{3}{5} \div 3$ c. $\frac{1}{2} \div 8$ d. $\frac{5}{6} \div 2$

What Is Your Answer?

5. **IN YOUR OWN WORDS** How can you use estimation to check that your answer is reasonable? Give some examples.

Practice Use what you learned about estimation to complete Exercises 7–14 on page 48.

2.1 Lesson

EXAMPLE 1 — Estimating Products

Key Vocabulary
underestimate, *p. 47*
overestimate, *p. 47*
compatible numbers, *p. 47*

Estimate the product by rounding to $0, \frac{1}{2}$, or 1.

a. $\dfrac{3}{8} \times \dfrac{11}{12}$

b. $\dfrac{4}{5} \times \dfrac{1}{6}$

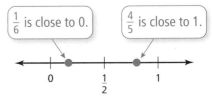

$\dfrac{3}{8} \times \dfrac{11}{12} \approx \dfrac{1}{2} \times 1 = \dfrac{1}{2}$

$\dfrac{4}{5} \times \dfrac{1}{6} \approx 1 \times 0 = 0$

∴ $\dfrac{3}{8} \times \dfrac{11}{12}$ is about $\dfrac{1}{2}$.

∴ $\dfrac{4}{5} \times \dfrac{1}{6}$ is about 0.

EXAMPLE 2 — Estimating with Mixed Numbers

Estimate the product or quotient by rounding each mixed number to the nearest whole number.

a. $5\dfrac{1}{4} \times 3\dfrac{9}{10}$

b. $11\dfrac{5}{6} \div 2\dfrac{2}{3}$

$5\dfrac{1}{4}$ is closer to 5 than to 6.

$11\dfrac{5}{6}$ is closer to 12 than to 11.

Reading

The symbol ≈ means "approximately equal to."

$5\dfrac{1}{4} \times 3\dfrac{9}{10} \approx 5 \times 4 = 20$

$11\dfrac{5}{6} \div 2\dfrac{2}{3} \approx 12 \div 3 = 4$

$3\dfrac{9}{10}$ is closer to 4 than to 3.

$2\dfrac{2}{3}$ is closer to 3 than to 2.

∴ $5\dfrac{1}{4} \times 3\dfrac{9}{10}$ is about 20.

∴ $11\dfrac{5}{6} \div 2\dfrac{2}{3}$ is about 4.

On Your Own

Now You're Ready
Exercises 7–22

Estimate the product or quotient.

1. $\dfrac{1}{9} \times \dfrac{4}{5}$

2. $\dfrac{9}{10} \times \dfrac{5}{12}$

3. $2\dfrac{7}{8} \times 6\dfrac{1}{3}$

4. $24\dfrac{1}{5} \div 3\dfrac{1}{2}$

Laurie's Notes

Introduction

Connect
- **Yesterday:** Students used visual models to estimate the sums, differences, products, and quotients of fractions.
- **Today:** Students will use rounding strategies to estimate products and quotients involving fractions and mixed numbers.

Motivate
- This introductory activity will help students develop their ability to estimate fractions and mixed numbers as close to 0, close to $\frac{1}{2}$, close to 1, etc.
- Write the numbers "0" and "1" on index cards. In another color, write proper fractions and mixed numbers between 1 and 2 on index cards.
- Have two students hold a clothesline to represent a number line. Use clothespins, tape, or paper clips to place the numbers 0 and 1 on the clothesline, as if you were placing them on a number line.

- Ask volunteers to locate the fractions and mixed numbers on the number line. Ask them to explain how they determined the locations. In talking about their estimates, listen for language such as $\frac{3}{8}$ is close to $\frac{4}{8}$, which is $\frac{1}{2}$.

Lesson Notes

Words of Wisdom
- Students have not learned the formal rules for multiplying and dividing fractions and mixed numbers. Tell students that today they will simply round fractions to a benchmark value (0, $\frac{1}{2}$, or 1) and will round mixed numbers to a whole number before they multiply or divide.

Example 1
- Work through both parts of the example. The number line is used as a visual reference to help students estimate the size of the fraction (close to 0, $\frac{1}{2}$, 1, or greater than 1). Note that the numbers are rounded, and then the operation (\times or \div) is performed.
- Remember that \approx means "approximately equal to."

Example 2
- Students are asked to round the mixed numbers to the nearest whole number before finding the product or quotient. Work through each example.
- A number line could be used as a visual for rounding.

On Your Own
- **Neighbor Check:** Have students work independently and then have their neighbor check their work. Have students discuss any discrepancies.

Goal Today's lesson is estimating products and quotients involving fractions and mixed numbers.

Lesson Materials
Introduction
• clothesline (6' long)
• index cards
• clothespins or tape

Start Thinking! and Warm Up

Lesson 2.1 Warm Up For use before Lesson 2.1

Lesson 2.1 Start Thinking! For use before Lesson 2.1
What are some advantages of using an estimate instead of finding an exact answer?

Extra Example 1
Estimate the product by rounding to 0, $\frac{1}{2}$, or 1.

a. $\frac{1}{8} \times \frac{7}{8}$ 0

b. $\frac{9}{10} \times \frac{6}{7}$ 1

Extra Example 2
Estimate the product or quotient by rounding each mixed number to the nearest whole number.

a. $8\frac{1}{3} \times 11\frac{2}{3}$ 96

b. $35\frac{3}{4} \div 6\frac{3}{8}$ 6

On Your Own

1. 0
2. $\frac{1}{2}$
3. 18
4. 6

English Language Learners
Vocabulary

Discuss the meaning of *compatible numbers* as *numbers that are easy to evaluate using mental math*. For example, 25 is a good substitute for 26 in $\frac{3}{5} \times 26$ because 25 is evenly divisible by 5. Ask students for compatible numbers for 18, 36, and 40. Write the compatible numbers on the board and have students explain why the numbers are compatible.

Extra Example 3
One gallon of stain covers 400 square feet. Is one gallon enough to cover a rectangular deck with a width of $15\frac{1}{2}$ feet and a length of $23\frac{3}{4}$ feet? Explain. *Yes, because 384 is an overestimate and is less than 400.*

On Your Own
5. yes; The overestimate for the area of the wall is 650 square feet and the two gallons of paint will cover 700 square feet.

Extra Example 4
A glass holds $1\frac{3}{4}$ cups of water. Estimate the number of glasses needed to fill a pitcher that holds 11 cups of water. *about 6 glasses*

On Your Own
6. about 50 panthers

Laurie's Notes

Words of Wisdom
- Before doing Example 3, talk about rounding numbers in a context. For example, when you use rounding while shopping, you typically round prices up, producing an **overestimate**. You want to make sure that you have enough money.
- ? "Can you think of other contexts involving fractions or decimals where you round up?"

Example 3
- ? "How do you find the area of a rectangle?" *Multiply the length and width.*
- ? "Do you think you want to *overestimate* or *underestimate* the amount of paint you need? Why?" *You want to overestimate the area of the wall so that you have enough paint. It's better to have too much paint than too little.*
- **Common Error:** Students often read too quickly and fail to answer the question being asked. Although they need to find the area of the rectangle, that is not the question being asked. Caution students to read carefully.

Example 4
- Discuss the meaning of the word *range* in this context. A similar context is the range a car can drive on a full tank of gas. The word range can also be used to describe the difference between data values, such as the range of students' heights, or the range of the prices of a product in a store.
- Review compatible numbers. Students may recall using compatible numbers when estimating the quotient of $176 \div 32$; $176 \approx 180$ and $32 \approx 30$, so $180 \div 30 = 6$.

On Your Own
- **Think-Pair-Share:** Students should read each question independently and then work with a partner to answer the questions. When they have answered the questions, the pair should compare their answers with another group and discuss any discrepancies.

Closure
- **Writing Prompt:** To estimate the product of two fractions... *Round each fraction to 0, $\frac{1}{2}$, or 1 and then multiply.*

The Dynamic Planning Tool
Editable Teacher's Resources at *BigIdeasMath.com*

An **underestimate** is an estimate that is less than the exact answer while an **overestimate** is greater than the exact answer.

EXAMPLE 3 Using an Overestimate

One gallon of paint covers 350 square feet. Is 1 gallon of paint enough to cover the rectangular ceiling? Explain.

$A = \ell w$ Write the formula for the area of a rectangle.

$= 27\frac{5}{6} \times 10\frac{1}{4}$ Substitute for ℓ and w.

$\approx 28 \times 11$ Round $27\frac{5}{6}$ up to 28. Round $10\frac{1}{4}$ up to 11.

$= 308$ Multiply.

∴ Because 308 is an overestimate and is less than 350, 1 gallon of paint is enough.

On Your Own

5. In Example 3, a hallway wall is $9\frac{3}{4}$ feet by $64\frac{1}{3}$ feet. Are 2 gallons of paint enough to cover the wall? Explain.

Compatible numbers are numbers that are easy to compute mentally.

EXAMPLE 4 Using a Compatible Number

Reading
The term "range" refers to the region where a Florida panther lives.

The range of a male Florida panther is about $3\frac{3}{4}$ times the range of a female Florida panther. The range of a male is about 275 square miles. Estimate the range of a female Florida panther.

$275 \div 3\frac{3}{4} \approx 275 \div 4$ Round $3\frac{3}{4}$ to the nearest whole number, 4.

275 is not evenly divisible by 4.

$\approx 280 \div 4$ 280 is close to 275 and is divisible by 4.

$= 70$ Divide.

∴ So, the range of a female Florida panther is about 70 square miles.

On Your Own

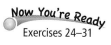
Exercises 24–31

6. There are about 100 Florida panthers in South Florida. A scientist wants $\frac{3}{8}$ of the panthers fitted with tracking collars. Estimate the number of panthers to be fitted with collars.

2.1 Exercises

Check It Out
Help with Homework
BigIdeasMath.com

Vocabulary and Concept Check

Tell whether you would use *rounding* or *compatible numbers* to estimate the product or quotient. Explain your reasoning.

1. $2\dfrac{1}{6} \times 5\dfrac{11}{12}$
2. $7\dfrac{3}{4} \div 1\dfrac{7}{9}$
3. $\dfrac{2}{5} \times \dfrac{7}{8}$
4. $34 \div 8\dfrac{2}{3}$

5. Copy and complete the table to estimate the quotient $77 \div 4\dfrac{2}{5}$.

How to Round	Estimate
Round 77 to the nearest hundred.	▨ ÷ 4 = ▨
Round 77 to the nearest ten.	▨ ÷ 4 = ▨
Round 77 to the nearest compatible number.	▨ ÷ 4 = ▨

6. **NUMBER SENSE** In Exercise 5, the quotient $77 \div 4\dfrac{2}{5}$ equals $17\dfrac{1}{2}$. What do you notice about the estimates in the table?

Practice and Problem Solving

Estimate the product or quotient.

7. $\dfrac{4}{7} \times \dfrac{1}{6}$
8. $\dfrac{9}{10} \times \dfrac{5}{9}$
9. $\dfrac{1}{5} \times \dfrac{7}{8}$
10. $\dfrac{8}{15} \times \dfrac{5}{6}$

11. $\dfrac{3}{4} \times \dfrac{1}{3}$
12. $\dfrac{2}{3} \times \dfrac{1}{7}$
13. $\dfrac{11}{8} \div 3$
14. $\dfrac{7}{9} \div 2$

15. $\dfrac{5}{13} \times \dfrac{4}{5}$
16. $\dfrac{1}{10} \times \dfrac{3}{16}$
17. $\dfrac{5}{6} \times \dfrac{7}{12}$
18. $\dfrac{3}{4} \times \dfrac{7}{9}$

19. $8\dfrac{3}{4} \times 2\dfrac{1}{2}$
20. $14\dfrac{11}{15} \times 4\dfrac{3}{7}$
21. $42\dfrac{2}{9} \div 6\dfrac{6}{7}$
22. $19\dfrac{1}{2} \div 4\dfrac{7}{8}$

23. **ERROR ANALYSIS** Describe and correct the error in estimating the product.

$\dfrac{5}{12} \times \dfrac{9}{10} \approx 0 \times 1 = 0$

Use compatible numbers to estimate the product or quotient.

24. $61 \div 4\dfrac{3}{8}$
25. $48 \div 6\dfrac{7}{12}$
26. $151 \times \dfrac{2}{5}$
27. $203 \times \dfrac{6}{7}$

28. $152 \div 6\dfrac{3}{11}$
29. $135 \div 19\dfrac{7}{10}$
30. $155 \div 7\dfrac{2}{9}$
31. $177 \div 8\dfrac{5}{6}$

32. **FLOWERS** You plant 25 flower bulbs in a garden. About $\dfrac{3}{4}$ of the flowers bloom. Estimate the number of flowers that bloom.

48 Chapter 2 Multiplying and Dividing Fractions

Assignment Guide and Homework Check

Level	Day 1 Activity Assignment	Day 2 Lesson Assignment	Homework Check
Basic	7–14, 42–46	1–6, 15–31 odd, 32, 33	5, 17, 21, 23, 25, 33
Average	7–14, 42–46	1–6, 15–31 odd, 34, 35, 37	17, 21, 23, 25, 34
Advanced	7–14, 42–46	1–6, 19–23, 28, 30, 36, 38, 40, 41	21, 23, 30, 38, 40

Common Errors

- **Exercises 7–22** Students may struggle finding an accurate estimate of some fractions and will get unreasonable answers. Tell them to draw a number line and place the fraction on the number line (as in the activity in the *Introduction* and Example 1). This will help the students visualize the fraction.
- **Exercises 24–31** Students may only round the fraction to a compatible number and try to estimate the expression. Remind them that *both* numbers should be easy to use mentally, and that they will most likely need to round both numbers.
- **Exercises 37–39** Students may need to be reminded that in the order of operations, multiplication and division are evaluated left to right. In Exercise 39, evaluating inside the parentheses comes first, so the multiplication would be done first even though it is to the right of the division.

Technology For the Teacher
Answer Presentation Tool
Quiz*Show*

2.1 Record and Practice Journal

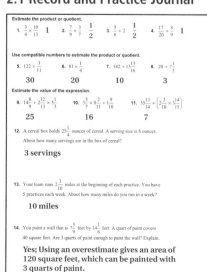

Vocabulary and Concept Check

1. rounding; The product will be easier to compute.
2. rounding; The quotient will be easier to compute.
3. rounding; The product will be easier to compute.
4. compatible numbers; The quotient will be easier to compute mentally.
5. $100 \div 4 = 25$
 $80 \div 4 = 20$
 $76 \div 4 = 19$
6. As you round 77 to closer numbers, the estimate gets closer to the actual answer.

Practice and Problem Solving

7. 0
8. $\frac{1}{2}$
9. 0
10. $\frac{1}{2}$
11. $\frac{1}{2}$
12. 0
13. $\frac{1}{3}$
14. $\frac{1}{2}$
15. $\frac{1}{2}$
16. 0
17. $\frac{1}{2}$
18. 1
19. 27
20. 60
21. 6
22. 4
23. $\frac{5}{12}$ is closer to $\frac{1}{2}$ than to 0.
 $\frac{5}{12} \times \frac{9}{10} \approx \frac{1}{2} \times 1 = \frac{1}{2}$
24. 15
25. 8
26. 75
27. 203
28. 25
29. 7
30. 20
31. 20
32. about 25 flowers

T-48

Practice and Problem Solving

33. See *Taking Math Deeper*.
34. about 8 loaves
35–36. Sample answers are given.
35. 27 in.²; underestimate
36. 30 ft²; overestimate
37. 18 38. 18
39. 5
40. yes; The overestimate for the area of the wall is 162 square feet and the 3 rolls of wallpaper will cover 180 square feet.
41. *Sample answer:*
 low estimate: 234 in.²
 high estimate: 320 in.²
 To find a low estimate, round the dimensions down. To find a high estimate, round the dimensions up.

Fair Game Review

42. 12 43. 36
44. $11\frac{2}{3}$ 45. $4\frac{1}{2}$
46. B

Mini-Assessment
Estimate the product or quotient.

1. $\frac{5}{6} \times \frac{3}{7}$ $\frac{1}{2}$
2. $3 \div 1\frac{1}{8}$ 3
3. $\frac{7}{8} \times \frac{14}{15}$ 1
4. $5\frac{8}{9} \div 3\frac{1}{3}$ 2
5. You cover a wall that is $10\frac{5}{6}$ feet by $20\frac{1}{4}$ feet with wallpaper. One roll of wallpaper covers 50 square feet. Are 4 rolls of wallpaper enough to cover the wall? no; You will need 5 rolls.

Taking Math Deeper

Exercise 33
An important problem-solving strategy is to learn to *simplify the question*. Racecars are 3-dimensional objects, with length, height, and width. But, students are only asked to compare their heights.

Racecar is 3-D

① Build intuition by seeing.
 Cut out paper that is $46\frac{7}{8}$ inches.
 Cut out paper that is $2\frac{7}{9}$ inches.
 How many times taller?

② Build intuition by doing.
 Mark off how many of the smaller pieces will fit onto the larger piece.
 About 16 or 17.

③ Use compatible numbers to estimate.
 $46\frac{7}{8}$ is about 48.
 $2\frac{7}{9}$ is about 3.
 $48 \div 3 = 16$

With a calculator, the answer is about 16.9. So, the height of the racecar is about 16 times greater than the height of the model.

$46\frac{7}{8}$ in.

2 yard sticks

$2\frac{7}{9}$ in.

Project
When is it important to use a model of a real life object? Design your own racecar. Make a sketch of your racecar and label the dimensions.

Reteaching and Enrichment Strategies

If students need help...	If students got it...
Resources by Chapter • Practice A and Practice B • Puzzle Time Record and Practice Journal Practice Differentiating the Lesson Lesson Tutorials Skills Review Handbook	Resources by Chapter • Enrichment and Extension Start the next section

33. **RACECAR** The height of a racecar is $46\frac{7}{8}$ inches. A model of the racecar is $2\frac{7}{9}$ inches tall. About how many times greater is the height of the racecar than the height of the model?

34. **BREAD** A recipe for a loaf of bread calls for $3\frac{1}{4}$ cups of flour. About how many loaves of bread can you make with 25 cups of flour?

GEOMETRY Estimate the area of the rectangle or parallelogram. Did you overestimate or underestimate the area? Explain.

35. $3\frac{3}{10}$ in. by $9\frac{1}{8}$ in.

36.
$4\frac{5}{12}$ ft, $5\frac{3}{5}$ ft

Estimate the value of the expression.

37. $6\frac{1}{4} \times 9\frac{3}{7} \div 2\frac{7}{8}$

38. $11\frac{2}{3} \div 3\frac{7}{12} \times 6\frac{2}{5}$

39. $100\frac{3}{8} \div \left(3\frac{5}{6} \times 5\frac{2}{9}\right)$

40. **WALLPAPER** You cover a wall that is $8\frac{5}{8}$ feet by $17\frac{1}{4}$ feet with wallpaper. One roll of wallpaper covers 60 square feet. Are 3 rolls of wallpaper enough to cover the wall? Explain.

41. **Geometry** Find a low estimate and a high estimate for the surface area of the jewelry box. Explain how you found your answers.

$3\frac{3}{4}$ in., $6\frac{7}{8}$ in., $11\frac{1}{4}$ in.

Fair Game Review What you learned in previous grades & lessons

Evaluate the expression. *(Skills Review Handbook)*

42. $\dfrac{2 \times 18}{3}$

43. $\dfrac{4 \times 45}{5}$

44. $\dfrac{5 \times 14}{6}$

45. $\dfrac{3 \times 12}{8}$

46. **MULTIPLE CHOICE** Which expression does *not* need the Commutative Property of Addition or the Commutative Property of Multiplication to simplify? *(Section 1.3)*

 Ⓐ $18 + (x + 3)$ **Ⓑ** $6(9x)$

 Ⓒ $(4 \cdot x) \cdot 11$ **Ⓓ** $5 + 10x + 7$

2.2 Multiplying Fractions and Whole Numbers

COMMON CORE STATE STANDARDS
6.NS.1

Essential Question What does it mean when a whole number is multiplied by a fraction? Will the product be *greater than* or *less than* the whole number?

1 EXAMPLE: Multiplying a Fraction and a Whole Number

You have 3 gallons of paint. You use $\frac{3}{4}$ of the paint. How many gallons did you use?

THINK ABOUT THE QUESTION: One way to think about this question is to rewrite the question.

Words: What is $\frac{3}{4}$ of 3? **Numbers:** $\frac{3}{4} \times 3 = ?$

Here is one way to get the answer.

- **Draw** a length of 3.

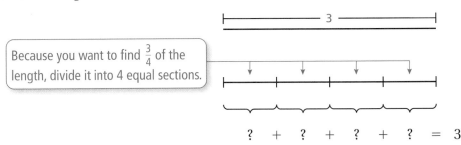

Because you want to find $\frac{3}{4}$ of the length, divide it into 4 equal sections.

$? + ? + ? + ? = 3$

Now, you need to think of a way to divide 3 into 4 equal parts.

- **Rewrite** the number 3 as a fraction whose numerator is divisible by 4.

Because the length is divided into 4 equal sections, multiply the numerator and denominator by 4.

$\frac{3}{1} = \frac{3 \times 4}{1 \times 4} = \frac{12}{4}$

In this form, you see that $\frac{12}{4}$ can be divided into four equal parts of $\frac{3}{4}$.

- Each part is $\frac{3}{4}$ gallon and you used three of them. Written as multiplication, you have

$\frac{3}{4} \times 3 = \frac{9}{4}.$

So, you used $\frac{9}{4}$ gallons of paint.

Laurie's Notes

Introduction

For the Teacher
- **Goal:** The goal of this activity is for students to develop an understanding of how to multiply a fraction and a whole number. There are two approaches, and each one helps students focus on the answer to the essential question.

Motivate
- Display 12 items (i.e., paper clips, bingo chips, pennies) to the class in a random fashion. Ask a student volunteer to show what $\frac{1}{2}$ of the collection would look like. Ask additional students to show the following fractional parts of the collection: $\frac{1}{4}, \frac{1}{3}, \frac{2}{3}, \frac{3}{4}, \frac{1}{6}$. In each case, be sure to phrase the problem as: "show ___ of 12".
- In each case, when you take a fractional part of the 12 items, you end up with a number less than 12 (connecting to the essential question).
- ? "Could you take $\frac{1}{5}$ of 12?" Yes, but because 12 is not divisible by 5, the result is a mixed number.
- As students solve the problems, record their work on the board.
 Examples: $\frac{1}{2}$ of 12 is 6 $\frac{1}{4}$ of 12 is 3 $\frac{1}{3}$ of 12 is 4

Words of Wisdom
- It is helpful for students to use concrete materials and visual models to represent the multiplication of a fraction and a whole number.

Activity Notes

Example 1
- Instead of modeling the three gallons of paint with an area model, a line segment with a length of three is used.
- Divide the 3 into 4 equal parts and shade three of the parts. Think of 3 as $\frac{12}{4}$. Then the shaded region is $\frac{3}{4}$ of $\frac{12}{4}$, which is $\frac{9}{4}$.
- **Connection:** Help students notice the connection to the opening activity. $\frac{3}{4}$ of 12 is 9 which is similar to $\frac{3}{4}$ of $\frac{12}{4}$ is $\frac{9}{4}$.
- ? After you represent 3 as $\frac{12}{4}$, ask "How can we represent $\frac{3}{4}$ of this amount?" Listen for the idea to multiply $\frac{3}{4}$ by $\frac{12}{4}$.
- ? "Are there any other ways in which you might solve this problem?" Listen for different methods.
- **Connection:** Some students might think of money. Divide $3 into 4 equal parts, which would be $0.75 in each part. Then 3 of the 4 parts is $0.75 + $0.75 + $0.75 = $2.25.

Common Core State Standards
6.NS.1 Interpret and compute quotients of fractions, and solve word problems involving division of fractions by fractions.

Activity Materials
Introduction
• 12 paper clips, pennies, or other items

Start Thinking! and Warm Up

2.2 Record and Practice Journal

T-50

English Language Learners

Vocabulary

Discuss with the class the connection between multiplication and the word *of*. For example: 3 boxes *of* 12 golf balls can be represented as 3×12 golf balls.

Similarly, $\frac{3}{4}$ *of* 12 golf balls can be represented as $\frac{3}{4} \times 12$ golf balls.

Encourage students to look for this wording in application problems.

2.2 Record and Practice Journal

Laurie's Notes

Example 2

- In this example students are multiplying a whole number by a fraction (notice the order).
- A common approach to multiplying a whole number by a fraction is to think of the multiplication as repeated addition.
- Say "If 4×2 is 4 groups of 2, then $4 \times \frac{2}{5}$ should be 4 groups of $\frac{2}{5}$."
- **?** Draw a number line on the board. Work together with students to count 4 groups of $\frac{2}{5}$. "What are 4 groups of $\frac{2}{5}$?" $\frac{8}{5}$ or $1\frac{3}{5}$
- **Common Error:** The classic error is to multiply 4 by the numerator AND denominator and say $4 \times \frac{2}{5} = \frac{8}{20}$ instead of $\frac{8}{5}$.
- **?** "To find $4 \times \frac{2}{5}$, could you have divided 4 into 5 equal parts and then taken 2 of the parts?" yes
- **Big Idea:** Students should start to realize that a whole number multiplied by a proper fraction results in a product that is less than the whole number.

Inductive Reasoning

- After students have worked on these problems with a partner, have them share their answers and explain their thinking about the process.
- **Note:** Students should not rewrite their answers as mixed numbers.
- **?** "Did you notice any patterns in your answers?" Listen for students to mention that you multiply the whole number by the numerator and put the resulting product over the denominator.
- **Extension:** Have students look at a pattern of products: $\frac{1}{4} \times 12, \frac{2}{4} \times 12, \frac{3}{4} \times 12, \frac{4}{4} \times 12, \frac{5}{4} \times 12$, etc.

What Is Your Answer?

- Students should be able to answer Questions 9 and 10 independently. Ask volunteers to share their answers with the class.

Closure

- It can be challenging for students to write real-life contexts. Have students share ideas to practice with the class.

The Dynamic Planning Tool
Editable Teacher's Resources at **BigIdeasMath.com**

② EXAMPLE: Multiplying a Whole Number and a Fraction

Find $4 \times \frac{2}{5}$.

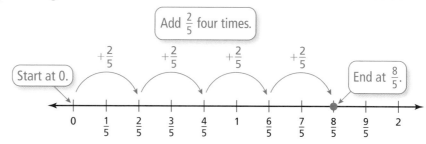

∴ So, $4 \times \frac{2}{5} = \frac{8}{5}$, or $1\frac{3}{5}$.

Inductive Reasoning

Work with a partner. Complete the table using a number line.

Exercise	Repeated Addition
① 3. $\frac{3}{4} \times 3$	$\frac{3}{4} + \frac{3}{4} + \frac{3}{4} = \frac{9}{4}$
② 4. $4 \times \frac{2}{5}$	$\frac{2}{5} + \frac{2}{5} + \frac{2}{5} + \frac{2}{5} = \frac{8}{5}$
5. $\frac{7}{6} \times 5$	
6. $3 \times \frac{9}{5}$	
7. $\frac{1}{3} \times 12$	

What Is Your Answer?

8. **a.** Write a real-life problem that is related to the product $\frac{2}{3} \times 5$.

 b. Write a different real-life problem that is related to the product $5 \times \frac{2}{3}$.

 c. Are the two products equal? How is your answer related to the Commutative Property of Multiplication?

9. **IN YOUR OWN WORDS** What does it mean when a whole number is multiplied by a fraction? Will the product be *greater than* or *less than* the whole number?

10. Write a general rule for multiplying fractions and whole numbers.

Practice — Use what you learned about multiplying fractions and whole numbers to complete Exercises 4–11 on page 54.

Section 2.2 Multiplying Fractions and Whole Numbers 51

2.2 Lesson

Key Idea

Multiplying Fractions and Whole Numbers

Words Multiply the numerator of the fraction by the whole number. Then write the product over the denominator.

Numbers $2 \times \dfrac{4}{9} = \dfrac{2 \times 4}{9} = \dfrac{8}{9}$

Algebra $a \cdot \dfrac{b}{c} = \dfrac{a \cdot b}{c}$, where $c \neq 0$

EXAMPLE 1 **Multiplying a Whole Number and a Fraction**

Find $3 \times \dfrac{7}{8}$. Estimate $3 \times 1 = 3$

$3 \times \dfrac{7}{8} = \dfrac{3 \times 7}{8}$ ← Multiply the numerator and whole number.
 ← Write the product over the denominator.

$= \dfrac{21}{8}$, or $2\dfrac{5}{8}$ Simplify.

∴ So, the product is $2\dfrac{5}{8}$. **Reasonable?** $2\dfrac{5}{8} \approx 3$ ✓

EXAMPLE 2 **Multiplying a Fraction and a Whole Number**

Find $\dfrac{11}{12} \times 6$. Estimate $1 \times 6 = 6$

Remember

A **common factor** is a factor that is shared by two or more whole numbers. For example, 3 and 9 share a common factor of 3.

$\dfrac{11}{12} \times 6 = \dfrac{11 \times 6}{12}$ ← Multiply the numerator and whole number.
 ← Write the product over the denominator.

$= \dfrac{11 \times \overset{1}{\cancel{6}}}{\underset{2}{\cancel{12}}}$ Divide out the common factor, 6, from 6 and 12.

$= \dfrac{11}{2}$, or $5\dfrac{1}{2}$ Simplify.

∴ So, the product is $5\dfrac{1}{2}$. **Reasonable?** $5\dfrac{1}{2} \approx 6$ ✓

On Your Own

Now You're Ready
Exercises 4–19

Multiply. Write the answer in simplest form.

1. $4 \times \dfrac{1}{5}$ 2. $\dfrac{3}{8} \times 7$ 3. $9 \times \dfrac{2}{3}$ 4. $\dfrac{7}{10} \times 5$

Laurie's Notes

Introduction

Connect
- **Yesterday:** Students used two different approaches to multiply a fraction and a whole number.
- **Today:** Students will develop a greater understanding of how to multiply fractions and whole numbers using a formal process.

Motivate
- Divide each of two cardboard circles into three equal parts. On one circle, shade one part to represent $\frac{1}{3}$. On the other, shade two parts to represent $\frac{2}{3}$.
- Ask two students to stand at the front of the room. Hand one student the model of $\frac{1}{3}$ and the other the model of $\frac{2}{3}$.
- ? "Imagine these circles are pizzas. How much does each student have? Who has more? How much more does he/she have?" $\frac{1}{3}$ and $\frac{2}{3}$; $\frac{2}{3}$ is two times more than $\frac{1}{3}$.
- Use a marker to shade $\frac{3}{12}$ and $\frac{9}{12}$ in the bottom of 2 egg cartons. Hand each student an egg carton.
- ? "If the shaded spaces were eggs, what fraction of a dozen does each student have? Who has more? How much more?" $\frac{3}{12}$ and $\frac{9}{12}$ OR $\frac{1}{4}$ and $\frac{3}{4}$; $\frac{3}{4}$ is more; $\frac{3}{4}$ is three times more than $\frac{1}{4}$.
- Record and discuss student observations.

Lesson Notes

Key Idea
- Make it very clear to students that only the numerator is being multiplied by the whole number, not the denominator.

Example 1
? "Would $\frac{7}{8} \times 3$ also be $2\frac{5}{8}$? Why?" Yes, because the answer is $\frac{21}{8}$, which is also $2\frac{5}{8}$.

Example 2
- Model the process of dividing out a common factor, as shown in the text.
- Reassure students that they will have time to practice this technique in future lessons. A few students may prefer to simplify after they multiply, and that's okay. As the factors increase in number and size, they may find it easier to divide out the common factors before multiplying.

On Your Own
- **Neighbor Check:** Have students work independently and then have their neighbor check their work. Have students discuss any discrepancies.

Goal Today's lesson is using a formal process to multiply fractions and whole numbers.

Lesson Materials
Introduction
• 2 cardboard circles
• 2 egg cartons

Start Thinking! and Warm Up

Lesson 2.2 Warm Up For use before Lesson 2.2

Lesson 2.2 Start Thinking! For use before Lesson 2.2
When is the product of a whole number and a fraction less than the whole number? greater than the whole number?

Hint: Try multiplying several different fractions by the same whole number to help you find the answer.

Extra Example 1
Find $8 \times \frac{3}{5}$. $4\frac{4}{5}$

Extra Example 2
Find $\frac{2}{9} \times 10$. $2\frac{2}{9}$

On Your Own
1. $\frac{4}{5}$
2. $2\frac{5}{8}$
3. 6
4. $3\frac{1}{2}$

T-52

Extra Example 3

Find the value of $s \cdot t - \dfrac{4}{5}$ when $s = 2$ and $t = \dfrac{3}{4}$. $\dfrac{7}{10}$

Extra Example 4

A dessert recipe calls for $\dfrac{2}{3}$ cup of oats per batch. How many cups of oats do you need to make 6 batches? 4 cups

On Your Own

5. 2

6. $13\dfrac{1}{2}$ pounds

Differentiated Instruction

Kinesthetic

Have students use fraction tiles to model multiplication of fractions and whole numbers. The hands-on practice allows students to understand how to find the products before they move on to using an algorithm.

T-53

Laurie's Notes

Example 3

? Ask a few questions about the example.
- "What operations are involved in this problem?" multiplication and addition; You can also divide out common factors.
- "What operation is performed first? Why?" multiplication; order of operations
- "How do we begin the problem?" substitute for the variables
- **Note:** Do not allow sloppy language. You *substitute* values; you do not *plug in* values.

Example 4

- Work through the example. Students need to read the scale to know that the watermelon weighs 20 pounds.
- **FYI:** Here are some fun facts about watermelons:
 - Early explorers used watermelons as canteens.
 - The first cookbook published in the U.S. in 1776 contained a recipe for watermelon rind pickles.
 - A 268.8-pound watermelon was grown in Arkansas in 2005.
 - Watermelon is thought to have originated in the Kalahari Desert of Africa.
 - Florida, Georgia, and Texas plant the most acreage of watermelon.

On Your Own

- **Think-Pair-Share:** Students should read each question independently and then work with a partner to answer the questions. When they have answered the questions, the pair should compare their answers with another group and discuss any discrepancies.

Words of Wisdom

- It is important to integrate algebra in daily lessons. Question 5 could have been presented as: Find the value of $\dfrac{5}{8} \cdot 4 - \dfrac{1}{2}$. If we want students to become comfortable with the use of variables to represent expressions, we need to look for opportunities to practice. This is an intentional feature of the textbook.

Closure

- **Writing Prompt:** If n is a whole number, explain how to multiply $n \cdot \dfrac{2}{5}$. Multiply n and 2, and then divide the product by 5.

The Dynamic Planning Tool
Editable Teacher's Resources at *BigIdeasMath.com*

EXAMPLE 3 **Standardized Test Practice**

What is the value of $\frac{1}{3} + x \cdot y$ when $x = 3$ and $y = \frac{2}{9}$?

Ⓐ $\frac{2}{3}$ **Ⓑ** $\frac{20}{27}$ **Ⓒ** 1 **Ⓓ** $1\frac{2}{3}$

> **Remember**
> Be sure to use the order of operations when evaluating numerical expressions.

$$\frac{1}{3} + x \cdot y = \frac{1}{3} + 3 \cdot \frac{2}{9}$$ Substitute 3 for x and $\frac{2}{9}$ for y.

$$= \frac{1}{3} + \frac{3 \cdot 2}{9}$$ Multiply 3 and $\frac{2}{9}$.

$$= \frac{1}{3} + \frac{\overset{1}{\cancel{3}} \cdot 2}{\underset{3}{\cancel{9}}}$$ Divide out the common factor 3.

$$= \frac{1}{3} + \frac{2}{3}$$ Simplify.

$$= 1$$ Add.

∴ The correct answer is **Ⓒ**.

EXAMPLE 4 **Real-Life Application**

About $\frac{9}{10}$ of the weight of a watermelon is water. How many pounds of water are in the watermelon?

To find $\frac{9}{10}$ of 20, multiply. **Estimate** $1 \times 20 = 20$

$$\frac{9}{10} \times 20 = \frac{9 \times 20}{10}$$

$$= \frac{9 \times \overset{2}{\cancel{20}}}{\underset{1}{\cancel{10}}}$$ Divide out the common factor 10.

$$= 18$$

∴ So, 18 pounds of water are in the watermelon.

Reasonable? $18 \approx 20$ ✓

● **On Your Own**

Now You're Ready
Exercises 32–39

5. Find the value of $a \cdot b - \frac{1}{2}$ when $a = \frac{5}{8}$ and $b = 4$.

6. **WHAT IF?** In Example 4, a watermelon weighs 15 pounds. How many pounds of water are in the watermelon?

2.2 Exercises

Vocabulary and Concept Check

1. **WRITING** Describe how to multiply a fraction by a whole number.

2. **NUMBER SENSE** Use repeated addition to find the product $6 \times \dfrac{7}{8}$.

3. **NUMBER SENSE** Without multiplying, which is greater, $\dfrac{1}{3} \times 24$ or $\dfrac{1}{4} \times 24$? Explain.

Practice and Problem Solving

Multiply. Write the answer in simplest form.

4. $3 \times \dfrac{1}{7}$
5. $\dfrac{1}{8} \times 5$
6. $\dfrac{3}{4} \times 7$
7. $4 \times \dfrac{4}{9}$

8. $\dfrac{6}{5} \times 8$
9. $9 \times \dfrac{5}{3}$
10. $\dfrac{5}{6} \times 8$
11. $25 \times \dfrac{7}{10}$

12. $\dfrac{2}{3} \times 5$
13. $7 \times \dfrac{3}{10}$
14. $15 \times \dfrac{2}{3}$
15. $\dfrac{13}{14} \times 28$

16. $\dfrac{3}{20} \times 5$
17. $\dfrac{3}{8} \times 24$
18. $9 \times \dfrac{13}{24}$
19. $18 \times \dfrac{3}{4}$

ERROR ANALYSIS Describe and correct the error in finding the product.

20.

21.

22. **CDs** Your friend has 12 CDs and $\dfrac{3}{4}$ of them are pop music. How many CDs are pop music?

23. **OATMEAL MUFFINS** A batch of oatmeal muffins calls for $\dfrac{2}{3}$ cup of oats. How many cups of oats do you need to make four batches of muffins?

24. **RAIN** There are 365 days in a year and rain falls on $\dfrac{2}{5}$ of the days. How many days does it rain during the year?

Use compatible numbers to estimate the product.

25. $\dfrac{1}{3} \times 17$
26. $35 \times \dfrac{1}{8}$
27. $22 \times \dfrac{5}{6}$
28. $\dfrac{4}{5} \times 7$

Assignment Guide and Homework Check

Level	Day 1 Activity Assignment	Day 2 Lesson Assignment	Homework Check
Basic	4–11, 43–47	1–3, 13–19 odd, 20–24, 29, 33, 35	2, 15, 19, 24, 29, 35
Average	4–11, 43–47	1–3, 12–20 even, 21, 24, 33–39 odd, 40	14, 18, 24, 37, 40
Advanced	4–11, 43–47	1–3, 20, 21, 30–38 even, 41, 42	30, 36, 38, 41

Common Errors

- **Exercises 4–19** Students may multiply the whole number by the numerator *and* the denominator, instead of just the numerator. Remind them of the Key Idea.
- **Exercises 4–19** Students may find a common factor between the numerator and the whole number. Explain to them that a common factor is divided out of the numerator *and* denominator to simplify the expression.
- **Exercises 32–39** Students may forget to use the order of operations. Remind them of the order of operations and that addition is evaluated after multiplication.

2.2 Record and Practice Journal

Technology For the Teacher
Answer Presentation Tool
QuizShow

 Vocabulary and Concept Check

1. Multiply the numerator of the fraction by the whole number. Then write the product over the denominator.
2. $5\frac{1}{4}$
3. $\frac{1}{3} \times 24$; because $\frac{1}{3} > \frac{1}{4}$

 Practice and Problem Solving

4. $\frac{3}{7}$
5. $\frac{5}{8}$
6. $5\frac{1}{4}$
7. $1\frac{7}{9}$
8. $9\frac{3}{5}$
9. 15
10. $6\frac{2}{3}$
11. $17\frac{1}{2}$
12. $3\frac{1}{3}$
13. $2\frac{1}{10}$
14. 10
15. 26
16. $\frac{3}{4}$
17. 9
18. $4\frac{7}{8}$
19. $13\frac{1}{2}$
20. The 8 and 2 should be multiplied, not added. $\frac{2}{5} \times 8 = \frac{2 \times 8}{5} = \frac{16}{5}$, or $3\frac{1}{5}$
21. 9 should be multiplied by 3, not 7. $9 \times \frac{3}{7} = \frac{9 \times 3}{7} = \frac{27}{7}$, or $3\frac{6}{7}$
22. 9 CDs
23. $2\frac{2}{3}$ cups
24. 146 days
25. 6
26. 4
27. 20
28. 4

T-54

Practice and Problem Solving

29. Multiply $25 \times \frac{2}{5}$ first by the Comm. Prop. of Mult.; 60

30. Multiply $\frac{5}{9} \times 18$ first by the Assoc. Prop. of Mult.; 80

31. Multiply $\frac{3}{7} \times 14$ first by the Comm. Prop. of Mult.; 78

32. $4\frac{1}{2}$ 33. $1\frac{1}{2}$

34. 14 35. $2\frac{2}{3}$

36. 48 37. $22\frac{1}{2}$

38. $1\frac{1}{2}$ 39. $4\frac{1}{6}$

40. 20 ft²

41. See Additional Answers.

42. a. See *Taking Math Deeper*.
 b. $32\frac{2}{5}$ cm
 c. no, You would need 7 more bugle beads.

Fair Game Review

43. $1\frac{7}{8}$ 44. $3\frac{3}{5}$

45. $\frac{32}{175}$ 46. $3\frac{31}{33}$

47. D

Mini-Assessment

Multiply. Write the answer in simplest form.

1. $2 \times \frac{1}{4}$ $\frac{1}{2}$
2. $4 \times \frac{5}{6}$ $3\frac{1}{3}$
3. $\frac{3}{8} \times 12$ $4\frac{1}{2}$
4. $\frac{2}{5} \times 25$ 10
5. Your friend has 36 DVDs and $\frac{2}{3}$ of them are comedies. How many DVDs are comedies? 24

Taking Math Deeper

Exercise 42(a)

Although this problem might seem straightforward, it has 9 numbers that need to be used to answer parts (a) and (b). This requires some organization and a *lot* of fraction work! A table can help organize the work.

 Organize the work for part (a) with a table.

Type	Length	Number	Multiply	Simplify
Clasp	$\frac{9}{10}$	1	$\frac{9}{10}$	$\frac{9}{10}$
Bugle	$\frac{9}{20}$	48	$\frac{432}{20}$	$21\frac{6}{10}$
Crow	$\frac{9}{10}$			
Tube	$\frac{2}{5}$	24	$\frac{48}{5}$	$9\frac{6}{10}$

 Find the total length by adding.

a. Total Length = $30 + \left(\frac{9}{10} + \frac{6}{10} + \frac{6}{10}\right)$

$= 30 + \frac{21}{10}$

$= 32\frac{1}{10}$ cm

About 32 cm

 Tactile: Make or draw the necklace as a check.

Project

Design your own necklace. Include in your design a picture of the beads that you would use, their lengths, and the pattern that you are using. How long will your necklace be? How much will it cost to make your necklace?

Reteaching and Enrichment Strategies

If students need help...	If students got it...
Resources by Chapter • Practice A and Practice B • Puzzle Time Record and Practice Journal Practice Differentiating the Lesson Lesson Tutorials Skills Review Handbook	Resources by Chapter • Enrichment and Extension Start the next section

Tell how the Commutative and Associative Properties of Multiplication can help find the product mentally. Then find the product.

29. $25 \times 6 \times \dfrac{2}{5}$
30. $\left(8 \times \dfrac{5}{9}\right) \times 18$
31. $\dfrac{3}{7} \times 13 \times 14$

ALGEBRA Evaluate the expression when $x = 6$, $y = \dfrac{3}{16}$, and $z = 20$.

③ 32. $\dfrac{3}{4} \cdot x$
33. $8 \cdot y$
34. $\dfrac{7}{10} \cdot z$
35. $x \cdot \dfrac{4}{9}$

36. $\dfrac{2}{5} xz$
37. xyz
38. $\dfrac{3}{8} + xy$
39. $\dfrac{5}{12} + yz$

40. AREA A rectangular picnic shelter is 75 feet long by 60 feet wide. A model of the shelter is $\dfrac{1}{15}$ as wide and $\dfrac{1}{15}$ as long as the actual shelter. What is the area, in square feet, of the model of the shelter?

41. REASONING You spend $\dfrac{1}{3}$ of your money and your friend spends $\dfrac{1}{2}$ of her money shopping. Is it possible that you spend more money than your friend? Explain your reasoning.

42. NECKLACES You make bead necklaces using the beads shown in the table. Each necklace has a clasp that is $\dfrac{9}{10}$ centimeter long.

Bead		Length
Bugle		$\dfrac{9}{20}$ cm
Crow		$\dfrac{9}{10}$ cm
Tube		$\dfrac{2}{5}$ cm

 a. How long is a necklace with 48 bugle beads and 24 tube beads?

 b. How long is a necklace with 18 bugle beads, 18 crow beads, and 18 tube beads?

 c. **Reasoning** You want to make a necklace 38 centimeters or longer. You have 26 bugle beads, 18 crow beads, and 16 tube beads. Do you have enough beads? If not, what is the least number of bugle beads you need to add to make the necklace? Explain how you found your answer.

Fair Game Review What you learned in previous grades & lessons

Evaluate the expression. *(Skills Review Handbook)*

43. $\dfrac{3 \times 5}{2 \times 4}$
44. $\dfrac{9 \times 2}{1 \times 5}$
45. $\dfrac{4 \times 8}{7 \times 25}$
46. $\dfrac{10 \times 13}{3 \times 11}$

47. MULTIPLE CHOICE What is the area of a parallelogram with a base of 12 inches and a height of 4 inches? *(Section 1.5)*

 Ⓐ 24 in. Ⓑ 48 in. Ⓒ 24 in.² Ⓓ 48 in.²

2.3 Multiplying Fractions

COMMON CORE STATE STANDARDS
6.NS.1

Essential Question What does it mean to multiply fractions?

1 EXAMPLE: Multiplying Fractions

A bottle of water is $\frac{4}{5}$ full. You drink $\frac{2}{3}$ of the water. How much do you drink?

THINK ABOUT THE QUESTION: To help you think about this question, rewrite the question.

Words: What is $\frac{2}{3}$ of $\frac{4}{5}$? **Numbers:** $\frac{2}{3} \times \frac{4}{5} = ?$

Here is one way to get the answer.

- **Draw** a length of $\frac{4}{5}$.

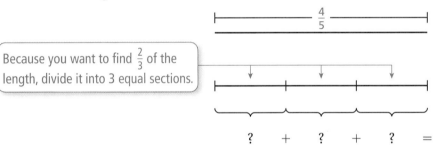

Because you want to find $\frac{2}{3}$ of the length, divide it into 3 equal sections.

$? + ? + ? = \frac{4}{5}$

Now, you need to think of a way to divide $\frac{4}{5}$ into three equal parts.

- **Rewrite** $\frac{4}{5}$ as a fraction whose numerator is divisible by 3.

Because the length is divided into 3 equal sections, multiply the numerator and denominator by 3.

$\frac{4}{5} = \frac{4 \times 3}{5 \times 3} = \frac{12}{15}$

In this form, you see that $\frac{12}{15}$ can be divided into three equal parts of $\frac{4}{15}$.

- Each part is $\frac{4}{15}$ of the water and you drank two of them. Written as multiplication, you have

$\frac{2}{3} \times \frac{4}{5} = \frac{8}{15}$.

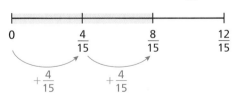

∴ So, you drank $\frac{8}{15}$ of the water.

56 Chapter 2 Multiplying and Dividing Fractions

Laurie's Notes

Introduction

For the Teacher
- **Goal:** The goal of this activity is for students to develop an understanding of how to multiply fractions.
- There are two different models: linear and area. In each case, a connection is made between the language "a fraction *of* a fraction" and "a fraction \times a fraction."
- In today's activities, students will find "a fraction of a fraction." It's important to remember that the answer represents a fraction of *the whole*.

Motivate
- Have the students try the following problems with a partner.

 What is $\frac{1}{2}$ of 20? What is $\frac{1}{2}$ of 6? What is $\frac{1}{2}$ of 2?

 What is $\frac{1}{2}$ of 1? What is $\frac{1}{2}$ of $\frac{1}{2}$? What is $\frac{1}{2}$ of $\frac{1}{4}$?

 What is $\frac{1}{2}$ of $\frac{1}{3}$? What is $\frac{1}{2}$ of $\frac{3}{4}$?

- Students may need to think a bit longer to get the last few answers. If so, give them a hint: "What would $\frac{1}{2}$ of $\frac{1}{4}$ of a sandwich look like?" "What would $\frac{1}{2}$ of $\frac{1}{3}$ of a pizza look like?" etc.
- Drawing pictures of the represented fractions may help students.

Activity Notes

Example 1
- Divide the bottle into fifths. Use a marker to label every $\frac{1}{5}$ interval.
- Fill the bottle with water until it is $\frac{4}{5}$ full. (The more cylindrical the bottle, the better. Also, a drop of food coloring may help the students in the back of the classroom to see the water.)
- **?** "How much water is in the bottle?" $\frac{4}{5}$
- **?** "I want to drink $\frac{2}{3}$ of this water. How much water will that be, and how do you know?" About $\frac{1}{2}$ $\left(\text{actually } \frac{8}{15}\right)$; Answers will vary.
- Discuss the various approaches students may have.
- Work through the linear model approach in the book.
- **Connection:** This method of multiplying fractions uses equivalent fractions. $\frac{2}{3}$ of $\frac{4}{5} = \frac{2}{3} \times \frac{4}{5} = \frac{2}{3} \times \frac{12}{15}$. If $\frac{1}{3}$ of $\frac{12}{15}$ is $\frac{4}{15}$, then $\frac{2}{3}$ of $\frac{12}{15}$ is $\frac{8}{15}$. By using number sense, students develop a sense of how fractions are multiplied (multiply the numerators and multiply the denominators).
- You want $\frac{2}{3}$ of a quantity that does not appear to be divisible by 3. By renaming the quantity $\left(\frac{4}{5} = \frac{12}{15}\right)$, then it is divisible by 3.

Common Core State Standards

6.NS.1 Interpret and compute quotients of fractions, and solve word problems involving division of fractions by fractions.

Activity Materials
Textbook
- clear plastic water bottle
- marker
- water |

Start Thinking! and Warm Up

2.3 Record and Practice Journal

T-56

English Language Learners
Analyzing Word Problems
Give students a copy of the word problems that are triple-spaced and have wide margins. The additional space gives students plenty of room to add notes. Demonstrate how to underline key words, phrases, and numbers; write down equivalent words in English (or their native language); and draw lines between elements to make the meaning clearer.

2.3 Record and Practice Journal

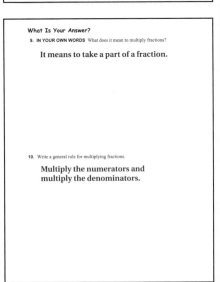

Laurie's Notes

Example 2
- In this activity, the students use an area model to multiply fractions. You are finding the area of a rectangle (length × width).
- Hand students a piece of paper and say "This rectangle represents a park. $\frac{3}{4}$ of its width and $\frac{4}{5}$ of its length is a playground!"
- Have students fold the paper as shown in the text. This may be difficult, so it is a good idea to have students sketch the fifths before they fold the paper.
- **Discuss:** In order to name the amount of paper that is shaded (representing the playground), students should think of a name for each small piece (twentieths) and count how many are shaded (12).

Inductive Reasoning
- Students should work with a partner to complete the remaining problems.
- Remind students that they should sketch a visual model for each problem. Most students will likely favor some sort of area model.
- Have students share their work for the visual model at the board, including both the problem and the answer. For example: $\frac{2}{3} \times \frac{5}{6} = \frac{10}{18} = \frac{5}{9}$.
- **?** "Do you notice any patterns in your answers?" Listen for students to describe multiplying the numerators and multiplying the denominators.

What Is Your Answer?
- **Think-Pair-Share:** Students should read each question independently and then work with a partner to answer the questions. When they have answered the questions, the pair should compare their answers with another group and discuss any discrepancies.

Closure
- Draw a picture to show $\frac{1}{2}$ of $\frac{3}{4}$, then use your picture to find $\frac{1}{2} \times \frac{3}{4}$.

 $\frac{3}{8}$; Sample picture:

The Dynamic Planning Tool
Editable Teacher's Resources at *BigIdeasMath.com*

② EXAMPLE: Multiplying Fractions

A park has a playground that is $\frac{3}{4}$ of its width and $\frac{4}{5}$ of its length.

What fraction of the park is covered by the playground?

Fold a piece of paper horizontally into fourths and shade three of the fourths to represent $\frac{3}{4}$.

Fold the paper vertically into fifths and shade $\frac{4}{5}$ of the paper another color.

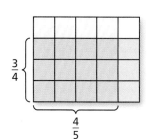

Count the total number of squares. This number is the denominator. The numerator is the number of squares shaded with both colors.

∴ $\frac{3}{4} \times \frac{4}{5} = \frac{12}{20} = \frac{3}{5}$. So, $\frac{3}{5}$ of the park is covered by the playground.

Inductive Reasoning

Work with a partner. Complete the table using a model or paper folding.

Exercise	Verbal Expression	Answer
① 3. $\frac{2}{3} \times \frac{4}{5}$	$\frac{2}{3}$ of $\frac{4}{5}$	$\frac{8}{15}$
② 4. $\frac{3}{4} \times \frac{4}{5}$	$\frac{3}{4}$ of $\frac{4}{5}$	$\frac{3}{5}$
5. $\frac{2}{3} \times \frac{5}{6}$		
6. $\frac{1}{6} \times \frac{1}{4}$		
7. $\frac{2}{5} \times \frac{1}{2}$		
8. $\frac{5}{8} \times \frac{4}{5}$		

What Is Your Answer?

9. **IN YOUR OWN WORDS** What does it mean to multiply fractions?

10. Write a general rule for multiplying fractions.

Practice — Use what you learned about multiplying fractions to complete Exercises 5–12 on page 60.

2.3 Lesson

🔑 Key Idea

Multiplying Fractions

Words Multiply the numerators and multiply the denominators.

Numbers $\dfrac{3}{7} \times \dfrac{1}{2} = \dfrac{3 \times 1}{7 \times 2} = \dfrac{3}{14}$

Algebra $\dfrac{a}{b} \cdot \dfrac{c}{d} = \dfrac{a \cdot c}{b \cdot d}$, where $b, d \neq 0$

EXAMPLE 1 Multiplying Fractions

Find $\dfrac{1}{5} \times \dfrac{1}{3}$.

$\dfrac{1}{5} \times \dfrac{1}{3} = \dfrac{1 \times 1}{5 \times 3}$ ← Multiply the numerators.
 ← Multiply the denominators.

$= \dfrac{1}{15}$ Simplify.

EXAMPLE 2 Multiplying Fractions with Common Factors

Find $\dfrac{8}{9} \times \dfrac{3}{4}$. Estimate $1 \times \dfrac{1}{2} = \dfrac{1}{2}$

$\dfrac{8}{9} \times \dfrac{3}{4} = \dfrac{8 \times 3}{9 \times 4}$ ← Multiply the numerators.
 ← Multiply the denominators.

$= \dfrac{\overset{2}{\cancel{8}} \times \overset{1}{\cancel{3}}}{\underset{3}{\cancel{9}} \times \underset{1}{\cancel{4}}}$ Divide out common factors.

$= \dfrac{2}{3}$ Simplify.

∴ So, the product is $\dfrac{2}{3}$. **Reasonable?** $\dfrac{2}{3} \approx \dfrac{1}{2}$ ✓

Study Tip

When the numerator of one fraction is the same as the denominator of another fraction, you can use mental math to multiply. For example, $\dfrac{4}{5} \times \dfrac{5}{9} = \dfrac{4}{9}$ because you can divide out the common factor 5.

🟢 On Your Own

Now You're Ready
Exercises 5–20

Multiply. Write the answer in simplest form.

1. $\dfrac{1}{2} \times \dfrac{5}{6}$ 2. $\dfrac{7}{8} \times \dfrac{1}{4}$ 3. $\dfrac{3}{7} \times \dfrac{2}{3}$ 4. $\dfrac{4}{9} \times \dfrac{3}{10}$

Laurie's Notes

Introduction

Connect
- **Yesterday:** Students used a visual model to multiply fractions.
- **Today:** Students will formalize the process for multiplying fractions.

Motivate
- **FYI:** The New York City Public School District is the largest public school district in the nation. About $\frac{1}{20}$ of the New York population attends school in this district. The population of New York is about $\frac{1}{15}$ of the population of the United States.
- "What is $\frac{1}{20} \times \frac{1}{15}$, and what does the answer mean in the context of the problem given?" $\frac{1}{300}$; $\frac{1}{20} \times \frac{1}{15}$ represents the fraction of the United States population that attends school in the New York City Public School District. Another way to interpret this answer is that for every 300 United States residents, 1 of them attends school in the New York City Public School District.
- Students should have a sense of how to multiply the fractions based upon yesterday's investigations.

Lesson Notes

Key Idea
- **Connections:** Previously, when a fraction was multiplied by a whole number, the whole number was written as a fraction with a denominator of 1. This rule still works!

Example 1
- Work through the example. It may help to show a visual model.

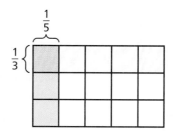

- Discuss the reasonableness of the answer: The product of $\frac{1}{5}$ and $\frac{1}{3}$ should be fairly small. Think of $\frac{1}{3}$ of the class. Now take $\frac{1}{5}$ of that group. It's not a lot!

Example 2
- Work through the example. Note the Study Tip.
- **Connection:** If the common factors are not divided out, the product is $\frac{24}{36}$. Notice that the GCF of 24 and 36 is 12.

On Your Own
- **Neighbor Check:** Have students work independently and then have their neighbor check their work. Have students discuss any discrepancies.

Goal Today's lesson is multiplying a fraction by a fraction.

Start Thinking! and Warm Up

Extra Example 1
Find $\frac{2}{7} \times \frac{3}{5}$. $\frac{6}{35}$

Extra Example 2
Find $\frac{2}{25} \times \frac{5}{6}$. $\frac{1}{15}$

On Your Own
1. $\frac{5}{12}$ 2. $\frac{7}{32}$
3. $\frac{2}{7}$ 4. $\frac{2}{15}$

Extra Example 3

Find the value of $u + r \cdot \dfrac{3}{10}$ when $u = \dfrac{1}{2}$ and $r = \dfrac{5}{12}$. $\dfrac{5}{8}$

Extra Example 4

You have $\dfrac{5}{8}$ of a large bottle of shampoo. You pour $\dfrac{3}{10}$ of the shampoo into a smaller bottle. How much of the entire larger bottle did you pour into the smaller bottle?

$\dfrac{3}{16}$ of the entire larger bottle

On Your Own

5. $\dfrac{8}{9}$

6. $\dfrac{1}{6}$ of the entire bag

Differentiated Instruction

Inclusion

Some students may confuse the concepts of adding fractions and multiplying fractions. Remind students when adding fractions, you must have a common denominator.

Like denominators:

$\dfrac{a}{b} + \dfrac{c}{b} = \dfrac{a+c}{b}$, where $b \neq 0$

Unlike denominators:

$\dfrac{a}{b} + \dfrac{c}{d} = \dfrac{ad}{bd} + \dfrac{bc}{bd}$

$= \dfrac{ad + bc}{bd}$, where $b, d \neq 0$

When multiplying fractions, you multiply the numerators and multiply the denominators.

$\dfrac{a}{b} \cdot \dfrac{c}{d} = \dfrac{ac}{bd}$, where $b, d \neq 0$

Laurie's Notes

Example 3

- Write the problem for the class. Ask the following questions.
 - "What operations are involved in this problem?" multiplication and subtraction
 - "What operation is performed first? Why?" Multiplication, because of the order of operations.
 - "How do you begin the problem?" substitute for the variables
- This is another example where students can practice dividing out a common factor.
- **Potential Error:** In this problem, make sure that students do not try to divide out the 4 in the denominator of $\dfrac{1}{4}$. This is NOT a factor. Factor means the number is being multiplied by another number. This 4 is part of a number that is being subtracted from a fraction. Make sure students can tell the difference.

Example 4

- Work through the example. The visual model should be familiar.
- **FYI:** An empanada is a Spanish word coming from *empanar* meaning *bread*, or *to bake pastry*. Although there are family-sized versions, it is basically an individual turnover with a pastry crust that is filled with chicken, meat, seafood, vegetables, or fruit.
- **Extension:** Some students find dividing out common factors to be pretty cool. If they want additional problems to try, here are two:

 $\dfrac{7}{12} \times \dfrac{3}{14}$ $\dfrac{10}{39} \times \dfrac{13}{15}$

- **Extension:** "Find the product of $\dfrac{2}{3} \times \dfrac{3}{4} \times \dfrac{4}{5} \times \dfrac{5}{6}$." $\dfrac{1}{3}$

On Your Own

- **Think-Pair-Share:** Students should read each question independently and then work with a partner to answer the questions. When they have answered the questions, the pair should compare their answers with another group and discuss any discrepancies.

Closure

- **Exit Ticket:** If you multiply two proper fractions, what do you know about the product? The product will be less than the greater fraction.

The Dynamic Planning Tool
Editable Teacher's Resources at *BigIdeasMath.com*

EXAMPLE 3 Standardized Test Practice

What is the value of $p \cdot \dfrac{7}{8} - q$ when $p = \dfrac{4}{5}$ and $q = \dfrac{1}{4}$?

Ⓐ $\dfrac{1}{4}$ **Ⓑ** $\dfrac{9}{20}$ **Ⓒ** $\dfrac{1}{2}$ **Ⓓ** 1

$p \cdot \dfrac{7}{8} - q = \dfrac{4}{5} \cdot \dfrac{7}{8} - \dfrac{1}{4}$ Substitute $\dfrac{4}{5}$ for p and $\dfrac{1}{4}$ for q.

$= \dfrac{\cancel{4}^1 \cdot 7}{5 \cdot \cancel{8}_2} - \dfrac{1}{4}$ Multiply. Divide out the common factor 4.

$= \dfrac{7}{10} - \dfrac{1}{4}$ Simplify.

$= \dfrac{14}{20} - \dfrac{5}{20} = \dfrac{9}{20}$ Subtract.

∴ The correct answer is **Ⓑ**.

EXAMPLE 4 Real-Life Application

You have $\dfrac{2}{3}$ of a bag of flour. You use $\dfrac{3}{4}$ of the flour to make empanada dough. How much of the entire bag do you use to make the dough?

Method 1: Use a model. Six of the 12 squares have both types of shading.

∴ So, you use $\dfrac{6}{12} = \dfrac{1}{2}$ of the entire bag.

Method 2: To find $\dfrac{3}{4}$ of $\dfrac{2}{3}$, multiply.

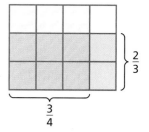

$\dfrac{3}{4} \times \dfrac{2}{3} = \dfrac{\cancel{3}^1 \times \cancel{2}^1}{\cancel{4}_2 \times \cancel{3}_1}$ Multiply the numerators and the denominators. Divide out common factors.

$= \dfrac{1}{2}$ Simplify.

∴ So, you use $\dfrac{1}{2}$ of the entire bag.

On Your Own

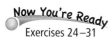
Exercises 24–31

5. Evaluate $a + b \cdot \dfrac{1}{12}$ when $a = \dfrac{5}{6}$ and $b = \dfrac{2}{3}$.

6. **WHAT IF?** In Example 4, you use $\dfrac{1}{4}$ of the flour to make the dough. How much of the entire bag do you use to make the dough?

Section 2.3 Multiplying Fractions

2.3 Exercises

Vocabulary and Concept Check

1. **WRITING** Explain how to multiply two fractions.

2. **OPEN-ENDED** Give three different sets of two fractions each having the same product.

3. **REASONING** Name the missing denominator.

 $\dfrac{3}{7} \times \dfrac{1}{\square} = \dfrac{3}{28}$

4. **NUMBER SENSE** Is $\dfrac{2}{3} \times \dfrac{5}{8}$ the same as $\dfrac{5}{8} \times \dfrac{2}{3}$? Explain.

Practice and Problem Solving

Multiply. Write the answer in simplest form.

5. $\dfrac{1}{7} \times \dfrac{2}{3}$ 6. $\dfrac{5}{8} \times \dfrac{1}{2}$ 7. $\dfrac{1}{4} \times \dfrac{2}{5}$ 8. $\dfrac{3}{7} \times \dfrac{1}{4}$

9. $\dfrac{2}{3} \times \dfrac{4}{5}$ 10. $\dfrac{5}{7} \times \dfrac{7}{8}$ 11. $\dfrac{3}{8} \times \dfrac{1}{9}$ 12. $\dfrac{5}{6} \times \dfrac{2}{5}$

13. $\dfrac{5}{12} \times 10$ 14. $6 \times \dfrac{7}{8}$ 15. $\dfrac{3}{4} \times \dfrac{8}{15}$ 16. $\dfrac{4}{9} \times \dfrac{4}{5}$

17. $\dfrac{3}{7} \times \dfrac{3}{7}$ 18. $\dfrac{5}{6} \times \dfrac{2}{9}$ 19. $\dfrac{13}{18} \times \dfrac{6}{7}$ 20. $\dfrac{7}{9} \times \dfrac{21}{10}$

21. **ERROR ANALYSIS** Describe and correct the error in finding the product.

22. **AQUARIUM** In an aquarium, $\dfrac{2}{5}$ of the fish are surgeonfish. Of these, $\dfrac{3}{4}$ are yellow tangs. What fraction of all fish in the aquarium are yellow tangs?

23. **JUMP ROPE** You exercise for $\dfrac{3}{4}$ of an hour. You jump rope for $\dfrac{1}{3}$ of that time. What fraction of the hour do you spend jumping rope?

ALGEBRA Evaluate the expression when $a = \dfrac{3}{4}$, $b = \dfrac{1}{6}$, and $c = \dfrac{2}{5}$.

24. $a \cdot \dfrac{5}{12}$ 25. $\dfrac{4}{7} \cdot b$ 26. $a \cdot b$ 27. $c \cdot a$

28. $\dfrac{5}{9}a + \dfrac{1}{9}$ 29. $\dfrac{14}{15} - \dfrac{7}{12}c$ 30. $bc + \dfrac{2}{3}$ 31. $ab + c$

Assignment Guide and Homework Check

Level	Day 1 Activity Assignment	Day 2 Lesson Assignment	Homework Check
Basic	5–12, 46–50	1–4, 13–21 odd, 22, 23–35 odd	4, 15, 22, 27, 35
Average	5–12, 46–50	1–4, 17–27 odd, 32–38 even, 41, 44	4, 19, 27, 34, 44
Advanced	5–12, 46–50	1–4, 21, 24–42 even, 43–45	30, 38, 43, 44

Common Errors

- **Exercises 5–20** Students may try to cross multiply instead of multiplying the numerators and denominators. Remind them that you cross multiply when an equal sign is between the fractions, but that you multiply the numerators and denominators when there is a multiplication symbol between the fractions.
- **Exercises 5–20** Students may try to get a common denominator and then multiply the fractions. Tell them this is not needed when multiplying, but is necessary for addition and subtraction. (Students will still end up with the same product, but the numbers will be greater before they simplify.)
- **Exercises 32–39** Students may not remember the definition of exponent and will multiply the fraction by the exponent. Remind them of the definition of exponent and also that exponents are evaluated before multiplication in the order of operations.

Technology For the Teacher
Answer Presentation Tool
QuizShow

2.3 Record and Practice Journal

Vocabulary and Concept Check

1. Multiply numerators and multiply denominators, then simplify the fraction.
2. Sample answer: $\frac{2}{7} \times \frac{7}{10}, \frac{2}{5} \times \frac{1}{2}, \frac{9}{10} \times \frac{2}{9}$
3. 4
4. yes; The Commutative Property of Multiplication applies to fractions.

Practice and Problem Solving

5. $\frac{2}{21}$ 6. $\frac{5}{16}$
7. $\frac{1}{10}$ 8. $\frac{3}{28}$
9. $\frac{8}{15}$ 10. $\frac{5}{8}$
11. $\frac{1}{24}$ 12. $\frac{1}{3}$
13. $4\frac{1}{6}$ 14. $5\frac{1}{4}$
15. $\frac{2}{5}$ 16. $\frac{16}{45}$
17. $\frac{9}{49}$ 18. $\frac{5}{27}$
19. $\frac{13}{21}$ 20. $1\frac{19}{30}$
21. You did not multiply the denominators.
$\frac{2}{5} \times \frac{3}{10} = \frac{2 \times 3}{5 \times 10} = \frac{\overset{1}{\cancel{2}} \times 3}{5 \times \underset{5}{\cancel{10}}} = \frac{3}{25}$
22. $\frac{3}{10}$ 23. $\frac{1}{4}$
24. $\frac{5}{16}$ 25. $\frac{2}{21}$
26. $\frac{1}{8}$ 27. $\frac{3}{10}$
28. $\frac{19}{36}$ 29. $\frac{7}{10}$
30. $\frac{11}{15}$ 31. $\frac{21}{40}$

T-60

Practice and Problem Solving

32. $\frac{2}{15}$ 33. $\frac{9}{80}$

34. $\frac{3}{14}$ 35. $\frac{7}{45}$

36. $\frac{81}{100}$ 37. $\frac{27}{125}$

38. $\frac{9}{25}$ 39. $\frac{25}{196}$

40. $\frac{4}{7} > \left(\frac{9}{10} \times \frac{4}{7}\right)$; Because $\frac{9}{10} < 1$, the product will be less than $\frac{4}{7}$.

41. $\left(\frac{5}{8} \times \frac{22}{15}\right) > \frac{5}{8}$; Because $\frac{22}{15} > 1$, the product will be greater than $\frac{5}{8}$.

42. See Additional Answers.

43. *Sample answer:* $\frac{1}{3}$

44. See *Taking Math Deeper*.

45. a. $\frac{3}{50}$ b. 45 people

Fair Game Review

46. $\frac{28}{3}$ 47. $\frac{35}{8}$

48. $\frac{31}{4}$ 49. $\frac{23}{6}$

50. B

Mini-Assessment

Multiply. Write the answer in simplest form.

1. $\frac{1}{6} \times \frac{1}{3}$ $\frac{1}{18}$
2. $\frac{2}{3} \times \frac{3}{4}$ $\frac{1}{2}$
3. $\frac{5}{6} \times \frac{6}{5}$ 1
4. $\frac{7}{8} \times \frac{4}{5}$ $\frac{7}{10}$

5. In an aquarium, $\frac{5}{6}$ of the fish are barbs. Of these, $\frac{1}{3}$ are tiger barbs. What fraction of all the fish in the aquarium are tiger barbs? $\frac{5}{18}$

T-61

Taking Math Deeper

Exercise 44

This type of problem is difficult because the information is given in an order that is not convenient. Remind students that a problem does not have to be solved using the information in the order it is given.

① Draw the given information on a number line.

② Work backwards. Start with the third sentence.

At Checkpoint 2, you are $\frac{1}{4}$ of the 40 miles.

$\frac{1}{4} \cdot 40 = 10$ mi Distance from S to C2

③ Now use the second sentence.

At Checkpoint 1, you are $\frac{2}{5}$ of the 10 miles to Checkpoint 2.

$\frac{2}{5} \cdot 10 = 4$ mi Distance from S to C1

Project

Rewrite Exercise 44 so that the information is given in order. Does this make the problem easier? Explain.

Reteaching and Enrichment Strategies

If students need help...	If students got it...
Resources by Chapter • Practice A and Practice B • Puzzle Time Record and Practice Journal Practice Differentiating the Lesson Lesson Tutorials Skills Review Handbook	Resources by Chapter • Enrichment and Extension • School-to-Work Start the next section

Multiply. Write the answer in simplest form.

32. $\dfrac{1}{2} \times \dfrac{3}{5} \times \dfrac{4}{9}$
33. $\dfrac{3}{4} \times \dfrac{5}{8} \times \dfrac{6}{25}$
34. $\dfrac{4}{7} \times \dfrac{2}{3} \times \dfrac{9}{16}$
35. $\dfrac{5}{6} \times \dfrac{4}{15} \times \dfrac{7}{10}$

36. $\left(\dfrac{9}{10}\right)^2$
37. $\left(\dfrac{3}{5}\right)^3$
38. $\left(\dfrac{4}{5}\right)^2 \times \left(\dfrac{3}{4}\right)^2$
39. $\left(\dfrac{5}{6}\right)^2 \times \left(\dfrac{3}{7}\right)^2$

Without finding the product, copy and complete the statement using <, >, or =. Explain your reasoning.

40. $\dfrac{4}{7}$ ▢ $\left(\dfrac{9}{10} \times \dfrac{4}{7}\right)$
41. $\left(\dfrac{5}{8} \times \dfrac{22}{15}\right)$ ▢ $\dfrac{5}{8}$
42. $\dfrac{5}{6}$ ▢ $\left(\dfrac{5}{6} \times \dfrac{7}{7}\right)$

43. **OPEN-ENDED** Find a fraction that when multiplied by $\dfrac{1}{2}$ is less than $\dfrac{1}{4}$.

44. **DISTANCES** You are in a bike race. When you get to the first checkpoint, you are $\dfrac{2}{5}$ the distance to the second checkpoint. When you get to the second checkpoint, you are $\dfrac{1}{4}$ the distance to the finish. What is the distance from the start to the first checkpoint?

45. **PETS** You ask 150 people about their pets. The results show that $\dfrac{9}{25}$ of the people own a dog. Of the people that own a dog, $\dfrac{1}{6}$ of them also own a cat.

 a. What fraction of the people own a dog and a cat?

 b. **Reasoning** How many people own a dog, but not a cat? Explain.

Fair Game Review *What you learned in previous grades & lessons*

Write the mixed number as an improper fraction. *(Skills Review Handbook)*

46. $9\dfrac{1}{3}$
47. $4\dfrac{3}{8}$
48. $7\dfrac{3}{4}$
49. $3\dfrac{5}{6}$

50. **MULTIPLE CHOICE** A science experiment calls for $\dfrac{3}{4}$ cup of baking powder. You have $\dfrac{1}{3}$ cup of baking powder. How much more baking powder do you need? *(Skills Review Handbook)*

 Ⓐ $\dfrac{1}{4}$ cup
 Ⓑ $\dfrac{5}{12}$ cup
 Ⓒ $\dfrac{4}{7}$ cup
 Ⓓ $1\dfrac{1}{12}$ cups

2.4 Multiplying Mixed Numbers

Essential Question How do you multiply a mixed number by a fraction?

COMMON CORE STATE STANDARDS
6.NS.1

1 ACTIVITY: Multiplying a Mixed Number and a Fraction

Work with a partner. Use a diagram to find the product.

a. Sample: $1\frac{1}{2} \times \frac{2}{3}$

What is two-thirds of $1\frac{1}{2}$?

Think of $1\frac{1}{2}$ as three halves.

Two of the halves is 1.

∴ So, $1\frac{1}{2} \times \frac{2}{3} = 1$.

b. $\frac{4}{9} \times 2\frac{1}{4}$

c. $2\frac{1}{4} \times \frac{1}{2}$

d. $\frac{1}{3} \times 3\frac{3}{4}$

2 ACTIVITY: Multiplying a Mixed Number and a Fraction

Work with a partner. How many square feet are in the piece of fabric?

$1\frac{1}{4}$ ft

$\frac{1}{2}$ ft

a. Use the Distributive Property and find the sum of the two pieces.

b. Rewrite the mixed number as an improper fraction and multiply.

Laurie's Notes

Introduction

For the Teacher
- **Goal:** The goal of this activity is for students to develop an understanding of how to multiply a mixed number by a fraction.

Motivate
- Pose two questions, each connected to the first activity:
- ❓ "What is $\frac{2}{3} \times 1$? What is $\frac{2}{3} \times 2$?" $\frac{2}{3}, \frac{4}{3}$
- ❓ "Use the last two answers to find $\frac{2}{3} \times 1\frac{1}{2}$." $\frac{3}{3}$ or 1
- **Discuss:** How do you multiply a mixed number and a fraction? Some problems can be solved using number sense, like the $\frac{2}{3} \times 1\frac{1}{2}$ example. An area model will also be used, as was done in the previous section.

Activity Notes

Activity 1
- Write the example and draw the model. Remind students that multiplying by $\frac{2}{3}$ means that you want $\frac{2}{3}$ of the model representing $1\frac{1}{2}$. Because $1\frac{1}{2}$ has three equal parts, two of those three parts is two halves, or 1.
- ❓ "What does it mean to take $\frac{2}{3}$ of something?" Listen for the idea of dividing the quantity into 3 equal parts, and then taking 2 of the 3 equal parts.
- In addition to writing the solution $1\frac{1}{2} \times \frac{2}{3} = 1$, you might also write $\frac{3}{2} \times \frac{2}{3} = 1$ as a check.
- The Commutative Property is used to rewrite the expression in part (a) so that the question becomes "What is two-thirds of one and a half?"
- Students will find it easier to diagram the mixed number and then divide the diagram into fractional parts. The goal is for students to find the number of fractional parts in the greater number.

Activity 2
- This example uses the area model to help develop the idea that when you multiply a fraction by a mixed number, you must multiply the fraction by the whole number and the fraction by the fraction. The underlying concept is an application of the Distributive Property. The pink region represents the fraction times the whole number $\left(\frac{1}{2} \times 1 = \frac{1}{2}\right)$ and the green region represents the fraction times the fraction $\left(\frac{1}{2} \times \frac{1}{4} = \frac{1}{8}\right)$.
- The model on the right and the model on the left represent the same shaded region, but some additional lines have been drawn on the model on the right.
- ❓ "How does the model on the right help you to find the answer?" There are 5 shaded pieces and each is $\frac{1}{8}$, so the area is $\frac{5}{8}$.

Common Core State Standards
6.NS.1 Interpret and compute quotients of fractions, and solve word problems involving division of fractions by fractions.

Previous Learning
Students should know how to multiply fractions, multiply a fraction and a whole number, and use the Distributive Property.

Start Thinking! and Warm Up

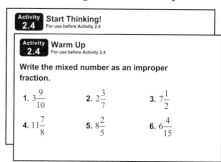

2.4 Record and Practice Journal

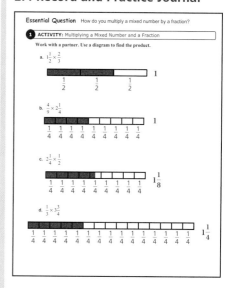

Differentiated Instruction

Visual

It is important for visual learners to complete Activities 1 and 2. The opportunity to create and examine several models for multiplying fractions and mixed numbers will help them gain a concrete understanding of the concept.

2.4 Record and Practice Journal

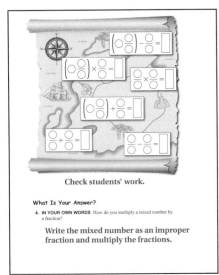

Laurie's Notes

Activity 3

- Students have the opportunity to review addition and subtraction of fractions and mixed numbers. They also will practice a new skill, multiplying a mixed number and a fraction. Using the sample from Activity 1, students should recognize that the first step in multiplying a mixed number and a fraction is to write the mixed number as an improper fraction.
- **Connections:** When trying to form the greatest possible value for any location, students need to consider the operation first. You might begin with a general discussion of a strategy using questions like those below.
- ? "When adding a mixed number and a fraction, how do you make the greatest sum?" Listen for making the whole number part of the mixed number the greatest possible value of the available numbers.
- ? "When subtracting fractions, how do you make the greatest difference?" Listen for making the first fraction the greatest possible and the second fraction the least possible.
- Each student will do 3 problems and add the answers. Students should check one another's work.

What Is Your Answer?

- **Think-Pair-Share:** Students should read each question independently and then work with a partner to answer the questions. When they have answered the questions, the pair should compare their answers with another group and discuss any discrepancies.

Closure

- Draw a model to find $\frac{1}{4} \times 1\frac{1}{2}$.

 Sample answer:

Technology For the Teacher

The Dynamic Planning Tool
Editable Teacher's Resources at *BigIdeasMath.com*

3 ACTIVITY: Buried Treasure Game

Number of Players: 2

Taking turns, each player will
- choose a treasure location.
- roll a number cube as many times as there are blanks in the expression.
- place the numbers in the blanks to form the greatest possible value for that treasure location.
- check each other's work.

Players then total the values of their treasures (sum of the three expressions). The player with the greater total wins the game.

What Is Your Answer?

4. IN YOUR OWN WORDS How do you multiply a mixed number by a fraction?

Use what you learned about multiplying mixed numbers to complete Exercises 8–15 on page 66.

2.4 Lesson

Check It Out
Lesson Tutorials
BigIdeasMath.com

EXAMPLE 1 Using the Distributive Property

Use the Distributive Property to find $\frac{1}{2} \times 2\frac{3}{4}$.

$\frac{1}{2} \times 2\frac{3}{4} = \frac{1}{2} \times \left(2 + \frac{3}{4}\right)$ Rewrite $2\frac{3}{4}$ as the sum $2 + \frac{3}{4}$.

$= \left(\frac{1}{2} \times 2\right) + \left(\frac{1}{2} \times \frac{3}{4}\right)$ Use the Distributive Property.

$= 1 + \frac{3}{8}$ Multiply.

$= 1\frac{3}{8}$ Add.

On Your Own

Now You're Ready
Exercises 4–7

Use the Distributive Property to find the product.

1. $\frac{2}{3} \times 1\frac{1}{2}$ 2. $\frac{3}{5} \times 2\frac{1}{6}$ 3. $\frac{1}{4} \times 4\frac{1}{5}$ 4. $\frac{2}{7} \times 3\frac{3}{4}$

Key Idea

Multiplying Mixed Numbers

Write each mixed number as an improper fraction. Then multiply as you would with fractions.

EXAMPLE 2 Multiplying a Fraction and a Mixed Number

Find $\frac{1}{2} \times 2\frac{3}{4}$. Estimate $\frac{1}{2} \times 3 = 1\frac{1}{2}$

$\frac{1}{2} \times 2\frac{3}{4} = \frac{1}{2} \times \frac{11}{4}$ Write $2\frac{3}{4}$ as the improper fraction $\frac{11}{4}$.

$= \frac{1 \times 11}{2 \times 4}$ Multiply the numerators and the denominators.

$= \frac{11}{8}$, or $1\frac{3}{8}$ Simplify.

∴ So, the product is $1\frac{3}{8}$. **Reasonable?** $1\frac{3}{8} \approx 1\frac{1}{2}$ ✓

64 Chapter 2 Multiplying and Dividing Fractions

Laurie's Notes

Introduction

Connect
- **Yesterday:** Students used a visual model to multiply a mixed number and a fraction.
- **Today:** Students will use the Distributive Property to multiply a fraction and a mixed number. This is generalized to multiply mixed numbers.

Motivate
- Play the *Multiplication Wizard*! Tell students that you can multiply a one-digit number by a two-digit number in your head. Have students give you these numbers.
- **FYI:** Use the Distributive Property to multiply the numbers in your head. Example: $8 \times 42 = 8(40 + 2) = 320 + 16 = 336$.
- Keep track of the first product while multiplying the second product. If this is difficult, write the two products (320 and 16) on the board and add aloud. Some students may recognize that you are using the Distributive Property.

Lesson Notes

Example 1
- Write the example on the board.
- ? "How can the Distributive Property be used to help you multiply $\frac{1}{2} \times 2\frac{3}{4}$?" Think of $2\frac{3}{4}$ as $2 + \frac{3}{4}$.
- Finish working through the example.

On Your Own
- The technique of using the Distributive Property to multiply a fraction and a mixed number can be used for each of these problems. The Distributive Property is reinforced because the fraction must be multiplied by both the whole number AND the fraction.
- **Common Error:** It is very common for students to multiply the fractions and then *add on* the whole number. Example: $\frac{1}{2} \times 2\frac{3}{4} \neq 2\frac{3}{8}$.
- ? "Can you think of a way to make sure that the fraction is being multiplied by both the whole number and the fraction without using the Distributive Property?" Listen for students to mention writing the mixed number as an improper fraction.

Key Idea
- **Discuss:** When the mixed number is written as an improper fraction, the entire number is being multiplied by the simple fraction. Connect this idea to Example 2.

Example 2
- Work through the example. (**Note:** It is the same problem as Example 1.)

Goal
Today's lesson is multiplying mixed numbers by fractions, fractions by mixed numbers, and mixed numbers by mixed numbers.

Start Thinking! and Warm Up

Lesson 2.4 Warm Up For use before Lesson 2.4

Lesson 2.4 Start Thinking! For use before Lesson 2.4

The distance around a track is $\frac{1}{4}$ mile.

Write a word problem using the track that involves multiplying a fraction and a mixed number.

Exchange word problems with a classmate and solve your classmate's problem.

Extra Example 1
Use the Distributive Property to find $\frac{5}{6} \times 3\frac{1}{10}$. $2\frac{7}{12}$

On Your Own
1. 1
2. $1\frac{3}{10}$
3. $1\frac{1}{20}$
4. $1\frac{1}{14}$

Extra Example 2
Find $6\frac{1}{3} \times \frac{3}{4}$. $4\frac{3}{4}$

T-64

Extra Example 3

Find $2\frac{3}{5} \times 3\frac{1}{3}$. $8\frac{2}{3}$

On Your Own

5. $\frac{7}{18}$
6. $1\frac{5}{9}$
7. $4\frac{1}{2}$
8. 12

Extra Example 4

A grass practice field is $80\frac{1}{2}$ yards by $30\frac{3}{4}$ yards. Find the area of the practice field. $2475\frac{3}{8}$ square yards

On Your Own

9. $36\frac{3}{32}$ ft²

English Language Learners

Vocabulary
Make sure that students understand the meanings of the phrases *mixed number* and *improper fraction*. Suggest that they add these phrases and examples to their notes.

Laurie's Notes

Example 3

- **Common Error:** Students may multiply the whole numbers and then multiply the fractions as shown: $1\frac{4}{5} \times 3\frac{2}{3} \neq 3\frac{8}{15}$.
- Write the problem. Estimate the product.
- "What is the first step in multiplying these mixed numbers?" Write the mixed numbers as improper fractions.
- As you work through this example, point out that by writing each of the mixed numbers as improper fractions, it ensures that all of the first number is being multiplied by all of the second number.
- Finish working through the problem. Note that the answer is more reasonable than $3\frac{8}{15}$.

On Your Own

- Have students work with a partner on these problems. Encourage them to write an estimate first, reinforcing the concept of rounding. Finish with judging the reasonableness of the answer.

Example 4

- Students have been using an area model in this chapter, so the formula is familiar. Notice that an estimate is made.
- Work through the problem. Students should appreciate the technique of dividing out common factors in the problem.
- "Why is the answer labeled with square meters?" *Sample answer:* Area is measured in square units.
- "If the basketball court was the same size but measured in yards instead of meters, would the answer be *more than* or *less than* 288 square yards? Explain." more; Meters are greater than yards.

On Your Own

- **Neighbor Check:** Have students work independently and then have their neighbor check their work. Have students discuss any discrepancies.

Closure

- **Exit Ticket:** If you multiply two mixed numbers, what do you know about the product? *Sample answer:* It will be greater than the product of the whole number portions of the mixed numbers.

Technology for the Teacher

The Dynamic Planning Tool
Editable Teacher's Resources at *BigIdeasMath.com*

EXAMPLE 3 Multiplying Mixed Numbers

Find $1\frac{4}{5} \times 3\frac{2}{3}$. Estimate $2 \times 4 = 8$

$1\frac{4}{5} \times 3\frac{2}{3} = \frac{9}{5} \times \frac{11}{3}$ Write $1\frac{4}{5}$ and $3\frac{2}{3}$ as improper fractions.

$= \frac{\overset{3}{\cancel{9}} \times 11}{5 \times \underset{1}{\cancel{3}}}$ Multiply fractions. Divide out the common factor 3.

$= \frac{33}{5}$, or $6\frac{3}{5}$ Simplify.

So, the product is $6\frac{3}{5}$. Reasonable? $6\frac{3}{5} \approx 8$

On Your Own

Now You're Ready Exercises 8–23

Multiply. Write the answer in simplest form.

5. $\frac{1}{3} \times 1\frac{1}{6}$ 6. $3\frac{1}{2} \times \frac{4}{9}$ 7. $1\frac{7}{8} \times 2\frac{2}{5}$ 8. $5\frac{5}{7} \times 2\frac{1}{10}$

EXAMPLE 4 Real-Life Application

A city is resurfacing a basketball court. Find the area of the court.

Estimate $21 \times 14 = 294$

$A = \ell w$ Write the formula for the area of a rectangle.

$= 21\frac{1}{3} \times 13\frac{1}{2}$ Substitute $21\frac{1}{3}$ for ℓ and $13\frac{1}{2}$ for w.

$= \frac{64}{3} \times \frac{27}{2}$ Write $21\frac{1}{3}$ and $13\frac{1}{2}$ as improper fractions.

$= \frac{\overset{32}{\cancel{64}} \times \overset{9}{\cancel{27}}}{\underset{1}{\cancel{3}} \times \underset{1}{\cancel{2}}}$ Multiply fractions. Divide out common factors.

$= 288$ Simplify.

The area of the court is 288 square meters.

Reasonable? $288 \approx 294$

On Your Own

9. Find the area of a rectangular air hockey table that is $8\frac{1}{4}$ feet by $4\frac{3}{8}$ feet.

2.4 Exercises

Check It Out
Help with Homework
BigIdeasMath.com

Vocabulary and Concept Check

1. **VOCABULARY** What is an improper fraction?
2. **WRITING** Describe how to multiply two mixed numbers.
3. **OPEN-ENDED** Write two mixed numbers between 3 and 4 that have a product between 9 and 12.

Practice and Problem Solving

Use the Distributive Property to find the product.

4. $\frac{1}{4} \times 2\frac{2}{7}$
5. $\frac{5}{6} \times 2\frac{2}{5}$
6. $\frac{5}{9} \times 4\frac{1}{2}$
7. $\frac{2}{15} \times 5\frac{5}{8}$

Multiply. Write the answer in simplest form.

8. $1\frac{1}{3} \times \frac{2}{3}$
9. $6\frac{2}{3} \times \frac{3}{10}$
10. $2\frac{1}{2} \times \frac{4}{5}$
11. $\frac{3}{5} \times 3\frac{1}{3}$

12. $7\frac{1}{2} \times \frac{2}{3}$
13. $\frac{5}{9} \times 3\frac{3}{5}$
14. $\frac{3}{4} \times 1\frac{1}{3}$
15. $3\frac{3}{4} \times \frac{2}{5}$

16. $4\frac{3}{8} \times \frac{4}{5}$
17. $\frac{3}{7} \times 2\frac{5}{6}$
18. $1\frac{3}{10} \times 18$
19. $15 \times 2\frac{4}{9}$

20. $1\frac{1}{6} \times 6\frac{3}{4}$
21. $2\frac{5}{12} \times 2\frac{2}{3}$
22. $5\frac{5}{7} \times 3\frac{1}{8}$
23. $2\frac{4}{5} \times 4\frac{1}{16}$

ERROR ANALYSIS Describe and correct the error in finding the product.

24. $4 \times 3\frac{7}{10} = 12\frac{7}{10}$

25. $2\frac{1}{2} \times 7\frac{4}{5} = (2 \times 7) + \left(\frac{1}{2} \times \frac{4}{5}\right)$
$= 14 + \frac{2}{5} = 14\frac{2}{5}$

26. **VITAMIN C** A vitamin C tablet contains $\frac{1}{40}$ gram of vitamin C. You take $1\frac{1}{2}$ tablets every day. How many grams of vitamin C do you take every day?

27. **SCHOOL BANNER** A banner is made for a football rally.
 a. What is the area of the banner?
 b. A $\frac{1}{4}$-foot border is added on each side. What is the new area of the banner?

GO PANTHERS! $1\frac{1}{2}$ ft
$4\frac{2}{3}$ ft

Assignment Guide and Homework Check

Level	Day 1 Activity Assignment	Day 2 Lesson Assignment	Homework Check
Basic	8–15, 40–48	1–3, 5, 7, 17–23 odd, 24–27, 29–33 odd	5, 17, 24, 27, 31
Average	8–15, 40–48	1–3, 5, 7, 17–27 odd, 28–36 even, 37	5, 17, 27, 30, 36
Advanced	8–15, 40–48	1–3, 17–25 odd, 28–36 even, 37–39	17, 30, 34, 38

Common Errors

- **Exercises 4–7** Students may struggle with the concept that a mixed number is the whole number plus the fraction. Tell them to write the whole number as a fraction with the same denominator as the fraction, then add. Compare this with the improper fraction.
- **Exercises 8–23** Students may forget to change the mixed number to an improper fraction. Do an example using improper fractions and using the Distributive Property with the mixed numbers. Show students visually that it is necessary to change the mixed number into an improper fraction.
- **Exercises 8–23** Students may write $6\frac{2}{3}$ as $\frac{3 \times 6 \times 2}{3}$ instead of as $\frac{3 \times 6 + 2}{3}$. Make sure they have a good understanding of how to convert mixed numbers to improper fractions and vice versa.

2.4 Record and Practice Journal

Technology For the Teacher
Answer Presentation Tool
QuizShow

Vocabulary and Concept Check

1. a fraction with a numerator that is greater than or equal to the denominator
2. Write each mixed number as an improper fraction, then multiply as you would with fractions.
3. *Sample answer:* $3\frac{1}{2} \times 3\frac{1}{7} = 11$

Practice and Problem Solving

4. $\frac{4}{7}$ 5. 2
6. $2\frac{1}{2}$ 7. $\frac{3}{4}$
8. $\frac{8}{9}$ 9. 2
10. 2 11. 2
12. 5 13. 2
14. 1 15. $1\frac{1}{2}$
16. $3\frac{1}{2}$ 17. $1\frac{3}{14}$
18. $23\frac{2}{5}$ 19. $36\frac{2}{3}$
20. $7\frac{7}{8}$ 21. $6\frac{4}{9}$
22. $17\frac{6}{7}$ 23. $11\frac{3}{8}$

24. You must first rewrite the mixed number as an improper fraction and then multiply.
$$4 \times 3\frac{7}{10} = 4 \times \frac{37}{10}$$
$$= \frac{\cancel{4}^{2} \times 37}{\cancel{10}_{5}}$$
$$= \frac{74}{5}, \text{ or } 14\frac{4}{5}$$

25. See Additional Answers.
26. $\frac{3}{80}$ gram

T-66

Practice and Problem Solving

27. a. 7 ft^2

 b. $10\frac{1}{3} \text{ ft}^2$

28. $6\frac{3}{4}$ **29.** $13\frac{3}{4}$

30. $6\frac{1}{14}$ **31.** $8\frac{13}{24}$

32. $2\frac{1}{12}$ **33.** 155

34–37. See Additional Answers.

38. See *Taking Math Deeper*.

39. a. $5\frac{1}{10}$ hours

 b. 11:39 A.M.

Fair Game Review

40. $\frac{6}{35}$ **41.** $\frac{1}{6}$

42. $\frac{4}{9}$ **43.** $\frac{7}{36}$

44. $5x + 30$ **45.** $9x - 27$

46. $28 + 4x$ **47.** $12x - 96$

48. D

Mini-Assessment

Multiply. Write the answer in simplest form.

1. $2\frac{3}{4} \times \frac{4}{5}$ $2\frac{1}{5}$

2. $3\frac{1}{2} \times \frac{3}{5}$ $2\frac{1}{10}$

3. $12 \times 2\frac{5}{9}$ $30\frac{2}{3}$

4. $1\frac{1}{3} \times 2\frac{7}{8}$ $3\frac{5}{6}$

5. You want to install carpet in your room. Your room is $10\frac{3}{4}$ feet by $16\frac{1}{2}$ feet. Find the area of your room. $177\frac{3}{8} \text{ ft}^2$

T-67

Taking Math Deeper

Exercise 38

This problem asks students to look for two ways to solve the problem. The object of this problem is to see that *changing units* can sometimes result in a lot less work with messy fractions.

① Draw a diagram. Identify the area to be found.

a. I want to find the area of the green region.

② **b.** Use feet or use inches.

Find the area using feet (involves fractions).

Area of Green = Area of Garden − Area of Fountain

$$= 6\frac{3}{4} \cdot 9\frac{1}{6} \quad - \quad 3\frac{1}{3} \cdot 5\frac{1}{4}$$

$$= \frac{27}{4} \cdot \frac{55}{6} - \frac{10}{3} \cdot \frac{21}{4}$$

$$= \frac{\overset{9}{\cancel{27}}}{4} \cdot \frac{55}{\underset{2}{\cancel{6}}} - \frac{\overset{5}{\cancel{10}}}{\underset{1}{\cancel{3}}} \cdot \frac{\overset{7}{\cancel{21}}}{\underset{2}{\cancel{4}}}$$

$$= \frac{495}{8} - \frac{35}{2}$$

$$= \frac{495}{8} - \frac{140}{8}$$

$$= \frac{355}{8} \text{ or } 44\frac{3}{8} \text{ ft}^2$$

③ Find the area using inches.

Area of Green = Area of Garden − Area of Fountain

$= 81 \cdot 110 \quad - \quad 40 \cdot 63$

$= 8910 - 2520$

$= 6390 \text{ in.}^2$

$= 44\frac{3}{8} \text{ ft}^2$ Divide 6390 by 144.

c.

This is easier.

Project

You double the length and width of your garden. Now you can add paths through the garden, a flower bed and other features. Draw a diagram of your new garden with your added features. Be sure to include dimensions.

Reteaching and Enrichment Strategies

If students need help...	If students got it...
Resources by Chapter • Practice A and Practice B • Puzzle Time Record and Practice Journal Practice Differentiating the Lesson Lesson Tutorials Skills Review Handbook	Resources by Chapter • Enrichment and Extension • School-to-Work Start the next section

ALGEBRA Evaluate the expression when $x = 5\frac{5}{8}$, $y = 2\frac{4}{9}$, and $z = 1\frac{5}{16}$.

28. $5\frac{1}{7} \cdot z$ **29.** xy **30.** $x \cdot 2\frac{2}{9} - 6\frac{3}{7}$ **31.** $yz + 5\frac{1}{3}$

Multiply. Write the answer in simplest form.

32. $\frac{4}{7} \times 4\frac{3}{8} \times \frac{5}{6}$

33. $3\frac{3}{4} \times 5\frac{1}{6} \times 8$

34. $2\frac{1}{10} \times \frac{5}{9} \times 4\frac{2}{3}$

35. $1\frac{1}{15} \times 5\frac{2}{5} \times 4\frac{7}{12}$

36. PICTURES Three pictures hang side by side on a wall. What is the total area of the wall covered by the pictures?

37. LOGIC Is the product of two positive mixed numbers ever less than 1? Explain.

38. GARDEN You plan to add a fountain to your garden.

a. Draw a diagram of the fountain in the garden. Label the dimensions.

b. Describe 2 methods for finding the area of the garden that surrounds the fountain.

c. Find the area. Which method did you use, and why?

39. The cooking time for a ham is $\frac{2}{5}$ hour for each pound.

a. How long should you cook a ham that weighs $12\frac{3}{4}$ pounds?

b. Dinner time is 4:45 P.M. What time should you start cooking the ham?

Fair Game Review What you learned in previous grades & lessons

Multiply. Write the answer in simplest form. *(Section 2.3)*

40. $\frac{2}{5} \times \frac{3}{7}$ **41.** $\frac{1}{4} \times \frac{2}{3}$ **42.** $\frac{5}{6} \times \frac{8}{15}$ **43.** $\frac{7}{8} \times \frac{2}{9}$

Use the Distributive Property to rewrite the expression. *(Section 1.4)*

44. $5(x + 6)$ **45.** $9(x - 3)$ **46.** $4(7 + x)$ **47.** $12(x - 8)$

48. MULTIPLE CHOICE How many inches are in 5 yards? *(Skills Review Handbook)*

Ⓐ 15 Ⓑ 60 Ⓒ 120 Ⓓ 180

2 Study Help

Check It Out
Graphic Organizer
BigIdeasMath.com

You can use a **notetaking organizer** to write notes, vocabulary and questions about a topic. Here is an example of a notetaking organizer for multiplying fractions and whole numbers.

Write important vocabulary or formulas in this space.

$$a \cdot \frac{b}{c} = \frac{a \cdot b}{c}$$
$$(c \neq 0)$$

Multiplying fractions and whole numbers

Multiply the numerator of the fraction by the whole number. Then write the product over the denominator.

Example:
$$2 \times \frac{4}{9} = \frac{2 \times 4}{9} = \frac{8}{9}$$

Write your notes about the topic in this space.

How can you multiply a fraction by another fraction?

Write your questions about the topic in this space.

On Your Own

Make a notetaking organizer to help you study these topics.

1. estimating with mixed numbers
2. multiplying two fractions
3. multiplying mixed numbers

After you complete this chapter, make notetaking organizers for the following topics.

4. dividing by a fraction
5. dividing mixed numbers
6. writing decimals as fractions
7. writing fractions as decimals

"The notetaking organizer in my math class gave me an idea of how to organize my doggy biscuits."

Sample Answers

1.

Underestimate: an estimate that is less than the exact answer.	Estimating with mixed numbers
	You may want to round mixed numbers up or down regardless of their fractional part.
Overestimate: an estimate that is greater than the exact answer.	Examples:
	Underestimate: $9\frac{1}{4} \times 4\frac{3}{4} \approx 9 \times 4 = 36$
Compatible Numbers: numbers that are easy to compute mentally.	Overestimate: $9\frac{1}{4} \times 4\frac{3}{4} \approx 10 \times 5 = 50$

How do you estimate with decimals?

2.

$\frac{a}{b} \cdot \frac{c}{d} = \frac{a \cdot c}{b \cdot d}$	Multiplying two fractions
($b \neq 0$ and $d \neq 0$)	Multiply the numerators and multiply the denominators. The denominators cannot be 0.
	Example:
	$\frac{3}{4} \times \frac{5}{8} = \frac{3 \times 5}{4 \times 8} = \frac{15}{32}$

How do you multiply two mixed numbers?

3. Available at *BigIdeasMath.com*.

List of Organizers
Available at *BigIdeasMath.com*

Comparison Chart
Concept Circle
Definition (Idea) and Example Chart
Example and Non-Example Chart
Formula Triangle
Four Square
Information Frame
Information Wheel
Notetaking Organizer
Process Diagram
Summary Triangle
Word Magnet
Y Chart

About this Organizer

A **Notetaking Organizer** can be used to write notes, vocabulary, and questions about a topic. In the space on the left, students write important vocabulary or formulas. In the space on the right, students write their notes about the topic. In the space at the bottom, students write their questions about the topic. A notetaking organizer can also be used as an assessment tool, in which blanks are left for students to complete.

Technology For the Teacher

Vocabulary Puzzle Builder

Answers

1. 10
2. 40
3. $\dfrac{4}{9}$
4. 3
5. $\dfrac{3}{28}$
6. $\dfrac{3}{5}$
7. $\dfrac{11}{12}$
8. $\dfrac{161}{60}$ or $2\dfrac{41}{60}$
9. $\dfrac{7}{15}$
10. $\dfrac{399}{20}$ or $19\dfrac{19}{20}$
11. 12 stores
12. $\dfrac{60{,}853}{8}$ or $7606\dfrac{5}{8}$ ft^2
13. about 5 times
14. $312.50

Alternative Quiz Ideas

100% Quiz — Math Log
Error Notebook — Notebook Quiz
Group Quiz — Partner Quiz
Homework Quiz — **Pass the Paper**

Pass the Paper

- Work in groups of four. The first student copies the problem and completes the first step, explaining his or her work.
- The paper is passed and the second student works through the next step, also explaining his or her work.
- This process continues until the problem is completed.
- The second member of the group starts the next problem. Students should be allowed to question and debate as they are working through the quiz.
- Student groups can be selected by the teacher, by students, through a random process, or any way that works for your class.
- The teacher walks around the classroom listening to the groups and asks questions to ensure understanding.

Assessment Book

Reteaching and Enrichment Strategies

If students need help...	If students got it...
Resources by Chapter • Study Help • Practice A and Practice B • Puzzle Time Lesson Tutorials BigIdeasMath.com Practice Quiz Practice from the Test Generator	Resources by Chapter • Enrichment and Extension • School-to-Work Game Closet at *BigIdeasMath.com* Start the next section

Technology For the Teacher

Answer Presentation Tool
Big Ideas Test Generator

T-69

2.1–2.4 Quiz

Use compatible numbers to estimate the product or quotient. *(Section 2.1)*

1. $52 \div 4\frac{2}{3}$
2. $21 \times 1\frac{7}{8}$

Multiply. Write the answer in simplest form. *(Section 2.2 and Section 2.3)*

3. $4 \times \frac{1}{9}$
4. $5 \times \frac{3}{5}$
5. $\frac{3}{7} \times \frac{1}{4}$
6. $\frac{9}{10} \times \frac{2}{3}$

Use the Distributive Property to find the product. *(Section 2.4)*

7. $\frac{1}{3} \times 2\frac{3}{4}$
8. $\frac{7}{10} \times 3\frac{5}{6}$

Multiply. Write the answer in simplest form. *(Section 2.4)*

9. $1\frac{1}{6} \times \frac{2}{5}$
10. $3\frac{1}{2} \times 5\frac{7}{10}$

11. **MALL** In a mall, $\frac{1}{15}$ of the stores sell shoes. There are 180 stores in the mall. How many of the stores sell shoes? *(Section 2.2)*

12. **CONCERT FLOOR** The floor of a concert venue is $100\frac{3}{4}$ feet by $75\frac{1}{2}$ feet. What is the area of the floor? *(Section 2.4)*

13. **PRESENTATION** You are practicing a presentation that is $15\frac{3}{4}$ minutes long. You have $1\frac{1}{3}$ hours before you give the presentation. Estimate how many times you can practice the presentation. *(Section 2.1)*

14. **DRAMA** The drama club has $500 to put on a play. The club can spend up to $\frac{3}{8}$ of the money on new costumes. How much money can they spend on other supplies? *(Section 2.2)*

2.5 Dividing Fractions

Essential Question How do you divide by a fraction?

COMMON CORE STATE STANDARDS
6.NS.1

1 ACTIVITY: Dividing by a Fraction

Work with a partner.

a. Describe the pattern of the blue numbers.

b. Describe the pattern of the red numbers. Use the pattern to complete the table.

c. The division $8 \div \frac{1}{2}$ can be read as "How many halves are in 8?" Use the completed table to answer this question. Then draw a model that shows your answer.

d. Use the pattern in the table to complete the following.

$8 \div \frac{1}{2} = 16 = 8 \times \frac{2}{1}$ Invert $\frac{1}{2}$ and multiply.

$8 \div \frac{1}{4} = 32 = $ _____ _____

$8 \div \frac{1}{8} = 64 = $ _____ _____

$8 \div 16$	$\frac{1}{2}$
$8 \div 8$	1
$8 \div 4$	2
$8 \div 2$	4
$8 \div 1$	8
$8 \div \frac{1}{2}$	
$8 \div \frac{1}{4}$	
$8 \div \frac{1}{8}$	

2 ACTIVITY: Dividing by a Fraction

Work with a partner.

a. Draw a model for $3 \div \frac{2}{3}$. Use the model to answer the question "How many two-thirds are in 3?"

b. Complete the table in two ways. First use the model. Then use the "invert and multiply" rule that you found in Activity 1. Compare your answers.

$3 \div \frac{2}{3}$	
$6 \div \frac{2}{3}$	
$9 \div \frac{2}{3}$	
$12 \div \frac{2}{3}$	

Chapter 2 Multiplying and Dividing Fractions

Laurie's Notes

Common Core State Standards

6.NS.1 Interpret and compute quotients of fractions, and solve word problems involving division of fractions by fractions.

Introduction

For the Teacher

- **Goal:** The goal of this activity is for students to develop an understanding of how to divide by a fraction.
- The activity shows division as a partition model. For example, 20 ÷ 5 means "How many sets of 5 can you make from 20?" In the activity, eight is partitioned into sets that keep getting smaller!

Activity Materials	
Introduction	Textbook (optional)
• 12 strips of paper • scissors	• one strip of paper for each student

Motivate

- ❓ "How many pieces of licorice would I need if I were to give each of you $\frac{1}{2}$ of a piece?" The answer is half of the number of students in your class. For 20 students, you need 10 pieces; for 21 students, you need $10\frac{1}{2}$ pieces.
- **Quick Model:** Show students six strips of paper. Say "These are pieces of licorice. How many halves are there in 6?" You may want to cut the pieces of paper in half. Then write $6 \div \frac{1}{2} = 12$.
- ❓ "How many pieces of licorice would I need if I were to give you each $\frac{1}{3}$ of a piece?" The answer is one-third the number of students.
- **Quick Model:** Take 6 strips of paper. Say "These are pieces of licorice. How many thirds are there in 6?" You may want to cut the pieces of paper into thirds. Then write: $6 \div \frac{1}{3} = 18$.

Start Thinking! and Warm Up

Activity Notes

Activity 1

- ❓ Students may have a question about the first entry. Use language such as "if a team is composed of 16 people, how many teams could be made with 8 people?" none; you only have $\frac{1}{2}$ a team.
- Leave the definition of reciprocal until tomorrow. For now, just tell students to invert the fraction and then multiply.

Activity 2

- You may want to give each student a strip of paper. Have them fold it to form 3 equal parts. Say "This is a model of 3. How many $\frac{2}{3}$ are in 3?"
- **Explanation:** Have students use a pencil to divide each whole into thirds. Then count the number of $\frac{2}{3}$.

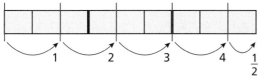

- There are 4 pieces of length $\frac{2}{3}$, and a remaining piece that is $\frac{1}{2}$ the length of $\frac{2}{3}$. So, there are a total of $4\frac{1}{2}$ pieces of length $\frac{2}{3}$ in 3.

2.5 Record and Practice Journal

Differentiated Instruction

Inclusion

Some students may have difficulty dividing fractions when the divisor or the dividend is a whole number. When this occurs, suggest that the student write the whole number as a fraction for the first step.

Example 1

$$8 \div \frac{1}{2} = \frac{8}{1} \div \frac{1}{2}$$
$$= \frac{8}{1} \times \frac{2}{1}$$
$$= 16$$

Example 2

$$\frac{3}{4} \div 9 = \frac{3}{4} \div \frac{9}{1}$$
$$= \frac{3}{4} \times \frac{1}{9}$$
$$= \frac{1}{12}$$

2.5 Record and Practice Journal

Laurie's Notes

Activity 3

- This is a good activity to help students think about the meaning of division. Before beginning, make sure students understand the model. The blue rectangles represent whole units and the dotted lines are parts of a whole. Encourage them to use the model and make additional lines as needed.
- For part (c), students may wish to show additional lines in order to see all of the fourths, although the quarters above the model are helpful.
- When students have finished, allow adequate time for them to share their approaches to the answers. Projecting this page for the class will help facilitate the discussion.
- **FYI:** Some students may simply *invert and multiply* and not try to make sense of what fraction division means and why the rule works. Pick one problem to explore in depth to make the connection. A good sample to pick is part (d).
- **Connection to the Model:** When the 8 units are divided into fifths, note that there are $\frac{40}{5}$ in 8. The question is asking how many $\frac{4}{5}$ are in 8. If groups of 4 are circled, this would happen 10 times to get to the end of the model for 8. So when you *invert and multiply*, the product tells you how many fifths there are in 8. You divide by 4 because you want to take 4 at a time.
- The explanation above is similar to saying the following:

$$8 \div \frac{4}{5} = \frac{40}{5} \div \frac{4}{5} = 40 \div 4 = 10.$$

What Is Your Answer?

- Invert the fraction and then multiply. The quotient (answer) is the number of fractional parts that are in the dividend.

Closure

- **Exit Ticket:** How many thirds are in 2? In other words, what is $2 \div \frac{1}{3}$? Draw a picture to explain your answer. 6

The Dynamic Planning Tool
Editable Teacher's Resources at *BigIdeasMath.com*

3 ACTIVITY: Dividing by a Fraction

Work with a partner. Write the division problem and answer it using a model.

a. How many halves are in five halves?

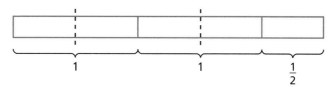

b. How many sixths are in three halves?

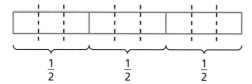

c. How many three-fourths are in 3?

d. How many four-fifths are in 8?

e. How many three-tenths are in 6?

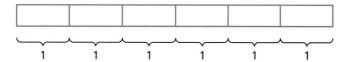

f. How many halves are in a fourth?

What Is Your Answer?

4. **IN YOUR OWN WORDS** How do you divide by a fraction? Give an example.

Practice — Use what you learned about dividing fractions to complete Exercises 11–18 on page 75.

2.5 Lesson

Key Vocabulary
reciprocals, *p. 72*

Two numbers whose product is 1 are **reciprocals**. To write a reciprocal of a number, write the number as a fraction. Then invert the fraction.

The Meaning of a Word • Invert

When you **invert** a glass, you turn it over.

EXAMPLE 1 Writing Reciprocals

Study Tip
When any number is multiplied by 0, the product is 0. So, the number 0 does not have a reciprocal.

	Original Number	Fraction	Reciprocal	Check
a.	$\dfrac{3}{5}$	$\dfrac{3}{5}$	$\dfrac{5}{3}$	$\dfrac{3}{5} \times \dfrac{5}{3} = 1$
b.	$\dfrac{9}{5}$	$\dfrac{9}{5}$	$\dfrac{5}{9}$	$\dfrac{9}{5} \times \dfrac{5}{9} = 1$
c.	2	$\dfrac{2}{1}$	$\dfrac{1}{2}$	$\dfrac{2}{1} \times \dfrac{1}{2} = 1$

● **On Your Own**

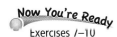
Exercises 7–10

Write the reciprocal of the number.

1. $\dfrac{3}{4}$ 2. 5 3. $\dfrac{7}{2}$ 4. $\dfrac{4}{9}$

Key Idea

Dividing Fractions

Words To divide a number by a fraction, multiply the number by the reciprocal of the fraction.

Numbers $\dfrac{1}{5} \div \dfrac{3}{4} = \dfrac{1}{5} \times \dfrac{4}{3}$

Algebra $\dfrac{a}{b} \div \dfrac{c}{d} = \dfrac{a}{b} \times \dfrac{d}{c}$, where b, c, and $d \neq 0$

Laurie's Notes

Introduction

Connect
- **Yesterday:** Students used a visual model to divide by a fraction.
- **Today:** Students will use a formal rule to divide by a fraction.

Motivate
- Hold up 3 quarters.
- ? "How many nickels are there in 3 quarters?" 15
- ? "How could I write that question as a division problem with fractions?"
 Think of what part of a dollar each coin represents ▶ $\frac{3}{4} \div \frac{1}{20} = 15$
- ? "Does our technique from yesterday, *invert and multiply*, work for this problem?" yes: $\frac{3}{4} \div \frac{1}{20} = \frac{3}{4} \times \frac{20}{1} = 15$
- **Discuss:** This problem is a fraction divided by a fraction. Most of the problems yesterday were a whole number divided by a fraction. Today you will look at many types of division problems with fractions.
- Introduce reciprocal as shown in the text.

Lesson Notes

Example 1
- To write the reciprocal of a whole number, you begin by representing the whole number over one. Then switch the numerator and denominator.
- **Common Error:** When you ask students to write a number and its reciprocal, they frequently write: $\frac{3}{4} = \frac{4}{3}$. Reciprocals are not equal (except for 1). Students should write $\frac{3}{4}, \frac{4}{3}$ or "the reciprocal of $\frac{3}{4}$ is $\frac{4}{3}$."
- **Check:** The product of the **reciprocals** should equal 1.

Key Idea
- Write the Key Idea on the board.
- Note that the definition says "to divide a number *by a fraction*," Students frequently ask, even two years later, which fraction to take the reciprocal of. Read this portion of the definition again. We are dividing a number (it could be a whole number or a fraction) *by a fraction* (Remind students that a whole number could also be written as a fraction).
- **Connection:** Notice that this definition can be used for dividing whole numbers, too: $8 \div 2 = 8 \div \frac{2}{1} = 8 \times \frac{1}{2} = \frac{8}{2} = 4$.

Goal Today's lesson is dividing by a fraction.

Lesson Materials
Introduction
• 3 quarters 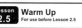

Start Thinking! and Warm Up

Lesson 2.5 Warm Up For use before Lesson 2.5

Lesson 2.5 Start Thinking! For use before Lesson 2.5
Come up with an easy way to remember how to find the reciprocal of a fraction.

Extra Example 1
Write the reciprocal of the number.
a. $\frac{7}{8}$ $\frac{8}{7}$
b. $\frac{3}{2}$ $\frac{2}{3}$
c. 9 $\frac{1}{9}$

On Your Own
1. $\frac{4}{3}$
2. $\frac{1}{5}$
3. $\frac{2}{7}$
4. $\frac{9}{4}$

T-72

Extra Example 2

Find $\dfrac{3}{4} \div \dfrac{9}{10}$. $\dfrac{5}{6}$

Extra Example 3

A length of rope is 6 meters long. How many $\dfrac{3}{5}$-meter pieces can be cut from the length of rope? ten $\dfrac{3}{5}$-meter pieces

On Your Own

5. $\dfrac{6}{7}$ 6. 4

7. $1\dfrac{1}{2}$ 8. $1\dfrac{1}{3}$

9. 14

English Language Learners
Vocabulary
For English learners, discuss the meaning of the word *reciprocal* in nonmathematical and mathematical contexts. A *reciprocal agreement* between two states is an exchange of privileges. For example, a Washington driver's license is honored in Pennsylvania as long as Washington honors a Pennsylvania driver's license. In math, a reciprocal of a fraction exchanges the numerator and the denominator.

Laurie's Notes

Example 2
- Estimate the quotient first.
- This question is asking how many $\dfrac{2}{3}$ are in $\dfrac{1}{6}$. Because $\dfrac{1}{6}$ is much less than $\dfrac{2}{3}$, there aren't many! The answer must be very small, close to 0.
- Work through the example. Check the reasonableness of the answer.

Example 3
- Write the problem and sketch the model. Work through each method.
- ? "Do you recognize this problem?" It is the same problem as part (c) of Activity 3 that was done with the quarters.
- Summarize with students what they have done so far: learned what a reciprocal is, divided two fractions, and divided a whole number by a fraction. Now they will look at dividing a fraction by a whole number.

Words of Wisdom
- It is important to have students think about the answer (estimate the answer) before they start working the problem. They should have a sense of the relative size of the answer before they begin.

On Your Own
- Before they begin, have students decide if the quotient is *less than* 1 or *greater than* 1. This is meant to tell if students can compare the fractions they are dividing.

T-73

EXAMPLE 2: Dividing a Fraction by a Fraction

Find $\dfrac{1}{6} \div \dfrac{2}{3}$.

$$\dfrac{1}{6} \div \dfrac{2}{3} = \dfrac{1}{6} \times \dfrac{3}{2}$$ Multiply by the reciprocal of $\dfrac{2}{3}$, which is $\dfrac{3}{2}$.

$$= \dfrac{1 \times \cancel{3}^{1}}{\cancel{6}_{2} \times 2}$$ Multiply fractions. Divide out the common factor 3.

$$= \dfrac{1}{4}$$ Simplify.

EXAMPLE 3: Dividing a Whole Number by a Fraction

A piece of wood is 3 feet long. How many $\dfrac{3}{4}$-foot pieces can be cut from the piece of wood?

Method 1: Draw a diagram. Mark each foot on the diagram. Then divide each foot into $\dfrac{1}{4}$-foot sections.

Count the number of $\dfrac{3}{4}$-foot pieces of wood. There are four.

∴ So, four $\dfrac{3}{4}$-foot pieces can be cut from the piece of wood.

Method 2: Divide 3 by $\dfrac{3}{4}$ to find the number of $\dfrac{3}{4}$-foot pieces.

$$3 \div \dfrac{3}{4} = 3 \times \dfrac{4}{3}$$ Multiply by the reciprocal of $\dfrac{3}{4}$, which is $\dfrac{4}{3}$.

$$= \dfrac{\cancel{3}^{1} \times 4}{\cancel{3}_{1}}$$ Multiply. Divide out the common factor 3.

$$= 4$$ Simplify.

∴ So, four $\dfrac{3}{4}$-foot pieces can be cut from the piece of wood.

On Your Own

Now You're Ready
Exercises 11–26

Divide. Write the answer in simplest form.

5. $\dfrac{2}{7} \div \dfrac{1}{3}$ 6. $\dfrac{1}{2} \div \dfrac{1}{8}$ 7. $\dfrac{3}{8} \div \dfrac{1}{4}$ 8. $\dfrac{2}{5} \div \dfrac{3}{10}$

9. How many $\dfrac{1}{2}$-foot pieces can be cut from a 7-foot piece of wood?

Section 2.5 Dividing Fractions 73

EXAMPLE 4 **Evaluating an Algebraic Expression**

Evaluate $a \div b$ when $a = \frac{4}{5}$ and $b = 2$.

$a \div b = \frac{4}{5} \div 2$ Substitute $\frac{4}{5}$ for a and 2 for b.

$= \frac{4}{5} \times \frac{1}{2}$ Multiply by the reciprocal of 2, which is $\frac{1}{2}$.

$= \frac{\overset{2}{\cancel{4}} \times 1}{5 \times \underset{1}{\cancel{2}}}$ Multiply fractions. Divide out the common factor 2.

$= \frac{2}{5}$ Simplify.

● **On Your Own**

Now You're Ready
Exercises 32–35

Evaluate the expression $x \div y$ for the given values of x and y.

10. $x = \frac{1}{2}, y = 3$ **11.** $x = \frac{2}{3}, y = 10$

12. $x = \frac{5}{8}, y = 4$ **13.** $x = 4, y = \frac{1}{3}$

EXAMPLE 5 **Using Order of Operations**

Evaluate $\frac{3}{8} + \frac{5}{6} \div 5$.

$\frac{3}{8} + \frac{5}{6} \div 5 = \frac{3}{8} + \frac{5}{6} \times \frac{1}{5}$ Multiply by the reciprocal of 5, which is $\frac{1}{5}$.

$= \frac{3}{8} + \frac{\overset{1}{\cancel{5}} \times 1}{6 \times \underset{1}{\cancel{5}}}$ Multiply $\frac{5}{6}$ and $\frac{1}{5}$. Divide out the common factor 5.

$= \frac{3}{8} + \frac{1}{6}$ Simplify.

$= \frac{9}{24} + \frac{4}{24}$ Rewrite fractions using the LCD 24.

$= \frac{13}{24}$ Add.

● **On Your Own**

Now You're Ready
Exercises 47–55

Evaluate the expression.

14. $\frac{4}{5} + \frac{2}{5} \div 4$ **15.** $\frac{3}{8} \div \frac{3}{4} - \frac{1}{6}$ **16.** $\frac{8}{9} \div 2 \div 8$

Laurie's Notes

Example 4

? "What do you know about the quotient when you divide a fraction by a whole number?" The quotient will be less than 1. Note: This would not necessarily be the case if the fraction was improper.

- If students seem to get lost in this problem because of the variables, try a visual diagram. The problem ultimately says that two numbers are being divided.

$$\boxed{} \div \boxed{} \rightarrow \boxed{a} \div \boxed{b} \rightarrow \boxed{\tfrac{4}{5}} \div \boxed{2}$$

This may also help students work the problems in the *On Your Own*.

On Your Own

- Have pairs of students present their work at the board.

Example 5

- In this example, students must pull all of the operations together. After spending a good deal of time on multiplication and division, students start to forget about needing a common denominator when adding and subtracting fractions. They may also forget about the order of operations. It is good to continually review and integrate skills.

? Write the problem on the board, and ask a few questions.
- "What operations are involved in this problem?" division and addition
- "Which operation is performed first? Why?" division; order of operations
- "How do you divide by 5?" Multiply by $\tfrac{1}{5}$.

- **Common Error:** Students try to divide out what they think is a common factor of 3 because there is a 6 in the denominator. Remind students that the numbers must be *factors*.

On Your Own

- If students are able to complete these problems correctly, they are ready for the assignment. Have pairs of students show their work at the board.

Closure

- Hold up 3 quarters. "How many dimes are in 3 quarters? Write and solve a division problem using fractions to represent this situation."
$$7\tfrac{1}{2};\ \tfrac{3}{4} \div \tfrac{1}{10} = \tfrac{3}{4} \cdot \tfrac{10}{1} = \tfrac{30}{4} = 7\tfrac{1}{2}$$

Technology For the Teacher

The Dynamic Planning Tool
Editable Teacher's Resources at *BigIdeasMath.com*

Extra Example 4

Evaluate $c \div d$ when $c = \tfrac{2}{3}$ and $d = 6$. $\tfrac{1}{9}$

On Your Own

10. $\tfrac{1}{6}$ 11. $\tfrac{1}{15}$
12. $\tfrac{5}{32}$ 13. 12

Extra Example 5

Evaluate $\tfrac{5}{8} - \tfrac{7}{8} \div 5$. $\tfrac{9}{20}$

On Your Own

14. $\tfrac{9}{10}$ 15. $\tfrac{1}{3}$
16. $\tfrac{1}{18}$

Differentiated Instruction

Kinesthetic

Give each student nine cards numbered 1 through 9. On a sheet of paper, have students draw a horizontal line halfway down the page. Ask one student to create a fraction using two of the cards. Have the rest of the class form the same fraction. Then ask the students how to find the reciprocal of the fraction. Have students make the reciprocal of the fraction with their cards and write a sentence naming the fraction and its reciprocal.

T-74

Vocabulary and Concept Check

1. Sample answer: $\frac{2}{5}, \frac{5}{2}$
2. $\frac{2}{9}$ does not belong because its reciprocal is not a whole number.
3. B
4. D
5. A
6. C

Practice and Problem Solving

7. $\frac{1}{8}$
8. $\frac{7}{6}$
9. $\frac{5}{2}$
10. $\frac{11}{8}$
11. $\frac{1}{2}$
12. $2\frac{11}{12}$
13. 16
14. 20
15. $\frac{1}{14}$
16. $\frac{3}{25}$
17. $\frac{1}{3}$
18. $\frac{2}{3}$
19. 3
20. $1\frac{13}{15}$
21. $\frac{2}{27}$
22. $\frac{1}{24}$
23. $\frac{27}{28}$
24. $\frac{26}{75}$
25. $20\frac{1}{4}$
26. 24

27. You need to invert the second fraction before you multiply. $\frac{4}{7} \div \frac{13}{28} = \frac{4}{7} \times \frac{28}{13} = \frac{4 \times \overset{4}{\cancel{28}}}{\underset{1}{\cancel{7}} \times 13} = \frac{16}{13}$, or $1\frac{3}{13}$

28. You invert the second fraction, not the first. $\frac{2}{5} \div \frac{8}{9} = \frac{2}{5} \times \frac{9}{8} = \frac{\overset{1}{\cancel{2}} \times 9}{5 \times \underset{4}{\cancel{8}}} = \frac{9}{20}$

Assignment Guide and Homework Check

Level	Day 1 Activity Assignment	Day 2 Lesson Assignment	Homework Check
Basic	11–18, 61–65	1–6, 19–25 odd, 27–33, 47–55 odd	19, 25, 28, 30, 32, 51
Average	11–18, 61–65	1–6, 19–25 odd, 28–34 even, 47–55 odd, 57, 58	19, 25, 28, 32, 51, 58
Advanced	11–18, 61–65	1–6, 24–28 even, 32–54 even, 59, 60	26, 28, 32, 52, 59

Common Errors

- **Exercise 7** Students may have difficulty rewriting a whole number as a fraction. Remind them that a whole number is a fraction with a denominator of 1.
- **Exercises 11–26** Students may find the reciprocal of the dividend instead of the divisor, or the reciprocal of both. Remind them that they only need to find the reciprocal of the divisor. Demonstrate this concept with a calculator. Enter the problem as a division problem, and then enter it as a multiplication problem with the reciprocal. Ask students to compare the answers. This comparison can also be made with the incorrect methods to demonstrate that they will not work.
- **Exercises 11–26** Students may find the reciprocal but forget to change the division to multiplication. Remind students that inverting the number is the *opposite*, so it requires the *opposite* operation.

2.5 Record and Practice Journal

Technology For the Teacher
Answer Presentation Tool
QuizShow

2.5 Exercises

Vocabulary and Concept Check

1. **OPEN-ENDED** Write a fraction and its reciprocal.

2. **WHICH ONE DOESN'T BELONG?** Which of the following does *not* belong with the other three? Explain your reasoning.

$$\frac{1}{3} \qquad \frac{1}{6} \qquad \frac{2}{9} \qquad \frac{1}{8}$$

MATCHING Match the expression with its value.

3. $\frac{2}{5} \div \frac{8}{15}$
4. $\frac{8}{15} \div \frac{2}{5}$
5. $\frac{2}{15} \div \frac{8}{5}$
6. $\frac{8}{5} \div \frac{2}{15}$

A. $\frac{1}{12}$
B. $\frac{3}{4}$
C. 12
D. $1\frac{1}{3}$

Practice and Problem Solving

Write the reciprocal of the number.

 7. 8
8. $\frac{6}{7}$
9. $\frac{2}{5}$
10. $\frac{8}{11}$

Divide. Write the answer in simplest form.

 11. $\frac{1}{8} \div \frac{1}{4}$
12. $\frac{5}{6} \div \frac{2}{7}$
13. $12 \div \frac{3}{4}$
14. $8 \div \frac{2}{5}$

15. $\frac{3}{7} \div 6$
16. $\frac{12}{25} \div 4$
17. $\frac{2}{9} \div \frac{2}{3}$
18. $\frac{8}{15} \div \frac{4}{5}$

19. $\frac{1}{3} \div \frac{1}{9}$
20. $\frac{7}{10} \div \frac{3}{8}$
21. $\frac{14}{27} \div 7$
22. $\frac{5}{8} \div 15$

23. $\frac{27}{32} \div \frac{7}{8}$
24. $\frac{4}{15} \div \frac{10}{13}$
25. $9 \div \frac{4}{9}$
26. $10 \div \frac{5}{12}$

ERROR ANALYSIS Describe and correct the error in finding the quotient.

27.

28. $\frac{2}{5} \div \frac{8}{9} = \frac{5}{2} \times \frac{8}{9}$

$= \frac{5 \times \overset{4}{\cancel{8}}}{\underset{1}{\cancel{2}} \times 9}$

$= \frac{20}{9}$

29. **REASONING** How can you use estimation to show that the quotient in Exercise 28 is incorrect?

30. APPLE PIE You have $\frac{3}{5}$ of an apple pie. You divide the remaining pie into five equal slices. What fraction of the original pie is each slice?

31. ANIMALS How many times longer is the baby alligator than the baby gecko?

$\frac{3}{4}$ ft

$\frac{2}{15}$ ft

Evaluate the expression when $a = \frac{1}{4}$, $b = \frac{5}{8}$, and $c = 2$.

④ 32. $a \div b$ **33.** $b \div c$ **34.** $c \div a$ **35.** $b \div a$

Determine whether the numbers are reciprocals. If not, write the reciprocal of each number.

36. $9, \frac{1}{9}$ **37.** $\frac{4}{5}, \frac{10}{8}$ **38.** $\frac{5}{6}, \frac{15}{18}$ **39.** $\frac{6}{5}, \frac{5}{6}$

Copy and complete the statement.

40. $\frac{5}{12} \times \boxed{} = 1$ **41.** $3 \times \boxed{} = 1$ **42.** $7 \div \boxed{} = 56$

Without finding the quotient, copy and complete the statement using <, >, or =. Explain your reasoning.

43. $5 \div \frac{7}{9} \boxed{} 5$ **44.** $\frac{3}{7} \div 1 \boxed{} \frac{3}{7}$

45. $8 \div \frac{3}{4} \boxed{} 8$ **46.** $\frac{5}{6} \div \frac{7}{8} \boxed{} \frac{5}{6}$

Evaluate the expression.

⑤ 47. $\frac{1}{6} \div 6 \div 6$ **48.** $\frac{7}{12} \div 14 \div 6$ **49.** $\frac{3}{5} \div \frac{4}{7} \div \frac{9}{10}$

50. $4 \div \frac{8}{9} - \frac{1}{2}$ **51.** $\frac{3}{4} + \frac{5}{6} \div \frac{2}{3}$ **52.** $\frac{7}{8} - \frac{3}{8} \div 9$

53. $\frac{9}{16} \div \frac{3}{4} \cdot \frac{2}{13}$ **54.** $\frac{3}{14} \cdot \frac{2}{5} \div \frac{6}{7}$ **55.** $\frac{10}{27} \cdot \left(\frac{3}{8} \div \frac{5}{24}\right)$

56. REASONING Use a model to evaluate the quotient $\frac{1}{2} \div \frac{1}{6}$. Explain.

Common Errors

- **Exercises 32–35** Because these problems require the extra step of substituting for the variable, students may get frustrated and not attempt these problems. Tell them to take the problem in small steps. Rewrite the problem as in the Lesson Notes for Example 4. Substitute for the variables. Then solve like any other division problem with fractions.
- **Exercises 36–39** Unless the fraction is obviously the reciprocal, students may state that the fraction is not a reciprocal. Remind them that if the product of the two fractions is 1, then the fractions are reciprocals of one another.
- **Exercise 42** Students may read the problem quickly and think that it is a multiplication problem instead of a division problem and that the answer is 8. The answer is the reciprocal of 8. Tell them to think of the problem backwards. "50 multiplied by what number equals 7?" This should help them to see that the answer must be a fraction.
- **Exercises 47–55** Students may try to apply the Associative and Commutative Properties. Remind them that the Associative and Commutative Properties apply only to multiplication and addition. Also remind students to follow the order of operations for all the problems in this section.

Practice and Problem Solving

29. Round $\frac{2}{5}$ to $\frac{1}{2}$ and $\frac{8}{9}$ to 1. $\frac{1}{2} \div 1 = \frac{1}{2}$, which is not close to the incorrect answer of $\frac{20}{9}$.

30. $\frac{3}{25}$ 31. $5\frac{5}{8}$ times

32. $\frac{2}{5}$ 33. $\frac{5}{16}$

34. 8 35. $2\frac{1}{2}$

36. yes 37. yes

38. no; Both fractions are equal to $\frac{5}{6}$. The reciprocal of $\frac{5}{6}$ is $\frac{6}{5}$.

39. yes 40. $\frac{12}{5}$

41. $\frac{1}{3}$ 42. $\frac{1}{8}$

43. >; When you divide a number by a fraction less than 1, the quotient is greater than the number.

44. =; When you divide a fraction by 1, the quotient is the fraction.

45. >; When you divide a number by a fraction less than 1, the quotient is greater than the number.

46. >; When you divide a fraction by a fraction less than 1, the quotient is greater than the fraction.

47. $\frac{1}{216}$ 48. $\frac{1}{144}$

49. $1\frac{1}{6}$ 50. 4

51. 2 52. $\frac{5}{6}$

53. $\frac{3}{26}$ 54. $\frac{1}{10}$

55. $\frac{2}{3}$ 56. 3

Practice and Problem Solving

57. when the fraction has, or can be simplified to have, a 1 in the numerator; The reciprocal will have, or can be simplified to have, a 1 in the denominator, so it is a whole number.

58. a. $3\frac{3}{4}$ times

 b. $3\frac{1}{3}$ times

 c. $\frac{1}{4}$

59. a. See *Taking Math Deeper*.

 b. 2 plates; $\frac{1}{2}$ pint

 c. See *Taking Math Deeper*.

60. a. $67\frac{1}{5}$ gallons

 b. $8\frac{2}{5}$ gallons

 c. $33\frac{3}{5}$ gallons

Fair Game Review

61. 3 **62.** 9
63. 10 **64.** 6
65. C

Mini-Assessment

Divide. Write the answer in simplest form.

1. $\frac{1}{3} \div \frac{1}{9}$ 3
2. $\frac{4}{5} \div \frac{2}{3}$ $1\frac{1}{5}$
3. $\frac{6}{10} \div \frac{1}{2}$ $1\frac{1}{5}$
4. $\frac{12}{15} \div 6$ $\frac{2}{15}$

5. You have $\frac{3}{4}$ of a pumpkin pie remaining. You divide the remaining pie into six equal slices. What fraction of the original pie is each slice? $\frac{1}{8}$

T-77

Taking Math Deeper

Exercise 59, parts (a) and (c)

Part (c) of this problem is a nice opportunity to show students how "making a table" can be a good problem-solving strategy.

 Organize the given information.

$\frac{7}{8}$ pint to glaze a bowl

$\frac{9}{16}$ pint to glaze a plate

6 pints of glaze

② **a.** I can glaze $6 \div \frac{7}{8}$ bowls.

$6 \div \frac{7}{8} = 6 \cdot \frac{8}{7} = \frac{48}{7} = 6\frac{6}{7}$ or about 6 bowls.

I can glaze $6 \div \frac{9}{16}$ plates.

$6 \div \frac{9}{16} = 6 \cdot \frac{16}{9} = \frac{32}{3} = 10\frac{2}{3}$ or about 10 plates.

③ Make a table for bowls and for plates.

Bowls	1	2	3	4	5	6
Glaze	$\frac{7}{8}$	$1\frac{3}{4}$	$2\frac{5}{8}$	$3\frac{1}{2}$	$4\frac{3}{8}$	$5\frac{1}{4}$

Plates	1	2	3	4	5	6
Glaze	$\frac{9}{16}$	$1\frac{1}{8}$	$1\frac{11}{16}$	$2\frac{1}{4}$	$2\frac{13}{16}$	$3\frac{3}{8}$

Use *Guess, Check, and Revise* to find two glaze amounts that add to 6 pints.

c. $2\frac{5}{8} + 3\frac{3}{8} = 6$

 3 bowls 6 plates

Reteaching and Enrichment Strategies

If students need help...	If students got it...
Resources by Chapter • Practice A and Practice B • Puzzle Time **Record and Practice Journal Practice** **Differentiating the Lesson** **Lesson Tutorials** **Skills Review Handbook**	**Resources by Chapter** • Enrichment and Extension • School-to-Work • Financial Literacy **Start the next section**

57. NUMBER SENSE When is the reciprocal of a fraction a whole number? Explain.

58. BUDGETS The table shows the portions of a family budget that are spent on several expenses.

 a. How many times more is the expense for housing than for automobiles?

 b. How many times more is the expense for food than for recreation?

 c. The expense for automobile fuel is $\frac{1}{60}$ of the total expenses. What fraction of the automobile expense is spent on fuel?

Expense	Portion of Budget
Housing	$\frac{1}{4}$
Food	$\frac{1}{12}$
Automobiles	$\frac{1}{15}$
Recreation	$\frac{1}{40}$

59. PROBLEM SOLVING You have 6 pints of glaze. It takes $\frac{7}{8}$ pint to glaze a bowl and $\frac{9}{16}$ pint to glaze a plate.

 a. How many bowls could you glaze? How many plates could you glaze?

 b. You want to glaze 5 bowls and then use the rest for plates. How many plates can you glaze? How much glaze will be left over?

 c. How many of each object could you glaze so that there is no glaze left over? Explain how you found your answer.

60. Reasoning A water tank is $\frac{1}{8}$ full. The tank is $\frac{3}{4}$ full when 42 gallons of water are added to the tank.

 a. How much water can the tank hold?

 b. How much water was originally in the tank?

 c. How much water is in the tank when it is $\frac{1}{2}$ full?

Fair Game Review What you learned in previous grades & lessons

Estimate the quotient. *(Section 2.1)*

61. $12\frac{1}{9} \div 3\frac{4}{5}$ **62.** $71\frac{2}{3} \div 8\frac{1}{4}$ **63.** $90\frac{2}{7} \div 9\frac{3}{8}$ **64.** $47\frac{3}{4} \div 7\frac{5}{6}$

65. MULTIPLE CHOICE The expression $3m$ represents the cost of renting m movies. What is the cost of renting four movies? *(Section 1.1)*

 Ⓐ $7 Ⓑ $9 Ⓒ $12 Ⓓ $27

2.6 Dividing Mixed Numbers

COMMON CORE STATE STANDARDS

6.NS.1

Essential Question How can you use division by a mixed number as part of a story?

1 EXAMPLE: Writing a Story

Write a story that uses the division problem $6 \div 1\frac{1}{2}$. Draw pictures for your story.

There are many possible stories. Here is one about a camping trip.

Joe goes on a camping trip with his aunt, his uncle, and three cousins. They leave at 5:00 P.M. and drive 2 hours to the campground.

Joe helps his uncle put up three tents. His aunt cooks hamburgers on a grill that is over a fire.

In the morning, Joe tells his aunt that he is making pancakes for everyone. He decides to triple the recipe so there will be plenty of pancakes for everyone. A single recipe uses 2 cups of water, so he needs a total of 6 cups.

Pancake Mix
Recipe:
2 cups water
2 cups pancake mix
1/4 cup oil
1 egg
1/4 teaspoon salt

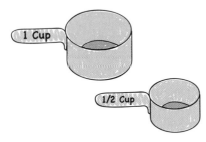

Joe's aunt has a 1-cup measuring cup and a ½-cup measuring cup. The water faucet is about 50 yards from the campsite. Joe tells his cousins that he can get 6 cups of water in only 4 trips.

When his cousins ask him how he knows that, he uses a stick to draw a diagram in the dirt. Joe says "This diagram shows that there are four 1½'s in 6." In other words,

$$6 \div 1\frac{1}{2} = 4.$$

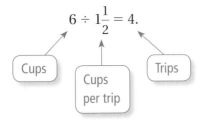

Cups Cups per trip Trips

Laurie's Notes

Introduction

For the Teacher
- **Goal:** The goal of this activity is for students to develop an understanding of how to divide by a mixed number.
- As part of writing the story problems, students need to pay attention to the units associated with each number. This is a subtle introduction to dimensional analysis, something the science teachers will be happy to see.
- This is a very different activity, one which we use in each book of the series. The goal is to have students use their literacy skills, putting mathematical skills in a context. Students need practice with writing stories, and many will find the first attempt challenging.

Motivate
- Share a colorful story while holding a travel coffee mug. The essence of the story should result in the following facts: a coffee pot holds 12 cups and the mug holds about $1\frac{1}{2}$ cups. How many times can the travel mug be filled?
- ? "What expression can you use to answer this question? What is the answer?" $12 \div 1\frac{1}{2}$; Listen for methods to find the solution.

Activity Notes

Example 1
- Ask a student to read the sample story to the class. Discuss the story.
- ? "Have any of you gone on a camping trip?"
- Discuss the pictures.
- ? "Would you prefer to read just the text, or do you enjoy having pictures that relate to the text?"
- Draw the diagram at the bottom of the page. Ask a student to explain the solution.
- **Big Idea:** Notice that the division problem at the bottom of the page is labeled with units: cups, cups *per* trip, and trips. As students start to think about writing a story, they need to consider a context that involves a rate (e.g., miles per hour, songs per minute, cookies per team, tons per truckload). The rate helps to define the context.
- **Common Error:** Students might decide on a rate where a mixed number does not make sense in terms of trying to write a story. For instance, they may want to work with "cars per group". You can have 6 cars and 4 groups, but you would not have $1\frac{1}{2}$ cars per group.
- Once students have decided on the context, allow 20–30 minutes of writing time.
- Allow a few students to share their short stories.

Common Core State Standards

6.NS.1 Interpret and compute quotients of fractions, and solve word problems involving division of fractions by fractions.

Previous Learning
Students should know how to divide a number by a fraction.

Activity Materials
Introduction
• travel coffee mug

Start Thinking! and Warm Up

2.6 Record and Practice Journal

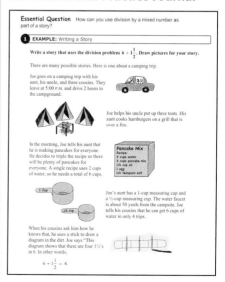

T-78

English Language Learners
Pair Activity
For Activity 3, pair English learners with English speakers to write a story based on the English learners' culture. Encourage students to share how to incorporate math into the story.

2.6 Record and Practice Journal

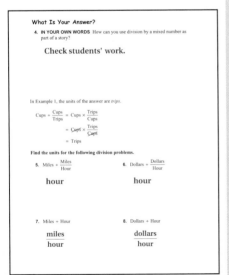

Laurie's Notes

Example 2
- The example in the text uses the math from the story in Example 1.
- Work through the mathematical solution to the problem which everyone used for their story.
- ? "How do you divide by a fraction?" multiply by the reciprocal
- ? "How do you take the reciprocal of a mixed number?" Write the mixed number as an improper fraction. Then find the reciprocal of the improper fraction.

Activity 3
- To help students with this open-ended question, have them brainstorm ideas for a rate associated with the mixed number. This is the first step, and unlike Example 1, students may use any numbers.
- Note that the dividend is not limited to a whole number, although students may be more comfortable with having it be a whole number.
- Allow 20–30 minutes of writing time, followed by time for sharing.

What Is Your Answer?
- Questions 5–8 focus on dimensional analysis. Students should follow the sample shown to determine the units for the answers.

Closure
- **Writing Prompt:** Dividing by a mixed number is like dividing by a fraction because... You can write a mixed number as an improper fraction.

The Dynamic Planning Tool
Editable Teacher's Resources at *BigIdeasMath.com*

T-79

2 EXAMPLE: Dividing by a Mixed Number

Show how Joe solves the division problem in Example 1.

$6 \div 1\frac{1}{2} = \frac{6}{1} \div \frac{3}{2}$ Rewrite 6 as $\frac{6}{1}$ and $1\frac{1}{2}$ as $\frac{3}{2}$.

$= \frac{6}{1} \times \frac{2}{3}$ Multiply by the reciprocal of $\frac{3}{2}$, which is $\frac{2}{3}$.

$= \frac{6 \times 2}{1 \times 3}$ Multiply fractions.

$= \frac{12}{3}$, or 4 Simplify.

3 ACTIVITY: Writing a Story

Work with a partner. Think of a story that uses division by a mixed number.

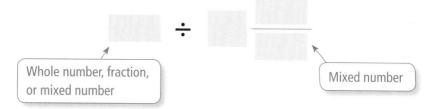

Whole number, fraction, or mixed number

Mixed number

a. Write your story. Then draw pictures for your story.

b. Solve the division problem and use the answer in your story. Include a diagram of the division problem.

What Is Your Answer?

4. IN YOUR OWN WORDS How can you use division by a mixed number as part of a story?

In Example 1, the units of the answer are *trips*.

$$\text{Cups} \div \frac{\text{Cups}}{\text{Trips}} = \text{Cups} \times \frac{\text{Trips}}{\text{Cups}}$$

$$= \cancel{\text{Cups}} \times \frac{\text{Trips}}{\cancel{\text{Cups}}} = \text{Trips}$$

Find the units for the following division problems.

5. Miles $\div \dfrac{\text{Miles}}{\text{Hour}}$

6. Dollars $\div \dfrac{\text{Dollars}}{\text{Hour}}$

7. Miles \div Hour

8. Dollars \div Hour

Practice — Use what you learned about dividing mixed numbers to complete Exercises 5–12 on page 82.

Section 2.6 Dividing Mixed Numbers

2.6 Lesson

Key Idea

Dividing Mixed Numbers

Write each mixed number as an improper fraction. Then divide as you would with proper fractions.

EXAMPLE 1 Dividing a Mixed Number by a Fraction

Find $4\frac{1}{2} \div \frac{3}{8}$. Estimate $5 \div \frac{1}{2} = 10$

$$4\frac{1}{2} \div \frac{3}{8} = \frac{9}{2} \div \frac{3}{8}$$ Write $4\frac{1}{2}$ as the improper fraction $\frac{9}{2}$.

$$= \frac{9}{2} \times \frac{8}{3}$$ Multiply by the reciprocal of $\frac{3}{8}$, which is $\frac{8}{3}$.

$$= \frac{\overset{3}{\cancel{9}} \times \overset{4}{\cancel{8}}}{\underset{1}{\cancel{2}} \times \underset{1}{\cancel{3}}}$$ Multiply fractions. Divide out common factors.

$$= 12$$ Simplify.

∴ So, the quotient is 12. Reasonable? $12 \approx 10$ ✓

EXAMPLE 2 Dividing Mixed Numbers

Find $3\frac{5}{6} \div 1\frac{2}{3}$. Estimate $4 \div 2 = 2$

$$3\frac{5}{6} \div 1\frac{2}{3} = \frac{23}{6} \div \frac{5}{3}$$ Write each mixed number as an improper fraction.

$$= \frac{23}{6} \times \frac{3}{5}$$ Multiply by the reciprocal of $\frac{5}{3}$, which is $\frac{3}{5}$.

$$= \frac{23 \times \overset{1}{\cancel{3}}}{\underset{2}{\cancel{6}} \times 5}$$ Multiply fractions. Divide out common factors.

$$= \frac{23}{10}, \text{ or } 2\frac{3}{10}$$ Simplify.

∴ So, the quotient is $2\frac{3}{10}$. Reasonable? $2\frac{3}{10} \approx 2$ ✓

On Your Own

Now You're Ready
Exercises 5–20

Divide. Write the answer in simplest form.

1. $1\frac{3}{7} \div \frac{2}{3}$ 2. $2\frac{1}{6} \div \frac{3}{4}$ 3. $8\frac{1}{4} \div 1\frac{1}{2}$ 4. $6\frac{4}{5} \div 2\frac{1}{8}$

Laurie's Notes

Introduction

Connect
- **Yesterday:** Students wrote story problems related to dividing with a mixed number.
- **Today:** Students will use a formal rule to divide with mixed numbers.

Motivate
- Do a warm-up with students where you ask them to give examples of different types of numbers. The goal is to review vocabulary and check on number sense.
- Examples: a fraction less than $\frac{1}{2}$, an improper fraction, a mixed number between 2 and 3, an improper fraction with an odd denominator, two fractions whose product is 1, a division problem involving fractions whose quotient is greater than 1, and a division problem involving fractions whose quotient is less than 1.

Lesson Notes

Key Idea
- **?** "How do you write a mixed number as an improper fraction? Give an example." *Sample answer:* Multiply the denominator of the fraction by the whole number part and then add the numerator. Put this quantity over the denominator of the fraction. $2\frac{3}{5} = \frac{5 \cdot 2 + 3}{5} = \frac{13}{5}$
- **?** "How do you write a whole number as a fraction? Give an example." Write the number as the numerator of the fraction with a denominator of 1.

Example 1
- Start by estimating the quotient. Note that $\frac{3}{8}$ was rounded to $\frac{1}{2}$, not 0.
 Students should be able to use mental math to divide by $\frac{1}{2}$.
- **Common Error:** Whenever students take the reciprocal of the first number instead of the second, the (incorrect) quotient will be the reciprocal of the actual quotient. Watch for this error, as it is easily corrected.

Example 2
- Estimate the quotient first.

On Your Own
- Before they begin, have students decide if the quotient is *greater than 1* or *less than 1*.
- Students should work with a partner on these problems. Have pairs of students share their work at the board. Check students' confidence with dividing out common factors.

Goal Today's lesson is dividing mixed numbers.

Start Thinking! and Warm Up

Warm Up For use before Lesson 2.6

Start Thinking! For use before Lesson 2.6

A beaded necklace has broken and the beads have fallen all over the floor. The diameter of each bead is $1\frac{1}{4}$ centimeters. Using the string from the broken necklace, how can you find out how many beads were on the necklace without counting them?

Extra Example 1
Find $5\frac{1}{5} \div \frac{1}{2}$. $10\frac{2}{5}$

Extra Example 2
Find $7\frac{3}{4} \div 2\frac{2}{3}$. $2\frac{29}{32}$

On Your Own
1. $2\frac{1}{7}$
2. $2\frac{8}{9}$
3. $5\frac{1}{2}$
4. $3\frac{1}{5}$

T-80

Extra Example 3

Evaluate $\frac{3}{4} - 3\frac{1}{2} \div 8\frac{1}{6}$. $\frac{9}{28}$

Extra Example 4

One serving of French onion soup is $1\frac{3}{4}$ cups. A restaurant cook makes 75 cups of soup. Is there enough to serve 40 people? Yes, because $42\frac{6}{7}$ is the number of $1\frac{3}{4}$-cup servings you have in 75 cups of soup. Because $42\frac{6}{7}$ is greater than 40, there is enough soup to serve 40 people.

On Your Own

5. $8\frac{1}{8}$
6. $4\frac{8}{9}$
7. 2
8. $\frac{1}{3}$
9. yes; There are 18 servings of soup, so you can serve 15 people.

Differentiated Instruction

Visual

Students that have difficulty dividing out common factors may benefit from a color-coding system. Have students circle pairs of numbers that share a common factor with the same color. For instance, in Example 3, 21 and 9 could be circled in blue because they share a common factor of three. The factors 8 and 4 could be circled in red, because they share a common factor of 4. To divide out common factors, the students only need to look at one color at a time.

Laurie's Notes

Example 3

- Write the problem on the board.
- ? "How many operations are involved in this problem? What operation is performed first?" division and subtraction; division
- **FYI:** Students do not always see the common factors in this problem. They sometimes believe that you need to divide out the entire number (written in the numerator or denominator), instead of a factor of the number. In this case, 9 and 21 have the common factor of 3, which is divided out.

Example 4

- Do the problem together as a class.
- ? "What does the quotient represent when you divide 50 by $1\frac{2}{3}$?" the number of servings from 50 cups of soup
- It is important to ask this question so that students will know how to interpret the answers.
- ? **Alternative Method:** Ask questions about solving the problem differently.
 - "What does the product of 35 and $1\frac{2}{3}$ represent?" the amount of soup needed to serve 35 people
 - "Is either approach $\left(50 \div 1\frac{2}{3} \text{ or } 35 \times 1\frac{2}{3}\right)$ sufficient to answer the question?" Yes, both approaches work.

On Your Own

- These problems require a bit more time. Have students work the problems individually first, and then check with a neighbor. There are many skills involved with each. Although all four problems involve the order of operations, Questions 7 and 8 are not solved from left to right.

Closure

- **Connection:** How is division of mixed numbers like multiplication of mixed numbers? In each case, you must first rewrite the mixed numbers as improper fractions.

Technology for the Teacher

The Dynamic Planning Tool
Editable Teacher's Resources at *BigIdeasMath.com*

EXAMPLE 3 Using Order of Operations

Evaluate $5\frac{1}{4} \div 1\frac{1}{8} - \frac{2}{3}$.

$5\frac{1}{4} \div 1\frac{1}{8} - \frac{2}{3} = \frac{21}{4} \div \frac{9}{8} - \frac{2}{3}$ Write each mixed number as an improper fraction.

$= \frac{21}{4} \times \frac{8}{9} - \frac{2}{3}$ Multiply by the reciprocal of $\frac{9}{8}$, which is $\frac{8}{9}$.

$= \frac{\overset{7}{\cancel{21}} \times \overset{2}{\cancel{8}}}{\underset{1}{\cancel{4}} \times \underset{3}{\cancel{9}}} - \frac{2}{3}$ Multiply $\frac{21}{4}$ and $\frac{8}{9}$. Divide out common factors.

$= \frac{14}{3} - \frac{2}{3}$ Simplify.

$= \frac{12}{3}$, or 4 Subtract.

Remember

Be sure to check your answers whenever possible. In Example 3, you can use estimation to check that your answer is reasonable.

$5\frac{1}{4} \div 1\frac{1}{8} - \frac{2}{3}$
$\approx 5 \div 1 - 1$
$= 5 - 1$
$= 4$ ✓

EXAMPLE 4 Real-Life Application

One serving of tortilla soup is $1\frac{2}{3}$ cups. A restaurant cook makes 50 cups of soup. Is there enough to serve 35 people? Explain.

Divide 50 by $1\frac{2}{3}$ to find the number of available servings.

$50 \div 1\frac{2}{3} = \frac{50}{1} \div \frac{5}{3}$ Rewrite each number as an improper fraction.

$= \frac{50}{1} \cdot \frac{3}{5}$ Multiply by the reciprocal of $\frac{5}{3}$, which is $\frac{3}{5}$.

$= \frac{\overset{10}{\cancel{50}} \cdot 3}{1 \cdot \underset{1}{\cancel{5}}}$ Multiply fractions. Divide out common factors.

$= 30$ Simplify.

∴ No. Because 30 is less than 35, there is not enough soup to serve 35 people.

On Your Own

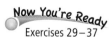
Exercises 29–37

Evaluate the expression.

5. $1\frac{1}{2} \div \frac{1}{6} - \frac{7}{8}$

6. $3\frac{1}{3} \div \frac{5}{6} + \frac{8}{9}$

7. $\frac{2}{5} + 2\frac{4}{5} \div 1\frac{3}{4}$

8. $\frac{2}{3} - 1\frac{4}{7} \div 4\frac{5}{7}$

9. In Example 4, can 30 cups of tortilla soup serve 15 people? Explain.

2.6 Exercises

Vocabulary and Concept Check

1. **VOCABULARY** What is the reciprocal of $7\frac{1}{3}$?

2. **NUMBER SENSE** Is $5\frac{1}{4} \div 3\frac{1}{2}$ the same as $3\frac{1}{2} \div 5\frac{1}{4}$? Explain.

3. **NUMBER SENSE** Is the reciprocal of an improper fraction *sometimes*, *always*, or *never* a proper fraction? Explain.

4. **DIFFERENT WORDS, SAME QUESTION** Which is different? Find "both" answers.

 What is $5\frac{1}{2}$ divided by $\frac{1}{8}$? Find the quotient of $5\frac{1}{2}$ and $\frac{1}{8}$.

 What is $5\frac{1}{2}$ times 8? Find the product of $5\frac{1}{2}$ and $\frac{1}{8}$.

Practice and Problem Solving

Divide. Write the answer in simplest form.

5. $2\frac{1}{4} \div \frac{3}{4}$ 6. $3\frac{4}{5} \div \frac{2}{5}$ 7. $8\frac{1}{8} \div \frac{5}{6}$ 8. $7\frac{5}{9} \div \frac{4}{7}$

9. $7\frac{1}{2} \div 1\frac{9}{10}$ 10. $3\frac{3}{4} \div 2\frac{1}{12}$ 11. $7\frac{1}{5} \div 8$ 12. $8\frac{4}{7} \div 15$

13. $8\frac{1}{3} \div \frac{2}{3}$ 14. $9\frac{1}{6} \div \frac{5}{6}$ 15. $13 \div 10\frac{5}{6}$ 16. $12 \div 5\frac{9}{11}$

17. $\frac{7}{8} \div 3\frac{1}{16}$ 18. $\frac{4}{9} \div 1\frac{7}{15}$ 19. $4\frac{5}{16} \div 3\frac{3}{8}$ 20. $6\frac{2}{9} \div 5\frac{5}{6}$

21. **ERROR ANALYSIS** Describe and correct the error in finding the quotient.

 $3\frac{1}{2} \div 1\frac{2}{3} = 3\frac{1}{2} \times 1\frac{3}{2} = \frac{7}{2} \times \frac{5}{2} = \frac{35}{4} = 8\frac{3}{4}$

22. **DOG FOOD** A bag contains 42 cups of dog food. Your dog eats $2\frac{1}{3}$ cups of dog food each day. How many days does the bag of dog food last?

23. **HAMBURGERS** How many $\frac{1}{4}$-pound hamburgers can be made from $3\frac{1}{2}$ pounds of ground beef?

24. **BOOKS** How many $1\frac{3}{5}$-inch thick books can fit on a $14\frac{1}{2}$-inch long bookshelf?

Assignment Guide and Homework Check

Level	Day 1 Activity Assignment	Day 2 Lesson Assignment	Homework Check
Basic	5–12, 41–45	1–4, 13–21 odd, 22–24, 29–33 odd	2, 15, 19, 24, 33
Average	5–12, 41–45	1–4, 13–23 odd, 30–38 even	15, 19, 30, 36, 38
Advanced	5–12, 41–45	1–4, 21–27 odd, 30–38 even, 39, 40	27, 30, 36, 39

Common Errors

- **Exercises 5–20** Students may try to divide the whole numbers and the fractions separately. Remind them to rewrite the mixed numbers as improper fractions and to think of each mixed number as one number.
- **Exercises 5–20** Students may find the reciprocal before writing the mixed number as an improper fraction. Remind them that the order of the steps in solving the problem does matter. Tell students to write out each step so that they can keep track of the correct order.
- **Exercises 29–37** Students may forget to use the order of operations. Remind them about the order of operations. Have them use parentheses to help remember which parts to evaluate first.

Technology For the Teacher
Answer Presentation Tool
QuizShow

2.6 Record and Practice Journal

Divide. Write the answer in simplest form.

1. $4\frac{1}{6} \div 5 \quad \frac{5}{6}$
2. $\frac{5}{8} \div 5\frac{3}{4} \quad \frac{5}{46}$
3. $8\frac{1}{6} \div 2\frac{1}{24} \quad 4$

Evaluate the expression when $x = 3\frac{3}{5}$ and $y = 6\frac{6}{7}$.

4. $2\frac{3}{10} \div x \quad \frac{23}{36}$
5. $y \div x \quad 1\frac{19}{21}$
6. $x \div y \quad \frac{21}{40}$

Evaluate the expression.

7. $4\frac{7}{12} \div \frac{3}{4} \times \frac{3}{11} \quad 1\frac{2}{3}$
8. $9 + 8\frac{1}{10} - \frac{5}{9} \quad \frac{5}{9}$
9. $5\frac{7}{8} \times \left(2\frac{4}{5} + 7\right) \quad 2\frac{7}{20}$

10. At a road race, you have $60\frac{3}{4}$ feet available for a water station. Your tables are $6\frac{3}{4}$ feet long. How many tables can you line up for the water station?
9 tables

11. A recipe calls for $2\frac{2}{3}$ teaspoons of salt. You can only find three of your measuring spoons: a $\frac{1}{2}$ teaspoon, a $\frac{1}{8}$ teaspoon, and a $\frac{1}{6}$ teaspoon.
 a. What measuring spoon(s) would you use to measure the salt?
 Sample answer: $\frac{1}{6}$ teaspoon
 b. How many scoops of each measuring spoon would you need?
 Sample answer: 16

Vocabulary and Concept Check

1. $\frac{3}{22}$
2. no; Division is not commutative.
3. sometimes; The reciprocal of $\frac{2}{2}$ is $\frac{2}{2}$, which is improper.
4. Find the product of $5\frac{1}{2}$ and $\frac{1}{8}$.; $\frac{11}{16}$; 44

Practice and Problem Solving

5. 3
6. $9\frac{1}{2}$
7. $9\frac{3}{4}$
8. $13\frac{2}{9}$
9. $3\frac{18}{19}$
10. $1\frac{4}{5}$
11. $\frac{9}{10}$
12. $\frac{4}{7}$
13. $12\frac{1}{2}$
14. 11
15. $1\frac{1}{5}$
16. $2\frac{1}{16}$
17. $\frac{2}{7}$
18. $\frac{10}{33}$
19. $1\frac{5}{18}$
20. $1\frac{1}{15}$

21. The mixed number $1\frac{2}{3}$ was not written as an improper fraction before inverting.

$$3\frac{1}{2} \div 1\frac{2}{3} = \frac{7}{2} \div \frac{5}{3}$$
$$= \frac{7}{2} \times \frac{3}{5}$$
$$= \frac{7 \times 3}{2 \times 5}$$
$$= \frac{21}{10}, \text{ or } 2\frac{1}{10}$$

22. 18 days

T-82

Practice and Problem Solving

23. 14 hamburgers
24. 9 books
25. $\frac{8}{15}$
26. $2\frac{17}{32}$
27. $\frac{22}{27}$
28. $\frac{32}{81}$
29. $1\frac{1}{3}$
30. $6\frac{9}{26}$
31. $5\frac{1}{6}$
32. $13\frac{1}{5}$
33. $\frac{7}{54}$
34. $3\frac{1}{3}$
35. $12\frac{1}{2}$
36. $\frac{7}{108}$
37. $\frac{22}{35}$
38. See *Taking Math Deeper*.
39. **a.** 6 ramps; *Sample answer:* The estimate is reasonable because $12\frac{1}{2}$ was rounded down.
 b. 6 ramps; $1\frac{1}{4}$ feet left over
40. $1\frac{1}{3}$ times greater

Fair Game Review
41. 0.43
42. 0.013
43. 3.8
44. 7.009
45. C

Mini-Assessment
Divide. Write the answer in simplest form.
1. $6\frac{1}{2} \div 2\frac{3}{4}$ $2\frac{4}{11}$
2. $4\frac{2}{3} \div 1\frac{1}{6}$ 4
3. $8\frac{4}{5} \div \frac{4}{10}$ 22
4. $6 \div 2\frac{3}{5}$ $2\frac{4}{13}$
5. A bag contains 13 cups of cat food. Your cats eat $1\frac{1}{2}$ cups of food each day. How many full days does the bag of cat food last? 8 days

T-83

Taking Math Deeper

Exercise 38
This problem is a nice opportunity to show how "making a table" can be a good problem-solving strategy. In this case, notice how it is convenient to make a table with 3 rows.

① Organize the given information.

8½ cups of peanuts

12 cups of granola

Trail Mix
$2\frac{3}{4}$ cups granola
$1\frac{1}{3}$ cups peanuts

② Make a table.

Batches	1	2	3	4	5
Granola	$2\frac{3}{4}$	$5\frac{1}{2}$	$8\frac{1}{4}$	11	$13\frac{3}{4}$
Peanuts	$1\frac{1}{3}$	$2\frac{2}{3}$	4	$5\frac{1}{3}$	$6\frac{2}{3}$

③ Use the table to make a conclusion.

It's 4 batches.

Batches	1	2	3	4	5
Granola	$2\frac{3}{4}$	$5\frac{1}{2}$	$8\frac{1}{4}$	11	$13\frac{3}{4}$
Peanuts	$1\frac{1}{3}$	$2\frac{2}{3}$	4	$5\frac{1}{3}$	$6\frac{2}{3}$

Five batches needs more granola than you have.

So, you can make 4 full batches of trail mix.

Project
Write your own recipe for trail mix. Be sure to include the amount of each ingredient, a serving size, and how many people your recipe serves. Give your recipe a name that will make people want to try it!

Reteaching and Enrichment Strategies

If students need help...	If students got it...
Resources by Chapter • Practice A and Practice B • Puzzle Time Record and Practice Journal Practice Differentiating the Lesson Lesson Tutorials Skills Review Handbook	Resources by Chapter • Enrichment and Extension • School-to-Work • Financial Literacy • Technology Connection Start the next section

ALGEBRA Evaluate the expression when $x = 5\frac{5}{8}$ and $y = 2\frac{2}{9}$.

25. $y \div 4\frac{1}{6}$
26. $x \div y$
27. $4\frac{7}{12} \div x$
28. $y \div x$

Evaluate the expression.

❸ 29. $5\frac{5}{6} \div 3\frac{3}{4} - \frac{2}{9}$
30. $6\frac{1}{2} - \frac{7}{8} \div 5\frac{11}{16}$
31. $9\frac{1}{6} \div 5 + 3\frac{1}{3}$

32. $3\frac{3}{5} + 4\frac{4}{15} \div \frac{4}{9}$
33. $\frac{3}{5} \times \frac{7}{12} \div 2\frac{7}{10}$
34. $4\frac{3}{8} \div \frac{3}{4} \times \frac{4}{7}$

35. $1\frac{9}{11} \times 4\frac{7}{12} \div \frac{2}{3}$
36. $3\frac{4}{15} \div \left(8 \times 6\frac{3}{10}\right)$
37. $2\frac{5}{14} \div \left(2\frac{5}{8} \times 1\frac{3}{7}\right)$

38. **TRAIL MIX** You have 12 cups of granola and $8\frac{1}{2}$ cups of peanuts to make trail mix. What is the greatest number of full batches of trail mix you can make? Explain how you found your answer.

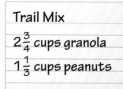

Trail Mix
$2\frac{3}{4}$ cups granola
$1\frac{1}{3}$ cups peanuts

$1\frac{7}{8}$ ft

$1\frac{7}{8}$ ft

39. **RAMPS** You make skateboard ramps by cutting pieces from a board that is $12\frac{1}{2}$ feet long.

 a. Estimate how many ramps you can cut from the board. Is your estimate reasonable? Explain.

 b. How many ramps can you cut from the board? How much wood is left over?

40. **Reasoning** At a track meet, the longest shot put throw by a boy is 25 feet and 8 inches. The longest shot put throw by a girl is 19 feet and 3 inches. How many times greater is the longest shot put throw by a boy than by a girl?

Fair Game Review What you learned in previous grades & lessons

Write the number as a decimal. *(Skills Review Handbook)*

41. forty-three hundredths
42. thirteen thousandths
43. three and eight tenths
44. seven and nine thousandths

45. **MULTIPLE CHOICE** The winner in a vote for class president received $\frac{3}{4}$ of the 240 votes. How many votes did the winner receive? *(Section 2.2)*

 Ⓐ 60 Ⓑ 150 Ⓒ 180 Ⓓ 320

2.7 Writing Decimals as Fractions

COMMON CORE STATE STANDARDS
6.NS.1

Essential Question When you write a terminating decimal as a fraction, what type of denominator do you get?

The Meaning of a Word • Terminate

When you **terminate** a phone call, you end the conversation.

A decimal that ends is called a **terminating decimal**. To convert a terminating decimal to a fraction, express the decimal in words.

0.4 → in words → four tenths → as a fraction → $\frac{4}{10}$

0.15 → in words → fifteen hundredths → as a fraction → $\frac{15}{100}$

1 ACTIVITY: Writing Common Decimals as Fractions

Work with a partner. Copy and complete the table.

Decimal	Write as a Fraction	Simplify
0.04	Four hundredths = $\frac{4}{100}$	$\frac{4}{100} = \frac{1 \cdot 4}{4 \cdot 25} = \frac{1 \cdot \cancel{4}}{\cancel{4} \cdot 25} = \frac{1}{25}$
0.125	One hundred twenty-five thousandths = $\frac{125}{1000}$	$\frac{125}{1000} = \frac{1 \cdot 125}{8 \cdot 125} = \frac{1 \cdot \cancel{125}}{8 \cdot \cancel{125}} = \frac{1}{8}$
0.25		
0.3		
0.5		
0.625		
0.7		
0.75		
0.875		

84 Chapter 2 Multiplying and Dividing Fractions

Laurie's Notes

Introduction

For the Teacher
- **Goal:** The goal of this activity is for students to review writing common decimals as fractions, and to expand this skill to include all terminating decimals. Students will discover that fractions which terminate will have something in common with their denominators.
- Students have only worked with terminating decimals in the past and they were referred to only as decimals. It may seem odd to students to now refer to these same decimals as *terminating*; what else would they do? Without going into great detail, allude to the idea that in the next lesson they will investigate *repeating* decimals.

Motivate
- Hold up a quarter.
- ❓ "How is the value of this coin represented as a decimal? as a fraction?" $0.25; \frac{1}{4}$ dollar
- Repeat this with a dime, nickel, penny, and half-dollar. All of these are examples of how decimals can be represented as fractions.
- **Common Error:** Make sure that when students represent the values of the coins as decimals, they use the $ symbol and not the ¢ symbol.

Activity Notes

The Meaning of a Word
- Define terminating decimals. Review place values. Write several decimals on the board and ask the value of the digit in each place value. (e.g., 0.35 → 3 is in the tenths place and 5 is in the hundredths place)
- **Connection:** Relate the name of the decimal to the fraction that is written.

Activity 1
- Discuss the number line model. Placing decimals on a number line is not an obvious skill for all students. If time permits, draw a number line from zero to one on the board with the tenths marked. Ask volunteers to come to the board to place the nine decimals shown in the activity.
- **Common Error:** Make sure that students pronounce the place values correctly and spell them correctly. They sometimes forget the "ths" at the end of the word. To assist students, record the correct spellings of the place values on the board.
- Remind students to simplify their answers. Check students' work.
- ❓ "Do you observe anything about the denominators of the fractions?" If you don't want to give away the answer, offer a hint—write the factors of each denominator. $25 = 5^2$, $8 = 2^3$, $4 = 2^2$, $10 = 2 \times 5, \ldots$ Students may see that the denominators have only factors of 2 and/or 5.

Common Core State Standards

6.NS.1 Interpret and compute quotients of fractions, and solve word problems involving division of fractions by fractions.

Previous Learning

Students have represented halves, fourths, and tenths as fractions and decimals. Students have simplified fractions by dividing the numerator and denominator by the Greatest Common Factor.

Activity Materials
Introduction
• coins with different values

Start Thinking! and Warm Up

2.7 Record and Practice Journal

T-84

Differentiated Instruction

Auditory

Writing terminating decimals as rational numbers is easier if the students read the decimal using place value as opposed to reading the digits.

Terminating decimal: 0.57

Read as place value:
fifty-seven hundredths

Don't read as digits:
zero point five seven

2.7 Record and Practice Journal

Laurie's Notes

Activity 2

- Review the results of Activity 1 with students. They now have a list of nine numbers that have been written as decimals and fractions. Each pair of numbers is equivalent, written in a different form. They will use these pairs of numbers to fill out the game card in this activity.
- This activity helps to develop fluency with common fraction-decimal equivalencies.
- Because there are 16 squares and 9 pairs of numbers, students may omit one of the pairs, or omit an equivalent form of two different numbers. They need to get four in a row, so students may experiment to determine which strategy is better.
- Clarify the directions. Both players cross out any numbers equal to the number called out. Of course, you may wish to change or modify the rules for your class.
- **Extension:** Add additional pairs of numbers to the list and/or make the grid smaller or larger.

Activity 3

? "Are there additional words that might be added to this list?"
 e.g., decathlon, decagon, decibel, etc.
- This same activity can be done with other prefixes such as octa-, quad-, and penta-.

Inductive Reasoning

- This section of the lesson allows students to summarize the three activities.

What Is Your Answer?

- Students should see that the denominator contains factors of 2, 5, or both 2 and 5.

Closure

- **Writing Prompt:** To write a decimal as a fraction, you . . . Write the digits of the decimal in the numerator. Then use the place value of the last digit in the decimal to write the denominator.

The Dynamic Planning Tool
Editable Teacher's Resources at *BigIdeasMath.com*

2 ACTIVITY: 4 in a Row

Number of Players: 2

Each player

- draws a 4 by 4 square (play card).
- fills the card with decimals or simplified fractions from Activity 1. (Don't show your play card to the other player.)
- takes turns calling out a fraction or decimal.

Both players cross out any number that is equal to the called out number.

The first person to get four numbers in the same row, column, or diagonal wins. Check each other's work.

3 ACTIVITY: Vocabulary Patterns

Work with a partner. Match each word on the left with its description on the right.

Decimal	Originally the 10th month
Deciliter	A number system with base 10
December	Celebration of a 10th anniversary
Decennial	Ten years
Decade	One-tenth of a liter

Inductive Reasoning

4. **a.** Make a list of all the denominators you found in Activity 1.

 b. Factor each denominator into prime factors.

 c. What similarities do you see in the denominators?

 d. Relate your answer to the decimal system.

5. Describe a good strategy you could use to win the game in Activity 2.

6. What does the prefix "dec" mean?

What Is Your Answer?

7. **IN YOUR OWN WORDS** When you write a terminating decimal as a fraction, what type of denominator do you get?

Use what you learned about writing decimals as fractions to complete Exercises 7–14 on page 88.

Section 2.7 Writing Decimals as Fractions 85

2.7 Lesson

Key Vocabulary 🔊
terminating decimal, p. 84

Every **terminating decimal** can be written as a fraction.

 Key Idea

Writing Decimals as Fractions

Words Write the digits of the decimal in the numerator. Use the place value of the last digit of the decimal to write the denominator.

Numbers

Write the digit 3 in the numerator. $0.3 = \frac{3}{10}$ The place value of the last digit is **tenths**.

EXAMPLE 1 Writing Decimals as Fractions

a. **Write 0.28 as a fraction in simplest form.**

Write the digits in the numerator.

$0.28 = \dfrac{28}{100}$ 8 is in the **hundredths'** place, so **100** is the denominator.

$= \dfrac{\cancel{28}\,^7}{\cancel{100}\,_{25}}$ Divide out the common factor 4.

$= \dfrac{7}{25}$ Simplify.

b. **Write 0.475 as a fraction in simplest form.**

Write the digits in the numerator.

$0.475 = \dfrac{475}{1000}$ 5 is in the **thousandths'** place, so **1000** is the denominator.

$= \dfrac{\cancel{475}\,^{19}}{\cancel{1000}\,_{40}}$ Divide out the common factor 25.

$= \dfrac{19}{40}$ Simplify.

● On Your Own

Now You're Ready
Exercises 7–18

Write the decimal as a fraction in simplest form.

1. 0.9
2. 0.52
3. 0.06
4. 0.135

Laurie's Notes

Introduction

Connect
- **Yesterday:** Students explored patterns in writing common decimals as fractions.
- **Today:** Students will use a formal rule to write any terminating decimal as a fraction.

Motivate
- Have students match the decimals in the first column with the descriptions in the second column:

 1) 0.24 A. four and two-tenths
 2) 0.42 B. two hundred four thousandths
 3) 0.204 C. twenty-four hundredths
 4) 4.2 D. two and four tenths
 5) 2.4 E. two and four hundredths
 6) 2.04 F. forty-two hundredths

 1C, 2F, 3B, 4A, 5D, 6E

- This quick activity will help students focus on place values and how decimals are read. If students can read the decimals correctly, they will be able to write them as fractions.

Lesson Notes

Key Idea
- Write the Key Idea on the board.
- **FYI:** This rule also works for decimal numbers greater than 1.

Example 1
- Work through the example together as a class.
- Students may take several steps to simplify the fractions.
- **?** For part (a), ask "28 and 100 are both even so they are at least divisible by what number?" *2*
- **?** For part (b), ask "475 and 1000 are both divisible by what number?" *25*

On Your Own
- **Common Error:** The decimal 0.06 equals $\frac{6}{100}$, not $\frac{6}{10}$.

Goal
Today's lesson is writing decimals as fractions or mixed numbers.

Start Thinking! and Warm Up

In gymnastics, 0.8 of a point is deducted from a gymnast's score if the gymnast falls off the apparatus during a routine. How can you write 0.8 as a fraction in simplest form?

All deductions in gymnastics scoring are given in decimal form. Why do you think that decimals are used instead of fractions when calculating gymnastics scores?

Extra Example 1
Write the decimal as a fraction in simplest form.

a. 0.48 $\frac{12}{25}$

b. 0.685 $\frac{137}{200}$

On Your Own

1. $\frac{9}{10}$ 2. $\frac{13}{25}$

3. $\frac{3}{50}$ 4. $\frac{27}{200}$

Extra Example 2

Write 9.125 as a mixed number in simplest form. $9\dfrac{1}{8}$

Extra Example 3

A male bird weighs $2\dfrac{1}{4}$ ounces and a female bird weighs 1.75 ounces. How many times greater is the weight of the male bird than the weight of the female bird? $1\dfrac{2}{7}$ times greater

On Your Own

5. $1\dfrac{7}{10}$
6. $3\dfrac{1}{4}$
7. $6\dfrac{17}{50}$
8. $10\dfrac{19}{25}$
9. $43\dfrac{17}{25}$ times greater

English Language Learners
Vocabulary

Explain to students that the word *decimal* comes from the Latin root "dec," which means "ten." That is why the first step in writing a fraction as a decimal is to rewrite the fraction with a denominator that is a power of ten.

Laurie's Notes

Example 2

? "How is the decimal 3.025 read?" 3 and 25-thousandths
- Tell students that the first step is to write 3.025 as the sum of a whole number and a decimal. This will help students read the decimal correctly.
? "What is the greatest common factor of 25 and 1000?" 25
? **Extension:** Ask a few questions to extend the problem:
 - "What is 3.25 as a fraction?" $3\dfrac{1}{4}$
 - "Which is greater, *3.025* or *3.25*?" 3.25

Example 3

? "Have you ever seen a whooping crane or a piping plover?"
? Give the weight information for each species, and then pose this question: "How many times greater is the weight of the whooping crane than the weight of the piping plover?" $131\dfrac{1}{3}$ times greater
- **Common Error:** Students sometimes read too quickly and think "How much greater?" instead of "How many *times* greater?" Ask students the difference between these two questions.
- Work through the problem. Students will need time to find the common factors to divide out.
? **Extension:** "What is the approximate weight of the whopping crane in pounds?" about 15.4 pounds
- **FYI:** The term *protected species* encompasses any plant or animal that is legally protected because it is endangered, threatened to become endangered, or one of special concern.
- A full-grown adult whooping crane stands 5 feet tall and has a wingspan of 7 to 8 feet. The adult piping plover has a length of 7 inches and a wingspan of 15 inches.

On Your Own

- These problems are a quick check of the basic skills for this lesson. Question 9 keeps the skill of dividing mixed numbers fresh. The computation in Question 9 is a bit messy, but provides a good challenge.

Closure

- **Exit Ticket:** Write each decimal as a fraction or mixed number in simplest form.
 1) 0.24 2) 0.42 3) 0.204 4) 4.2
 $\dfrac{6}{25}$ $\dfrac{21}{50}$ $\dfrac{51}{250}$ $4\dfrac{1}{5}$

Technology For the Teacher

The Dynamic Planning Tool
Editable Teacher's Resources at *BigIdeasMath.com*

EXAMPLE 2 Writing Decimals as Mixed Numbers

Write 3.025 as a mixed number in simplest form.

$$3.025 = 3 + 0.025 \quad \text{Write as a sum.}$$

$$= 3 + \frac{25}{1000} \quad \text{Write "twenty-five thousandths" as a fraction.}$$

$$= 3 + \frac{\cancel{25}^{\,1}}{\underset{40}{\cancel{1000}}} \quad \text{Simplify fraction.}$$

$$= 3\frac{1}{40} \quad \text{Write as a mixed number.}$$

EXAMPLE 3 Real-Life Application

Two protected bird species are the whooping crane and the piping plover. How many times greater is the weight of a whooping crane than the weight of the piping plover?

Write each weight as a mixed number.

Whooping Crane: $246.25 = 246 + 0.25 = 246 + \frac{25}{100} = 246\frac{1}{4}$

Piping Plover: $1\frac{7}{8}$

Divide the weights. **Estimate** $246 \div 2 = 123$

Whooping Crane
246.25 ounces

Piping Plover
$1\frac{7}{8}$ ounces

$$246\frac{1}{4} \div 1\frac{7}{8} = \frac{985}{4} \div \frac{15}{8} \quad \text{Write as improper fractions.}$$

$$= \frac{985}{4} \times \frac{8}{15} \quad \text{Multiply by the reciprocal } \frac{8}{15}.$$

$$= \frac{\overset{197}{\cancel{985}} \times \overset{2}{\cancel{8}}}{\underset{1}{\cancel{4}} \times \underset{3}{\cancel{15}}} \quad \text{Divide out common factors.}$$

$$= \frac{394}{3}, \text{ or } 131\frac{1}{3} \quad \text{Simplify.}$$

∴ So, the weight of the whooping crane is $131\frac{1}{3}$ times greater.

Reasonable? $131\frac{1}{3} \approx 123$ ✓

On Your Own

Now You're Ready
Exercises 19–22

Write the decimal as a mixed number in simplest form.

5. 1.7 **6.** 3.25 **7.** 6.34 **8.** 10.76

9. A wood stork weighs 81.9 ounces. How many times greater is the weight of a wood stork than the weight of the piping plover in Example 3?

2.7 Exercises

Vocabulary and Concept Check

1. **OPEN-ENDED** Find a decimal with a value between 0.25 and 0.5. Write both the decimal and the equivalent fraction.

2. **OPEN-ENDED** Describe a real-life example where you write a decimal as a mixed number.

Fill in the blank to make the statement true.

3. $6.7 = \boxed{}\dfrac{7}{10}$

4. $2.33 = 2\dfrac{33}{\boxed{}}$

5. $7.9 = 7\dfrac{\boxed{}}{10}$

6. $1.622 = 1\dfrac{622}{\boxed{}}$

Practice and Problem Solving

Write the decimal as a fraction or mixed number in simplest form.

7. 0.9	8. 0.6	9. 0.4	10. 0.8
11. 0.64	12. 0.18	13. 0.375	14. 0.655
15. 0.11	16. 0.28	17. 0.435	18. 0.812
19. 2.75	20. 3.06	21. 8.113	22. 4.764

23. **ERROR ANALYSIS** Describe and correct the error in writing 0.073 as a fraction.

24. **CELL PHONE** The thickness of a cell phone is 0.46 inch. Write the thickness as a fraction in simplest form.

25. **MOVIE** A movie is 2.3 hours long. Write the length of the movie as a mixed number in simplest form.

26. **RUNNING** A person runs 26.22 miles in a marathon. Write the number of miles as a mixed number in simplest form.

Write the decimal as an improper fraction in simplest form.

27. 7.2 28. 3.35 29. 6.82 30. 2.265
31. 4.75 32. 1.17 33. 5.52 34. 8.125

Write the value of the money as a decimal and as a fraction of a dollar.

35. one nickel

36. one quarter

37. eight pennies

38. one dollar and six dimes

Assignment Guide and Homework Check

Level	Day 1 Activity Assignment	Day 2 Lesson Assignment	Homework Check
Basic	7–14, 45–53	1–6, 15–23 odd, 24–26, 39, 41	15, 23, 26, 39
Average	7–14, 45–53	1–6, 15–33 odd, 40, 42	15, 23, 33, 40
Advanced	7–14, 45–53	1–6, 23, 32–42 even, 43, 44	23, 34, 40, 42

Common Errors

- **Exercises 7–18** Students may use the wrong denominator in writing the decimal as a fraction. For example, in Exercise 13, they may think 0.375 means "375 hundredths" instead of "375 thousandths."
 Remind students of the name of each decimal place and encourage them to write out the decimal in words. Tell students that when the decimal is less than 1, the denominator must be greater than the numerator.
- **Exercises 19–22** Students may try to write the decimal as an improper fraction instead of as a mixed number. Remind them that the decimal point separates the whole number from the fraction.

2.7 Record and Practice Journal

Write the decimal as a fraction or mixed number in simplest form.
1. 0.7 $\frac{7}{10}$ 2. 0.282 $\frac{141}{500}$ 3. 4.66 $4\frac{33}{50}$ 4. 2.125 $2\frac{1}{8}$

Write the decimal as an improper fraction in simplest form.
5. 7.25 $\frac{29}{4}$ 6. 5.32 $\frac{133}{25}$ 7. 9.545 $\frac{1909}{200}$ 8. 4.118 $\frac{2059}{500}$

9. You have 3.28 yards of fabric. The instructions for a project require $3\frac{2}{5}$ yards of fabric. Do you have enough fabric for the project? Explain.
 No; You need $3\frac{2}{5}$ yards of fabric. You only have 3.28, or $3\frac{7}{25}$, yards.

10. You have to finish a race in $\frac{7}{8}$ of the time you ran in the semifinals in order to win. Your semifinal time was 25.6 seconds.
 a. Write your semifinal time as a mixed number in simplest form.
 $25\frac{3}{5}$ seconds
 b. What does your time need to be to win the race?
 $22\frac{2}{5}$ seconds

Technology For the Teacher
Answer Presentation Tool
QuizShow

Vocabulary and Concept Check

1. *Sample answer:* $0.3 = \frac{3}{10}$
2. *Sample answer:* While traveling, you may convert some mileage to mixed numbers to perform calculations.
3. $6\frac{7}{10}$ 4. $2\frac{33}{100}$
5. $7\frac{9}{10}$ 6. $1\frac{622}{1000}$

Practice and Problem Solving

7. $\frac{9}{10}$ 8. $\frac{3}{5}$
9. $\frac{2}{5}$ 10. $\frac{4}{5}$
11. $\frac{16}{25}$ 12. $\frac{9}{50}$
13. $\frac{3}{8}$ 14. $\frac{131}{200}$
15. $\frac{11}{100}$ 16. $\frac{7}{25}$
17. $\frac{87}{200}$ 18. $\frac{203}{250}$
19. $2\frac{3}{4}$ 20. $3\frac{3}{50}$
21. $8\frac{113}{1000}$ 22. $4\frac{191}{250}$
23. Because 3 is in the thousandths' place, the denominator should be 1000.
 $0.073 = \frac{73}{1000}$
24. $\frac{23}{50}$ inch
25. $2\frac{3}{10}$ hours
26. $26\frac{11}{50}$ miles
27. $\frac{36}{5}$ 28. $\frac{67}{20}$
29. $\frac{341}{50}$ 30. $\frac{453}{200}$

T-88

Practice and Problem Solving

31. $\dfrac{19}{4}$ **32.** $\dfrac{117}{100}$

33. $\dfrac{138}{25}$ **34.** $\dfrac{65}{8}$

35. $0.05; \dfrac{1}{20}$ **36.** $0.25; \dfrac{1}{4}$

37. $0.08; \dfrac{2}{25}$ **38.** $1.60; 1\dfrac{3}{5}$

39. $0.32; \dfrac{8}{25}$ **40.** $0.84; \dfrac{21}{25}$

41. $0.55; \dfrac{11}{20}$ **42.** 292 cars

43. See *Taking Math Deeper*.

44. a *Sample answer:* The player could have shot 25 3-point shots and made 9 of them.

 b. 126 shots

Fair Game Review

45. 5 **46.** 16

47. 17 **48.** 9

49. $\dfrac{5}{7}$ **50.** $1\dfrac{7}{11}$

51. $\dfrac{2}{7}$ **52.** $\dfrac{20}{27}$

53. B

Mini-Assessment

Write the decimal as a fraction or mixed number in simplest form.

1. 0.3 $\dfrac{3}{10}$ **2.** 0.625 $\dfrac{5}{8}$

3. 2.07 $2\dfrac{7}{100}$ **4.** 1.65 $1\dfrac{13}{20}$

5. A movie is 2.6 hours long. Write the length of the movie as a mixed number in simplest form. $2\dfrac{3}{5}$ hours

Taking Math Deeper

Exercise 43

Like many problems with a lot of numbers, a table can help organize the work.

1 Organize the information in a table.

a.

Exhibit	Distance	
	Decimal	Fraction
	0.16 mi	$\dfrac{16}{100} = \dfrac{4}{25}$ mi
	0.35 mi	$\dfrac{35}{100} = \dfrac{7}{20}$ mi
	0.78 mi	$\dfrac{78}{100} = \dfrac{39}{50}$ mi

2 b. How many times farther is than ?

$\dfrac{39}{50} \div \dfrac{4}{25} = \dfrac{39}{\cancel{50}_2} \cdot \dfrac{\cancel{25}^1}{4}$

$= \dfrac{39}{8}$ or $4\dfrac{7}{8}$ times farther

About 5 times

3 c. How far is the manatee exhibit?

$3 \cdot \dfrac{7}{20} = \dfrac{21}{20} = 1\dfrac{1}{20}$ mi or 1.05 mi

Project

Have students make a poster for the zoo. Include the distance to the manatee exhibit. Include art to help visitors from other countries understand where the exhibits are located. Include both miles and kilometers. Write the names of the animals in both English and Spanish.

Reteaching and Enrichment Strategies

If students need help...	If students got it...
Resources by Chapter • Practice A and Practice B • Puzzle Time Record and Practice Journal Practice Differentiating the Lesson Lesson Tutorials Skills Review Handbook	Resources by Chapter • Enrichment and Extension • School-to-Work • Financial Literacy • Technology Connection • Life Connections Start the next section

Write the shaded part of the unit square as a decimal. Then write the decimal as a fraction.

39. 40. 41.

42. **TRAFFIC JAM** You are in a traffic jam that is $\frac{7}{8}$ mile long. The cars are in one lane bumper to bumper. The average length of a car is 0.003 mile. About how many cars are in the traffic jam?

Monkeys 0.16 mile →
Tigers 0.35 mile ←
Elephants 0.78 mile ↙

43. **ZOO** The sign shows the distances to several zoo exhibits.

 a. Write the distances as fractions.
 b. How many times farther is the elephant exhibit than the monkey exhibit?
 c. The manatee exhibit is 3 times farther than the tiger exhibit. How far is the manatee exhibit?

44. **Reasoning** A basketball player's 3-point average is the number of 3-point shots made divided by the number of attempts.

 a. A player's 3-point average is 0.36. How could this happen?
 b. If the player in part (a) attempted 350 shots, how many shots did the player make?

Fair Game Review *What you learned in previous grades & lessons*

Divide. *(Skills Review Handbook)*

45. $30 \div 6$ 46. $64 \div 4$ 47. $85 \div 5$ 48. $81 \div 9$

Divide. Write the answer in simplest form. *(Section 2.5)*

49. $\frac{1}{7} \div \frac{1}{5}$ 50. $\frac{6}{11} \div \frac{1}{3}$ 51. $\frac{1}{6} \div \frac{7}{12}$ 52. $\frac{8}{15} \div \frac{18}{25}$

53. **MULTIPLE CHOICE** What is the area of the triangle? *(Section 1.5)*

 Ⓐ 15 ft^2 Ⓑ 27 ft^2
 Ⓒ 54 ft^2 Ⓓ 90 ft^2

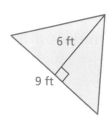

2.8 Writing Fractions as Decimals

COMMON CORE STATE STANDARDS
6.NS.2

Essential Question How can you tell from the denominator of a fraction if its decimal form is terminating or repeating?

1 EXAMPLE: Writing a Fraction as a Decimal

Write the fraction as a decimal. Is it terminating or repeating?

a. $\frac{3}{8}$

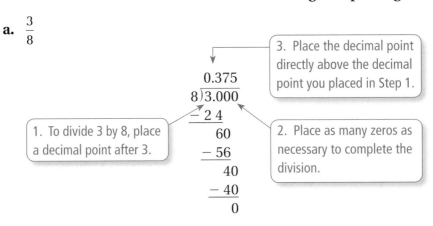

The division terminates. So, $\frac{3}{8} = 0.375$ is a terminating decimal.

b. $\frac{2}{3}$

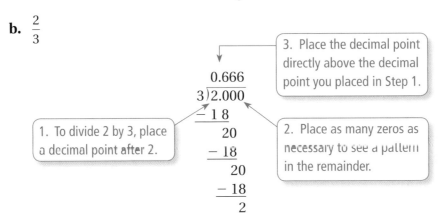

The division does not end. So, $\frac{2}{3} = 0.666\ldots$ is a **repeating decimal**.

Inductive Reasoning

Write the fraction as a decimal. Is it terminating or repeating?

2. $\frac{3}{4}$ 3. $\frac{1}{16}$ 4. $\frac{1}{6}$ 5. $\frac{2}{9}$

6. $\frac{7}{20}$ 7. $\frac{7}{8}$ 8. $\frac{5}{12}$ 9. $\frac{21}{40}$

90 Chapter 2 Multiplying and Dividing Fractions

Laurie's Notes

Introduction

For the Teacher
- **Goal:** These activities have two goals. The first goal is for students to write fractions as decimals by dividing the numerator by the denominator. The second goal is to recognize that when the denominator has only 2 and/or 5 as prime factors, the decimal will terminate. Denominators that have prime factors other than 2 and/or 5 will repeat.

Motivate
- **Story Time!** This morning I was listening to the radio as I ate my breakfast. The morning trivia question was about fractions. I had a good feeling! The question was "Which fraction is greater, $\frac{7}{8}$ or $\frac{8}{9}$, and why?" I dialed into the radio station as fast as I could, but I did not get through.
- ❓ "How would you have answered the question?"
- Write $\frac{7}{8}$ and $\frac{8}{9}$ on the board. See if students have an explanation as to which is greater.
- ❓ "Do you remember the decimal equivalent for $\frac{7}{8}$?" Some students may remember.
- Tell students that today's investigation is about writing fractions as decimals. Tell students that you'll return to this question at the end of class.

Activity Notes

Example 1
- Two problems are presented here, one which terminates and one which repeats.
- ❓ "How does $\frac{3}{8}$ compare to $\frac{1}{2}$? How do you know?" $\frac{3}{8} < \frac{1}{2}; \frac{1}{2} = \frac{4}{8}$ and $\frac{3}{8} < \frac{4}{8}$
- Tell students that because $\frac{1}{2} = 0.5$, $\frac{3}{8}$ should be less than 0.5. To write $\frac{3}{8}$ as a decimal, you divide 3 by 8.
- Work through the division of each problem.
- Discuss the number of zeros that are added to the right of the decimal point. Initially you make a guess. It's okay to add more than what is needed. For example: In part (b), you need to carry the division through enough place values until students can recognize the repeating pattern.
- ❓ In part (b) "Do you think the remainder will ever be zero?"
- Once students are convinced that the pattern keeps repeating, have them describe the part that repeats. State that the notation of using three dots means that the pattern continues. The bar notation will be introduced tomorrow.
- **FYI:** Student errors occur when they are not neat with lining up their digits. Some students have greater success doing division on grid paper.

Common Core State Standards

6.NS.2 Fluently divide multi-digit numbers using the standard algorithm.

Previous Learning
Students have represented common fractions—halves, fourths, and tenths—as decimals. Students should know that a decimal that ends is a *terminating decimal*.

Start Thinking! and Warm Up

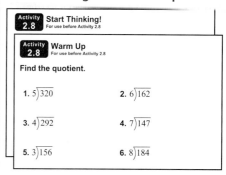

2.8 Record and Practice Journal

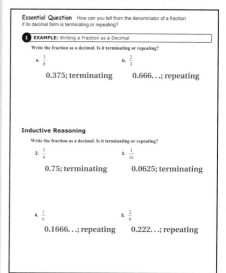

T-90

Differentiated Instruction

Auditory

Write the numbers $\frac{71}{100}$ and 0.71 on the board. Explain to students that both numbers are read in the same way, "seventy-one hundredths." Decimal numbers greater than one and mixed numbers are read the same way. Auditory learners may be confused as to which form they should use when solving a word problem. Explain that they should look for clues in the text to determine whether they should use a fraction or a decimal.

2.8 Record and Practice Journal

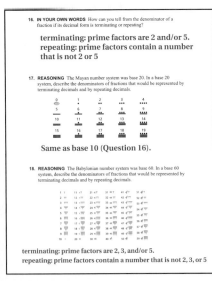

Laurie's Notes

Inductive Reasoning

- **Neighbor Check:** Have students work independently and then have their neighbor check their work. Have students discuss any discrepancies.
- **Common Error:** This is the first time that students have used the technique of dividing the numerator by the denominator in order to write a fraction as a decimal. A common error later on will be to divide the denominator by the numerator, producing a quotient greater than 1. Remind students to always divide the numerator by the denominator in order to write the fraction as a decimal.
- Check to see that students are dividing correctly. Also make sure that students are placing the decimal point correctly and deciding on the first digit of the quotient. These are usually the most difficult steps for students.
- **Connection:** When teachers return tests or quizzes, the score is often written as a fraction. If students want to know how they did, they need to divide the number of points they got (the numerator) by the total number of points that were possible (the denominator).
- Have students do Questions 10–13 with a partner and then discuss these as a class. Have students explain their reasoning.
- If you are short on time, assign Question 14 as part of the homework.

What Is Your Answer?

- To assist students with Question 15, make a table with two columns: one column to list the fractions that terminate and the other column to list the fractions that repeat.
- **Alternative Activity:** For Questions 17 and 18, if you do not want your students to explore Mayan and Babylonian number systems, you could have them use calculators to explore the patterns generated when certain sets of fractions are written as decimals. Particular denominators to explore include thirds, sixths, ninths, elevenths, and twenty-seconds. Avoid certain denominators (7, 13, 19) unless students are using calculators with a large display window.

Closure

- Have students refer back to the radio story used to motivate the lesson. "Compare $\frac{7}{8}$ and $\frac{8}{9}$ by writing each as a decimal." $\frac{7}{8} = 0.875$; $\frac{8}{9} = 0.888\ldots$; $0.875 < 0.888$, so $\frac{7}{8} < \frac{8}{9}$.

The Dynamic Planning Tool
Editable Teacher's Resources at *BigIdeasMath.com*

Use estimation to match the fraction with its decimal. Then use a calculator to check your answer.

10. $\dfrac{5}{6}$ **11.** $\dfrac{1}{3}$ **12.** $\dfrac{5}{8}$ **13.** $\dfrac{3}{16}$

A. 0.625 **B.** 0.1875 **C.** 0.333... **D.** 0.83333...

14. In $\dfrac{1}{7} = 0.1428571428571428571\ldots$, what are the repeating digits? Can you find another fraction that has at least six digits that repeat?

What Is Your Answer?

15. a. Describe the denominators of fractions that can be written as terminating decimals.

b. Describe the denominators of fractions that can be written as repeating decimals.

16. IN YOUR OWN WORDS How can you tell from the denominator of a fraction if its decimal form is terminating or repeating?

17. Reasoning The Mayan number system was base 20. In a base 20 system, describe the denominators of fractions that would be represented by terminating decimals and by repeating decimals.

18. Reasoning The Babylonian number system was base 60. In a base 60 system, describe the denominators of fractions that would be represented by terminating decimals and by repeating decimals.

Practice Use what you learned about writing fractions as decimals to complete Exercises 11–18 on page 94.

Section 2.8 Writing Fractions as Decimals 91

2.8 Lesson

Key Vocabulary
repeating decimal, p. 92

When writing a fraction as a decimal, your result is a terminating or repeating decimal. A **repeating decimal** *repeats* a pattern of one or more digits.

Terminating decimal: $\dfrac{1}{2} = 0.5$

Repeating decimal: $\dfrac{2}{3} = 0.666\ldots = 0.\overline{6}$

Show that a decimal repeats by using three dots or placing a bar over the digit that repeats.

Key Idea

Method 1: Writing Fractions as Decimals
To write a fraction as a decimal, divide the numerator by the denominator.

EXAMPLE 1 Writing Fractions as Decimals

a. Write $\dfrac{1}{8}$ as a decimal.

b. Write $\dfrac{5}{6}$ as a decimal.

$$\therefore \ \dfrac{1}{8} = 1 \div 8 = 0.125$$

$$\therefore \ \dfrac{5}{6} = 5 \div 6 = 0.833\ldots \text{ or } 0.8\overline{3}$$

Reading
One way to say the repeating decimal $0.8\overline{3}$ is "zero point eight three, with the three repeating."

On Your Own

Write the fraction as a decimal. Is it terminating or repeating?

1. $\dfrac{3}{5}$
2. $\dfrac{1}{9}$
3. $\dfrac{9}{20}$
4. $\dfrac{8}{5}$

Laurie's Notes

Introduction

Connect
- **Yesterday:** Students explored patterns in writing common fractions as decimals.
- **Today:** Students will use two different methods for writing a fraction as a decimal. Given a fraction, it is important for students to distinguish which method is most efficient.

Motivate
- Confess to an exciting life and say "I was playing around with fractions last night and look at what I found."
- Write the following on the board:

 $\frac{5}{9} = 0.555\ldots$ $\frac{7}{11} = 0.636363\ldots$ $\frac{8}{27} = 0.296296296\ldots$

- **"What do you notice?"** Listen for students describing the number of digits that repeat for each fraction.
- **"Can we predict if a fraction will repeat or terminate?"** yes; If a denominator has only 2 and/or 5 as prime factors, the fraction will terminate.
- **Discuss: Repeating decimals** can have one or more digits that repeat. In addition to using the three dots to show that a pattern repeats, you can also use bar notation. Write the bar notation for each of the previous decimals.
 $0.555\ldots = 0.\overline{5}$ $0.636363\ldots = 0.\overline{63}$ $0.296296296\ldots = 0.\overline{296}$

Lesson Notes

Key Idea
- Write the key idea on the board.

Example 1
- Work through each part.
- It is common for students to ask how they will know if they need to keep dividing, meaning will the decimal repeat or terminate. Remind them that if the denominator has only prime factors of 2 and/or 5, the fraction will terminate as a decimal. This helps students who might make an error in dividing.
- Discuss the bar notation for $\frac{5}{6}$. Only the digit 3 repeats, hence the bar is written over the 3 alone and not the 8. Make sure it is clear to students that $0.8\overline{3} \neq 0.\overline{83}$.

On Your Own
- **"Can you predict which fractions will be repeating decimals?"** yes
- **"When you write $\frac{8}{5}$ as a decimal, will it be *greater than* or *less than* 1?"** It will be greater than 1.
- **Think-Pair-Share:** Students should read each question independently and then work with a partner to answer the questions. When they have answered the questions, the pair should compare their answers with another group and discuss any discrepancies.

Goal Today's lesson is writing fractions as either a **repeating** or terminating decimals.

Start Thinking! and Warm Up

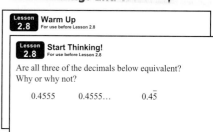

Extra Example 1
Write the fraction as a decimal. Is it terminating or repeating?
a. $\frac{17}{20}$ 0.85, terminating
b. $\frac{9}{11}$ $0.\overline{81}$, repeating

On Your Own
1. 0.6; terminating
2. $0.\overline{1}$; repeating
3. 0.45; terminating
4. 1.6; terminating

T-92

English Language Learners
Vocabulary
Let English language learners know that the pronunciation of rational numbers that end in *-th*, such as *hundredths* and *thousandths*, is sometimes difficult for native English speakers.

Extra Example 2
Write $\frac{6}{25}$ as a decimal. 0.24

Extra Example 3
You ran a 5-mile race in 44.6 minutes. Your friend ran the race in $44\frac{11}{20}$ minutes. Who was faster and by how much? Your friend was faster by 0.05 minute.

On Your Own
5. 0.8
6. 0.55
7. 0.14
8. 0.188
9. yes; Pedro ran it in 6.74 seconds, which is the fastest time.

Laurie's Notes

Key Idea
- Write the Key Idea on the board.
- ? "What fractions can be written as equivalent fractions whose denominators are 10, 100, or 1000?" *listen for halves, fourths, fifths, eighths, and tenths*
- If time permits, look back at fractions previously written as decimals by dividing. Ask students how they would write fractions such as $\frac{3}{5}, \frac{9}{20}, \frac{8}{5}$, and $\frac{1}{8}$ as decimals without dividing.

Example 2
- ? Ask a few questions about this example.
 - "What is $\frac{1}{10}$ as a decimal? How do you know?" 0.1; The fraction $\frac{1}{10}$ is "one tenth", the same as 0.1.
 - "What is $\frac{1}{4}$ as a decimal? How do you know?" 0.25; $\frac{1}{4} = \frac{25}{100} = 0.25$
 - "What fractions can be written as a decimal without dividing?" If students need a hint, have them look at the decimals that terminated on the previous page. Give students sufficient time to think.

Example 3
- Write $6\frac{21}{25}$ on the board.
- ? "How is this fraction read?" *listen for 6 and twenty one twenty-fifths*
- Explain that there is no need to write a mixed number as an improper fraction in order to write it as a decimal. This is a common misconception for students.
- ? Write $6\frac{21}{25} = 6.$ _____ on the board. Say, "You know that $6\frac{21}{25}$ will be written as 6. _____ as a decimal. How would you write $\frac{21}{25}$ as a fraction with a denominator of 100? How do you write $6\frac{21}{25}$ as a decimal?" $\frac{21}{25} = 0.84$; $6\frac{21}{25} = 6.84$

Closure
- **Connection:** You looked at two methods for writing a fraction as a decimal. How will you decide which method to use? If the denominator of the fraction can easily be written as 10, 100, or 1000, use equivalent fractions to find the decimal. Otherwise, divide the numerator by the denominator.

Technology for the Teacher

The Dynamic Planning Tool
Editable Teacher's Resources at *BigIdeasMath.com*

Key Idea

Method 2: Writing Fractions as Decimals

To write a fraction as a decimal, write an equivalent fraction (if possible) whose denominator is 10, 100, or 1000.

EXAMPLE 2 Writing a Fraction as a Decimal

Write $\frac{3}{20}$ as a decimal.

20 is a factor of 100.

$$\frac{3}{20} = \frac{3 \times 5}{20 \times 5}$$ Because $20 \times 5 = 100$, multiply the numerator and denominator by 5.

$$= \frac{15}{100}$$ Multiply.

$$= 0.15$$ $\frac{15}{100}$ is read as "fifteen hundredths."

EXAMPLE 3 Real-Life Application

You run the 40-yard dash in $6\frac{21}{25}$ seconds. Your teammate runs it in the time shown. Who is faster and by how much?

Write $6\frac{21}{25}$ as a decimal and compare it to 6.9.

$$6\frac{21}{25} = 6 + \frac{21}{25}$$ Write the mixed number as a sum.

$$= 6 + \frac{21 \times 4}{25 \times 4}$$ Because $25 \times 4 = 100$, multiply the numerator and denominator by 4.

$$= 6 + \frac{84}{100}$$ Multiply.

$$= 6 + 0.84 = 6.84$$ $\frac{84}{100}$ is read as "eighty-four hundredths."

$6.84 < 6.9$. The difference is $6.9 - 6.84 = 0.06$ second.

∴ So, you are faster by 0.06 second.

On Your Own

Now You're Ready
Exercises 11–26

Write the fraction as a decimal.

5. $\frac{4}{5}$ 6. $\frac{11}{20}$ 7. $\frac{7}{50}$ 8. $\frac{47}{250}$

9. In Example 3, Pedro runs the 40-yard dash in $6\frac{37}{50}$ seconds. Is he the fastest? Explain.

2.8 Exercises

Vocabulary and Concept Check

1. **NUMBER SENSE** When using division to write a fraction as a decimal, when do you stop dividing?

2. **WHICH ONE DOESN'T BELONG?** Which fraction does *not* belong with the other three? Explain your reasoning.

$$\frac{1}{4} \qquad \frac{1}{5} \qquad \frac{1}{6} \qquad \frac{1}{8}$$

Tell whether the decimal is repeating or terminating.

3. 0.625
4. $0.1\overline{3}$
5. $7.\overline{4}$
6. 0.470

Rewrite the repeating decimal using bar notation.

7. $0.1111\ldots$
8. $3.4444\ldots$
9. $0.5333\ldots$
10. $0.1666\ldots$

Practice and Problem Solving

Write the fraction as a decimal.

11. $\frac{7}{10}$
12. $\frac{5}{6}$
13. $\frac{31}{50}$
14. $\frac{2}{15}$

15. $\frac{17}{18}$
16. $\frac{23}{40}$
17. $\frac{21}{25}$
18. $\frac{11}{45}$

19. $\frac{3}{20}$
20. $\frac{7}{18}$
21. $\frac{3}{8}$
22. $\frac{6}{25}$

23. $\frac{19}{30}$
24. $\frac{51}{40}$
25. $\frac{114}{25}$
26. $\frac{22}{15}$

27. **ERROR ANALYSIS** Describe and correct the error in writing $\frac{8}{15}$ as a decimal.

28. **GOLDFISH** The length of a goldfish is $\frac{5}{16}$ foot. Write the length of the goldfish as a decimal.

29. **REPEATED REASONING** Your batting average for a baseball season is $\frac{11}{15}$. Write your batting average as a decimal.

Copy and complete the statement using <, >, or =.

30. $\frac{7}{8}$ ⬚ 0.85
31. $\frac{5}{12}$ ⬚ 0.4
32. 1.95 ⬚ $\frac{39}{20}$

94 Chapter 2 Multiplying and Dividing Fractions

Assignment Guide and Homework Check

Level	Day 1 Activity Assignment	Day 2 Lesson Assignment	Homework Check
Basic	11–18, 44–48	1–10, 19–35 odd, 28	2, 21, 28, 31
Average	11–18, 44–48	1–10, 19–37 odd, 38	2, 21, 31, 37
Advanced	11–18, 44–48	1–10, 27, 30–36 even, 38–43	2, 27, 30, 36

Common Errors

- **Exercises 11–26** Students may use the denominator as the dividend and the numerator as the divisor. Remind them to estimate their answer ahead of time and check the reasonableness of their answer. For example, if the fraction is less than 1, then the decimal should be less than 1 as well.
- **Exercises 11–26** Students may try to write an equivalent fraction with a denominator of 10, 100, or 1000, and get frustrated that they cannot write the fraction as a decimal. Remind them that you will not always be able to write an equivalent fraction with a denominator of 10, 100, or 1000.
- **Exercise 38** Students may forget to add the whole number onto the decimal. Remind them about the whole number, or have them write the mixed number as an improper fraction before writing it as a decimal.

Technology For the Teacher
Answer Presentation Tool
Quiz*Show*

2.8 Record and Practice Journal

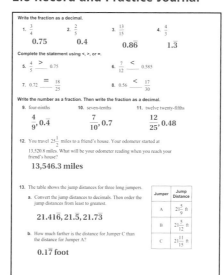

Vocabulary and Concept Check

1. When the remainder is 0 or the decimal begins repeating
2. $\frac{1}{6}$ because it is the only repeating decimal
3. terminating
4. repeating
5. repeating
6. terminating
7. $0.\overline{1}$ 8. $3.\overline{4}$
9. $0.5\overline{3}$ 10. $0.1\overline{6}$

Practice and Problem Solving

11. 0.7 12. $0.8\overline{3}$
13. 0.62 14. $0.1\overline{3}$
15. $0.9\overline{4}$ 16. 0.575
17. 0.84 18. $0.2\overline{4}$
19. 0.15 20. $0.3\overline{8}$
21. 0.375 22. 0.24
23. $0.6\overline{3}$ 24. 1.275
25. 4.56 26. $1.4\overline{6}$
27. The digit 5 does not repeat; $\frac{8}{15} = 0.5\overline{3}$.
28. 0.3125 foot
29. $0.7\overline{3}$
30. $>$ 31. $>$
32. $=$

 Practice and Problem Solving

33. $\frac{5}{8}$; 0.625

34. $\frac{11}{12}$, $0.91\overline{6}$

35. $\frac{9}{14}$, $0.6\overline{428571}$

36. $\frac{63}{20}$, 3.15

37. Sample answer: $\frac{3}{5}$; 0.6

38. See Taking Math Deeper.

39. a. $\frac{3}{9} = 0.333...$; $\frac{4}{9} = 0.444...$;
$\frac{5}{9} = 0.555...$; $\frac{6}{9} = 0.666...$;
$\frac{7}{9} = 0.777...$; $\frac{8}{9} = 0.888...$;
$\frac{9}{9} = 0.999...$

b. 1; $\frac{9}{9} = 1$

40. $0.\overline{345}$ 41. $0.00\overline{345}$

42. $0.\overline{125125}$

43. $0.00\overline{125125}$

 Fair Game Review

44. 11 45. 7
46. 9 47. 12
48. C

Mini-Assessment

Write the fraction as a decimal.

1. $\frac{9}{10}$ 0.9 2. $\frac{5}{6}$ $0.8\overline{3}$

3. $\frac{6}{5}$ 1.2 4. $\frac{7}{8}$ 0.875

5. An alligator is $\frac{51}{4}$ feet long. Write the length of the alligator as a decimal.
12.75 ft

T-95

Taking Math Deeper

Exercise 38

In this problem, students can observe how much easier it is to order numbers as decimals than as fractions. Notice that the question does not tell students how many decimal places to round to. When working the problem, students need to see that the decimals should be rounded to three places to compare 2.45 and 2.454545…

 a. Organize the work with a table.

Sea Turtle		Shell Length	
		Fraction	Decimal
Green		$2\frac{9}{20}$ ft	2.450 ft
Hawksbill		$2\frac{1}{9}$ ft	≈ 2.111 ft
Kemp's Ridley		$2\frac{1}{10}$ ft	2.100 ft
Leatherback		$2\frac{5}{11}$ ft	≈ 2.455 ft

Use a number line to order the shell lengths.

 How much longer is 🐢 than 🐢?

b. 2.455 − 2.450 = 0.005 ft

"How many times longer" would be a different question.
2.455 ÷ 2.450 ≈ 1.002

Reteaching and Enrichment Strategies

If students need help…	If students got it…
Resources by Chapter • Practice A and Practice B • Puzzle Time Record and Practice Journal Practice Differentiating the Lesson Lesson Tutorials Skills Review Handbook	Resources by Chapter • Enrichment and Extension • School-to-Work • Financial Literacy • Technology Connection • Life Connections • Stories in History Start the next section

Write the number as a fraction. Then write the fraction as a decimal.

33. five-eighths

34. eleven-twelfths

35. nine-fourteenths

36. sixty-three twentieths

37. **OPEN-ENDED** Find a fraction whose decimal value is between 0.5 and 0.65. Write the fraction and its decimal equivalent.

38. **ENDANGERED SPECIES** The table shows shell lengths of four sea turtles that are on the endangered species list.

Sea Turtle	Shell Length
Green	$2\frac{9}{20}$ ft
Hawksbill	$2\frac{1}{9}$ ft
Kemp's Ridley	$2\frac{1}{10}$ ft
Leatherback	$2\frac{5}{11}$ ft

 a. Convert the shell lengths to decimals. Then order the shell lengths from least to greatest.

 b. How much longer is the leatherback sea turtle than the green sea turtle?

39. **PATTERNS** Use the following pattern.

 $\frac{0}{9} = 0.000000...,$ $\frac{1}{9} = 0.111111...,$ $\frac{2}{9} = 0.222222...$

 a. Complete the pattern for $\frac{3}{9}$ through $\frac{9}{9}$.

 b. Two of the decimals in the pattern have terminating decimal forms. One is $0.000000...$, which is simply 0. The other is $0.999999....$ What is the terminating form of this decimal? Explain your reasoning.

 Write the fraction as a decimal.

40. $\frac{115}{333}$

41. $\frac{115}{33,300}$

42. $\frac{1001}{8000}$

43. $\frac{1001}{800,000}$

 Fair Game Review What you learned in previous grades & lessons

Use rounding to estimate the sum or difference. *(Skills Review Handbook)*

44. $4.8 + 6.4$

45. $10.7 - 3.8$

46. $2.16 + 7.44$

47. $16.58 - 5.26$

48. **MULTIPLE CHOICE** Which expression is equivalent to $9(3 + x + 5)$? *(Section 1.4)*

 Ⓐ $x + 32$ Ⓑ $9x + 32$ Ⓒ $9x + 72$ Ⓓ $27x + 45$

2.5–2.8 Quiz

Divide. Write the answer in simplest form. *(Section 2.5 and Section 2.6)*

1. $\dfrac{1}{9} \div \dfrac{1}{3}$

2. $7 \div \dfrac{5}{8}$

3. $4\dfrac{7}{8} \div \dfrac{1}{8}$

4. $7\dfrac{2}{3} \div 1\dfrac{1}{9}$

Write the decimal as a fraction or mixed number in simplest form. *(Section 2.7)*

5. 0.2

6. 8.3

Write the fraction as a decimal. *(Section 2.8)*

7. $\dfrac{9}{10}$

8. $\dfrac{2}{3}$

Evaluate the expression. *(Section 2.5)*

9. $5 \div \dfrac{1}{4} + 2$

10. $\dfrac{1}{11} + 8\dfrac{3}{5} \div 2\dfrac{1}{5}$

11. **BAND** Band members make $\dfrac{2}{3}$ of their profit from selling concert tickets. They make $\dfrac{1}{5}$ of their profit from selling band merchandise at the concerts. How many times more profit do they make from ticket sales than from merchandise sales? *(Section 2.5)*

12. **SKATEBOARDS** You are cutting as many $32\dfrac{1}{4}$-inch sections as you can out of the board to make skateboards. How many skateboards can you make? *(Section 2.6)*

13. **JAVELIN** The table shows the distances of javelin throws during a meet. Which throw was the longest? *(Section 2.8)*

Throw	Distance (meters)
1	96.72
2	$96\dfrac{5}{8}$
3	96.56
4	$96\dfrac{7}{12}$

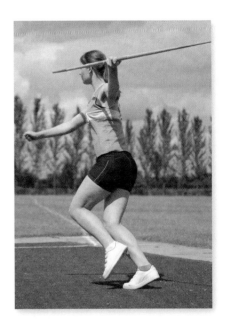

Alternative Assessment Options

Math Chat Student Reflective Focus Question
Structured Interview Writing Prompt

Math Chat
- Work in groups of four. Two members work together to discuss the similarities and differences of multiplying two whole numbers compared to multiplying two fractions. The other pair does the same for division. When they are finished, they explain their findings to the other members of their group.
- The teacher should walk around the classroom listening to the groups and asking questions to ensure understanding.

Study Help Sample Answers

Remind students to complete Graphic Organizers for the rest of the chapter.

4.

$\frac{a}{b} \div \frac{c}{d} = \frac{a}{b} \times \frac{d}{c}$ (b, c, and d ≠ 0)	**Dividing by a fraction** Multiply the number by the reciprocal of the fraction. To write a reciprocal, switch the numerator and denominator. **Examples:** $\frac{4}{7} \div \frac{3}{5} = \frac{4}{7} \times \frac{5}{3} = \frac{4 \times 5}{7 \times 3} = \frac{20}{21}$ $8 \div \frac{2}{3} = 8 \times \frac{3}{2} = \frac{\overset{4}{\cancel{8}} \times 3}{\cancel{2}_1} = 12$
	How do you divide by a mixed number?

5–7. Available at *BigIdeasMath.com*

Reteaching and Enrichment Strategies

If students need help...	If students got it...
Resources by Chapter • Study Help • Practice A and Practice B • Puzzle Time Lesson Tutorials *BigIdeasMath.com* Practice Quiz Practice from the Test Generator	Resources by Chapter • Enrichment and Extension • School-to-Work Game Closet at *BigIdeasMath.com* Start the Chapter Review

Technology For the Teacher
Answer Presentation Tool

Answers

1. $\frac{1}{3}$
2. $\frac{56}{5}$ or $11\frac{1}{5}$
3. 39
4. $\frac{69}{10}$ or $6\frac{9}{10}$
5. $\frac{1}{5}$
6. $8\frac{3}{10}$
7. 0.9
8. $0.\overline{6}$
9. 22
10. 4
11. $3\frac{1}{3}$ times more profit
12. 4 skateboards
13. throw 1

Assessment Book

For the Teacher
Additional Review Options
- **Quiz**Show
- Big Ideas Test Generator
- Game Closet at *BigIdeasMath.com*
- Vocabulary Puzzle Builder
- Resources by Chapter
 Puzzle Time
 Study Help

Answers

1. 1
2. 0
3. 2
4. 3
5. $\frac{3}{5}$
6. $\frac{16}{3}$ or $5\frac{1}{3}$
7. $\frac{24}{7}$ or $3\frac{3}{7}$
8. $\frac{15}{2}$ or $7\frac{1}{2}$
9. $\frac{5}{56}$
10. $\frac{3}{10}$
11. $\frac{1}{6}$
12. $\frac{6}{25}$

Review of Common Errors

Exercises 1–4
- Students may struggle with determining an accurate estimate of certain fractions and will get unreasonable answers.

Exercises 5–8
- Students may multiply the whole number by the denominator instead of the numerator.

Exercises 9–12
- Students may cross multiply instead of multiplying the numerators and denominators, or try to find a common denominator before multiplying.

Exercises 13–16
- Students may not change the mixed number into an improper fraction.

Exercises 17–20
- Students may find the reciprocal of the dividend instead of the divisor, the reciprocal of both, or forget to switch from division to multiplication.

Exercises 21–24
- Students may not change the mixed number into an improper fraction.

Exercises 25–28
- Students may use the wrong denominator in writing the decimal as a fraction.

Exercises 29–33
- Students may mix up the denominator and the numerator when dividing.

2 Chapter Review

Review Key Vocabulary

underestimate, *p. 47*
overestimate, *p. 47*
compatible numbers, *p. 47*
reciprocals, *p. 72*
terminating decimal, *p. 84*
repeating decimal, *p. 92*

Review Examples and Exercises

2.1 Fractions and Estimation (pp. 44–49)

$$3\frac{1}{5} \times 4\frac{7}{8} \approx 3 \times 5 = 15 \qquad \text{Round } 3\frac{1}{5} \text{ to 3 and round } 4\frac{7}{8} \text{ to 5.}$$

Exercises

Estimate the product or quotient.

1. $\frac{4}{5} \times \frac{2}{3}$
2. $\frac{1}{8} \times \frac{6}{11}$
3. $6\frac{4}{15} \div 2\frac{7}{8}$
4. $14\frac{5}{6} \div 4\frac{10}{13}$

2.2 Multiplying Fractions and Whole Numbers (pp. 50–55)

$4 \times \frac{5}{6} = \frac{4 \times 5}{6}$ Multiply the numerator and whole number.
Write the product over the denominator.

$= \frac{20}{6} = \frac{10}{3}$, or $3\frac{1}{3}$ Simplify.

Exercises

Multiply. Write the answer in simplest form.

5. $3 \times \frac{1}{5}$
6. $8 \times \frac{2}{3}$
7. $\frac{4}{7} \times 6$
8. $\frac{5}{6} \times 9$

2.3 Multiplying Fractions (pp. 56–61)

$\frac{1}{4} \times \frac{3}{5} = \frac{1 \times 3}{4 \times 5} = \frac{3}{20}$ Multiply the numerators and the denominators.

Exercises

Multiply. Write the answer in simplest form.

9. $\frac{1}{8} \times \frac{5}{7}$
10. $\frac{3}{5} \times \frac{1}{2}$
11. $\frac{2}{9} \times \frac{3}{4}$
12. $\frac{3}{10} \times \frac{4}{5}$

2.4 Multiplying Mixed Numbers (pp. 62–67)

$$\frac{3}{5} \times 1\frac{1}{8} = \frac{3}{5} \times \frac{9}{8}$$ Write $1\frac{1}{8}$ as the improper fraction $\frac{9}{8}$.

$$= \frac{3 \times 9}{5 \times 8} = \frac{27}{40}$$ Multiply the numerators and denominators.

Exercises

Multiply. Write the answer in simplest form.

13. $2\frac{2}{3} \times \frac{4}{5}$
14. $\frac{2}{7} \times 4\frac{4}{9}$
15. $1\frac{5}{6} \times 2\frac{3}{8}$
16. $2\frac{3}{10} \times 5\frac{1}{3}$

2.5 Dividing Fractions (pp. 70–77)

$$\frac{3}{7} \div \frac{5}{8} = \frac{3}{7} \times \frac{8}{5}$$ Multiply by the reciprocal of $\frac{5}{8}$, which is $\frac{8}{5}$.

$$= \frac{3 \times 8}{7 \times 5} = \frac{24}{35}$$ Multiply fractions and simplify.

Exercises

Divide. Write the answer in simplest form.

17. $\frac{1}{9} \div \frac{2}{5}$
18. $\frac{3}{4} \div \frac{5}{6}$
19. $5 \div \frac{1}{3}$
20. $\frac{8}{9} \div \frac{3}{10}$

2.6 Dividing Mixed Numbers (pp. 78–83)

$$3\frac{3}{4} \div 1\frac{1}{2} = \frac{15}{4} \div \frac{3}{2}$$ Write each mixed number as an improper fraction.

$$= \frac{15}{4} \times \frac{2}{3}$$ Multiply by the reciprocal of $\frac{3}{2}$, which is $\frac{2}{3}$.

$$= \frac{\overset{5}{\cancel{15}} \times \overset{1}{\cancel{2}}}{\underset{2}{\cancel{4}} \times \underset{1}{\cancel{3}}}$$ Multiply fractions. Divide out common factors.

$$= \frac{5}{2}, \text{ or } 2\frac{1}{2}$$ Simplify.

Exercises

Divide. Write the answer in simplest form.

21. $1\frac{2}{5} \div \frac{4}{7}$
22. $2\frac{3}{8} \div \frac{3}{5}$
23. $4\frac{1}{8} \div 2\frac{1}{4}$
24. $5\frac{5}{8} \div 1\frac{2}{9}$

Review Game
Operations with Fractions

Materials
- 10 cards numbered 1 through 10

- 4 cards labeled $+$, $-$, \div, and \times

Players: 2

Directions
One player shuffles the numbered cards and deals four cards, face down, to the other player and then himself or herself. The other player places the operation cards face down, mixes them, selects one, and turns it over.

Players turn over their numbered cards, arrange them into two fractions, and perform the selected operation. Then the players compare their answers. The player whose answer is greater wins the round and receives 1 point. (*Variation:* The player whose answer is lesser wins the round and receives 1 point.) After each round, players reshuffle the numbered cards and remix the operation cards.

Who Wins?
Students can play for a set number of minutes, with the higher score winning, or they can play to a set number of points.

For the Student
Additional Practice
- Lesson Tutorials
- Study Help (textbook)
- Student Website
 Multi-Language Glossary
 Practice Assessments

Answers

13. $\dfrac{32}{15}$ or $2\dfrac{2}{15}$ 14. $\dfrac{80}{63}$ or $1\dfrac{17}{63}$

15. $\dfrac{209}{48}$ or $4\dfrac{17}{48}$ 16. $\dfrac{184}{15}$ or $12\dfrac{4}{15}$

17. $\dfrac{5}{18}$ 18. $\dfrac{9}{10}$

19. 15 20. $\dfrac{80}{27}$ or $2\dfrac{26}{27}$

21. $\dfrac{49}{20}$ or $2\dfrac{9}{20}$ 22. $\dfrac{95}{24}$ or $3\dfrac{23}{24}$

23. $\dfrac{11}{6}$ or $1\dfrac{5}{6}$ 24. $\dfrac{405}{88}$ or $4\dfrac{53}{88}$

25. $\dfrac{3}{5}$ 26. $\dfrac{6}{25}$

27. $\dfrac{2}{25}$ 28. $3\dfrac{31}{200}$

29. 0.6; terminating

30. 0.875; terminating

31. $0.\overline{2}$; repeating

32. $1.1\overline{6}$; repeating

33. $0.6\overline{1}$

My Thoughts on the Chapter

What worked. . .

> **Teacher Tip**
> Not allowed to write in your teaching edition? Use sticky notes to record your thoughts.

What did not work. . .

What I would do differently. . .

2.7 Writing Decimals as Fractions (pp. 84–89)

Write 0.56 as a fraction in simplest form.

Write the digits in the numerator.

$0.56 = \dfrac{56}{100}$ ← 6 is in the hundredths' place, so 100 is the denominator.

$= \dfrac{\overset{14}{\cancel{56}}}{\underset{25}{\cancel{100}}}$ Divide out common factor of 4.

$= \dfrac{14}{25}$ Simplify.

Exercises

Write the decimal as a fraction or mixed number in simplest form.

25. 0.6 **26.** 0.24 **27.** 0.08 **28.** 3.155

2.8 Writing Fractions as Decimals (pp. 90–95)

Write $\dfrac{1}{6}$ as a decimal.

```
       0.166
    6)1.000   ← Place the decimal point.
     − 6
       40
     − 36     ← Place zeros to complete the division.
        40
      − 36
         4    ← The remainder 4 keeps repeating.
```

∴ So, $\dfrac{1}{6} = 1 \div 6 = 0.166\ldots$, or $0.1\overline{6}$.

Exercises

Write the fraction as a decimal. Is it terminating or repeating?

29. $\dfrac{3}{5}$ **30.** $\dfrac{7}{8}$ **31.** $\dfrac{2}{9}$ **32.** $\dfrac{7}{6}$

33. MINIATURE GOLF During a round of miniature golf, you made par on $\dfrac{11}{18}$ of the holes. Write this fraction as a decimal.

2 Chapter Test

Estimate the product or quotient.

1. $\dfrac{3}{8} \times \dfrac{7}{9}$

2. $2\dfrac{4}{5} \div \dfrac{3}{4}$

Multiply. Write the answer in simplest form.

3. $9 \times \dfrac{1}{4}$

4. $\dfrac{1}{10} \times \dfrac{5}{6}$

5. $1\dfrac{3}{7} \times 6\dfrac{7}{10}$

Divide. Write the answer in simplest form.

6. $\dfrac{1}{6} \div \dfrac{1}{3}$

7. $10 \div \dfrac{2}{5}$

8. $8\dfrac{3}{4} \div 2\dfrac{7}{8}$

Write the fraction as a decimal.

9. $\dfrac{4}{5}$

10. $\dfrac{7}{8}$

11. $\dfrac{9}{25}$

Write the decimal as a fraction or mixed number in simplest form.

12. 0.23

13. 0.125

14. 0.35

Evaluate the expression.

15. $\dfrac{1}{3} + 8 \div \dfrac{3}{4}$

16. $4\dfrac{1}{4} \div 1\dfrac{2}{3} + 1\dfrac{5}{8}$

17. **FRESHWATER FISH** Convert the lengths of both fish to decimals. Which fish is longer?

Florida gar

$2\dfrac{3}{8}$ feet

Chain pickerel

$2\dfrac{5}{14}$ feet

18. **BLOG** You spend $2\dfrac{1}{2}$ hours online. You spend $\dfrac{1}{5}$ of that time writing a blog. How long do you spend writing your blog?

19. **PIT STOP** You are in a car race. During a 10-second pit stop, the other cars gain $\dfrac{1}{4}$ mile on you. How much would the other cars gain on you during a one-minute pit stop?

20. **GARDEN** You border the edge of a garden with bricks that are $12\dfrac{1}{2}$ inches long. The length of the garden is $15\dfrac{1}{2}$ feet.

 a. How many bricks do you need to border the length of the garden? Explain your answer.

 b. A brick costs $3. You have $50. Do you have enough money to buy the bricks you need to border the length of the garden?

Test Item References

Chapter Test Questions	Section to Review	Common Core State Standards
1, 2	2.1	6.NS.1
3–5, 19	2.2	6.NS.1
3–5	2.3	6.NS.1
3–5, 18	2.4	6.NS.1
6–8	2.5	6.NS.1
6–8, 15, 16, 20	2.6	6.NS.1
12–14	2.7	6.NS.1
9–11, 17	2.8	6.NS.2

Test-Taking Strategies

Remind students to quickly look over the entire test before they start so that they can budget their time. When students hurry they may make unintentional mistakes, such as writing a mixed number as an improper fraction incorrectly or inverting the wrong fraction in a division problem. Remind students to **Stop** and **Think** before they write their answers.

Common Assessment Errors

- **Exercises 3–5** Students may try to get a common denominator when multiplying fractions.
- **Exercises 6–8** Students might find the reciprocal of the dividend instead of the divisor.
- **Exercise 10** Students may have difficulty writing an equivalent fraction. Remind students that they can use long division.
- **Exercises 12–14** Students may use the wrong power of 10 when rewriting the decimal. Remind students of the name of each decimal place.
- **Exercises 15 and 16** Students may forget to use the order of operations. Remind students that addition is evaluated after division.

Reteaching and Enrichment Strategies

If students need help...	If students got it...
Resources by Chapter • Practice A and Practice B • Puzzle Time Record and Practice Journal Practice Differentiating the Lesson Lesson Tutorials Practice from the Test Generator Skills Review Handbook	Resources by Chapter • Enrichment and Extension • School-to-Work • Financial Literacy • Life Connections • Stories in History Game Closet at *BigIdeasMath.com* Start Standardized Test Practice

Answers

1. $\frac{1}{2}$
2. 3
3. $\frac{9}{4}$ or $2\frac{1}{4}$
4. $\frac{1}{12}$
5. $\frac{67}{7}$ or $9\frac{4}{7}$
6. $\frac{1}{2}$
7. 25
8. $\frac{70}{23}$ or $3\frac{1}{23}$
9. 0.8
10. 0.875
11. 0.36
12. $\frac{23}{100}$
13. $\frac{1}{8}$
14. $\frac{7}{20}$
15. 11
16. $\frac{167}{40}$ or $4\frac{7}{40}$
17. Florida gar: 2.375 ft; Chain pickerel: $2.3\overline{571428}$ ft; Florida gar
18. $\frac{1}{2}$ hour
19. $1\frac{1}{2}$ miles
20. a. about 15 bricks; When you divide, you get $14\frac{22}{25}$ bricks, round up so you have enough bricks for the length of the garden.
 b. yes

Assessment Book

Test-Taking Strategies
Available at *BigIdeasMath.com*

After Answering Easy Questions, Relax
Answer Easy Questions First
Estimate the Answer
Read All Choices before Answering
Read Question before Answering
Solve Directly or Eliminate Choices
Solve Problem before Looking at Choices
Use Intelligent Guessing
Work Backwards

About this Strategy

When taking a multiple choice test, be sure to read each question carefully and thoroughly. After reading the question, estimate the answer before trying to solve it.

Answers

1. D
2. H
3. 90 points
4. C
5. G

Technology For the Teacher
Big Ideas Test Generator

T-101

Item Analysis

1. **A.** The student estimates both mixed numbers by rounding down and then adds the two estimates.

 B. The student estimates both mixed numbers by rounding down and then multiplies the two estimates.

 C. The student multiplies the two whole numbers and then adds this product to the sum of the two fractions.

 D. Correct answer

2. **F.** The student makes an order of operations error, adding $3c$ to $5b$ and then subtracting this sum from $8a$.

 G. The student substitutes 5 for c and 4 for b.

 H. Correct answer

 I. The student makes two-digit numbers out of the coefficients and the variables, such as 43 instead of 4×3.

3. **Gridded Response:** Correct answer: 90 points

 Common Error: The student multiplies the number of numerical response questions by the point value of the multiple choice questions and the number of multiple choice questions by the point value of the numerical response questions to get an answer of 120 points.

4. **A.** The student multiplies the numerator and the denominator and uses this product as the decimal part of the number.

 B. The student uses the numerator and denominator as digits in the decimal part of the number.

 C. Correct answer

 D. The student confuses $\frac{5}{8} = 0.625$ with $\frac{3}{8} = 0.375$.

5. **F.** The student finds the fraction of the garden that was dug up rather than the fraction that still contains tomato plants.

 G. Correct answer

 H. The student divides the number of tomato plant sections by the fraction destroyed, and then divides this quotient by the total number of sections.

 I. The student multiplies the number of tomato plant sections by the fraction destroyed and does not divide by the total number of sections.

6. **A.** The student confuses the intended division algorithm with an addition/subtraction algorithm.

 B. The student finds the reciprocal of the dividend rather than the reciprocal of the divisor.

 C. Correct answer

 D. The student renames the first number of the expression as if the student were going to follow a subtraction algorithm.

2 Standardized Test Practice

1. Which integer is closest to the value of the expression below? *(6.NS.1)*

 $$3\frac{1}{2} \times 2\frac{1}{2}$$

 A. 5 **C.** 7

 B. 6 **D.** 9

2. What is the value of the expression below when $a = 6$, $b = 5$, and $c = 4$? *(6.EE.2c)*

 $$8a - 3c + 5b$$

 F. 11 **H.** 61

 G. 53 **I.** 107

3. The formula $p = 5n + 2m$ can be used to determine the total number of points p on an exam with n numerical response questions and m multiple choice questions. What is the total number of points for an exam with 10 numerical response questions and 20 multiple choice questions? *(6.EE.2c)*

4. Which decimal is equivalent to $\frac{5}{8}$? *(6.NS.2)*

 A. 0.40 **C.** 0.625

 B. 0.58 **D.** 0.675

5. Your garden is divided into 12 sections of equal size.

 In 5 of these sections, you planted tomato plants. An armadillo dug up and destroyed $\frac{1}{3}$ of the tomato plants. What fraction of the garden now contains tomato plants? *(6.NS.1)*

 F. $\frac{5}{36}$ **H.** $\frac{5}{4}$

 G. $\frac{5}{18}$ **I.** $\frac{5}{3}$

6. The steps your friend took to divide two mixed numbers are shown below. What should your friend change in order to divide the two mixed numbers correctly? *(6.NS.1)*

$$3\frac{3}{5} \div 1\frac{1}{2} = \frac{18}{5} \times \frac{3}{2}$$
$$= \frac{27}{5}$$
$$= 5\frac{2}{5}$$

A. Find a common denominator of 5 and 2.

B. Multiply by the reciprocal of $\frac{18}{5}$.

C. Multiply by the reciprocal of $\frac{3}{2}$.

D. Rename $3\frac{3}{5}$ as $2\frac{8}{5}$.

7. On Saturday, you earned $35 mowing lawns. This was x dollars more than you earned on Thursday. Which expression represents the amount, in dollars, you earned mowing lawns on Thursday? *(6.EE.6)*

F. $35x$

G. $x + 35$

H. $x - 35$

I. $35 - x$

8. Tom and Erlene both purchased an MP3 player like the one shown. Tom received $\frac{1}{3}$ off the price of the MP3 player, and Erlene received $\frac{1}{5}$ off. How much more money, in dollars, did Erlene pay than Tom? *(6.NS.1)*

9. Which property is illustrated by the statement below? *(6.EE.3)*

$$4 + (6 + n) = (4 + 6) + n$$

A. Associative Property of Addition

B. Commutative Property of Addition

C. Associative Property of Multiplication

D. Identity Property of Addition

Item Analysis (continued)

7. **F.** The student misinterprets "more than" as meaning multiplication.

 G. The student misinterprets the problem, thinking that you received more on Thursday than Saturday.

 H. The student subtracts in the wrong order and writes an expression that represents $35 less than the difference of the amounts earned on Thursday and Saturday.

 I. Correct answer

8. **Gridded Response:** Correct answer: $20

 Common Error: The student thinks that the difference of $\frac{1}{3}$ and $\frac{1}{5}$ is $\frac{1}{2}$ and then takes $\frac{1}{2}$ of 150, which is 75.

9. **A.** Correct answer

 B. The student misidentifies the property being used as the Commutative Property of Addition.

 C. The student misidentifies the property being used as the Associative Property of Multiplication.

 D. The student misidentifies the property being used as the Identity Property of Addition.

10. **F.** The student subtracts $\frac{2}{3}$ from 1, and then multiplies $\frac{1}{3} \times \frac{3}{4} \times 16$.

 G. Correct answer

 H. The student approximates $\frac{2}{3}$ of 16 and ignores the $\frac{3}{4}$.

 I. The student approximates $\frac{3}{4} \times 16$ and ignores the $\frac{2}{3}$.

11. **4 points** The student demonstrates a thorough understanding of multiplying and dividing mixed numbers and whole numbers. The student calculates the length, width, and cost of the walkway correctly and gets answers of $112\frac{3}{4}$ inches for Part A and $68.06 for Part B. The student shows accurate, complete work for both parts and provides clear and complete explanations.

 3 points The student demonstrates an understanding of multiplying and dividing mixed numbers and whole numbers, but the student's work and explanations demonstrate an essential but less than thorough understanding.

 2 points The student demonstrates a partial understanding of multiplying and dividing mixed numbers and whole numbers. The student's work and explanations demonstrate a lack of essential understanding.

 1 point The student demonstrates very limited understanding of multiplying and dividing mixed numbers and whole numbers. The student's response is incomplete and exhibits many flaws.

 0 points The student provided no response, a completely incorrect or incomprehensible response, or a response that demonstrates insufficient understanding of multiplying and dividing mixed numbers and whole numbers.

Answers

6. C
7. I
8. $20
9. A

Answers

10. G
11. Part A $112\frac{3}{4}$ inches
 Part B $68.06
12. A
13. F

Answers for Extra Examples

1. **A.** In evaluating 5x, the student finds the product of 5 and 3 and then adds $\frac{1}{2}$.
 B. In evaluating 5x, the student finds the product of 5 and 3 and then adds $\frac{1}{2}$. The student then subtracts $\frac{2}{3} - \frac{1}{2}$ in the wrong order to avoid renaming.
 C. Correct answer
 D. The student subtracts $\frac{2}{3} - \frac{1}{2}$ in the wrong order to avoid renaming.

2. **F.** The student finds the reciprocal of $\frac{5}{4}$ and then finds $\frac{4}{5} \div 10$.
 G. Correct answer
 H. The student finds the reciprocal of $\frac{5}{4}$ and then finds $\frac{4}{5} \times 10$.
 I. The student subtracts $\frac{5}{4}$ from 10.

Item Analysis (continued)

12. **A.** Correct answer
 B. The student incorrectly rounds down and obtains an insufficient number of jars.
 C. The student finds $17\frac{3}{4} \times \frac{1}{2}$ and rounds this product up.
 D. The student finds $17\frac{3}{4} \times \frac{1}{2}$ and rounds this product down.

13. **F.** Correct answer
 G. The student does not recognize that $\frac{8}{5}$ is equivalent to 1.6.
 H. The student does not recognize that $1\frac{3}{5}$ is equivalent to 1.6.
 I. The student does not recognize that $1\frac{6}{10}$ is equivalent to 1.6.

Extra Examples for Standardized Test Practice

1. What is the value of the expression below when $x = 3\frac{1}{2}$ and $y = 2\frac{2}{3}$?

 $$5x - y$$

 A. $12\frac{5}{6}$ **C.** $14\frac{5}{6}$

 B. $13\frac{1}{6}$ **D.** $15\frac{1}{6}$

2. What number belongs in the box to make the equation true?

 $$\frac{5}{4} \div \Box = 10$$

 F. $\frac{2}{25}$ **H.** 8

 G. $\frac{1}{8}$ **I.** $8\frac{3}{4}$

10. There are 16 pet carriers at an animal show. Three-fourths of the pet carriers contain a dog and $\frac{2}{3}$ of the dogs are spotted. How many of the pet carriers contain spotted dogs? *(6.NS.1)*

F. 4

G. 8

H. 11

I. 12

11. A walkway was built using identical concrete blocks. *(6.NS.1)*

Part A How much longer, in inches, is the length of the walkway than the width of the walkway? Show your work and explain your reasoning.

Part B The cost of the walkway was $0.02 per square inch. What was the total cost, in dollars to the nearest cent, of the walkway? Show your work and explain your reasoning.

12. You made $17\frac{3}{4}$ pints of orange marmalade. You want to store the marmalade in $\frac{1}{2}$-pint jars. What is the minimum number of jars you will need? *(6.NS.1)*

A. 36

B. 35

C. 9

D. 8

13. Which number is *not* equivalent to 1.6? *(6.NS.1)*

F. $\frac{1}{6}$

G. $\frac{8}{5}$

H. $1\frac{3}{5}$

I. $1\frac{6}{10}$

3 Multiplying and Dividing Decimals

3.1 **Decimals and Estimation**

3.2 **Multiplying Decimals and Whole Numbers**

3.3 **Multiplying Decimals**

3.4 **Dividing Decimals by Whole Numbers**

3.5 **Dividing Decimals**

"Hey, I just noticed that my dog license is a palindrome."

"825 434 528 ... Forwards or backwards, it's the same."

"I entered a contest for dog biscuits."

"I was notified that the number of biscuits I won was in the 3 digit range."

Connections to Previous Learning

- Describe decimals as an extension of the base-ten number system.
- Write common fractions as decimals and decimals as fractions.
- Compare and order decimals.

- Represent addition and subtraction of decimals using models and place value.
- Add and subtract decimals efficiently, and verify reasonableness.
- Estimate sums and differences of decimals by rounding.

- Explain and justify procedures for multiplication and division of decimals.
- Multiply and divide decimals efficiently, including real-world problems.
- Estimate the results of computations with decimals, and verify reasonableness.

Math in History

The decimal point is still not used universally to denote the transition between whole numbers and parts of a whole. Many countries use a period to denote the transition. Many others use a comma.

★ Countries that use a period, as in 3.1416:
Australia, Brunei, Botswana, Canada, Hong Kong, India, Ireland, Israel, Japan, Kenya, Korea, Malaysia, Mexico, New Zealand, Nigeria, Pakistan, China, Philippines, Singapore, Taiwan, Tanzania, Thailand, Uganda, United Kingdom, United States, and Zimbabwe.

★ Countries that use a comma, as in 3,1416:
Albania, Argentina, Austria, Belgium, Bolivia, Bosnia, Brazil, Bulgaria, Costa Rica, Chile, Colombia, Cuba, Czech Republic, Denmark, Dominican Republic, Ecuador, El Salvador, Finland, France, Germany, Greece, Guatemala, Honduras, Hungary, Indonesia, Iceland, Italy, Norway, Nicaragua, Panama, Paraguay, Peru, Poland, Portugal, Romania, Russia, Spain, Sweden, Switzerland, Turkey, Ukraine, Uruguay, Venezuela, and Vietnam.

Pacing Guide for Chapter 3

Chapter Opener	1 Day
Section 1 Activity	1 Day
Lesson	1 Day
Section 2 Activity	1 Day
Lesson	1 Day
Section 3 Activity	1 Day
Lesson	1 Day
Study Help / Quiz	1 Day
Section 4 Activity	1 Day
Lesson	1 Day
Section 5 Activity	1 Day
Lesson	1 Day
Quiz / Chapter Review	1 Day
Chapter Test	1 Day
Standardized Test Practice	1 Day
Total Chapter 3	15 Days
Year-to-Date	52 Days

Check Your Resources for

- Record and Practice Journal
- Resources by Chapter
- Skills Review Handbook
- Assessment Book
- Worked-Out Solutions

The Dynamic Planning Tool
Editable Teacher's Resources at
BigIdeasMath.com

Common Core State Standards

5.NBT.5 Fluently multiply multi-digit whole numbers using the standard algorithm.
5.NBT.6 Find whole-number quotients of whole numbers with up to four-digit dividends and two-digit divisors,
5.NBT.7 Add, subtract, multiply, and divide decimals to hundredths,

Additional Topics for Review
- Place value
- Estimating products and quotients

Try It Yourself

1. 3825
2. 12,978
3. 17,371
4. 6036
5. 40
6. 107
7. 67
8. 113
9. 1.634
10. 0.44
11. 3.621
12. 2.115
13. 5.221
14. 1.208
15. 4.001
16. 0.153

Record and Practice Journal

1. 5265
2. 5049
3. 37
4. 158
5. 95
6. $8632
7. 1.537
8. 11.46
9. 3.207
10. 13.92
11. 1.79
12. 0.56
13. 7.117
14. 6.572
15. $34.50

Math Background Notes

Vocabulary Review
- Product
- Quotient
- Divisor
- Evaluate

Multiplying and Dividing Whole Numbers
- Students have been multiplying and dividing whole numbers.
- The review of multiplication and division skills with whole numbers will help students transition smoothly into the new material being presented in the chapter.
- You may want to review the multiplication and division algorithm with students.
- **Teaching Tip:** This section could be difficult for students who are still struggling to master their basic multiplication facts. You may consider allowing these students to use a multiplication chart to aid them through the problems, while encouraging them to master the facts.
- **Common Error:** Students will often make errors using the division algorithm in problems that contain a middle zero, such as Exercise 7. Remind students to record a zero in the quotient before carrying the 2 down.

Adding and Subtracting Decimals
- Students should know how to add and subtract decimals.
- Remind students that to add and subtract decimals, you must align the decimal points. This will be an important distinction as students learn the rules to multiply and divide decimals.
- **Common Error:** Students may forget to insert zeros into blank columns. Example 4 requires students to expand 1.8 to 1.80 before finding the difference. Encourage students to insert the zero and align the columns before finding the sum or difference.

Reteaching and Enrichment Strategies

If students need help. . .	If students got it. . .
Record and Practice Journal • Fair Game Review Skills Review Handbook Lesson Tutorials	Game Closet at *BigIdeasMath.com* Start the next section

What You Learned Before

"Uh Oh! I thought they wanted my phone number. I think I just ordered 9,045,551,212 dog biscuits."

● **Multiplying and Dividing Whole Numbers** (5.NBT.5, 5.NBT.6)

Example 1 Find 356 × 21.

$$\begin{array}{r} \overset{1\,1}{356} \\ \times\ 21 \\ \hline 356 \\ +\ 7120 \\ \hline 7476 \end{array}$$

Example 2 Find 765 ÷ 3.

$$\begin{array}{r} 255 \\ 3\overline{)765} \\ -\ 6\downarrow \\ \hline 16 \\ -\ 15\downarrow \\ \hline 15 \\ -\ 15 \\ \hline 0 \end{array}$$

Try It Yourself

Find the product or quotient.

1. 425 × 9
2. 721 × 18
3. 599 × 29
4. 503 × 12

5. 7)280
6. 4)428
7. 6)402
8. 4)452

● **Adding and Subtracting Decimals** (5.NBT.7)

Example 3 Find 1.57 + 2.003.

$$\begin{array}{r} 1.570 \\ +\ 2.003 \\ \hline 3.573 \end{array}$$

Insert a zero so the numbers have the same number of decimal places.

Example 4 Find 1.8 − 0.59.

$$\begin{array}{r} \overset{7\ 10}{1.8\cancel{0}} \\ -\ 0.59 \\ \hline 1.21 \end{array}$$

Try It Yourself

Find the sum or difference.

9. 1.63 + 0.004
10. 1.5 − 1.06
11. 3.62 + 0.001
12. 2.025 + 0.09

13. 5.23 − 0.009
14. 4.21 − 3.002
15. 1.701 + 2.3
16. 2.053 − 1.9

3.1 Decimals and Estimation

COMMON CORE STATE STANDARDS
6.NS.3

Essential Question How can you use estimation to check that your answer is reasonable?

The newspaper ad shows the weekly specials at a grocery store.

1 ACTIVITY: Estimating a Decimal Sum

Work with a partner. You are buying the items on your shopping list.

- Find the *exact* total cost.
- *Estimate* the total cost.
- Use your estimate to check that your total is reasonable.

a. Sample:

Shopping List
- Hot Dogs
- Bread
- Potatoes
- Cereal
- Apples
- Water

Hot Dogs	$2.52	→	3
Bread	$1.99	→	2
Potatoes	$2.99	→	3
Cereal	$3.15	→	3
Apples	$3.99	→	4
Water	+ $6.59	→	+ 7
	$21.23		22

Exact Estimate

Your estimate is close to the exact total, so your answer is reasonable.

b. Shopping List
- Crackers
- Potatoes
- Milk
- Ketchup
- Orange Juice

c. Shopping List
- Cereal
- Bread
- Eggs
- Syrup
- Apples

d. Shopping List
- Baby Food
- Soup
- Water
- Ketchup
- Hot Dogs

106 Chapter 3 Multiplying and Dividing Decimals

Laurie's Notes

Introduction

For the Teacher
- **Goal:** The goal of this activity is for students to use estimation to judge if a grocery bill seems reasonable.
- It is often helpful to relate decimal operations to money.

Motivate
- Play the *Target Challenge*, described below.
- Write the numbers 0–9 on slips of paper or on the sides of a 10-sided die. You may want to use a calculator with a random number generator.
- Have students make a quick game card, shown below.

$$\square . \square\square + \square . \square\square = \boxed{\text{TARGET}}$$

- Tell students the target number (e.g., $8.00). Explain that the goal is to get as close to the target as possible. Decide in advance if the goal is to be the closest to the target, or to be the closest without going over.
- Determine the first number using the slips of paper, 10-sided die, or number generator. Have students record the number in any open box. Continue generating numbers until all six boxes are full. The choice of number placement uses some number sense.
- Have students add the numbers and determine who is closest to the target.
- Play a few quick games, changing the target number.

Activity Notes

Activity 1
- Discuss why you would want to estimate your total when you do any shopping. Students should realize it's helpful to know if you have enough money and whether the purchases are added up correctly at the register.
- **?** "What do you need to remember to do with the decimal points when you add decimals?" line up the decimal points
- **?** "Why is it necessary to line up decimal points when adding decimals?" to ensure that the correct place values are added together
- Work through the sample problem together as a class.
- **?** "How do you decide what whole number to round the price to?" Listen for standard rules of rounding. Some students might correctly say that you should round up if you want to make sure you have enough money.
- **?** "Would it be okay to round $2.52 to $2.50 and $6.59 to $6.50? Why?" Yes, the estimate would be closer, but less than the actual cost.
- After Activity 1, have students check their work.
- **?** "If you were purchasing 6 cans of soup, how would you find the exact total?" Listen for "Multiply $1.29 by 6."

Common Core State Standards
6.NS.3 Fluently add, subtract, multiply, and divide multi-digit decimals using the standard algorithm for each operation.

Previous Learning
Students should know how to add and subtract decimals, round a decimal to a whole number, make reasonable estimates, and verify the reasonableness of results.

Activity Materials
Introduction
• 10 pieces of paper, a 10-sided die, or a random number generator

Start Thinking! and Warm Up

Activity 3.1 Start Thinking!
For use before Activity 3.1

Activity 3.1 Warm Up
For use before Activity 3.1

Round the decimal to the nearest whole number.

1. 16.3 2. 98.9 3. 7.878
4. 1.02 5. 6.592 6. 0.24

3.1 Record and Practice Journal

T-106

Differentiated Instruction

Visual

Help students that have difficulty with rounding by breaking the process down into steps.

Round 3.548 to the nearest tenth.

3.548 **Step 1** Underscore the place value to which you are rounding.

3.5④8 **Step 2** Circle the number to the right.

3.5 **Step 3** If the value of the circled number is 0 to 4, keep the underscored value as is. If the value of the circled number is 5 to 9, increase the value of the underscored number by 1.

3.1 Record and Practice Journal

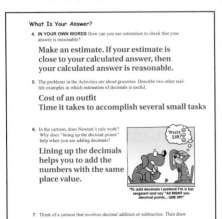

Laurie's Notes

Activity 2

- Discuss the problem. While $8 is a good mathematical estimate, in this context, you would have too little money.
- "Could you use estimation and multiplication to approximate the cost of 6 cans of soup? How?" Yes, $1.29 × 6 ≈ $6. If students say 6 × $2, ask how they might get closer to the exact answer. Listen for 6 × $1.25 or perhaps 6 × $1.50.
- "Would you prefer to have an overestimate or an underestimate in this context? Why?" an overestimate; You want to have enough money.

Activity 3

- **Writing:** This activity presents another opportunity for students to write. Encourage students to be creative and add interesting details to the story.

What Is Your Answer?

- Question 4 relates to the *Essential Question* at the start of the lesson. You may want to discuss it with the class.
- Have students share their real-life examples for Question 5.

Closure

- If time permits, play the *Target Challenge* game again, but this time allow students to hear all 6 digits before placing any of them in the boxes.
- Are you able to get closer to the target? Why? Yes. You know which numbers you can use ahead of time, so you can make better choices for the place value of each number and get closer to the target.

The Dynamic Planning Tool
Editable Teacher's Resources at BigIdeasMath.com

2 ACTIVITY: Estimating Decimal Products

You get home and realize you forgot to buy four boxes of crackers. Your friend says the crackers cost about $2.00 a box, so you should take $8.00.

a. Fill in the blanks to show how your friend got the estimate.

$$2.09 \longrightarrow \boxed{}$$
$$\underline{\times \quad 4} \longrightarrow \underline{\times \boxed{}}$$
$$\boxed{}$$

b. Find the actual total.

$$\begin{aligned}&\$2.09\\&\$2.09\\&\$2.09\\+\;&\$2.09\\\hline\end{aligned}$$

c. What is wrong with your friend's estimate?

3 ACTIVITY: Writing a Story

Work with a partner. Write a story about the shopping list. As part of your story, imagine that the grocery clerk told you that the total was $137.56 and you used estimation to decide that the total was way too much.

Shopping List
2 Gallons Milk
3 Cartons Eggs
3 Boxes Cereal
Syrup
6 Cans Soup
2 Boxes Crackers

What Is Your Answer?

4. **IN YOUR OWN WORDS** How can you use estimation to check that your answer is reasonable?

5. The problems on this page are about groceries. Describe two other real-life examples in which estimation of decimals is useful.

6. In the cartoon, does Newton's rule work? Why does "lining up the decimal points" help when you are adding decimals?

7. Think of a cartoon that involves decimal addition or subtraction. Then draw the cartoon.

"To add decimals I pretend I'm a top sergeant and say "All RIGHT you decimal points... LINE UP!"

Practice

Use what you learned about estimation to complete Exercises 8–15 on page 110.

3.1 Lesson

To estimate decimal products and quotients, you can round each factor to the nearest whole number.

EXAMPLE 1 Estimating Decimal Products and Quotients

a. **Estimate 12.3 × 4.8 by rounding to the nearest whole number.**

$$12.3 \times 4.8 \approx 12 \times 5 \qquad \text{12.3 rounds down to 12. 4.8 rounds up to 5.}$$
$$= 60 \qquad \text{Multiply.}$$

∴ So, 12.3 × 4.8 is about 60.

b. **Estimate 62.9 ÷ 7.48 by rounding to the nearest whole number.**

$$62.9 \div 7.48 \approx 63 \div 7 \qquad \text{62.9 rounds up to 63. 7.48 rounds down to 7.}$$
$$= 9 \qquad \text{Divide.}$$

∴ So, 62.9 ÷ 7.48 is about 9.

Remember
When rounding, identify the place value being rounded. Then look at the digit to the right. If the digit is 5 or greater, round *up*. If it is less than 5, round *down*.

 On Your Own

Now You're Ready
Exercises 8–23

Estimate by rounding each factor to the nearest whole number.

1. 11.2 × 5.7 **2.** 15.81 × 3.1 **3.** 26.5 ÷ 2.5 **4.** 75.22 ÷ 24.61

When rounding, you do not always get numbers that are easy to use. In this case, use *compatible numbers*.

EXAMPLE 2 Using Compatible Numbers

a. **Use compatible numbers to estimate 48.97 × 3.91.**

$$48.97 \times 3.91 \approx 50 \times 4 \qquad \text{50 and 4 are compatible numbers.}$$
$$= 200 \qquad \text{Multiply.}$$

∴ So, 48.97 × 3.91 is about 200.

Study Tip
Compatible numbers are numbers that are easy to compute mentally.

b. **Use compatible numbers to estimate 62.45 ÷ 11.72.**

$$62.45 \div 11.72 \approx 60 \div 12 \qquad \text{60 and 12 are compatible numbers.}$$
$$= 5 \qquad \text{Divide.}$$

∴ So, 62.45 ÷ 11.72 is about 5.

108 Chapter 3 Multiplying and Dividing Decimals

Laurie's Notes

Introduction

Connect
- **Yesterday:** Students used estimation to judge the reasonableness of grocery purchases.
- **Today:** Students will use rounding strategies to estimate products and quotients with decimals.

Motivate
- Share some world records or facts with the students. Ask the students to round the decimals to a specific place value. Examples:
 - largest carved sapphire = 35.49 lb
 - smallest living dog (length) = 15.2 cm
 - longest skateboard = 9.17 m
- Ask these questions to review and assess rounding of decimals. You may also want to ask students about situations when exact measurements are needed and situations when an estimate would be appropriate.

Lesson Notes

Example 1
- **FYI:** Students have not learned the formal rules for multiplying and dividing decimals. Tell students that today they will simply round decimals to a whole number in order to get an estimated product or quotient.
- "How do you round a decimal? For example, how would you round 12.46 to the nearest tenth?" Listen for rules described in the textbook and also for students to recognize that 12.46 is closer to 12.5 than to 12.4.
- **Visual:** If students only quote rounding rules without understanding, draw a number line like the one shown below and ask "Between what two tenths is 12.46?" Then have a student graph 12.46.

12.? 12.?

- "The number 12.46 is between what two whole numbers?" 12 and 13
- Work through both parts of the example.

On Your Own
- Ask volunteers to share their answers with the class.

Example 2
- "Which is easier, 5 × 39 or 5 × 40? Why?" Students will likely say 5 × 40 because multiplying by multiples of 10 is easier for them.
- "Which is easier, 48 ÷ 6 or 48 ÷ 7? Why?" Students will likely say 48 ÷ 6 because it divides evenly.
- Describe compatible numbers. Then work through the two examples.
- Be sure that students recognize that compatible numbers are easier to work with, even though the estimate may not be as close as it could be.

Goal Today's lesson is estimating products and quotients of decimals by using rounding strategies.

Start Thinking! and Warm Up

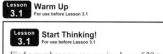

Lesson 3.1 Warm Up For use before Lesson 3.1

Lesson 3.1 Start Thinking! For use before Lesson 3.1

Find a number you can use in place of 29 to estimate the quotient in each division problem below. Tell what number you used.

29 ÷ 3 29 ÷ 4 29 ÷ 8 29 ÷ 9

Then write a division problem in which the number used in place for 29 is 33.

Extra Example 1

Estimate by rounding to the nearest whole number.

a. 15.4 × 6.8 105
b. 74.8 ÷ 15.2 5

On Your Own

1. 66 2. 48
3. 9 4. 3

Extra Example 2

Use compatible numbers to estimate the product or quotient.

a. 41.58 × 5.15 *Sample answer:* 200
b. 78.05 ÷ 4.25 *Sample answer:* 20

T-108

On Your Own

5–8. Sample answers are given.

5. 400 6. 150
7. 6 8. 7

Extra Example 3

The money collected from tickets sales is $3612.50. The cost of a ticket is $4.25. Estimate the number of tickets sold. about 900 tickets

Extra Example 4

A river is 12 inches below its flood stage. A storm causes the river to rise 1.75 inches each hour. Will the river overflow its banks in 5 hours? Explain. No. In five hours, the river will rise about $5 \times 2 = 10$ inches. This is an overestimate and is less than 12, so the river will not overflow its banks.

On Your Own

9–10. Sample answers are given.

9. 600 tickets

10. Yes; An underestimate of the beach erosion in 10 years is 110 feet. Because 110 > 100, the water will reach the home.

English Language Learners

Vocabulary
English learners may not be fluent with the base-ten system. A place value chart can be used to visualize and organize information.

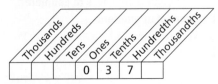

0.37
thirty-seven hundredths
$\dfrac{37}{100}$

T-109

Laurie's Notes

On Your Own

- **Neighbor Check:** Have students work independently and then share their answers with a neighbor. Have them share what they used for compatible numbers as well as the product.

Example 3

- Have a student read the problem. He or she will also need to read the information contained in the graphic.
- "How can you estimate the number of tickets sold?" Divide the amount collected by the cost of a ticket.
- "What compatible numbers can you use?" $2800 \div 4$
- "Would it be okay to use $3000 \div 5$?" yes
- "How does the second estimate compare to the first?" It is not as accurate, but it is still a reasonable estimate.

Example 4

- Ask students if they are familiar with beach erosion. Some facts about erosion are provided below.
 - A study of 450 miles of sandy shoreline in California showed that 66% are experiencing erosion.
 - About 299 of the 825 miles of sandy beach in Florida are experiencing "critical erosion," which is a level of erosion that threatens development, recreation, and the environment.
- "Why do you want an overestimate in this problem?" If the answer to the overestimate is less than 75 feet, then the water won't reach the shop in 9 years. If you used an underestimate, you would not be sure.
- "If the problem 6.374×9 had no context, how would you estimate the product?" $6 \times 9 = 54$

On Your Own

- **Think-Pair-Share:** Students should read each question independently and then work with a partner to answer the questions. When they have answered the questions, the pair should compare their answers with another group and discuss any discrepancies.

Closure

- What are compatible numbers and how are they used to find a product or quotient? Compatible numbers are numbers that are easy to simplify using mental math. You can use compatible numbers when you are estimating products or quotients.

Technology For the Teacher

The Dynamic Planning Tool
Editable Teacher's Resources at BigIdeasMath.com

On Your Own

Now You're Ready
Exercises 24–31

Use compatible numbers to estimate the product or quotient.

5. 38.79×9.8
6. 26.025×6.12
7. $17.34 \div 2.91$
8. $136.9 \div 22.83$

EXAMPLE 3 Real-Life Application

The money collected from ticket sales for the talent show is $2692.50. Estimate the number of tickets sold.

The number of tickets sold is $2692.50 \div 3.75$. You can estimate this quotient using compatible numbers.

$$3.75\overline{)2692.50} \longrightarrow 4\overline{)2800}^{700}$$

Round 3.75 up to 4.

A number that is close to 2692.50 and easily divisible by 4 is 2800.

∴ So, about 700 tickets were sold.

EXAMPLE 4 Real-Life Application

A surf shop sits 75 feet from a shoreline. Each year 6.375 feet of the beach is lost due to erosion. Your friend says the water will reach the shop within 9 years. Is your friend correct? Explain.

To find how much beach will be lost in 9 years, estimate 6.375×9. If an overestimate is less than 75, you can be sure the water will not reach the shop.

$6.375 \times 9 \approx 7 \times 9$ Round 6.375 *up* to 7.

$ = 63$ Multiply. This is an overestimate, so the actual amount is less than 63.

∴ No. Because 63 is less than 75, the water will not reach the shop.

On Your Own

9. Each ticket for a school play costs $3.25. The money collected from ticket sales is $1774.50. Estimate the number of tickets sold.

10. A home sits 100 feet from the shoreline. Each year, 11.25 feet of the beach is lost due to erosion. Will the water reach the home in the next decade? Explain.

3.1 Exercises

Vocabulary and Concept Check

1. **VOCABULARY** Describe a real-life example of decimal estimation.
2. **OPEN-ENDED** Fill in the blanks for three prices: $ ___.95 + $ ___.95 + $ ___.95 ≈ $10.00.
3. **NUMBER SENSE** Which three quotients have an estimate of 4?

 $3.9 \div 1.1$ $3\frac{1}{2} \div 1\frac{1}{2}$ $7\frac{3}{5} \div 1\frac{7}{8}$ $11.3 \div 2.9$

Tell whether you would use *rounding* or *compatible numbers* to estimate the product or quotient. Explain your reasoning.

4. $23.6 \div 4.2$
5. 33.72×18.11
6. $46.18 \div 7.57$
7. 8.5×7.83

Practice and Problem Solving

Estimate by rounding each factor to the nearest whole number.

① 8. 3.21×8 9. 7.06×3 10. 2.8×7 11. 4.57×5
12. 7.9×12 13. 5.42×6 14. 6.11×10 15. 9.7×9
16. 5.6×7.1 17. 2.5×9.4 18. 19.8×5.7 19. $9.1 \div 1.1$
20. 10.83×7.91 21. 7.74×9.18 22. $35.71 \div 6.15$ 23. $40.31 \div 19.55$

Use compatible numbers to estimate the product or quotient.

② 24. 57.3×8.62 25. 18.6×32.7 26. $42.17 \div 3.84$ 27. $179.6 \div 32.1$
28. 4.35×33.41 29. 8.73×46.9 30. $73.55 \div 2.23$ 31. $87.16 \div 17.43$

32. **ANOTHER WAY** Show another way to estimate the quotient. Explain your steps.

 $42.3 \div 6.459 \approx 42 \div 7$
 $= 6$

33. **SAND** You use 38 buckets of sand to bury your friend at the beach. Your bucket holds 8.2 pounds of sand. Estimate how many pounds of sand you use. Is your estimate *too little* or *too much*?

34. **GEOMETRY** You want to estimate the amount of bedding you need for your hamster cage.

 a. Estimate the area A of the hamster cage floor. Is your estimate *too little* or *too much*?

 b. Use the expression $2A$ to estimate how much bedding (in cubic inches) is needed for a 2-inch layer.

28.3 in.
17.3 in.

110 Chapter 3 Multiplying and Dividing Decimals

Assignment Guide and Homework Check

Level	Day 1 Activity Assignment	Day 2 Lesson Assignment	Homework Check
Basic	8–15, 44–52	1–7, 17–23 odd, 24–27, 33–35	21, 24, 26, 34
Average	8–15, 44–52	1–7, 20–27, 32, 33, 35, 38, 41	21, 24, 26, 32
Advanced	8–15, 44–52	1–7, 28–32, 36–40, 42, 43	28, 30, 36, 39

Common Errors

- **Exercises 8–23** Students may not round correctly. Have students circle the tenths places. If the number circled is 4 or less, they leave the whole number as it is. If the number circled is 5 or more, they need to increase the whole number by one.
- **Exercises 24–31** Students may just round the numbers up or down instead of finding compatible numbers. While this strategy is not wrong, the problem can be easier to evaluate if the numbers are compatible, such as multiples of 5. Have students think of math-friendly numbers.
- **Exercise 39** Students may try to evaluate the multiplication part first. Remind them of the order of operations and that multiplication and division are evaluated left to right.

Technology For the Teacher
Answer Presentation Tool
*Quiz*Show

3.1 Record and Practice Journal

Estimate by rounding each factor to the nearest whole number.
1. 6×5.8 — 36
2. 9.07×3.33 — 27
3. $21.11 \div 6.73$ — 3

Use compatible numbers to estimate the product or quotient. *Sample answers are given.*
4. 19.57×4.86 — 100
5. $47.84 \div 11.99$ — 4
6. $65.58 \div 10.71$ — 6

Estimate the value of the expression.
7. $4.32 \times 9.44 \div 2.16$ — 18
8. $44.98 \div 4.86 \times 5.23$ — 45
9. $71.94 \div (1.85 \times 4.33)$ — 9

10. You and five friends go to a haunted house. Admission is $16.95 per person. Estimate how much tickets will cost for all of you. Is your estimate *too little* or *too much*? Explain.
$102; *Sample answer:* My estimate is too much because I rounded 16.95 up.

11. You need four pieces of string that are 2.25 feet long to make a friendship bracelet. You have 20 feet of string.
 a. Estimate how much string you need to make one bracelet.
 8 feet
 b. Estimate how many bracelets you can make from the spool of string. Is your estimate *too little* or *too much*? Explain.
 2 bracelets; *Sample answer:* My estimate is too little because I rounded down.

Vocabulary and Concept Check

1. *Sample answer:* You use an estimate to approximate the price of an item after a 25% discount.
2. *Sample answer:* $1.95, $3.95, $3.95
3. $3.9 \div 1.1$, $7\frac{3}{5} \div 1\frac{7}{8}$, $11.3 \div 2.9$
4. Rounding, because it gives numbers that are easy to use.
5. Compatible numbers, because rounding does not give numbers that are easy to use.
6. Compatible numbers, because rounding does not give numbers that are easy to use.
7. Rounding, because it gives numbers that are easy to use.

Practice and Problem Solving

8. 24		9. 21	
10. 21		11. 25	
12. 96		13. 30	
14. 60		15. 90	
16. 42		17. 27	
18. 120		19. 9	
20. 88		21. 72	
22. 6		23. 2	

24–43. Sample answers are given.

24. 600		25. 600	
26. 10		27. 6	
28. 150		29. 450	
30. 35		31. 6	

T-110

Practice and Problem Solving

32. $42 \div 6 = 7$; Round to the nearest whole number.

33. 320 lb; too much

34. **a.** 600 in.2; too much
 b. 1200 in.3

35. 35 m^2 36. 36 in.2

37. 96 ft^2 38. 18

39. 36 40. 6

41. See *Taking Math Deeper*.

42. **a.** 30 gallons
 b. *Sample answer:* $3.65
 c. $120

43. *Sample answer:* Your cousins, from out of town, leave at 9:00 A.M. to visit you. To be sure you are home when they arrive, you underestimate their travel time.

Fair Game Review

44. 294 45. 192
46. 308 47. 195
48. 9.78 49. 8.92
50. 11.625 51. 10.004
52. D

Mini-Assessment

Estimate by rounding each factor to the nearest whole number.

1. 2.4×3.97 8
2. 8.75×6.36 54
3. $11.8 \div 4.19$ 3
4. $19.77 \div 4.80$ 4
5. Your friend's car gets 25.3 miles per gallon of gasoline. You and your friend are driving to a beach which is 375 miles away. Estimate how many gallons you will need to drive to the beach. 15 gallons

Taking Math Deeper

Exercise 41

Notice that the "90 pounds" is not specifically used in the problem. However, students might realize that a person who weighs more burns more calories.

This problem demonstrates that it can be useful to make a table that has much more information than is needed to answer the questions. Once the table is made, it is easier to answer almost any question about the data.

 Make a reference table to answer any question.

Activity	Actual 5 min	Estimated Calories Burned			
		5 min	30 min	45 min	60 min
Basketball	27.2	25	6 • 25	9 • 25	12 • 25
Inline Skating	42.6	40	6 • 40	9 • 40	12 • 40
Swimming	20.4	20	6 • 20	9 • 20	12 • 20
Tennis	23.8	25	6 • 25	9 • 25	12 • 25
Walking	21.4	20	6 • 20	9 • 20	12 • 20

 Use the reference table.
a. Basketball for 30 min: $6 \cdot 25 = 150$ calories
b. Tennis for 45 min: $9 \cdot 25 = 225$ calories

 Compare inline skating and walking.
c. Inline skating for 45 min:
$9 \cdot 40 = 360$ calories
Walking for 60 min:
$12 \cdot 20 = 240$ calories
$360 - 240 = 120$ calories

Project

Make a poster that shows the number of calories a person burns while doing different exercises. Be sure to make the poster so that it encourages participation!

Reteaching and Enrichment Strategies

If students need help...	If students got it...
Resources by Chapter • Practice A and Practice B • Puzzle Time Record and Practice Journal Practice Differentiating the Lesson Lesson Tutorials Skills Review Handbook	Resources by Chapter • Enrichment and Extension Start the next section

Estimate the area of the figure by rounding to the nearest whole number.

35. **36.** **37.**

Estimate the value of the expression.

38. 6.16 × 8.8 ÷ 2.91

39. 24.45 ÷ 3.62 × 5.9

40. 35.7 ÷ (3.33 × 2.37)

41. FITNESS The table shows the number of calories burned in 5 minutes by a 90-pound person during various activities.

Activity	Calories Burned
Basketball	27.2
Inline Skating	42.6
Swimming	20.4
Tennis	23.8
Walking	21.4

 a. Estimate the number of calories burned by playing basketball for 30 minutes.

 b. Estimate the number of calories burned by playing tennis for 45 minutes.

 c. Estimate how many more calories are burned by inline skating for 45 minutes than by walking for 60 minutes.

42. TRAVEL Your family is driving to visit friends who live about 420 miles away. Your car gets 27.3 miles per gallon of gasoline.

 a. Estimate how many gallons you will use driving there *and* back.

 b. RESEARCH Find the cost of gasoline in your neighborhood.

 c. Estimate the total cost of gasoline for the trip.

43. Describe a real-life situation where you would want to underestimate.

Fair Game Review What you learned in previous grades & lessons

Multiply. *(Skills Review Handbook)*

44. 21 × 14 **45.** 16 × 12 **46.** 28 × 11 **47.** 15 × 13

Add. *(Skills Review Handbook)*

48. 2.4 + 7.38 **49.** 5.62 + 3.3 **50.** 4.71 + 6.915 **51.** 8.324 + 1.68

52. MULTIPLE CHOICE A rectangular postcard measures $5\frac{3}{4}$ inches by $4\frac{1}{2}$ inches. What is the area of the postcard? *(Section 2.4)*

 Ⓐ $14\frac{1}{4}$ in.² **Ⓑ** $14\frac{7}{8}$ in.² **Ⓒ** $20\frac{3}{8}$ in.² **Ⓓ** $25\frac{7}{8}$ in.²

3.2 Multiplying Decimals and Whole Numbers

COMMON CORE STATE STANDARDS
6.NS.3

Essential Question What happens to the decimal point when you multiply a whole number by a decimal?

1 ACTIVITY: Multiplying by Powers of 10

Work with a partner. Copy and complete the table. Then describe how to multiply by a power of 10.

Exponent	Power of 10	Product	Evaluate
1	$10^1 = 10$	10×0.825	8.25
2	$10^2 = 100$	100×0.825	82.5
3	$10^3 = 1000$	1000×0.825	825
4	$10^4 = 10{,}000$		
5	$10^5 = 100{,}000$		

2 ACTIVITY: Multiplying a Decimal by a Whole Number

Work with a partner. Your school is selling tickets to the school carnival.

a. Copy and complete the table.

Number of Tickets	Price per Ticket	Find the Total	Total Cost
3	0.25	$0.25 + 0.25 + 0.25$	$0.75
4	0.25		
5	0.25		
15	0.25		
100	0.25		

b. The example in the table shows how to find the total cost using addition. This works for small numbers of tickets. How did you find the total cost for the last two rows?

Laurie's Notes

Introduction

For the Teacher
- **Goal:** The goal of this activity is to develop a sense of what happens to the decimal point when a decimal is multiplied by a whole number.
- It is often helpful to relate decimals to money.

Motivate
- ❓ Show the class the coins. Ask students to tell you the value of *x* coins, where *x* is a power of 10. Record the value of the coins.
 "What is the value of 10 quarters?" Write $10 \times \$0.25 = \2.50 on the board.
 "What is the value of 20 quarters?" Write $20 \times \$0.25 = \5.00 on the board.
 "What is the value of 100 quarters?" Write $100 \times \$0.25 = \25.00 on the board.
- Tell students that today's lesson is about multiplying a decimal by a whole number, similar to what they just did with the coins.

Activity Notes

Activity 1
- The table in this activity has been started. Students are expected to use the pattern observed to help them finish completing the table.
- Encourage students to use the term "decimal point" instead of just "point."
- It's okay for students to not have the language to describe the pattern yet.

Activity 2
- This activity connects multiplication to repeated addition.
- ❓ "How did you decide the cost for 15 tickets? 100 tickets?" Listen for a variety of strategies, such as multiplying 15 by 25 and placing the decimal point or adding the total for 5 tickets three times.
- ❓ If no one uses the strategy of multiplying 15 by 25 and using reasonableness to position the decimal point, ask the following questions:
 - "Could I find the cost of 15 tickets by multiplying 15 times 25? How would I know where to put the decimal point?" yes; You know the total is under $15, so place the decimal point so that the number is under 15.
 - "If I wanted to find the cost of 54 tickets, could I multiply 54 times 25? How would I know where to put the decimal point?" yes; *Sample Answer:* Put the decimal point two places from the end so it looks like money.
- **Common Error:** Students may write $0.75 or 75¢, but make sure they don't write both the decimal and the cent symbols: 75 cents ≠ 0.75¢.
- ❓ **Connection:** If no one mentions fractions during this activity, make the connection before moving on.
 - "What fraction is equivalent to 0.25?" $\frac{1}{4}$
 - "Explain another way to find the cost of 15 tickets."
 $15 \times 0.25 = 15 \times \frac{1}{4} = \frac{15}{4} = 3\frac{3}{4} = 3.75$

Common Core State Standards
6.NS.3 Fluently add, subtract, multiply, and divide multi-digit decimals using the standard algorithm for each operation.

Previous Learning
Students should know powers of 10, multiples of 10, and factors.

Activity Materials
Introduction
• sample of U.S. or Mexican coins

Start Thinking! and Warm Up

3.2 Record and Practice Journal

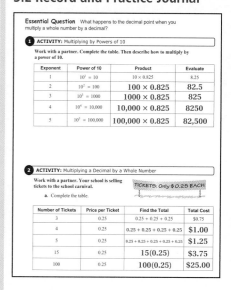

T-112

Differentiated Instruction

Kinesthetic

Have students use base ten blocks to model decimal and whole number multiplication. Assign the products below and ask students to describe any patterns they see.

2 × 3	2 × 0.3	2 × 0.03
2 × 4	2 × 0.4	2 × 0.04
2 × 5	2 × 0.5	2 × 0.05
2 × 6	2 × 0.6	2 × 0.06

3.2 Record and Practice Journal

Check students' work.

What Is Your Answer?

4. **IN YOUR OWN WORDS** What happens to the decimal point when you multiply a whole number by a decimal?

Place the decimal point in the product so it has the same number of decimal places as the decimal factor.

Laurie's Notes

Activity 3

- If there is an odd number of students in the class, a team of three will also work.
- The context of money will help students locate the decimal point correctly.
- Have students record the quantity and price of the items they "purchase" with each turn. Decide ahead of time if you want to allow students to buy an item again on another turn. (For example, do you want to allow students to buy three binders on one turn and three more binders on another turn?)
- Have students keep a running total of all their purchases.
- You may want to tell students to compare their totals after each person purchases three items.
- You can increase the difficulty of the game by asking students to compare their totals after two purchases, or even after six or ten purchases.

What Is Your Answer?

- Most students should have a general sense that the decimal point in the product is related to the number of decimal places in the decimal factor.
- If students are struggling with language, guide them to find the products and the patterns of questions like the ones below.

$$2 \times 80 = 160$$
$$2 \times 8 = 16$$
$$2 \times 0.8 = 1.6$$
$$2 \times 0.08 = 0.16$$

Closure

- Describe how the problems 12 × 12 and 12 × 1.2 are alike and how they are different. *They are alike because you multiply 12 by 12 in both. They are different because 12 × 12 is a whole number greater than 100 (actually 144) and 12 × 1.2 is a decimal close to 12 (actually 14.4).*

The Dynamic Planning Tool
Editable Teacher's Resources at *BigIdeasMath.com*

3 ACTIVITY: Back to School Shopping

Game Rules

- Take turns with your partner.
- When it is your turn, choose one item from the list. Decide whether you want to buy 1, 2, or 3 of your item.
- The person who comes closest to $30 without going over, wins.

Back to School List

Item	Price	Item	Price
Ink Pens	$1.41	Paper Clips	$3.49
Pencils	$0.33	Markers	$3.29
Erasers	$0.24	Colored Pencils	$0.89
Poster Board	$0.64	Tissues	$2.29
Rulers	$1.99	Rubber Bands	$3.49
Protractors	$2.29	Notebook Paper	$3.98
Pocket Folders	$0.33	Graph Paper	$3.52
Sticky Notes	$0.99	Stapler	$12.63
Spirals	$1.15	Staples	$2.99
3-ring Binders	$4.26	Stickers	$0.99
Index Cards	$1.99	Calculator	$10.98
Scissors	$3.99	Book Covers	$1.15

What Is Your Answer?

4. **IN YOUR OWN WORDS** What happens to the decimal point when you multiply a whole number by a decimal?

Use what you learned about multiplying decimals and whole numbers to complete Exercises 8–15 on page 116.

3.2 Lesson

Key Idea

Multiplying Decimals by Whole Numbers

Words Multiply as you would with whole numbers. Then count the number of decimal places in the decimal factor. The product has the same number of decimal places.

Numbers

13.91 → 2 decimal places
× 7
97.37

6.218 → 3 decimal places
× 4
24.872

EXAMPLE 1 Using Estimation to Find a Product

Find 15.8×4. Estimate $16 \times 4 = 64$

$$\begin{array}{r} \overset{2\;3}{15.8} \\ \times\;\;4 \\ \hline 63.2 \end{array}$$

Multiply as you would with whole numbers.

The estimate is 64. So, place the decimal point after the 3.

EXAMPLE 2 Multiplying Decimals and Whole Numbers

a. Find 8.7×5.

Estimate $9 \times 5 = 45$

$$\begin{array}{r} \overset{3}{8.7} \\ \times\;\;5 \\ \hline 43.5 \end{array}$$

8.7 ← One decimal place

43.5 ← Count one decimal place from right to left.

∴ So, $8.7 \times 5 = 43.5$.

Reasonable? $43.5 \approx 45$ ✓

b. Find 6×0.91.

Estimate $6 \times 1 = 6$

$$\begin{array}{r} \overset{5}{0.91} \\ \times\;\;\;6 \\ \hline 5.46 \end{array}$$

0.91 ← Two decimal places

5.46 ← Count two decimal places from right to left.

∴ So, $6 \times 0.91 = 5.46$.

Reasonable? $5.46 \approx 6$ ✓

On Your Own

Exercises 4–23

Find the product. Use an estimate to place the decimal point.

1. 2.1×3
2. 24.3×7
3. 5.9×11
4. 19.6×2

Multiply. Use estimation to check your answer.

5. 12.3×8
6. 5×14.51
7. 0.7×4
8. 0.88×9

Laurie's Notes

Introduction

Connect
- **Yesterday:** Students developed an understanding of how the location of the decimal point in a product is related to the number of decimal places in a factor.
- **Today:** Students will use a formal rule to find the product of a whole number and a decimal.

Motivate
? The first question below relates to the student's understanding of yesterday's activity. The second question assesses how well students can apply this understanding.
- "A sheet of paper is about 0.008 millimeter thick. How thick would a stack of 10 sheets be?" 0.08 mm
- "How many sheets of paper would you need to have a stack 8 millimeters thick?" 1000

Lesson Notes

Example 1
- In the first two examples, students are reminded that they should still use estimation to approximate an answer. Estimation can now be used to check their work and make sure their answer is reasonable.
? "Would the answer to 15.8 × 4 be greater than 100? How do you know?" no; 16 × 4 = 64, and this is an overestimate.
? "Would the answer to 15.8 × 4 be less than 10? How do you know?" no; 15 × 4 = 60, and this is an underestimate.
- Work through the example. The first step is to multiply as you would whole numbers. Then use the estimate to help determine where to place the decimal point.

Example 2
- Encourage students to estimate the product before beginning.
- Work through the two examples. In order to place the decimal point correctly, students should count from the right in the product.
- **Common Error:** A common error is for students to count the number of decimal places from left to right when placing the decimal point in the product.

On Your Own
- Students should be able to use mental math to find the estimate in each problem.
- **Neighbor Check:** Have students work independently and then have their neighbor check their work. Have students discuss any discrepancies.

Goal
Today's lesson is finding the product of a whole number and a decimal.

Start Thinking! and Warm Up

Extra Example 1
Find 16.3 × 5. 81.5

Extra Example 2
Multiply.
a. 4.8 × 3 14.4
b. 7 × 0.86 6.02

On Your Own
1. 6.3
2. 170.1
3. 64.9
4. 39.2
5. 98.4
6. 72.55
7. 2.8
8. 7.92

Extra Example 3
Find 0.0045×5. 0.0225

Extra Example 4
Use the Distributive Property to rewrite $5(x + 0.0013)$. $5x + 0.0065$

On Your Own
9. 0.096 10. 0.009
11. $4p + 0.0084$

Extra Example 5
A half-dollar is 2.15 millimeters thick. How high is a stack of 100 half-dollars? 215 mm

On Your Own
12. 1750 mm

English Language Learners
Numbers and Words
English learners are more comfortable with numbers than with words. Provide students with the opportunity to improve their English. On a piece of paper, give students decimal numbers written in numbers and words. Ask the students to match the number with the words.

Numbers	Words
4.35	twelve and five thousandths
12.005	six and seven tenths
0.83	four and thirty-five hundredths
6.7	eighty-three hundredths

Have students copy the number and then write the corresponding words next to the number. Encourage students to say the number out loud.

Laurie's Notes

Example 3
- **Connection:** When the decimal is very small, it may be helpful to have students write the number as a fraction and then multiply.
$$3 \times \frac{16}{1000} = \frac{48}{1000} = 0.048$$
- Remind students that this is another strategy they can use, like estimating, to check their answer when they multiply decimals.

Example 4
- **Connection:** The decimal 0.0013 is very small. Write 0.0013 as a fraction.
$$7 \times 0.0013 = 7 \times \frac{13}{10,000} = \frac{91}{10,000} = 0.0091, \text{ so Choice C is correct.}$$
- This is a good exercise to use to show students the test-taking strategy of *Eliminate Choices*. Students need to distribute the 7 to both terms. Choice A can be eliminated because the *n* did not get multiplied by 7. Some students may recognize that the product 7×0.0013 will have four decimal places. Of the remaining three answers, only Choice C has four decimal places.

Example 5
- Work through this example together as a class. Have students estimate the answer to 1.35×100. Then do Method 1.
- **?** After completing Method 1, ask students if they can think of how to do the problem using mental math. See if anyone describes the process in Method 2. If not, review Method 2.
- **?** Ask students which method they prefer. Some may prefer the accuracy of Method 1. Most will probably like to use mental math, as in Method 2.

On Your Own
- Have students use Method 2 first and then check their mental math by finding the product using Method 1.
- Have two students come up to the board to do the problem while the other students work at their desks. Both students should write down their mental math calculations and then find the product using Method 1.

Closure
- **Exit Ticket:** Multiply 5×0.0212. 0.106
- Students might include the last zero of the calculation in the answer and write 0.1060. This is correct, but not really the simplest form. Explain that 1.5 and 1.50 are equivalent, but we often don't write the last zero unless we are trying to be very precise, as in science class.

The Dynamic Planning Tool
Editable Teacher's Resources at *BigIdeasMath.com*

EXAMPLE 3 Inserting Zeros in the Product

Study Tip
When multiplying, you may not have enough decimal places in the product. In this case, insert one or more zeros in the product.

Find 3×0.016.

$$
\begin{array}{r}
1 \\
0.016 \\
\times \quad 3 \\
\hline
0.048
\end{array}
$$

← Three decimal places

← To have three decimal places, insert zeros to the left of 48.

So, $3 \times 0.016 = 0.048$.

EXAMPLE 4 Standardized Test Practice

Which expression is equivalent to $7(n + 0.0013)$?

A $n + 0.0091$ **B** $7n + 0.091$
C $7n + 0.0091$ **D** $7n + 0.91$

$7(n + 0.0013) = 7(n) + 7(0.0013)$ Distributive Property
$\quad\quad\quad\quad\quad\quad = 7n + 0.0091$ Multiply 7 and 0.0013.

The correct answer is **C**.

On Your Own

Now You're Ready
Exercises 24–31 and 41–46

9. Find 0.012×8. **10.** Find 0.003×3.

11. Use the Distributive Property to rewrite $4(p + 0.0021)$.

EXAMPLE 5 Use Mental Math

How high is a stack of 100 dimes?

1.35 millimeters

Method 1: Multiply 1.35 by 100.

$$
\begin{array}{r}
1.35 \\
\times\; 1\,00 \\
\hline
0\,00 \\
00\,0 \\
135\quad \\
\hline
135.00
\end{array}
$$

Two decimal places

Method 2: You are multiplying by a power of 10. Use mental math.

There are two zeros in 100. So, move the decimal point in 1.35 two places to the right.

$1.35 \times 100 = 135. = 135$

So, a stack of 100 dimes is 135 millimeters high.

On Your Own

12. A quarter is 1.75 millimeters thick. How high is a stack of 1000 quarters? Solve using both methods.

Section 3.2 Multiplying Decimals and Whole Numbers

3.2 Exercises

Vocabulary and Concept Check

1. **NUMBER SENSE** How many decimal places are in the product of 0.0087 and 23? Explain.

2. **NUMBER SENSE** Is the product 1.23 × 8 greater than or less than 8? Explain.

3. **DIFFERENT WORDS, SAME QUESTION** Which is different? Find "both" answers.

 What is the product of 7.6 and 3? How much is 7.6 times 3?

 7.6 is how much more than 3? Multiply 7.6 and 3.

Practice and Problem Solving

Find the product. Use an estimate to place the decimal point.

4. 0.2 × 6
5. 0.7 × 8
6. 4.6 × 3
7. 3.1 × 13

Multiply. Use estimation to check your answer.

8. 2.8 × 7
9. 6.1 × 9
10. 0.36 × 18
11. 7.43 × 3

12. 1.7 × 12
13. 9.5 × 5
14. 4.08 × 8
15. 1.24 × 2

16. 4.8 × 7
17. 6.3 × 5
18. 7.19 × 16
19. 0.87 × 21

20. 1.95 × 11
21. 5.89 × 5
22. 3.472 × 4
23. 8.188 × 12

Multiply.

24. 0.024 × 3
25. 0.013 × 7
26. 0.083 × 2
27. 0.017 × 5

28. 7 × 0.007
29. 19 × 0.004
30. 0.0038 × 9
31. 0.0093 × 6

ERROR ANALYSIS Describe and correct the error in the solution.

32.
```
    0.0045
  ×      9
    4.05
```

33.
```
    0.32
  ×    5
   0.160
```

34. **MOON** The weight of an object on the moon is about 0.167 of its weight on Earth. How much does a 180-pound astronaut weigh on the moon?

35. **BAMBOO** A bamboo plant grows about 1.25 feet each day. Find the growth in one week. Use an estimate to place the decimal point.

36. **NAILS** A fingernail grows about 0.1 millimeter each day. How much does a fingernail grow in 30 days? 90 days?

116 Chapter 3 Multiplying and Dividing Decimals

Assignment Guide and Homework Check

Level	Day 1 Activity Assignment	Day 2 Lesson Assignment	Homework Check
Basic	8–15, 54–58	1–3, 5, 7, 17–33 odd, 34–36, 41, 43	2, 17, 25, 36, 41
Average	8–15, 54–58	1–3, 5, 7, 17–33 odd, 36, 41, 43, 51, 52	2, 17, 25, 36, 41
Advanced	8–15, 54–58	1–3, 28–33, 36–50 even, 51, 53	28, 36, 42, 48

Common Errors

- **Exercises 8–23** Students might count the decimal places to the left of the decimal point instead of the right. Remind them that they are looking to the right of the decimal point, just like they did when writing decimals as fractions. Stress that estimating the product first will help.
- **Exercises 8–23** When determining where to place the decimal point, students might count left to right instead of right to left. Tell students to count from the least to greatest place value, so that means they must start counting decimal places from the right.
- **Exercises 24–31** Students might ignore the zeros when counting the number of decimal places, or forget to add them when finding the answer. Remind them that zeros are placeholders and are important for determining the correct answer. For example, 0.024 is not the same number as 0.24. Students may be more successful if they write these decimals as fractions and then multiply.

3.2 Record and Practice Journal

Find the product. Use an estimate to place the decimal point.
1. $\begin{array}{r} 0.5 \\ \times\ 4 \\ \hline 2 \end{array}$ 2. $\begin{array}{r} 0.9 \\ \times\ .7 \\ \hline 6.3 \end{array}$ 3. $\begin{array}{r} 3.8 \\ \times\ 6 \\ \hline 22.8 \end{array}$ 4. $\begin{array}{r} 2.1 \\ \times\ 11 \\ \hline 23.1 \end{array}$

Multiply. Use estimation to check your answer.
5. 0.32×5 — 1.6
6. 7.81×2 — 15.62
7. 0.0046×9 — 0.0414

Use the Distributive Property to rewrite the expression.
8. $3(a + 0.21)$ — $3a + 0.63$
9. $4(x - 0.55)$ — $4x - 2.2$
10. $12(k + 0.097)$ — $12k + 1.164$

11. You use a microscope to look at bacteria that is 0.0034 millimeter long. The microscope magnifies the bacteria 430 times. How long does the bacteria appear to be when you look at it through the microscope?
1.462 millimeters

12. Each weekday a bus travels 17.6 miles to school. How many miles does the bus travel to school and back in one week?
176 miles

Technology For the Teacher
Answer Presentation Tool
QuizShow

Vocabulary and Concept Check

1. 4; The decimal factor has 4 decimal places.
2. Greater than 8; Because 8 is being multiplied by a number greater than 1.
3. 7.6 is how much more than 3?; 4.6; 22.8

Practice and Problem Solving

4. 1.2
5. 5.6
6. 13.8
7. 40.3
8. 19.6
9. 54.9
10. 6.48
11. 22.29
12. 20.4
13. 47.5
14. 32.64
15. 2.48
16. 33.6
17. 31.5
18. 115.04
19. 18.27
20. 21.45
21. 29.45
22. 13.888
23. 98.256
24. 0.072
25. 0.091
26. 0.166
27. 0.085
28. 0.049
29. 0.076
30. 0.0342
31. 0.0558
32. The decimal is in the wrong place. $0.0045 \times 9 = 0.0405$
33. The decimal is in the wrong place. $0.32 \times 5 = 1.60$
34. 30.06 lb
35. 8.75 ft
36. 3 mm; 9 mm
37. 28.3
38. 710
39. 9670
40. 33,200
41. $5x + 1.55$
42. $3y - 0.51$
43. $7b + 0.364$

T-116

Practice and Problem Solving

44. $8w + 0.052$
45. $12w + 0.0264$
46. $9r - 0.00738$
47. 28.4
48. 23.7
49. 282.24
50. 852
51. $7.12 \times 8.22 \times 100$
 $= 7.12 \times 822 = 5852.64$
52. See *Taking Math Deeper*.
53. *Answers should include, but are not limited to:*
 a. menu with main items, desserts, beverages, and prices
 b. guest check for 5 people showing items, prices, and subtotal
 c. tax and total with tax are shown
 d. amount rounded to nearest dollar, 20% tip, and total amount paid are shown

Fair Game Review

54. 2480
55. 18,272
56. 44,044
57. 47,107
58. A

Mini-Assessment
Multiply.
1. 7.2×4 28.8
2. 9.6×5 48
3. 2.47×3 7.41
4. 4.321×2 8.642
5. The subtotal of your restaurant bill is $12. Multiply the subtotal by 0.06 to find the tax. What is the total of your bill? $12.72

T-117

Taking Math Deeper

Exercise 52
This problem provides a nice opportunity to include some geography in your class. You could display a world map and ask students to name the different continents. Which continent is missing in the table? Why is it missing? Antarctica, it has no tall buildings.

Put the data in geographic perspective.

Find the heights in feet.
223 m ≈ 731.4 ft 509 m ≈ 1669.5 ft
268 m ≈ 879.0 ft 442 m ≈ 1449.8 ft
323 m ≈ 1059.4 ft 221 m ≈ 724.9 ft
For buildings, round to the nearest tenth of a foot.

Visualize the different heights.

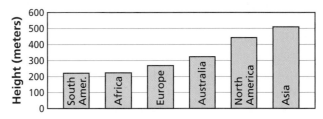

Note: The bar graph would show the same relative heights if the *y*-axis was measured in feet.

Project
Make a collage of the buildings. You could cut and paste photos, do hand drawings, or use other representations. Be sure to make it clear which buildings are taller than others. Use the Internet to research which building is shown in the picture in the textbook. Taipei 101

Reteaching and Enrichment Strategies

If students need help...	If students got it...
Resources by Chapter • Practice A and Practice B • Puzzle Time Record and Practice Journal Practice Differentiating the Lesson Lesson Tutorials Skills Review Handbook	Resources by Chapter • Enrichment and Extension Start the next section

MENTAL MATH Use mental math to find the product.

37. 2.83×10 **38.** 7.1×100

39. 9.67×1000 **40.** $0.332 \times 100{,}000$

Use the Distributive Property to rewrite the expression.

4 41. $5(x + 0.31)$ **42.** $3(y - 0.17)$ **43.** $7(b + 0.052)$

44. $8(w + 0.0065)$ **45.** $12(w + 0.0022)$ **46.** $9(r - 0.00082)$

Evaluate the expression.

47. $3.4 \times 6 + 8$ **48.** $3 + 9 \times 2.3$ **49.** $8.82 \times 4 \times 8$ **50.** $2.84 \times 3 \times 100$

51. REASONING Show how to evaluate $7.12 \times 8.22 \times 100$ without multiplying the two decimals.

52. BUILDING HEIGHTS One meter is approximately 3.28 feet. Find the height of each building in feet by multiplying its height in meters by 3.28.

Continent	Tallest Building	Height
Africa	Carlton Centre Office Tower	223 m
Asia	Taipei 101	509 m
Europe	Naberezhnaya Tower C	268 m
North America	Sears Tower	442 m
Australia	Q1 Tower	323 m
South America	Parque Central Torre Este	221 m

53. Open-Ended You and four friends have dinner at a restaurant.

 a. Draw a restaurant menu that has main items, desserts, and beverages, with their prices.

 b. Write a guest check that shows what each of you ate. Find the subtotal.

 c. Multiply by 0.07 to find the tax. Then find the total.

 d. Round the total to the nearest whole number. Multiply by 0.20 to estimate a tip. Including the tip, how much did you spend?

Fair Game Review What you learned in previous grades & lessons

Multiply. *(Skills Review Handbook)*

54. 124×20 **55.** 571×32 **56.** 364×121 **57.** 289×163

58. MULTIPLE CHOICE You buy a shirt for $9.99 and a hat for $6.60. You give the cashier a $20 bill. How much change do you receive? *(Skills Review Handbook)*

 Ⓐ $3.41 **Ⓑ** $4.41 **Ⓒ** $9.35 **Ⓓ** $11.33

3.3 Multiplying Decimals

COMMON CORE STATE STANDARDS
6.NS.3

Essential Question When multiplying decimals, how do you know where to place the decimal point in the product?

1 EXAMPLE: Multiplying Decimals

Find 0.2×0.3.

$0.2 \times 0.3 = \dfrac{2}{10} \times \dfrac{3}{10}$ Write as fractions.

$= \dfrac{6}{100}$ Multiply the fractions.

$= \dfrac{6}{10^2}$ Rewrite the denominator as a power of 10.

$= 0.06$ Rewrite the fraction as a decimal.

∴ So, $0.2 \times 0.3 = 0.06$.

2 ACTIVITY: Multiplying Decimals Using Powers of 10

Work with a partner.

a. Copy and complete the table. Use Example 1 as a model.

Problem	Rewrite as Fractions	Product	Denominator as Base 10	Rewrite as Decimal
0.2×3	$\dfrac{2}{10} \times \dfrac{3}{1}$	$\dfrac{6}{10}$	$\dfrac{6}{10^1}$	0.6
0.2×0.3	$\dfrac{2}{10} \times \dfrac{3}{10}$	$\dfrac{6}{100}$	$\dfrac{6}{10^2}$	0.06
0.2×0.03				
0.2×0.003				
0.2×0.0003				
0.2×0.00003				

b. Describe the connection between the first and last columns.

Laurie's Notes

Introduction

For the Teacher
- **Goal:** The goal of this activity is for students to observe the relationship between the number of decimal places in the factors and the number of decimal places in the product.

Motivate
- Tell a story similar to this one: "At lunch yesterday, I brought half of a pizza into the faculty room. I ate half of the pizza I brought."
- ❓ "How much of the pizza is left?" one-fourth
- Continue the story: "One of the other teachers entered the room and said he had forgotten his lunch. I offered him half of the remaining pizza."
- ❓ "Now, how much of the pizza is left?" one-eighth
- Record what happened in the story, but use decimals instead of fractions.

Amount of pizza to start	×	Amount of pizza I ate		Amount left after I ate	×	Amount the other teacher ate

$$0.5 \times 0.5 = 0.25 \quad \text{and} \quad 0.25 \times 0.5 = 0.125$$

- ❓ "Do you remember how to write 0.125 as a fraction? Is it equivalent to $\frac{1}{8}$?"
 Listen for students to say to write the number over the place value.
 $0.125 = \frac{125}{1000} = \frac{1 \times 125}{8 \times 125} = \frac{1}{8}$; 0.125 is equivalent to $\frac{1}{8}$.

Activity Notes

Example 1
- **Connection:** This example is a review of the last lesson. The significant difference is the extra step of writing the denominator of the fraction as a power of 10. The reason this is done is so students will see the connection between the exponent and the sum of the place values in the factors.

Activity 2
- ❓ "How do you multiply a fraction by a whole number?" Write the whole number as a fraction with a denominator of 1. Then write the product of the numerators over the product of the denominators.
- ❓ "How do you multiply two fractions?" Multiply the numerators and multiply the denominators.
- Have students work with a partner to complete the table.
- ❓ "What is the pattern in the first column?" Listen for the key ideas that the first factor is constant 0.2 and the second factor is decreasing (the 3 is moving to the right and 0s are being added).
- Ask for volunteers to describe the connection between the first and last columns of the table.
- ❓ "In the last column, is the product increasing or decreasing?" decreasing

Common Core State Standards

6.NS.3 Fluently add, subtract, multiply, and divide multi-digit decimals using the standard algorithm for each operation.

Previous Learning

Students should know how to multiply fractions and write decimals as fractions. Students should know that powers of 10 are written as 10^x, where x is a counting number.

Start Thinking! and Warm Up

3.3 Record and Practice Journal

T-118

Differentiated Instruction

Auditory

Remind students of the rule for placing the decimal point: *The number of decimal places in the product must equal the sum of the number of decimal places in the factors.*

3.3 Record and Practice Journal

b. Describe the connection between the first and last columns of the table in part (a).

The sum of the numbers of decimal places of the two factors is the same as the number of decimal places in the product.

ACTIVITY 3: Multiplying Decimals Using Powers of 10

Complete the table. Use Example 1 as a model.

Problem	Rewrite as Fractions	Product	Denominator as Base 10	Rewrite as Decimal
2×0.3	$\frac{2}{1} \times \frac{3}{10}$	$\frac{6}{10}$	$\frac{6}{10^1}$	0.6
0.2×0.3	$\frac{2}{10} \times \frac{3}{10}$	$\frac{6}{100}$	$\frac{6}{10^2}$	0.06
0.02×0.3	$\frac{2}{100} \times \frac{3}{10}$	$\frac{6}{1000}$	$\frac{6}{10^3}$	0.006
0.002×0.3	$\frac{2}{1000} \times \frac{3}{10}$	$\frac{6}{10,000}$	$\frac{6}{10^4}$	0.0006
0.0002×0.3	$\frac{2}{10,000} \times \frac{3}{10}$	$\frac{6}{100,000}$	$\frac{6}{10^5}$	0.00006

What Is Your Answer?

4. a. What differences do you notice between the tables in Activities 2 and 3?

The factor that is divided by increasing powers of 10 is 3 in Activity 2 and 2 in Activity 3.

b. What similarities do you notice?

The last three columns are the same.

5. IN YOUR OWN WORDS When multiplying decimals, how do you know where to place the decimal point in the product? Give examples in your description.

Add the numbers of decimal places in the factors. Use this number of decimal places in the product.

6. Write a general rule for multiplying two decimals. Give examples with your rule.

Multiply as if the numbers were whole numbers. Add the numbers of decimal places in the factors. Use this number of decimal places in the product.

7. How many products can you find in the circle maze? List each product.

$3 \times 0.3 = 0.9$, $0.3 \times 3 = 0.9$, $0.07 \times 0.6 = 0.042$,
$0.2 \times 2 = 0.4$, $10 \times 0.06 = 0.6$, $2 \times 0.4 = 0.8$,
$0.1 \times 0.6 = 0.06$, $0.6 \times 0.1 = 0.06$, $10 \times 0.07 = 0.7$

Laurie's Notes

Activity 3
- This is the same concept as Activity 2.
- Have students work with a partner to complete the table.

Words of Wisdom
- Completing the tables in Activities 2 and 3 reinforces several skills. Students convert between decimals and fractions, write powers of 10 using exponents, and review multiplication of fractions.

What Is Your Answer?
- Give students adequate time to answer these questions.
- **Neighbor Check:** Have students work independently on Question 4 and then have their neighbor check their work. Have students discuss any discrepancies.
- **Class Review:** Ask volunteers to share their answers to Questions 5 and 6.
- Question 7 gives students practice counting place values in the factors and product.

Closure
- **Exit ticket:**

 $0.08 \times 0.9 = ?$ $0.4 \times ? = 0.00028$

 0.072 0.0007

Technology for the Teacher

The Dynamic Planning Tool
Editable Teacher's Resources at *BigIdeasMath.com*

3 ACTIVITY: Multiplying Decimals Using Powers of 10

Copy and complete the table. Use Example 1 as a model.

Problem	Rewrite as Fractions	Product	Denominator as Base 10	Rewrite as Decimal
2×0.3	$\dfrac{2}{1} \times \dfrac{3}{10}$	$\dfrac{6}{10}$	$\dfrac{6}{10^1}$	0.6
0.2×0.3				
0.02×0.3				
0.002×0.3				
0.0002×0.3				

What Is Your Answer?

4. **a.** What differences do you notice between the tables in Activities 2 and 3?
 b. What similarities do you notice?

5. **IN YOUR OWN WORDS** When multiplying decimals, how do you know where to place the decimal point in the product? Give examples in your description.

6. Write a general rule for multiplying two decimals. Give examples with your rule.

7. How many products can you find in the circle maze? List each product.

 Sample: $3 \times 0.3 = 0.9$

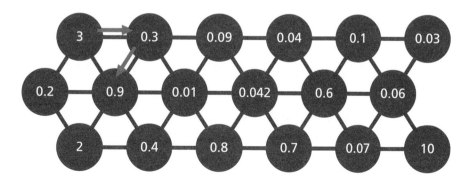

Practice

Use what you learned about multiplying decimals to complete Exercises 8–15 on page 122.

3.3 Lesson

The rule for multiplying two decimals is similar to the rule for multiplying a decimal by a whole number.

🔑 Key Idea

Multiplying Decimals by Decimals

Words Multiply as you would with whole numbers. Then add the number of decimal places in the factors. This sum gives you the number of decimal places in the product.

Numbers
```
  4.7 1 6  ← 3 decimal places
×    0.2  ← + 1 decimal place
  0.9 4 3 2  ← 4 decimal places
```

EXAMPLE 1 Multiplying Decimals

a. Find 4.8 × 7.2. **Estimate** 5 × 7 = 35

```
    4.8    ← 1 decimal place
  × 7.2    ← + 1 decimal place
    9 6
  3 3 6
  3 4.5 6  ← 2 decimal places
```

∴ So, 4.8 × 7.2 = 34.56. **Reasonable?** 34.56 ≈ 35 ✓

b. Find 3.1 × 0.05. **Estimate** 3 × 0 = 0

```
     3.1    ← 1 decimal place
  × 0.0 5   ← + 2 decimal places
    0.1 5 5 ← 3 decimal places
```

∴ So, 3.1 × 0.05 = 0.155. **Reasonable?** 0.155 ≈ 0 ✓

● On Your Own

Multiply. Use estimation to check your answer.

Now You're Ready
Exercises 8–23

1. 8.1 × 5.6
2. 2.7 × 9.04
3. 6.32 × 0.09
4. 1.785 × 0.2

Laurie's Notes

Introduction

Connect
- **Yesterday:** Students explored multiplication of decimals by rewriting decimals as fractions and then rewriting the product as a decimal.
- **Today:** Students will use a formal rule to multiply decimals.

Motivate
- ❓ "Is the product of two numbers always greater than either of the two factors?" No, only if both factors are greater than 1.
- ❓ "Can you give an example of a multiplication problem where the product is greater than one factor, but less than the other factor?" $4 \times 0.5 = 2$
- ❓ "Can you give an example of a multiplication problem where the product is less than both factors?" $0.4 \times 0.2 = 0.08$

Lesson Notes

Key Idea
- Students generally have little difficulty at this stage with the rule for multiplying decimals. You want to be sure that they have an understanding of why the rule works.
- Connect place value and fraction multiplication.
- **FYI:** If no multiplication error is made, locating the decimal point in the product is usually not difficult for students.
- The concepts students find most difficult are often related to trailing zeros in the product or inserting zeros in the product.

Example 1
- Work through both parts of the example.
- Begin each problem by estimating the product. The last step should always be to judge the reasonableness of the solution compared to your estimate. This is especially important in part (b), where the product is much less than the first factor.
- In part (b), it is more difficult to estimate the size of the answer. You may want to remind students they can also use fractions to estimate the answer.

Words of Wisdom
- Ask students to record the number of decimal places in each factor. They should do this in all of the problems at this point.

 Example: 3.1 (1) ← Number of decimal places
 × 0.05 (2)
 0.155 (3)

On Your Own
- **Neighbor Check:** Have students work independently and then have their neighbor check their work. Have students discuss any discrepancies.
- Watch for multiplication errors in Question 2 due to the 0 in 9.04.

Goal Today's lesson is finding the product of two decimals.

Start Thinking! and Warm Up

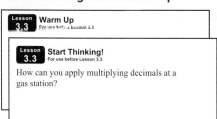

Extra Example 1
Multiply.
a. 1.3×3.5 4.55
b. 5.8×0.02 0.116

On Your Own
1. 45.36
2. 24.408
3. 0.5688
4. 0.3570

T-120

Extra Example 2

Evaluate the expression 1.8y for the given value of y.
a. y = 6.75 12.15
b. y = 0.029 0.0522

On Your Own
5. 8.435
6. 64.40
7. 3.7170
8. 0.0245

Extra Example 3

You buy 4.5 pounds of bananas at $0.49 a pound. You hand the cashier a $20 bill. How much change will you get back? $17.79

On Your Own
9. $0.52

English Language Learners

Vocabulary

Students should be warned not to drop the *trailing zeros* before placing the decimal point. Students need to include the zeros in the number of decimal places to be counted. Trailing zeros may be dropped after the decimal point has been placed.

$$\begin{array}{r} 3.25 \\ \times\, 0.8 \\ \hline 2.600 \end{array}$$

$3.25 \times 0.8 = 2.6$

Laurie's Notes

Example 2
- ❓ "What does the expression 2.4x mean?" Multiply 2.4 by the value of x.
- Remind students of why you generally do not use the × symbol or a dot to represent multiplication. Now in a chapter on decimals, it makes even more sense to avoid the dot for multiplication.

On Your Own
- **Neighbor Check:** Have students work independently and then have their neighbor check their work. Have students discuss any discrepancies.
- **Representation:** Students may ask if they can write .5 instead of 0.5. The initial zero makes it easier to see that the number is a decimal.

Example 3
- ❓ It may have been some time since students multiplied two 3-digit numbers. This problem may prompt questions about multiplication in general.
 - "What is a reasonable estimate for the cost of 2.75 pounds at $1.89 a pound?" $3 \times \$2 = \6
 - "What is 2.75 written as a fraction?" $2\frac{3}{4}$; This question is simply to review this benchmark fraction. It should be automatic at this point.
 - "Why is the exact answer (5.1975) written as $5.20?" You need to round the answer to the nearest hundredth, because money is only written with two decimal places to represent the cents.

Closure
- If $4321 \times 56 = 241{,}976$ then $4.321 \times \,? = 2.41976$. Explain your reasoning. 0.56; The number of decimal places in the product is the sum of the number of decimal places in the factors.

The Dynamic Planning Tool
Editable Teacher's Resources at *BigIdeasMath.com*

EXAMPLE 2 Evaluating Expressions

Evaluate the expression 2.4x for the given value of x.

a. x = 3.95

2.4x = 2.4(3.95) Substitute.

$$\begin{array}{r} 3.95 \\ \times\ 2.4 \\ \hline 1580 \\ 790 \\ \hline 9.480 \end{array}$$ ← 2 decimal places
← + 1 decimal place

← 3 decimal places

∴ So, 2.4x = 9.40 when x = 3.95.

b. x = 0.016

2.4x = 2.4(0.016) Substitute.

$$\begin{array}{r} 0.016 \\ \times\ 2.4 \\ \hline 64 \\ 32 \\ \hline 0.0384 \end{array}$$ ← 3 decimal places
← + 1 decimal place

← 4 decimal places

∴ So, 2.4x = 0.0384 when x = 0.016.

On Your Own

Now You're Ready
Exercises 28–35

Evaluate the expression 3.5x for the given value of x.

5. x = 2.41 **6.** x = 18.4 **7.** x = 1.062 **8.** x = 0.007

EXAMPLE 3 Real-Life Application

You buy 2.75 pounds of tomatoes. You hand the cashier a $10 bill. How much change will you get back?

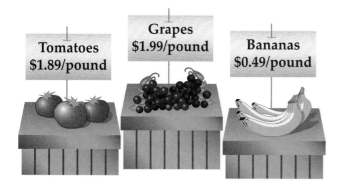

Tomatoes $1.89/pound
Grapes $1.99/pound
Bananas $0.49/pound

Step 1: Multiply 1.89 by 2.75.

$$\begin{array}{r} 1.89 \\ \times\ 2.75 \\ \hline 945 \\ 1323 \\ 378 \\ \hline 5.1975 \end{array}$$ ← 2 decimal places
← + 2 decimal places

← 4 decimal places

The cost of 2.75 pounds of tomatoes is $5.20.

Step 2: Subtract $5.20 from $10.

$$\begin{array}{r} 10.00 \\ -\ 5.20 \\ \hline 4.80 \end{array}$$

∴ So, you will get $4.80 back.

On Your Own

9. You buy 2.25 pounds of grapes. You hand the cashier a $5 bill. How much change will you get back?

Section 3.3 Multiplying Decimals 121

3.3 Exercises

Vocabulary and Concept Check

1. **NUMBER SENSE** If you know $12 \times 24 = 288$, how can you find 1.2×2.4?

Copy the problem and place the decimal point in the product.

2. 1.7 8
 × 4.9
 8 7 2 2

3. 9.2 4
 × 0.6 8
 6 2 8 3 2

4. 3.7 5
 × 5.2 2
 1 9 5 7 5 0

How many decimal places are in the product?

5. 6.17×8.2 6. 1.684×10.2 7. 0.053×2.78

Practice and Problem Solving

Multiply. Use estimation to check your product.

8. 0.7
 × 0.2

9. 0.08
 × 0.3

10. 0.007
 × 0.03

11. 0.0008
 × 0.09

12. 0.004
 × 0.9

13. 0.06
 × 0.5

14. 0.0008
 × 0.004

15. 0.0002
 × 0.06

16. 12.4×0.2 17. 18.6×5.9 18. 7.91×0.72 19. 1.16×3.35

20. 6.478×18.21 21. 1.064×7.216 22. 0.0021×18.2 23. 6.109×8.407

24. **ERROR ANALYSIS** Describe and correct the error in the solution.

25. **TAKEOUT** A Chinese restaurant offers buffet takeout for $4.99 per pound. How much does your takeout meal cost?

26. **CROPLAND** Alabama has about 2.51 million acres of cropland. Florida has about 1.15 times as much cropland as Alabama. How much cropland does Florida have?

27. **GOLD** On a tour of an old gold mine, you find a nugget containing 0.82 ounce of gold. Gold is worth $904.62 per ounce. How much is your nugget worth?

ALGEBRA Evaluate the expression when $x = 3.7$, $y = 6.19$, and $z = 0.072$.

28. $5x$ 29. $8z$ 30. $2.21y$ 31. $0.006x$

32. xy 33. xz 34. $3.6y + 3.2$ 35. $2.7x - 3.79$

Assignment Guide and Homework Check

Level	Day 1 Activity Assignment	Day 2 Lesson Assignment	Homework Check
Basic	8–15, 48–52	1–7, 17–35 odd, 24, 26	17, 21, 26, 31
Average	8–15, 48–52	1–7, 17–23 odd, 24, 26, 29–39 odd, 44, 45	17, 21, 26, 31, 37
Advanced	8–15, 48–52	1–7, 24, 28–44 even, 46, 47	30, 36, 40, 46

Common Errors

- **Exercises 8–23** Students may only count the number of decimal places in one number instead of two. Remind them that they need to count the number of decimal places in both numbers.
- **Exercises 8–23** Suggest to students who are struggling with the concept to write the decimals as fractions. They can also write the number of decimal places to the right of the problem, as shown in the example, so that they remember where to place the decimal point.
- **Exercises 28–35** Students may forget that the coefficient in front of the variable is the number being multiplied by x. It may help some students to write the problem with a multiplication symbol between the coefficient and the variable before evaluating.

Technology
For the Teacher
Answer Presentation Tool
QuizShow

3.3 Record and Practice Journal

 Vocabulary and Concept Check

1. Place the decimal point so that there are two decimal places. $1.2 \times 2.4 = 2.88$
2. 8.722
3. 6.2832
4. 19.5750
5. 3
6. 4
7. 5

 Practice and Problem Solving

8. 0.14
9. 0.024
10. 0.00021
11. 0.000072
12. 0.0036
13. 0.03
14. 0.0000032
15. 0.000012
16. 2.48
17. 109.74
18. 5.6952
19. 3.886
20. 117.96438
21. 7.677824
22. 0.03822
23. 51.358363
24. The product should have two decimal places.

$$\begin{array}{r} 4.9 \\ \times\ 3.8 \\ \hline 18.62 \end{array}$$

25. $3.24
26. 2.8865 million acres
27. $741.79
28. 18.5
29. 0.576
30. 13.6799
31. 0.0222
32. 22.903
33. 0.2664
34. 25.484
35. 6.2

T-122

Practice and Problem Solving

36. Each number is 0.6 times the previous number; 0.1296, 0.07776, 0.046656

37. Each number is 0.1 times the previous number; 0.0015, 0.00015, 0.000015

38. Each number is 0.5 times the previous number; 0.0025, 0.00125, 0.000625

39. Each number is 1.5 times the previous number; 25.3125, 37.96875, 56.953125

40. $0.875 \times 0.6 = 0.525$

41. $0.45 \times 0.75 = 0.3375$

42. $1.0625 \times 3.74 = 3.97375$

43. $5.525 \times 6.64 = 36.686$

44. $14.03

45. a. 190.06 miles
 b. 91.29 miles

46. 2; 3; 4

47. See *Taking Math Deeper*.

Fair Game Review

48. 26 **49.** 5
50. 3 **51.** 7
52. D

Mini-Assessment
Multiply.
1. 14.5×0.4 5.8
2. 20.25×0.6 12.15
3. 2.18×3.24 7.0632
4. 5.426×6.2 33.6412
5. Apples are on sale for $1.29 per pound. You buy 4.25 pounds of apples. How much will it cost to buy the apples? $5.49

T-123

Taking Math Deeper

Exercise 47
This problem gives students a chance to factor with decimals. The key to part (a) is to write 952 as the product of two whole number factors. Then insert a decimal point to get the factors for 9.52.

 a. Factor 9.52.

3.4 ft
2.8 ft
5.6 ft
1.7 ft
6.8 ft
1.4 ft

 Find the perimeter of each painting. The painting with the shortest perimeter is the least expensive to frame.

b. $P = 2(3.4) + 2(2.8) = 12.4$ ft
$P = 2(5.6) + 2(1.7) = 14.6$ ft
$P = 2(6.8) + 2(1.4) = 16.4$ ft

Draw a diagram. Remember to include the mitered corners.

$1 + 2.8 + 1$
$1 + 3.4 + 1$
$P = 20.4$ ft
$C = 5(20.4) = \$102$

$0.2 + 2.8 + 0.2$
$0.2 + 3.4 + 0.2$
$P = 14.0$ ft
$C = 1(14) = \$14$

c. $5(14) = \$70; \$102 > \$70$
So, the fancy frame is more than 5 times the cost.

Project
Make a drawing that you would like to frame. What type of frame would you use? Why?

Reteaching and Enrichment Strategies

If students need help...	If students got it...
Resources by Chapter • Practice A and Practice B • Puzzle Time Record and Practice Journal Practice Differentiating the Lesson Lesson Tutorials Skills Review Handbook	Resources by Chapter • Enrichment and Extension • School-to-Work Start the next section

Describe the pattern. Find the next three numbers.

36. 1, 0.6, 0.36, 0.216, . . .

37. 15, 1.5, 0.15, 0.015, . . .

38. 0.04, 0.02, 0.01, 0.005, . . .

39. 5, 7.5, 11.25, 16.875, . . .

Convert the fractions or mixed numbers to decimals. Then multiply.

40. $\dfrac{7}{8} \times \dfrac{3}{5}$

41. $\dfrac{9}{20} \times \dfrac{3}{4}$

42. $1\dfrac{1}{16} \times 3\dfrac{37}{50}$

43. $5\dfrac{21}{40} \times 6\dfrac{16}{25}$

44. FOOD You buy 2.6 pounds of apples and 1.475 pounds of peaches. You hand the cashier a $20 bill. How much change will you get back?

Apples $1.23/pound

Peaches $1.88/pound

45. MILEAGE A car can travel 22.36 miles on one gallon of gasoline.

 a. How far can the car travel on 8.5 gallons of gasoline?

 b. A hybrid car can travel 33.1 miles on one gallon of gasoline. How much farther can the hybrid car travel on 8.5 gallons of gasoline?

46. REASONING Without multiplying, how many decimal places does 3.4^2 have? 3.4^3? 3.4^4? Explain your reasoning.

47. Geometry A rectangular painting has an area of 9.52 square feet.

 a. Draw three different ways in which this can happen.

 b. The cost of a frame depends on the perimeter of a painting. Which of your drawings from part (a) is the least expensive to frame? Explain your reasoning.

 c. The thin black framing costs $1 per foot. The fancy framing costs $5 per foot. Will the fancy framing cost five times as much as the black framing? Explain why or why not.

Fair Game Review *What you learned in previous grades & lessons*

Divide. *(Skills Review Handbook)*

48. 78 ÷ 3

49. 65 ÷ 13

50. 57 ÷ 19

51. 84 ÷ 12

52. MULTIPLE CHOICE How many edges does the rectangular prism at the right have? *(Skills Review Handbook)*

 Ⓐ 4
 Ⓑ 6
 Ⓒ 8
 Ⓓ 12

3 Study Help

You can use a **process diagram** to show the steps involved in a procedure. Here is an example of a process diagram for multiplying decimals.

On Your Own

Make a process diagram with an example to help you study these topics.

1. multiplying decimals and whole numbers
2. estimating with decimals

After you complete this chapter, make process diagrams with examples for the following topics.

3. dividing decimals by whole numbers
4. dividing decimals
5. Describe the steps in the process that are similar for multiplying decimals and dividing decimals.

"Descartes, you should use my process diagram when you eat your treats."

Sample Answers

1.

2.

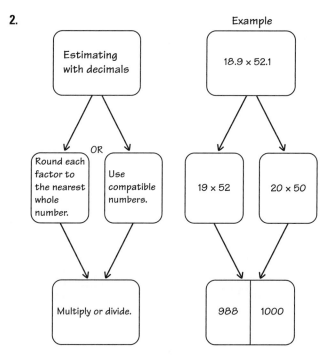

List of Organizers
Available at *BigIdeasMath.com*

Comparison Chart
Concept Circle
Definition (Idea) and Example Chart
Example and Non-Example Chart
Formula Triangle
Four Square
Information Frame
Information Wheel
Notetaking Organizer
Process Diagram
Summary Triangle
Word Magnet
Y Chart

About this Organizer

A **Process Diagram** can be used to show the steps involved in a procedure. Process diagrams are particularly useful for illustrating procedures with two or more steps, and they can have one or more branches. As shown, students' process diagrams can have two parallel parts, in which the procedure is stepped out in one part and an example illustrating each step is shown in the other part. Or, the diagram can be made up of just one part, with example(s) included in the last "bubble" to illustrate the steps that precede it.

Technology For the Teacher
Vocabulary Puzzle Builder

Answers

1. 20
2. 5
3. 45
4. 8
5. 38
6. 10.54
7. 76.32
8. 0.666
9. 0.05
10. 0.00024
11. 0.486
12. 2.912
13. *Sample answer:* 13 pizzas; too much
14. 8.4 qt
15. $334.40
16. 7.8324 m^2

Assessment Book

Alternative Quiz Ideas

100% Quiz Math Log
Error Notebook Notebook Quiz
Group Quiz Partner Quiz
Homework Quiz Pass the Paper

Error Notebook

- An error notebook provides an opportunity for students to analyze and learn from their errors. Have students make an error notebook for this chapter. They should work in their notebook a little each day. Give students the following directions.
- Use a notebook and divide the page into three columns.
- Label the first column *problem,* second column *error,* and third column *correction*.
- In the first column, write down the problem in which errors were made. Record the source of the problem (homework, quiz, in-class assignment).
- The second column should show the exact error that was made. Include a statement of why you think the error was made. Statements might include, "I didn't pay close enough attention to the decimal point." This is where the learning takes place, so it is helpful to use a different color ink for the work in this column.
- The last column contains the corrected problems and comments that will help with future work.
- Separate each problem with horizontal lines.

Reteaching and Enrichment Strategies

If students need help...	If students got it...
Resources by Chapter • Study Help • Practice A and Practice B • Puzzle Time Lesson Tutorials *BigIdeasMath.com* Practice Quiz Practice from the Test Generator	Resources by Chapter • Enrichment and Extension • School-to-Work Game Closet at *BigIdeasMath.com* Start the next section

Technology For the Teacher

Answer Presentation Tool
Big Ideas Test Generator

3.1–3.3 Quiz

Estimate by rounding each factor to the nearest whole number.
(Section 3.1)

1. 3.8×4.9
2. $15.2 \div 2.6$

Use compatible numbers to estimate the product or quotient. *(Section 3.1)*

3. 14.9×3.02
4. $48.24 \div 5.76$

Multiply. Use estimation to check your answer. *(Section 3.2)*

5. 7.6
 $\times5$

6. 0.62
 $\times17$

7. 9.54
 $\times8$

8. 0.074
 $\times9$

Multiply. Use estimation to check your answer. *(Section 3.3)*

9. 0.5
 $\times0.1$

10. 0.004
 $\times0.06$

11. 0.54
 $\times0.9$

12. 4.16
 $\times0.7$

13. **PIZZA** A school budgets $125.50 to buy pizza for a dance. Each pizza costs $11.29. Estimate how many pizzas the school can buy. Is your estimate *too little* or *too much*? *(Section 3.1)*

14. **LEMONADE** A two-liter bottle contains 2.1 quarts of lemonade. You buy four bottles. How many quarts do you have? *(Section 3.2)*

15. **WHITEWATER RAFTING** A whitewater rafting trip costs $41.80 per person. A family of eight goes on a rafting trip. What is the total cost of the trip? *(Section 3.2)*

3.66 m

2.14 m

16. **GEOMETRY** Find the area of the mouth of the field hockey goal. *(Section 3.3)*

3.4 Dividing Decimals by Whole Numbers

COMMON CORE STATE STANDARDS
6.NS.2
6.NS.3

Essential Question How is dividing a decimal by a whole number similar to dividing a whole number by a whole number?

1 ACTIVITY: Dividing a Decimal by a Whole Number

Work with a partner. Use base ten blocks to model the division.

a. **Sample:** 3.2 ÷ 4

Begin by modeling 3.2.

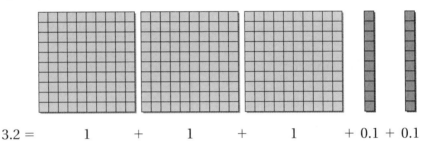

3.2 = 1 + 1 + 1 + 0.1 + 0.1

Next, think of the division 3.2 ÷ 4 as dividing 3.2 into four parts.

Replace the ones blocks with tenths blocks. You have a total of 32 tenths blocks.

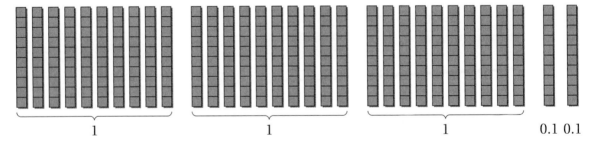

Separate the blocks into four groups.

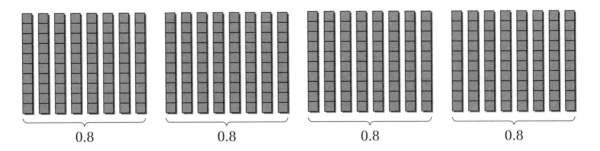

There are four groups of 0.8. So, 3.2 ÷ 4 = 0.8.

b. 1.2 ÷ 3 c. 2.8 ÷ 4 d. 2.6 ÷ 2 e. 3.5 ÷ 5

Laurie's Notes

Introduction

For the Teacher
- **Goal:** The goal of this activity is for students to begin to develop an understanding of where to place the decimal point in the quotient of a division problem.

Motivate
- Show $0.64 using 2 quarters, 1 dime, and 4 pennies. Students should be able to distinguish their value from the size.
- ? "How much money is here?" $0.64
- Ask a volunteer to demonstrate how this money could be shared equally between two people. Have additional coins nearby for exchanging.
- Record on the board: $0.64 ÷ 2 = $0.32.
- ? Move the money back to one group and ask "How could you share this money equally between four people?"
- Ask a volunteer to show this, again having coins nearby for exchanging.
- Record on the board: $0.64 ÷ 4 = $0.16.
- The coin activity models the process of exchanging one amount for lesser, but equal, amounts. This will help students when they need to exchange base ten ones blocks for tenths blocks.

Activity Notes

Words of Wisdom
- The process of dividing decimals is similar to dividing whole numbers. The challenge for students is in deciding where to place the decimal point.
- These activities include quotients with compatible numbers and a connection to fraction division. This helps students start to understand how the placement of the decimal point in the quotient might be determined.

Activity 1
- Begin with a review of base ten blocks.
- ? "If the large square represents 1, what is the value of this piece?" Hold up a tenths piece (the thin rectangle). 0.1
- ? "What is the value of the smallest piece?" Hold up the small square. 0.01
- If students need help finding these values, ask them how many rectangles and how many little squares fit on the large square. 10 rectangles and 100 little squares. What number represents one out of 10? 0.1 What number represents one out of 100? 0.01
- ? Ask a student to use the base ten blocks to model the number 3.2. "How could you divide the tiles representing 3.2 into 4 equal groups?" Ask a student to show this to the class. The process is shown in the text.
- Use numbers to represent the base ten blocks: 3.2 ÷ 4 = 0.8.
- **Summarize:** "Three and two-tenths divided by four is eight-tenths. This could also be read as thirty-two tenths divided by four is eight-tenths."

Common Core State Standards
6.NS.2 Fluently divide multi-digit numbers using the standard algorithm.
6.NS.3 Fluently add, subtract, multiply, and divide multi-digit decimals using the standard algorithm for each operation.

Activity Materials	
Introduction	Textbook
• 2 quarters • 4 dimes • 4 nickels • 4 pennies	• base 10 blocks

Start Thinking! and Warm Up

3.4 Record and Practice Journal

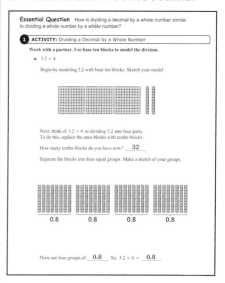

T-126

English Language Learners
Vocabulary
Show students the following problems to reinforce the key words of division. Have students make a copy of these in their notebooks.

dividend ÷ divisor = quotient

$$\text{divisor} \overline{)\text{dividend}}^{\text{quotient}}$$

3.4 Record and Practice Journal

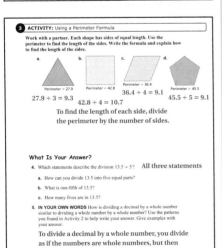

Laurie's Notes

Activity 2
- In this activity, students use pattern recognition and number sense to find the quotients.
- These questions help students think about how the quotient changes.
- ? When students have finished, ask them what patterns they observed and how they determined the quotient. Listen for students to describe the decimal moving one space to the left every time the decimal in the dividend moved one space to the left. In parts (d) and (e), they should notice that as the divisor increases by a factor of 10, the quotient decreases by a factor of 10 (the decimal moves to the left).
- ? How can you check your answers? Students may use estimation or may multiply the quotient and the divisor to see if they get the dividend.

Activity 3
- ? "What does perimeter mean? How do you find the perimeter of a rectangle?" The distance around a figure. The sum of all of the side lengths.
- ? As a review, ask students to name the four shapes seen in this activity. triangle, square, rhombus, and pentagon
- You may want to help students with part (a). Note that the numbers are very compatible.

$P = s + s + s$	Write a formula for the perimeter of a triangle.
$27.9 = 3s$	Substitute the value for the perimeter.
$9.3 = s$	Divide each side by 3.

- Part (b) is a bit more challenging, so students need to be careful.

Closure
- **Exit ticket:** Find 84.08 ÷ 4. Explain your thinking. 21.02; The dividend has two decimal places, so the answer needs to have two decimal places.

The Dynamic Planning Tool
Editable Teacher's Resources at *BigIdeasMath.com*

2 ACTIVITY: Where Does the Decimal Go?

Work with a partner. Use a pattern to complete each row. Use estimation to check that your answer is reasonable.

a. 1236 ÷ 3 = 412 123.6 ÷ 3 = 41.2 12.36 ÷ 3 = 1.236 ÷ 3 =

b. 5120 ÷ 10 = 512 512.0 ÷ 10 = 51.2 51.20 ÷ 10 = 5.120 ÷ 10 =

c. 4 ÷ 2 = 2 0.4 ÷ 2 = 0.2 0.04 ÷ 2 = 0.004 ÷ 2 =

d. 482.5 ÷ 1 = 482.5 482.5 ÷ 10 = 48.25 482.5 ÷ 100 = 482.5 ÷ 1000 =

e. 10 ÷ 5 = 2 10 ÷ 50 = 0.2 10 ÷ 500 = 10 ÷ 5000 =

3 ACTIVITY: Using a Perimeter Formula

Work with a partner. Each shape has sides of equal length. Use the perimeter to find the length of the sides. Write the formula and explain how to find the length of the sides.

a.
Perimeter = 27.9

b.
Perimeter = 42.8

c.
Perimeter = 36.4

d.
Perimeter = 45.5

What Is Your Answer?

4. Which statements describe the division 13.5 ÷ 5?
 - How can you divide 13.5 into five equal parts?
 - What is one-fifth of 13.5?
 - How many fives are in 13.5?

5. **IN YOUR OWN WORDS** How is dividing a decimal by a whole number similar to dividing a whole number by a whole number? Use the patterns you found in Activity 2 to help write your answer. Give examples with your answer.

Use what you learned about dividing decimals by whole numbers to complete Exercises 9–16 on page 130.

3.4 Lesson

Key Idea

Dividing Decimals by Whole Numbers

Words Place the decimal point in the quotient above the decimal point in the dividend. Then divide as you would with whole numbers.

Numbers
$$4 \overline{)7.32} = 1.83$$

Place the decimal point in the quotient above the decimal point in the dividend.

EXAMPLE 1 — Dividing Decimals by Whole Numbers

a. Find $7.6 \div 4$. Estimate $8 \div 4 = 2$

$$\begin{array}{r} 1.9 \\ 4\overline{)7.6} \\ -4 \\ \hline 3\,6 \\ -3\,6 \\ \hline 0 \end{array}$$

Place the decimal point in the quotient above the decimal point in the dividend.

∴ So, $7.6 \div 4 = 1.9$. Reasonable? $1.9 \approx 2$ ✓

Remember

Another way to check your answer is to multiply the quotient by the divisor. Example 1(b):

$$\begin{array}{r} 6.35 \\ \times 5 \\ \hline 31.75 \end{array}$$ ✓

The product is the dividend.

b. Find $31.75 \div 5$. Estimate $30 \div 5 = 6$

$$\begin{array}{r} 6.35 \\ 5\overline{)31.75} \\ -30 \\ \hline 1\,7 \\ -1\,5 \\ \hline 25 \\ -25 \\ \hline 0 \end{array}$$

Place the decimal point in the quotient above the decimal point in the dividend.

∴ So, $31.75 \div 5 = 6.35$. Reasonable? $6.35 \approx 6$ ✓

On Your Own

Divide. Use estimation to check your answer.

1. $36.4 \div 2$
2. $22.2 \div 6$
3. $59.64 \div 7$
4. $43.26 \div 14$

Laurie's Notes

Introduction

Connection
- **Yesterday:** Students explored dividing a decimal by a whole number using base-ten blocks and patterns.
- **Today:** Students will use a formal rule to divide a decimal by a whole number.

Motivate
- Give students some contexts for dividing a dollar amount among three or four people. Have students estimate the whole dollar amount of the quotient using compatible numbers.
- ? "Three siblings purchase a gift for their parents that costs $22.98. About how much does each sibling pay?" round $22.98 to $24; $24 ÷ 3 = $8
- ? "Four friends share the cost of a pizza and drinks that total $18.75. About how much does each friend owe?" round $18.75 to $20; $20 ÷ 4 = $5
- ? "You pay $4.32 for three pens. Were the pens more than $2 each?" No, the cost for each pen is somewhere between $1 and $2.

Lesson Notes

Words of Wisdom
- Placing the decimal point in the quotient above the decimal point in the dividend makes sense to the students.
- **FYI:** Some students may have difficulty lining up the columns when they do long division. Have them do their work on $\frac{1}{4}$-inch graph paper.

Example 1
- Work through both examples. Begin by estimating the quotient and end with judging the reasonableness of your answer.
- Students often ask where the decimal point is written in the work below the dividend. "Why don't you bring down the decimal point?" In theory, you could bring the decimal point down. But, once the decimal point is placed in the quotient, you treat the problem as if you were dividing whole numbers.

On Your Own
- Question 4 is a little more challenging due to the two-digit divisor and the zero in the quotient.
- **Neighbor Check:** Have students work independently and then have their neighbor check their work. Have students discuss any discrepancies. Make sure students check each other's estimates to reinforce the use of this strategy.
- Ask for volunteers to share their work at the board.

Goal Today's lesson is dividing a decimal by a whole number.

Start Thinking! and Warm Up

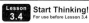

Lesson 3.4 Warm Up
For use before Lesson 3.4

Lesson 3.4 Start Thinking!
For use before Lesson 3.4

How could dividing a decimal by a whole number apply to going to a rock concert?

Extra Example 1
Divide.
a. 3.5 ÷ 7 0.5
b. 58.68 ÷ 9 6.52

On Your Own
1. 18.2 2. 3.7
3. 8.52 4. 3.09

T-128

Extra Example 2

Find 6.92 ÷ 8. 0.865

On Your Own

5. 1.55 6. 7.35
7. 0.195 8. 1.707

Extra Example 3

A can of three tennis balls costs $2.69. A can that contains four tennis balls is $4.19. Which can of tennis balls is the better buy? Explain. The can with three tennis balls, because the price per tennis ball is $0.90 compared to $1.05 for the can with four tennis balls.

On Your Own

9. Yes; the price per bottle is $0.91 for the 12-pack.

Differentiated Instruction

Inclusion

Remind students that one difference between multiplying decimals and dividing decimals is the placement of the decimal point. In multiplication, the decimal point is placed after the decimals are multiplied. In division, the placement of the decimal point is determined before dividing.

Laurie's Notes

Example 2

- ❔ "If the divisor is greater than the dividend, what do you know about the quotient?" It will be less than 1.
- You may want to share whole number examples to illustrate the above question. For example, use 8 ÷ 4 and 4 ÷ 8.
- ❔ When you start Example 2, place the decimal point in the quotient. Then ask "Does 12 go into 4?" no "Does 12 go into 43?" yes
- ❔ In this problem, it is necessary to add a zero to the dividend in order to continue the division. Ask, "Is 4.38 equivalent to 4.380? 4.3800?" yes; yes
- Students want to know how many zeros to add. Tell students that you need to continue to add zeros until the division ends with a zero remainder, or until a pattern occurs, as it did when they studied repeating decimals.

On Your Own

- **Neighbor Check:** Have students work independently and then have their neighbor check their work. Have students discuss any discrepancies.

Example 3

- Ask a student to read the problem and suggest a method for answering the question.
- Dividing the cost by the number of bottles will give the *unit* cost for one bottle. The pack of sports drinks that has the lower unit cost is the better buy.
- ❔ "Why is the division carried out to only 3 decimal places?" In this problem, the cost per bottle is rounded to the nearest penny.

Closure

- **Writing:** Have students describe how to divide a decimal by a whole number. *Sample answer:* Place the decimal point in the quotient above the decimal point in the dividend. Then divide as you would with whole numbers. Check your answer by rounding the decimal to a number compatible with the divisor and dividing. Confirm that your answer is reasonable by comparing it to your estimate.

The Dynamic Planning Tool
Editable Teacher's Resources at *BigIdeasMath.com*

EXAMPLE 2 Inserting a Zero in the Dividend

Find 4.38 ÷ 12.

```
     0.365
12)4.380
   -3 6
     78
    -72
     60
    -60
      0
```

Place the decimal point in the quotient above the decimal point in the dividend.

Insert a zero and continue to divide.

So, 4.38 ÷ 12 = 0.365. Check 0.365 × 12 = 4.38 ✓

On Your Own

Now You're Ready
Exercises 9–24

Evaluate the expression.

5. 6.2 ÷ 4
6. 34.8 ÷ 8 + 3
7. 3.12 ÷ 16
8. 1 + 14.14 ÷ 20

EXAMPLE 3 Real-Life Application

Which pack of sports drinks is the better buy? Explain.

To find the better buy, divide each price by the number of bottles. Continue dividing until the quotient can be rounded to the nearest cent.

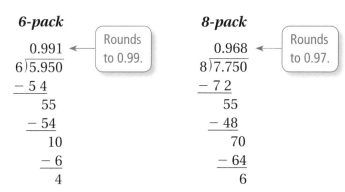

6-pack
```
     0.991
6)5.950
  -5 4
    55
   -54
    10
    -6
     4
```
Rounds to 0.99.

8-pack
```
     0.968
8)7.750
  -7 2
    55
   -48
    70
   -64
     6
```
Rounds to 0.97.

The price per bottle is $0.99 for the 6-pack and $0.97 for the 8-pack. So, the 8-pack is the better buy.

On Your Own

9. **WHAT IF?** In Example 3, a 12-pack costs $10.95. Is it the best buy? Explain.

3.4 Exercises

Vocabulary and Concept Check

1. **NUMBER SENSE** Which number is the quotient? dividend? divisor?

 $$3\overline{)12.9} = 4.3$$

2. **NUMBER SENSE** Fix the one that is not correct.

 $$4\overline{)24.4} = 6.1 \qquad 4\overline{)244} = 61 \qquad 4\overline{)2.44} = 6.1$$

Copy the problem and place the decimal point in the correct location.

3. $18.6 \div 4 = 465$
4. $6.38 \div 11 = 58$
5. $88.27 \div 7 = 1261$
6. $19.2 \div 3 = 64$
7. $43.254 \div 9 = 4806$
8. $3.24 \div 12 = 27$

Practice and Problem Solving

Divide. Check your answer.

9. $6\overline{)25.2}$
10. $5\overline{)33.5}$
11. $7\overline{)3.5}$
12. $8\overline{)10.4}$
13. $38.7 \div 9$
14. $37.6 \div 4$
15. $43.4 \div 7$
16. $25.6 \div 8$
17. $44.64 \div 8$
18. $0.294 \div 3$
19. $3.6 \div 24$
20. $64.26 \div 18$
21. $34.64 \div 16$
22. $24.2 \div 11$
23. $61.299 \div 7$
24. $12.25 \div 14$

ERROR ANALYSIS Describe and correct the error in finding the quotient.

25.

26.

27. **TEXT MESSAGING** You send 40 text messages in one month. The total cost is $4.80. How much does each text message cost?

28. **SUNTAN LOTION** A 5-ounce bottle of suntan lotion costs $10.29. A 4-ounce bottle costs $8.49. Which is the better buy? Explain.

ORDER OF OPERATIONS Evaluate the expression.

29. $7.68 + 3.18 \div 12$
30. $10.56 \div 3 - 1.9$
31. $19.6 \div 7 \times 9$
32. $5.5 \times 16.56 \div 9$
33. $35.25 \div 5 \div 3$
34. $13.41 \times (5.4 \div 9)$

130 Chapter 3 Multiplying and Dividing Decimals

Assignment Guide and Homework Check

Level	Day 1 Activity Assignment	Day 2 Lesson Assignment	Homework Check
Basic	9–16, 39–43	1–8, 17–33 odd, 28	17, 21, 28, 29
Average	9–16, 39–43	1–8, 17–25 odd, 28–31, 35, 36	17, 21, 28, 29
Advanced	9–16, 39–43	1–8, 25, 26, 30–34 even, 35–38	25, 26, 30, 35

Common Errors

- **Exercises 9–24** Students may divide without putting the decimal point in the quotient. Tell them to place the decimal point immediately after they put the numbers in long division form. Remind students to use estimation to check the reasonableness of their answers.
- **Exercises 9–24** Students may have difficulty keeping their columns organized and will forget to include placeholder zeros or put in extra placeholders. Have them use graph paper as an organizational tool to help line up the numbers. Remind students to use estimation to check the reasonableness of their answers.
- **Exercise 36** Students may automatically divide $35.95 by four without taking into account that $35.95 is the cost for *each* pair of jeans (except the free pair!). Remind them to read the problem carefully. Have students write out the cost for each of the four pairs of jeans and find the total. Then they can find the cost per pair. This type of question is common on state assessments, so take time to go over this question as a class.

3.4 Record and Practice Journal

Divide. Check your answer.
1. $3 \overline{)18.6}$ — 6.2
2. $6 \overline{)46.8}$ — 7.8
3. $4 \overline{)7.6}$ — 1.9
4. $24.5 \div 7$ — 3.5
5. $0.096 \div 8$ — 0.012
6. $15.65 \div 5$ — 3.13

Evaluate the expression.
7. $15.54 + 22.2 \div 6$ — 19.24
8. $59.85 \div 9 \times 8.23$ — 54.7295
9. $10.02 \times (45.408 \div 3)$ — 151.66272

10. You and three friends buy a birthday gift for a classmate. The gift costs $50.24. Each of you pays an equal amount for the gift. How much do you pay for the gift? — $12.56

11. It costs $859.32 to have a school dance. Each ticket costs $8.
 a. How many tickets must be sold to cover the cost? — 108 tickets
 b. How many tickets must be sold to make a $980.68 profit? — 230 tickets

Technology For the Teacher
Answer Presentation Tool
QuizShow

Vocabulary and Concept Check

1. 4.3; 12.9; 3
2. $4 \overline{)2.44}$ with 0.61 above
3. $18.6 \div 4 = 4.65$
4. $6.38 \div 11 = 0.58$
5. $88.27 \div 7 = 12.61$
6. $19.2 \div 3 = 6.4$
7. $43.254 \div 9 = 4.806$
8. $3.24 \div 12 = 0.27$

Practice and Problem Solving

9. 4.2
10. 6.7
11. 0.5
12. 1.3
13. 4.3
14. 9.4
15. 6.2
16. 3.2
17. 5.58
18. 0.098
19. 0.15
20. 3.57
21. 2.165
22. 2.2
23. 8.757
24. 0.875

25. They brought down 2 zeros instead of 1.

$9 \overline{)28.008}$ with 3.112 above
```
  3.112
9)28.008
  27
  ──
   10
    9
   ──
   10
    9
   ──
   18
   18
   ──
    0
```

26. A zero should be placed before the 8 in the quotient.

```
  0.086
6)0.516
   48
   ──
   36
   36
   ──
    0
```

27. $0.12

T-130

 Practice and Problem Solving

28. the 5-ounce bottle; The price per ounce is $2.06 for the 5-ounce bottle and $2.12 for the 4-ounce bottle.

29. 7.945 **30.** 1.62

31. 25.2 **32.** 10.12

33. 2.35 **34.** 8.046

35. the 12-pack; The price per unit is $0.74 for the 4-pack, $0.72 for the 12-pack, and $0.73 for the 24-pack. So, the 12-pack is the best buy.

36. $26.96

37. See *Taking Math Deeper*.

38. a. 9 weeks

b. 3 more weeks; *Sample answer:* Use the same method as part (a) to find the total number of weeks needed, then subtract the number of weeks in part (a).

 Fair Game Review

39. 6 **40.** 6
41. 7 **42.** 12
43. A

Mini-Assessment
Divide.
1. $6.8 \div 2$ 3.4
2. $15.4 \div 4$ 3.85
3. $27.6 \div 5$ 5.52
4. $0.963 \div 3$ 0.321
5. You send 15 pictures from your cell phone in one month. The total cost is $3.75. How much does it cost to send each picture? $0.25

T-131

Taking Math Deeper

Exercise 37
Part of the goal of this problem is to help students understand how competitive many timed sports are. In this problem, the U.S. team would have taken the gold medal if each swimmer on the team had finished one-quarter second quicker.

 a. Divide the time for each team by 4.

Country	Total Time (seconds)	Time per Swimmer (seconds)
Australia	215.94	$215.94 \div 4 = 53.985$
U.S.	216.39	$216.39 \div 4 = 54.098$
Netherlands	217.59	$217.59 \div 4 = 54.398$

 Subtract 0.25 second from each U.S. swimmer.
$$54.098 - 0.25 = 53.848 \text{ sec}$$
Multiply by 4 to get the team total.
$$4(53.848) = 215.392 \text{ sec}$$

b. 215.392 seconds is faster than 215.94 seconds. So, the U.S. team would have won the gold medal.

 Help me understand it.

It takes about 0.3 second for your eyes to blink.

A quarter of a second, the amount of time each swimmer would have had to make up to win the gold medal, is *faster* than the time it takes your eyes to blink.

The swimmers on the U.S. team lost the gold medal by the blink of an eye!

Reteaching and Enrichment Strategies

If students need help. . .	If students got it. . .
Resources by Chapter • Practice A and Practice B • Puzzle Time Record and Practice Journal Practice Differentiating the Lesson Lesson Tutorials Skills Review Handbook	Resources by Chapter • Enrichment and Extension • School-to-Work Start the next section

35. **FRUIT PUNCH** Which pack of fruit punch is the best buy? Explain.

36. **SALE** You buy three pairs of jeans for $35.95 each and get a fourth pair for free. What is your cost per pair of jeans?

37. **SWIMMING** The table shows the top three times in a swimming event at the Summer Olympics. A team consists of four women swimming 100 meters each.

 a. For each team, suppose the times of all four swimmers were the same. Find how long it took a swimmer from each team to swim 100 meters.

 b. If each swimmer on the U.S. team swam a quarter second faster, would the U.S. team have won the gold medal? Explain your reasoning.

Women's 4 × 100 Freestyle Relay		
Medal	Country	Time (seconds)
Gold	Australia	215.94
Silver	United States	216.39
Bronze	Netherlands	217.59

38. **Analyze** You are saving money to buy a new bicycle that costs $155.75. You have $30 and plan to save $5 each week. Your aunt decides to give you an additional $10 each week.

 a. How many weeks will you have to save until you have enough money to buy the bicycle?

 b. How many more weeks would you have to save to buy a new bicycle that costs $203.89? Explain how you found your answer.

 Fair Game Review What you learned in previous grades & lessons

Divide. *(Skills Review Handbook)*

39. $84 \div 14$ 40. $102 \div 17$ 41. $161 \div 23$ 42. $372 \div 31$

43. **MULTIPLE CHOICE** Which is the best estimate of how many $3\frac{3}{4}$-inch pieces of string can be cut from a string that is $32\frac{1}{8}$ inches long? *(Section 2.1)*

 Ⓐ 8 Ⓑ 10 Ⓒ 11 Ⓓ 28

3.5 Dividing Decimals

Essential Question How can you use base ten blocks to model decimal division?

COMMON CORE STATE STANDARDS

6.NS.2
6.NS.3

1 ACTIVITY: Dividing Decimals

Work with a partner. Use base ten blocks to model the division.

a. **Sample:** 2.4 ÷ 0.6

Begin by modeling 2.4.

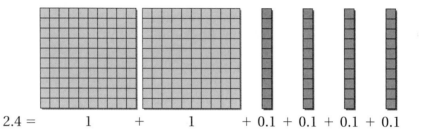

2.4 = 1 + 1 + 0.1 + 0.1 + 0.1 + 0.1

Next, think of the division 2.4 ÷ 0.6 as the question "How many 0.6's are in 2.4?" To answer this, divide 2.4 into groups of 0.6 each.

Replace the ones blocks with tenths blocks. You have a total of 24 tenths blocks.

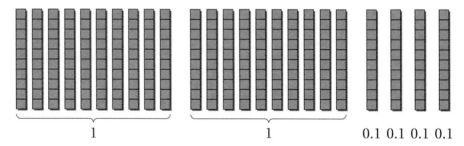

 1 1 0.1 0.1 0.1 0.1

Group the blocks into groups of 0.6 each.

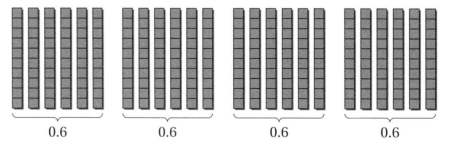

 0.6 0.6 0.6 0.6

 There are four groups of 0.6. So, 2.4 ÷ 0.6 = 4.

b. 1.6 ÷ 0.8 c. 2.8 ÷ 0.7 d. 3.2 ÷ 0.4 e. 3.6 ÷ 0.9

Laurie's Notes

Introduction

For the Teacher
- **Goal:** The goal of this activity is for students to develop a visual understanding of decimal division.

Motivate
- Show students the equations below, and have them complete and describe the pattern.

 $16.4 \div 20 = 0.82$
 $16.4 \div 2 = 8.2$
 $16.4 \div 0.2 = 82$
 $16.4 \div 0.02 = ?$

 820; The decimal place of the quotient moves one place to the right every time the decimal point in the divisor moves one place to the left.

Activity Notes

Words of Wisdom
- **Connections:** In Activity 1, there are two possible ways to model $2.4 \div 0.6$.
- Think of 2.4 as the improper fraction, 24-tenths.
- Think of 2.4 as the mixed number, 2 and 4-tenths.

Activity 1
- Have students model 2.4 with their base ten blocks. You want to divide this amount by 0.6. Read it as "sixth-tenths," not "point six."
- ? "How do you model 2.4 so that only tenths are used?" Exchange 2 whole units for 20-tenths.
- ? "What does it mean to divide 24-tenths by 6-tenths?" It means "How many groups of 6-tenths can you make from 24-tenths?"
- Ask a student to show how 24-tenths can be divided into groups of 6-tenths. Record the problem and the quotient on the board: $2.4 \div 0.6 = 4$.
- **Connection:** The rule for dividing decimals is to multiply the divisor and dividend by a power of 10. This is accomplished using base ten blocks, when 2 wholes are exchanged for 20-tenths. You now have 24-tenths ÷ 6-tenths, which is equivalent to $24 \div 6$.
- Have students work in pairs on parts (b)–(e). Remind them to model the dividend with base ten pieces first, and then exchange the whole number portion for the equivalent number of tenths.
- Ask for volunteers to record the problem and quotients on the board.

Common Core State Standards
6.NS.2 Fluently divide multi-digit numbers using the standard algorithm.
6.NS.3 Fluently add, subtract, multiply, and divide multi-digit decimals using the standard algorithm for each operation.

Previous Learning
Students have used base ten blocks to model operations with decimals.

Activity Materials
Textbook
• base 10 blocks

Start Thinking! and Warm Up

Activity 3.5 Start Thinking! For use before Activity 3.5

Activity 3.5 Warm Up For use before Activity 3.5

Use base ten blocks to complete the statement.
1. 3 tenths = __?__ hundredths
2. 5 ones = __?__ tenths
3. 2 ones = __?__ hundredths
4. 20 hundredths = __?__ tenths

3.5 Record and Practice Journal

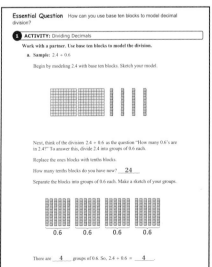

T-132

English Language Learners
Graphic Organizer
Have students copy the table in their notes as a reference. Encourage them to add their own examples using other powers of ten.

Multiplying by Whole Number Powers of 10	
Move decimal point right.	$10 \times 2.5 = 25$
Multiplying by Decimal Powers of 10	
Move decimal point left.	$0.1 \times 2.5 = 0.25$
Dividing by Whole Number Powers of 10	
Move decimal point left.	$2.5 \div 10 = 0.25$
Dividing by Decimal Powers of 10	
Move decimal point right.	$2.5 \div 0.1 = 25$

3.5 Record and Practice Journal

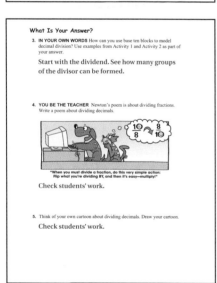

Laurie's Notes

Activity 2
- ❓ "Is $0.3 \div 0.06$ *greater than* 1 or *less than* 1? How do you know?" greater than 1; The dividend (0.3) is greater than the divisor (0.06).
- Work through the sample with students as they work in groups.
- When students exchange tenths for hundredths, it would be helpful for them to say or write "two-tenths is equivalent to twenty-hundredths."
- Note: Parts (d) and (e) already have like units (16-hundredths divided by 8-hundredths, and 28-hundredths divided by 7-hundredths).

Words of Wisdom
- The base ten blocks are used in the activities so that students can develop a visual model for division.
- The base ten pieces work easily because the numbers are compatible *and* because the quotients are all whole numbers. Notice that none of the quotients are decimals.
- To help students make the connection to the former lesson, ask them to look at the following.

 $0.16 \div 0.08 = 2$ (16-hundredths divided by 8-hundredths)
 $1.6 \div 0.8 = 2$ (16-tenths divided by 8-tenths)
 $16 \div 8 = 2$ (16 ones divided by 8 ones)
 $160 \div 80 = 2$ (16 tens divided by 8 tens)

This concept is a big idea in the lesson that follows. Notice that when the dividend and divisor are both multiplied by 10, the quotient stays the same.

What Is Your Answer?
- When students explain how base ten blocks are used to model decimal division, listen for references to like-sized pieces (the same units) being divided into groups of the same size.
- ❓ "You can use base ten blocks to model $0.8 \div 0.4$. Can you model $0.4 \div 0.8$?" You can't model $0.4 \div 0.8$ using base ten blocks, but you can model $\frac{1}{2}$ of the group that you need.

In tomorrow's lesson, problems of this type will be evaluated.
- ❓ "Could you use fractions to find $0.4 \div 0.8$?" yes; $\frac{4}{10} \div \frac{8}{10} = \frac{4}{8} = \frac{1}{2} = 0.5$

Closure
- Write each problem as a decimal division problem and solve.
 a. 54-hundredths divided by 9-hundredths $0.54 \div 0.09 = 6$
 b. 8-tenths divided by 5-hundredths $0.8 \div 0.05 = 16$

The Dynamic Planning Tool
Editable Teacher's Resources at *BigIdeasMath.com*

2 ACTIVITY: Dividing Decimals

Work with a partner. Use base ten blocks to model the division.

a. **Sample:** 0.3 ÷ 0.06

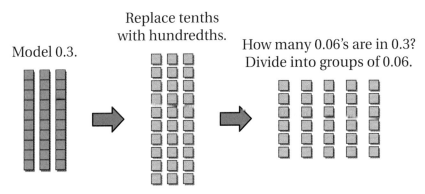

Model 0.3. Replace tenths with hundredths. How many 0.06's are in 0.3? Divide into groups of 0.06.

There are 5 groups of 0.06. So, 0.3 ÷ 0.06 = 5.

b. 0.2 ÷ 0.04 c. 0.6 ÷ 0.01
d. 0.16 ÷ 0.08 e. 0.28 ÷ 0.07

What Is Your Answer?

3. **IN YOUR OWN WORDS** How can you use base ten blocks to model decimal division? Use examples from Activity 1 and Activity 2 as part of your answer.

4. **YOU BE THE TEACHER** Newton's poem is about dividing fractions. Write a poem about dividing decimals.

"When you must divide a fraction, do this very simple action: Flip what you're dividing BY, and then it's easy—multiply!"

5. Think of your own cartoon about dividing decimals. Draw your cartoon.

Practice Use what you learned about dividing decimals to complete Exercises 9–12 on page 136.

3.5 Lesson

Key Idea

Dividing Decimals by Decimals

Words Multiply the divisor *and* dividend by 10, 100, or 1000 to make the divisor a whole number. Then place the decimal point in the quotient and divide as you would with whole numbers.

Numbers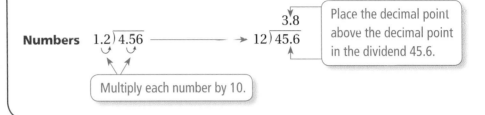

Multiply each number by 10.

Place the decimal point above the decimal point in the dividend 45.6.

EXAMPLE 1 Dividing Decimals

a. Find $18.2 \div 1.4$.

Study Tip

Multiplying the divisor and dividend by 10, 100, or 1000 does not change the quotient.

For example:
$18.2 \div 1.4 = 13$
$182 \div 14 = 13$
$1820 \div 140 = 13$

$$1.4\overline{)18.2} \longrightarrow 14\overline{)182.}$$

Multiply each number by 10.

$$\begin{array}{r} 13. \\ 14\overline{)182.} \\ -14 \\ \hline 42 \\ -42 \\ \hline 0 \end{array}$$

Place the decimal point above the decimal point in the dividend 182.

∴ So, $18.2 \div 1.4 = 13$. **Check** $13 \times 1.4 = 18.2$ ✓

b. Find $0.273 \div 0.39$.

$$0.39\overline{)0.273} \longrightarrow 39\overline{)27.3}$$

Multiply each number by 100.

$$\begin{array}{r} 0.7 \\ 39\overline{)27.3} \\ -27.3 \\ \hline 0 \end{array}$$

∴ So, $0.273 \div 0.39 = 0.7$. **Check** $0.7 \times 0.39 = 0.273$ ✓

On Your Own

Divide. Check your answer.

1. $1.2\overline{)9.6}$
2. $3.4\overline{)57.8}$
3. $21.643 \div 2.3$
4. $0.459 \div 0.51$

Laurie's Notes

Introduction

Connect
- **Yesterday:** Students used base ten blocks to explore decimal division.
- **Today:** Students will use a formal rule to divide decimals.

Motivate
- Play *Lucky Seven*, which gets its name because seven lucky students are each asked to hold a card with a division problem written on it.
- You can have multiple groups of seven in different areas of your classroom. If the number of students is not evenly divisible by seven, ask the remaining students to sort a set of seven cards on the desktop.
- Create a set of index cards for each group. Each set of cards should include the division problems shown below and a reference letter.

 A: $510 \div 170$ B: $5.1 \div 0.17$ C: $51 \div 0.17$
 D: $51 \div 17$ E: $51 \div 1.7$ F: $5.1 \div 1.7$
 G: $510 \div 17$

- Students should find the quotient of the problem on the card. They can work with their group to find the quotient, if needed. Cards A, D, and F equal 3, Cards B, E, and G equal 30, and Card C equals 300.
- Students should group themselves within their group by quotient.
- When the groups have finished, ask one group to list the quotients they found and discuss the strategies they used to determine the quotients. Use the letters to reference the cards.
- All quotients involve a 3, so knowing where to locate the decimal point in the quotient is the challenge.

Lesson Notes

Key Idea
- Discuss why the rule works: When you multiply both the dividend and the divisor by a power of ten, the quotient is not changed. You may want to remind students of the activity from the motivation of the lesson.
- Remind students that they need to multiply by a large enough power of ten to make the divisor a whole number. Only when you divide by a whole number can you place the decimal point in the quotient.

Example 1
- Work through parts (a) and (b). The original problem in part (a) can be read "182-tenths ÷ 14-tenths." Multiply the divisor and the dividend by 10 to get the equivalent problem $182 \div 14$.
- Do the check as part of each example to model this good habit for students.

On Your Own
- Note the two representations of division in these four problems.
- Ask students to estimate an answer before they begin. Do this as a class.

Goal Today's lesson is dividing decimals.

Lesson Materials
Introduction
- index cards

Start Thinking! and Warm Up

Lesson 3.5 **Warm Up** For use before Lesson 3.5

Lesson 3.5 **Start Thinking!** For use before Lesson 3.5
List all of the ways that you can check your answer for a division problem.

Extra Example 1
Divide.
a. $68.4 \div 5.7$ 12
b. $0.336 \div 0.42$ 0.8

On Your Own
1. 8
2. 17
3. 9.41
4. 0.9

T-134

Extra Example 2

Find $1.76 \div 0.004$. 440

 On Your Own

5. 23.75 6. 20
7. 150 8. 60

Extra Example 3

In 2008, there were 6.4 thousand commercial FM radio stations. In 1990, there were 4.4 thousand. How many times more commercial FM radio stations were there in 2008 than in 1990? Round your answer to the nearest tenth. about 1.5 times more

 On Your Own

9. about 6 times more

Differentiated Instruction

Visual

When dividing decimals, students may forget which number to make the whole number (the *divisor*) or they may forget to multiply the dividend by the same power of 10. Students may be careless in placing the decimal point in the quotient. The decimal point should be placed above the new decimal point in the dividend. Create a bulletin board display as a visual reminder.

whole number $0.05\overline{)2.655}$

Laurie's Notes

Example 2

- ❓ "How can you estimate the answer to $2.45 \div 0.007$?" Listen for students to explain that the quotient has to be greater than 2.45.
- Sometimes students focus on the process and forget about the mathematics behind the process. For example, they may talk about moving decimal points but lose sight of why doing so doesn't change the problem.
- To help students understand (and remind them of) the mathematics, write 2.4×1000 and 0.007×1000 off to the side.
- Remind them that $2.45 \div 0.007$ and $2450 \div 7$ will have the same quotient.
- ❓ "Is $0.03 \div 0.004$ *greater than one* or *less than one*? Why?" greater than one; The dividend is greater than the divisor.
- ❓ "What number do you need to multiply by to make the divisor a whole number?" 1000

On Your Own

- **Neighbor Check:** Have students work independently and then compare their answers with their neighbor. Have students discuss any differences.

Example 3

- ❓ **Review:** Ask questions about the graph: How many cell phone subscribers were there in 1994? In what year were there almost 70 million cell phone subscribers? Describe the general trend of the graph. 24.1; 1998; The graph is increasing.
- **Common Error:** Students often read this type of question as "How many more?" versus "How many *times* more?" Relate this to a question about themselves that they might understand, such as "How much older is your mom than you?" versus "How many times older is your mom than you?"
- ❓ "If you want to round to the nearest whole number, what place value do you need to look at?" the tenths
- Rounding to the nearest whole number may be difficult for students due to the decimal point. Remind students that rounding to the nearest whole number is still done in the same way as any other place value.

On Your Own

- Have students estimate first. Then work through the solution to the question with the class.

 Closure

- **Writing:** Use 8.4 and 0.42 to write an example whose quotient is greater than 1 and an example whose quotient is less than 1. Solve each problem. $8.4 \div 0.42 = 20$; $0.42 \div 8.4 = 0.05$

Technology For the Teacher

The Dynamic Planning Tool
Editable Teacher's Resources at *BigIdeasMath.com*

EXAMPLE 2 Inserting Zeros in the Dividend and Quotient

Find $2.45 \div 0.007$.

$0.007\overline{)2.450} \longrightarrow 7\overline{)2450}$

Multiply each number by 1000. Insert a zero in the dividend.

$$\begin{array}{r} 350 \\ 7\overline{)2450} \\ -21 \\ \hline 35 \\ -35 \\ \hline 00 \end{array}$$

$0 \div 7 = 0$. So, insert a zero in the quotient.

Study Tip
Remember to check your answer by multiplying the quotient by the divisor.

So, $2.45 \div 0.007 = 350$.

On Your Own

Divide. Check your answer.

Now You're Ready Exercises 13–24

5. $3.8 \div 0.16$
6. $15.6 \div 0.78$
7. $7.2 \div 0.048$
8. $0.18 \div 0.003$

EXAMPLE 3 Real-Life Application

How many times more cellular phone subscribers were there in 2006 than in 1990? Round to the nearest whole number.

From the graph, there were 233.04 million subscribers in 2006 and 5.3 million in 1990. So, divide 233.04 by 5.3.

Estimate $230 \div 5 = 46$

$5.3\overline{)233.04} \longrightarrow$

$$\begin{array}{r} 43.9 \\ 53\overline{)2330.4} \\ -212 \\ \hline 210 \\ -159 \\ \hline 51\ 4 \\ -47\ 7 \\ \hline 3\ 7 \end{array}$$

Rounds to 44.

Cellular Phone Subscribers

233.04
140.77
69.21
24.1
5.3

1990 1994 1998 2002 2006
Year

So, there were about 44 times more subscribers in 2006 than in 1990.

Reasonable? $44 \approx 46$ ✓

On Your Own

9. How many times more subscribers were there in 2002 than in 1994? Round your answer to the nearest whole number.

3.5 Exercises

Vocabulary and Concept Check

1. **NUMBER SENSE** Estimate $62.4 \div 6.5$.

Rewrite the problem so that the divisor is a whole number.

2. $4.7\overline{)13.6}$
3. $0.21\overline{)17.66}$
4. $2.16\overline{)18.5}$
5. $15.6 \div 4.7$
6. $9.28 \div 0.3$
7. $5.59 \div 0.647$

8. **DIFFERENT WORDS, SAME QUESTION** Which is different? Find "both" answers.

 9.6 is how many times greater than 1.6?

 What is the quotient of 9.6 and 1.6?

 1.6 is how much less than 9.6?

 What is 9.6 divided by 1.6?

Practice and Problem Solving

Divide.

9. $3.6 \div 0.3$
10. $2.6 \div 0.2$
11. $0.72 \div 0.06$
12. $0.36 \div 0.04$

Divide. Check your answer.

① ②
13. $2.1\overline{)25.2}$
14. $3.8\overline{)34.2}$
15. $0.4\overline{)14.6}$
16. $8.75\overline{)35.35}$
17. $78.54 \div 4.2$
18. $9.52 \div 1.19$
19. $36.47 \div 0.7$
20. $0.984 \div 12.3$
21. $4.23 \div 1.2$
22. $0.52 \div 0.0013$
23. $95.04 \div 13.2$
24. $32.2 \div 0.07$

25. **ERROR ANALYSIS** Describe and correct the error in rewriting the problem.

26. **TICKETS** Tickets to the school musical cost $6.25. The amount received from ticket sales is $706.25. How many tickets were sold?

27. **HEIGHT** A person's running stride is about 1.14 times his or her height. Your friend's stride is 5.472 feet. How tall is your friend?

28. **MP3 PLAYER** You have 3.4 gigabytes available on your MP3 player. Each song is about 0.004 gigabyte. How many more songs can you download onto your MP3 player?

Divide. Round to the nearest hundredth if necessary.

29. $44.8 \div 1.25$
30. $80.88 \div 8.425$
31. $9.9 \div 3.18$
32. $0.8 \div 0.6$
33. $38.9 \div 6.44$
34. $11.6 \div 0.95$
35. $0.35 \div 0.45$
36. $87.6 \div 9.24$

Assignment Guide and Homework Check

Level	Day 1 Activity Assignment	Day 2 Lesson Assignment	Homework Check
Basic	9–12, 45–49	1–8, 13–25 odd, 26–28	13, 17, 21, 28
Average	9–12, 45–49	1–8, 17–25 odd, 28–32, 41, 42	17, 21, 28, 32
Advanced	9–12, 45–49	1–8, 25, 28–40 even, 43, 44	28, 32, 38, 43

Common Errors
Exercises 13–24, 29–36
- Students may try to multiply by a power of 10 to make the dividend a whole number, instead of the divisor. While this method will still provide the same answer, it is easier for students to divide when the divisor is the smallest whole number possible. Remind them to multiply by whatever power of 10 will make the divisor a whole number.
- Some students may forget to multiply the dividend by the same multiple of 10 that they used to multiply the divisor. Remind them that they must multiply every number in the problem by the same multiple of 10.
- Students may place the decimal point in the wrong place as a result of sloppiness. Encourage them to use graph paper to keep their work organized.
- Remind students that they should check their answers by multiplying the quotient by the divisor and confirm that the product is equal to the dividend.

3.5 Record and Practice Journal

Technology For the Teacher
Answer Presentation Tool
QuizShow

Vocabulary and Concept Check

1. 10
2. $47\overline{)136}$
3. $21\overline{)1766}$
4. $216\overline{)1850}$
5. $156 \div 47$
6. $92.8 \div 3$
7. $5590 \div 647$
8. 1.6 is how much less than 9.6?; 8; 6

Practice and Problem Solving

9. 12
10. 13
11. 12
12. 9
13. 12
14. 9
15. 36.5
16. 4.04
17. 18.7
18. 8
19. 52.1
20. 0.08
21. 3.525
22. 400
23. 7.2
24. 460
25. The decimal point in the dividend should be moved to the right instead of the left; $0.32\overline{)146.4} \rightarrow 32\overline{)14,640}$
26. 113 tickets
27. 4.8 ft
28. 850 songs
29. 35.84
30. 9.6
31. 3.11
32. about 1.33
33. about 6.04
34. about 12.21
35. about 0.78
36. about 9.48

T-136

Practice and Problem Solving

37. = **38.** >

39. < **40.** <

41. 5357 bees

42. 1.7 times

43. See *Taking Math Deeper*.

44. **a.** 3 times

b. 9 times

c. no; The answers are different because perimeter is measured in units and area is measured in square units.

d. The answers become (a) 1.5 times and (b) 2.25 times.

Fair Game Review

45. 25% **46.** 70%

47. 21% **48.** 57%

49. B

Taking Math Deeper

Exercise 43

This is a good opportunity to help students practice unit analysis. With addition, subtraction, and multiplication, unit analysis is straightforward. With division, you need to "invert and multiply" in the same way you do when dividing fractions so that units divide out.

 Organize the given information.

24 oz
$2.63

3.9 oz
$0.52

 How many bowls of applesauce fit into the jar?

$$24 \frac{oz}{jar} \div 3.9 \frac{oz}{bowl}$$

$$= 24 \frac{oz}{jar} \cdot \frac{1}{3.9} \frac{bowls}{oz}$$

a. ≈ 6.15 bowls per jar "Bowls per jar" is a check that I did it right.

 b. and c. Which is the better buy?

- One way: Multiply. At the bowl price, you pay more for a jar.

$$0.52 \frac{\$}{bowl} \cdot 6.15 \frac{bowls}{jar} = \$3.20 \text{ per jar}$$

Buy the jar.

- 2nd way: Find the cost per ounce.

$$\frac{2.63}{24} \frac{\$}{oz} \approx \$0.11 \text{ per oz} \quad \text{Jar}$$

$$\frac{0.52}{3.9} \frac{\$}{oz} \approx \$0.13 \text{ per oz} \quad \text{Bowl}$$

Project

Find a recipe for applesauce. How many apples would it take to make a 24-ounce jar of applesauce? How many apples are in a 3.9-ounce bowl?

Mini-Assessment

Divide.

1. 7.6 ÷ 3.2 2.375
2. 15.47 ÷ 2.6 5.95
3. 7.25 ÷ 2.50 2.9
4. 0.75 ÷ 0.015 50
5. Cookies for a school fundraiser cost $3.75 per box. The total amount received from the cookie sale is $2718.75. How many boxes of cookies were sold? 725 boxes

Reteaching and Enrichment Strategies

If students need help…	If students got it…
Resources by Chapter • Practice A and Practice B • Puzzle Time Record and Practice Journal Practice Differentiating the Lesson Lesson Tutorials Skills Review Handbook	Resources by Chapter • Enrichment and Extension • School-to-Work • Financial Literacy Start the next section

T-137

Without finding the quotient, copy and complete the statement using <, >, or =.

37. 6.66 ÷ 0.74 ___ 66.6 ÷ 7.4 **38.** 32.2 ÷ 0.7 ___ 3.22 ÷ 7

39. 160.72 ÷ 16.4 ___ 160.72 ÷ 1.64 **40.** 75.6 ÷ 63 ___ 7.56 ÷ 0.63

41. BEES The formula $N = \dfrac{3x}{0.014}$ approximates the number N of bees in a hive based on the number x of bees seen leaving the hive in one minute. You see 25 bees leave the hive in one minute. About how many bees are in the hive?

42. ORANGES How many times more expensive was one pound of oranges in 2006 than in 1990? Round your answer to the nearest tenth.

43. PRECISION A store sells applesauce in two sizes.

 a. How many *bowls* of applesauce fit in a *jar*? Round your answer to the nearest hundredth.

 b. Explain two ways to find the better buy.

 c. What is the better buy?

Applesauce
3.9-ounce bowl $0.52
24-ounce jar $2.63

44. Geometry The large rectangle's dimensions are three times the dimensions of the small rectangle.

23.1 ft
49.2 ft

 a. How many times greater is the perimeter of the larger rectangle compared to the perimeter of the smaller rectangle?

 b. How many times greater is the area of the larger rectangle compared to the area of the smaller rectangle?

 c. Are the answers to parts (a) and (b) the same? *Explain* why or why not.

 d. What happens in parts (a) and (b) if the dimensions of the small rectangle are doubled?

Fair Game Review What you learned in previous grades & lessons

Write the fraction as a percent. *(Skills Review Handbook)*

45. $\dfrac{1}{4}$ **46.** $\dfrac{7}{10}$ **47.** $\dfrac{21}{100}$ **48.** $\dfrac{57}{100}$

49. MULTIPLE CHOICE Melissa earns $7.40 an hour working at a grocery store. She works 14.25 hours this week. How much will she earn? *(Section 3.3)*

 Ⓐ $83.13 Ⓑ $105.45 Ⓒ $156.75 Ⓓ $1054.50

3.4–3.5 Quiz

Divide. Use estimation to check your answer. *(Section 3.4)*

1. $4.75 \div 5$
2. $84.2 \div 4$
3. $5\overline{)8.4}$
4. $6\overline{)6.48}$

Divide. Round to the nearest hundredth if necessary. *(Section 3.5)*

5. $5.6 \div 0.7$
6. $1.8 \div 0.03$
7. $0.4\overline{)5.16}$
8. $0.16\overline{)4.6}$

Evaluate the expression. *(Section 3.4)*

9. $5.19 + 6.27 \div 3$
10. $8.49 \div 6 - 1.04$

Without finding the quotient, copy and complete the statement using <, >, or =. *(Section 3.5)*

11. $5.18 \div 0.48$ ▭ $51.8 \div 4.8$
12. $65.4 \div 0.8$ ▭ $6.54 \div 8$

13. **BROADWAY** The bar graph shows the yearly attendance at traveling Broadway shows. *(Section 3.4 and Section 3.5)*

 a. Suppose the attendance was the same each month in 2005. How many people attended each month?

 b. How many times more people attended shows in 2005 than in 2002?

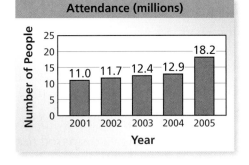

14. **RACE** You and three friends form a team to run a 3.1-mile relay race. Each of you runs the same distance. How far does each person run? *(Section 3.4)*

15. **RECYCLING** At a recycling center, you receive $1.30 for cans. Each can is worth $0.05. How many cans did you recycle? *(Section 3.5)*

16. **TILE** A floor is 10.6 feet by 12 feet. A tile covers 0.2 square feet. How many tiles will cover the floor? *(Section 3.4)*

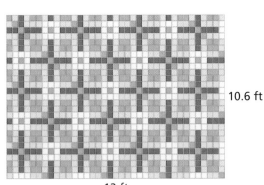

Alternative Assessment Options

Math Chat Student Reflective Focus Question
Structured Interview Writing Prompt

Structured Interview

Interviews can occur formally or informally. Ask a student to perform a task and to explain it as they work. Have them describe their thought process. Probe the student for more information. Do not ask leading questions. Keep a rubric or notes.

Teacher Prompts	Student Answers	Teacher Notes
Tell me a story about having a dog's teeth cleaned. Include this number sentence. 42 times 1.3 = ?	I took my dog to the vet to have his teeth cleaned. The vet charged me $1.30 per tooth for each of his 42 teeth. So the total cost was $54.60.	Student can correctly understand and apply the algorithm for multiplication of decimals and whole numbers.
Add to your story using this number sentence. 54.60 divided by 2.5 = ?	I asked if I could make $2\frac{1}{2}$ payments of $21.84 each.	Student can correctly understand and apply the algorithm for division of decimals by decimals.

Study Help Sample Answers

Remind students to complete Graphic Organizers for the rest of the chapter.

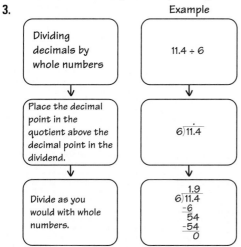

4–5. Available at *BigIdeasMath.com*

Reteaching and Enrichment Strategies

If students need help...	If students got it...
Resources by Chapter • Study Help • Practice A and Practice B • Puzzle Time Lesson Tutorials *BigIdeasMath.com* Practice Quiz Practice from the Test Generator	Resources by Chapter • Enrichment and Extension • School-to-Work Game Closet at *BigIdeasMath.com* Start the Chapter Review

Answers

1. 0.95
2. 21.05
3. 1.68
4. 1.08
5. 8
6. 60
7. 12.9
8. 28.75
9. 7.28
10. 0.375
11. =
12. >
13. a. $1.51\overline{6}$ million
 b. $1.\overline{5}$ times
14. 0.775 mile
15. 26 cans
16. 636 tiles

Technology For the Teacher
Answer Presentation Tool

Assessment Book

For the Teacher
Additional Review Options
- **Quiz**Show
- Big Ideas Test Generator
- Game Closet at *BigIdeasMath.com*
- Vocabulary Puzzle Builder
- Resources by Chapter
 Puzzle Time
 Study Help

Review of Common Errors

Exercises 1–6
- Remind students that for decimals less than 0.5, they round down. For decimals greater than or equal to 0.5, they round up.

Exercises 7–13
- Students may count the decimal places to the left of the decimal point instead of the right.
- Students may ignore the zeros when counting the number of decimal places, or forget to add them when finding the answer.

Exercises 14–20
- Students may not count the decimal places of both decimals factors.

Exercises 21–27
- Students may divide without putting the decimal point in the quotient.
- Students may have difficulty keeping their numbers organized and will forget to include placeholder zeros or put in extra placeholders.

Exercises 28–33
- Some students may struggle with determining the multiple of 10 to multiply by to get a whole number divisor.
- Students may try to multiply by the number of decimal places in the dividend instead of the divisor.

Answers

1. 32
2. 120
3. 8
4–6. Sample answers are given.
4. 69
5. 8
6. 8

3 Chapter Review

Review Examples and Exercises

3.1 Decimals and Estimation (pp. 106–111)

Estimate by rounding to the nearest whole number.

a. 9.6×16.2

$$9.6 \times 16.2 \approx 10 \times 16 = 160$$

9.6 rounds up to 10.

16.2 rounds down to 16.

∴ So, 9.6×16.2 is about 160.

b. $55.7 \div 8.43$

$$55.7 \div 8.43 \approx 56 \div 8 = 7$$

55.7 rounds up to 56.

8.43 rounds down to 8.

∴ So, $55.7 \div 8.43$ is about 7.

Use compatible numbers to estimate the product or quotient.

a. 32.21×4.87

$32.21 \times 4.87 \approx 30 \times 5$ 30 and 5 are compatible numbers.

$= 150$ Multiply.

∴ So, 32.21×4.87 is about 150.

b. $73.63 \div 24.88$

$73.63 \div 24.88 \approx 75 \div 25$ 75 and 25 are compatible numbers.

$= 3$ Divide.

∴ So, $73.63 \div 24.88$ is about 3.

Exercises

Estimate by rounding each factor to the nearest whole number.

1. 15.8×2.1
2. 24.45×5.27
3. $31.98 \div 4.21$

Use compatible numbers to estimate the product or quotient.

4. 22.9×3.02
5. $47.89 \div 6.99$
6. $87.57 \div 10.97$

3.2 Multiplying Decimals and Whole Numbers (pp. 112–117)

Find 3.6 × 4.

$$\begin{array}{r} 3.6 \\ \times\ \ 4 \\ \hline 14.4 \end{array}$$

3.6 ← One decimal place

14.4 ← Count one decimal place from right to left

Exercises

Multiply. Use estimation to check your answer.

7. 5.3 × 8
8. 6.1 × 7
9. 4.68 × 3
10. 8.194 × 9
11. 0.052 × 4
12. 0.302 × 5

13. **BIKING** Your friend lives 1.3 miles from you. One week, you rode your bike to your friend's house and back home three times. How many miles did you ride?

3.3 Multiplying Decimals (pp. 118–123)

Find 7.5 × 5.3.

$$\begin{array}{r} 7.5 \\ \times\ 5.3 \\ \hline 2\ 2\ 5 \\ +\ 3\ 7\ 5\ \ \\ \hline 3\ 9.7\ 5 \end{array}$$

7.5 ← 1 decimal place
× 5.3 ← + 1 decimal place
39.75 ← 2 decimal places

So, 7.5 × 5.3 = 39.75.

Exercises

Multiply. Use estimation to check your answer.

14. 4.8 × 1.9
15. 6.2 × 5.18
16. 5.45 × 3.57
17. 9.475 × 8.03
18. 0.27 × 4.42
19. 0.051 × 0.244

20. **AREA** Find the area of the computer screen.

13.8 in.
10.4 in.

Review Game
Multiplying and Dividing Decimals

Big Ideas Game Closet

Materials per Group
- 2 number cubes
- 2 coins (or integer counters)
- paper
- pencil

Players: groups of 4

Directions
- Each group of four divides into teams of two. Each team competes with the other team.
- Each team tosses a coin. One coin determines whether multiplication (heads) or division (tails) will be used by the group. The other coin determines whether estimation (heads) or exact calculation (tails) will be used by the group.
- Each team rolls a number cube twice to determine the numbers to be used. For example, if one team rolls a 5 and then a 3, the first number is 5.3. If the other team rolls a 4 and then a 2, the other number is 4.2.

Now the race is on!

Using the determined numbers and operations, each team estimates or calculates the answer. The team that finishes first has their work checked by the challenging team. If the challenging team accepts the answer, the team that finishes first wins the round. If the challenging team finds an error, the challenging team wins the round.

If the operation is carried out with estimation, the winning team receives one point. If the operation is carried out with exact calculation, the winning team receives two points.

Who Wins?
After a set amount of time, the team with the most points wins.

For the Student
Additional Practice
- Lesson Tutorials
- Study Help (textbook)
- Student Website
 Multi-Language Glossary
 Practice Assessments

Answers

7. 42.4
8. 42.7
9. 14.04
10. 73.746
11. 0.208
12. 1.51
13. 7.8 miles
14. 9.12
15. 32.116
16. 19.4565
17. 76.08425
18. 1.1934
19. 0.012444
20. 143.52 in.2
21. 1.7
22. 6.2
23. 7.1
24. 0.46
25. 0.35
26. 7.25
27. $30.12
28. 3
29. 0.13
30. 6.48
31. 30
32. 0.05
33. 5.7

My Thoughts on the Chapter

What worked...

Teacher Tip
Not allowed to write in your teaching edition? Use sticky notes to record your thoughts.

What did not work...

What I would do differently...

3.4 Dividing Decimals by Whole Numbers (pp. 126–131)

Find 0.64 ÷ 5.

```
      0.128
  5 ) 0.640
     − 5
       14
     − 10
        40
      − 40
         0
```

Place the decimal point in the quotient above the decimal point in the dividend.

Insert a zero and continue to divide.

∴ So, 0.64 ÷ 5 = 0.128.

Exercises

Evaluate the expression.

21. 6.8 ÷ 4
22. 13.2 ÷ 6 + 4
23. 49.7 ÷ 7
24. 3.68 ÷ 8
25. 3.15 ÷ 9
26. 46.25 ÷ 5 − 2

27. **SOCCER LEAGUE** It costs $542.24 for a team to play in an indoor soccer league. There are 18 players on the team and they split the cost evenly. How much does each player pay?

3.5 Dividing Decimals (pp. 132–137)

Find 22.8 ÷ 1.2.

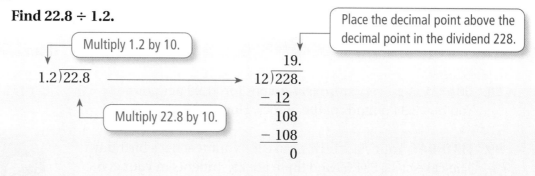

∴ So, 22.8 ÷ 1.2 = 19.

Exercises

Divide.

28. 1.2 ÷ 0.4
29. 0.195 ÷ 1.5
30. 34.992 ÷ 5.4
31. 0.12) 3.6
32. 2.5) 0.125
33. 3.9) 22.23

3 Chapter Test

Estimate by rounding each factor to the nearest whole number.

1. 6.3×7.5
2. $54.2 \div 8.6$

Use compatible numbers to estimate the product or quotient.

3. 26.32×5.54
4. $87.84 \div 9.45$

Multiply. Use estimation to check your answer.

5. $\begin{array}{r} 6.7 \\ \times\ \ 8 \\ \hline \end{array}$
6. $\begin{array}{r} 0.4 \\ \times\ 0.7 \\ \hline \end{array}$
7. 0.023×3
8. 4.87×7.23

Divide. Check your answer by multiplying.

9. $5.6 \div 7$
10. $2.6 \div 0.02$
11. $4\overline{)9.32}$
12. $0.25\overline{)5.46}$

Evaluate the expression.

13. $28.75 \div 5 \times 0.4$
14. $8.253 \div 0.9 + 8.65$
15. $1.25 \times 3.2 \div 0.8$
16. $17.3 \times (7.96 \div 4)$

17. **DVD SALE** Which deal is the better buy?

18. **GRAPES** A grocery store sells grapes for $1.99 per pound. You buy 2.34 pounds of the grapes. How much do you pay?

19. **BIRTHDAY** Your class buys a cake for your teacher's birthday. The cake costs $34.50 and there are 25 students in your class. How much does each student pay?

20. **PHOTOGRAPHY** A motocross rider is in the air for 2.5 seconds. Your camera can take a picture every 0.125 second. Your friend's camera can take a picture every 0.15 second.

 a. How many times faster is your camera than your friend's camera?

 b. How many more pictures can you take of the rider?

Test Item References

Chapter Test Questions	Section to Review	Common Core State Standards
1–4	3.1	6.NS.3
5–8	3.2	6.NS.3
5–8, 18, 20	3.3	6.NS.3
9–16, 17, 20	3.4	6.NS.2, 6.NS.3
9–12, 19, 20	3.5	6.NS.2, 6.NS.3

Test-Taking Strategies

Remind students to quickly look over the entire test before they start so that they can budget their time. When working through the test, students should mentally read the numbers to ensure correct decimal point placement. Remind students to use **Stop** and **Think** strategies to ensure that they understand what is being asked. Estimating the answer will also help ensure correct decimal point placement.

Common Assessment Errors

- **Exercises 1 and 2** Students may not round correctly. It may help to have students circle the tenths place before rounding to the nearest whole number.
- **Exercises 6 and 8** Students may only count the number of decimal places in one number instead of two. Remind them that they need to count *both* sets of decimal places. Stress estimation as a check for reasonableness.
- **Exercise 7** Students may forget to count the zero to the right of the decimal point when placing the decimal point in the answer. Remind students that zeros are placeholders and are important for determining the correct answer.
- **Exercises 9–12** Students may place the decimal point in the wrong place. Encourage them to use grid paper to keep their work organized. Remind students to check their answers by multiplying.
- **Exercises 10 and 12** Students may forget to multiply the dividend by the same multiple of 10 that they used to multiply the divisor. Remind them that they must multiply *both* numbers by the same multiple of 10.
- **Exercises 13 and 14** Students may try to evaluate the multiplication or addition first. Remind students of the order of operations.

Reteaching and Enrichment Strategies

If students need help...	If students got it...
Resources by Chapter • Practice A and Practice B • Puzzle Time Record and Practice Journal Practice Differentiating the Lesson Lesson Tutorials Practice from the Test Generator Skills Review Handbook	Resources by Chapter • Enrichment and Extension • School-to-Work • Financial Literacy Game Closet at *BigIdeasMath.com* Start Standardized Test Practice

Answers

1. 48
2. 6

3–4. Sample answers are given.

3. 150
4. 10
5. 53.6
6. 0.28
7. 0.069
8. 35.2101
9. 0.8
10. 130
11. 2.33
12. 21.84
13. 2.3
14. 17.82
15. 5
16. 34.427
17. 5 for $58.99
18. $4.66
19. $1.38
20. a. 1.2 times faster
 b. 4 more pictures

Assessment Book

Test-Taking Strategies
Available at *BigIdeasMath.com*

After Answering Easy Questions, Relax
Answer Easy Questions First
Estimate the Answer
Read All Choices before Answering
Read Question before Answering
Solve Directly or Eliminate Choices
Solve Problem before Looking at Choices
Use Intelligent Guessing
Work Backwards

About this Strategy

When taking a multiple choice test, be sure to read each question carefully and thoroughly. It is also very important to read each answer choice carefully. Do not pick the first answer that you think is correct! If two answer choices are the same, eliminate them both. There can only be one correct answer.

Answers

1. B
2. H
3. $0.186 per ounce
4. B

Technology For the Teacher
Big Ideas Test Generator

T-143

Item Analysis

1. **A.** The student ignores the $\frac{1}{2}$ and finds the quotient of 2 and 10.
 B. Correct answer
 C. The student recognizes the quotient of $2\frac{1}{2}$ and 10 as 0.25, and then uses the digits of the decimal part of the number as the numerator and denominator of the fraction.
 D. The student finds the quotient of $2\frac{1}{2}$ and 10, and then subtracts this quotient from 1.

2. **F.** The student ignores the decimal points and finds $8 \times 5 \times 6$.
 G. The student thinks $0.8 \times 0.5 = 4$ and multiplies this product by 6.
 H. Correct answer
 I. The student thinks $0.8 \times 0.5 = 0.04$ and multiplies this product by 6.

3. **Gridded Response:** Correct answer: $0.186 per ounce
 Common Error: The student thinks that the quotient has two decimal places because the dividend has two decimal places, therefore getting 1.86.

4. **A.** The student multiplies the dividend and the divisor and uses this product as the decimal part of the number.
 B. Correct answer
 C. The student divides in the wrong order and rounds the quotient to the nearest hundredth.
 D. The student uses the dividend as the ones digit and the divisor as the tenths digit.

5. **F.** The student puts five digits in the answer because there are five digits total in *a*, *b*, and *c*.
 G. The student thinks that $0.6 \div 0.03 = 0.02$ by aligning the decimal points and then multiplies this quotient by 0.02.
 H. The student thinks that $0.6 \div 0.03 = 0.02$ by aligning the decimal points and then thinks that $0.02 \times 0.02 = 0.04$ by aligning the decimal points.
 I. Correct answer

6. **A.** The student moves the decimal point in the dividend two places, but in the wrong direction, and does not move the decimal point in the divisor.
 B. The student moves the decimal point in the dividend correctly but does not move the decimal point in the divisor.
 C. The student moves the decimal point in the divisor correctly but moves the decimal point two places in the wrong direction.
 D. Correct answer

3 Standardized Test Practice

1. At a party, 10 people equally shared $2\frac{1}{2}$ gallons of ice cream. How much ice cream did each person eat? *(6.NS.1)*

 A. $\frac{1}{5}$ gal

 B. $\frac{1}{4}$ gal

 C. $\frac{2}{5}$ gal

 D. $\frac{3}{4}$ gal

Test-Taking Strategy
Read All Choices Before Answering

"Reading all choices before answering can get you a lot more yummy treats!"

2. Stan draws sketches on rectangular sheets of paper. The dimensions of each sheet of paper are shown below.

 0.5 m

 0.8 m

 Stan will hang 6 of these sketches on the wall. If they do not overlap, what is the total area they will cover? *(6.NS.3)*

 F. 240 m²

 G. 24 m²

 H. 2.4 m²

 I. 0.24 m²

3. A jar of peanut butter costs $2.79. What is the *exact* cost, in dollars per ounce, of the peanut butter? *(6.NS.2)*

4. Which number is equivalent to 7 ÷ 8? *(6.NS.2)*

 A. 0.56

 B. 0.875

 C. 1.14

 D. 7.8

5. What is the value of the expression below when $a = 0.02$, $b = 0.6$, and $c = 0.03$? *(6.NS.3)*

$$a(b \div c)$$

F. 0.00004

G. 0.0004

H. 0.04

I. 0.4

6. Your friend divided two decimal numbers. Her work is shown in the box below. What should your friend change in order to divide the two decimal numbers correctly? *(6.NS.3)*

$$0.07\overline{)14.56} \rightarrow 7\overline{)14.56}^{2.08}$$

A. Rewrite the problem as $0.07\overline{)0.1456}$.

B. Rewrite the problem as $0.07\overline{)1456}$.

C. Rewrite the problem as $7\overline{)0.1456}$.

D. Rewrite the problem as $7\overline{)1456}$.

7. What is the value of the expression below? *(6.NS.1)*

$$8\frac{1}{2} \times 5\frac{3}{4}$$

F. $40\frac{3}{8}$

G. $41\frac{1}{4}$

H. $48\frac{1}{4}$

I. $48\frac{7}{8}$

8. David's mother flew on an airplane for a business trip. The airplane averaged 500 miles per hour for the entire 1250 mile flight. The airplane spent 45 minutes on the runway. How many hours did David's mother spend on the airplane? *(6.NS.1)*

9. Which expression is equivalent to the expression below? *(6.EE.3)*

$$2(m + n)$$

A. $2m \times 2n$

B. $2m + 2n$

C. $(2 + m) \times (2 + n)$

D. $(2 + m) + (2 + n)$

Chapter 3 Multiplying and Dividing Decimals

Item Analysis (continued)

7. **F.** The student multiplies the whole numbers together and the fractions together, and then uses these two products to make a mixed number.

 G. The student multiplies the whole numbers together, and then adds the two fractions to this product.

 H. The student rounds $8\frac{1}{2}$ down to 8 and rounds $5\frac{3}{4}$ up to 6, and then multiplies these two rounded numbers. The student then adds $8\frac{1}{2} - 8 = \frac{1}{2}$ to and subtracts $6 - 5\frac{3}{4} = \frac{1}{4}$ from this product.

 I. Correct answer

8. **Gridded Response:** Correct answer: $\frac{13}{4}$ or 3.25 hours

 Common Error: The student thinks that 45 minutes equals 0.45 hour and adds 0.45 to 2.5 to get 2.95.

9. **A.** The student distributes the multiplication by 2 correctly, but then multiplies the resulting terms instead of adding them.

 B. Correct answer

 C. The student incorrectly distributes the multiplication by 2 by using addition, and then multiplies the resulting sums.

 D. The student incorrectly distributes the multiplication by 2 by using addition.

10. **F.** The student represents the situation with division and finds $2.6 \div 5.2 = 0.50$.

 G. The student represents the situation with division and finds $2.6 \div 5.2 = 0.50$. The student then aligns the decimal point in the quotient with the decimal point in the dividend and divisor.

 H. The student represents the situation with addition and finds $2.6 + 5.2 = 7.80$.

 I. Correct answer

Answers

5. I
6. D
7. I
8. $\frac{13}{4}$ or 3.25 hours
9. B

Answers

10. I
11. C
12. F
13. $2.92 or $2.93

Answers for Extra Examples

1. **A.** The student incorrectly finds the number of quarters in $80 and/or the number of pennies in $4 (and may make other errors).

 B. The student incorrectly finds the number of dimes in $25 and/or the number of pennies in $4 (and may make other errors).

 C. The student incorrectly finds the number of nickels in $15 and/or the number of pennies in $4 (and may make other errors).

 D. Correct answer

2. **F.** The student puts two zeros in the product because there are two total digits in the factors.

 G. Correct answer

 H. The student puts four digits in the product because there are four total zeros in the factors.

 I. The student puts four digits in the product because each of the factors has four digits.

3. **Gridded Response:** Correct answer: $19.14

 Common Error: The student multiplies the whole numbers together and the decimal parts together and finds $1 \times 15 + 2 \times 0.95$, which equals 16.90.

Item Analysis (continued)

11. **A.** The student ignores the decimal point of the divisor and approximates $100 \div 19$.

 B. The student thinks that 0.19 is approximately 0.02, then ignores the decimal point in the approximated divisor, and finds $100 \div 2$.

 C. Correct answer

 D. The student thinks that 0.19 is approximately 0.02 and then finds $100 \div 0.02$.

12. **F.** Correct answer

 G. The student does not find the reciprocal of $\frac{10}{3}$.

 H. The student does not change division to multiplication.

 I. The students incorrectly rewrites $3\frac{1}{3}$ as $\frac{7}{3}$ by adding 3, 3, and 1 to get 7.

13. **2 points** The student demonstrates a thorough understanding of multiplication and division of decimal numbers and whole numbers. The student correctly finds the unit cost for an ounce of oats and an ounce of sugar, and then uses these unit costs to find the total cost of $2.92 (or $2.93).

 1 point The student demonstrates a partial understanding of multiplication and division of decimal numbers and whole numbers. The student makes some correct computations but does not successfully answer the question.

 0 points The student demonstrates insufficient understanding of multiplication and division of decimal numbers and whole numbers. The student makes many errors and shows little or no ability to compute with decimal numbers.

Extra Examples for Standardized Test Practice

1. Which of the following quantities is the greatest?

 A. the number of quarters in $80 **C.** the number of nickels in $15
 B. the number of dimes in $25 **D.** the number of pennies in $4

2. Which expression is equivalent to 0.0028?

 F. 0.7×0.4 **H.** 0.007×0.004
 G. 0.07×0.04 **I.** 0.0007×0.0004

3. You bought 1.2 pounds of chocolates for $15.95 per pound. How much did you pay?

10. You bought some grapes at a farm stand. You paid $2.60 per pound.

What was the total amount that you paid for the grapes? *(6.NS.3)*

 F. $0.50 **H.** $7.80

 G. $5.00 **I.** $13.52

11. It costs approximately $0.19 to dry a load of laundry in a clothes dryer that runs on natural gas. Which of the following is closest to the number of loads that can be dried for $100? *(6.NS.3)*

 A. 5 **C.** 500

 B. 50 **D.** 5000

12. Which expression is equivalent to the expression below? *(6.NS.1)*

$$k \div 3\frac{1}{3}$$

 F. $k \cdot \dfrac{3}{10}$ **H.** $k \div \dfrac{3}{10}$

 G. $k \cdot \dfrac{10}{3}$ **I.** $k \div \dfrac{7}{3}$

13. You need 22 ounces of oats and 12 ounces of sugar for a cookie recipe.

What is the total cost, in dollars to the nearest cent, of the oats and sugar for your recipe? Show your work and explain your reasoning. (1 pound = 16 ounces) *(6.NS.3)*

4 Fractions, Decimals, and Percents

- **4.1 Percents and Fractions**
- **4.2 Percents and Decimals**
- **4.3 Comparing and Ordering Fractions, Decimals, and Percents**
- **4.4 Finding the Percent of a Number**
- **4.5 Percents and Estimation**

"At 4 a day, I have chewed 17,536 dog biscuits. At only 99.9% pure, that means that..."

"I hope the same logic doesn't apply for mice!"

"I have swallowed seventeen and a half contaminated dog biscuits during the past twelve years."

"It says 75% tomatoes, 15% sugar, 5% vinegar, 4% water, and 1% salt."

"See... no cats in catsup."

"That's a relief!"

Connections to Previous Learning

- Write common fractions as decimals and percents, only include halves, fourths, tenths, and hundredths.
- Locate fractions on a number line.
- Decimals may have an equivalent fraction where the denominator is (or can be) a power of 10.
- Compare and order decimals.

- Define the benchmark fractions.

- Use equivalent forms of fractions, decimals, and percents.
- Compare and order fractions, decimals, and percents, including using a number line.
- Estimate the results of computations with percents, and verify reasonableness.

Math in History

Most of what ancient people knew about mathematics is lost. However, occasionally, a fabulous record is found.

★ The Rhind papyrus (about 1650 B.C.) is an Egyptian textbook about arithmetic and geometry. The book gives area formulas and methods for multiplication, division, and prime numbers.

★ The Surya Siddhanta (about 400 A.D.) is an Indian textbook that gives the length of a year as 365.2563627 days, which is only 1.4 seconds longer than the modern value of 365.25636305 days. This book was translated to Arabic and Latin during the Middle Ages.

Pacing Guide for Chapter 4

Chapter Opener	1 Day
Section 1	
Activity	1 Day
Lesson	1 Day
Section 2	
Activity	1 Day
Lesson	1 Day
Section 3	
Activity	1 Day
Lesson	1 Day
Study Help / Quiz	1 Day
Section 4	
Activity	1 Day
Lesson	1 Day
Section 5	
Activity	1 Day
Lesson	1 Day
Quiz / Chapter Review	1 Day
Chapter Test	1 Day
Standardized Test Practice	1 Day
Total Chapter 4	15 Days
Year-to-Date	67 Days

Check Your Resources

- Record and Practice Journal
- Resources by Chapter
- Skills Review Handbook
- Assessment Book
- Worked-Out Solutions

The Dynamic Planning Tool
Editable Teacher's Resources at
BigIdeasMath.com

Common Core State Standards

5.NBT.3a Read and write decimals to thousandths using base-ten numerals, number names, and expanded form,

4.NF.7 Compare two decimals to hundredths by reasoning about their size. . . . Record the results . . . and justify the conclusions . . . by using a visual model.

6.NS.2 Fluently divide multi-digit numbers using the standard algorithm.

Additional Topics for Review
- Place value
- The Division Algorithm
- Estimation with decimals

Try It Yourself
1. two hundred one ten thousandths
2. seven tenths
3. eight hundred thousandths
4. one and twenty-three thousandths
5. 0.0102
6. 0.676
7. 0.01, 0.2, 0.42, 0.5
8. 0.75, 0.95, 1.01, 1.05
9. $\frac{41}{100}$
10. $\frac{521}{1000}$
11. 0.9
12. 0.625

Record and Practice Journal
1. eight tenths
2. one hundred eighty-five thousandths
3. one and sixty-seven hundredths
4. five hundred six ten thousandths
5. 0.38
6. 0.724
7. 0.0009
8. 2.5
9. fifteen and eighty-seven hundredths
10. 0.2, 0.4, 0.54, 0.61
11. 1.02, 1.24, 1.33, 1.7
12. 0.87, 0.9, 0.98, 1.23
13. 0.003, 0.03, 0.031, 0.033
14. $\frac{11}{20}$
15. $\frac{121}{500}$
16. 0.4
17. 0.6875
18. $\frac{29}{50}$

T-147

Math Background Notes

Vocabulary Review
- Divisor
- Numerator
- Denominator

Naming Decimals
- Students should know how to name decimals and work with place value.
- Encourage students to use correct punctuation and capitalization when they are writing decimals in words.
- **Teaching Tip:** This section can be difficult for English Language Learners. Help them to see the difference between twenty-five thousand (25,000) and twenty-five thousandths (0.025). Reading the decimals out loud before writing them in words can also be helpful.
- **Common Error:** Students will often forget to insert the zero to the right of the decimal. This zero holds the place so that the last digit in the number appears in the thousandths place.

Ordering Decimals
- Students should know how to order decimals.

Writing Decimals and Fractions
- Students learned to convert between decimals and fractions. Because this is a fairly recent skill, students may still need practice to achieve mastery.
- Naming decimals provides a good review of place value. This should help students convert decimals to fractions.
- You may wish to review the division algorithm prior to Example 4.
- **Common Error:** Students may use the numerator of the given fraction as the divisor in the corresponding division problem. Remind students that the fraction bar can be read as "divided by." Encourage students to read the problem out loud from top to bottom as they write their division problem. If the fraction is less than one, the equivalent decimal should be less than one.

Reteaching and Enrichment Strategies

If students need help. . .	If students got it. . .
Record and Practice Journal • Fair Game Review Skills Review Handbook Lesson Tutorials	Game Closet at *BigIdeasMath.com* Start the next section

What You Learned Before

- ## Naming Decimals (5.NBT.3a)

 Example 1

 a. Write 0.251 in words. **∴** Two hundred fifty-one thousandths

 b. Write *twenty-five thousandths* as a decimal. **∴** 0.025

 ### Try It Yourself
 Write the decimal in words.

 1. 0.0201 **2.** 0.7 **3.** 0.00008 **4.** 1.023

 Write the number as a decimal.

 5. One hundred two ten thousandths **6.** Six hundred seventy-six thousandths

- ## Ordering Decimals (4.NF.7)

 Example 2 Use a number line to order 0.25, 1.15, 0.2, and 0.34 from least to greatest.

 ### Try It Yourself
 Use a number line to order the numbers from least to greatest.

 7. 0.01, 0.42, 0.2, 0.5 **8.** 1.05, 0.95, 0.75, 1.01

- ## Writing Decimals and Fractions (6.NS.2)

 Example 3 Write 0.21 as a fraction. **Example 4** Write $\frac{6}{25}$ as a decimal.

 $$0.21 = \frac{21}{100}$$ $$\frac{6}{25} = 25\overline{)6} = 0.24$$

 ### Try It Yourself
 Write the decimal as a fraction or the fraction as a decimal.

 9. 0.41 **10.** 0.521 **11.** $\frac{9}{10}$ **12.** $\frac{5}{8}$

4.1 Percents and Fractions

COMMON CORE STATE STANDARDS
6.RP.3c

Essential Question How can you use a model to write a percent as a fraction or write a fraction as a percent?

The Meaning of a Word • Percent

A century is 100 years.

A cent is one hundredth of a dollar.

In Mexico, a centavo is one hundredth of a peso.

"Cent" means one hundred, so "percent" means per one hundred. The symbol for percent is **%**.

Sample: $30\% = \dfrac{30}{100}$ ← per ← cent

1 ACTIVITY: Writing Percents as Fractions

Work with a partner. Write the percent shown by the model. Write the percent as a fraction with a denominator of 100. Simplify the fraction.

a. Sample: b. c.

$30\% = \dfrac{30}{100} = \dfrac{3}{10}$

2 ACTIVITY: Writing Percents as Fractions

Work with a partner. Draw a model to represent the percent. Write the percent as a fraction with a denominator of 100. Simplify the fraction.

a. 60% b. 5% c. 85% d. 28%

148 Chapter 4 Fractions, Decimals, and Percents

Laurie's Notes

Introduction

For the Teacher
- **Goal:** The goal of this activity is for students to use visual models to represent a percent.
- Relating percents to money (part of a dollar) is often a helpful technique.

Motivate
- Shake a few coins in your hand to draw attention to the money.
- ? Ask students to write each U.S. coin as a fractional amount of $1.00.
 penny $\frac{1}{100}$, nickel $\frac{5}{100}$, dime $\frac{10}{100}$, quarter $\frac{25}{100}$, half-dollar $\frac{50}{100}$, dollar $\frac{100}{100}$
- Be sure to include the presidential $1 coins.

Activity Notes

Discuss
- The word *percent* is used in common language. Discuss applications and contexts.
- ? "Where do you hear the word *percent* used? Give an example of how it is used." *Sample answers:* 80% on a test; 50% chance of rain; 25% off on a sale item; 6% interest rate
- For each example given, ask students to explain.
- Discuss the examples in the book (century, cent, and centavo) and the notation.

Activity 1
- Make sure students understand the 100-grid model. Ten rows and ten columns make 100 parts.
- ? "Where have you seen this model before?" decimals; Some students may say it looks like Base 10 pieces for whole numbers—flats, strips, units.
- Check answers to each problem.
- ? "Does it matter where the squares are shaded?" No; Could be checkerboard style or random.
- **Summarize:** Be sure to tell students that when the percent is given, for instance 40%, the percent symbol (%) means writing the number as a fraction with a denominator of 100. Write $\frac{40}{100}$, then say "to complete, be sure to simplify the fraction to $\frac{2}{5}$."

Activity 2
- Students often like to be creative when sketching a model. Discuss why it is hard to get a quick read or estimate of a percent model when the shading is not a compact rectangle.

Common Core State Standards

6.RP.3c Find a percent of a quantity as a rate per 100; solve problems involving finding the whole, given a part and the percent.

Previous Learning

Students should know how to write halves, fourths, tenths, and hundredths as percents. Students also need to be able to convert between fractions and decimals, and to simplify and write equivalent fractions.

Activity Materials	
Introduction	Textbook
• Sample of U.S. coins	• 100 grids

Start Thinking! and Warm Up

Activity 4.1 Start Thinking! For use before Activity 4.1

Activity 4.1 Warm Up For use before Activity 4.1

Rewrite the fraction with a denominator of 100.

1. $\frac{12}{20}$ 2. $\frac{1}{4}$ 3. $\frac{15}{25}$

4. $\frac{33}{50}$ 5. $\frac{27}{20}$ 6. $\frac{54}{10}$

4.1 Record and Practice Journal

T-148

English Language Learners
Visual
In their notebooks, have students make a table showing the fractional equivalents of common percents.

$10\% = \frac{1}{10}$	$20\% = \frac{1}{5}$	$25\% = \frac{1}{4}$
$30\% = \frac{3}{10}$	$40\% = \frac{2}{5}$	$50\% = \frac{1}{2}$
$60\% = \frac{3}{5}$	$70\% = \frac{7}{10}$	$75\% = \frac{3}{4}$
$80\% = \frac{4}{5}$	$90\% = \frac{9}{10}$	$100\% = 1$

4.1 Record and Practice Journal

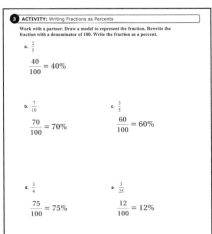

Laurie's Notes

Activity 3

- This activity reverses the process of what students did in Activity 2. In Activity 2, they wrote a percent as a fraction and now they are going to write a fraction as a percent. Have students sketch a model of the fraction. Then have them write an equivalent fraction with a denominator of 100. Then write the percent.

- **FYI:** Sketch fractions on a transparency using a square model that is the same size as a square 100 grid. After shading the fraction $\left(\text{say } \frac{3}{5}\right)$, overlay the transparency that has the 100 grid. Students can see that the area shaded has not changed, only the size of the pieces.

? Extension: "You sketched models for $\frac{3}{5}, \frac{3}{4},$ and $\frac{3}{25}$, and were able to write them as percents. If you wanted to write a fraction as a percent, which would be easier, $\frac{3}{10}$ or $\frac{3}{7}$ and why?" $\frac{3}{10}$; 10 is a factor of 100 and 7 is not.

What Is Your Answer?

- Sketching models to answer questions similar to Question 7 is a useful skill.
- **Extension:** Show a model. Ask students what questions could be answered by looking at the model.

$1.25	$1.25	$1.25	$1.25

30 points	30 points	30 points	30 points	30 points

Sample answers:
$\frac{1}{4}$ or 25% of $5 is $1.25. $\frac{2}{5}$ or 40% of 150 points is 60 points.

Closure

- Write this list on the board and ask students to put the numbers in order from least to greatest.

$\frac{1}{2}$, 60%, $\frac{2}{5}$, 75%, $\frac{3}{10}$, 90%, $\frac{3}{20}$, 25% $\frac{3}{20}$, 25%, $\frac{3}{10}$, $\frac{2}{5}$, $\frac{1}{2}$, 60%, 75%, 90%

Technology For the Teacher

Dynamic Classroom

The Dynamic Planning Tool
Editable Teacher's Resources at *BigIdeasMath.com*

3 ACTIVITY: Writing Fractions as Percents

Work with a partner. Draw a model to represent the fraction. Rewrite the fraction with a denominator of 100. Write the fraction as a percent.

a. Sample: $\dfrac{2}{5} = \dfrac{40}{100} = 40\%$

b. $\dfrac{7}{10}$ c. $\dfrac{3}{5}$ d. $\dfrac{3}{4}$ e. $\dfrac{3}{25}$

What Is Your Answer?

4. **IN YOUR OWN WORDS** How can you use a model to write a percent as a fraction or write a fraction as a percent? Give an example with your answer.

5. Fractions that are terminating decimals are easier to write as percents than fractions that are repeating decimals. Write the percent shown by the model as a fraction. Explain your reasoning.

$33\dfrac{1}{3}\%$		

6. One way to answer a question about a percent is to write the percent as a fraction.

 a. Write the following question using a fraction.

 "How much is 50% of $2.00?"

 b. Use what you know about fractions to answer the question.

"Dear Sir, you could save a letter in writing 50% OFF by simply writing 50% ON."

7. A notebook has an original price of $8.00. The notebook is on sale for 75% of the original price. Use a model to determine how much you will pay for the notebook.

Practice — Use what you learned about percents and fractions to complete Exercises 4–7 and 17–20 on page 152.

Section 4.1 Percents and Fractions 149

4.1 Lesson

Check It Out
Lesson Tutorials
BigIdeasMath com

Key Vocabulary
percent, p. 150

A **percent** is the number of parts per one hundred.

$$60\% = 60 \text{ out of } 100 = \frac{60}{100}$$
(parts per one hundred)

Key Idea

Writing Percents as Fractions

Words Write the percent as a fraction with a denominator of 100. Then simplify if possible.

Numbers $30\% = \dfrac{30}{100} = \dfrac{3}{10}$ **Algebra** $n\% = \dfrac{n}{100}$

EXAMPLE 1 — Writing Percents as Fractions

a. Write 35% as a fraction in simplest form.

$35\% = \dfrac{35}{100}$ Write as a fraction with a denominator of 100.

$= \dfrac{7}{20}$ Simplify.

b. Write 174% as a mixed number in simplest form.

$174\% = \dfrac{174}{100}$ Write as a fraction with a denominator of 100.

$= \dfrac{87}{50}$, or $1\dfrac{37}{50}$ Simplify.

∴ So, $174\% = 1\dfrac{37}{50}$.

On Your Own

Now You're Ready
Exercises 4–15

Write the percent as a fraction or mixed number in simplest form.

1. 15%
2. 168%
3. 36%
4. 83%

Chapter 4 Fractions, Decimals, and Percents Multi-Language Glossary at BigIdeasMath com.

Laurie's Notes

Introduction

Connect
- **Yesterday:** Students used visual models to represent percents, and then wrote the percents as fractions.
- **Today:** Students will convert between percents and fractions.

Motivate
- Review *benchmark fractions* and *percents*.
- Students should start to memorize, and have quick recall of benchmark fractions and percents: halves, fourths, fifths, and tenths.
- Students have little difficulty writing a percent as a fraction. Writing fractions as percents is more challenging.
- As a motivator today, use any game format *Matching, Memory, I Have, Who Has?*, and so on, to review these benchmarks.

Lesson Notes

Key Idea
- Write the Key Idea.
- Discuss the algebraic representation.

Example 1
- Work through both parts of the example.
- ? "What is 100% as a fraction?" 1
- ? "If the percent is greater than 100%, what do you know about the equivalent fraction?" It will be greater than 1.
- ? "Can you think of a context where the percent would be greater than 100%?" *Sample answer:* When you give 110%, it means that you give more than expected.

On Your Own
- **Think-Pair-Share:** Students should read each question independently and then work with a partner to answer the questions. When they have answered the questions, the pair should compare their answers with another group and discuss any discrepancies.

Goal Today's lesson is converting between **percents** and fractions.

Start Thinking! and Warm Up

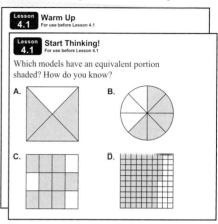

Extra Example 1
a. Write 76% as a fraction in simplest form. $\frac{19}{25}$
b. Write 118% as a mixed number in simplest form. $1\frac{9}{50}$

On Your Own
1. $\frac{3}{20}$
2. $1\frac{17}{25}$
3. $\frac{9}{25}$
4. $\frac{83}{100}$

T-150

Differentiated Instruction

Auditory

Give students fractions that have denominators of 100 or any fraction with a denominator that is a factor of 100, such as $\frac{3}{50}$ or $\frac{3}{10}$. Ask students to give the percent equivalents of the fractions.

Extra Example 2

Write $\frac{13}{20}$ as a percent. 65%

Extra Example 3

You delete $\frac{3}{16}$ of the pictures on your digital camera. What percent of the pictures did you delete? 18.75%

On Your Own

5. 62% 6. 28%
7. 95% 8. 22.5%
9. 12.5%

Laurie's Notes

Key Idea

? "What are equivalent fractions? For instance, $\frac{2}{5}$ and $\frac{4}{10}$ are equivalent fractions. $\frac{2}{5} \cdot \frac{20}{20} = \frac{40}{100}$ and $\frac{4}{10} \cdot \frac{10}{10} = \frac{40}{100}$." two fractions that represent the same amount

? "How do you write equivalent fractions with a denominator of 100?" Find a number you can multiply the denominator by so that it equals 100. Multiply the numerator by the same number.

Example 2

- **Connection:** The visual model for $\frac{3}{50}$ and $\frac{6}{100}$ would be equivalent. Use a simpler problem to demonstrate this. $\frac{1}{4} = \frac{4}{16}$.

 =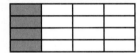

? "How many hundredths would $\frac{1}{4}$ equal?" 25

Example 3

- **FYI:** Most students do not know that 8 divides evenly into 1000. The divisibility rule for 8 states that if 8 divides evenly into the last 3 digits of a number then the number is divisible by 8.
- ? "How do you convert a fraction to a decimal?" Divide the numerator of the fraction by the denominator.
- ? "How do you convert a decimal to a fraction?" Write the decimal as a fraction. For example, 0.875 is read as 875 thousandths, so that is $\frac{875}{1000}$.
- Discuss the step of dividing the numerator and denominator by 10. Discuss that the denominator is 1000 and you need it to be 100. What number can be divided into 1000 to get 100?

On Your Own

- Question 9 can be written with a denominator of 1000, or begin by writing the fraction as a decimal.

Closure

- **Writing prompts:**
 To write a percent as a fraction, . . .
 To write a fraction as a percent, . . .

Technology For the Teacher

The Dynamic Planning Tool
Editable Teacher's Resources at BigIdeasMath.com

T-151

Key Idea

Writing Fractions as Percents

Words Write an equivalent fraction with a denominator of 100.

Numbers $\dfrac{1}{4} \xrightarrow{\times 25} \dfrac{25}{100} = 25\%$

EXAMPLE 2 **Writing a Fraction as a Percent**

Write $\dfrac{3}{50}$ as a percent.

$\dfrac{3}{50} \xrightarrow{\times 2} \dfrac{6}{100} = 6\%$

$50 \times 2 = 100$. So, multiply the numerator and denominator by 2. Write the numerator with a percent symbol.

EXAMPLE 3 **Real-Life Application**

You delete $\dfrac{7}{8}$ of the pictures on a digital camera. What percent of the pictures did you delete?

$\dfrac{7}{8} = 0.875$ Write $\dfrac{7}{8}$ as a decimal.

$= \dfrac{875}{1000}$ 0.875 is *875 thousandths*.

$= \dfrac{87.5}{100}$ ← $\dfrac{875 \div 10}{1000 \div 10} = \dfrac{87.5}{100}$

$= 87.5\%$ Write the numerator with a percent symbol.

∴ So, you deleted 87.5% of your pictures.

On Your Own

Now You're Ready
Exercises 17–28

Write the fraction as a percent.

5. $\dfrac{31}{50}$ 6. $\dfrac{7}{25}$ 7. $\dfrac{19}{20}$ 8. $\dfrac{9}{40}$

9. You delete $\dfrac{1}{8}$ of the messages in your email inbox. What percent of the messages did you delete?

Section 4.1 Percents and Fractions

4.1 Exercises

Vocabulary and Concept Check

1. **WHICH ONE DOESN'T BELONG?** Which one does *not* have the same value as the other three? Explain your reasoning.

 $\dfrac{4}{25}$ 16% $\dfrac{2}{12}$ $\dfrac{16}{100}$

2. **OPEN-ENDED** Write three different fractions that are less than 40%.

3. **NUMBER SENSE** Can $1\dfrac{1}{4}$ be written as a percent? Explain.

Practice and Problem Solving

Write the percent as a fraction or mixed number in simplest form.

 4. 45% 5. 90% 6. 15% 7. 77.5%

8. 34% 9. 79% 10. 23.9% 11. 188%

12. 0.25% 13. 224% 14. 146.8% 15. 0.4%

16. **ERROR ANALYSIS** Describe and correct the error in writing 225% as a fraction.

 ✗ $225\% = \dfrac{225}{1000} = \dfrac{9}{40}$

Write the fraction as a percent. Which method did you use?

 17. $\dfrac{1}{10}$ 18. $\dfrac{1}{5}$ 19. $\dfrac{11}{20}$ 20. $\dfrac{2}{25}$

21. $\dfrac{27}{50}$ 22. $\dfrac{18}{25}$ 23. $\dfrac{3}{8}$ 24. $\dfrac{13}{16}$

25. $\dfrac{17}{20}$ 26. $\dfrac{9}{16}$ 27. $\dfrac{33}{40}$ 28. $\dfrac{3}{50}$

29. **ERROR ANALYSIS** Describe and correct the error in writing $\dfrac{14}{25}$ as a percent.

 ✗ $\dfrac{14}{25} = \dfrac{14 \times 4}{25 \times 4} = \dfrac{56}{100} = 0.56\%$

30. **LEFT-HANDED** Of the students in your class, 12% are left-handed. What *fraction* of the students are left-handed?

31. **BASKETBALL** A basketball player's made shots (•) and missed shots (✗) are shown. What percent of shots did the player make?

Assignment Guide and Homework Check

Level	Day 1 Activity Assignment	Day 2 Lesson Assignment	Homework Check
Basic	4–7, 17–20, 42–46	1–3, 9, 16, 21–33 odd, 30	9, 16, 17, 23, 30
Average	4–7, 17–20, 42–46	1–3, 9–15 odd, 16, 23–31 odd, 32–38 even, 39	9, 16, 23, 32
Advanced	4–7, 17–20, 42–46	1–3, 12–16 even, 23–29 odd, 32–38 even, 39–41	12, 16, 23, 32

Common Errors

- **Exercises 4–15** Students may write the percent as a fraction using the wrong denominator, or try to remove the decimal point when writing it over 100. For example, a student may write $\frac{775}{100}$ instead of $\frac{77.5}{100}$. Ask students how many percents are in one whole (100). Remind students that the denominator represents the whole and the denominator will always be 100. Also tell students that the number to the left of the percent symbol will be the numerator.
- **Exercises 17–28** Students may not know when it is appropriate to use an equivalent fraction to write a fraction as a percent. While there is no error in writing the fraction as a decimal first, sometimes it is easier to just write an equivalent fraction. Brainstorm some denominators that are easily rewritten as denominators of 100 (such as 4, 5, 10, 20, 25, etc.).

Technology For the Teacher
Answer Presentation Tool
QuizShow

4.1 Record and Practice Journal

Write the percent as a fraction or mixed number in simplest form.
1. 55% $\frac{11}{20}$
2. 140% $\frac{7}{5}$ or $1\frac{2}{5}$
3. 12.5% $\frac{1}{8}$
4. 0.6% $\frac{3}{500}$

Write the fraction or mixed number as a percent. Which method did you use?
5. $\frac{3}{4}$ 75%
6. $\frac{23}{40}$ 57.5%
7. $\frac{53}{200}$ 26.5%
8. $2\frac{7}{10}$ 270%
9. $8\frac{9}{25}$ 836%
10. $3\frac{11}{20}$ 355%

11. You answered 85% of the questions on the quiz correctly. What fraction of the questions did you answer correctly?
$\frac{17}{20}$

12. The states with the greatest number of electoral votes are shown in the table. There are 538 total electoral votes.
 a. Each state's electoral votes represent what percent of the total electoral votes? Round your answers to the nearest tenth.
 California: 10.2%, Texas: 6.3%, Indiana: 2.0%

State	Electoral Votes
California	55
Texas	34
Indiana	11

 b. California has about how many times more votes than Indiana?
 California has 5 times as many votes as Indiana.

Vocabulary and Concept Check

1. $\frac{2}{12}$, because it is equal to $0.1\overline{6}$, and the others are equal to 0.16.
2. *Sample answer:* $\frac{3}{20}, \frac{23}{100}, \frac{1}{8}$
3. yes; You can rewrite $1\frac{1}{4}$ as the improper fraction $\frac{125}{100}$, which is equal to 125%.

Practice and Problem Solving

4. $\frac{9}{20}$
5. $\frac{9}{10}$
6. $\frac{3}{20}$
7. $\frac{31}{40}$
8. $\frac{17}{50}$
9. $\frac{79}{100}$
10. $\frac{239}{1000}$
11. $1\frac{22}{25}$
12. $\frac{1}{400}$
13. $2\frac{6}{25}$
14. $1\frac{117}{250}$
15. $\frac{1}{250}$
16. The 225 should be over 100. $225\% = \frac{225}{100} = \frac{9}{4} = 2\frac{1}{4}$
17. 10%; method of Example 2
18. 20%; method of Example 2
19. 55%; method of Example 2
20. 8%; method of Example 2
21. 54%; method of Example 2
22. 72%; method of Example 2
23. 37.5%; method of Example 3
24. 81.25%; method of Example 3
25. 85%; method of Example 2
26. 56.25%; method of Example 3
27. 82.5%; method of Example 3

T-152

Practice and Problem Solving

28. 6%; method of Example 2

29. The decimal point should not have been added to the percent expression.
$\frac{14}{25} = \frac{14 \times 4}{25 \times 4} = \frac{56}{100} = 56\%$

30. $\frac{3}{25}$ 31. 40%

32. $\frac{3}{10} = 30\%$ 33. $\frac{13}{25} = 52\%$

34. $\frac{3}{4} = 75\%$ 35. 294%

36. 615% 37. 392%

38. 418%

39. The answer depends on the goals for each year.

40. See *Taking Math Deeper*.

41. $8\frac{1}{3}\%$ or $8.\overline{3}\%$; *Sample answer:* Multiply the numerator and the denominator by $\frac{25}{3}$. Then write the result as a mixed number.

Fair Game Review

42. 0.85 43. 0.225

44. $0.4\overline{6}$ 45. 0.9375

46. C

Taking Math Deeper

Exercise 40

Have you ever tried to find a U.S. map that shows Alaska in proportion to the lower 48 states? Most maps show Alaska much smaller than it is.

c.

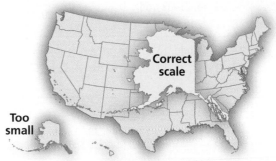

① a. Make a table.

State	Area as Portion of Total U.S. Area	
	as a Percent	as a Fraction
Alaska	17.5%	$\frac{17.5}{100} = \frac{175}{1000} = \frac{7}{40}$
Florida	1.7%	$\frac{1.7}{100} = \frac{17}{1000}$
Hawaii	0.288%	$\frac{0.288}{100} = \frac{288}{100,000} = \frac{9}{3125}$
Illinois	1.5%	$\frac{1.5}{100} = \frac{15}{1000} = \frac{3}{200}$

About 5 times

② Use division to compare Illinois to Hawaii.

b. $\frac{3}{200} \div \frac{9}{3125} = \frac{\cancel{3}^{1}}{\cancel{200}_{8}} \cdot \frac{\cancel{3125}^{125}}{\cancel{9}_{3}} = \frac{125}{24} \approx 5.2$ times larger

③ d. 24 states are larger than Illinois. In order: AK, TX, CA, MT, NM, AZ, NV, CO, OR, WY, MI, MN, UT, ID, KS, NE, SD, WA, ND, OK, MO, FL, WI, GA

Project

How was the western border of Illinois determined? Draw a poster showing how far the borders of Illinois would need to be moved so that the area of Illinois would be approximately equal to the area of Alaska.

Mini-Assessment

Write the percent as a fraction in simplest form.

1. 54% $\frac{27}{50}$ 2. 62.5% $\frac{5}{8}$

Write the fraction as a percent.

3. $\frac{3}{10}$ 30% 4. $\frac{4}{25}$ 16%

5. You ate $\frac{4}{5}$ of a loaf of bread. What percent of the bread did you eat? 80%

Reteaching and Enrichment Strategies

If students need help...	If students got it...
Resources by Chapter • Practice A and Practice B • Puzzle Time Record and Practice Journal Practice Differentiating the Lesson Lesson Tutorials Skills Review Handbook	Resources by Chapter • Enrichment and Extension Start the next section

Write a fraction and a percent to represent the shaded portion of the model.

32.
33.
34.

Write the mixed number as a percent.

35. $2\frac{47}{50}$
36. $6\frac{3}{20}$
37. $3\frac{23}{25}$
38. $4\frac{9}{50}$

39. **FUNDRAISER** A school fundraiser raised 120% of its goal last year and 125% of its goal this year. Did the fundraiser raise more money this year? Explain your reasoning.

40. **GEOGRAPHY** The percent of the area of the United States that is in each of four states is shown.

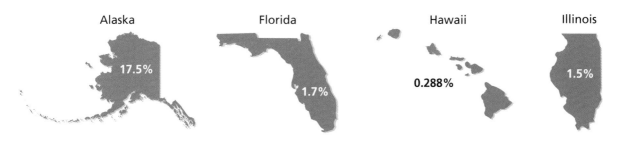

 a. Write the percents as fractions in simplest form.
 b. How many times larger is Illinois than Hawaii?
 c. Compared to the map of Florida, is the map of Alaska the correct size? Explain your reasoning.
 d. **RESEARCH** Which of the fifty states are larger than Illinois?

41. Write $\frac{1}{12}$ as a percent. Explain how you found your answer.

Fair Game Review *What you learned in previous grades & lessons*

Write the fraction as a decimal. *(Section 2.8)*

42. $\frac{17}{20}$
43. $\frac{9}{40}$
44. $\frac{7}{15}$
45. $\frac{15}{16}$

46. **MULTIPLE CHOICE** Twelve tickets to a concert cost $523.56. What is the cost of one ticket? *(Section 3.4)*

 A $31.36 **B** $43.51 **C** $43.63 **D** $44.38

4.2 Percents and Decimals

Essential Question How does the decimal point move when you rewrite a percent as a decimal and when you rewrite a decimal as a percent?

COMMON CORE STATE STANDARDS
6.RP.3c

1 ACTIVITY: Writing Percents as Decimals

Work with a partner. Write the percent shown by the model. Write the percent as a decimal.

a. Sample:

$$30\% = \frac{30}{100} \quad \begin{matrix} \leftarrow 30 \\ \leftarrow \text{per} \\ \leftarrow \text{cent} \end{matrix}$$

$$= \frac{\overset{3}{\cancel{30}}}{\underset{10}{\cancel{100}}} \qquad \text{Divide out common factor of 10.}$$

$$= \frac{3}{10} \qquad \text{Simplify.}$$

$$= 0.3 \qquad \text{Write fraction as a decimal.}$$

b.

c.

d.

e.

f.

g.

154 Chapter 4 Fractions, Decimals, and Percents

Laurie's Notes

Introduction

For the Teacher
- **Goal:** The goal of this activity is for students to use a visual model to represent a percent.
- Students have converted between fractions and decimals and between fractions and percents. This lesson completes the triangle by converting between decimals and percents.

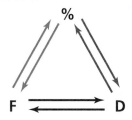

- Relating percents to money (part of a dollar) is often a helpful technique.

Motivate
- Share a fictional story about collecting student homework on a USB drive and the need to purchase a new drive with greater capacity—for all of their homework! Work the following nonfictional facts into the story.
 - Most common USB drives hold 2 GB (gigabytes) or 4 GB.
 - An 8 GB USB drive holds 4 or 400% more data than the 2 GB, and 2 or 200% more data than the 4 GB.

Activity Notes

Activity 1
- Work through the sample problem. There are 3 steps: write the percent; write the percent as a fraction; write the fraction as a decimal.
- A common question from students is whether they can leave the answer as 0.30 versus 0.3. The two decimals are equivalent. However, explain that you generally write the simplified version 0.3 for the same reason you simplify fractions.
- Students should work with a partner to answer the remaining parts.
- Probe what strategies were used to count the shaded squares.
- **Common Error:** In part (b), students often write "100% = 100" instead of "100% = 1."

Common Core State Standards
6.RP.3c Find a percent of a quantity as a rate per 100; solve problems involving finding the whole, given a part and the percent.

Previous Learning
Students should know how to write halves, fourths, tenths, and hundredths as percents. Students also need to be able to convert between fractions and decimals, and multiply and divide decimals by powers of 10.

Start Thinking! and Warm Up

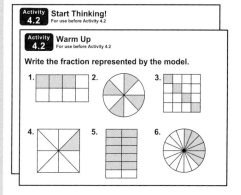

4.2 Record and Practice Journal

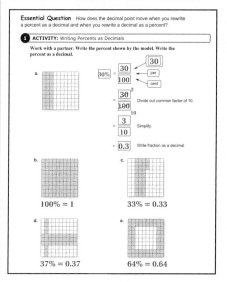

English Language Learners

Have students make drawings to visualize percents. Have them use grid paper and draw 50%, 5%, and $\frac{5}{10}$% so that they can see the difference between the percents.

Laurie's Notes

Activity 2

- **Discuss:** Percents do not need to involve whole numbers, and in fact often do not.
- **?** "What is different about the visual model from what you have seen previously?" *One-half of a square is shaded.*
- **FYI:** Students initially find it odd to have a decimal as part of a fraction. Multiplying the numerator and denominator by a power of 10 makes sense to them.
- **Common Error:** When percents are less than 1%, students often just remove the percent symbol and leave the decimal. In part (d), 0.5% does not equal 0.5.

Activity 3

- This example reverses the process students have just completed. After they sketch a model of the decimal, they write the decimal as a fraction with a denominator of 100, and then write the percent.

What Is Your Answer?

- Students should recognize that the decimal point has moved two decimal places left or right, although they may not have a good understanding of why.
- Listen for explanations about the percent symbol and its meaning.
- **?** Probe about the location of the decimal point for percents involving whole numbers, such as 25%. "Where is the decimal point located?" *after the 5*

Closure

- **Exit Ticket:** Write each percent as a decimal. 60%, 6%, and 0.6% *0.6, 0.06, 0.006*

4.2 Record and Practice Journal

The Dynamic Planning Tool
Editable Teacher's Resources at *BigIdeasMath.com*

2 ACTIVITY: Writing Percents as Decimals

Work with a partner. Write the percent as a decimal.

a. **Sample:** 13.5%

$$13.5\% = \frac{13.5}{100}$$

- 13.5 ← per
- cent

$$= \frac{13.5 \cdot 10}{100 \cdot 10}$$ Multiply numerator and denominator by 10.

$$= \frac{135}{1000}$$ One hundred thirty-five thousandths

$$= 0.135$$ Write fraction as a decimal.

b. 12.5% c. 3.8% d. 0.5%

3 ACTIVITY: Writing Decimals as Percents

Work with a partner. Draw a model to represent the decimal. Write the decimal as a percent.

a. **Sample:** 0.1

$$0.1 \quad = \quad 0.10 = \frac{10}{100} \quad = \quad 10\%$$

One Tenth Ten Hundredths Ten Percent

b. 0.24 c. 0.58 d. 0.05

What Is Your Answer?

4. **IN YOUR OWN WORDS** How does the decimal point move when you rewrite a percent as a decimal and when you rewrite a decimal as a percent?

Practice — Use what you learned about percents and decimals to complete Exercises 7–12 and 19–24 on page 158.

Section 4.2 Percents and Decimals 155

On Your Own

5. 94% 6. 120%
7. 31.6% 8. 0.5%

Extra Example 3

On a science test, you get 85 out of a possible 100 points. Write your score as a percent, a fraction, and a decimal.

85%, $\frac{17}{20}$, 0.85

Extra Example 4

In Example 4, how many times more UV rays are reflected by water than by grass? 28 times more

On Your Own

9. 18% = $\frac{9}{50}$ = 0.18
10. 5.6

Differentiated Instruction

Kinesthetic

Some students may struggle with writing decimals for very small percents (i.e., 0.0075% as 0.000075) or writing percents for very small decimals (i.e., 0.0075 as 0.75%). To improve their number sense, have students draw a 10 by 10 grid with sides of 20 centimeters. Divide one of the squares to make a smaller 10 by 10 grid. Have students compare 75% and 0.75% by shading squares to represent each percent. Then have students write the decimal equivalents of the percents.

T-157

Laurie's Notes

On Your Own

- **Think-Pair-Share:** Students should read each question independently and then work with a partner to answer the questions. When they have answered the questions, the pair should compare their answers with another group and discuss any discrepancies.

Words of Wisdom

- Students quickly recognize that the decimal point moves 2 places left or right depending on the type of conversion.
- Students generally do well with 2-digit problems: 24%, or 0.24.
- Students have difficulty with percents greater than 100% or less than 1% (i.e., 250%, 0.025%). They want to quickly move the decimal point without thinking first.
- Reinforce the meaning behind moving the decimal point.
 - 25% is 25 per one hundred or 0.25.
 - 250% is 250 per one hundred or 2.5.
 - 0.25% is 25 hundredths per one hundred or 0.0025.
 - 0.025% is 25 thousandths per one hundred or 0.00025.
- When you feel that students are comfortable with percents between 1 and 100, practice converting with greater and lesser percents.

Example 3

- Note the test-taking strategy of eliminating choices.

Example 4

- **FYI:** UV rays are necessary for our bodies to produce vitamin D, a substance that helps strengthen bones. The downside: reflection of UV rays off of snow and sand are enough to cause photokeratitis, sunburn of the cornea.
- This example reviews all three forms: percents, decimals, and fractions, as well as division of decimals.
- ? "What is the key question being asked?" This is a language connection. "How many times more...."

On Your Own

- Students should estimate an answer for Question 10 before computing the answer.

Closure

- **Writing prompts:**
 To write a percent as a decimal, ...
 To write a decimal as a percent, ...

The Dynamic Planning Tool
Editable Teacher's Resources at *BigIdeasMath.com*

● **On Your Own**

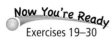

Write the decimal as a percent. Use a model to check your answer.

5. 0.94 **6.** 1.2 **7.** 0.316 **8.** 0.005

EXAMPLE 3 **Standardized Test Practice**

On a math test, you get 92 out of a possible 100 points. Which of the following is *not* another way of expressing 92 out of 100?

Ⓐ $\dfrac{23}{25}$ Ⓑ 92% Ⓒ $\dfrac{17}{20}$ Ⓓ 0.92

$$92 \text{ out of } 100 = \dfrac{92}{100} \begin{cases} = 92\% & \text{Eliminate choice B.} \\ = \dfrac{23}{25} & \text{Eliminate choice A.} \\ = 0.92 & \text{Eliminate choice D.} \end{cases}$$

∴ The correct answer is Ⓒ.

EXAMPLE 4 **Real-Life Application**

The figure shows the portions of ultraviolet (UV) rays reflected by four different surfaces. How many times more UV rays are reflected by water than by sea foam?

Write 25% and $\dfrac{21}{25}$ as decimals.

Sea Foam: 25% = 25.% = 0.25 **Water:** $\dfrac{21}{25} = \dfrac{84}{100} = 0.84$

Divide 0.84 by 0.25: $0.25\overline{)0.84}$ → $25\overline{)84.00}$ = 3.36

∴ So, water reflects about 3.4 times more UV rays than sea foam.

● **On Your Own**

9. Write "18 out of 100" as a percent, fraction, and decimal.

10. In Example 4, how many times more UV rays are reflected by water than by sand?

4.2 Exercises

Vocabulary and Concept Check

MATCHING Match the decimal with its equivalent percent.

1. 0.42
2. 4.02
3. 0.042
4. 0.0402

A. 4.02%
B. 42%
C. 4.2%
D. 402%

5. **OPEN-ENDED** Write three different decimals that are between 10% and 20%.

6. **WHICH ONE DOESN'T BELONG?** Which one does *not* belong with the other three? Explain your reasoning.

 70% 0.7 $\frac{7}{10}$ 0.07

Practice and Problem Solving

Write the percent as a decimal.

7. 78%
8. 55%
9. 18.5%
10. 57.4%
11. 33%
12. 9%
13. 47.63%
14. 91.25%
15. 166%
16. 217%
17. 0.06%
18. 0.034%

Write the decimal as a percent.

19. 0.74
20. 0.52
21. 0.89
22. 0.768
23. 0.99
24. 0.49
25. 0.487
26. 0.128
27. 3.68
28. 5.12
29. 0.0371
30. 0.046

31. **ERROR ANALYSIS** Describe and correct the error in writing 0.86 as a percent.

 ✗ 0.86 = 00.86 = 0.0086%

32. **MUSIC** Thirty-six percent of the songs on your MP3 player are pop songs. Write this percent as a decimal.

33. **CAT** About 0.34 of the length of a cat is its tail. Write this decimal as a percent.

34. **COMPUTER** Write the percent of free space on the computer as a decimal.

Volume	Capacity	Free Space	% Free Space
(C:)	149 GB	133 GB	89 %

Write each percent as a fraction in simplest form and as a decimal.

35. 36%
36. 23.5%
37. 16.24%

Assignment Guide and Homework Check

Level	Day 1 Activity Assignment	Day 2 Lesson Assignment	Homework Check
Basic	7–12, 19–24, 42–50	1–6, 13, 15, 17, 25–37 odd, 32, 34	13, 25, 32, 35
Average	7–12, 19–24, 42–50	1–6, 16–18, 28–31, 34–38 even, 39	16, 28, 34, 36
Advanced	7–12, 19–24, 42–50	1–6, 18, 30, 31, 34, 35–41	18, 30, 34, 36

Common Errors

- **Exercises 7–30** Students may move the decimal point the wrong way, forget to place zeros as placeholders, or move the decimal point too many places (especially when the percent is greater than 100). As a class, gather some helpful information for remembering how to convert decimals and percents. For example, when the percent is greater than 100%, the decimal equivalent will be greater than 1. Because there are two zeros in 100, you need to move the decimal point two places. When converting from percents to decimals, you move the decimal point to the left because D is to the left of P in the alphabet. When converting from decimals to percents, you move the decimal point to the right because P is to the right of D in the alphabet.
- **Exercises 35–37** Students may move the decimal point to the right instead of to the left. Remind them how to get rid of the decimal point in the percent when writing a fraction.

4.2 Record and Practice Journal

Write the percent as a decimal.
1. 35% 2. 160% 3. 74.8% 4. 0.3%
 0.35 1.6 0.748 0.003

Write the decimal as a percent.
5. 1.23 6. 0.49 7. 0.024 8. 0.881
 123% 49% 2.4% 88.1%

Write each percent as a fraction in simplest form and as a decimal.
9. 48% 10. 15.5% 11. 84.95%
 $\frac{12}{25}$; 0.48 $\frac{31}{200}$; 0.155 $\frac{1699}{2000}$; 0.8495

12. People with severe hearing loss were given a sentence and word recognition test six months after they got implants in their ears. The patients scored an average of 82% on the test. Write this percent as a decimal.
 0.82

13. Out of the first 15 serves in a volleyball game, 0.6 of them were in bounds.
 a. What percent of the serves were in bounds?
 60%
 b. What percent of the serves were not in bounds?
 40%

Technology For the Teacher
Answer Presentation Tool
QuizShow

Vocabulary and Concept Check

1. B 2. D
3. C 4. A
5. Sample answer: 0.11, 0.13, 0.19
6. 0.07 because it represents 7% instead of 70%.

Practice and Problem Solving

7. 0.78 8. 0.55
9. 0.185 10. 0.574
11. 0.33 12. 0.09
13. 0.4763 14. 0.9125
15. 1.66 16. 2.17
17. 0.0006 18. 0.00034
19. 74% 20. 52%
21. 89% 22. 76.8%
23. 99% 24. 49%
25. 48.7% 26. 12.8%
27. 368% 28. 512%
29. 3.71% 30. 4.6%
31. The decimal point was moved in the wrong direction. 0.86 = 0.86 = 86%
32. 0.36
33. 34%
34. 0.89
35. $\frac{9}{25} = 0.36$
36. $\frac{47}{200} = 0.235$
37. $\frac{203}{1250} = 0.1624$

T-158

Practice and Problem Solving

38. See Additional Answers.
39. 40%
40. See *Taking Math Deeper*.
41. a. $16.\overline{6}\%$ or $16\frac{2}{3}\%$
 b. $\frac{5}{6}$

Fair Game Review

42. $\frac{23}{50}$
43. $\frac{31}{100}$
44. $2\frac{1}{5}$
45. $4\frac{8}{25}$
46. $9 + (4 + x) = (9 + 4) + x$
 Assoc. Prop. of Add.
 $= 13 + x$
 Add 9 and 4.
47. $(7 + y) + 14 = (y + 7) + 14$
 Comm. Prop. of Add.
 $= y + (7 + 14)$
 Assoc. of Add.
 $= y + 21$
 Add 7 and 14.
48. $7(8w) = (7 \cdot 8)w$
 Assoc. Prop. of Mult.
 $= 56w$
 Multiply 7 and 8.
49. $6 \cdot c \cdot 5 = 6 \cdot 5 \cdot c$
 Comm. Prop. of Mult.
 $= 30c$
 Multiply 6 and 5.
50. B

Taking Math Deeper

Exercise 40

This is a good review that all regions of a circle graph must add up to 1 or 100%. Begin by making a table to summarize the given information.

Color	Portion Who Said this Color	
	Percent	Decimal
●	26%	0.26
●	40%	0.40
●	4%	0.04
●		
●	14%	0.14

① What % said red, blue, or yellow? "Or" means to add the percents.

a. $26\% + 40\% + 4\% = 70\%$

② "How many times more" means to divide.

b. $0.26 \div 0.04 = 6.5$ times more.

③ Add the 4 given percents. Then subtract from 100%.

$26\% + 40\% + 4\% + 14\% = 84\%$

c. $100\% - 84\% = 16\% = 0.16$

Project

Conduct a survey concerning favorite colors in your class or with your friends. Do their answers agree with the results of Exercise 40?

Mini-Assessment

Write the percent as a decimal.
1. 12% 0.12
2. 130% 1.3

Write the decimal as a percent.
3. 0.98 98%
4. 2.25 225%
5. Forty-two percent of your cell phone ringtones are pop songs. Write this percent as a decimal. 0.42

Reteaching and Enrichment Strategies

If students need help...	If students got it...
Resources by Chapter • Practice A and Practice B • Puzzle Time Record and Practice Journal Practice Differentiating the Lesson Lesson Tutorials Skills Review Handbook	Resources by Chapter • Enrichment and Extension Start the next section

38. SCHOOL The percent of students that travel to school by car, bus, and bicycle is shown for a school of 825 students.

Car 20%

School bus 48%

Bicycle 8%

 a. Write the percents as decimals.
 b. Write the percents as fractions.
 c. What percent of students use another method to travel to school?
 d. **RESEARCH** Make a bar graph that represents how the students in your class travel to school.

39. ELECTIONS In an election, the winning candidate received 60% of the votes. What percent of the votes did the other candidate receive?

40. COLORS Students in a class were asked to tell their favorite color.

 a. What percent said red, blue, or yellow?
 b. How many times more students said red than yellow?
 c. What percent of students said green? Write this percent as a decimal.

41. In the first 42 Super Bowls, $0.1\overline{6}$ of the MVPs (most valuable players) were running backs.

 a. What percent of the MVPs were running backs?
 b. What fraction of the MVPs were *not* running backs?

Fair Game Review What you learned in previous grades & lessons

Write the decimal as a fraction or mixed number in simplest form. *(Section 2.7)*

42. 0.46 **43.** 0.31 **44.** 2.2 **45.** 4.32

Simplify the expression. Explain each step. *(Section 1.3)*

46. $9 + (4 + x)$ **47.** $(7 + y) + 14$

48. $7(8w)$ **49.** $6 \cdot c \cdot 5$

50. MULTIPLE CHOICE Ham costs $4.48 per pound. Cheese costs $6.36 per pound. You buy 1.5 pounds of ham and 0.75 pound of cheese. How much more do you pay for the ham? *(Section 3.3)*

 Ⓐ $1.41 Ⓑ $1.95 Ⓒ $4.77 Ⓓ $6.18

4.3 Comparing and Ordering Fractions, Decimals, and Percents

COMMON CORE STATE STANDARDS
6.NS.7a
6.NS.7b

Essential Question How can you order numbers that are written as fractions, decimals, and percents?

1 ACTIVITY: Ordering Numbers

Work with a partner to order the following numbers.

$\frac{1}{8}$ 11% $\frac{3}{20}$ 0.172 0.32 43% 7% 0.7 $\frac{5}{6}$

a. Decide on a strategy for ordering the numbers. Will you write them all as fractions, decimals, or percents?

b. Use your strategy and a number line to order the numbers from least to greatest. (Note: Label the number line appropriately.)

2 ACTIVITY: Using Fractions, Decimals, and Percents

Work with a partner. Decide which number form (fraction, decimal, or percent) is more common. Then find which is greater.

a. 7% sales tax or $\frac{1}{20}$ sales tax

b. 0.37 cup of flour or $\frac{1}{3}$ cup of flour

c. $3\frac{5}{8}$-inch wrench or 3.69-inch wrench

d. $12\frac{2}{3}$ dollars or 12.56 dollars

e. 93% test score or $\frac{7}{8}$ test score

f. $5\frac{5}{6}$ fluid ounces or 5.6 fluid ounces

Laurie's Notes

Introduction

For the Teacher
- **Goal:** The goal of this activity is for students to develop greater fluency with common benchmark fractions, decimals, and percents.

Motivate
- Write each number on an index card: $\frac{1}{10}, \frac{13}{20}, \frac{4}{5}$, 0.18, 0.45, 0.5, 35%, 60%, 85%.
- Hand out the 9 cards. Ask the 3 students holding fraction cards to stand in order (least to greatest). Then ask the 3 students holding decimal cards and the 3 students holding percent cards to do the same.
- Have each group of students describe the strategies for ordering themselves.
- Now ask all 9 students to order themselves.
- Discuss strategies used when all 3 forms are used in the same problem.

Activity Notes

Activity 1
- Remind students of the conversion triangle. Today they will work with all 3 forms.
- Have students work in groups on this activity, with each student sharing suggestions for how to order the numbers. Listen for key understanding, such as 0.7 and $\frac{5}{6}$ are the only two numbers greater than $\frac{1}{2}$.
- Note that the number line has scaling for tenths and only one number will be graphed in a tenth scale mark.
- In completing this activity, many skills are reviewed. This is an opportunity for informal assessment to guide instruction in the sections ahead.
- **Communication:** One group should present their answers to the class. They should discuss their results and the strategies they used.

Activity 2
- Students may continue in groups or perhaps only with a partner for this activity.
- The context should be familiar to most students.
- Remind students that there are two parts to each question.
- Discuss results as a class.
- **?** "Which fractions were repeating decimals?" Listen for knowledge that thirds and sixths are repeating decimals.
- **Common Error:** Some students, who do not have a good understanding of how to convert a fraction to a decimal, will take simple fractions such as $\frac{1}{20}$ and write 1.20 or 0.120. For $\frac{1}{3}$, they write 1.3 or 0.13.

Common Core State Standards

6.NS.7a Interpret statements of inequality as statements about the relative position of two numbers on a number line diagram.

6.NS.7b Write, interpret, and explain statements of order for rational numbers in real-world contexts.

Previous Learning

Students should know how to write halves, fourths, tenths, and hundredths as percents. Students also need to be able to convert between fractions and decimals, and multiply and divide decimals by powers of 10.

Activity Materials	
Introduction	Textbook
• index cards	• index cards

Start Thinking! and Warm Up

Activity 4.3 Start Thinking! For use before Activity 4.3

Activity 4.3 Warm Up For use before Activity 4.3

Order the decimals from least to greatest.

1. 1.4, 1.44, 4.1, 4.3, 3.3
2. 0.5, 1.75, 0.7, 1.57, 1.5
3. 0.44, 0.02, 0.2, 0.04, 0.24
4. 6.6, 0.6, 6.63, 6.06, 0.06
5. 0.2, 0.17, 0.02, 0.72, 2.27
6. 8.1, 0.81, 0.18, 1.8, 1.88

4.3 Record and Practice Journal

T-160

English Language Learners
Auditory
Students may struggle with the idea that three different forms of a number are useful. Have a classroom discussion where students brainstorm when to use fractions (construction, recipes), when to use decimals (money, chemistry), and when to use percents (surveys, chance of rain).

4.3 Record and Practice Journal

Laurie's Notes

Activity 3
- To preserve cards for multiple uses, make cards on colored card stock and store individual sets in plastic zipper bags.
- The card game *War* is common to most students. The question asked for each play is, "Which number is greater?" The player with the greater value collects both cards.
- If the cards have equivalent values (i.e., 75% and $\frac{3}{4}$), there is a tie. As stated in the text, each player lays 3 cards face down and 1 card face up. The player with the card of greater value collects all of the cards in play.
- **Comparing Cards:** The learning component of this activity is when students actually compare two numbers. Listen to students discuss how they are comparing the numbers.
- To start the play, give students the opportunity to preview the cards. Explain the rules and let students begin.
- If one group finishes early, they should shuffle the cards and play again.
- **Extension:** You can use the same set of cards to do a matching activity with the cards either face up (easier) or down. If they make a 3-card match, such as 75%, $\frac{3}{4}$, and 0.75, they get a certain number of points. A 2-card match, such as 0 and 0%, is worth fewer points.

? What cards have no match?

What Is Your Answer?
- Listen for the big idea, namely that the farther to the right the number is on the number line, the greater the number.

Closure
- **Exit Ticket:**
 - Write a percent between 25% and 40%. Write the equivalent fraction and decimal. *Sample answer:* 30%, $\frac{3}{10}$, 0.3
 - Write a percent between 75% and 90%. Write the equivalent fraction and decimal. *Sample answer:* 85%, $\frac{17}{20}$, 0.85

What Is Your Answer?
4. **IN YOUR OWN WORDS** How can you order numbers that are written as fractions, decimals, and percents? Give an example with your answer.
Convert all numbers to one form then order them.

5. All but one U.S. coin has a name that is related to its value. Which one is it? How are the names of the others related to their values?

Nickel; the name nickel comes from the metal the coin was originally made from; 1 cent = $\frac{1}{100}$; dime comes from the French word "disme" meaning tenth-part; quarter = $\frac{1}{4}$ dollar; half dollar = $\frac{1}{2}$ dollar

The Dynamic Planning Tool
Editable Teacher's Resources at **BigIdeasMath.com**

3 ACTIVITY: The Game of Math Card War

Preparation:
- Cut index cards to make 40 playing cards.
- Write each number in the table on a card.

To Play:
- Play with a partner.
- Deal 20 cards to each player face-down.
- Each player turns one card face-up. The player with the greater number wins. The winner collects both cards and places them at the bottom of his or her cards.
- Suppose there is a tie. Each player lays three cards face-down, then a new card face-up. The player with the greater of these new cards wins. The winner collects all ten cards and places them at the bottom of his or her cards.
- Continue playing until one player has all the cards. This player wins the game.

75%	$\frac{3}{4}$	$\frac{1}{3}$	$\frac{3}{10}$	0.3	25%	0.4	0.25	100%	0.27
0.75	$66\frac{2}{3}\%$	12.5%	40%	$\frac{1}{4}$	4%	0.5%	0.04	$\frac{1}{100}$	$\frac{2}{3}$
0	30%	5%	$\frac{27}{100}$	0.05	$33\frac{1}{3}\%$	$\frac{2}{5}$	0.333...	27%	1%
1	0.01	$\frac{1}{20}$	$\frac{1}{8}$	0.125	$\frac{1}{25}$	$\frac{1}{200}$	0.005	0.666...	0%

What Is Your Answer?

4. **IN YOUR OWN WORDS** How can you order numbers that are written as fractions, decimals, and percents? Give an example with your answer.

5. All but one U.S. coin has a name that is related to its value. Which one is it? How are the names of the others related to their values?

Practice

Use what you learned about ordering numbers to complete Exercises 4–7, 16, and 17 on page 164.

4.3 Lesson

When comparing and ordering fractions, decimals, and percents write the numbers as all fractions, all decimals, or all percents.

EXAMPLE 1 Comparing Fractions, Decimals, and Percents

a. **Which is greater, $\frac{3}{20}$ or 16%?**

> **Study Tip**
> It is usually easier to order decimals or percents.

Write $\frac{3}{20}$ as a percent: $\frac{3}{20} = \frac{15}{100} = 15\%$ (× 5)

∴ 15% is less than 16%. So, 16% is the greater number.

b. **Which is greater, 79% or 0.08?**

Write 79% as a decimal: $79\% = 79.\% = 0.79$

∴ 0.79 is greater than 0.08. So, 79% is the greater number.

On Your Own

Exercises 4–15

1. Which is greater, 25% or $\frac{7}{25}$?
2. Which is greater, 0.49 or 94%?

EXAMPLE 2 Real-Life Application

You, your sister, and a friend each take the same number of shots at a soccer goal. You make 72% of your shots, your sister makes $\frac{19}{25}$ of her shots, and your friend makes 0.67 of his shots. Who made the fewest shots?

Write 72% and $\frac{19}{25}$ as decimals.

Remember
To order numbers from least to greatest, write them as they appear on a number line from left to right.

You: $72\% = 72.\% = 0.72$ **Sister:** $\frac{19}{25} = \frac{76}{100} = 0.76$ (× 4)

Graph the decimals on a number line.

Friend: 0.67 You: 72% = 0.72 Sister: $\frac{19}{25}$ = 0.76

0.66 0.68 0.70 0.72 0.74 0.76 0.78

∴ 0.67 is the least number. So, your friend made the fewest shots.

162 Chapter 4 Fractions, Decimals, and Percents

Laurie's Notes

Introduction

Connect
- **Yesterday:** Students compared and ordered common percents, decimals, and fractions.
- **Today:** Students will compare and order less common percents, decimals, and fractions.

Motivate
- Read the 5 statistics about the United States and have students order the percents from least to greatest.
 1) 90% of the states joined the United States before 1900.
 2) The U.S. population represents about 4.5% of the world population.
 3) About 20% of the U.S. population is under 15 years old.
 4) 62% of the states are entirely east of the Mississippi River.
 5) Water area represents about 6.9% of the total area of the United States.
- ❓ "Why is it usually easier to order decimals or percents than fractions?" Listen for an understanding of unlike denominators being harder to order than place value.

Lesson Notes

Example 1
- ❓ "To compare $\frac{3}{20}$ and 16%, should you write $\frac{3}{20}$ as a percent or 16% as a fraction and why?" Discuss both options and perhaps work through both options, if time permits.
- ❓ "Could you compare $\frac{3}{20}$ and 16% by writing each as a decimal?" yes; Depending on time, show this option as well.
- ❓ Work through part (b). "What is 0.08 as a percent?" 8%
- **Common Error:** Students might say 80% because they are comparing it to 79%.

On Your Own
- **Think-Pair-Share:** Students should read each question independently and then work with a partner to answer the questions. When they have answered the questions, the pair should compare their answers with another group and discuss any discrepancies.
- Discuss methods used by students to make comparisons.

Example 2
- Note the scale on the number line. A common misconception is that the scale must be in units of 1, 5, or 10. Even digits are not as common.
- ❓ "Why is scaling by even digits helpful in this problem? Could we use a scale of 10?" Even digits show the distance between the 3 numbers in an accurate display. You could use 10, but it would be less accurate.
- ❓ "What other strategies could be used to compare these three numbers?" Answers will vary.

Goal Today's lesson is comparing and ordering less common fractions, decimals, and percents.

Start Thinking! and Warm Up

Extra Example 1
a. Which is greater, $\frac{17}{20}$ or 80%? $\frac{17}{20}$
b. Which is greater, 28% or 0.29? 0.29

On Your Own
1. $\frac{7}{25}$ 2. 94%

Extra Example 2
You, your sister, and your friend each take the same number of shots at a soccer goal. You make 0.67 of your shots, your sister makes 68% of her shots, and your friend makes $\frac{17}{20}$ of her shots. Who made the fewest shots? You made the fewest shots.

T-162

On Your Own

3. You made the most shots.

Extra Example 3

The table shows the portions of the population of Rhode Island that live in each county. List the counties in order by population from least to greatest.

Counties	Fraction	Decimal	Percent
Bristol	$\frac{1}{20}$		
Kent		0.16	
Newport	$\frac{2}{25}$		
Providence			59%
Washington	$\frac{3}{25}$		

Bristol, Newport, Washington, Kent, Providence

On Your Own

4. See Additional Answers.

Differentiated Instruction

Kinesthetic

Some students have difficulty seeing that 60% is not only equal to "60 out of 100," but is also equal to "6 out of 10" and "3 out of 5." Give students three 100-grid squares. In the first grid, have students shade 60 out of 100 squares. With the second grid, have students draw thick lines around groups of 10 squares and then shade 6 of the groups. With the third grid, have students draw thick lines around groups of 20 squares and then shade 3 of the groups. When students compare all three grids, they should see that the same amount of 60 squares have been shaded on each of the grids. This can be seen easily if the students have grouped the squares by rows or columns.

T-163

Laurie's Notes

On Your Own

- **Think-Pair-Share:** Students should read the question independently and then work with a partner to answer the question. When they have answered the question, the pair should compare their answer with another group and discuss any discrepancies.
- Have students graph the three numbers on a scaled number line.

Example 3

- Give students time to read the information in the problem.
- ? "Interpret what it means for $\frac{1}{50}$ of the U.S. population to live in Washington." Listen for an understanding that one in every 50 U.S. residents lives in Washington.
- ? **Extension:** "If there were 100 U.S. residents, how many would live in Washington?" 2 "If there were 1000 U.S. residents, how many would live in Washington?" 20
- Discuss the columns of the table. Note that the decimals and percents are easier to compare than the fractions.

On Your Own

- This is a good assessment of a student's readiness for the homework.

Closure

- Complete this table of benchmark numbers.

Fraction	$\frac{1}{4}$	$\frac{3}{10}$	$\frac{1}{2}$	$\frac{2}{5}$	$\frac{1}{20}$	$\frac{3}{20}$	$\frac{7}{10}$	$\frac{3}{4}$	$\frac{1}{5}$	1
Decimal	0.25	0.3	0.5	0.4	0.05	0.15	0.7	0.75	0.2	1
Percent	25%	30%	50%	40%	5%	15%	70%	75%	20%	100%

The Dynamic Planning Tool
Editable Teacher's Resources at *BigIdeasMath.com*

On Your Own

3. WHAT IF? You make 75% of your shots, your sister makes $\frac{13}{20}$ of her shots, and your friend makes 0.7 of his shots. Who made the most shots?

EXAMPLE 3 Real-Life Application

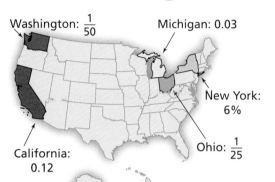

The map shows the portions of the U.S. population that live in five states.

List the five states in order by population from least to greatest.

Begin by writing each portion as a fraction, decimal, and percent.

State	Fraction	Decimal	Percent
Michigan	$\frac{3}{100}$	0.03	3%
New York	$\frac{6}{100}$	0.06	6%
Washington	$\frac{1}{50}$	0.02	2%
California	$\frac{12}{100}$	0.12	12%
Ohio	$\frac{1}{25}$	0.04	4%

Graph the percent for each state on a number line.

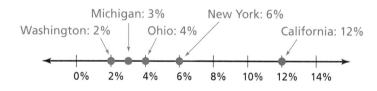

∴ The states in order from least to greatest are Washington, Michigan, Ohio, New York, and California.

On Your Own

4. The portion of the U.S. population that lives in Texas is $\frac{2}{25}$. The portion that lives in Illinois is 0.043. Reorder the states in Example 3 including Texas and Illinois.

4.3 Exercises

Check It Out
Help with Homework
BigIdeasMath.com

Vocabulary and Concept Check

1. **NUMBER SENSE** Copy and complete the table.

2. **NUMBER SENSE** How would you decide whether $\frac{3}{5}$ or 59% is greater? Explain.

3. **WHICH ONE DOESN'T BELONG?** Which one does *not* belong with the other three? Explain your reasoning.

 40% $\frac{2}{5}$ 0.4 0.04

Fraction	Decimal	Percent
$\frac{18}{25}$	0.72	
$\frac{17}{20}$		85%
$\frac{13}{50}$		
	0.62	
		45%

Practice and Problem Solving

Tell which number is greater.

4. 0.9, 95%
5. 20%, 0.02
6. $\frac{37}{50}$, 37%
7. 50%, $\frac{13}{25}$
8. 0.086, 86%
9. 76%, 0.67
10. 60%, $\frac{5}{8}$
11. 0.12, 1.2%
12. 17%, $\frac{4}{25}$
13. 140%, 0.14
14. $\frac{1}{3}$, 30%
15. 80%, $\frac{7}{9}$

Use a number line to order the numbers from least to greatest.

16. 38%, $\frac{8}{25}$, 0.41
17. 68%, 0.63, $\frac{13}{20}$
18. $\frac{43}{50}$, 0.91, $\frac{7}{8}$, 84%
19. 0.15%, $\frac{3}{20}$, 0.015
20. 2.62, $2\frac{2}{5}$, 26.8%, 2.26, 271%
21. $\frac{87}{200}$, 0.44, 43.7%, $\frac{21}{50}$

22. **TEST** You answer 21 out of 25 questions correctly on a test. Did you reach your goal of getting 80% or better?

23. **POPULATION** The table shows the portions of the world population that live in four countries. Order the countries by population from least to greatest.

Country	Brazil	China	Japan	United States
Portion of World Population	3%	$\frac{1}{5}$	$\frac{1}{50}$	0.046

164 Chapter 4 Fractions, Decimals, and Percents

Assignment Guide and Homework Check

Level	Day 1 Activity Assignment	Day 2 Lesson Assignment	Homework Check
Basic	4–7, 16, 17, 33–37	1–3, 9–15 odd, 19–25 odd, 22	9, 13, 19, 22, 25
Average	4–7, 16, 17, 33–37	1–3, 12–15, 20–22, 25–29 odd, 30	13, 20, 22, 25, 27
Advanced	4–7, 16, 17, 33–37	1–3, 18–32 even, 31	18, 22, 24, 26

Common Errors

- **Exercises 4–21** Students may try to order the numbers without converting them, or will only convert them mentally and do so incorrectly. Tell them that it is necessary to convert all the numbers to one form and that they should write out the steps to make sure that they are converting correctly.
- **Exercises 24 and 25** Students may try to round the numbers that have repeating decimals and incorrectly order the numbers. Remind them that even though you often round repeating decimals, the decimal is actually less than or greater than the rounded decimal. For example, $0.\overline{6}$ is less than 0.667, but greater than 0.666.

4.3 Record and Practice Journal

Technology For the Teacher
Answer Presentation Tool
QuizShow

Vocabulary and Concept Check

1.

$\frac{18}{25}$	0.72	72%
$\frac{17}{20}$	0.85	85%
$\frac{13}{50}$	0.26	26%
$\frac{31}{50}$	0.62	62%
$\frac{9}{20}$	0.45	45%

2. *Sample answer*: Write $\frac{3}{5}$ as a percent, then compare 59% to determine which is greater.

3. 0.04; 0.04 = 4%, but 40%, $\frac{2}{5}$, and 0.4 are all equal to 40%.

Practice and Problem Solving

4. 95% 5. 20%
6. $\frac{37}{50}$ 7. $\frac{13}{25}$
8. 86% 9. 76%
10. $\frac{5}{8}$ 11. 0.12
12. 17% 13. 140%
14. $\frac{1}{3}$ 15. 80%

16.

17.

18.

T-164

Practice and Problem Solving

19–21. See Additional Answers.

22. yes

23. Japan, Brazil, United States, China

24. $0.66, 66.1\%, \frac{2}{3}, 0.667$

25. $21\%, 0.2\overline{1}, \frac{11}{50}, \frac{2}{9}$

26. A **27.** D

28. B **29.** C

30. Stage 21, Stage 7, Stage 8, Stage 1, Stage 17

31. See *Taking Math Deeper*.

32. a. 7

 b. There is none. $\frac{1}{a}$ is less than 33% when a is greater than 3, but when a is greater than 3, $\frac{a}{8}$ is greater than 33%.

Fair Game Review

33. 4.8 **34.** 16.5

35. 6.66 **36.** 26.28

37. C

Mini-Assessment

Tell which number is greater.

1. 27%, 0.48 0.48
2. $\frac{2}{5}, 0.125$ $\frac{2}{5}$
3. $\frac{3}{4}, 76\%$ 76%
4. $\frac{5}{8}, 60\%$ $\frac{5}{8}$
5. On a quiz, you answer 7 out of 10 questions correctly. Did you reach your goal of getting 80% or better? no

Taking Math Deeper

Exercise 31

The problem doesn't tell students which form to use to order the numbers. It is up to the students to decide when percents or decimals are easier to use than fractions.

 Use a table to organize the information.

Animal	Portion of Day Sleeping		
	Percent	Decimal	Fraction
Dolphin	43.3%	0.433	
Lion	56.3%	0.563	
Rabbit	47.5%	0.475	$\frac{19}{40}$
Squirrel	62.0%	0.620	$\frac{31}{50}$
Tiger	65.8%	0.658	

 Use a number line to order the decimals.
a.

 Using an estimate of 8 hours a day, the portion of the day a teen sleeps is

 b $\frac{8}{24} = \frac{1}{3} = 33.\overline{3}\%$.

 c. On the ordered list, the teen would be first!

Project

Use a chart to keep track of the number of hours you sleep each day for a week. Compare your chart with three of your classmates. Find the average number of hours each of you sleep each day. Which day seems to be the sleepiest day of the week?

Reteaching and Enrichment Strategies

If students need help...	If students got it...
Resources by Chapter • Practice A and Practice B • Puzzle Time Record and Practice Journal Practice Differentiating the Lesson Lesson Tutorials Skills Review Handbook	Resources by Chapter • Enrichment and Extension • School-to-Work Start the next section

Order the numbers from least to greatest.

24. 66.1%, 0.66, $\frac{2}{3}$, 0.667

25. $\frac{2}{9}$, 21%, $0.2\overline{1}$, $\frac{11}{50}$

Tell which letter shows the graph of the number.

26. $\frac{2}{5}$ 27. 45.2% 28. 0.435 29. $\frac{4}{9}$

30. **TOUR DE FRANCE** The Tour de France is a bicycle road race. The whole race is made up of 21 small races called stages. The table shows how several stages compare to the whole Tour de France. Order the stages from shortest to longest.

Stage	1	7	8	17	21
Portion of Total Distance	$\frac{11}{200}$	0.044	$\frac{6}{125}$	0.06	4%

31. **SLEEP** The table shows the portions of the day that several animals sleep.

 a. Order the animals by sleep time from least to greatest.
 b. Estimate the portion of the day that you sleep.
 c. Where do you fit on the ordered list?

Animal	Portion of Day Sleeping
Dolphin	0.433
Lion	56.3%
Rabbit	$\frac{19}{40}$
Squirrel	$\frac{31}{50}$
Tiger	65.8%

32. **Number Sense** Tell what whole number you can substitute for a in each list so the numbers are ordered from least to greatest. If there is none, explain why.

 a. $\frac{2}{a}, \frac{a}{22}, 33\%$ b. $\frac{1}{a}, \frac{a}{8}, 33\%$

Fair Game Review What you learned in previous grades & lessons

Multiply. *(Section 3.2)*

33. 0.6×8 34. 3.3×5 35. 0.74×9 36. 2.19×12

37. **MULTIPLE CHOICE** A flute is 27 inches long. A piccolo is $12\frac{1}{2}$ inches long. How many times greater is the length of the flute than the length of the piccolo? *(Section 2.6)*

 Ⓐ $\frac{25}{54}$ Ⓑ $1\frac{1}{5}$ Ⓒ $2\frac{4}{25}$ Ⓓ 6

4 Study Help

You can use a **summary triangle** to explain a concept. Here is an example of a summary triangle for writing a percent as a fraction.

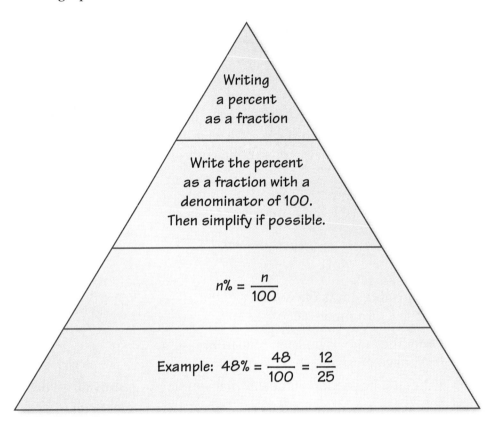

On Your Own

Make a summary triangle to help you study these topics.

1. writing a fraction as a percent
2. writing a percent as a decimal
3. writing a decimal as a percent
4. comparing and ordering fractions, decimals, and percents

After you complete this chapter, make summary triangles for the following topics.

5. finding the percent of a number
6. estimating the percent of a number

"I found this great summary triangle in my *Beautiful Beagle Magazine*."

Sample Answers

1.

2.

3.
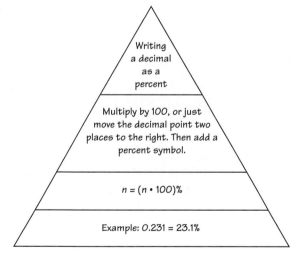

4. Available at *BigIdeasMath.com*.

List of Organizers
Available at *BigIdeasMath.com*

Comparison Chart
Concept Circle
Definition (Idea) and Example Chart
Example and Non-Example Chart
Formula Triangle
Four Square
Information Frame
Information Wheel
Notetaking Organizer
Process Diagram
Summary Triangle
Word Magnet
Y Chart

About this Organizer

A **Summary Triangle** can be used to explain a concept. Typically, the summary triangle is divided into 3 or 4 parts. In the top part, students write the concept being explained. In the middle part(s), students write any procedure, explanation, description, definition, theorem, and/or formula(s). In the bottom part, students write an example to illustrate the concept. A summary triangle can be used as an assessment tool, in which blanks are left for students to complete. Also, students can place their summary triangles on note cards to use as a quick study reference.

Technology For the Teacher
Vocabulary Puzzle Builder

T-166

Answers

1. 15%
2. 137.5%
3. 6.25%
4. 0.34
5. 0.0012
6. 0.625
7. 67%
8. 535%
9. 68.5%
10. 74%
11. 0.3

12. $\frac{6}{5} = 1.2$, 1.22, 125% = 1.25

13. 42% = 0.42, $\frac{17}{40} = 0.425$, 0.43

14. $\frac{19}{50}$
15. 0.62

16. Yes; The goalie saved 12 out of 18 shots, or $66.\overline{6}\%$, which is greater than 65%.

17. **a.** Team 5, Team 1, Team 6, Team 3, Team 7, Team 2, Team 8, Team 4

 b. Teams 2, 8, and 4

Alternative Quiz Ideas

100% Quiz Math Log
Error Notebook Notebook Quiz
Group Quiz **Partner Quiz**
Homework Quiz Pass the Paper

Partner Quiz
- Students should work in pairs. Each pair should have a small white board.
- The teacher selects certain problems from the quiz and writes one on the board.
- The pairs work together to solve the problem and write their answer on the white board.
- Students show their answers and, as a class, discuss any differences.
- Repeat for as many problems as the teacher chooses.
- For the word problems, teachers may choose to have students read them out of the book.

Assessment Book

Reteaching and Enrichment Strategies

If students need help...	If students got it...
Resources by Chapter • Study Help • Practice A and Practice B • Puzzle Time Lesson Tutorials *BigIdeasMath.com* Practice Quiz Practice from the Test Generator	Resources by Chapter • Enrichment and Extension • School-to-Work Game Closet at *BigIdeasMath.com* Start the next section

Technology For the Teacher
Answer Presentation Tool
Big Ideas Test Generator

T-167

4.1–4.3 Quiz

Write the fraction as a percent. *(Section 4.1)*

1. $\dfrac{3}{20}$
2. $\dfrac{11}{8}$
3. $\dfrac{1}{16}$

Write the percent as a decimal. *(Section 4.2)*

4. 34%
5. 0.12%
6. 62.5%

Write the decimal as a percent. *(Section 4.2)*

7. 0.67
8. 5.35
9. 0.685

Tell which number is greater. *(Section 4.3)*

10. $\dfrac{11}{15}$, 74%
11. 3%, 0.3

Use a number line to order the numbers from least to greatest. *(Section 4.3)*

12. 125%, $\dfrac{6}{5}$, 1.22
13. 42%, 0.43, $\dfrac{17}{40}$

14. **MUSEUMS** Of the museums in the United States, 38% are history museums. What *fraction* of the museums are history museums? *(Section 4.1)*

15. **ANATOMY** About 62% of the human body is composed of water. Write this percent as a decimal. *(Section 4.2)*

16. **SAVES** A goalie's saves (•) and goals scored against (×) are shown. Did the goalie save at least 65% of the shots on goal? Explain. *(Section 4.3)*

17. **SCAVENGER HUNT** The table shows the results of 8 teams competing in a scavenger hunt. *(Section 4.3)*

Team	1	2	3	4	5	6	7	8
Portion Collected	$\dfrac{3}{4}$	0.8	77.5%	0.825	$\dfrac{29}{40}$	76.25%	$\dfrac{63}{80}$	81.25%

a. Order the teams by portion collected from least to greatest.

b. The top three teams move on to the final round. Which teams are in the final round?

4.4 Finding the Percent of a Number

COMMON CORE STATE STANDARDS
6.RP.3c

Essential Question How can you use mental math to find the percent of a number?

"I have a secret way for finding 21% of 80."

"10% is 8 and 1% is 0.8."

"So 21% is 8 + 8 + 0.8 = 16.8."

1 EXAMPLE: Finding 10% of a Number

a. How did Newton know that 10% of 80 is 8?

Write 10% as a fraction.

$$10\% = \frac{10}{100} = \frac{1}{10}$$

(10 per cent)

Method 1: Using a Model

| 0% | 10% | 20% | 30% | 40% | 50% | 60% | 70% | 80% | 90% | 100% |
| 0 | 8 | 16 | 24 | 32 | 40 | 48 | 56 | 64 | 72 | 80 |

Method 2: Using Multiplication

$$10\% \text{ of } 80 = \frac{1}{10} \text{ of } 80 = \frac{1}{10} \times 80 = \frac{80}{10} = 8$$

b. How do you move the decimal point to find 10% of a number?

Move the decimal point one place to the left. **10% of 80. = 8.0**

2 ACTIVITY: Finding 1% of a Number

Work with a partner.

a. How did Newton know that 1% of 80 is 0.8?

b. How do you move the decimal point to find 1% of a number?

Laurie's Notes

Common Core State Standards

6.RP.3c Find a percent of a quantity as a rate per 100; solve problems involving finding the whole, given a part and the percent.

Introduction

For the Teacher
- **Goal:** The goal of this activity is for students to develop an understanding of how to find 10% and 1% of a number using mental math.

Motivate
- Draw a 24-tower showing different ways 24 can be divided into groups of equal size.
- ❓ Use this model to ask questions such as:
 "What is $\frac{1}{2}$ of 24? What is $\frac{3}{4}$ of 24?
 What is $\frac{5}{6}$ of 24?" 12; 18; 20

Previous Learning
Students should know how to write halves, fourths, tenths, and hundredths as percents. Students should also be able to multiply decimals and fractions, and write percents as fractions or decimals.

Start Thinking! and Warm Up

Activity 4.4 Start Thinking! For use before Activity 4.4

Activity 4.4 Warm Up For use before Activity 4.4

Multiply.

1. $\frac{1}{10} \times 120$ 2. $\frac{2}{5} \times 150$
3. $\frac{1}{2} \times 26$ 4. $\frac{3}{10} \times 100$
5. $\frac{1}{4} \times 20$ 6. $\frac{1}{100} \times 920$

Activity Notes

Example 1
❓ This example uses a percent bar model to find 10% of 80. Introduce the idea with a series of questions.
- "If 10 friends share $80, how much will each person receive?" $8
- "What is $\frac{1}{10}$ of $80? What is 10% of $80?" $8; $8
- The percent bar model displays multiples of 8 on the bottom and increments of 10% on the top. It is important for students to see that each of the small rectangles in the model are 8. There are ten 8s, or 80 total.

Activity 2
- Help students develop an understanding of the relationship between 1%, 10%, and 100%.
- ❓ "Finding $\frac{1}{10}$ of a number moves the decimal point one place to the left, so finding $\frac{1}{100}$ of a number should move the decimal point how many places to the left? Why?"

 2; $\frac{1}{100}$ is $\frac{1}{10} \cdot \frac{1}{10}$ or $\frac{1}{10}$ times itself.

- ❓ Students may find these questions difficult because of the decimal answer.
 - "If 100 friends share $80, how much will each person receive?" $0.80
 - "What is $\frac{1}{100}$ of $80?" $0.80
 - "What is 1% of $80?" $0.80
- **Connection:** Percents of the same number can be added or subtracted. For example, 12% of 30 could be found by finding 10% of 30 = 3 and 1% of 30 = 0.3 and adding 3 + 0.3 + 0.3 = 3.6. Discuss this concept with the students, because it is not obvious to all students.

4.4 Record and Practice Journal

T-168

English Language Learners

Vocabulary

English learners may be unfamiliar with the concept of commission in Activity 4. Explain that the three most common ways for a salesperson to be compensated are salary, commission, and base salary with commission.

Examples

Salary:
 $12 per hour, $500 per week

Commission:
 17% of the dollar amount sold

Base salary with commission:
 $300 per week plus 5% of the dollar amount sold

4.4 Record and Practice Journal

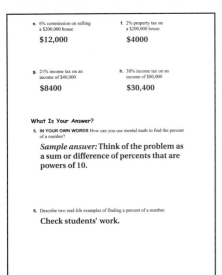

Laurie's Notes

Example 3

? "What is 10% of 40? What is 1% of 40?" 4; 0.4
- Some students may need to write down each percent as they find it instead of trying to hold more than one answer in their heads.
- Work through each example.

? **Extension:** Try to help students develop number sense with percents by asking the following questions.
 - "What is the relationship between 10% of 40 and 10% of 80?" 4 is half of 8.
 - "10% of 40 is 4. What is 5% of 40?" half of 4; 2
 - "19% of 50 is 9.5. What is 19% of 100?" two times 9.5; 19
 - "19% of 50 is 9.5. What is 50% of 19? half of 19; 9.5

Activity 4

- **Vocabulary:** Commission, property tax, and income tax may be unfamiliar concepts, so discuss their meanings before students begin.
- Students will use mental math to compute the percents, but will need to record answers as they are done.
- **Example:** Find 38% of $80,000.

 Find 10% of $80,000. $8,000
 Quadruple $8,000, which is 40% of $80,000. $32,000
 Find 1% of $80,000. $800
 Double $800, which is 2% of $80,000. $1600
 Find 38%, which is 40% − 2%. $32,000 − $1600
 Answer: So, 38% of $80,000 is $30,400.

- Students should work in pairs. When everyone has finished, ask for volunteers to explain their work.
- **Common Difficulty:** When numbers become fairly large, a student's ability to use mental math seems to diminish.

Closure

- Use mental math to find the following percents of 120: 10%, 15%, and 20%. 12, 18, 24

The Dynamic Planning Tool
Editable Teacher's Resources at *BigIdeasMath.com*

3 EXAMPLE: Using Mental Math

Use mental math to find each percent of a number.

a. 12% of 40

Think: 12% = 10% + 1% + 1%

> 10% of 40 = 4 1% of 40 = 0.4

$$4 + 0.4 + 0.4 = 4.8$$

b. 19% of 50

Think: 19% = 10% + 10% − 1%

> 10% of 50 = 5 1% of 50 = 0.5

$$5 + 5 - 0.5 = 9.5$$

4 ACTIVITY: Using Mental Math

Work with a partner. Use mental math to find each percent of a number.

a. 20% tip for a $30 meal

b. 18% tip for a $30 meal

c. 6% sales tax on a $20 shirt

d. 9% sales tax on a $20 shirt

e. 6% commission on selling a $200,000 house

f. 2% property tax on a $200,000 house

g. 21% income tax on an income of $40,000

h. 38% income tax on an income of $80,000

What Is Your Answer?

5. **IN YOUR OWN WORDS** How can you use mental math to find the percent of a number?

6. Describe two real-life examples of finding a percent of a number.

Use what you learned about finding the percent of a number to complete Exercises 3–10 on page 172.

Section 4.4 Finding the Percent of a Number

4.4 Lesson

Check It Out
Lesson Tutorials
BigIdeasMath.com

Key Idea

Finding the Percent of a Number

Words Write the percent as a fraction or decimal. Then multiply.

Numbers 20% of 60 is 12.

$$\frac{1}{5} \times 60 = 12$$

$$0.2 \times 60 = 12$$

Model

0% 20% 40% 60% 80% 100%
0 12 24 36 48 60

EXAMPLE 1 — Finding the Percent of a Number

Use a fraction to find the percent of the number.

a. Find 25% of 40.

$$25\% \text{ of } 40 = \frac{1}{4} \times 40$$

$$= \frac{1 \times \overset{10}{\cancel{40}}}{\cancel{4}_1}$$

$$= 10$$

∴ So, 25% of 40 is 10.

0% 25% 50% 75% 100%
0 10 20 30 40

b. Find 60% of 150.

$$60\% \text{ of } 150 = \frac{3}{5} \times 150$$

$$= \frac{3 \times \overset{30}{\cancel{150}}}{\cancel{5}_1}$$

$$= 90$$

∴ So, 60% of 150 is 90.

0% 20% 40% 60% 80% 100%
0 30 60 90 120 150

On Your Own

Use a fraction to find the percent of the number.

1. 90% of 20
2. 75% of 32
3. 10% of 110
4. 30% of 75

EXAMPLE 2 — Standardized Test Practice

Summer Vacation
Did you go on a vacation this past summer?

| Yes | 48% |
| No | 52% |

Note: 200 students surveyed

How many students went on vacation?

Ⓐ 48 Ⓑ 96 Ⓒ 100 Ⓓ 104

From the survey, you can see that 48% out of 200 students said yes.

48% of 200 = 0.48 × 200 Write 48% as a decimal.

= 96 Multiply.

∴ So, 96 students went on vacation. The correct answer is Ⓑ.

170 Chapter 4 Fractions, Decimals, and Percents

Laurie's Notes

Introduction

Connect
- **Yesterday:** Students used mental math to find the percent of a number.
- **Today:** Students will use multiplication to find the percent of a number.

Motivate
- Play a game of *Find Your Partner*. Make cards where one-half of the cards have a fraction problem and one-half of the cards have a percent problem.
- Each pair of cards has the same solution. Use mental math to solve.
- Depending on class size, you may want all or some of your students to have a card.

Sample:

$\frac{1}{2}$ of 48	$\frac{1}{4}$ of 120	$\frac{1}{3}$ of 45	$\frac{1}{5}$ of 45
10% of 240	50% of 60	30% of 50	75% of 12

- Ask each pair to explain how they found the value of their card. Connect yesterday's investigation to today's introduction.

Lesson Notes

Key Idea
- **Connection:** Connect this idea to multiplying a fraction by a whole number. Instead of "find $\frac{1}{2}$ of 48," the problem is "find 50% of 48."

Example 1
- ❓ "What is 25% as a fraction? How do you find $\frac{1}{4}$ of 40?" $\frac{1}{4}$; You multiply.
- Discuss the visual model. Each part, $\frac{1}{4}$ or 25%, represents 10.
- ❓ "What is 60% as a fraction?" $\frac{6}{10}$ or $\frac{3}{5}$
- ❓ "How do you find $\frac{3}{5}$ of 150?" Multiply $\frac{3}{5} \times 150 = \frac{(3 \times 30 \times 5)}{5} = 90$
- Discuss why a common factor needs to be divided out in each problem.

On Your Own
- **Think-Pair-Share:** Students should read each question independently and then work with a partner to answer the questions. When they have answered the questions, the pair should compare their answers with another group and discuss any discrepancies.

Example 2
- Give students time to read the information in the graphic display.
- Probe the reasoning involved in an alternative approach.
- ❓ "What is 48% of 100? What is 48% of 200?" 48; 96

Goal Today's lesson is finding the percent of a number.

Lesson Materials
Introduction
• index cards

Start Thinking! and Warm Up

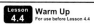

> **Lesson 4.4 Warm Up** For use before Lesson 4.4
>
> **Lesson 4.4 Start Thinking!** For use before Lesson 4.4
>
> Research the sales tax rates in various U.S. states.
>
> Write a word problem that compares the cost of buying an item in one state with buying the same priced item in a state with a different sales tax rate.
>
> Exchange word problems with a classmate and solve your classmate's problem.

Extra Example 1
Use a fraction to find the percent of a number.
a. Find 80% of 75. 60
b. Find 40% of 200. 80

On Your Own
1. 18
2. 24
3. 11
4. $22\frac{1}{2}$

Extra Example 2
Twenty-five percent of 120 students went on vacation. How many students went on vacation? 30 students

T-170

On Your Own

5. 6
6. 117
7. 5.6
8. 2.8

Extra Example 3
You leave a 20% tip on a restaurant bill of $35. How much is the tip? $7

On Your Own
9. $270

Extra Example 4
The length of a garden is 225% of its width. If the width of the garden is 4 meters, what is the area of the garden? 36 m²

On Your Own
10. 7920 ft²

Differentiated Instruction
Kinesthetic
Use a 100 grid to illustrate 80% of 60. Have students shade 60 of the 100 squares. Ask students for strategies for finding 80% or $\frac{4}{5}$ of the shaded squares by just counting. One strategy would be to repeatedly shade 4 squares with a second color and skip 1 square.

T-171

Laurie's Notes

On Your Own
- Students will likely write the percent as a decimal and multiply by the whole number. Probe to see if Question 5 was computed by finding 10% + 5%.
- **Common Error:** When multiplying a decimal by a whole number that ends in a zero (Questions 6 and 8), it is common to be off by a place value. Encourage students to estimate their answer before they begin. *Example:* 4% of 70 is less than 10% of 70, so the answer should be less than 7.

Example 3
- "What is 100% of any number?" the number
- "What is 50% of any number?" half of the number
- "What is 150% of any number?" The number with half of the number added on.
- **FYI:** Students need plenty of practice with "nice" numbers when developing fluency for finding the percent of a number using mental math.
- Ask students to identify the key phrase in the problem. 150% of the current bid
- Share a third method. Write the percent as a fraction.
 150% of 120 = $\frac{3}{2} \times 120 = 180$.

On Your Own
- Probe different strategies for solving this problem.

Example 4
- Be sure that students have a basic understanding that if the width is 80% of the length, the width is shorter than the length. This may not be obvious to all students.
- The percent bar model visually shows the 5 equal parts of 15. Students could use this model to find 20%, 40%, 60%, or 80% of 15.
- Note that the question asks for the area. The answer is stated in terms of square feet.

Closure
- Draw a percent bar model to find 75% of 64.

0%	25%	50%	75%	100%
0	16	32	48	64

- Use a fraction to find 75% of 48. $\frac{3}{4} \times 48 = 36$
- Use a decimal to find 75% of 40. $0.75 \times 40 = 30$

The Dynamic Planning Tool
Editable Teacher's Resources at *BigIdeasMath.com*

● **On Your Own**

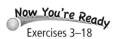
Exercises 3–18

Use a decimal to find the percent of the number.

5. 15% of 40 6. 78% of 150 7. 35% of 16 8. 4% of 70

EXAMPLE 3 Using Mental Math

Current bid: US $120.00
Time remaining: 1 hour 45 min

Your friend is bidding online for concert tickets. The current bid is shown. The winning bid is 150% of the current bid. How much is the winning bid?

Method 1: Write 150% as a decimal and multiply.

150% of 120 = 1.5 × 120

$\phantom{150\% \text{ of } 120} = 180$

Method 2: Using mental math, think 150% = 100% + 50%.

100% of 120 = 1 × 120 = 120

50% of 120 = $\frac{1}{2}$ × 120 = 60

Add: 120 + 60 = 180

∴ So, the winning bid is $180.

● **On Your Own**

Exercises 23–30

9. **WHAT IF?** In Example 3, the winning bid is 225% of the current bid. How much is the winning bid?

EXAMPLE 4 Real-Life Application

The width of a rectangular room is 80% of its length. What is the area of the room?

Find 80% of 15 feet.

15 ft

$80\% \text{ of } 15 = \dfrac{4}{5} \times 15$

$= \dfrac{4 \times \cancel{15}^{3}}{\cancel{5}_{1}}$

$= 12$ The width is 12 feet.

0%	20%	40%	60%	80%	100%
0	3	6	9	12	15

Use the formula for the area A of a rectangle.

$A = 15 \times 12 = 180$

∴ So, the area of the room is 180 square feet.

● **On Your Own**

10. The width of a rectangular stage is 55% of its length. The stage is 120 feet long. What is the area?

4.4 Exercises

Check It Out
Help with Homework
BigIdeasMath.com

Vocabulary and Concept Check

1. **DIFFERENT WORDS, SAME QUESTION** Which is different? Find "both" answers.

 What is twenty percent of 30? What is one-fifth of 30?

 What is 20 multiplied by 30? What is 0.2 times 30?

2. **REASONING** If 52 is 130% of a number, is the number greater or less than 52? Explain.

Practice and Problem Solving

Find the percent of the number.

3. 20% of 60	4. 10% of 40	5. 18% of 70	6. 32% of 30
7. 8% of 90	8. 14% of 20	9. 26% of 50	10. 3% of 60
11. 30% of 70	12. 75% of 48	13. 45% of 45	14. 92% of 19
15. 40% of 60	16. 38% of 22	17. 70% of 20	18. 87% of 55

19. **ERROR ANALYSIS** Describe and correct the error in finding 40% of 75.

 40% of 75 = 40% × 75 = 3000

20. **MANGROVES** Lake Worth, near West Palm Beach, had about 2120 acres of mangrove trees 40 years ago. Only about 13% of the mangrove trees remain. How many acres of mangrove trees remain?

21. **SPIDER MONKEY** The tail of the spider monkey is 64% of its length. What is the length of its tail?

 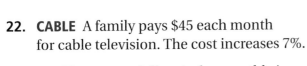
 55 in.

22. **CABLE** A family pays $45 each month for cable television. The cost increases 7%.

 a. How many dollars is the monthly increase?
 b. What is the new monthly cost?

Find the percent of the number.

23. 140% of 60	24. 120% of 33	25. 175% of 54	26. 250% of 146
27. 4.5% of 50	28. 0.7% of 40	29. 2.8% of 150	30. 7.2% of 235

Assignment Guide and Homework Check

Level	Day 1 Activity Assignment	Day 2 Lesson Assignment	Homework Check
Basic	3–10, 42–46	1, 2, 11–35 odd, 20, 22	11, 20, 23, 27, 31
Average	3–10, 42–46	1, 2, 11–19 odd, 22–34 even, 35–38	11, 22, 24, 28, 32
Advanced	3–10, 42–46	1, 2, 15–19, 24–36 even, 39–41	16, 24, 28, 32, 36

Common Errors

- **Exercises 3–18** If the number ends in a zero and students multiply by a decimal, they may be off by a place value. This is because they have forgotten about the zero at the end. Encourage students to estimate their answer first. For example, 8% of 90 should be less than 10% of 90, so the answer should be less than 9.
- **Exercises 23–30** Students often struggle with percents greater than 100% and percents with decimals. They may find a number significantly greater than or less than what they should. Remind students of the methods of converting percents to decimals and how to move the decimal point. Encourage students to estimate their answer. Review that percents greater than 100% will have answers that are greater than the original number.
- **Exercises 35 and 36** Students may forget to convert the units after they have found the percent. Remind them to read the question carefully.

4.4 Record and Practice Journal

Find the percent of the number.
1. 30% of 50 — **15**
2. 12% of 85 — **10.2**
3. 2% of 96 — **1.92**
4. 150% of 66 — **99**
5. 7.5% of 120 — **9**
6. 0.3% of 15 — **0.045**

Complete the statement using <, >, or =.
7. 70% of 80 __=__ 80% of 70
8. 92% of 30 __<__ 48% of 75

9. You make homemade lip balm. About 11% of the lip balm is made from beeswax. You make $4\frac{2}{5}$ teaspoons of the lip balm. About how many teaspoons of beeswax do you need? Round your answer to the nearest tenth.
0.5 teaspoon

10. You score $\frac{63}{75}$ on your first test. After your second test, your teacher tells you your test average is 88%.
 a. The second test had 75 points possible. What percent did you receive on the second test?
 92%
 b. How many total points have you earned in the class?
 132

Technology For the Teacher
Answer Presentation Tool
QuizShow

Vocabulary and Concept Check

1. What is 20 multiplied by 30?; 600; 6
2. less than 52; If you take more than 100% of a number, you get a greater number. So, 52 is greater than the original number.

Practice and Problem Solving

3. 12
4. 4
5. 12.6
6. 9.6
7. 7.2
8. 2.8
9. 13
10. 1.8
11. 21
12. 36
13. 20.25
14. 17.48
15. 24
16. 8.36
17. 14
18. 47.85
19. The percent was not written as a decimal or fraction before multiplying: 40% × 75 = 0.40 × 75 = 30
20. about 275.6 acres
21. 35.2 in.
22. a. $3.15
 b. $48.15
23. 84
24. 39.6
25. 94.5
26. 365
27. 2.25
28. 0.28
29. 4.2
30. 16.92

Practice and Problem Solving

31. = **32.** <
33. > **34.** =
35. 48 min
36. 28.08 in.
37. a. 432 in.²
 b. 37.5%
38. yes; To pass inspection, the ball must bounce back to between 68% and 75% of the starting height, or between 4.08 feet and 4.5 feet. It bounced back to 4.08$\overline{3}$ feet, so it passes.
39. *Sample answer:* Because 30% of n is equal to 2 times 15% of n and 45% of n is equal to 3 times 15% of n, you can write 30% of $n = 2 \times 12 = 24$ and 45% of $n = 3 \times 12 = 36$.
40. See *Taking Math Deeper*.
41. a. 243 points
 b. 97.2%

Fair Game Review

42. $2\frac{2}{3}$ **43.** $1\frac{1}{2}$
44. $3\frac{3}{5}$ **45.** 10
46. D

Mini-Assessment
Find the percent of the number.
1. 15% of 70 10.5
2. 65% of 50 32.5
3. 32% of 88 28.16
4. 6% of 32 1.92
5. How many inches is 40% of 5 feet? 24 in.

Taking Math Deeper

Exercise 40
Sale items can be discounted in different ways. In this problem, students can learn that sequential discounting can produce a lesser discount than if each discount is taken off the original price.

 Organize the given information.

Regular Price $150

Coupon 10% off Sale price

Sale: 30% off

② Find the sale price and the price after using the coupon.
30% of $150 = 0.3(150) = $45
 a. So, the sale price = 150 − 45 = $105.
 10% of $105 = 0.1(105) = $10.5
 b. Therefore, the after coupon price = 105 − 10.5 = $94.50.

③ Compare the two discount plans.
40% of $150 = 0.4(150) = $60
 c. Sale price = 150 − 60 = $90

Discounts and markups will be taught in a future course. However, it is good preparation for students to think about the various ways that discounts can occur.

Project
Prepare a budget for a surfing trip or other outing. Include all expenses that might be incurred. Give some ideas as to how you might earn the money needed for the trip.

Reteaching and Enrichment Strategies

If students need help...	If students got it...
Resources by Chapter • Practice A and Practice B • Puzzle Time Record and Practice Journal Practice Differentiating the Lesson Lesson Tutorials Skills Review Handbook	Resources by Chapter • Enrichment and Extension • School-to-Work Start the next section

Copy and complete the statement using <, >, or =.

31. 80% of 60 ⬚ 60% of 80
32. 20% of 30 ⬚ 30% of 40
33. 120% of 5 ⬚ 0.8% of 250
34. 85% of 40 ⬚ 25% of 136

35. **TIME** How many minutes is 40% of 2 hours?

36. **LENGTH** How many inches is 78% of 3 feet?

37. **GEOMETRY** The width of the rectangle is 75% of its length.

 a. What is the area of the rectangle?
 b. The length of the rectangle is doubled. What percent of the length is the width now? Explain your reasoning.

24 in.

38. **BASKETBALL** To pass inspection, a new basketball should bounce back to between 68% and 75% of the starting height. A new ball is dropped from 6 feet and bounces back 4 feet 1 inch. Does the ball pass inspection? Explain.

39. **REASONING** You know that 15% of a number n is 12. How can you use this to find 30% of n? 45% of n? Explain.

40. **SURFBOARD** You have a coupon for 10% off the sale price of a surfboard.

 a. What is the sale price of the surfboard?
 b. What is the price of the surfboard after using the coupon?
 c. Is taking 40% off the regular price the same as taking 30% off the regular price and then 10% off the sale price? Explain your reasoning.

41. **Number Sense** On three geography tests, you earned grades of 88%, 94%, and 90%. Each test was worth 150 points.

 a. The final exam is worth 250 points. How many points do you need on the final exam to earn 93% of the total points on tests?
 b. What *percent* do you need on the final?

 Fair Game Review What you learned in previous grades & lessons

Multiply. Write the answer in simplest form. *(Section 2.2)*

42. $\frac{2}{3} \times 4$
43. $\frac{3}{8} \times 4$
44. $6 \times \frac{3}{5}$
45. $12 \times \frac{5}{6}$

46. **MULTIPLE CHOICE** What is the quotient of 7.5 and 2.4? *(Section 3.5)*

 Ⓐ 0.0032 Ⓑ 0.03125 Ⓒ 0.32 Ⓓ 3.125

4.5 Percents and Estimation

COMMON CORE STATE STANDARDS

6.RP.3c

Essential Question How can you use mental math and estimation to help solve real-life problems?

1 ACTIVITY: Estimating a Percent

Work with a partner. In the U.S. Constitution, the nation's capitol, Washington, D.C., was not allowed to exceed 10 miles square. After the capitol was built, it ended up having less than the maximum allowed area.

a. What was the maximum area allowed by the Constitution?

b. Use the grid to estimate the area of Washington, D.C. Explain your reasoning.

c. What percent of the maximum allowed area did the capitol use?

$$\frac{\text{Actual Area}}{\text{Maximum Area Allowed}}$$

174 **Chapter 4** Fractions, Decimals, and Percents

Laurie's Notes

Introduction

For the Teacher
- **Goal:** The goal of this activity is for students to use their rounding skills to estimate the answer to a real-life problem.

Motivate
- **History:** Share a little history with students related to the first activity. President Washington hired Pierre Charles L'Enfant, a French artist-architect, to design the American capitol, which was to be built between the states of Maryland and Virginia. L'Enfant designed the layout of the city, but was fired by President Washington a year after he was hired because he didn't listen to instructions and spent too much money. Although there is some disagreement as to who recreated L'Enfant's plans after he took his designs back to Paris, it is agreed that Benjamin Banneker, an African American mathematician, astronomer, and publisher, surveyed the land for the national capital. Recently discovered records of the survey show Banneker was paid $60 for three months of work.

Activity Notes

Activity 1
- **FYI:** Information for Activity 1 can be found in Article I.8 of the U.S. Constitution.
- Students are asked to estimate a percent, which is a new concept. The set-up of the problem is given and the denominator is 100, thus making this new concept a very reasonable problem for students to try.
- Also, students are estimating area by counting squares, which is a good method to reinforce the area concept.
- Have a student read the introduction. If time permits, ask who has visited Washington, D.C.
- Explain that "10 miles square" means a square with side lengths of 10 miles.
- ❓ "What is the area of a square with side lengths of 10 miles?" 100 mi^2
- Have students work with a partner to complete this activity.
- Compare answers. Record the range of answers to part (b). Have a few students explain how they estimated the area. Having a transparency of the map will help facilitate the discussion.
- As a prelude to the next chapter when students will calculate mean, median, and mode, discuss features of the data collected for part (b).

Common Core State Standards

6.RP.3c Find a percent of a quantity as a rate per 100; solve problems involving finding the whole, given a part and the percent.

Previous Learning

Students need to know how to round numbers and how to use compatible numbers to estimate. Students should also know how to write a percent as a fraction or decimal.

Start Thinking! and Warm Up

Activity 4.5 **Start Thinking!** For use before Activity 4.5

Activity 4.5 **Warm Up** For use before Activity 4.5

Find the percent of the number.
1. 10% of 50
2. 12% of 60
3. 2% of 45
4. 30% of 90
5. 15% of 40
6. 5% of 200

4.5 Record and Practice Journal

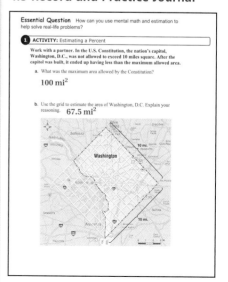

Essential Question How can you use mental math and estimation to help solve real-life problems?

1 ACTIVITY: Estimating a Percent

Work with a partner. In the U.S. Constitution, the nation's capitol, Washington, D.C., was not allowed to exceed 10 miles square. After the capitol was built, it ended up having less than the maximum allowed area.

a. What was the maximum area allowed by the Constitution?

100 mi^2

b. Use the grid to estimate the area of Washington, D.C. Explain your reasoning. 67.5 mi^2

T-174

Differentiated Instruction

Auditory

Students who learn best by listening will benefit from hearing the process of estimating out loud. For example, to estimate the 20% tip for a $29.45 meal say, "$29.45 is close to $30.00. 10% of $30.00 is $3.00. 20% is 2 times 10%, so the tip is about 2 times $3.00, or $6.00."

4.5 Record and Practice Journal

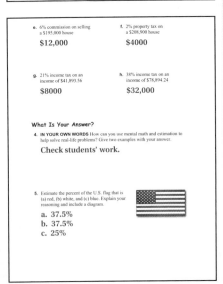

Laurie's Notes

Example 2
- This example is similar to the example in Activity 4.4, except that the numbers used are not whole numbers. Students will need to round the numbers in order to estimate the answer.
- Discuss rounding dollar amounts. Often the context influences how much rounding is done. For instance, if you are trying to determine if you have enough money to purchase an item, rounding up is wise.
- Discuss rounding $38.57 to $39 versus $40.

Activity 3
- This activity uses the same contexts that were estimated in Activity 4.4, except now the numbers are not whole numbers.
- Review vocabulary, if necessary.
- Students work with a partner. Ask students to record their work so that you can see how they rounded the numbers to get estimates. If students check their estimates with a calculator, demonstrate how it is done.
- Discuss the estimates found, as answers will likely vary. For example, 9% of $21.89 might be 10% of $20 or 10% of $22.
- **Common Error:** Students may be off by a decimal place when finding 2% of $208,900. *Think:* 10% is about $20,000, so 1% is about $2000. Double to find 2%.

What Is Your Answer?
- Question 5 would be a good problem for students to work with a partner.

Closure
- **Exit Ticket:** Estimate 9% of $44.98. *Sample answer:* 10% of 45 = 4.5

The Dynamic Planning Tool
Editable Teacher's Resources at *BigIdeasMath.com*

② EXAMPLE: Using Mental Math

Use mental math to estimate each percent of a number.

a. 10% of $38.57
Round $38.57 to $40.
10% of $40 is $4.

So, 10% of $38.57 is about $4.

b. 19% of $71.33
Round 19% up to 20%.
Round $71.33 down to $70.
20% of $70 is $14.

So, 19% of $71.33 is about $14.

③ ACTIVITY: Using Mental Math

Work with a partner. Use mental math to estimate each percent of a number. Use a calculator to check your estimate.

a. 20% tip for a $29.45 meal

b. 18% tip for a $29.45 meal

c. 6% sales tax on a $21.89 shirt

d. 9% sales tax on a $21.89 shirt

e. 6% commission on selling a $195,000 house

f. 2% property tax on a $208,900 house

g. 21% income tax on an income of $41,893.56

h. 38% income tax on an income of $78,894.24

What Is Your Answer?

4. **IN YOUR OWN WORDS** How can you use mental math and estimation to help solve real-life problems? Give two examples with your answer.

5. Estimate the percent of the U.S. flag that is (a) red, (b) white, and (c) blue. Explain your reasoning and include a diagram.

Practice → Use what you learned about percents and estimation to complete Exercises 7–12 on page 178.

Section 4.5 Percents and Estimation 175

4.5 Lesson

Check It Out
Lesson Tutorials
BigIdeasMath.com

In many real-life problems, you do not need an exact answer. To estimate a percent of a number, use common percents that are easy to work with.

Common Percent-to-Fraction Conversions

$10\% = \frac{1}{10}$	$20\% = \frac{1}{5}$	$30\% = \frac{3}{10}$	$40\% = \frac{2}{5}$	$50\% = \frac{1}{2}$	$60\% = \frac{3}{5}$
$70\% = \frac{7}{10}$	$80\% = \frac{4}{5}$	$90\% = \frac{9}{10}$	$100\% = 1$	$25\% = \frac{1}{4}$	$75\% = \frac{3}{4}$

EXAMPLE 1 — Estimating the Percent of a Number

An inflatable pool contains 800 gallons of water. The pool loses 74% of its water through a leak. Estimate the amount of water lost.

74% is close to 75%, or $\frac{3}{4}$. So, find 75% of 800.

$$75\% \text{ of } 800 = \frac{3}{4} \times 800$$
$$= \frac{3 \times \overset{200}{\cancel{800}}}{\cancel{4}_{1}}$$
$$= 600$$

So, about 600 gallons of water are lost.

On Your Own

Now You're Ready
Exercises 7–18

Estimate the percent of the number.

1. 27% of 40
2. 8% of 50
3. 61% of 125
4. 99% of 230

EXAMPLE 2 — Using Compatible Numbers

a. Estimate 46% of 177.

46% is close to 50%, or $\frac{1}{2}$. For 177, use the compatible number 180.

46% of 177
↓ ↓
$50\% \text{ of } 180 = \frac{1}{2} \times 180$
$= 90$

Remember
Compatible numbers are numbers that are easy to compute mentally.

So, 46% of 177 is about 90.

176 Chapter 4 Fractions, Decimals, and Percents

Laurie's Notes

Introduction

Connect
- **Yesterday:** Students used estimation and mental math to find the percent of a number.
- **Today:** Students will estimate percents using common fractional equivalents.

Motivate
❓ About 80% of American lobsters caught in the U.S. come from Maine. If 7 million pounds were caught in the U.S. in one month, about how many pounds were caught in Maine? 5.6 million pounds

Lesson Notes

Example 1
- Just as students are fluent with multiplication facts to ease computations, students should be fluent with common percent-fraction conversions. Tell students that you expect them to be able to recall these easily.
- **Discuss:** Estimating with percents is helpful in answering many real-life problems. Sometimes an exact number is not needed. That is when you can estimate. Being able to round a percent with a common fraction conversion allows us to use mental math to find the answer.
- Discuss the technique of dividing out a common factor.
- ❓ Get students to think about how much water was lost in the leak by asking, "About how much is 800 gallons? Would 800 gallons fit in this room?" Have students think about a 50-gallon trash can or a 200-gallon hot water tank. When they have something to compare it to, students should be able to tell that 800 gallons would fit in a classroom.
- **Extension:** Use the percent bar model to find 25% of 800.

On Your Own
- **Think-Pair-Share:** Students should read each question independently and then work with a partner to answer the questions. When they have answered the questions, the pair should compare their answers with another group and discuss any discrepancies.

Example 2
- **Vocabulary:** Review compatible numbers.
- ❓ "Is the estimate in part (a) greater than or less than the exact value? Explain." greater
- Students may ask if 177 could be rounded to other numbers, such as 200.
- ❓ "How is the estimate affected when 177 is rounded to 200 versus 180?" Less accurate, but it may be close enough for the context.
- ❓ "Is the estimate in part (b) greater than or less than the exact value? Explain." less
- ❓ "Why is 36 rounded to 35?" Listen for an understanding that because $80\% = \frac{4}{5}$, you want to round to a number that is divisible by five. You could also round to 40.

Goal Today's lesson is estimating the percent of a number.

Start Thinking! and Warm Up

Warm Up For use before Lesson 4.5

Start Thinking! For use before Lesson 4.5

When solving real-life problems involving percents, sometimes an exact answer is needed and other times an estimate is sufficient.

In which of the following situations is an estimate sufficient?
- Calculating the tip at a restaurant
- Comparing basketball players' field goal percentages
- Comparing student survey results

Extra Example 1
Estimate 38% of 60.
Sample answer: 24 (40% of 60)

On Your Own
1–4. Sample answers are given.
1. 10
2. 5
3. 75
4. 230

Extra Example 2
a. Estimate 79% of 155.
 Sample answer: 128 (80% of 160)
b. Estimate 92% of 28.
 Sample answer: 27 (90% of 30)

On Your Own

5–8. Sample answers are given.

5. 15
6. 10
7. 225
8. 45

Extra Example 3

In Example 3, how many of the 913 students are likely to say hanging out with friends is their favorite summer activity? *Sample answer:* about 225 students (25% of 900)

On Your Own

9. *Sample answer:* about 400 students

English Language Learners

Class Activity

Materials needed for this activity: poster board, markers, and pictures of clothing cut from magazines.

Have small groups of students work together to make a clothing sale sign that shows the clothing, the price tag, and the percent off the price. The prices should not be whole dollar amounts and the signs should have wording, such as "Take 25% off," or "Save 60%." Once all the signs are complete, have students walk around the room and "shop" at each of the locations. Have them see how quickly they can estimate the cost of each item.

Laurie's Notes

On Your Own

- **Neighbor Check:** Have students work independently and then have their neighbor check their work. Have students discuss any discrepancies.
- Have students compare their strategies. The example 24% of 63 might be done as 20% of 60 or 25% of 60.
- **Big Idea:** Share an easy strategy for dividing by 4. Divide by 2 twice. *Example:* $60 \div 4 = 15$. Also, $60 \div 2 = 30$ and $30 \div 2 = 15$. Students can generally divide by 2 in their head, so dividing by 4 means they must do the division twice.
- **FYI:** Dividing by 8 is the same as dividing by 2 three times.

Example 3

- **Discuss:** Surveys are often used to make predictions about a larger population. Ask students to give examples of surveys that they have read or heard about.
- Discuss how random samples are needed to make predictions.
- For additional practice, have students estimate the number of students that would choose one of the other summer activities.
- ❓ "What is the sum of the percents in the table? Does this make sense?" 100%; Unless those surveyed could select more than one activity, the percents should add to approximately 100%.

Closure

- **Exit Ticket:** Estimate the savings, if a $27.95 item is marked 25% off. *Sample answer:* $\frac{1}{4}$ of 28 is 7.

The Dynamic Planning Tool
Editable Teacher's Resources at *BigIdeasMath.com*

b. Estimate 81% of 36.

81% is close to 80%, or $\frac{4}{5}$. For 36, use the compatible number 35.

$$81\% \text{ of } 36$$
$$\downarrow \quad \downarrow$$
$$80\% \text{ of } 35 = \frac{4}{5} \times 35$$
$$= 28$$

∴ So, 81% of 36 is about 28.

On Your Own

Now You're Ready
Exercises 22–25

Use compatible numbers to estimate the percent of the number.

5. 24% of 63 6. 17% of 49 7. 76% of 297 8. 52% of 91

EXAMPLE 3 Real-Life Application

The circle graph shows the results of a survey of several students at a school. The school has 913 students. How many of them are likely to say spending time at the beach is their favorite summer activity?

Favorite Summer Activity

From the circle graph, 39% chose spending time at the beach. Use this percent to estimate the number from the school.

39% is close to 40%, or $\frac{2}{5}$. For 913, use the compatible number 900.

$$40\% \text{ of } 900 = \frac{2}{5} \times 900 \quad \text{Write 40\% as a fraction.}$$
$$= 360 \quad \text{Multiply.}$$

∴ So, about 360 students would say spending time at the beach is their favorite summer activity.

On Your Own

9. The bar graph shows the results of a survey of several students at a school. The school has 1038 students. How many of them are likely to say pizza is their favorite cafeteria food?

4.5 Exercises

Vocabulary and Concept Check

Round to a common percent.

1. 27%
2. 63%
3. 38%
4. 93%

5. **WHICH ONE DOESN'T BELONG?** Which one does *not* belong with the other three? Explain your reasoning.

 | 48% of 16 | 64% of 37 | 22% of 39 | 9% of 83 |

6. **OPEN-ENDED** Write a real-world problem that can be solved by estimating 22% of 60.

Practice and Problem Solving

Estimate the percent of the number.

7. 28% of 52
8. 71% of 126
9. 17% of 23
10. 12% of 47
11. 87% of 233
12. 74% of 31
13. 22% of 60
14. 33% of 200
15. 24% of 180
16. 96% of 66
17. 4% of 20
18. 6% of 120

19. **RESTAURANT** The daily special at a restaurant costs $10. About how much more does the daily special cost when the restaurant increases its prices 17%?

20. **GOLF** About 6% of the golf courses in the United States are in Florida. In 2008, there were 17,151 golf courses in the United States. About how many of them were in Florida?

21. **SCHOOL CLUBS** A middle school has 722 students.

 a. About how many students are not members of a club?
 b. About how many students are members of at least two clubs?

Number of Clubs	Percent
0	22
1	42
2	29
3	7

Use compatible numbers to estimate the percent of the number.

22. 70% of 38
23. 43% of 202
24. 13% of 80
25. 24% of 120

Estimate the percent of the number.

26. 142% of 50
27. 223% of 80
28. 296% of 33
29. 114% of 67

178 Chapter 4 Fractions, Decimals, and Percents

Assignment Guide and Homework Check

Level	Day 1 Activity Assignment	Day 2 Lesson Assignment	Homework Check
Basic	7–12, 36–40	1–6, 13–25 odd, 20	13, 17, 20, 23
Average	7–12, 36–40	1–6, 13–25 odd, 31–33	13, 17, 23, 32
Advanced	7–12, 36–40	1–6, 22–28 even, 30–35	22, 26, 30, 32

Common Errors

- **Exercises 7–18** Students might round up or down to the next percent and use the method from the previous section. While there is nothing wrong with this method, the goal of this lesson is for the student to use common percents to estimate the answer. Remind students of the table and ask them to round only to those percents.
- **Exercises 22–29** Even though they have used compatible numbers before, students might struggle with finding compatible numbers. Explain to students that finding compatible numbers is similar to rounding, but without the strict rules. Students are finding numbers that are easy to compute mentally. Tell students to write the original problem and then write the new problem with the compatible numbers underneath.

Technology For the Teacher
Answer Presentation Tool
QuizShow

4.5 Record and Practice Journal

Vocabulary and Concept Check

1–4. Sample answers are given.

1. 25% 2. 60%
3. 40% 4. 90%

5. 64% of 37 does not belong because it is about 24, and the others are about 8.

6. *Sample answer:* You decide to leave a 22% tip for a meal that costs $60. How much of a tip do you leave?

Practice and Problem Solving

7–29. Sample answers are given.

7. 15 8. 90
9. 4 10. 5
11. 207 12. 24
13. 12 14. 60
15. 45 16. 66
17. 1 18. 6
19. about $2
20. about 1000
21. a. 150
 b. 280
22. 28 23. 80
24. 10 25. 30
26. 75 27. 180
28. 90 29. 70
30. always true; Because both factors, the percent and the number, are rounded down, the estimate will be less than the actual answers.

T-178

Practice and Problem Solving

31. sometimes true; It depends on how much the percent is rounded down compared to how much the number is rounded up.

32. See *Taking Math Deeper*.

33. *Sample answer:*
Northeast: 100 children
Midwest: 50 children
South: 80 children
West: 160 children

34. *Sample answer:*

 a. 40

 b. 32 students; Subtract the sum of all known percents from 100% to find the percent who chose other. Then use the percent to estimate the number out of 388 who chose other.

35. Your friend's estimate is closer, because $\frac{1}{3} = 0.3\overline{3}$, which is closer to 33% (0.33) than 0.3.

Fair Game Review

36. $\frac{4}{5}$ **37.** $\frac{5}{6}$

38. $\frac{7}{8}$ **39.** $\frac{2}{7}$

40. B

Mini-Assessment

Estimate the percent of the number.

1. 24% of 68 17
2. 51% of 36 18
3. 77% of 56 42
4. 9% of 90 9
5. You want to get 89% of the questions on a math test correct. If there are 50 questions on the test, about how many do you need to answer correctly? 45 questions

T-179

Taking Math Deeper

Exercise 32

This problem looks forward to the concept of ratio. It also foreshadows the study of proportion that will be taught in a future course.

① Organize the given information.

Package	Weight	Price
Small	18 oz	$4.49
Large	?	$6.29

② Find the weight of the large bag.
33% of 18 is about $\frac{1}{3}$ of 18.

 a. So, the new bag weighs about $6 + 18 = 24$ oz.

③ Compare the percent increases of price and weight.
For weight, the percent increase = 33%.
For price, the percent increase is

 b. $\frac{1.80}{4.49} \approx 40\%$.

Consumers tend to think that bigger packages have lesser unit prices. This example shows that is *not* always true.

Project

Create a newspaper ad that the *Trail Mix Company* can use to sell the new size bags.

Reteaching and Enrichment Strategies

If students need help...	If students got it...
Resources by Chapter • Practice A and Practice B • Puzzle Time Record and Practice Journal Practice Differentiating the Lesson Lesson Tutorials Skills Review Handbook	Resources by Chapter • Enrichment and Extension • School-to-Work • Financial Literacy Start the next section

Determine whether the statement is *sometimes, always,* or *never* true. Explain your reasoning.

30. If both the percent and the number are rounded down, then the estimate will be less than the actual answer.

31. If the percent is rounded down and the number is rounded up, then the estimate will be less than the actual answer.

32. **TRAIL MIX** A company increases the size of a bag of trail mix.

 a. About how many ounces are in the new bag?
 b. The new bag costs $1.80 more. Did the cost increase by the same percent as the size? Does the new cost seem *fair*? Explain.

33. **LANGUAGE** The bar graph gives information about different regions of the United States. Out of 500 children from each region, estimate the number of children that speak another language at home.

34. **SPORTS** The circle graph shows the results of a survey of 388 students.

 a. Estimate how many more students preferred soccer than baseball and tennis combined.
 b. Estimate how many students chose *other*. Explain how you found your answer.

35. **Reasoning** A pair of shoes that costs $90 is discounted by 33%. To estimate the amount of the discount, you multiply the price by 0.3. Your friend multiplies by $\frac{1}{3}$. Which estimate is closer to the actual amount of the discount? Explain.

Fair Game Review What you learned in previous grades & lessons

Write the fraction in simplest form. *(Skills Review Handbook)*

36. $\frac{20}{25}$ 37. $\frac{15}{18}$ 38. $\frac{42}{48}$ 39. $\frac{8}{28}$

40. **MULTIPLE CHOICE** Which number is not equal to $\frac{36}{100}$? *(Section 4.1 and Section 4.2)*

 Ⓐ $\frac{18}{50}$ Ⓑ 0.36% Ⓒ $\frac{9}{25}$ Ⓓ 0.36

4.4–4.5 Quiz

Find the percent of the number. *(Section 4.4)*

1. 25% of 40
2. 15% of 80
3. 24% of 60
4. 12% of 35
5. 235% of 14
6. 120% of 88

Estimate the percent of the number. *(Section 4.5)*

7. 51% of 83
8. 9% of 79
9. 24% of 45
10. 96% of 66
11. 82% of 94
12. 77% of 220

13. **SHOPPING** You went to the mall with $80. You spent 25% of your money on a pair of shorts and 65% of the remainder on sandals. How much did you spend on the sandals? Explain how you found your answer. *(Section 4.4)*

14. **TAXI** You want to give a taxi driver a 20% tip. About how much should you tip the driver? *(Section 4.5)*

15. **TOURNAMENT** Your volleyball team pays a tournament entry fee of $350. When your team withdraws before the start of the tournament, 75% of the entry fee is refunded. How much money does your team get back? *(Section 4.4)*

16. **SCHOOL DANCE** The circle graph shows the results of a vote of 774 students on what type of school dance should be held. *(Section 4.5)*

 a. Estimate how many more students preferred a costume party than a formal dance. Explain how you found your answer.

 b. Estimate how many students chose *other*. Explain how you found your answer.

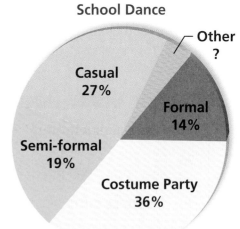

180 Chapter 4 Fractions, Decimals, and Percents

Alternative Assessment Options

Math Chat
Structured Interview
Student Reflective Focus Question
Writing Prompt

Writing Prompt

Ask students to draw figures that represent 100%, one less than 100%, and one greater than 100%. Remind them to label their drawings. Then have students explain their drawings in writing.

Study Help Sample Answers

Remind students to complete Graphic Organizers for the rest of the chapter.

5.

6.

Reteaching and Enrichment Strategies

If students need help...	If students got it...
Resources by Chapter • Study Help • Practice A and Practice B • Puzzle Time Lesson Tutorials *BigIdeasMath.com* Practice Quiz Practice from the Test Generator	Resources by Chapter • Enrichment and Extension • School-to-Work Game Closet at *BigIdeasMath.com* Start the Chapter Review

Answers

1. 10 2. 12
3. 14.4 4. 4.2
5. 32.9 6. 105.6

7–12. Sample answers are given.

7. 40 8. 8
9. 10 10. 66
11. 80 12. 165

13. $39; To find the answer, find the amount spent on shorts, subtract this from the amount you took, and multiply the result by 65%.

14. *Sample answer:* $5

15. $262.50

16. *Sample answer:*
 a. 156
 b. 39

Technology For the Teacher
Answer Presentation Tool

Assessment Book

T-180

For the Teacher
Additional Review Options
- Quiz*Show*
- Big Ideas Test Generator
- Game Closet at *BigIdeasMath.com*
- Vocabulary Puzzle Builder
- Resources by Chapter
 Puzzle Time
 Study Help

Answers

1. $\frac{3}{25}$
2. $\frac{22}{25}$
3. $\frac{1}{250}$
4. 80%
5. 172%
6. 42%
7. 0.76
8. 0.06
9. 3.34
10. 15%
11. 124%
12. 9.7%

Review of Common Errors

Exercises 1–6
- Students may put the percent over the wrong denominator, or try to remove the decimal point when writing it over 100.
- Students may struggle with knowing when it is appropriate to use the equivalent fraction method of writing a fraction as a percent.

Exercises 7–12
- Students may move the decimal point the wrong way, forget to insert zeros as placeholders, or move the decimal too many places (especially when the percent is greater than 100).

Exercises 13–21
- Students may try to order the numbers without converting them, or will only convert them mentally and do so incorrectly.

Exercises 22–25
- When the number ends in a zero, students may be off by a place value because they forgot about the zero at the end. Encourage students to estimate their answer first.

Exercises 26–30
- Students may round up or down to the next percent and use the method from the previous section. While there is nothing wrong with this method, the goal is for the student to use common percents to estimate the answer. Remind students of the table and ask them to round only to those percents.

4 Chapter Review

Review Key Vocabulary

percent, *p. 150*

Review Examples and Exercises

4.1 Percents and Fractions (pp. 148–153)

Write $\frac{3}{20}$ as a percent.

$$\frac{3}{20} = \frac{15}{100}$$ (× 5)

$20 \times 5 = 100$. So, multiply the numerator and denominator by 5.

$= 15\%$ Write the numerator with a percent symbol.

∴ So, $\frac{3}{20}$ is 15%.

Exercises

Write the percent as a fraction or mixed number in simplest form.

1. 12%
2. 88%
3. 0.4%

Write the fraction as a percent.

4. $\frac{4}{5}$
5. $\frac{43}{25}$
6. $\frac{21}{50}$

4.2 Percents and Decimals (pp. 154–159)

a. Write 64% as a decimal.

$64\% = 64.\% = 0.64$

b. Write 0.023 as a percent.

$0.023 = 0.023 = 2.3\%$

Exercises

Write the percent as a decimal. Use a model to check your answer.

7. 76%
8. 6%
9. 334%

Write the decimal as a percent. Use a model to check your answer.

10. 0.15
11. 1.24
12. 0.097

4.3 Comparing and Ordering Fractions, Decimals, and Percents
(pp. 160–165)

a. Which is greater, $\frac{9}{10}$ or 88%?

Write $\frac{9}{10}$ as a percent: $\frac{9}{10} = \frac{90}{100} = 90\%$

∴ 88% is less than 90%. So, $\frac{9}{10}$ is the greater number.

b. Use a number line to order 0.39, 31%, and $\frac{8}{25}$ from least to greatest.

Write 31% and $\frac{8}{25}$ as decimals.

31% = 31.% = 0.31

∴ From least to greatest, the order is 31%, $\frac{8}{25}$, and 0.39.

Exercises

Tell which number is greater.

13. $\frac{1}{2}$, 52%

14. $\frac{12}{5}$, 245%

15. 0.46, 43%

16. 0.023, 22%

Use a number line to order the numbers from least to greatest.

17. $\frac{41}{50}$, 0.83, 80%

18. $\frac{9}{4}$, 220%, 2.15

19. 0.67, 66%, $\frac{2}{3}$

20. 0.88, $\frac{7}{8}$, 90%

21. **FRUIT** Students are asked to name their favorite fruit. Of the responses, 0.32 are apple, 25% are banana, $\frac{7}{20}$ are orange, and 0.08 are *other*. What fruit is the students' favorite?

Review Game
Fractions, Decimals, and Percents

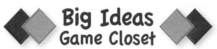
Big Ideas Game Closet

For the Student
Additional Practice
- Lesson Tutorials
- Study Help (textbook)
- Student Website
 Multi-Language Glossary
 Practice Assessments

Materials
- One set of fraction, decimal, and percent cards showing the following fractions and their decimal and percent equivalents.

$$\frac{1}{4}, \frac{1}{2}, \frac{1}{8}, \frac{3}{4}, \frac{2}{3}, \frac{1}{5}, \frac{5}{6}, \frac{1}{6}, \frac{2}{5}, \frac{3}{5},$$

$$\frac{4}{5}, \frac{3}{8}, \frac{5}{8}, \frac{7}{8}, \frac{3}{10}, \frac{7}{10}, \frac{9}{10}, \frac{1}{3}, \frac{1}{10}$$

Players: 2

Directions
This game is like the card game *War*. The deck of cards is shuffled. One player gets half of the cards and the other player gets the other half. Both players turn a card over. The player whose card has the greater value takes both cards. In the event of a tie, both players turn another card over and continue if necessary until one player has a card with a greater value. That player takes all of the cards that were turned over. Play continues until one player has all of the cards.

Who Wins?
The player with all of the cards wins.

Answers
13. 52%

14. 245%

15. 0.46 **16.** 22%

17.

18.

19.

20.

21. oranges **22.** 48

23. 44 **24.** 18.5

25. 22.75

26–29. *Sample answers are given.*

26. 15 **27.** 36

28. 35 **29.** 16

30. *Sample answer:* 24

My Thoughts on the Chapter

What worked...

Teacher Tip
Not allowed to write in your teaching edition? Use sticky notes to record your thoughts.

What did not work...

What I would do differently...

4.4 Finding the Percent of a Number (pp. 168–173)

a. **Find 75% of 80. Use a fraction.**

$$75\% \text{ of } 80 = \frac{3}{4} \times 80$$

$$= \frac{3 \times \cancel{80}^{20}}{\cancel{4}_{1}}$$

$$= 60$$

∴ 75% of 80 is 60.

b. **Find 30% of 90. Use a decimal.**

$$30\% \text{ of } 90 = 0.3 \times 90 = 27$$

∴ 30% of 90 is 27.

Exercises

Find the percent of the number.

22. 60% of 80 **23.** 80% of 55 **24.** 74% of 25 **25.** 35% of 65

4.5 Percents and Estimation (pp. 174–179)

Estimate 24% of 58.

24% is close to 25%, or $\frac{1}{4}$. For 58, use the compatible number 60.

24% of 58
↓ ↓
25% of 60 = $\frac{1}{4} \times 60$

= 15

∴ So, 24% of 58 is about 15.

Exercises

Estimate the percent of the number.

26. 49% of 32 **27.** 41% of 89 **28.** 66% of 48 **29.** 89% of 17

30. PIZZA PARTY About 79% of your class voted to have a pizza party on the last day of school. Your class has 31 students. Estimate how many students voted to have a pizza party.

4 Chapter Test

Write the fraction as a percent.

1. $\frac{21}{25}$
2. $\frac{17}{16}$
3. $\frac{2}{5}$

Write the percent as a decimal.

4. 0.96%
5. 65%
6. 25.7%

Write the decimal as a percent.

7. 0.47
8. 7.88
9. 0.5854

Tell which number is greater.

10. $\frac{16}{25}$, 65%
11. 56%, 5.6

Use a number line to order the numbers from least to greatest.

12. 85%, $\frac{15}{18}$, 0.84
13. 58.3%, 0.58, $\frac{7}{12}$

Find the percent of the number.

14. 8% of 50
15. 120% of 75
16. 54% of 89

Estimate the percent of the number.

17. 21% of 42
18. 74% of 98
19. 12% of 55

20. **TENNIS** A tennis player faults 0.14 of her serves. What percent of the player's serves are faults?

21. **DOWNLOAD** Your computer displays the progress of a downloading video. What fraction of the video is downloaded? Write your answer in simplest form.

22. **BASKETBALL** You, your cousin, and a friend each take the same number of free throws at a basketball hoop. Who made the most free throws?

23. **TIP** Estimate an 18% tip on a meal costing $15.67.

Player	Made
You	$\frac{17}{20}$
Cousin	0.8
Friend	87.5%

24. **GLASSES** In a class of 20 students, 40% are boys. Twenty-five percent of the boys and 50% of the girls wear glasses. How many students wear glasses?

Test Item References

Chapter Test Questions	Section to Review	Common Core State Standards
1–3, 21	4.1	6.RP.3c
4–9, 20	4.2	6.RP.3c
10–13, 22	4.3	6.NS.7a, 6.NS.7b
14–16, 24	4.4	6.RP.3c
17–19, 23	4.5	6.RP.3c

Test-Taking Strategies

Remind students to quickly look over the entire test before they start so they can budget their time. Remind them that the test is on fractions, decimals, *and* percents and that they need to read the problems carefully. Students need to use the **Stop** and **Think** strategy before they answer a question. Students need to remember to think of the different representations of each number as they work through the test, such as 0.5, $\frac{1}{2}$, and 50%.

Common Assessment Errors

- **Exercises 1–3** Students may not know when it is appropriate to use an equivalent fraction to write a fraction as a percent. Point out that sometimes it is easier to write an equivalent fraction.
- **Exercises 4–9** Students may move the decimal point the wrong way, forget to insert zeros as placeholders, or move the decimal point too many places. Have them think about the following to help them remember which way to move the decimal point. When converting from percents to decimals, the decimal point moves to the left (D is to the left of P in the alphabet.) When converting from decimals to percents, the decimal point moves to the right (P is to the right of D in the alphabet.)
- **Exercises 10–13** Students may try to order the numbers without converting them, or will only convert them mentally and do so incorrectly. Tell students that it is necessary to convert all the numbers to one form.
- **Exercises 14–16** Students might find a number that is significantly greater or less than the correct answer. Encourage students to estimate their answers. Remind them that percents greater than 100 will have answers that are greater than the original number.

Reteaching and Enrichment Strategies

If students need help...	If students got it...
Resources by Chapter • Practice A and Practice B • Puzzle Time Record and Practice Journal Practice Differentiating the Lesson Lesson Tutorials Practice from the Test Generator Skills Review Handbook	Resources by Chapter • Enrichment and Extension • School-to-Work • Financial Literacy Game Closet at *BigIdeasMath.com* Start Standardized Test Practice

Answers

1. 84%
2. 106.25%
3. 40%
4. 0.0096
5. 0.65
6. 0.257
7. 47%
8. 788%
9. 58.54%
10. 65%
11. 5.6

12.

13.

14. 4
15. 90
16. 48.06

17–19. *Sample answers are given.*

17. 8
18. 75
19. 6
20. 14%
21. $\frac{23}{50}$
22. Your friend
23. *Sample answer:* $3
24. 8 students

Assessment Book

Test-Taking Strategies
Available at *BigIdeasMath.com*

After Answering Easy Questions, Relax
Answer Easy Questions First
Estimate the Answer
Read All Choices before Answering
Read Question before Answering
Solve Directly or Eliminate Choices
Solve Problem before Looking at Choices
Use Intelligent Guessing
Work Backwards

About this Strategy

When taking a multiple choice test, be sure to read each question carefully and thoroughly. Look closely for words that change the meaning of the question, such as *not*, *never*, *all*, *every*, and *always*.

Answers

1. C
2. F
3. 5.4
4. A
5. G

Item Analysis

1. A. The student multiplies the numerator and denominator.
 B. The student uses the digits from the numerator and denominator.
 C. Correct answer
 D. The student divides 5 by 4 and converts the quotient into a percent.

2. F. Correct answer
 G. The student reverses the values of k and p.
 H. The student uses the values of k and p as coefficients.
 I. The student incorrectly uses a decimal number division algorithm and thinks that $0.8 \times 0.5 = 4$ and that $0.5 \times 1.6 = 8$.

3. **Gridded Response:** Correct answer: 5.4
 Common Error: The student divides 36 by 0.15 and gets 240.

4. A. Correct answer
 B. The student thinks that 25% = $25 and subtracts $25 from $123.75.
 C. The student thinks that 25% = $0.25, and then either adds $0.25 to or subtracts $0.25 from $123.75.
 D. The student thinks that 25% = $25 and adds $25 to $123.75.

5. F. The student multiplies $\frac{5}{9}$ by 90 and then subtracts 32.
 G. Correct answer
 H. The student subtracts 32 from 90 and ignores the $\frac{5}{9}$.
 I. The student subtracts 32 from 90 and then changes $\frac{5}{9}$ to $\frac{9}{5}$ before multiplying.

Technology For the Teacher
Big Ideas Test Generator

4 Standardized Test Practice

1. Which percent is equivalent to $\frac{4}{5}$? *(6.RP.3c)*

 A. 20% C. 80%

 B. 45% D. 125%

2. Your friend evaluated an expression using $k = 0.5$ and $p = 1.6$ and got an answer of 12. Which expression did your friend evaluate? *(6.EE.2c)*

 F. $5p + 8k$ H. $0.5k + 1.6p$

 G. $8p + 5k$ I. $0.8k + 0.5p$

3. What is 15% of 36? *(6.RP.3c)*

4. A pair of running shoes is on sale for 25% off the original price.

 Which price is closest to the sale price of the running shoes? *(6.RP.3c)*

 A. $93 C. $124

 B. $99 D. $149

5. The expression below can be used to find the temperature in degrees Celsius when given F, the temperature in degrees Fahrenheit.

$$\frac{5}{9}(F - 32)$$

 What is the temperature in degrees Celsius, to the nearest degree, when the temperature in degrees Fahrenheit is 90°? *(6.EE.2c)*

 F. 18° H. 58°

 G. 32° I. 104°

6. For a party, you made a gelatin dessert in a rectangular pan and cut the dessert into equal-sized pieces as shown below.

The dessert consisted of seven layers of equal height. Each height was a different flavor, as shown below by a side view of the pan.

Your guests ate $\frac{5}{6}$ of the pieces of the dessert. Cherry gelatin was what fraction of the dessert eaten by your guests? *(6.NS.1)*

A. $\frac{37}{42}$

B. $\frac{5}{42}$

C. $\frac{5}{13}$

D. $\frac{1}{7}$

7. What property was used to simplify the expression? *(6.EE.3)*

$$12 \times 47 = 12 \times (40 + 7)$$
$$= 12 \times 40 + 12 \times 7$$
$$= 480 + 84$$
$$= 564$$

F. Distributive Property

G. Identity Property of Addition

H. Commutative Property of Addition

I. Associative Property of Multiplication

Item Analysis (continued)

Answers
6. B
7. F

6. **A.** The student finds the part of the entire dessert eaten by the guests that was NOT cherry.

 B. Correct answer

 C. The student adds the denominators when multiplying $\frac{5}{6}$ by $\frac{1}{7}$.

 D. The student finds the total part of the entire dessert that was cherry.

7. **F.** Correct answer

 G. The student incorrectly identifies the property being used as the Identity Property of Addition.

 H. The student incorrectly identifies the property being used as the Commutative Property of Addition.

 I. The student incorrectly identifies the property being used as the Associative Property of Multiplication.

8. **A.** The student thinks that 0.8 is less than 0.09 and does not know how to compare decimal numbers with fractions or percents.

 B. The student orders the decimal numbers and fractions correctly but thinks that 70% equals 70.

 C. The student orders the numbers using either the numerator or the leading digit.

 D. Correct answer

9. **Gridded Response:** Correct answer: $5.40

 Common Error: The student aligns the decimal points in the factors and gets a product of 0.54.

Answers

8. D
9. $5.40
10. G
11. *Part A* $371 at the hardware store; $355.20 at the online store

 Part B The hardware store offers the better final cost by $5.30.

Answers for Extra Examples

1. **A.** The student multiplies 1.2 by 0.4 and then aligns the decimal point of the product with the decimal points of the factors.
 B. The student misunderstands the inverse relationship between multiplication and division and divides 1.2 by 0.4.
 C. Correct answer
 D. The student misunderstands the inverse relationship between multiplication and division and divides 1.2 by 0.4. The student then aligns the decimal point of the quotient with the decimal points of the dividend and divisor.

2. **F.** Correct answer
 G. The student adds 0.3 to 0.08 and 0.01 to 0.005 and then finds the product of these two sums.
 H. The student finds $3 \times 8 \times 1 \times 5$ and then aligns the decimal point of the product with the decimal points of the factors.
 I. The student adds 0.3 to 0.08 and 0.01 to 0.005 and then finds the sum of these two sums.

Item Analysis (continued)

10. **F.** The student does not recognize that $\frac{1}{4} = \frac{25}{100} = 25\%$.
 G. Correct answer
 H. The student does not recognize that $\frac{5}{20} = \frac{25}{100} = 25\%$.
 I. The student does not recognize that $\frac{25}{100} = 25\%$.

11. **4 points** The student demonstrates a thorough understanding of solving problems involving finding percents of numbers. In Part A, the student correctly calculates the cost of the ladder at each store, getting $371 at the hardware store and $355.20 at the online store. In Part B, the student correctly subtracts 10% of the cost of the ladder at the hardware store before adding the tax. The student also correctly recalculates the cost of the ladder at the online store without the shipping and handling charge. The student then compares the final costs, showing that the hardware store offers the better final cost by $5.30. The student shows accurate, complete work for both parts and provides clear and complete explanations.

 3 points The student demonstrates an understanding of solving problems involving finding percents of numbers, but the student's work and explanations demonstrate an essential but less than thorough understanding.

 2 points The student demonstrates a partial understanding of solving problems involving finding percents of numbers. The student's work and explanations demonstrate a lack of essential understanding.

 1 point The student demonstrates very limited understanding of solving problems involving finding percents of numbers. The student's response is incomplete and exhibits many flaws.

 0 points The student provided no response, a completely incorrect or incomprehensible response, or a response that demonstrates insufficient understanding of solving problems involving finding percents of numbers.

Extra Examples for Standardized Test Practice

1. What number belongs in the box to make the equation true?

 $\square \div 0.4 = 1.2$

 A. 4.8
 B. 3
 C. 0.48
 D. 0.3

2. What is the value of the expression below when $t = 0.08$ and $u = 0.005$?

 $0.3t \cdot 0.01u$

 F. 0.0000012
 G. 0.0057
 H. 0.012
 I. 0.395

8. Which list of numbers is in order from least to greatest? *(6.NS.7a)*

 A. 0.8, $\frac{5}{8}$, 70%, 0.09

 B. 0.09, $\frac{5}{8}$, 0.8, 70%

 C. $\frac{5}{8}$, 70%, 0.8, 0.09

 D. 0.09, $\frac{5}{8}$, 70%, 0.8

9. You bought 0.8 kilogram of mixed nuts. What was the total cost, in dollars, of the mixed nuts that you bought? *(6.NS.3)*

MIXED NUTS
$6.75 per kilogram

10. Which fraction is *not* equivalent to 25%? *(6.RP.3c)*

 F. $\frac{1}{4}$

 G. $\frac{2}{5}$

 H. $\frac{5}{20}$

 I. $\frac{25}{100}$

11. You are comparing the costs of a certain model of ladder at a hardware store and at an online store. *(6.RP.3c)*

HARDWARE STORE
Ladders: $350/ea.
Sales tax: 6% of cost of ladder

BUY HARDWARE ONLINE
Ladder $320
Sales tax: 6% of cost of ladder
Shipping & handling: 5% of cost of ladder

Part A What is the cost of the ladder at each of the stores? Show your work and explain your reasoning.

Part B Suppose that the hardware store is offering 10% off the price of the ladder and that the online store is offering free shipping & handling. Which store offers the better final cost and by how much? Show your work and explain your reasoning.

5 Ratios, Rates, and Data Analysis

5.1 Ratios

5.2 Rates

5.3 Solving Rate Problems

5.4 Mean

5.5 Median, Mode, and Range

5.6 Analyzing Data Sets

"The speed limit is 18 miles per hour, but that's no problem because we're going to the hot dog stand, and that's only one-half mile away."

"By my records, I ate 1460 dog biscuits last year."

"So I calculated that my unit rate is 4 biscuits per day."

Connections to Previous Learning

- Construct and analyze frequency tables, bar graphs, pictographs, and line plots from data.

- Construct and analyze line graphs and double bar graphs.

- Interpret and compare ratio and rates.
- Use reasoning about multiplication and division to solve ratio and rate problems.
- Determine measures of central tendency including mean, median, mode, and range (variability).
- Select appropriate measures of central tendency to describe a data set.

Math in History

The concept of a prime number dates back to ancient Greek mathematics. The word "prime" is related to the words "primal" and "primitive."

★ The Sieve of Eratosthenes is a simple algorithm for finding all prime numbers up to a specified integer. Eratosthenes was a Greek mathematician. To create the sieve, he listed all numbers up to a given number such as 100.
- Cross out 1.
- Leave 2, cross out all multiples of 2.
- Leave 3, cross out all multiples of 3.
- 4 is already crossed out.
- Leave 5, cross out all multiples of 5.
- 6 is already crossed out.
- Leave 7, cross out all multiples of 7.
- All numbers not crossed out are prime.

★ Properties of prime numbers have been studied for hundreds of years. An ancient Chinese hypothesis stated that if an integer is prime, then the number $2^n - 2$ is divisible by n. For instance, $2^3 - 2 = 6$, which is divisible by 3.

Pacing Guide for Chapter 5

Chapter Opener	1 Day
Section 1	
Activity	1 Day
Lesson	1 Day
Section 2	
Activity	1 Day
Lesson	1 Day
Section 3	
Activity	1 Day
Lesson	1 Day
Study Help / Quiz	1 Day
Section 4	
Activity	1 Day
Lesson	1 Day
Section 5	
Activity	1 Day
Lesson	1 Day
Section 6	
Activity	2 Days
Lesson	1 Day
Lesson b	2 Days
Quiz / Chapter Review	1 Day
Chapter Test	1 Day
Standardized Test Practice	1 Day
Total Chapter 5	**20 Days**
Year-to-Date	**87 Days**

Check Your Resources

- Record and Practice Journal
- Resources by Chapter
- Skills Review Handbook
- Assessment Book
- Worked-Out Solutions

Technology For the Teacher

The Dynamic Planning Tool
Editable Teacher's Resources at
BigIdeasMath.com

T-188

Common Core State Standards

4.NF.1 Explain why a fraction a/b is equivalent to a fraction $(n \times a)/(n \times b)$ by using visual fraction models, with attention to how the number and size of the parts differ even though the two fractions themselves are the same size. Use this principle to recognize and generate equivalent fractions.

4.NBT.4 Fluently add and subtract multi-digit whole numbers using the standard algorithm.

Additional Topics for Review

- Equivalent fractions
- Greatest Common Factor

Try It Yourself

1. $\dfrac{1}{10}$
2. $\dfrac{4}{7}$
3. $\dfrac{5}{12}$
4. $\dfrac{31}{40}$
5. 2042 more females
6. *Sample answer:* There are more females and fewer males in 2004 than in 2000.

Record and Practice Journal

1. $\dfrac{2}{5}$
2. $\dfrac{1}{5}$
3. $\dfrac{1}{8}$
4. $\dfrac{2}{3}$
5. $\dfrac{5}{9}$
6. $\dfrac{7}{12}$
7. $\dfrac{19}{50}$
8. $122
9. pants
10. pants

T-189

Math Background Notes

Vocabulary Review
- Greatest Common Factor
- Numerator
- Denominator

Simplifying Fractions
- Students should know how to simplify fractions.
- Some students may have learned simplifying fractions as reducing fractions.
- It is often helpful to find the Greatest Common Factor of the numerator and the denominator before simplifying the fraction.
- Divide the numerator and denominator by the Greatest Common Factor to simplify the fraction.
- **Common Error:** Students may try to divide the numerator and denominator by different factors. Remind them that the factor in the numerator must match the factor in the denominator. Otherwise, the factors cannot be divided out.

Analyzing Double Bar Graphs
- Students should know how to read and draw double bar graphs. Because this is a fairly recent skill, students may still need practice to achieve mastery.
- In single bar graphs, each category contains one data set (one bar). Remind students that in a double bar graph each category contains two data sets (two bars).
- **Teaching Tip:** This is an excellent opportunity to review the parts of a graph with students. Sample questions could include: What is the title of the graph? Where is the legend? What does it tell you? What are the categories? What data sets are being compared? Or, what scale is being used to measure the axes?

Reteaching and Enrichment Strategies

If students need help...	If students got it...
Record and Practice Journal • Fair Game Review Skills Review Handbook Lesson Tutorials	Game Closet at *BigIdeasMath.com* Start the next section

What You Learned Before

"Our mean weight is 18 pounds."

Simplifying Fractions (4.NF.1)

Example 1 Simplify $\frac{3}{6}$.

$$\frac{3 \div 3}{6 \div 3} = \frac{1}{2}$$

Use the Greatest Common Factor to simplify the fraction.

Example 2 Simplify $\frac{28}{36}$.

$$\frac{28 \div 4}{36 \div 4} = \frac{7}{9}$$

Try It Yourself
Simplify the fraction.

1. $\frac{10}{100}$
2. $\frac{20}{35}$
3. $\frac{60}{144}$
4. $\frac{62}{80}$

Analyzing Double Bar Graphs (4.NBT.4)

Example 3 How many more male athletes than female athletes participated in the 1992 Summer Olympics?

$$6652 - 2704 = 3948$$

∴ 3948 more male athletes participated.

Example 4 How many athletes participated in the 2000 Summer Olympics.

∴ $6582 + 4069 = 10{,}651$ athletes participated in the 2000 Summer Olympics.

Try It Yourself

5. How many more female athletes participated in the 2008 Summer Olympics than in the 1992 Summer Olympics?

6. Describe the relationship between the number of athletes in the 2000 Summer Olympics and the number of athletes in the 2004 Summer Olympics.

5.1 Ratios

COMMON CORE STATE STANDARDS
6.RP.1

Essential Question How can you tell whether two recipes make the same mixture?

A **ratio** is a comparison of two quantities using division.

Ratios

$\dfrac{4 \text{ ft}}{2 \text{ ft}}$ $\dfrac{3 \text{ c}}{5 \text{ c}}$ $\dfrac{20 \text{ sec}}{45 \text{ sec}}$ $\dfrac{120 \text{ mi}}{80 \text{ mi}}$

1 ACTIVITY: Comparing Recipes

Work with a partner.

You are making some homemade hand lotion. You find three recipes.

Do the recipes make the same lotion? How can you tell?

Recipe 1

Melt these ingredients over low heat:
2/3 cup of apricot oil
1/3 cup of cocoa butter
1 teaspoon of lanolin
1/2 ounce of grated beeswax

When cool, add the following:
2/3 cup of rosewater
1/3 cup of aloe vera gel
2 drops of rose oil
1 Vitamin E capsule

Whip together until the mixture resembles lotion.

Recipe 2

Melt these ingredients over low heat:
1 cup of apricot oil
1/2 cup of cocoa butter
1 1/2 teaspoons of lanolin
3/4 ounce of grated beeswax

When cool, add the following:
1 cup of rosewater
1/2 cup of aloe vera gel
3 drops of rose oil
1 1/2 Vitamin E capsules

Whip together until the mixture resembles lotion.

Recipe 3

Melt these ingredients over low heat:
1 1/2 cups of apricot oil
2/3 cup of cocoa butter
2 teaspoons of lanolin
1 ounce of grated beeswax

When cool, add the following:
1 1/2 cups of rosewater
2/3 cup of aloe vera gel
4 drops of rose oil
2 Vitamin E capsules

Whip together until the mixture resembles lotion.

Laurie's Notes

Introduction

For the Teacher
- **Goal:** The goal of this activity is for students to use a context of recipes to explore equivalent ratios.

Motivate
- Use a portion of a page from a telephone book to quickly gather the following data: number of telephone numbers that end in 0, 1, 2, …, 9. Form different ratios from the data gathered: even to odd; 4s to 8s; primes to composites.

Activity Notes

Activity 1
- Students have heard *ratio* and *rate* before but may not be able to explain the difference. Generate examples of ratios and rates that are familiar to them. Ratio: 4 girls to 5 boys; Rate: 15 miles per gallon
- **?** Ask a few questions about baking:
 - "Have you ever followed a recipe when cooking?" Most respond yes.
 - "Have you ever adjusted a recipe to make more or less than the original?"
 - "Can someone explain how to adjust a recipe?" Listen for a concept of proportionality, some may use language such as "double all the ingredients, or use $\frac{1}{2}$ of what is called for."
 - "If a recipe calls for 2 eggs and 3 cups of flour, can you double it? Can you cut it in half?"
- It is important for students to understand that *all* the ingredients must be increased or decreased by the same factor.
- Students should initially observe that all three recipes contain the same ingredients. The amount of each ingredient differs.
- Give sufficient time for groups to discuss the recipes.
- Recipes 1 and 2 make the same lotion. You can multiply each ingredient in Recipe 2 by $\frac{2}{3}$. There are other equivalent explanations.
- Recipe 3 does not make the same lotion as the other two. There are several equivalent explanations.
- If no group suggests it, find the ratio of each ingredient from Recipes 1 and 2.

 Example: $\dfrac{\frac{2}{3} \text{ cup}}{1 \text{ cup}} = \dfrac{2}{3} \cdot \dfrac{\frac{1}{3} \text{ cup}}{\frac{1}{2} \text{ cup}} = \dfrac{1}{3} \div \dfrac{1}{2} = \dfrac{2}{3}$; continue for all ingredients.

- **?** Try this same technique using Recipe 3 and either of the other recipes. Are the ratios of the ingredients all the same or different? different
- This activity provides a good review of fraction operations.

Common Core State Standards
6.RP.1 Understand the concept of a ratio and use ratio language to describe a ratio relationship between two quantities.

Previous Learning
Students should know how to simplify fractions, write a fraction as a decimal, and write a decimal as a percent.

Activity Materials
Introduction
• page from a telephone book

Start Thinking! and Warm Up

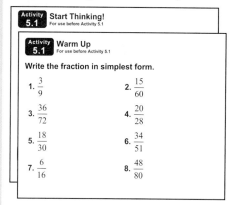

5.1 Record and Practice Journal

Differentiated Instruction

Visual

Students may see little difference between fractions and ratios. Fractions are one type of ratio, part-to-whole. Two other types of ratios are part-to-part and whole-to-whole.

Part-to-whole of Circle 1:

$$\frac{\text{unshaded parts}}{\text{whole number of parts}} = \frac{2}{4}$$

Part-to-part of Circle 2:

$$\frac{\text{shaded parts}}{\text{unshaded parts}} = \frac{5}{3}$$

Whole-to-whole:

$$\frac{\text{parts of Circle 1}}{\text{parts of Circle 2}} = \frac{4}{8}$$

Circle 1

Circle 2

5.1 Record and Practice Journal

2 ACTIVITY: Finding Equivalent Ratios

Work with a partner.

a. The ratios $\frac{1}{3}, \frac{2}{6}, \frac{3}{9}, \frac{4}{12}, \frac{5}{15}, \frac{6}{18}$ are all equivalent. Explain how you can use the multiplication table to show this.

The numerator and denominator are multiplied by the same number.

Use the multiplication table to find 10 more ratios that are equivalent to each ratio.

d. Explain why the strategy in parts (a), (b), and (c) works to produce equivalent ratios.

Multiplying the numerator and denominator by the same number, is equal to multiplying by one.

What Is Your Answer?

3. You and two friends are making cookies. You make the original recipe amount. One of your friends makes a "half batch." Your other friend makes a "double batch." If you taste a spoonful of cookie dough from each batch, will they all taste the same? Explain your reasoning.

Yes, because the ratios are the same.

4. IN YOUR OWN WORDS How can you tell whether two recipes make the same mixture? Give an example.

All of the ingredients have equivalent ratios.

Laurie's Notes

Activity 2

- **Big Idea:** Students are asked to use multiplication and division as they reason about equivalent ratios.
- **"What are equivalent fractions? How do you know if two fractions are equivalent?"** Fractions that have the same value.
- Students are familiar with equivalent fractions, so use this knowledge to introduce the activity. Suggest that they think of equivalent ratios in the same way. Use ratio language: 1 to 3 is the same as 2 to 6.
- **"What patterns do you observe in the multiplication table?"** There are many; listen for vocabulary such as multiples, factor, product, even, odd, square numbers.
- In this activity, students need to think about the relationship between the numbers found in a multiplication table.
- Students should work in small groups.
- Groups should be encouraged to suggest alternate strategies for finding equivalent ratios.
- **Big Idea:** Because the rows and columns represent the multiples of a particular number, dividing the corresponding entries in two rows or two columns will result in equivalent ratios.
- Give time for different groups to explain their strategy for finding equivalent ratios.

What Is Your Answer?

- Students should consider their work in Activity 1 to answer these questions.

Closure

- **Exit Ticket:** Conduct a survey in class and record how many students have brown, green, blue, and hazel eyes. What is the ratio of students with blue eyes to the total number of students? What is the ratio of students with brown eyes to the total number of students? What conclusions can you draw from these ratios? Answers will vary. Typically, more students will have brown eyes than blue eyes. Because the denominator of each ratio will be the same (each will be the total number of students in the class), students should be able to compare the fractions to determine that the ratio of brown eyes to total students is greater than the ratio of blue eyes to total students.

Technology For the Teacher

The Dynamic Planning Tool
Editable Teacher's Resources at *BigIdeasMath.com*

2 ACTIVITY: Finding Equivalent Ratios

Work with a partner.

a. The ratios $\frac{1}{3}, \frac{2}{6}, \frac{3}{9}, \frac{4}{12}, \frac{5}{15}, \frac{6}{18}$ are all equivalent. Explain how you can use the multiplication table to show this.

b. Use the multiplication table to find 11 ratios that are equivalent to $\frac{2}{7}$.

c. Use the multiplication table to find 11 ratios that are equivalent to $\frac{8}{3}$.

d. Explain why the strategy in parts (a), (b), and (c) works to produce equivalent ratios.

	1	2	3	4	5	6	7	8	9	10	11	12
1	1	2	3	4	5	6	7	8	9	10	11	12
2	2	4	6	8	10	12	14	16	18	20	22	24
3	3	6	9	12	15	18	21	24	27	30	33	36
4	4	8	12	16	20	24	28	32	36	40	44	48
5	5	10	15	20	25	30	35	40	45	50	55	60
6	6	12	18	24	30	36	42	48	54	60	66	72
7	7	14	21	28	35	42	49	56	63	70	77	84
8	8	16	24	32	40	48	56	64	72	80	88	96
9	9	18	27	36	45	54	63	72	81	90	99	108
10	10	20	30	40	50	60	70	80	90	100	110	120
11	11	22	33	44	55	66	77	88	99	110	121	132
12	12	24	36	48	60	72	84	96	108	120	132	144

What Is Your Answer?

3. You and two friends are making cookies. You make the original recipe amount. One of your friends makes a "half batch." Your other friend makes a "double batch." If you taste a spoonful of cookie dough from each batch, will they all taste the same? Explain your reasoning.

4. **IN YOUR OWN WORDS** How can you tell whether two recipes make the same mixture? Give an example.

Practice

Use what you learned about writing equivalent ratios to complete Exercises 12–15 on page 194.

5.1 Lesson

Check It Out
Lesson Tutorials
BigIdeasMathcom

Key Vocabulary
ratio, p. 192
equivalent ratios, p. 193

Key Idea

Ratio

Words A **ratio** is a comparison of two quantities using division.

Numbers A ratio of 2 red crayons to 5 blue crayons can be written in three ways.

$$\frac{2}{5}, 2 \text{ to } 5, \text{ or } 2:5$$

EXAMPLE 1 — Writing a Ratio

Write the ratio of pennies to quarters in three ways.

pennies → $\frac{5}{8}$
quarters

∴ The ratio of pennies to quarters is $\frac{5}{8}$, 5 to 8, or 5 : 8.

EXAMPLE 2 — Writing and Simplifying Ratios

a. Write the ratio of boys to girls at Oak Grove.

Oak Grove Middle School	
Boys	Girls
600	540

boys → $\frac{600}{540} = \frac{10}{9}$ Write in simplest form.
girls

∴ The ratio of boys to girls is $\frac{10}{9}$.

b. Write the ratio of girls to the total number of students at Oak Grove.

girls → $\frac{540}{1140} = \frac{9}{19}$ Write in simplest form.
total number of students

∴ The ratio of girls to the total number of students is $\frac{9}{19}$.

On Your Own

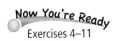
Now You're Ready
Exercises 4–11

1. In Example 1, write the ratio of quarters to the total number of coins in three ways.

2. Write the ratio of boys to the total number of students at Oak Grove in simplest form.

192 Chapter 5 Ratios, Rates, and Data Analysis Multi-Language Glossary at BigIdeasMathcom.

Laurie's Notes

Introduction

Connect
- **Yesterday:** Students explored equivalent ratios in the context of recipes and a standard multiplication table.
- **Today:** Students will write and simplify ratios.

Motivate
- Have 2 boys and 4 girls stand at the front of the room.
- Ask a few questions.
 - "What is the ratio of boys to girls?" accept 2 : 4 or 1 : 2
 - "What is the ratio of girls to boys?" 4 : 2 or 2 : 1; This helps to point out that order matters in a ratio.
 - "What is the ratio of girls to people standing?" 4 : 6 or 2 : 3
 - "How many boys need to stand so the ratio of boys to girls is 3 : 2?" total of 6 boys

Lesson Notes

Key Idea
- Students should be comfortable with simplifying fractions from earlier lessons. Simplifying ratios will be done in the same fashion.
- Write the Key Idea.
- Review the three ways in which ratios are written.

Example 1
- Work through the example.
- **Extension:** "What is the ratio of the value of the pennies to the value of the quarters?" Five pennies equals $0.05. Eight quarters equals $2.00. So the ratio is $0.05 to $2.00.

Example 2
- Review what it means to simplify a ratio: dividing out common factors in each term.
- "How is the ratio $\frac{600}{540}$ simplified?" Divide out a common factor of 60 from each term in the ratio.
- **Note:** Ratios should not be written as mixed numbers.
- **Discuss:** As with fractions, it is okay to simplify in stages, meaning dividing out a common factor of 10, followed by dividing out a common factor of 6.
- In part (b), students may read too quickly and forget to sum all students.

On Your Own
- **Neighbor Check:** Have students work independently and then have their neighbor check their work. Have students discuss any discrepancies.

Goal Today's lesson is writing and simplifying **ratios**.

Start Thinking! and Warm Up

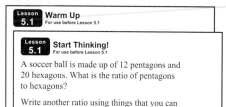

Lesson 5.1 Warm Up For use before Lesson 5.1

Lesson 5.1 Start Thinking! For use before Lesson 5.1

A soccer ball is made up of 12 pentagons and 20 hexagons. What is the ratio of pentagons to hexagons?

Write another ratio using things that you can find in your classroom.

Extra Example 1
You have 7 dimes and 10 nickels. Write the ratio of dimes to nickels in three ways. $\frac{7}{10}$, 7 to 10, or 7 : 10

Extra Example 2

Fairfield Middle School	
Boys	Girls
450	475

a. Write the ratio of girls to boys at Fairfield Middle School.
$\frac{19}{18}$

b. Write the ratio of boys to the total number of students at Fairfield Middle School.
$\frac{18}{37}$

On Your Own
1. $\frac{8}{13}$, 8 to 13, 8 : 13
2. $\frac{10}{19}$

T-192

Extra Example 3

Write two equivalent ratios for $\frac{8}{10}$.

Sample answer: $\frac{4}{5}, \frac{80}{100}$

Extra Example 4

You answer 16 out of 20 questions correctly on a test. Your friend answers 25 out of 30 questions correctly on a different test. Who has the better score? *Your friend*

On Your Own

3–5. Sample answers are given.

3. $\frac{1}{2}, \frac{2}{4}$

4. $\frac{6}{14}, \frac{9}{21}$

5. $\frac{1}{4}, \frac{10}{40}$

6. Your teammate has the greater catch ratio because $\frac{23}{30} = 0.7\overline{6} = 76.\overline{6}\%$, which is greater than $\frac{18}{25} = 0.72 = 72\%$.

English Language Learners

Vocabulary

Have students add a glossary to their math notebook. Key vocabulary words should be added as they are introduced. Illustrations next to vocabulary words will help in understanding and reinforcing the concept.

Laurie's Notes

Example 3

- Define equivalent ratios.
- Fractions do not have units associated with the numerator and denominator. Example: $\frac{3}{4}$ inch versus the ratio 3 inches : 4 inches.
- Fractions have a numerator and denominator. Ratios have two terms.
- Like fractions, a ratio can be written in simplest form.
- This example is similar to the activity used to motivate today's lesson. Instead of a collection of students (boys and girls), there is a collection of shapes (triangles and squares).
- **Discuss:** The order in which the ratio is written matters; 4 triangles to 6 squares is not the same as 6 squares to 4 triangles.
- ? "How do you simplify the ratio $\frac{4}{6}$?" Divide the numerator and denominator each by 2.
- Discuss the reading note. You could then ask, "If there are 24 triangles, how many squares would there be?" 36 squares

Example 4

? Begin with a few questions to get students thinking about what it means to compare two ratios.
- "Looking at the information, your friend answered 35 questions correctly and you only answered 24 correctly, so your friend has the better score. Do you agree?" No; you have to consider how many total questions there were.
- "Your friend missed 5 questions and you missed 6 questions so your friend did better. Do you agree?" No; you have to consider how many total questions there were.
- "How can you compare the ratios 24 : 30 and 35 : 40?" Simplify each ratio and then write it as a decimal.
- "How do you write the ratio 4 : 5 as a decimal?" Think of 4 : 5 as the fraction $\frac{4}{5}$; divide the numerator by the denominator.
- **Note:** Students can compare the decimals in order to decide who had the better score. Writing the decimals as percents is a good review.

Closure

- Explain what equivalent ratios are and give an example. *Sample answer:* Equivalent ratios describe the same relationship. For example, the ratio of eyes to toes (2 : 10 or 1 : 5) is equivalent to the ratio of ears to fingers (2 : 10 or 1 : 5).

The Dynamic Planning Tool
Editable Teacher's Resources at *BigIdeasMath.com*

Two ratios that describe the same relationship are **equivalent ratios**.

EXAMPLE 3 Writing Equivalent Ratios

Write two equivalent ratios for triangles to squares.

Reading

The ratio $\frac{2}{3}$ means that for every 2 triangles, there are 3 squares.

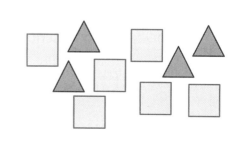

∴ Two equivalent ratios are $\frac{4}{6}$ and $\frac{2}{3}$.

EXAMPLE 4 Comparing Ratios

You answer 24 out of 30 questions correctly on a quiz. Your friend answers 35 out of 40 questions correctly on a different quiz. Who has the better score?

You: number correct → $\frac{24}{30} = \frac{4}{5}$ Write in simplest form.
number of questions

Your friend: number correct → $\frac{35}{40} = \frac{7}{8}$ Write in simplest form.
number of questions

Use percents to compare the scores.

You: $\frac{4}{5} = 0.8 = 80\%$

Your friend: $\frac{7}{8} = 0.875 = 87.5\%$

∴ Your friend has the better score.

On Your Own

Now You're Ready
Exercises 12–19

Write two equivalent ratios for the given ratio.

3. 4 : 8 **4.** $\frac{3}{7}$ **5.** 5 to 20

6. You catch 18 out of 25 passes. Your teammate catches 23 out of 30 passes. Who has the greater catch ratio? Explain.

Section 5.1 Ratios 193

5.1 Exercises

Vocabulary and Concept Check

1. **VOCABULARY** Write the ratio $\frac{3}{7}$ in two other ways.

2. **OPEN-ENDED** Describe the circles and triangles using four different ratios.

3. **OPEN-ENDED** Write the ratio of vowels to consonants in your last name.

Practice and Problem Solving

Write the ratio in three ways. Explain what the ratio means.

4. frogs to turtles

5. basketballs : soccer balls

6. $\dfrac{\text{calculators}}{\text{pencils}}$

7. $\dfrac{\text{shirts}}{\text{pants}}$

Write the ratio in simplest form.

8. $\dfrac{25}{30}$ 9. $\dfrac{12}{16}$ 10. $\dfrac{24}{8}$ 11. $\dfrac{21}{15}$

Write two equivalent ratios for the given ratio.

12. $\dfrac{4}{12}$ 13. $\dfrac{6}{18}$ 14. $\dfrac{7}{14}$ 15. $\dfrac{11}{33}$

16. $\dfrac{6}{27}$ 17. $\dfrac{9}{10}$ 18. $\dfrac{15}{35}$ 19. $\dfrac{12}{8}$

194 Chapter 5 Ratios, Rates, and Data Analysis

Assignment Guide and Homework Check

Level	Day 1 Activity Assignment	Day 2 Lesson Assignment	Homework Check
Basic	12–15, 28–32	1–3, 5–11 odd, 17–21 odd, 20, 22	5, 9, 17, 20
Average	12–15, 28–32	1–3, 5–11 odd, 16–20 even, 21, 23, 24	5, 9, 16, 20, 24
Advanced	12–15, 28–32	1–3, 8, 10, 16–20 even, 21, 24–27	8, 16, 20, 24

Common Errors

- **Exercises 4–7** Students may switch the order of the ratio when writing it in three different ways. For example, the student might write the ratio of frogs to turtles as: $2:5$, $\frac{5}{2}$, and 2 to 5. Remind students that the first quantity given will always be first or in the numerator, and the second will be second or in the denominator.
- **Exercises 8–11** Students may not reduce the ratio to the simplest form. Remind them that reducing ratios is just like reducing fractions and they need to reduce until there are no more common factors in the numerator and denominator.
- **Exercises 12–19** Students may only find one equivalent ratio by reducing and then state that there are no more equivalent ratios. Remind them of equivalent fractions and that to find them you can multiply the numerator and denominator by the same number.

Technology For the Teacher
Answer Presentation Tool
QuizShow

5.1 Record and Practice Journal

Vocabulary and Concept Check

1. 3 to 7, 3 : 7
2. *Sample answer:*
 circles to triangles = $\frac{4}{6}$,
 triangles to circles = $\frac{6}{4}$,
 circles to shapes = $\frac{4}{10}$,
 triangles to shapes = $\frac{6}{10}$
3. *Sample answer:* Smith: vowels to consonants = $\frac{1}{4}$.

Practice and Problem Solving

4. $\frac{2}{5}$, 2 : 5, 2 to 5; There are 2 frogs for every 5 turtles.
5. $\frac{6}{4}$, 6 : 4, 6 to 4 or $\frac{3}{2}$, 3 : 2, 3 to 2; There are 3 basketballs for every 2 soccer balls.
6. $\frac{2}{6}$, 2 : 6, 2 to 6 or $\frac{1}{3}$, 1 : 3, 1 to 3; There is 1 calculator for every 3 pencils.
7. $\frac{3}{7}$, 3 : 7, 3 to 7; There are 3 shirts for every 7 pants.
8. $\frac{5}{6}$
9. $\frac{3}{4}$
10. $\frac{3}{1}$
11. $\frac{7}{5}$

12–19. Sample answers are given.

12. $\frac{1}{3}, \frac{2}{6}$
13. $\frac{1}{3}, \frac{2}{6}$
14. $\frac{1}{2}, \frac{14}{28}$
15. $\frac{1}{3}, \frac{22}{66}$
16. $\frac{2}{9}, \frac{12}{54}$
17. $\frac{18}{20}, \frac{27}{30}$
18. $\frac{3}{7}, \frac{30}{70}$
19. $\frac{3}{2}, \frac{6}{4}$

T-194

Practice and Problem Solving

20. Adding the same number in the numerator and denominator does not produce an equivalent ratio, multiplying or dividing does. $\dfrac{2}{3} = \dfrac{2 \times 4}{3 \times 4} = \dfrac{8}{12}$

21. The numerator and denominator should be multiplied by the same number; $\dfrac{2}{3} = \dfrac{2 \times 2}{3 \times 2} = \dfrac{4}{6}$

22. 48 cups of dog food

23. 6 black pieces; The ratio of black to red is 3 : 5, so the ratio of black to all is 3 : 8. An equivalent ratio is 6 : 16.

24. 20 boys; The ratio of boys to girls is 5 : 7, so the ratio of boys to students is 5 : 12. An equivalent ratio is 20 : 48.

25. See *Taking Math Deeper*.

26–27. See Additional Answers.

Fair Game Review

28. 4.6 29. 3.29
30. 2.53 31. 1.478
32. B

Mini-Assessment

Write two equivalent ratios for the given ratio.

1. $\dfrac{1}{2}$ Sample answer: $\dfrac{2}{4}, \dfrac{3}{6}$

2. $\dfrac{10}{15}$ Sample answer: $\dfrac{20}{30}, \dfrac{30}{45}$

3. $\dfrac{36}{6}$ Sample answer: $\dfrac{18}{3}, \dfrac{12}{2}$

4. $\dfrac{12}{9}$ Sample answer: $\dfrac{8}{6}, \dfrac{16}{12}$

5. There are 32 students in the math club. The ratio of boys to girls is 1 : 3. How many boys are in the math club? 8 boys

T-195

Taking Math Deeper

Exercise 25

This problem introduces students to a fundamental concept regarding perimeter and area. As the dimensions of a figure are increased by a factor of k, the perimeter increases by a factor of k, but the area increases by a factor of k^2.

 Make a table.

Figure	2 ft × 3 ft	4 ft × 6 ft
Length	3 ft	6 ft
Width	2 ft	4 ft
Perimeter	10 ft	20 ft
Area	6 ft²	24 ft²

 Find ratios. a. b.

$\dfrac{\text{Blue Length}}{\text{Green Length}} = \dfrac{3 \text{ ft}}{6 \text{ ft}} = \dfrac{1}{2}$

$\dfrac{\text{Blue Width}}{\text{Green Width}} = \dfrac{2 \text{ ft}}{4 \text{ ft}} = \dfrac{1}{2}$

$\dfrac{\text{Blue Perimeter}}{\text{Green Perimeter}} = \dfrac{10 \text{ ft}}{20 \text{ ft}} = \dfrac{1}{2}$

$\dfrac{\text{Blue Area}}{\text{Green Area}} = \dfrac{6 \text{ ft}^2}{24 \text{ ft}^2} = \dfrac{1}{4}$

 Cut it out to help me see it.

4 of these = 1 of these

Reteaching and Enrichment Strategies

If students need help. . .	If students got it. . .
Resources by Chapter • Practice A and Practice B • Puzzle Time Record and Practice Journal Practice Differentiating the Lesson Lesson Tutorials Skills Review Handbook	Resources by Chapter • Enrichment and Extension Start the next section

ERROR ANALYSIS Describe and correct the error in writing an equivalent ratio.

20. ✗ $\dfrac{2}{3} = \dfrac{2+4}{3+4} = \dfrac{6}{7}$

21. ✗ $\dfrac{2}{3} = \dfrac{2 \times 2}{3 \times 3} = \dfrac{4}{9}$

22. **PUPPY** One way to make puppy food is to soften three parts dog food with one part water. How much dog food can be softened with 16 cups of water?

23. **CHECKERS** During a checkers game, there are 16 pieces left. The ratio of black to red is 3 : 5. How many black pieces are on the board? Explain how you found your answer.

24. **SCHOOL PLAY** There are 48 students in a school play. The ratio of boys to girls is 5 : 7. How many boys are in the school play? Explain how you found your answer.

25. **GEOMETRY** Use the blue and green rectangles.

 a. Find the ratio of the length of the blue rectangle to the length of the green rectangle. Repeat this for width, perimeter, and area.

 b. Compare and contrast your ratios in part (a).

Player	Hits	At Bats
1	30	48
2	16	40
3	20	32
4	36	60

26. **SOFTBALL** The table shows the number of hits and at bats for four softball players.

 a. Write each player's batting average as a ratio of hits to at bats.

 b. Did the player with the most hits have the best batting average? Explain.

27. **Reasoning** There are 12 boys and 10 girls in your gym class. If 6 boys joined the class, how many girls would need to join for the ratio to remain the same?

Fair Game Review *What you learned in previous grades & lessons*

Divide. *(Section 3.4)*

28. $13.8 \div 3$ 29. $16.45 \div 5$ 30. $53.13 \div 21$ 31. $19.214 \div 13$

32. **MULTIPLE CHOICE** Which numbers are ordered from least to greatest? *(Section 4.3)*

 Ⓐ $0.83, \dfrac{17}{20}, 86\%, \dfrac{7}{8}, 87.25\%$

 Ⓑ $0.83, \dfrac{17}{20}, 86\%, 87.25\%, \dfrac{7}{8}$

 Ⓒ $\dfrac{17}{20}, 0.83, 86\%, 87.25\%, \dfrac{7}{8}$

 Ⓓ $\dfrac{7}{8}, 87.25\%, 86\%, 0.83, \dfrac{17}{20}$

5.2 Rates

Essential Question How can you use rates to describe changes in real-life problems?

COMMON CORE STATE STANDARDS
6.RP.1
6.RP.2
6.RP.3b

1 ACTIVITY: Stories Without Words

Work with a partner. Each diagram shows a story problem.

- Describe the story problem in your own words.
- Write the rate indicated by the diagram. What are the units?
- Rewrite the rate so that the denominator is 1. (This is called a *unit rate*.)

a.

b.

c.

d.

January 2008
Length: 3 ft

January 2012
Length: 7 ft

196 Chapter 5 Ratios, Rates, and Data Analysis

Laurie's Notes

Introduction

For the Teacher
- **Goal:** The goal of this activity is for students to explore how to calculate rates and convert a rate to a unit rate.

Motivate
- Help students calculate their heart rate, the number of beats per unit of time.
- First have students practice finding their pulse, with wrist or neck.
- Count the number of beats in a 6 second period. Record the information as a rate. Example: $\dfrac{7 \text{ beats}}{6 \text{ seconds}}$.
- ? Ask what they would predict their answer to be if they found their heart rate for one minute. They should recognize that by multiplying both the beats and seconds by 10, the rate is $\dfrac{70 \text{ beats}}{60 \text{ seconds}}$ which is the same as 70 beats per minute.
- ? "If you had found the number of beats in 10 seconds, how would you find the beats per minute?" Multiply by a factor of 6.
- Discuss the idea that they could find their heart rate in beats per hour; however that is not a common rate.

Activity Notes

Activity 1
- Students may not know a formal definition for rates, but it is a familiar idea. They just found their heart rate and have heard rates such as 65 miles per hour, 32 miles per gallon, $7.50 per hour, and $1.99 per pound.
- Explain to students that they are going to work with a partner and write a story about information given in a diagram.
- Encourage students to be creative in writing the story problem. You want students to have fun and think about each context and what the diagram is telling them.
- Give sufficient time for students to write their story problems. Encourage students to use proper grammar, spelling, and sentence structure.
- Remind students to label the rate with appropriate units.
- If students have questions about writing the unit rate, remind them of the heart rate problem. Their initial rate was rewritten with a denominator of 1 minute.
- Have several groups share their stories and unit rates.
- **Summarize:** All four problems involve a unit of time as the second term in the rate.
- ? "Do all rates involve time? Explain." No, although they are common it could also be 32 miles per gallon.

Common Core State Standards

6.RP.1 Understand the concept of a ratio and use ratio language to describe a ratio relationship between two quantities.
6.RP.2 Understand the concept of a unit rate a/b associated with a ratio $a : b$ with $b \neq 0$, and use rate language in the context of a ratio relationship.
6.RP.3b Solve unit rate problems including those involving unit pricing and constant speed.

Previous Learning
Students know benchmark conversion facts, and how to simplify fractions.

Activity Materials
Introduction
• stopwatch or clock with second hand

Start Thinking! and Warm Up

Activity 5.2 **Start Thinking!** For use before Activity 5.2

Activity 5.2 **Warm Up** For use before Activity 5.2

Complete the statement.
1. 1 hour = ___?___ minutes
2. 1 week = ___?___ days
3. 1 foot = ___?___ yard
4. 1 foot = ___?___ inches
5. 1 year = ___?___ days

5.2 Record and Practice Journal

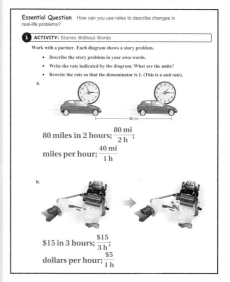

T-196

Differentiated Instruction

Auditory

Ask students to recall when they have heard or used expressions with the word *per*, such as dollars per gallon. Make a list of the expressions. Explain that these are all rates and have a denominator of 1. They are called *unit rates*. Ask students how they would find the unit rate from a rate such as $42.50 for 10 gallons of gasoline. Conversely, ask how they would use a unit rate of 25 miles per gallon to determine the number of miles traveled using 10 gallons. Motivate the students by posting the list and adding to it as you work through the chapter.

5.2 Record and Practice Journal

Laurie's Notes

Activity 2

- Introduce this activity by referring to the heart rate problem. Students rewrote their heart rates by multiplying each term by a factor of 10.
- **Multiple Representations:** $\dfrac{7 \text{ beats}}{6 \text{ seconds}} \times \dfrac{60 \text{ seconds}}{1 \text{ minute}} = \dfrac{70 \text{ beats}}{1 \text{ minute}}$.
- **Explain:** Students should recognize that 60 seconds = 1 minute. The fractions $\dfrac{60 \text{ seconds}}{1 \text{ minute}}$ and $\dfrac{1 \text{ minute}}{60 \text{ seconds}}$ are equivalent to 1. This is the definition of Magic One. Multiplying by 1 doesn't change the rate's value, only the units that are used to describe it.
- **Common Error:** Students multiply by the reciprocal of the "Magic One." Tell students to look at the units they have (i.e., dollars per hour) and the units they want (i.e., dollars per minute). Then ask in what order should they write the fraction so that the unwanted units are divided out and the unit they want is left. This is working backwards.
- **Connection:** When multiplying fractions, common factors are divided out. When rewriting a rate as a unit rate, common units are divided out. In some examples the numerical portion must be simplified as shown in the sample (120 ÷ 60 = 2).
- Have different pairs of students share their results at the board.

Closure

- **Exit Ticket:** Have students do jumping jacks or jog in place for 1 minute and then find their heart rate. Count the number of beats in 6 seconds and have them write the unit rate per minute. "How does this unit rate compare with the unit rate they found at the beginning of class?" *Sample answer:* Students should find that the unit heart rate after doing jumping jacks will be greater than the resting heart rate they calculated at the beginning of class.

The Dynamic Planning Tool
Editable Teacher's Resources at *BigIdeasMath.com*

2 ACTIVITY: Changing Units in a Rate

Work with a partner.

- Change the units of the rate by multiplying by a "Magic One." Show your work.
- Write your answer as a unit rate.

Original Rate *Magic One* *New Units* *Unit Rate*

a. Sample:

$$\frac{\$120}{\cancel{h}} \times \frac{1\,\cancel{h}}{60\text{ min}} = \frac{\$120}{60\text{ min}} = \frac{\$2}{1\text{ min}}$$

b. $\dfrac{\$3}{\text{min}} \times = = \dfrac{\$\square}{1\text{ h}}$

c. $\dfrac{36\text{ people}}{\text{yr}} \times = = \dfrac{\square\text{ people}}{1\text{ mo}}$

d. $\dfrac{12\text{ in.}}{\text{ft}} \times = = \dfrac{\square\text{ in.}}{1\text{ yd}}$

e. $\dfrac{60\text{ mi}}{\text{h}} \times = = \dfrac{\square\text{ mi}}{1\text{ min}}$

f. $\dfrac{2\text{ ft}}{\text{week}} \times = = \dfrac{\square\text{ ft}}{1\text{ yr}}$

What Is Your Answer?

3. One problem-solving strategy is called *Working Backwards*. What does this mean? How can this strategy be used to find the rates in Activity 2?

4. **IN YOUR OWN WORDS** How can you use rates to describe changes in real-life problems? Give two examples.

Practice — Use what you learned about rates to complete Exercises 11–14 on page 200.

5.2 Lesson

Key Vocabulary
rate, *p. 198*
unit rate, *p. 198*
unit cost, *p. 199*

Key Idea

Rate and Unit Rate

Words A **rate** is a ratio of two quantities using different units.

Numbers You pay $27 for 3 pizzas.

$$\text{rate} = \frac{\$27}{3 \text{ pizzas}} \leftarrow \text{dollars} \\ \leftarrow \text{pizzas}$$

Words A **unit rate** compares a quantity to one unit of the other quantity.

Numbers

Rate: $\dfrac{\$27}{3 \text{ pizzas}}$ Unit rate: $\dfrac{\$9}{1 \text{ pizza}}$

EXAMPLE 1 Writing a Rate

In a jalapeño pepper-eating contest, a contestant eats 70 peppers in 3 minutes. Write a rate that represents this situation.

$$\text{rate} = \frac{70 \text{ peppers}}{3 \text{ minutes}} \leftarrow \text{peppers} \\ \leftarrow \text{minutes}$$

∴ A rate is $\dfrac{70 \text{ peppers}}{3 \text{ minutes}}$.

On Your Own

Now You're Ready
Exercises 3–10

1. A contestant eats 55 peppers in 4 minutes. Write a rate that represents this situation.

EXAMPLE 2 Writing a Unit Rate

In orbit, a space shuttle travels 70,000 miles in 4 hours. How far does it travel in 1 hour?

$$\frac{70{,}000 \text{ miles}}{4 \text{ hours}} \xrightarrow{\div 4} \frac{17{,}500 \text{ miles}}{1 \text{ hour}}$$

To get a denominator of 1, divide each number by 4.

Reading
Dollars per hour and miles per gallon are commonly used unit rates.

∴ The space shuttle travels 17,500 miles in 1 hour.

198 Chapter 5 Ratios, Rates, and Data Analysis

Laurie's Notes

Introduction

Connect
- **Yesterday:** Students wrote unit rates from a diagram and converted unit rates.
- **Today:** Students will write rates to describe real-life situations.

Motivate
- Pair students; give them preliminary think time and then have them write as many rates as possible in two minutes. The rates need to have an *action verb* at the beginning of the rate and must be realistic.
- Example: *Sleep* 8 hours per night; *blink* 5 times per 20 seconds; *purchase* 1 ball per 4 people

Lesson Notes

Key Idea
- Write the Key Idea.
- When students are writing rates it is very important to write the related units. The units tell the context for the rate.
- Without the units the problem becomes $\frac{27}{3} = 9$. The 9 doesn't convey the context of $9 per pizza. Note that the word "per" is used in reading unit rates.

Example 1
- Work through the example.
- **FYI:** The record for eating jalapeno peppers is 247 pickled jalapeno peppers in 8 minutes.
- **Extension:** "How many peppers did they eat per minute?"

On Your Own
- **Reasoning:** Ask students to compare 70 peppers in 3 minutes to 55 peppers in 4 minutes. Which is the greater unit rate without converting? Explain.

Example 2
- **FYI:** *Columbia*, the first space shuttle, was launched April 12, 1981 at Kennedy Space Center. It was the first piloted reusable orbiter.
- **Connection:** Writing this rate as a **unit rate** is equivalent to simplifying the improper fraction $\frac{70,000}{4}$.
- **Extension:** "If the shuttle flies 17,500 miles in one hour, would it make it from Los Angeles to Philadelphia in more or less than 1 hour?" less than; The distance from Los Angeles to Philadelphia is 2400 miles. The circumference of Earth at the equator is approximately 25,000 miles.

Goal
Today's lesson is writing rates.

Start Thinking! and Warm Up

A grocery store has three options for your favorite hot breakfast cereal: single serving containers, a box of several single serving packets, and a large container. How can you decide which is the best buy for your money?

Extra Example 1
You pay $24 for 4 medium pizzas. Write a rate that represents this situation.

$\dfrac{\$24}{4 \text{ pizzas}}$

On Your Own

1. $\dfrac{55 \text{ peppers}}{4 \text{ min}}$

Extra Example 2
A car travels 275 miles in 5 hours. How far does it travel in 1 hour?
55 miles in 1 hour

Extra Example 3

You earn $5 for every hour you babysit. How much money do you earn for babysitting 3 hours? $15

 On Your Own

2. 186 mi
3. 4500 points

Extra Example 4

A 7-pound bag of cat food sells for $4.90. An 18-pound bag of cat food sells for $11.70. Which bag of cat food is the better buy? Explain. The 18-pound bag is the better buy, because $0.65 per pound is less than $0.70 per pound.

 On Your Own

4. The 32-pack is a better buy because its unit rate ($1.55 per roll) is less than the unit rate of the 30-pack ($1.61 per roll).

English Language Learners

Labels
English learners may recognize the fraction bar as division, but may not be familiar with its use in the concept of rate. The following unit rates are equivalent.

$\frac{3\,m}{1\,h}$, 3 m/h, 3 meters per hour

Each of these rates can be read as "three meters *for every* hour."

Laurie's Notes

Example 3
- Set up and work through the example.
- Students will ask why you can't just multiply 150 by 25 to answer the question. They may view the set up as extraneous when they know how to solve the problem.
- Explain that the set up is focusing on the units, a problem solving strategy that students need to develop.

On Your Own
- **Neighbor Check:** Have students work independently and then have their neighbor check their work. Have students discuss any discrepancies.
- **FYI:** *Shinkansen* is the Japanese name for the bullet train. Shinkansen runs at an average speed of 261 kilometers per hour or 163 miles per hour. All the lines combined carry 755,000 passengers a day. Since the Shinkansen was inaugurated 40 years ago, there have been no derailments or collision accidents and only one passenger fatality.

Example 4
- Ask students questions about the cost of common items. Answers may include text messages, CDs, or items of clothing.
- ? "What is the cost of a 16-ounce drink?" Answers will vary based on location. This size bottle is normally dispensed from a machine and costs a minimum of one dollar.
- ? "What is the cost of a music download?" Answers will vary.
- These are examples of unit costs. For the drink can, students may bring up different brands that differ in cost.
- A strategy in deciding which is a better buy is to find the unit cost as done in this example. Make sure students understand that in dividing by the number of pounds in each bag, the answer represents the cost *per* pound.
- ? "Is there another strategy you could use to find the better buy in this example?" Students may recognize that in purchasing two 20-pound bags you would spend $34.40, or $0.80 more than the 40-pound bag.
- Because 20 and 40 are compatible numbers, an alternate strategy was possible. This is not always the case.

On Your Own
- This is a good review of decimal division.

Closure
- **Exit Ticket:** Write the following as a unit rate in miles per minute and miles per hour: You run 2 miles in 20 minutes. 0.1 mile per minute; 6 miles per hour

The Dynamic Planning Tool
Editable Teacher's Resources at *BigIdeasMath.com*

EXAMPLE **Using a Unit Rate**

You earned 150 points for every note you successfully hit in a music video game. How many points did you earn?

∴ You earned 3750 points.

On Your Own

Exercises 11–20

2. A Japanese bullet train travels 558 miles in 3 hours. How far does the train travel in 1 hour?

3. **WHAT IF?** In Example 3, you successfully hit 30 notes. How many points did you earn?

A unit rate for cost per unit is a **unit cost**. A unit cost helps compare prices.

EXAMPLE 4 **Comparing Unit Costs**

Which bag of dog food is the better buy? Explain.

20-pound bag 40-pound bag

 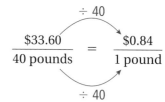

∴ Because $0.84 is less than $0.86, the 40-pound bag is the better buy.

On Your Own

Exercises 21 and 22

4. A 30-pack of paper towels costs $48.30. A 32-pack costs $49.60. Which is the better buy? Explain.

5.2 Exercises

Vocabulary and Concept Check

1. **WRITING** Describe a unit rate that you use in real life.

2. **DIFFERENT WORDS, SAME QUESTION** Which is different? Find "both" answers.

 What is the cost per bagel?

 What is the unit cost of a bagel?

 What is the cost per dozen bagels?

 How much does each bagel cost?

6 for $7.50

Practice and Problem Solving

Write a rate that represents the situation.

3. 105 words in 35 minutes
4. 12 feet in 5 seconds
5. 36 students for 16 computers
6. $1.40 for 4 apples
7. 3 baskets in 45 minutes
8. 2 inches in 8 years
9. $48 for 6 tickets
10. 150 gallons in 25 minutes

Write a unit rate for the situation.

11. $28 saved in 4 weeks
12. 18 necklaces in 3 hours
13. 270 miles in 6 hours
14. 228 students in 12 classes
15. 2520 kilobytes in 18 seconds
16. 880 calories in 8 servings
17. 1080 miles on 15 gallons
18. $12.50 for 5 ounces

19. **LIGHTNING** Lightning strikes Earth 100 times per second. What is the rate, in strikes per minute?

20. **HEART RATE** Your heart beats 240 times in 4 minutes. What is your heart rate, in beats per second?

21. **ENERGY BARS** Which pack of energy bars is the better buy? Explain.

22. **DEBATE** Do you think it is true that the bigger package is always the better buy? Give examples to support your decision.

Assignment Guide and Homework Check

Level	Day 1 Activity Assignment	Day 2 Lesson Assignment	Homework Check
Basic	11–14, 31–35	1, 2, 3–9 odd, 15–21 odd, 20, 22	3, 15, 20, 21
Average	11–14, 31–35	1, 2, 3–9 odd, 16–22 even, 23–27 odd	3, 16, 20, 23
Advanced	11–14, 31–35	1, 2, 16–26 even, 27–30	16, 20, 24, 28

Common Errors

- **Exercises 3–10** Students may not write units on the rates. Compare rates to ratios, but with different units in the numerator and denominator. Illustrate the importance of the units with an examination of the speed of a car. In the United States speed is measured in miles per hour, but in most of Europe, speed is measured in kilometers per hour. If you say that the speed of a car is 35, you do not know if it is miles or kilometers per hour.
- **Exercises 11–18** Students may try to make the numerator 1 instead of the denominator. Remind them of familiar unit rates like miles per hour and ask them to identify which part of the fraction has the 1.
- **Exercises 25 and 26** Students may compare the unit rates that have different units. Remind them that they must have the same units to compare the rates. Refer students to Activity 2 and the "Magic One" for the units.

5.2 Record and Practice Journal

Technology For the Teacher
Answer Presentation Tool
QuizShow

Vocabulary and Concept Check

1. *Sample answer:* You walk at a rate of $\frac{2 \text{ blocks}}{1 \text{ min}}$, so you walk 12 blocks in 6 minutes.
2. What is the cost per dozen bagels?; $15; $1.25.

Practice and Problem Solving

3. $\frac{105 \text{ words}}{35 \text{ min}}$
4. $\frac{12 \text{ ft}}{5 \text{ sec}}$
5. $\frac{36 \text{ students}}{16 \text{ computers}}$
6. $\frac{\$1.40}{4 \text{ apples}}$
7. $\frac{3 \text{ baskets}}{45 \text{ min}}$
8. $\frac{2 \text{ in.}}{8 \text{ yr}}$
9. $\frac{\$48}{6 \text{ tickets}}$
10. $\frac{150 \text{ gal}}{25 \text{ min}}$
11. $\frac{\$7}{1 \text{ week}}$
12. $\frac{6 \text{ necklaces}}{1 \text{ h}}$
13. $\frac{45 \text{ mi}}{1 \text{ h}}$
14. $\frac{19 \text{ students}}{1 \text{ class}}$
15. $\frac{140 \text{ kilobytes}}{1 \text{ sec}}$
16. $\frac{110 \text{ calories}}{1 \text{ serving}}$
17. $\frac{72 \text{ mi}}{1 \text{ gal}}$
18. $\frac{\$2.50}{1 \text{ oz}}$
19. 6000 strikes per minute
20. 1 beat per second
21. The 15-pack of energy bars is the better buy because it has a lesser unit rate ($1.28 per bar) than the 10-pack ($1.35 per bar).
22. *Answer should include, but is not limited to*: A decision that is supported by examples of unit price comparisons of same-brand products of different sizes.

T-200

Practice and Problem Solving

23. equivalent
24. equivalent
25. not equivalent
26. equivalent
27. See *Taking Math Deeper*.
28. a. about 0.12 mile per minute
 b. 8.02 minutes per mile
 c. The runner is talking about the rate in part (b) because "8-minute miles" is a way of talking about the rate in minutes per mile.
29. 66
30. 4.8

Fair Game Review

31–34. Sample answers are given.

31. $\dfrac{2}{6}, \dfrac{3}{9}$ 32. $\dfrac{10}{12}, \dfrac{20}{24}$

33. $\dfrac{4}{10}, \dfrac{8}{20}$ 34. $\dfrac{8}{18}, \dfrac{12}{27}$

35. B

Mini-Assessment

Write a unit rate for the situation.

1. 8 miles in 2 hours $\dfrac{4 \text{ miles}}{1 \text{ hour}}$
2. 104 students in 4 classes $\dfrac{26 \text{ students}}{1 \text{ class}}$
3. $8 for 5 ounces $\dfrac{\$1.60}{1 \text{ ounce}}$
4. $35 saved in 5 weeks $\dfrac{\$7}{1 \text{ week}}$
5. Your mom paid $27.60 for a package of 92 diapers. How much did your mom pay per diaper? $0.30

T-201

Taking Math Deeper

Exercise 27

The problem is interesting because it could be solved more simply using algebra. However, we don't know how to use algebra at this stage. So, we need to use another problem-solving strategy, like *Guess, Check, and Revise*.

 Make a table.

	Homeroom A	Homeroom B
Students	24	16
Canned Food	30	22
Dry Food	42	24
Additional	20	?

 Find a rate for Homeroom A.

$$\dfrac{30 + 42 + 20}{24} = 3.83 \text{ items per student}$$

③ *Guess, Check, and Revise* a rate for Homeroom B.

It's 16.

$\dfrac{22 + 24 + 20}{16} = 4.13$ items per student

$\dfrac{22 + 24 + \boxed{16}}{16} = 3.88$ items per student

$\dfrac{22 + 24 + 15}{16} = 3.81$ items per student

Project

Organize a food drive at your school. Determine the rates per student for each class. Make a poster to award the class with the highest rate.

Reteaching and Enrichment Strategies

If students need help. . .	If students got it. . .
Resources by Chapter • Practice A and Practice B • Puzzle Time Record and Practice Journal Practice Differentiating the Lesson Lesson Tutorials Skills Review Handbook	Resources by Chapter • Enrichment and Extension Start the next section

Decide whether the rates are equivalent.

23. $\dfrac{24 \text{ laps}}{6 \text{ minutes}}, \dfrac{72 \text{ laps}}{18 \text{ minutes}}$

24. $\dfrac{126 \text{ points}}{3 \text{ games}}, \dfrac{210 \text{ points}}{5 \text{ games}}$

25. $\dfrac{15 \text{ breaths}}{36 \text{ seconds}}, \dfrac{90 \text{ breaths}}{3 \text{ minutes}}$

26. $\dfrac{\$16}{4 \text{ pounds}}, \dfrac{\$1}{4 \text{ ounces}}$

27. **FOOD DRIVE** The table shows the amount of food collected by two homerooms. Homeroom A collects 20 additional items of food. How many more items does Homeroom B need to collect to have a higher rate of items per student?

	Homeroom A	Homeroom B
Students	24	16
Canned Food	30	22
Dry Food	42	24

28. **MARATHON** A runner completed his first 26.2-mile marathon in 210 minutes.

 a. Find the unit rate $\dfrac{\text{miles}}{\text{minute}}$.

 b. Find the unit rate $\dfrac{\text{minutes}}{\text{mile}}$.

 c. The runner says, "I averaged 8-minute miles in the marathon." Is he talking about the rate in part (a) or the rate in part (b)? Explain your reasoning.

 Convert the unit rate.

29. $\dfrac{45 \text{ miles}}{1 \text{ hour}} = \dfrac{\text{____ feet}}{1 \text{ second}}$

30. $\dfrac{18 \text{ gallons}}{1 \text{ hour}} = \dfrac{\text{____ cups}}{1 \text{ minute}}$

 Fair Game Review What you learned in previous grades & lessons

Write two fractions that are equivalent to the given fraction. *(Skills Review Handbook)*

31. $\dfrac{1}{3}$

32. $\dfrac{5}{6}$

33. $\dfrac{2}{5}$

34. $\dfrac{4}{9}$

35. **MULTIPLE CHOICE** A handheld video game system that normally costs $150 is on sale for 35% off. What is the sale price? *(Section 4.4)*

 Ⓐ $52.50 Ⓑ $97.50 Ⓒ $109.50 Ⓓ $375

5.3 Solving Rate Problems

Essential Question How can you use rates to help show how a country can save valuable natural resources?

COMMON CORE STATE STANDARDS
6.RP.2
6.RP.3b
6.RP.3d

For want of a nail the shoe was lost.
For want of a shoe the horse was lost.
For want of a horse the rider was lost.
For want of a rider the battle was lost.
For want of a battle the kingdom was lost.
And all for the want of a horseshoe nail.

1 ACTIVITY: Saving Water

The nursery rhyme above is an example of how a small problem can lead to a big problem.

Work with a partner. Here is an example about a leaky faucet that drips a drop of water every 2 seconds.

a. Copy and complete the table showing how many drops of water drip in different amounts of time. Write each entry in the table as a rate in drops per unit of time.

Drops	1					
Time	2 sec	1 min	1 h	1 d	1 wk	1 yr

b. How many gallons of water are wasted in a year? Show your work.

> 80 drops = 1 teaspoon
> 96 teaspoons = 1 pint
> 8 pints = 1 gallon

c. There are about 125 million homes and apartments in the United States. Suppose every one of them has a leaky faucet. How many gallons of water will be wasted each year? Explain your reasoning.

d. The swimming pool shown at the right holds about 15,000 gallons of water. How many times could this pool be filled by the amount of water you found in part (c)?

202 Chapter 5 Ratios, Rates, and Data Analysis

Laurie's Notes

Introduction

For the Teacher
- **Goal:** The goal of this activity is for students to recognize how one's perspective might change when rates are rewritten using different time units.

Motivate
- Given the theme of this lesson, share some *green facts*.
- It takes 3 liters of water to produce a plastic bottle that holds 1 liter of water. Americans throw away about 38 billion water bottles per year.
- The average American throws out 7.5 pounds of garbage a day.
- More than 10 million trees were planted in the Beijing area to help absorb carbon dioxide (CO_2) emissions that are produced by automobiles and factories. A single tree can absorb about 48 pounds of CO_2 per year.
- For additional information, check out *www.nrc-recycle.org*.

Activity Notes

Activity 1
- Ask students to read the rhyme. Discuss the meaning behind it.
- ? "Do your parents tell you to turn off the lights when you're not in a room, or turn off the water when you're brushing your teeth?" Your parents are trying to conserve our natural resources, and save money as well.
- Introduce the activity.
- In part (a), multiplying the daily rate by 365 and the weekly rate by 52 will give different answers because there are $52\frac{1}{7}$ weeks per year. Perhaps you could have two groups explore the difference that $\frac{1}{7}$ makes.
- In part (b), the answer is approximately 260 gallons per year.
- **FYI:** The average 5-minute shower uses 12.5–50 gallons of water depending upon the shower head.
- **Extension:** Students could talk with their parents about the cost of water in their home.
- The dimensions of the pool shown in part (d) are only for the purpose of giving students a context for the actual size. The key information is that the pool holds 15,000 gallons.
- Have groups share their findings at the conclusion of the activity. Listen for units when numbers are reported.

Common Core State Standards

6.RP.2 Understand the concept of a unit rate a/b associated with a ratio $a : b$ with $b \neq 0$, and use rate language in the context of a ratio relationship.

6.RP.3b Solve unit rate problems including those involving unit pricing and constant speed.

6.RP.3d Use ratio reasoning to convert measurement units; manipulate and transform units appropriately when multiplying or dividing quantities.

Previous Learning
Students know basic conversion facts.

Start Thinking! and Warm Up

Activity 5.3 Start Thinking!
For use before Activity 5.3

Activity 5.3 Warm Up
For use before Activity 5.3

Complete the statement.
1. 5 weeks = __?__ days
2. 2.5 feet = __?__ inches
3. 18 yards = __?__ feet
4. 48 hours = __?__ days
5. 11 years = __?__ months
6. 4 years = __?__ days

5.3 Record and Practice Journal

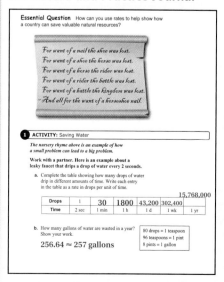

English Language Learners

Vocabulary
Some students may be used to hearing about gasoline purchased in Pounds per Liter, Pesos per Liter, Euros per Liter, or even Gourdes per Liter. Explain to students that gasoline is always purchased with a rate of currency per unit of volume.

Different words are used to describe rates. For example, $2 a bushel, $7.50 for every sheet, and 6 inches per year. Prepositions are often hard for the English learner to master. Encourage them to learn prepositions, but remind them that in this context they are often interchangeable.

5.3 Record and Practice Journal

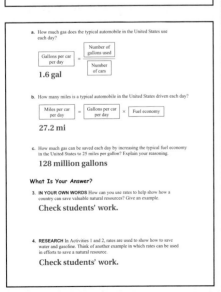

Laurie's Notes

Activity 2

- Discuss the meaning of fuel efficiency ratings as stated for each type of car and the size of the gas tank. You could begin with a few warm-up questions.
- ? "At $4.00 per gallon, how much money does it cost to fill an SUV with an empty gas tank?" $100
- ? "Approximately how many gallons does a full size car consume in driving 99 miles from Denver, Colorado to Cheyenne, Wyoming?" 3.4 gallons
- Part (a) asks students to describe the daily gasoline use of a typical car in the U.S. This is an average because there are different sizes of cars. The information students need is in the introduction and the graphic.
- Students should recognize that 400 million ÷ 250 million can be found by dividing 400 by 250 (common factor of millions are divided out), or simply 40 ÷ 25. This is a good example of how dividing out common factors first can simplify a division problem.
- **Multiple Representation:** For part (b) it may be helpful for students to see the units in this problem written in the following fashion:

$$\frac{\text{gallons}}{\text{car} \cdot \text{day}} \times \frac{\text{miles}}{\text{gallon}} = \frac{\frac{\text{miles}}{\text{car}}}{\text{day}} = \frac{\frac{\text{miles}}{\text{day}}}{\text{car}}.$$

The common unit of *gallons* divides out.
- Once students have done the computation, have them read it with the associated units. The typical car drives 27.2 miles per day.

What Is Your Answer?

- Water conservation and improved fuel economy are important topics. Students should have many other examples to share.

Closure

- **Exit Ticket:** Properly inflated tires can improve fuel economy. How significant is it to get 49 miles per gallon versus 48 miles per gallon? Explain. If the fuel economy is 49 miles per gallon, you can drive 1 mile farther than if the fuel economy is 48 miles per gallon. That's an extra 5280 feet!

The Dynamic Planning Tool
Editable Teacher's Resources at *BigIdeasMath.com*

2 ACTIVITY: Saving Gasoline

Work with a partner.

Drivers in the United States use about 400 million gallons of gasoline each day. There are about 250 million automobiles in the United States. The typical fuel economy of automobiles is about 17 miles per gallon.

a. How much gasoline does the typical automobile in the United States use each day?

$$\text{Gallons per car per day} = \frac{\text{Number of gallons used}}{\text{Number of cars}}$$

b. How many miles is a typical automobile in the United States driven each day?

$$\text{Miles per car per day} = \text{Gallons per car per day} \times \text{Fuel economy}$$

c. How much gasoline can be saved each day by increasing the typical fuel economy in the United States to 25 miles per gallon? Explain your reasoning.

What Is Your Answer?

3. **IN YOUR OWN WORDS** How can you use rates to help show how a country can save valuable natural resources? Give an example.

4. **RESEARCH** In Activities 1 and 2, rates are used to show how to save water and gasoline. Think of another example in which rates can be used in efforts to save a natural resource.

Use what you learned about solving ratio and rate problems to complete Exercises 11–14 on page 206.

Section 5.3 Solving Rate Problems 203

5.3 Lesson

Key Idea

Distance, Speed, and Time

Words To find the distance traveled d, multiply the speed r by the time t.

Algebra $d = rt$

Words To find the speed r, divide the distance d by the time t.

Algebra $r = \dfrac{d}{t}$

Remember
Speed is an example of a rate.

EXAMPLE 1 Finding a Distance

A peregrine falcon can dive at a speed of 80 meters per second. How far can it dive in 4 seconds?

$d = rt$ Write the formula for distance.

$= \dfrac{80 \text{ meters}}{1 \text{ second}} \times 4 \text{ seconds}$ Substitute the given values. The seconds divide out.

$= 320 \text{ meters}$ Multiply.

∴ The peregrine falcon can dive 320 meters in 4 seconds.

EXAMPLE 2 Finding a Speed

A great white shark swims 90 feet in 2.5 seconds. What is its speed?

$r = \dfrac{d}{t}$ Write the formula for speed.

$= \dfrac{90 \text{ feet}}{2.5 \text{ seconds}}$ Substitute the given values.

$= \dfrac{36 \text{ feet}}{1 \text{ second}}$ Divide.

∴ Its speed is 36 feet per second.

On Your Own

Now You're Ready
Exercises 3–10

1. **WHAT IF?** In Example 1, how far can the falcon dive in 7.5 seconds?

2. A grizzly bear runs 60 feet in 1.25 seconds. What is its speed?

Laurie's Notes

Introduction

Connect
- **Yesterday:** Students solved rate problems relating to natural resources, with increased awareness of the need to conserve.
- **Today:** Students will solve rate problems relating mostly to speed. The units of measure must be carefully noted.

Motivate
- Share a few interesting speed facts.
- Usain Bolt, of Jamaica, ran the 100 meters in 9.69 seconds at the Beijing Olympics. His average speed during the race was about 23.1 miles per hour.
- Many major league pitchers throw a fastball at more than 90 miles per hour.

Lesson Notes

Key Idea
- **?** "How is speed calculated?" distance traveled divided by the time involved
- **?** "What are some common units of speed?" miles per hour; meters per second; millimeters per day; yards per minute; inches per year
- Walk across the front of your classroom and ask students to estimate your speed.
- **?** "If I walk the same distance in less time, how does my speed change?"
- Write the Key Idea. Use your walking example or athletes to discuss the two formulas.
- **Connection:** Make sure that students understand that the speed formula is derived from the distance formula by dividing both sides of the equation by time.

Example 1
- Work through the problem. Dividing out the common units of *seconds* is a strategy called dimensional analysis.
- **Common Error:** Students may divide 80 by 4. Remind students that the question asks about a distance which is found by multiplying the rate by the time.

Example 2
- **?** "What are you solving for in this problem?" Speed
- **Multiple Representations:** Students don't always see these two representations as the same: $\frac{90 \text{ feet}}{2.5 \text{ seconds}} = \frac{90}{2.5}$ feet per second. Explain the equivalence.

On Your Own
- **?** **Neighbor Check:** Have students work independently and then have their neighbor check their work. Have students discuss any discrepancies.

Goal Today's lesson is solving rate problems.

Start Thinking! and Warm Up

Lesson 5.3 Warm Up
For use before Lesson 5.3

Lesson 5.3 Start Thinking!
For use before Lesson 5.3

How fast can you walk? Estimate your walking rate in feet per second.

Measure the length of a hallway in your school.

Have a classmate record the number of seconds it takes you to walk the entire length of the hallway.

Find your actual rate in feet per second. How close was your estimate?

Extra Example 1
A giraffe runs at 880 yards per minute. How far can it run in 3 minutes?
2640 yards

Extra Example 2
A lion runs 288 feet in 4 seconds. What is its speed? 72 feet per second

 On Your Own
1. 600 m
2. 48 ft/sec

Extra Example 3

A hot air balloon rises 8 meters every 4 seconds. How high does the hot air balloon rise in 20 seconds? 40 meters

Extra Example 4

Five cans of soup cost $6. How much do 7 cans of soup cost? $8.40

On Your Own

3. 90 m
4. $8.75

Differentiated Instruction

Visual

Have students create a chart to reinforce the relationship between distance, rate, and time.

distance = rate × time
$d = rt$, where d is in miles, kilometers, feet, and so on

time = $\dfrac{\text{distance}}{\text{rate}}$
$t = \dfrac{d}{r}$, where t is in seconds, minutes, hours, years, and so on

rate = $\dfrac{\text{distance}}{\text{time}}$
$r = \dfrac{d}{t}$, where r is in feet per second, kilometers per hour, miles per minute, and so on

Laurie's Notes

Example 3

- ? "Has anyone been on a hot air balloon ride?"
- **FYI:** Hot air balloons were invented in France in 1783. The first manned flight carried two men for 5.5 miles over Paris. The flight lasted 23 minutes.
- ? "What are you solving for in this problem?" Distance
- Discuss each method. Both methods divide out a factor of 3, along with the common unit of *seconds*.

Example 4

- This is the only problem not related to distance and speed in this lesson. The analysis, using units, is the same however.
- ? "How do you find the unit cost?" Divide the $7 cost by the number of bottles.
- ? "If you know the unit cost, how do you compute the cost of 9 bottles?" multiply unit cost by 9
- **Alternate Strategy:** Students might suggest that if 4 bottles cost $7, 8 bottles could cost $14. Then find the cost of one bottle (unit cost) and add that to $14.
- **Financial Literacy:** This type of problem builds financial and consumer literacy. This is a basic skill which students need to develop.

On Your Own

- Students can use mental math to see that the balloon rises twice as far in 30 seconds as in 15 seconds.

Closure

- **Exit Ticket:** The cheetah is the world's fastest land animal. It can reach speeds of 70 miles per hour. If a baby cheetah runs 20 miles per hour for 6 minutes, how far will it travel? *Hint:* 6 minutes = 0.1 hour; 2 miles

The Dynamic Planning Tool
Editable Teacher's Resources at *BigIdeasMath.com*

EXAMPLE 3 Solving a Rate Problem

How high does the hot air balloon rise in 15 seconds?

Method 1: Use the formula for distance.

$d = rt$ Write the formula for distance.

$= \dfrac{9 \text{ meters}}{\cancel{3}\, \text{seconds}} \times \cancel{15}^{\,5} \text{ seconds}$ Substitute the given values.

$= 45$ meters Multiply.

∴ The balloon rises 45 meters in 15 seconds.

Method 2: Use a unit rate.

Rises 9 meters every 3 seconds.

$\dfrac{9 \text{ meters}}{3 \text{ seconds}} \xrightarrow{\div 3} \dfrac{3 \text{ meters}}{1 \text{ second}}$ Find the unit rate.

$\dfrac{3 \text{ meters}}{1 \text{ second}} \times 15 \text{ seconds} = 45 \text{ meters}$ Multiply the unit rate by the time.

∴ The balloon rises 45 meters in 15 seconds.

EXAMPLE 4 Solving a Cost Problem

Four bottles of fruit juice cost $7. How much do nine bottles cost?

Find a unit rate.

$\dfrac{\$7}{4 \text{ bottles}} \xrightarrow{\div 4} \dfrac{\$1.75}{1 \text{ bottle}}$ ← Each bottle costs $1.75.

$\dfrac{\$1.75}{1 \text{ bottle}} \times 9 \text{ bottles} = \15.75 Multiply the unit rate by 9.

∴ Nine bottles cost $15.75.

On Your Own

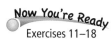

Exercises 11–18

3. **WHAT IF?** In Example 3, how high does the balloon rise in 30 seconds? Solve using both methods.

4. Four bottles of water cost $5. How much do seven bottles cost?

5.3 Exercises

Vocabulary and Concept Check

1. **FORMULA** Which is correct?

 | Distance equals speed times time. | Distance equals speed divided by time. |

2. **WRITING** What are the units for the product? $\dfrac{2 \text{ mi}}{1 \text{ h}} \cdot 2.5 \text{ h}$

Practice and Problem Solving

Find the distance.

3. $d =$ ____, $r = 35$ mi/h, $t = 3$ h
4. $d =$ ____, $r = 50$ m/sec, $t = 25$ sec
5. $d =$ ____, $r = 23$ ft/sec, $t = 12$ sec
6. $d =$ ____, $r = 6$ in./min, $t = 53$ min

Find the speed.

7. 425 inches in 85 seconds
8. 900 miles in 18 hours
9. 870 feet in 15 minutes
10. 3096 meters in 72 seconds

Find how far the object travels in the given amount of time.

11. 18 hours

Moves 2 meters every 3 hours.

12. 24 seconds

Rises 5 stories every 6 seconds.

13. 40 seconds

Falls 1810 feet every 10 seconds.

14. 20 minutes

Moves 960 kilometers every 4 minutes.

15. **YOGURT** Six containers of yogurt cost $4.50. How much do eight containers cost?

16. **CEREAL** Three boxes of cereal cost $9.57. How much do five boxes cost?

206 Chapter 5 Ratios, Rates, and Data Analysis

Assignment Guide and Homework Check

Level	Day 1 Activity Assignment	Day 2 Lesson Assignment	Homework Check
Basic	11–14, 23–26	1, 2, 3–9 odd, 15, 16, 17	2, 3, 7, 16
Average	11–14, 23–26	1, 2, 3–9 odd, 16, 17, 18, 20	2, 3, 7, 16
Advanced	11–14, 23–26	1, 2, 4–10 even, 17, 18, 21, 22	4, 8, 18, 21

Common Errors

- **Exercises 3–6** Students may forget to write the units with the answer. Remind them that without units the answer would not make sense because the answer could be in miles, feet, yards, meters, etc.
- **Exercises 7–10** Students may just substitute the values into the equation without simplifying to the unit rate. Remind them that speed is a unit rate which can be found using the formula.
- **Exercises 11–14** Students may take the distance part of the rate instead of finding the unit rate first to solve the problem. For example, the student might write 2 meters per hour multiplied by 18 hours instead of $\frac{2}{3}$ meter per hour multiplied by 18 hours. Students don't have to write the rate as a unit rate, but they cannot forget to include the time given. Encourage them to find the unit rate first to help them solve the problem without errors.

Vocabulary and Concept Check

1. Distance equals speed times time.
2. miles

Practice and Problem Solving

3. 105 mi
4. 1250 m
5. 276 ft
6. 318 in.
7. 5 in./sec
8. 50 mi/h
9. 58 ft/min
10. 43 m/sec
11. 12 m
12. 20 stories
13. 7240 ft
14. 4800 km
15. $6
16. $15.95

Technology For the Teacher
Answer Presentation Tool
Quiz*Show*

5.3 Record and Practice Journal

Practice and Problem Solving

17. 50 city blocks
18. 40 ounces
19. $18.\overline{6}$ in./min
20. *Sample answer:*
 a. 8 ounces
 b. 4 people
 c. 15 punch recipes
21. See *Taking Math Deeper*.
22. 10 students; Because out of 120 students, 18 are in concert band, out of 300 students, you expect 45 to be in concert band. Because 2 of 9 concert band members are in the jazz band, you expect 10 of 45 concert band members to be in the jazz band.

Fair Game Review

23. 86 24. 8.6
25. 141
26. B

Mini-Assessment

Find the distance.

1. $d = \boxed{}$, $r = 55$ mi/h, $t = 2$ h
 110 miles
2. $d = \boxed{}$, $r = 40$ mi/h, $t = 3$ h
 120 miles

Find the speed.

3. 800 feet in 20 minutes
 40 feet per minute
4. 1300 miles in 25 hours
 52 miles per hour
5. You walk 2 city blocks in 10 minutes. How many city blocks can you walk in 1 hour? 12 city blocks

T-207

Taking Math Deeper

Exercise 21

This problem keeps students on their toes regarding units. Each rate is given in terms of seconds, but each time is given as minutes. Moreover, with the distances ending up as thousands of feet, it makes sense to list the final distances in miles.

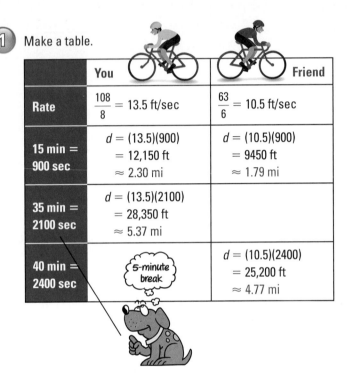

② After 15 minutes, your distance apart is about:
 a. $2.30 + 1.79 = 4.09$ mi

③ After 40 minutes, your distance apart is about:
 b. $5.37 + 4.77 = 10.14$ mi

Reteaching and Enrichment Strategies

If students need help...	If students got it...
Resources by Chapter • Practice A and Practice B • Puzzle Time Record and Practice Journal Practice Differentiating the Lesson Lesson Tutorials Skills Review Handbook	Resources by Chapter • Enrichment and Extension • School-to-Work Start the next section

17. **WALKING** You walk 5 city blocks in 12 minutes. How many city blocks can you walk in 2 hours?

18. **HAM** Eight ounces of ham cost $1.92. You bought $9.60 worth of ham. How many ounces did you buy?

19. **ANT** How fast should an ant run to go around the rectangle in three minutes?

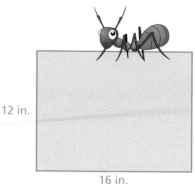

Fruit Punch
- 8 ounces orange juice
- 2 ounces lemon juice
- 6 ounces pineapple juice
- 6 ounces apple juice
- 12 ounces lemon-lime soda

20. **FRUIT PUNCH** You are using the recipe to make fruit punch.

 a. At a party, how many ounces of punch do you think each person will drink?

 b. Use your answer to part (a) to estimate how many people this recipe will serve.

 c. You invite 60 people to the party. How many punch recipes should you make?

21. **BIKING** You and a friend start biking in opposite directions from the same point. You travel 108 feet every 8 seconds. Your friend travels 63 feet every 6 seconds.

 a. How far apart are you and your friend after 15 minutes?

 b. After 20 minutes, you take a 5-minute rest, but your friend does not. How far apart are you and your friend after 40 minutes? Explain your reasoning.

22. **Critical Thinking** A survey of 120 students had the following results.
 - Eighteen are in the concert band.
 - Two out of every nine concert band members are in the jazz band.
 - Every jazz band member is in the concert band.

 If you increase the size of the survey to 300 students, how many would you expect to be in the jazz band? Explain your reasoning.

 Fair Game Review What you learned in previous grades & lessons

Add. *(Skills Review Handbook)*

23. $12 + 16 + 24 + 18 + 16$ 24. $2.6 + 3.5 + 1.7 + 0.8$ 25. $33 + 47 + 25 + 36$

26. **MULTIPLE CHOICE** What is the area of the trapezoid? *(Section 1.5)*

 Ⓐ 6 m^2 Ⓑ 36 m^2
 Ⓒ 60 m^2 Ⓓ 210 m^2

5 Study Help

You can use a **definition and example chart** to organize information about a concept. Here is an example of a definition and example chart for ratio.

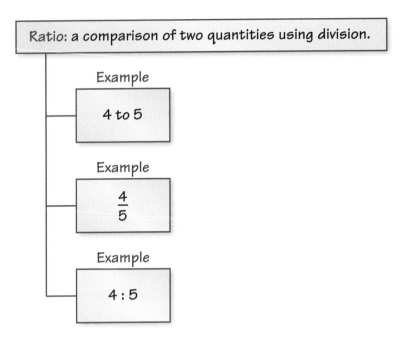

On Your Own

Make a definition and example chart to help you study these topics.

1. equivalent ratios
2. rate
3. unit rate
4. unit cost

After you complete this chapter, make definition and example charts for the following topics.

5. mean
6. outlier
7. measure of central tendency
8. median
9. mode
10. range

"My math teacher taught us how to make a definition and example chart."

208 Chapter 5 Ratios, Rates, and Data Analysis

Sample Answers

1.

Equivalent ratios: two ratios that describe the same relationship.

- Example: $\frac{3}{4}$ and $\frac{9}{12}$
- Example: 50 to 40 and 5 to 4
- Example: 1 : 12 and 5 to 60

2.

Rate: a ratio of two quantities using different units.

- Example: $\frac{\$135}{3 \text{ hours}}$
- Example: $\frac{63 \text{ feet}}{12 \text{ seconds}}$
- Example: $\frac{8 \text{ hits}}{25 \text{ at bats}}$

3.

Unit rate: a rate that compares a quantity to one unit of another quantity.

- Example: $\frac{60 \text{ miles}}{1 \text{ hour}}$
- Example: $\frac{5.25 \text{ feet}}{1 \text{ second}}$
- Example: $\frac{72 \text{ words}}{1 \text{ minute}}$

4. Available at *BigIdeasMath.com*.

List of Organizers

Available at *BigIdeasMath.com*

Comparison Chart
Concept Circle
Definition (Idea) and Example Chart
Example and Non-Example Chart
Formula Triangle
Four Square
Information Frame
Information Wheel
Notetaking Organizer
Process Diagram
Summary Triangle
Word Magnet
Y Chart

About this Organizer

A **Definition and Example Chart** can be used to organize information about a concept. Students fill in the top rectangle with a term and its definition or description. Students fill in the rectangles that follow with examples to illustrate the term. Each sample answer shows 3 examples, but students can show more or fewer examples. Definition and example charts are useful for concepts that can be illustrated with more than one type of example.

Technology For the Teacher
Vocabulary Puzzle Builder

Answers

1. $\dfrac{3}{2}$, 3 to 2, or 3 : 2; There are 3 tulips for every 2 lilies.

2. $\dfrac{2}{1}$, 2 to 1, or 2 : 1; There are 2 crayons for every marker.

3. *Sample answer:* $\dfrac{1}{3}, \dfrac{10}{30}$

4. *Sample answer:* $\dfrac{22}{16}, \dfrac{44}{32}$

5. $\dfrac{2 \text{ touchdowns}}{1 \text{ game}}$

6. $\dfrac{3 \text{ text messages}}{1 \text{ minute}}$

7. $\dfrac{20 \text{ entries}}{1 \text{ contest}}$ 8. $\dfrac{3 \text{ questions}}{1 \text{ minute}}$

9. 150 m 10. 28 ft

11. 24 ounces 12. $59.95

13. a. $\dfrac{17}{14}$, 17 to 14, or 17 : 14; The Celtics have won 17 championships to the 14 championships by the Lakers.

 b. $\dfrac{3}{4}$, 3 to 4, or 3 : 4; The Pistons have won 3 championships to the 4 championships by the Spurs.

 c. $\dfrac{3}{7}$, 3 to 7, or 3 : 7; The Bulls have won 3 championships to the 7 championships by the Lakers.

Assessment Book

T-209

Alternative Quiz Ideas

100% Quiz	Math Log
Error Notebook	**Notebook Quiz**
Group Quiz	Partner Quiz
Homework Quiz	Pass the Paper

Notebook Quiz

A notebook quiz is used to check students' notebooks. Students should be told at the beginning of the course what the expectations are for their notebooks: notes, class work, homework, date, problem number, goals, definitions, or anything else that you feel is important for your class. They also need to know that it is their responsibility to obtain the notes when they miss class.

1. On a certain day, what was the answer to the warm up question?
2. On a certain day, how was this vocabulary term defined?
3. In Section 5.2, what is the answer to On Your Own Question 1?
4. In Section 5.3, what is the answer to the Essential Question?
5. On a certain day, what was the homework assignment?

Give students 5 minutes to answer these questions.

Reteaching and Enrichment Strategies

If students need help...	If students got it...
Resources by Chapter • Study Help • Practice A and Practice B • Puzzle Time Lesson Tutorial *BigIdeasMath.com* Practice Quiz Practice from the Test Generator	Resources by Chapter • Enrichment and Extension • School-to-Work Game Closet at *BigIdeasMath.com* Start the next section

Technology For the Teacher

Answer Presentation Tool
Big Ideas Test Generator

5.1–5.3 Quiz

Write the ratio as a fraction in simplest form. Explain what the ratio means. *(Section 5.1)*

1. tulips : lilies

2. crayons : markers

Write two equivalent ratios for the given ratio. *(Section 5.1)*

3. $\dfrac{5}{15}$

4. $\dfrac{11}{8}$

Write a unit rate for the situation. *(Section 5.2)*

5. 12 touchdowns in 6 games

6. 15 text messages in 5 minutes

7. 80 entries in 4 contests

8. 75 questions in 25 minutes

Find how far the object travels in the given amount of time. *(Section 5.3)*

9. 30 seconds

Surfs 50 meters in 10 seconds.

10. 63 seconds

Rises 4 feet in 9 seconds.

11. **SHAMPOO** You can buy 18 ounces of shampoo for $3.99 or 24 ounces for $5.09. Which is the better buy? *(Section 5.2)*

12. **DOWNLOADS** Three album downloads cost $35.97. How much do five album downloads cost? *(Section 5.3)*

13. **NBA CHAMPIONSHIPS** Write each ratio in simplest form. Explain what the ratio means. *(Section 5.1)*

 a. Celtics championships to Lakers championships

 b. Pistons championships to Spurs championships

 c. Bulls championships to Lakers championships

5.4 Mean

COMMON CORE STATE STANDARDS
6.SP.3
6.SP.5c

Essential Question What is the meaning of the word "average?" How can you find the average of a collection of numbers?

1 ACTIVITY: Describing an Average

Work with a partner. A women's shoe store is analyzing its stock. The bar graph shows the percent of women's shoes in stock for each size.

a. What percent of the shoes are size $7\frac{1}{2}$, 8, or $8\frac{1}{2}$?

b. There are 200 pairs of shoes in stock. How many are size 7? Explain your reasoning.

c. What is the average shoe size for the shoes in stock? Explain.

2 ACTIVITY: Describing a Collection of Shoe Sizes

Work with a partner. A women's shoe store has 20 customers with the following sizes.

$6\frac{1}{2}$, 8, 9, 7, 7, 6, 10, 8, $8\frac{1}{2}$, 11

$7\frac{1}{2}$, $8\frac{1}{2}$, 8, 7, 8, $5\frac{1}{2}$, 6, 9, 8, $8\frac{1}{2}$

a. Use a table or a graph to organize the shoe sizes of the 20 customers.

b. Write a short paragraph describing the shoe sizes. In your paragraph, describe the "average shoe size" of the customers.

c. Is the entire stock in the shoe store, as shown in Activity 1, well represented by these 20 customers?

Chapter 5 Ratios, Rates, and Data Analysis

Laurie's Notes

Introduction

For the Teacher
- **Goal:** The goal of this activity is for students to develop an intuitive understanding about what the word *average* means.

Motivate
- Give each student a sticky note, with different colors for girls and boys. Ask students to write their shoe size on the note. Make a quick bar graph on the board representing shoe size for each gender.
- Ask students to make observations about each bar graph. Some vocabulary to listen for: *least, greatest, typical, most common,* and so on. Compare your graph to the given graph.
- If appropriate, ask, "What is the typical, or average, shoe size for girls? For boys?"

Activity Notes

Activity 1
- This lesson and the next discuss three types of *averages:* mean, median, and mode.
- **Common Misconception:** The arithmetic mean (the average) is the only type of average. Median and mode are valid averages for certain contexts.
- Listen for language that suggests an understanding or intuition about the different types of averages.
- ? "What do the numbers on the horizontal axis represent?" women's shoe sizes "the vertical axis?" percent of shoes in stock in that size
- ? "What do you notice about the bar graph?" symmetry; more than 50% between 7 and 9, inclusive
- Discuss reasoning for part (b). "What does "10% of the shoes in stock are size 7" mean in the context?" You might expect about 10% of the 200 to be size 7.
- **Common Error:** Students may read the vertical axis as *number* of shoes instead of the percent of shoes in stock. Discuss the difference.
- Discuss reasoning for part (c). Students may say that size 8 is the *average* because it is the middle (median), there are more size 8 shoes than any other (mode), or the shoe sizes in stock are balanced around size 8 (mean).

Activity 2
- Have several pairs of students show their work at the board, or on a transparency, to discuss various samples.
- **Common Error:** Students often have difficulty correctly scaling the axes. Check students' work.
- Have several pairs of students read their paragraphs.
- ? "Did all the groups use the same type of graph or table?"
- ? "How did each group describe the *average* shoe size?"

Common Core State Standards

6.SP.3 Recognize that a measure of center for a numerical data set summarizes all of its values with a single number,

6.SP.5c Summarize numerical data sets in relation to their context by giving quantitative measures of center (median and/or mean), . . . as well as describing any overall pattern and any striking deviations from the overall pattern with reference to the context in which the data were gathered.

Previous Learning

Students know benchmark conversion facts, and how to simplify fractions.

Start Thinking! and Warm Up

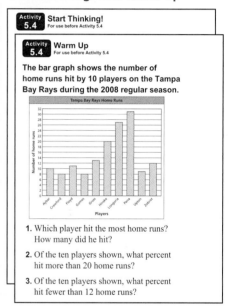

5.4 Record and Practice Journal

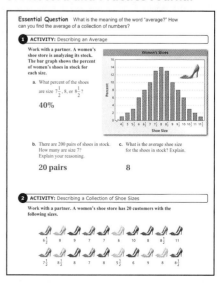

T-210

English Language Learners

Vocabulary

English learners may be confused by the word *average*. In math they may read phrases such as "the average rate" or "the average number of participants." Explain to students that the word *average* can be used to describe a common outcome or result. Looking up *average* in the dictionary and writing synonyms may help students to better understand the concept.

5.4 Record and Practice Journal

a. Use a table or a graph to organize the shoe sizes of the 20 customers.
Check students' work.

b. Write a short paragraph describing the shoe sizes.
Check students' work.

c. Is the entire stock in the shoe store, as shown in Activity 1, well represented by these 20 customers?
No.

3 ACTIVITY: Talking About Averages

Work with a partner. Talk about the statement. What type of survey or research do you think was done to write each statement?

a. The average height for men in the United States is 5 feet, 9 inches.
survey

b. The average annual income for a family in the United States is $52,000.
survey or research

c. The average fuel economy for a car in the United States is 17 miles per gallon.
survey or research

d. The average age of a person living in the United States is 36.4 years.
research

e. The average amount of dog food eaten by a dog in the United States is 1.2 pounds per day.
survey or research

What Is Your Answer?

4. **IN YOUR OWN WORDS** What is the meaning of the word "average"? How can you find the average of a collection of numbers? Give two examples of averages.
Sample answer: Add all numbers and divide by the number of values.

5. There are 5 students in the cartoon. Four of the students are 66 inches tall. One is 96 inches tall.

a. How do you think the students decided their average height is 6 feet?
Sample answer: Added all heights and divided by 5.

b. Does a height of 6 feet seem like a good representation of the average height of the 5 students? Explain why or why not.
No

Laurie's Notes

Activity 3

- Discuss how surveys and research are done in order to make statements about a particular segment of the population.
- ? "If you want to know the average number of songs a 12-year old downloads in one year, how would you find this out?" You would conduct a survey in different locations around the country, not just 12-year olds who have an account with an internet music site.
- Explain that students need to think about how the survey or research was done to gather the information reflected in each statement.
- When students have had sufficient time, have a class discussion.
- Students may not be familiar with the U.S. Census Bureau *(census.gov)* and the information gained from the Census that is taken every 10 years.

What Is Your Answer?

- For Question 5, chances are someone in the class knows that to find a mean you add all numbers and divide by the number of values.
- The bigger idea of this question is to help students develop a sense about outliers and how they affect the mean of a data set.

Closure

- **Exit Ticket:** How would you find the *average* time it takes students in your school to commute from home to school? *Sample answer:* You could conduct a survey to see how long it takes each student in the class to commute from home to school. Add all the commute times together and divide by the total number of students that responded to the survey.

Technology For the Teacher

Dynamic Classroom

The Dynamic Planning Tool
Editable Teacher's Resources at *BigIdeasMath.com*

3 ACTIVITY: Talking About Averages

Work with a partner. Talk about the statement. What type of survey or research do you think was done to write each statement?

a. The average height for men in the United States is 5 feet, 9 inches.

b. The average annual income for a family in the United States is $52,000.

c. The average fuel economy for a car in the United States is 17 miles per gallon.

d. The average age of a person living in the United States is 36.4 years.

e. The average amount of dog food eaten by a dog in the United States is 1.2 pounds per day.

What Is Your Answer?

4. **IN YOUR OWN WORDS** What is the meaning of the word "average?" How can you find the average of a collection of numbers? Give two examples of averages.

5. There are 5 students in the cartoon. Four of the students are 66 inches tall. One is 96 inches tall.

 a. How do you think the students decided that their average height is 6 feet?

 b. Does a height of 6 feet seem like a good representation of the average height of the 5 students? Explain why or why not.

"Yup, the average height in our class is 6 feet."

Practice

Use what you learned about averages to complete Exercises 4 and 5 on page 214.

Section 5.4 Mean **211**

5.4 Lesson

Key Vocabulary
mean, *p. 212*
outlier, *p. 213*

Key Idea

Mean

Words The **mean** of a data set is the sum of the data divided by the number of data values.

Numbers Data: 8, 5, 6, 9 (4 data values) Mean: $\dfrac{8+5+6+9}{4} = \dfrac{28}{4} = 7$

EXAMPLE 1 Standardized Test Practice

Text Messages Sent
Mark: 120
Laura: 95
Stacy: 101
Josh: 125
Kevin: 82
Maria: 108
Manny: 90

The table shows the number of text messages sent by a group of friends over one week. What is the mean number of messages sent?

Ⓐ 100 Ⓑ 102 Ⓒ 103 Ⓓ 104

$$\text{mean} = \dfrac{120 + 95 + 101 + 125 + 82 + 108 + 90}{7}$$

← Sum of the data
← Number of values

$$= \dfrac{721}{7}, \text{ or } 103 \quad \text{Simplify.}$$

∴ The mean number of text messages sent is 103. The correct answer is Ⓒ.

EXAMPLE 2 Comparing Means

A mean is a type of average.

The double bar graph shows the monthly rainfall amounts for two cities over a six-month period. Compare the mean monthly rainfalls.

City 1 mean: $\dfrac{3.5 + 2.2 + 1.9 + 2.1 + 2.5 + 3.4}{6} = \dfrac{15.6}{6}$, or 2.6

City 2 mean: $\dfrac{1.7 + 1.6 + 2.2 + 2.1 + 2.7 + 1.7}{6} = \dfrac{12}{6}$, or 2

∴ Because 2.6 is greater than 2, City 1 averaged more rainfall.

Laurie's Notes

Introduction

Connect
- **Yesterday:** Students developed an intuitive understanding about different types of averages.
- **Today:** Students will use the formal definition of a mean to answer real-life problems.

Motivate
- Use 24 pennies or counters and sort them into 3 stacks of height 3, 7, and 14. Ask how to find the *average* number of pennies in a stack.
- Ask a volunteer to show how it could be done. Perhaps they will "share" some pennies among the stacks. The stacks will even out to 8 per stack, which is the mean.
- **Big Idea:** Finding the mean of a data set is a sharing process. By summing the data set and dividing by the number of data values you even out the data, so they are all the same.
- The 24 pennies divided into 3 unequal stacks model the process that the sum (24) divided by the number of data values (3) gives you even stacks (8).

Lesson Notes

Key Idea
- There are three *measures of central tendency* or *averages*. Today's lesson is about the mean.
- Write the Key Idea.
- Discuss the example. For the data values of 5, 6, 8, and 9, the mean is somewhere between the low (5) and high (9) data values. After sharing some from the 9 to the 5 and the 8 to the 6, the mean is 7.

Example 1
- ❓ "What is the greatest value?" 125 "the least value?" 82
- ❓ "What do you know about the mean?" It is a number between 82 and 125.
- Work through the problem.
- ❓ "Is 103 a reasonable answer? How can you tell?" Yes, because 103 falls between 82 (the low) and 125 (the high).
- **Caution:** If students use a calculator and enter all the operations in one step, they must remember to keep the numerator in parentheses and then divide by the denominator: $\frac{(120 + 95 + 101 + 125 + 82 + 108 + 90)}{7} = 103$.

Example 2
- ❓ "Which city is represented by the dark green bar?" City 1
- ❓ "In what month was the rainfall the same in the two cities?" February
- Work through the computation for each city.

Goal Today's lesson is finding the **mean** of data sets and discovering the relationship between the mean and **outliers**.

Lesson Materials
Introduction
• 24 pennies or counters

Start Thinking! and Warm Up

Lesson 5.4 Warm Up
For use before Lesson 5.4

Lesson 5.4 Start Thinking!
For use before Lesson 5.4

What do you think is the average height of all of the people in your class?

Measure everyone in your class (including your teacher). Round to the nearest inch and record everyone's height in inches.

Find the average by adding all of the heights and dividing by the number of people.

Was your guess close to the average?

Extra Example 1
Find the mean of the data.
30, 81, 50, 24, 15, 64 44

Extra Example 2
Compare the mean monthly rainfall (in inches) for the two cities.
City A: 2.5, 4.3, 4.8, 2.7, 1.2
City B: 1.7, 4.1, 5.5, 3.2, 0.5
Because 3.1 is greater than 3, City A averaged more rainfall.

On Your Own

1. 57 2. 7.5

Extra Example 3

Kitten Weights (pounds)				
4.5	5.7	4.4	4.45	5.5
5.6	4.7	4.9	7.25	5

The table shows the weights of several kittens.
a. Identify the outlier. 7.25 pounds
b. Find the mean with and without the outlier. Mean with outlier: 5.2 pounds, mean without outlier: about 4.97 pounds
c. Describe how the outlier affects the mean. With the outlier, the mean is greater than 6 of the values (that's over half!). Without the outlier, the mean better represents the weights.

On Your Own

3. The weight of 8 pounds is the outlier, and it makes the mean less than all but two of the weights.

4. The price of $900 is the outlier, and it makes the mean greater than all of the other prices.

Differentiated Instruction

Vocabulary
Have students add a glossary to their math notebook. Key vocabulary words should be added as they are introduced. Mean, median, and mode are often confusing to students. Illustrations next to the vocabulary words will help in understanding and reinforcing their meanings.

Laurie's Notes

On Your Own

- **Neighbor Check:** Have students work independently and then have their neighbor check their work. Have students discuss any discrepancies.

Example 3

- Ask students to examine the height of each pony.
- ? "Is there one pony that is noticeably different in size from the others? Describe how you can tell." Yes. There is one pony that is 28 inches in height. This is much smaller compared to the other ponies.
- Define outlier. In the pony example, 28 inches is the outlier.
- **Multiple Representations:** Some students will be able to find the outlier simply by reading through the chart. Others may struggle to find the outlier. Have these students graph the heights on the number line as shown in part (a). This will visually reinforce why 28 inches is the outlier.
- Calculate the mean of the data.
- ? "Are any of the ponies exactly 37.9 inches tall? How many ponies are taller than the mean?" No. There are 7 ponies that are more than 37.9 inches tall.
- ? "Does the mean do a good job of describing the average pony height?" No.
- Ask students to calculate the mean *without* the outlier.
- **Common Error:** Some students will forget to subtract one from the total number of data values. Remind students that they are using one less value and so the denominator of their calculation should decrease by 1.
- ? "Are there ponies exactly 39 inches tall? How many ponies are taller than the mean? Shorter than the mean?" Yes. There are 4 ponies taller than 39 inches and 4 ponies shorter than 39 inches (ignoring the outlier).
- ? "Which mean better describes the average pony height?" 39 inches. Listen for: the heights are more balanced (4 greater than the mean, 4 less than the mean).
- **Big Idea:** An **outlier** that is very low compared to the rest of the data values will cause the mean to be too low. By eliminating the outlier, you can calculate a mean that better demonstrates the average of the data.
- **Discuss:** "What effect would an extremely tall pony have on the mean? What could you do to correct this?" It would cause the mean to be too high. You could eliminate the outlier and calculate the mean without it.

Closure

- **Exit Ticket:** Compute the mean of points scored by a football team in their first 5 games one season: 16, 21, 6, 30, 27 20 points

The Dynamic Planning Tool
Editable Teacher's Resources at *BigIdeasMath.com*

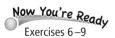

On Your Own

Find the mean of the data.

1. 49, 62, 52, 54, 61, 70, 55, 53
2. 7.2, 8.5, 7.0, 8.1, 6.7

An **outlier** is a data value that is much greater or much less than the other values. When included in a data set, it can affect the mean.

EXAMPLE 3 Finding a Mean With and Without an Outlier

Shetland Pony Heights (in.)				
40	37	39	40	42
38	38	37	28	40

The table shows the heights of several Shetland ponies.

a. Identify the outlier.
b. Find the mean with and without the outlier.
c. Describe how the outlier affects the mean.

a. Graph the heights on a number line.

The height of 28 inches is very low compared to the other heights. So, it is an outlier.

b. **Mean with outlier:**

$$\frac{40 + 37 + 39 + 40 + 42 + 38 + 38 + 37 + 28 + 40}{10} = \frac{379}{10}, \text{ or } 37.9$$

Mean without outlier:

$$\frac{40 + 37 + 39 + 40 + 42 + 38 + 38 + 37 + 40}{9} = \frac{351}{9}, \text{ or } 39$$

c. With the outlier, the mean is less than all but three of the heights. Without the outlier, the mean better represents the heights.

On Your Own

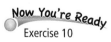

For each data set, identify the outlier. Then describe how it affects the mean.

3. Weights (in pounds) of dogs at a kennel

 48, 50, 55, 60, 8, 37, 50

4. Prices for flights from Miami, Florida to San Juan, Puerto Rico

 $456, $512, $516, $900, $436, $516

5.4 Exercises

Check It Out
Help with Homework
BigIdeasMath.com

Vocabulary and Concept Check

1. **NUMBER SENSE** Is the mean always equal to a value in the data set? Explain.
2. **NUMBER SENSE** How can you tell whether a data set has an outlier?
3. **REASONING** Arrange the words to explain how to find a mean.

 the data values divide by the number of data values add then

Practice and Problem Solving

Describe the "average" of the data.

4. Ages in a class: 11, 12, 12, 12, 12, 12, 13
5. Movies seen this week: 0, 0, 0, 1, 1, 2, 4

Find the mean of the data.

6.
Pets Owned	
Brandon	I
Jill	III
Mark	II
Nicole	IIII
Steve	0

7.
Brothers and Sisters	
Amanda	♀
Eve	♀♀♀♀♀
Joseph	♂♂♂♂
Michael	♂♂

8.

9.

10. **GOLF** The table shows where Tiger Woods finished in tournaments in a recent year.

 a. What was his mean finish?
 b. Describe two outliers for the data.

Tiger Woods Finishes							
1	1	2	1	1	12	6	2
15	37	1	2	1	22	9	1

11. **COMMERCIALS** The table shows the lengths of the commercial breaks during a one-hour television show. What is the mean length of a commercial break?

Time (minutes)	4.2	3.5	4.55	2.75	2.25

214 Chapter 5 Ratios, Rates, and Data Analysis

Assignment Guide and Homework Check

Level	Day 1 Activity Assignment	Day 2 Lesson Assignment	Homework Check
Basic	4, 5, 18–22	1–3, 7, 9, 10, 11, 12	2, 7, 10, 12
Average	4, 5, 18–22	1–3, 7, 9, 11–14	2, 7, 12, 13
Advanced	4, 5, 18–22	1–3, 6, 8, 13, 15–17	2, 6, 13, 16

Common Errors

- **Exercises 6–9** Students may forget to count the total to divide by but instead use the maximum number. Remind them that the definition of mean is an *average* so they must take into account the total number of items or numbers to get an average. Explain to students that it is as if they are dividing the total evenly among the number of groups that there are. For example, the mean number of SUVs in a dealership describes how many SUVs each dealership would have if they all had the same number of SUVs.
- **Exercise 10** Students may have difficulty finding the outliers just by looking at the table. Ask them to place all the values on a number line. This will help make the two outliers more obvious.

5.4 Record and Practice Journal

Technology For the Teacher
Answer Presentation Tool
QuizShow

Vocabulary and Concept Check

1. No; Dividing the sum of the data by the number of data values to find the mean does not necessarily result in one of the data values.

2. If a data value is much greater than or much less than the other data values, it is an outlier.

3. Add the data values then divide by the number of data values.

Practice and Problem Solving

4. The "average" age is 12 years old, because most of the ages are 12.

5. The "average" could be 0 if you consider the most common value, or $\frac{8}{7}$ if you consider the mean.

6. 2 pets

7. 3 brothers and sisters

8. 6.5 seats

9. 16 visits

10. a. 7.125
 b. 37 and 22 are outliers because they are much greater than the other values.

11. 3.45 minutes

T-214

Practice and Problem Solving

12. 3.245 inches

13. *Sample answer:*
20, 21, 21, 21, 21, 22
20, 20.5, 20.5, 21.5, 21.5, 22

14. **a.** The 288 minutes used in September is much less than the other values, so it is an outlier.

b. With outlier: 488.4
Without outlier: 538.5
The outlier caused the mean to be about 50 minutes lower.

c. *Sample answer:* School could have caused you to spend less time talking on your cell phone.

15. 3.9 inches; No, neither team has a height that is much shorter or taller than the other heights. So, you can say that the Tigers are taller than the Dolphins on average.

16. See *Taking Math Deeper*.

17. See Additional Answers.

Fair Game Review

18. 9 **19.** 30
20. 18.5 **21.** 15.5
22. B

Mini-Assessment
Find the mean of the data set.

1. 2, 5, 7, 2, 4 4
2. 22, 29, 18, 31 25
3. 11, 15, 8, 14, 18, 10, 9, 19 13
4. 35, 56, 41, 59, 37, 48, 53 47
5. In your class, 7 students do not receive a weekly allowance, 8 students receive $8, 6 students receive $5, and 10 students receive $3. What is the mean weekly allowance? $4

Taking Math Deeper

Exercise 16
This is an important type of problem in statistics. The key to understanding this problem is to realize that each of 5 different allowance amounts are represented more than once.

 Make a table.

Students	Allowances	Total
7	$0	7 • 0 = $0
5	$3	5 • 3 = $15
7	$5	7 • 5 = $35
3	$6	3 • 6 = $18
2	$8	2 • 8 = $16

24 Students $84

 Summarize the table.
So, 24 students receive a weekly total of $84.

Mean: $\frac{84}{24} = \$3.50$

③ According to National Longitudinal Survey of Youth, allowance for 12–17 year olds varies. 50% reported receiving no regular allowance. However, the remaining 50% reported a median allowance of $50 a week. In spite of this, we kept the amounts low to avoid making those with no or low allowances feel bad.

Project
Assume you receive an allowance of $7.00 per week. Keep a budget for one week. In one week, how much money did you spend on food? Calculate the average amount of money you spent on food per day.

Reteaching and Enrichment Strategies

If students need help...	If students got it...
Resources by Chapter • Practice A and Practice B • Puzzle Time Record and Practice Journal Practice Differentiating the Lesson Lesson Tutorials Skills Review Handbook	Resources by Chapter • Enrichment and Extension • School-to-Work Start the next section

Month	Rainfall (inches)	Month	Rainfall (inches)
Jan	2.22	Jul	3.27
Feb	1.51	Aug	5.40
Mar	1.86	Sep	5.45
Apr	2.06	Oct	4.34
May	3.48	Nov	2.64
Jun	4.57	Dec	2.14

12. RAINFALL The table shows the monthly rainfall at a measuring station. What is the mean monthly rainfall?

13. OPEN-ENDED Create two different sets of data that have six values and a mean of 21.

14. CELL PHONE The bar graph shows your cell phone usage for five months.

 a. Which data value is an outlier? Explain.
 b. Find the mean with and without the outlier. Then describe how the outlier affects the mean.
 c. Describe a situation that could have caused the outlier in this problem.

15. HEIGHT The table shows the heights of the basketball players from two schools. What is the difference between the mean heights of the two teams? Do outliers affect either mean? Explain.

Player Height (inches)															
Dolphins	59	65	53	56	58	61	64	68	51	56	54	57			
Tigers	63	68	66	58	54	55	61	62	53	70	64	64	62	67	69

16. PRECISION In your class, 7 students do not receive a weekly allowance, 5 students receive $3, 7 students receive $5, 3 students receive $6, and 2 students receive $8. What is the mean weekly allowance? Explain how you found your answer.

17. Reasoning A collection of 8 backpacks has a mean weight of 14 pounds. A different collection of 12 backpacks has a mean weight of 9 pounds. What is the mean weight of the 20 backpacks? Explain how you found your answer.

Fair Game Review What you learned in previous grades & lessons

Evaluate the expression. *(Skills Review Handbook)*

18. $\dfrac{8+10}{2}$ 19. $\dfrac{26+34}{2}$ 20. $\dfrac{18+19}{2}$ 21. $\dfrac{14+17}{2}$

22. MULTIPLE CHOICE What is the best estimate of 26% of 38? *(Section 4.5)*

 Ⓐ 8 Ⓑ 10 Ⓒ 12 Ⓓ 15

5.5 Median, Mode, and Range

COMMON CORE STATE STANDARDS
6.SP.3
6.SP.5c

Essential Question Describe situations in real life where the mean is not a good representation of the average.

1 ACTIVITY: Comparing Three Samples

Work with a partner. Surveys are taken in three grade 6–12 schools. Make up a story about the three surveys. Find the mean for each survey. Do you think the mean is a good way to represent the "average" of each group? Why or why not?

a.

b.

c.
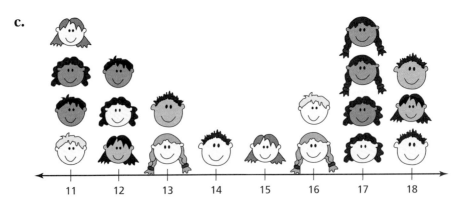

216 Chapter 5 Ratios, Rates, and Data Analysis

Laurie's Notes

Introduction

For the Teacher
- **Goal:** The goal of this activity is for students to develop an understanding of the need for other types of averages besides the mean.

Motivate
- ❓ "If you scored well on all of your tests except one, would you still want the mean to be reported as your average?" No. The test on which you scored poorly will act as an outlier and lower the mean.
- This question usually prompts students to say that the lowest grade should be dropped (never the highest grade though). This is a natural introduction to median and mode.

Activity Notes

Activity 1
- The last lesson presented the mean. Today students explore data sets where the mean is not a good choice in describing the data. The mean does not reflect the "typical" data value.
- This activity is open-ended. In each problem there are 20 data values ranging from 11 to 18. How the data is distributed over that range varies.
- ❓ "What affect did the outlier have on the mean?" pulled the mean towards the outlier
- **Discuss:** If an outlier is quite different from the rest of the data, the mean can be very different from all of the data values. It is no longer a "typical" piece of data, representative of the rest of the data.
- Be sure that students understand that each data set contains 20 values. To find the mean of each, sum the 20 values and divide by 20.
- If students need help thinking of a context for the data you could suggest one or two of the following:
 - age of students
 - number of push-ups done in one minute
 - number of movies watched since school began
 - length of hair (in inches) of 20 different girls
- Encourage students to work with a partner and think of interesting contexts for each data set.
- **Big Idea:** Students should recognize that the data is distributed differently in each problem. The first is somewhat bell-shaped and symmetric; the second is skewed to the right; the third has a U-shaped distribution.
- ❓ "What did you find for the mean of each?" 14.6, 15.65, 14.45
- ❓ "How well does the mean represent the data displayed?" In part (a), the mean works well. In parts (b) and (c), the mean does not work well.
- ❓ "Is it possible for the class average (mean) on a test to be around a C but everyone in the class did really well or really poorly on the test?" Yes
- Discuss the idea that sometimes the mean doesn't represent the typical value in a data set and that perhaps there are other types of averages.

Common Core State Standards

6.SP.3 Recognize that a measure of center for a numerical data set summarizes all of its values with a single number while a measure of variation describes how its values vary with a single number.

6.SP.5c Summarize numerical data sets in relation to their context by giving quantitative measures of center (median and/or mean) and variability (interquartile range and/or mean absolute deviation), as well as describing any overall pattern and any striking deviations from the overall pattern with reference to the context in which the data were gathered.

Previous Learning
Students should know how to find the mean.

Start Thinking! and Warm Up

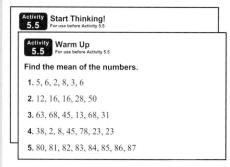

Activity 5.5 **Start Thinking!** For use before Activity 5.5

Activity 5.5 **Warm Up** For use before Activity 5.5

Find the mean of the numbers.
1. 5, 6, 2, 8, 3, 6
2. 12, 16, 16, 28, 50
3. 63, 68, 45, 13, 68, 31
4. 38, 2, 8, 45, 78, 23, 23
5. 80, 81, 82, 83, 84, 85, 86, 87

5.5 Record and Practice Journal

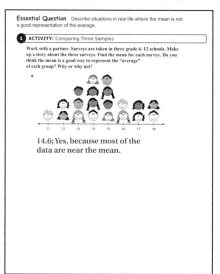

14.6; Yes, because most of the data are near the mean.

English Language Learners
Graphic Organizer
Provide small groups with different data sets and a copy of the organizer. Have each group find the mean, median, mode, and range of their data set and present their work to the class.

Data set:	
Mean	
Median	
Mode	
Range	

Laurie's Notes

Activity 2
- It is important for students to be critical readers, especially when statistics are quoted.
- The examples presented show the effect of an outlier on the mean. The calculations of the means are correct. The means do not reflect the typical data values. The statistics are misleading.
- The bar graphs are a visual model of the outliers.
- **Common Error:** Students may state the *number* of data values less than the mean instead of the *percent* less than the mean.
- Ask students if they have examples to share where the mean was misleading or it did not represent the data set well.

Closure
- Explain how an outlier can affect the mean. *Sample answer:* If the outlier is lower than the data, the mean will be lower. If the outlier is greater than the data, the mean will be higher.

5.5 Record and Practice Journal

15.65; No, most of the data are greater than the mean.

14.45; No, most values are not close to the mean.

2 ACTIVITY: When the Mean is Misleading

Work with a partner. Read and re-read each statement. Think of a better way to represent the "average" so that the statement is not so misleading.

a. Someone is trying to criticize a small high school by saying "Last year, the average age of the graduating class was 22 years old." When you look into the facts, you find that the class had a senior citizen who went back to school to earn a diploma. Here are the ages for the class.

18, 18, 18, 18, 18, 17, 18, 19, 18, 18, 18, 18, 18, 74

What percent of the ages are *below* the mean?

92.9%

b. There is a small town where most of the people are having a difficult time getting by because of low incomes. Someone is trying to ignore the problem and writes an article in the newspaper saying "It is not so bad in the town. The average income for a family is $52,000 a year." Here are the incomes.

$20,000, $20,000, $20,000, $20,000, $30,000, $30,000, $30,000, $30,000, $40,000, $40,000, $40,000, $50,000, $50,000, $50,000, $310,000

What percent of families have incomes *below* the mean?

93.3%

What Is Your Answer?

3. **IN YOUR OWN WORDS** Describe situations in real life where the mean is not a good representation of the average. What measures (other than mean) can you use to describe an average?

Check students' work.

The Dynamic Planning Tool
Editable Teacher's Resources at *BigIdeasMath.com*

2 ACTIVITY: When the Mean is Misleading

Work with a partner. Read and re-read each statement. Think of a better way to represent the "average" so that the statement is not so misleading.

a. Someone is trying to criticize a small high school by saying "Last year, the average age of the graduating class was 22 years old." When you look into the facts, you find that the class had a senior citizen who went back to school to earn a diploma. Here are the ages for the class.

18, 18, 18, 18, 18, 17, 18,
19, 18, 18, 18, 18, 18, 74

What percent of the ages are *below* the mean?

b. There is a small town where most of the people are having a difficult time getting by because of low incomes. Someone is trying to ignore the problem and writes an article in the newspaper saying "It is not so bad in the town. The average income for a family is $52,000 a year." Here are the incomes.

$20,000, $20,000, $20,000,
$20,000, $30,000, $30,000,
$30,000, $30,000, $40,000,
$40,000, $40,000, $50,000,
$50,000, $50,000, $310,000

What percent of the families have incomes *below* the mean?

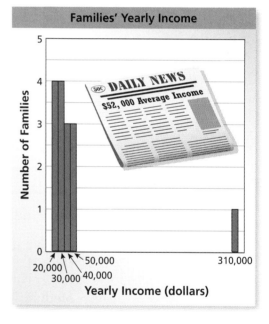

What Is Your Answer?

3. **IN YOUR OWN WORDS** Describe situations in real life where the mean is not a good representation of the average. What measures (other than mean) can you use to describe an average?

Practice — Use what you learned about the mean to complete Exercises 5 and 6 on page 220.

Section 5.5 Median, Mode, and Range

5.5 Lesson

Check It Out
Lesson Tutorials
BigIdeasMath*com*

A **measure of central tendency** is a measure that represents the *center* of a data set. The mean is one type of measure of central tendency. Here are two others.

Key Vocabulary
measure of central tendency, *p. 218*
median, *p. 218*
mode, *p. 218*
range, *p. 219*

🔑 Key Ideas

Median

Words Order the data. For a set with an odd number of values, the **median** is the middle value. For a set with an even number of values, the **median** is the mean of the two middle values.

Numbers Data: 5, 8, 9, 12, 14 The median is 9.

Data: 2, 3, 5, 7, 10, 11

The median is $\frac{5+7}{2}$, or 6.

Study Tip
The mode is the only measure of central tendency that can be used to describe a set of data that is *not* made up of numbers.

Mode

Words The **mode** of a data set is the value or values that occur most often. Data can have one mode, more than one mode, or no mode. When all values occur only once, there is no mode.

Numbers Data: 11, 13, 15, 15, 18, 21, 24, 24

The modes are 15 and 24.

EXAMPLE 1 **Finding the Median and Mode**

Bowling Scores

| 120 | 135 | 160 | 125 | 90 |
| 205 | 160 | 175 | 105 | 145 |

Find the median and mode of the bowling scores.

90, 105, 120, 125, 135, 145, 160, 160, 175, 205 Order the data.

Median: $\frac{135 + 145}{2} = \frac{280}{2}$, or 140 Add the two middle values and divide by 2.

Mode: 90, 105, 120, 125, 135, 145, 160, 160, 175, 205

The value 160 occurs most often.

∴ The median is 140. The mode is 160.

● On Your Own

Find the median and mode of the data.

1. 20, 4, 17, 8, 12, 9, 5, 20, 13
2. 100, 75, 90, 80, 110, 102

Laurie's Notes

Introduction

Connect
- **Yesterday:** Students developed an understanding that the mean is not always a good representative of a data set.
- **Today:** Students will find the median and mode for a data set.

Motivate
 Ask a few questions that don't necessarily need to be answered. The questions are intended to get students thinking about another type of average (mode).
- "What is the most common (average) color of car in the parking lot?"
- "What is the typical (average) eye color of students in the class?"
- "What is the typical (average) style of shoe worn by students in the class?"

Lesson Notes

Key Ideas
- Define **measure of central tendency**. The three types (mean, median, and mode) describe how the data *tends*, its center.
- Write the Key Ideas.
- Discuss the median and mode and how each is determined.
- **FYI:** The median housing value in the U.S. is $185,200 (U.S. Census Bureau 2006). Discuss with students that this means half of the houses have a value greater than the median and half have a value less than the median.
- To help students remember the difference between mean and median, think about the median on a highway.
- Note the Study Tip: The mode is the only measure of central tendency that can be used to describe non-numeric data sets such as the questions in the beginning of the lesson.

Example 1

 "What is the first step in finding the median?" order the data
- **Common Error:** Students forget to arrange the data in order before finding the median.
- **Note:** Sorting the data also makes it easier to determine the mode(s).
- **Connection:** Adding two middle values and dividing by 2 is the same as finding the mean of the middle two data values.
- If time permits, find the mean. Discuss the 3 measures of central tendency found and if all are representative of the data.
- **Extension:** "Is it possible for the median and the mode to be the same?" yes

On Your Own
- Ask students to explain how to find the median for a data set.

Goal Today's lesson is finding the **median** and **mode** and identifying which **measure of central tendency** best describes a data set.

Start Thinking! and Warm Up

Lesson 5.5 Warm Up
For use before Lesson 5.5

Lesson 5.5 Start Thinking!
For use before Lesson 5.5

The **mode** of a data set is the value that occurs most often.

Do you think it's possible for a set of data to have no mode? If so, give an example. If not, explain why not.

Do you think it's possible for a set of data to have more than one mode? If so, give an example. If not, explain why not.

Extra Example 1
Find the median and mode of the data.
95, 80, 85, 86, 89, 100, 95
median: 89, mode: 95

On Your Own
1. median: 12; mode: 20
2. median: 95; no mode

Extra Example 2

The list shows the favorite colors for students in a class. Organize the data in a frequency table. Then find the mode.

blue, green, purple, blue, green, pink, purple, blue, green, purple, blue, blue, green, pink, blue, blue, green, blue, green

Color	Tally	Frequency
Blue	JHT III	8
Green	JHT I	6
Pink	II	2
Purple	III	3

Mode: Blue

Extra Example 3

Five of the world's tallest buildings are Taipei 101 Tower (509 meters), Shanghai World Financial Center (492 meters), Petronas Towers (452 meters), Sears Tower (442 meters), and Jin Mao Building (421 meters). What is the range of the heights? 88 meters

On Your Own

3. Yes; Comedy and horror are both modes now that horror has the same number of votes as comedy.

4. 0.8 in.

Differentiated Instruction

Visual
Some students may incorrectly find the median by counting the number of values and dividing by 2. Have students write the numbers in increasing order. Then cross the numbers out in pairs, one on the left end and one on the right end, until 1 or 2 numbers remain. If there is one number remaining, the median is that number. If two numbers remain, then the median is the mean of those two numbers.

Laurie's Notes

Example 2

- **Common Error:** When making the frequency table students can forget to make a tally for the data on the first entry; for instance, in this example they would write comedy as the first type of movie and then forget to make a tally mark for that entry.

Example 3

- The three measures thus far describe the center of the data – its tendency. The **range** describes how spread out the data are.
- "Does the data have to be sorted?" No; however, it makes it easier to find the least and greatest!
- Work through the example.
- **Extension:** "Does the range have any affect on the median or the mode?" No.

On Your Own

- Students need to order and compare decimals.

Closure

- **Writing Prompt:** To determine the median of one dozen data values, ...
 Sample answer: Start by sorting the data values. Then find the mean of the two middle values by adding them and dividing by two.

The Dynamic Planning Tool
Editable Teacher's Resources at *BigIdeasMath.com*

EXAMPLE 2 Finding the Mode

The list shows the favorite types of movie for students in a class. Organize the data in a frequency table. Then find the mode.

Favorite Types of Movies

Comedy	Drama	Horror
Horror	Drama	Horror
Comedy	Comedy	Action
Action	Comedy	Action
Horror	Drama	Comedy
Comedy	Comedy	Horror
Horror	Comedy	Action
Horror	Action	Drama

Type	Tally	Frequency							
Action						5			
Comedy									8
Drama						4			
Horror								7	

The number of tally marks is the frequency.

Make a tally for each vote.

Comedy received the most votes.

∴ So, the mode is comedy.

On Your Own

Exercises 7–15

3. In Example 2, one member of the class was absent and ends up voting for horror. Does this change the mode? Explain.

The **range** of a data set is the difference between the greatest value and the least value. The range describes how spread out the data are.

EXAMPLE 3 Standardized Test Practice

The table shows the record lengths of six venomous snakes. What is the range of the lengths?

Snake	Record Length (inches)
Copperhead	53
Cottonmouth	74.5
Diamondback rattlesnake	96
Timber rattlesnake	74.5
Pygmy rattlesnake	31
Coral snake	47.5

Ⓐ 43.5 inches Ⓑ 48.5 inches

Ⓒ 65 inches Ⓓ 74.5 inches

31, 47.5, 53, 74.5, 74.5, 96 Order the data from least to greatest.

The least value is 31. The greatest value is 96. So, the range is 96 − 31, or 65 inches.

∴ The correct answer is Ⓒ.

On Your Own

4. The data set shows how many inches of rain fell in several cities during a storm.

3.9, 4.25, 4.1, 3.7, 3.8, 4.3, 3.8, 4.5

Find the range of the data.

5.5 Exercises

Vocabulary and Concept Check

1. **NUMBER SENSE** Give an example of a data set that has no mode.
2. **WRITING** Which is affected more by an outlier: the mean or the median? Explain.
3. **REASONING** What two data values do you need to know to find the range? Explain.
4. **NUMBER SENSE** A data set has a mean of 7, a median of 5, and a mode of 8. Which of the numbers 7, 5, and 8 *must* be in the data set? Explain.

Practice and Problem Solving

Find the mean of the data. Is the mean a good "average"? If not, what would be a better "average"? Explain.

5. 8, 29, 30, 30, 30
6. 16, 24, 13, 66, 22, 26, 22, 28

Find the median, mode(s), and range of the data.

7. 3, 5, 7, 9, 11, 3, 8
8. 14, 19, 16, 13, 16, 14
9. 93, 81, 94, 71, 89, 92, 94, 99
10. 44, 13, 36, 52, 19, 27, 33
11. 12, 33, 18, 28, 29, 12, 17, 4, 2
12. 55, 44, 40, 55, 48, 44, 58, 67

13. **ERROR ANALYSIS** Describe and correct the error in finding the median and range of the data.

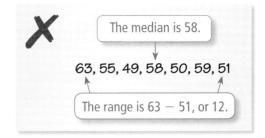

Find the mode(s) of the data.

14.
Shirt Color		
Black	Blue	Red
Pink	Black	Black
Gray	Green	Blue
Blue	Blue	Red
Yellow	Blue	Blue
Black	Orange	Black
Black		

15.
Talent Show Acts		
Singing	Dancing	Comedy
Singing	Singing	Dancing
Juggling	Dancing	Singing
Singing	Poetry	Dancing
Comedy	Magic	Dancing
Poetry	Singing	Singing

16. **REASONING** In Exercises 14 and 15, can you find the mean and median of the data? Explain.

17. **REASONING** How does an outlier affect the range of a data set? Explain.

220 Chapter 5 Ratios, Rates, and Data Analysis

Assignment Guide and Homework Check

Level	Day 1 Activity Assignment	Day 2 Lesson Assignment	Homework Check
Basic	5, 6, 28–30	1–4, 7–17 odd, 14, 16	2, 7, 14, 16
Average	5, 6, 28–30	1–4, 7–15 odd, 14, 16, 23–25	2, 7, 14, 23
Advanced	5, 6, 28–30	1–4, 13–16, 18–21, 25–27	2, 14, 19, 25

Common Errors

- **Exercises 7–12** Students may try to identify the median without ordering the data first. Remind them that it is essential to order the data first and then find the median. This makes finding the mode easier as well.
- **Exercises 14 and 15** Students may make an educated guess of which color or act appears the most instead of organizing the data. Remind them of the importance of organizing data and how it can help answer other questions, not just what is the mode, but something like "how many students wore black?"
- **Exercises 19–21** Students may have difficulties ordering the data before finding the median because the numbers are harder to work with. Remind them to take one step at a time. Tell students to write the numbers as they are and then cross out the numbers as they re-write them in order from least to greatest.

Technology For the Teacher
Answer Presentation Tool
Quiz*Show*

5.5 Record and Practice Journal

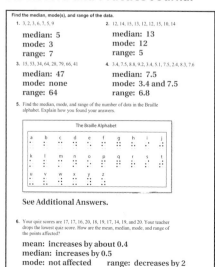

See Additional Answers.

Vocabulary and Concept Check

1. *Sample answer:* 1, 2, 3, 4, 5, 6

2. It depends on the data. The mean is often affected more by an outlier when the middle values of the ordered data are relatively close, but if they are not close, the median may be affected more.

3. The greatest and least data values. The range is the difference of these values.

4. The number 8 must be in the data set at least twice, because the mode is the data value that occurs most often.

Practice and Problem Solving

5. 25.4; No, the mode would be better because most of the data values are the mode of 30.

6. 27.125; No, either the median or mode would be better because each is closer to more of the data values.

7. median: 7; mode: 3; range: 8

8. median: 15; modes: 14, 16; range: 6

9. median: 92.5; mode: 94; range: 28

10. median: 33; no mode; range: 39

11. median: 17; mode: 12; range: 31

12. median: 51.5; modes: 44, 55; range: 27

13. The data was not ordered from least to greatest; The median is 55.
49, 50, 51, 55, 58, 59, 63
The range is 63 − 49 = 14.

T-220

Practice and Problem Solving

14. black, blue

15. singing

16. No, only the mode can describe a set of data that is not made up of numbers.

17. *Sample answer:* An outlier increases the range of a data set because there is a wider spread between the greatest and least value.

18. mean: 16 years; median: 17 years; no mode; range: 31 years

19. mean: 14.9; median: 17.5; modes: 5, 19; range: 28

20. mean: 76; median: 76; no mode; range: 15

21. mean: 7.61; median: 7.42; no mode; range: 4.94

22–24. See Additional Answers.

25. See *Taking Math Deeper*.

26–27. See Additional Answers.

Fair Game Review

28. Outlier: 83
 With outlier: 29
 Without outlier: 23

29. Outlier: 35
 With outlier: 62.125
 Without outlier: 66

30. C

Mini-Assessment

Find the median, mode(s), and range of the data.

1. 3, 14, 3, 17, 7 7, 3, 14

2. 5, 12, 4, 8, 5 5, 5, 8

3. 56, 27, 35, 27, 44, 25 31, 27, 31

4. 65, 2, 23, 41, 11, 7, 55 23, no mode, 63

5. The data are your test scores. Find the median, mode, and range of the data.
 92, 85, 95, 87, 82, 96 89.5, no mode, 14

Taking Math Deeper

Exercise 25

This problem shows how the change of a single number in a data set can affect the mean, median, mode, and range. Of these, the mode is the least likely to be affected.

 Measures with the 30-year-old:

14, 14, 17, 20, 21, 22, 25, 25, 26, 27, 29, 30

Mean: $\frac{270}{12} = 22.5$ years

Median: $\frac{22 + 25}{2} = 23.5$ years

Modes: 14 and 25

Range: $30 - 14 = 16$ years

 Measures with the 15-year-old:

14, 14, 15, 17, 20, 21, 22, 25, 25, 26, 27, 29

Mean: $\frac{255}{12} = 21.25$ years

Median: $\frac{21 + 22}{2} = 21.5$ years

Modes: 14 and 25

Range: $29 - 14 = 15$ years

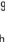 Compare the measures.

The mean, median, and range were lowered. The modes are unchanged.

Project

Research motocross racing in the United States. Write a report that shows when it started, costs involved, number of participants, and other interesting facts.

Reteaching and Enrichment Strategies

If students need help...	If students got it...
Resources by Chapter • Practice A and Practice B • Puzzle Time Record and Practice Journal Practice Differentiating the Lesson Lesson Tutorials Skills Review Handbook	Resources by Chapter • Enrichment and Extension • School-to-Work • Financial Literacy Start the next section

18. **SUPREME COURT** The data show the years of service of the justices of the Supreme Court as of 2008. Find the mean, median, mode(s), and range of the data.

 2, 14, 15, 20, 3, 22, 18, 33, 17

Find the mean, median, mode(s), and range of the data.

19. 5, 13, 24, 28, 5, 19, 20, 16, 0, 19

20. 77, 73, 75, 78, 79, 72, 74, 68, 81, 83, 76

21. 4.7, 8.51, 6.5, 7.42, 9.64, 7.2, 9.3

22. $8\frac{1}{2}, 6\frac{5}{8}, 3\frac{1}{8}, 5\frac{3}{4}, 6\frac{5}{8}, 5\frac{1}{4}, 10\frac{5}{8}, 4\frac{1}{2}$

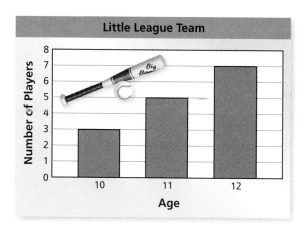

23. **BASEBALL** Find the median, mode, and range of the ages of players on the Little League team. Explain how you found your answers.

24. **HOMEWORK** Write down the number of minutes you spent on homework each day of the past week.

 a. Find the mean, median, mode(s), and range of the data.

 b. How do you think your success as a student is related to the median of your data? the mode? the range?

25. **MOTOCROSS** The ages of the racers in a bicycle motocross race are 14, 22, 20, 25, 26, 17, 21, 30, 27, 25, 14, and 29. The 30-year-old drops out of the race and is replaced with a 15-year-old. How are the mean, median, mode, and range of the ages affected?

26. **REASONING** The median of the data is 37 and the range is 29. What is the missing value? Explain how you found your answer.

 48, 56, 29, ___, 42, 27

27. Write a set of data with six values that has a mean of 28, a median of 29, and a range of 18.

Fair Game Review What you learned in previous grades & lessons

Identify the outlier. Find the mean of the data set with and without the outlier. *(Section 5.4)*

28. 22, 25, 17, 27, 29, 83, 26, 20, 20, 21

29. 64, 71, 58, 60, 65, 35, 70, 74

30. **MULTIPLE CHOICE** A parking lot contains 18 vans, 26 SUVs, and 24 cars. What is the ratio of vans to the total number of vehicles? *(Section 5.1)*

 Ⓐ $\frac{9}{25}$ Ⓑ $\frac{8}{17}$ Ⓒ $\frac{9}{34}$ Ⓓ $\frac{3}{4}$

5.6 Analyzing Data Sets

Essential Question How can you use tables and graphs to help organize data?

COMMON CORE STATE STANDARDS
6.SP.4
6.SP.5d

1 ACTIVITY: Conducting an Experiment

Work with a partner.

a. Roll a number cube 20 times. Record your results in a tally chart.

b. Make a bar graph of the totals.

c. Go to the board and enter your totals in the class tally chart.

d. Make a second bar graph showing the class totals. Compare and contrast the two bar graphs.

Tally Chart

1	
2	
3	
4	
5	
6	

Key: | = 1 ||||| = 5

2 ACTIVITY: Organizing Data

Work with a partner. You are judging a paper airplane contest. Each contestant flies his or her paper airplane 20 times. Make a tally chart and a graph of the distances.

Sample:
20.5 ft, 24.5 ft, 18.5 ft, 19.5 ft, 21.0 ft, 14.0 ft, 12.5 ft, 20.5 ft, 17.5 ft, 24.5 ft,
19.5 ft, 17.0 ft, 18.5 ft, 12.0 ft, 21.5 ft, 23.0 ft, 13.5 ft, 19.0 ft, 22.5 ft, 19.0 ft

Tally Chart

Interval	Tally	Total					
10.0–12.9	II	2					
13.0–15.9	II	2					
16.0–18.9	IIII	4					
19.0–21.9						III	8
22.0–24.9	IIII	4					

a. Make a different tally chart and graph of the distances using the following intervals.

10.0–11.9, 12.0–13.9, 14.0–15.9, 16.0–17.9, 18.0–19.9, 20.0–21.9, 22.0–23.9, 24.0–25.9

b. Which graph do you think represents the distances better? Explain.

Laurie's Notes

Introduction

For the Teacher
- **Goal:** The goal of this activity is for students to develop an understanding of how tables (charts) and graphs help to organize data so that features of the data become more apparent.

Motivate
- Ask if students have ever made a paper airplane. If so, how long did their plane usually stay in flight before crashing?
- Ken Blackburn is an aerospace engineer and the Guinness World Record holder for the longest paper airplane flight time (27.60 seconds).

Activity Notes

Activity 1
- ❓ "When you roll a standard number cube, what are the possible outcomes?" 1, 2, 3, 4, 5, 6
- Students should work with a partner. One could roll and the other record. Switch roles after 20 tallies.
- Discuss part (d). Depending upon the number of students, the class results should be close to the theoretical probability for this event, whereas each student's results may not be.
- ❓ "Was it helpful to use a chart and tally marks to keep track of the results from your experiment? Explain." Answers will vary.

Activity 2
- Discuss the context of this problem. Ask for the range of the data. Compare the range to the distances in the classroom to help students learn to judge distances.
- Ask students to compare the column of tally marks in the chart with the graph at the right known as a histogram. Some might observe that if the tally marks are rotated 90° counterclockwise, the shape resembles the histogram.
- ❓ "How are these intervals different from the sample problem?" Smaller; these intervals have a width of 1.9 versus 2.9.
- ❓ "How will the chart and graph be different?" It's ok for students not to have an answer – it's a big idea!
- **Big Idea:** When the interval is smaller, you have *more* intervals and the tallies become more *spread out*, meaning the number of tallies in each interval will be smaller.
- ❓ "Looking only at the tally chart or graph, did the person fly their airplane 20.0 feet?" cannot determine
- **Big Idea:** Once the tallies or graph have been made, the actual identity of the data is lost. Only in looking at the raw data can the question above be answered. The interval 19.0–21.9 contains 20.0 but you do not know if it is actually one of the data values.

Common Core State Standards
6.SP.4 Display numerical data in plots on a number line, including dot plots, histograms, and box plots.
6.SP.5d Summarize numerical data sets in relation to their context by relating the choice of measures of center and variability to the shape of the data distribution and the context in which the data were gathered.

Previous Learning
Students know how to find the mean, median, and mode and describe the range of a data set.

Activity Materials
Textbook
- scrap paper
- measuring tapes

Start Thinking! and Warm Up

5.6 Record and Practice Journal

T-222

Differentiated Instruction

Auditory

Some students may benefit by hearing how to construct a bar graph. Ask students questions and have them explain how they got their answers. Should you draw a horizontal or vertical bar graph? What increments should you use? What scale should you use and where should it start and end? What is the title of the bar graph?

5.6 Record and Practice Journal

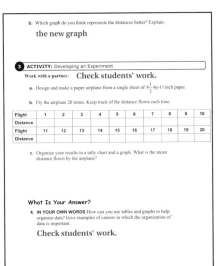

Laurie's Notes

Activity 3

- This activity allows students to collect their own data. It is always more engaging and interesting for students to work with data they generated.
- Use scrap paper and allow only one sheet per student. They need to think before they fold!
- **Management Tip:** Use tape measures and lay them out in a parallel fashion about 6 feet apart on the soccer field, or other school grounds.
- Two students at a time fly their planes. They must make a quick read of the distance to the nearest foot and call out their distance to their partner who records it.
- Review how to round to the nearest foot before going outside, if necessary.
- Students continue in pairs until all have completed 20 trials. The recorder and pilot switch roles and repeat the process.
- If the planes hook left or right too far and make a reading difficult, use your judgment to call out a quick measurement.
- **Safety:** Students should fly their planes in the same direction so that no one is accidentally hit in the face.
- **Management Tip:** You may choose to complete this activity inside. If so, it is helpful to mark the floor prior to the start of the activity. For instance, have students measure and mark distances 2 feet apart. When students test their planes, they only have to measure between the marks. This will allow students to review their measuring skills as well as make the activity more time efficient.
- Once the data is collected, complete part (c) inside. To save time, do not have students do any conversions to total inches. The intervals are generally a width of about 2 feet.
- Students have fun folding and flying the airplane, but don't forget the mathematics. Raw data does not tell much of a story until it is organized in some fashion.
- ? "Where are the mean, median, and mode located in the chart and graph?"
- **Extension:** You are the executive of a paper airplane business. Use the charts and graphs to decide which airplane you would put into production and explain the reasoning of your choice. Answers will vary.

Closure

- Explain how the intervals in a tally chart help you organize data. The interval with the most tallies helps you determine the most common answer (the mode).

Technology For the Teacher

The Dynamic Planning Tool
Editable Teacher's Resources at *BigIdeasMath.com*

3 ACTIVITY: Developing an Experiment

Work with a partner.

a. Design and make a paper airplane from a single sheet of $8\frac{1}{2}$-by-11-inch paper.

Sample:

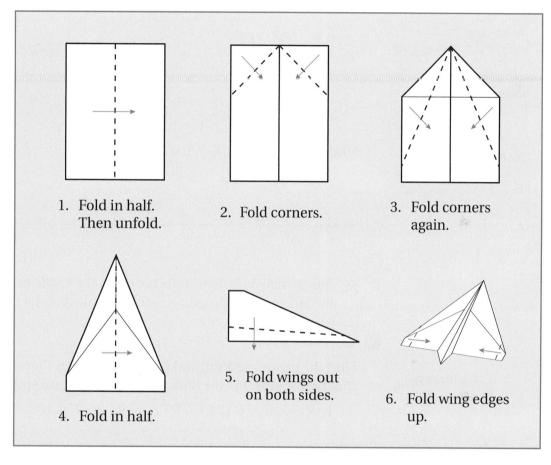

1. Fold in half. Then unfold.
2. Fold corners.
3. Fold corners again.
4. Fold in half.
5. Fold wings out on both sides.
6. Fold wing edges up.

b. Fly the airplane 20 times. Keep track of the distance flown each time.

c. Organize your results in a tally chart and a graph. What is the mean distance flown by the airplane?

What Is Your Answer?

4. **IN YOUR OWN WORDS** How can you use tables and graphs to help organize data? Give examples of careers in which the organization of data is important.

Practice

Use what you learned about organizing data to complete Exercises 4 and 5 on page 226.

5.6 Lesson

EXAMPLE 1 Choosing the Best Measure of Central Tendency

Find the mean, median, and mode of the sneaker prices. Which measure best represents the data?

Mean: $\dfrac{20 + 31 + 122 + 48 + 37 + 20 + 45 + 65}{8} = \dfrac{388}{8}$, or 48.5

Median: 20, 20, 31, 37, 45, 48, 65, 122 Order from least to greatest.

$$\dfrac{37 + 45}{2} = \dfrac{82}{2}, \text{ or } 41$$

Mode: 20, 20, 31, 37, 45, 48, 65, 122 The value 20 occurs most often.

∴ The median best represents the data. The mode is less than most of the data and the mean is greater than most of the data.

On Your Own

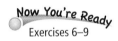
Exercises 6–9

Find the mean, median, and mode of the data. Choose the measure that best represents the data. Explain your reasoning.

1. 1, 93, 46, 48, 34, 194, 67, 55
2. 96, 150, 102, 87, 150, 75

EXAMPLE 2 Removing an Outlier

Identify the outlier in Example 1. Find the mean, median, and mode without the outlier. Which measure does the outlier affect the most?

The price of $122 is much greater than any other price. So, it is the outlier.

	Mean	Median	Mode
With Outlier (Example 1)	48.5	41	20
Without Outlier	38	37	20

∴ The mean is affected the most by the outlier.

Laurie's Notes

Introduction

Connect
- **Yesterday:** Students developed an understanding of how tables (charts) and graphs can help to organize data so that features of the data become more apparent.
- **Today:** Students will explore how to decide which measure of central tendency best represents a data set.

Motivate
- **FYI:** The average daily temperature for January in Cleveland, OH is approximately 26°F.
- ❓ "Could the temperature in Cincinnati, OH in January be 60°F?" yes "Would this be the typical January temperature?" no

Lesson Notes

Example 1
- **Discuss:** When you shop for sneakers, there is a range of prices in the store. How do you decide what the *typical* price of sneakers is?
- ❓ Work through the first example, asking the following questions.
 - "How do you find the mean for a set of data?" Add the data values and divide by the total number of data values.
 - "How do you find the median for a set of data?" Order the data and find the mean of the middle values.
 - "How do you find the mode?" Find the most frequent data value.
- Note that by sorting the data first, the mode is easier to identify.
- ❓ "What is the range for this data set?" Range = 122 − 20 = 102.
- In analyzing the data, ask why the mean is so high.
- ❓ "What affect did the outlier have on the data?" pulls the mean up

On Your Own
- **Neighbor Check:** Have students work independently and then have their neighbor check their work. Have students discuss any discrepancies.
- Have students share their reasoning about their choice.

Example 2
- ❓ "Which of the three measures does the outlier impact the most?" mean
- ❓ "If $122 is not the *typical* price for a pair of sneakers, would it be okay to eliminate it from the data set?" Yes. If you're trying to describe the typical price, eliminating the outlier is fine.
- Computing the mean, median, and mode should be efficient due to the work previously done.
- ❓ "Why does the median change?" There are an odd number of data values now and the middle two were not the same in the original set of data.
- **Big Idea:** The mean of a data set is affected the most by an outlier.
- ❓ "If the outlier of a data set is a least value, what effect does it have on the mean?" pulls the mean down

Goal Today's lesson is identifying which measure of central tendency best describes a data set.

Start Thinking! and Warm Up

Use the results from your survey question in the Start Thinking! for use before Activity 5.6. Calculate the mean, median, and mode of the data. Which measure of central tendency best describes the data? Explain why.

Extra Example 1
Find the mean, median, and mode of the data. Choose the measure that best represents the data. Explain your reasoning.
85, 70, 60, 10, 89, 85, 63
Mean: 66, median: 70, mode: 85
Sample: The mean or the median could be used to represent the data. The mode is greater than most of the data.

On Your Own
1. mean: 67.25; median: 51.5; no mode
 The median best represents the data. It is closer to more of the data because the mean is influenced by outliers.
2. mean: 110; median: 99; mode: 150
 The median best represents the data because the mode is the greatest data value and the mean is also greater than most of the data.

Extra Example 2
Identify the outlier in Extra Example 1. Find the mean, median, and mode without the outlier. Which measure does the outlier affect the most? outlier: 10; mean: $75.\overline{3}$, median: 77.5, mode: 85
The mean is affected most by the outlier.

 On Your Own

3. 45 is the outlier.
 With outlier: mean: 18; median: 12.5; mode: 10
 Without outlier: mean: 12.6; median: 10; mode: 10
 The mean is affected most by the outlier.

Extra Example 3

For six days you inline skate for 2 miles, 3 miles, 2 miles, 4 miles, 5 miles, and 3 miles. How many miles do you need to skate on the seventh day so that your mean for the week is 3 miles a day? 2 miles

On Your Own

4. 5 miles

English Language Learners

Vocabulary
English learners may need help understanding the word *scale*. There are several meanings in the English language.
 a series of musical notes
 the covering of a reptile
 a device for weighing
 a ratio
 to climb
In bar graphs the scale is a series of markings used for measuring. Most scales start at 0 and go to (at least) the greatest value of the data.

Laurie's Notes

On Your Own
- This question will take time for students to complete. They could work with a partner.

Example 3
- This example models a very common question for students: "What do I need to get on my test to have an A average?" The example also models the test taking strategy of *eliminating choices*.
- Discuss the reasoning aspect of the problem that is needed after the mean of the first four seasons is found.
- **Alternative Method:** Another way to solve this example would be to multiply the desired mean (400) by the number of seasons (5). You then subtract the number of free throws for the first four seasons $(277 + 356 + 390 + 529)$.

$$(400 \cdot 5) - (277 + 356 + 390 + 529) = 448$$

On Your Own
- Students need to read a pictograph and recognize the scale.
- They also need to reason that one week is 7 days and if you want the mean daily number of miles to be 5, you would need a total of 35 miles.

Closure
- **Exit Ticket:** Determine the mean, median, and mode for the data set 10, 20, 15, 15, 45. Which measure best describes the data and why? mean = 21; median = 15; mode = 15. The best answer will depend upon the context of the data set. Challenge students to create a scenario in which the median would be the best measure, and one in which the mean would be the best measure.

The Dynamic Planning Tool
Editable Teacher's Resources at *BigIdeasMath.com*

On Your Own

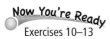
Exercises 10–13

3. The times (in minutes) it takes six students to travel to school are 8, 10, 10, 15, 20, and 45. Identify the outlier. Find the mean, median, and mode with and without the outlier. Which measure does the outlier affect the most?

EXAMPLE 3 Standardized Test Practice

The bar graph shows the free throws made by a basketball player in the last four seasons. How many free throws does he need to make in Season 5 to have a mean of 400 for the five seasons?

- **A** 388
- **B** 396
- **C** 428
- **D** 448

Find the mean of the four seasons.

Mean of four seasons: $\dfrac{277 + 356 + 390 + 529}{4} = \dfrac{1552}{4}$, or 388

A mean of 400 is an increase. So, he needs to make more than 400 free throws in Season 5. You can eliminate choices A and B.

Try choice C.

Choice C: $\dfrac{277 + 356 + 390 + 529 + 428}{5} = \dfrac{1980}{5}$, or 396 ✗

∴ You have eliminated choices A, B, and C. The correct answer is **D**.

On Your Own

4. The pictograph shows the number of miles you ride on your bike for 6 days. How many miles do you need to bike on Sunday so that your mean for the week is 5 miles per day?

5.6 Exercises

Vocabulary and Concept Check

1. **NUMBER SENSE** Which is most affected by an outlier: the mean, median, or mode? Explain.

2. **NUMBER SENSE** The median of a data set stays the same after an outlier is removed. How is this possible?

3. **OPEN-ENDED** Give an example of a data set that is best represented by the mode.

Practice and Problem Solving

Make a tally chart and a bar graph of the data.

4. 1, 1, 1, 4, 3, 1, 2, 2, 2, 4, 1, 2, 1

5. 7, 10, 9, 8, 9, 9, 7, 6, 9, 9, 8, 8, 9, 9, 10

Find the mean, median, and mode(s) of the data. Choose the measure that best represents the data. Explain your reasoning.

 6. 48, 12, 11, 45, 48, 48, 43, 32

7. 12, 13, 40, 95, 88, 7, 95

8. 2, 8, 10, 12, 56, 9, 5, 2, 4

9. 126, 62, 144, 81, 144, 103

Find the mean, median, and mode(s) of the data with and without the outlier. Describe the effect of the outlier on the measures of central tendency.

 10. 45, 52, 18, 63, 57

11. 65, 72, 180, 72, 85, 59, 64, 67

12. 6, 12, 5, 8, 15, 5, 4, 7, 16, 80

13. 85, 71, 210, 88, 91, 84, 85

14. **WEATHER** The weather forecast for a week is shown.

	Sun	Mon	Tue	Wed	Thu	Fri	Sat
High	90° F	91° F	89° F	97° F	101° F	99° F	91° F
Low	74° F	78° F	77° F	77° F	83° F	78° F	72° F

a. Find the mean, median, and mode(s) of the high temperatures. Which measure best represents the data? Explain your reasoning.

b. Repeat part (a) for the low temperatures.

15. **CEREAL** Go to a grocery store and find the unit cost of ten different kinds of cereal. Choose one cereal whose unit cost will be an outlier.

a. Find the mean, median, and mode(s) of the data. Which measure best represents the data? Explain your reasoning.

b. Identify the outlier in the data set. Find the mean, median, and mode(s) of the data set without the outlier. Which measure did the outlier affect the most?

Assignment Guide and Homework Check

Level	Day 1 Activity Assignment	Day 2 Lesson Assignment	Homework Check
Basic	4, 5, 20–24	1–3, 7–15 odd, 14, 16	7, 11, 14, 16
Average	4, 5, 20–24	1–3, 7, 9, 10–16 even, 17	7, 10, 14, 16
Advanced	4, 5, 20–24	1–3, 10–13, 16–19	10, 12, 16, 18

Common Errors

- **Exercises 6–9** Students may only find the mean, median, and mode instead of putting all the numbers on a number line before deciding which measure of central tendency is the best representative. Even if students are correct, encourage them to draw the number line to check their answer. Creating a visual of the data helps the student to understand and it is also a way for them to explain why that measure of central tendency is the best representative.
- **Exercises 10–13** When finding the mean for the data after eliminating the outlier, students may forget to subtract one from the denominator. Remind them that when the outlier is removed, there is one less data value so you must divide by a lesser number.

Technology For the Teacher
Answer Presentation Tool
QuizShow

5.6 Record and Practice Journal

Vocabulary and Concept Check

1. The mean is likely to be most affected unless the middle values are far apart, because then the median can be affected more by removing the outlier. The mode is unaffected unless the mode consists of outliers.

2. There are two identical values at the middle of the data.

3. *Sample answer*: 190, 190, 190, 190, 192, 192, 195, 195

Practice and Problem Solving

4.

Number	Tally	Total						
1								6
2						4		
3			1					
4				2				

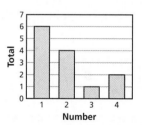

5. See Additional Answers.

6. mean: 35.875
 median: 44
 mode: 48
 Median is probably best, because it is close to most of the data. The mean is less than most of the data and the mode is the greatest value.

7–15. See Additional Answers.

T-226

Practice and Problem Solving

16. 12 hours; Use the mean of 10 hours and the numbers of hours in the first six weeks to work backwards.

17. a. No; The price is the mode, but it is the lowest price. Most cameras cost more.

 b. By advertising the lowest price, they are likely to draw more customers to the store.

 c. Knowing all the measures can help you to know whether the store has many models in your price range.

18. See *Taking Math Deeper*.

19. The mean, median, and mode increase by 2, but the range does not change.

Fair Game Review

20. 25.12 **21.** 5.2

22. 18.97 **23.** 55.2

24. B

Mini-Assessment

The data are the number of times you win playing a video game.

Mon	Tues	Wed	Thu	Fri	Sat
2	10	12	14	2	11

a. Find the mean, median, and mode(s) of the data. Choose the measure that best represents the data. Explain your reasoning. 8.5, 10.5, 2; The median best represents the data. The mean and mode are less than most of the data.

b. Suppose you win 33 times on Sunday. Find the mean, median, and mode(s) of the data with this outlier. Which measure does the outlier affect the most? 12, 11, 2; The mean is affected the most by the outlier.

T-227

Taking Math Deeper

Exercise 18

This problem gives students a chance to work with a large data set. From this experience, they can see that the median and mode of a large set is not so difficult to find. However, the mean is difficult because it requires adding a lot of numbers.

① List the data. Find the mode.

0, 0, 0, 0, 0, 0, 0, 0, 0, 0, 0, 0, 1, 1, 1, 1, 2, 2, 3, 3, 5, 9, 9, 15, 18, 28, 33, 35, 45, 46, 54, 58, 105, 107, 113, 127, 138, 155, 188, 212

 a. The mode is 0. It is not a good representation for the hits of team players. However, it is a good representation for the hits of the pitchers.

② Find the mean and median.

Use a calculator → Hits per season

$$\text{Mean} = \frac{1514}{40} = 37.85 \text{ hits}$$

$$\text{Median} = \frac{3+5}{2} = 4 \text{ hits}$$

 b. The mean is more useful because it is at least near some of the values that represent hits by non-pitchers.

③ c. Find the average hits per game.

$$\text{Mean} = \frac{1514 \text{ hits}}{162 \text{ games}} = 9.35 \text{ hits per game}$$

Project

Do research to determine the number of hits per player last year for two professional baseball teams. Which team had the greater number of hits per player? Justify your answer. Did the same team have the better win/loss record? What conclusions can you draw?

Reteaching and Enrichment Strategies

If students need help...	If students got it...
Resources by Chapter • Practice A and Practice B • Puzzle Time Record and Practice Journal Practice Differentiating the Lesson Lesson Tutorials Skills Review Handbook	Resources by Chapter • Enrichment and Extension • School-to-Work • Financial Literacy • Technology Connection Start the next section

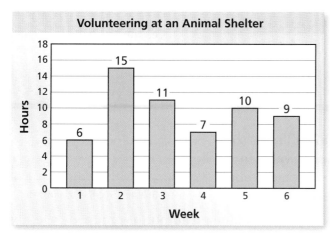

16. VOLUNTEERING The bar graph shows the numbers of hours you volunteered at an animal shelter. You wanted to work a mean of 10 hours for the 7 weeks. How many hours do you have to work in the 7th week to meet your goal? Explain how you found your answer.

17. CAMERAS The data are the prices of several digital cameras at a store.

$130 $170 $230 $130 $250 $275
$130 $185

 a. Does the price shown in the advertisement represent the prices well? Explain.
 b. Why might the store use this advertisement?
 c. In this situation, why might a person want to know the mean? the median? the mode? the range? Explain.

18. BASEBALL The data show the number of hits for each player on a baseball team in a recent season. The red numbers are hits for pitchers.

0, 0, 0, 0, 0, 0, 0, 0, 0, 0, 0, 0, 1, 1, 1, 1, 2, 2, 3, 3, 5, 9, 9, 15, 18,
28, 33, 35, 45, 46, 54, 58, 105, 107, 113, 127, 138, 155, 188, 212

 a. Explain why the mode is not useful for describing how the team hit. What does the mode describe well?
 b. Is the mean or the median useful for describing how the team hit? Explain.
 c. The team played 162 games. What is the average number of hits in a game for the team for the season? Explain how you found your answer.

19. Reasoning You increase every number in a data set by 2. What happens to the mean, median, mode, and range? Explain using an example.

Fair Game Review What you learned in previous grades & lessons

Multiply. *(Section 3.2)*

20. 3.14×8 **21.** 2.6×2 **22.** 2.71×7 **23.** $4.6 \times 3 \times 4$

24. MULTIPLE CHOICE It rains 7 inches in 4 hours. At this rate how much will it rain in 15 hours? *(Section 5.3)*

 Ⓐ 8.57 in. Ⓑ 26.25 in. Ⓒ 28 in. Ⓓ 105 in.

5.6b Data Displays

Key Idea

Line Plots

A **line plot** uses a number line to show the number of times each value in a data set occurs. An "×" is placed above the number line for each data value.

Line plots show how much data are spread out or grouped together. This is called the *distribution* of the data.

EXAMPLE 1 Making a Line Plot

Ages			
14	18	13	15
16	15	24	14
17	16	18	15
15	19	16	16
24	14	15	17

The table shows the ages of people in a school library. Display the data in a line plot. Describe the distribution of the data.

Step 1: Draw a number line that includes the least and greatest values in the data set. The least age is 13 and the greatest age is 24.

Step 2: Place an "×" above the number line for each data value.

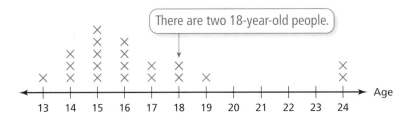

∴ Most of the people are in their teens. Two of the people are 24 years old.

Practice

Display the data in a line plot. Describe the distribution of the data.

1.

Number of Fouls					
2	1	2	0	0	3
2	1	5	0	3	0

2.

Camper Registrations				
21	25	19	22	21
23	24	27	25	16
24	21	22	25	22

Laurie's Notes

Introduction

Connect
- **Yesterday:** Students found the mean, median, and mode of a data set, and chose which measure best represents the data.
- **Today:** Students will display data using a line plot, a histogram, and a box-and-whisker plot.

Motivate
- Give each student a sticky note when class begins.
- **?** "How many times have you eaten pizza (or another popular food) in the last month? If you can't remember, an estimate is fine. Write your answer on the sticky note."
- Use student responses to make a line plot on the board. Draw a horizontal number line with a scale that fits the range of the data.
- Have students go to the board to place their sticky notes above the appropriate values.
- Tell students that this is one type of plot they will be making today. Ask questions about the completed plot.

Lesson Notes

Key Idea
- Students may say that a line plot is similar to a bar graph. They are similar, but the data in a line plot is numerical and a bar graph can include categorical data (i.e., favorite vegetable) as well as numerical data.
- Point out that line plots are not practical for large sets of data.

Example 1
- It is important to try to make each × the same size. Looking at the plot should give a quick visual sense of the relative occurrence of each value.
- Tell students that other ways of labeling the number line can be used. However, the number line must contain the least and greatest data values (i.e., the labels could be 12 to 25, or 10 to 30, etc).
- **?** "Does some of the data appear grouped, or clustered?" Yes; 13–19 or 14–18
- **?** "Is there a gap in the data?" Yes; between 20 and 23
- **Connection:** You can use a line plot to find the median of a data set. Ask students to find the median of this data set. 16

Practice
- **Extension:** Have students circle the median in each line plot.

Goal
Today's lesson is constructing and analyzing a **line plot**, a **histogram**, and a **box-and-whisker plot**.

Lesson Materials
Introduction
• sticky notes

Warm Up

Lesson 5.6b Warm Up For use before Lesson 5.6b

Find the mean, median, and mode(s) of the data.
1. 20, 11, 14, 16, 15, 16, 13
2. 4, 7, 6, 2, 1, 7, 5, 4
3. 42, 39, 41, 42, 43, 45
4. 22, 17, 16, 20, 21, 22, 18, 19
5. 87, 83, 89, 90, 84, 84, 85
6. 8, 9, 11, 8, 7, 10, 4, 12

Extra Example 1
The table shows the shoe sizes of the girls on a soccer team. Display the data in a line plot. Describe the distribution of the data.

Shoe Size				
8	7	4	8	7
9	6	6	7	8
6	7	8	7	6

Most girls on the team wear a size 6, 7, or 8.

Practice
1. See Additional Answers.
2. See Additional Answers.

Record and Practice Journal Practice
See Additional Answers.

T-227A

Extra Example 2

The frequency table shows the number of sweatshirts each person in a class owns. Display the data in a histogram.

Sweatshirts	Frequency
1–2	7
3–4	10
5–6	6
7–8	0
9–10	3

Practice

3.

4.

Laurie's Notes

Key Idea

- Explain that a **histogram** is a particular type of bar graph where the data is numerical, and the data is grouped into intervals of equal size.
- "Use the sample shown to ask questions, checking students' understanding of how to read a histogram.
 - "How many students were polled?" 25
 - "How many students polled can do 30 or more push-ups?" 7
 - "How many students polled can do 15 push-ups?" cannot determine
- Discuss with students how a line plot is related to a histogram. You can group the data in a line plot into equal intervals to resemble a histogram. Demonstrate this idea using the data from Example 1.

Example 2

- Use the Remember box to review the definition of a frequency table.
- The intervals for the histogram have been pre-determined. The size of each interval is 3. Students should notice the pattern in the right-end value of each interval: 3, 6, 9, 12, 15.
- Labeling the axes can often present a challenge for students. Students may write a number below the hash mark instead of writing an interval between the hash marks.
- Explain that no space is left between the bars because the intervals are continuous.
- Remind students that the axes are labeled and the histogram is given a title explaining what the data is about.
- Ask questions about the completed histogram.

Practice

- **Common Error:** Students may leave spaces between the bars when drawing the histograms.

Key Idea

Histograms

A **histogram** is a bar graph that shows the frequencies of data values in intervals of the same size.

The height of a bar represents the frequency of the values in the interval.

Remember

The *frequency* is the number of data values in an interval. A *frequency table* groups data values into intervals.

EXAMPLE 2 — Making a Histogram

The frequency table shows the number of points that each student in a class earns on a 15-point quiz. Display the data in a histogram.

Points	Frequency
1–3	2
4–6	0
7–9	6
10–12	10
13–15	9

Step 1: Draw and label the axes.

Step 2: Draw a bar to represent the frequency of each interval.

Include any interval with a frequency of 0. The bar height is 0.

There is no space between the bars of a histogram.

Practice

Display the data in a histogram.

3.
Heights of Plants	
Inches	Frequency
10–13	8
14–17	11
18–21	5
22–25	2
26–29	7

4.
Time on Hold	
Minutes	Frequency
0–4	11
5–9	19
10–14	5
15–19	0
20–24	3

Key Idea

Box-and-Whisker Plots

A **box-and-whisker plot** displays a data set along a number line using medians. **Quartiles** divide the data set into four equal parts. The median (second quartile) divides the data set into two halves. The median of the lower half is the first quartile. The median of the upper half is the third quartile.

Study Tip

A box-and-whisker plot shows the *variability* of a data set.

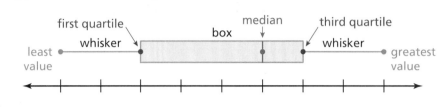

EXAMPLE 3 Making a Box-and-Whisker Plot

Make a box-and-whisker plot for the ages (in years) of the spider monkeys at a zoo.

15, 20, 14, 38, 30, 36, 30, 30, 27, 26, 33, 35

Step 1: Order the data. Find the median and the quartiles.

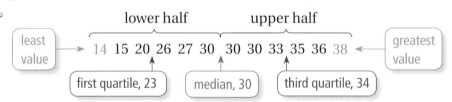

Step 2: Draw a number line that includes the least and greatest values. Graph points above the number line for the least value, greatest value, median, first quartile, and third quartile.

Step 3: Draw a box using the quartiles. Draw a line through the median. Draw whiskers from the box to the least and greatest values.

Practice

Make a box-and-whisker plot for the data.

5. Hours spent online:
1, 0, 2, 3, 3, 6, 1, 1, 2, 2

6. Ages of teachers:
30, 62, 45, 22, 28, 50, 42, 35

7. Quiz scores:
8, 12, 9, 12, 8, 5, 9, 7, 10

8. Weights (in pounds) of dogs:
10, 30, 15, 20, 25, 5, 20, 15, 35

Laurie's Notes

Key Idea
- Define a box-and-whisker plot. Draw the sample plot and discuss the process and vocabulary.
- **Discuss:** The box-and-whisker plot shows the *variability* of the data. Refer to this idea in each example done today.

Example 3
- **FYI:** Spider monkeys live in the rain forests of Central and South America. They do not have thumbs, yet they are able to grasp tree branches. They spend most of their time in trees.
- Point out that ordering the data from least to greatest makes it possible to find the least value, greatest value, and quartiles.
- ❓ "How many data values are there?" 12 "When there are 12 data values, how do you find the median?" Find the mean of the 6th and 7th data values.
- There are six data values in the lower half and six data values in the upper half. The first quartile and third quartile are the mean of the middle two data values in each half.
- **Note:** The quartiles divide the data set into four equal parts, each part having three data values.

Practice
- These are very small data sets so that they are manageable for students.
- **Common Error:** Students may forget to order the data before creating the box-and-whisker plot.

Extra Example 3
Make a box-and-whisker plot for the numbers of text messages received over several days.

32, 18, 11, 28, 42, 33, 40, 21, 37, 24

Practice

5.

6.

7.

8.

Extra Example 4
The box-and-whisker plot shows the times that the students in a gym class spent on an obstacle course.

a. What portion of the class spent 110 seconds or less on the obstacle course? $\frac{1}{4}$

b. Find and interpret the interquartile range of the data. 40; The middle half of the times vary by no more than 40 seconds.

 Practice

9. a. $\frac{3}{4}$

 b. 80; The middle half of the heights vary by no more than 80 feet.

Mini-Assessment
In Exercises 1–3, use the table below that shows the number of songs played at several concerts.

Songs					
20	19	15	13	8	15
12	13	20	15	21	14

1. Display the data in a line plot. Describe the distribution of the data.

Most of the concerts played 12–15 songs or 19–21 songs.

2. Display the data in a histogram using the intervals 7–9, 10–12, 13–15, 16–18, and 19–21. See Additional Answers.

3. Display the data in a box-and-whisker plot.

T-227D

Laurie's Notes

Discuss
- The five key values plotted in a box-and-whisker plot divide the data into four parts.
- ❓ "What portion of the data in the box are to the left of the median? to the right?" $\frac{1}{4}, \frac{1}{4}$
- Relate the definition of interquartile range to the box-and-whisker plot, indicating that it represents the "length" of the box.

Example 4
- This example does not state how many data values the plot represents. This is often true of box-and-whisker plots found in print materials.
- ❓ "Can someone read the 5 key values from the plot and describe what they mean in the context of the problem?" least value (shortest player) = 64 in.; first quartile = 72 in.; median = 74 in. (one-half of the players are shorter than 74 in. and one-half of the players are taller than 74 in.); third quartile = 76 in.; greatest value (tallest player) = 78 in.
- Work through the problem as shown.
- ❓ "75% of the players are at least what height?" 72 in.
- ❓ "Do you think the least value, 64, is an outlier?" Sample answer: Possibly because the left whisker is more spread out than the rest of the data.
- Explain what the lengths of the whiskers and the box tell you about the data.
- Stress that each of the four parts of the plot represent 25% or $\frac{1}{4}$ of the data.

Practice
- **Common Error:** Students may read the box-and-whisker plot incorrectly and find the range instead of the interquartile range.

Closure
- **Exit Ticket:** Create a box-and-whisker plot from the data used in the Motivate from today's lesson.

The Dynamic Planning Tool
Editable Teacher's Resources at *BigIdeasMath.com*

The figure shows how data are distributed in a box-and-whisker plot.

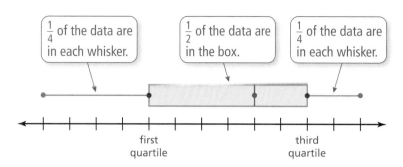

The **interquartile range** of a data set is the difference of the third quartile and the first quartile. It represents the range of the middle half of the data.

EXAMPLE 4 Analyzing a Box-and-Whisker Plot

The box-and-whisker plot shows the heights of the players on a basketball team.

a. **What portion of the players are at least 76 inches tall?**

The right whisker represents the players that are at least 76 inches tall.

∴ So, $\frac{1}{4}$ of the players are at least 76 inches tall.

b. **Find and interpret the interquartile range of the data.**

$$\text{interquartile range} = \text{third quartile} - \text{first quartile}$$
$$= 76 - 72$$
$$= 4$$

∴ So, the middle half of the heights vary by no more than 4 inches.

Practice

9. **ROLLER COASTERS** The box-and-whisker plot shows the heights of the roller coasters at an amusement park.

a. What portion of the roller coasters are between 120 feet tall and 220 feet tall?
b. Find and interpret the interquartile range of the data.

5.4–5.6 Quiz

Find the mean of the data. *(Section 5.4)*

1.
Tour Dates	
May	III
June	JHT JHT
July	JHT JHT IIII
August	JHT JHT JHT III
September	JHT

2.
Scores	
Judge 1	8.9
Judge 2	9.4
Judge 3	8.6
Judge 4	9.1

Find the median, mode(s), and range of the data. *(Section 5.5)*

3. 3, 5, 9, 11, 3

4. 24, 4, 37, 56, 6, 56, 45

5. 20, 54, 85, 14

6. 85, 100, 124, 54, 57, 68, 100, 68

Find the mean, median, and mode(s) of the data. Choose the measure that best represents the data. Explain your reasoning. *(Section 5.6)*

7. 3, 5, 15, 10, 24, 25, 100

8. 34, 57, 58, 56, 21

9. 47, 147, 24, 47, 38, 42

10. 52, 55, 58, 24, 97, 98

11. **EMAILS** The number of emails you received this week is shown. What is the mean number of emails you received per day? *(Section 5.4)*

Emails	
Monday	4
Tuesday	6
Wednesday	3
Thursday	6
Friday	7
Saturday	7
Sunday	2

12. **QUIZZES** The data are your quiz scores for a class. Find the mean, mode, and range of the data. *(Section 5.5)*

 18, 19, 17, 14, 20, 20, 15, 21

13. **MUSIC** The data are the lengths of the songs (in minutes) on your new CD. Which measure of central tendency best represents the data with and without the outlier? Explain. *(Section 5.6)*

 2.2, 2.2, 2.4, 2.6, 2.8, 3.0, 3.2, 3.4, 14.2

Alternative Assessment Options

Math Chat
Structured Interview
Student Reflective Focus Question
Writing Prompt

Math Chat
- Put students in pairs to discuss Exercises 7–10 from the Quiz. The discussion of the mean, median, and mode should include comparing and contrasting each measure of central tendency. It should also include choosing which of these measures best represents each data set. Have students add one or more numbers to the data set and explain how it affects the mean, median, mode, and range.
- The teacher should walk around the classroom listening to the pairs and ask questions to ensure understanding.

Study Help Sample Answers

Remind students to complete Graphic Organizers for the rest of the chapter.

5.

6.

7–10. Available at *BigIdeasMath.com*

Reteaching and Enrichment Strategies

If students need help...	If students got it...
Resources by Chapter • Study Help • Practice A and Practice B • Puzzle Time Lesson Tutorial *BigIdeasMath.com* Practice Quiz Practice from the Test Generator	Resources by Chapter • Enrichment and Extension • School-to-Work Game Closet at *BigIdeasMath.com* Start the Chapter Review

Answers

1. 10
2. 9
3. median: 5; mode: 3; range: 8
4. median: 37; mode: 56; range: 52
5. median: 37; mode: none; range: 71
6. median: 76.5; modes: 68, 100; range: 70
7. mean: 26; median: 15; mode: none; *Sample answer:* The median is best because there is no mode and the mean is greater than most of the data.
8. mean: 45.2; median: 56; mode: none; *Sample answer:* The median is best because there is no mode and the median is very close to 3 out of 5 values.
9. mean: 57.5; median: 44.5; mode: 47; *Sample answer:* The median is best because the mean and mode are greater than most of the data.
10–13. See Additional Answers.

Technology For the Teacher
Answer Presentation Tool

Assessment Book

T-228

For the Teacher
Additional Review Options
- Quiz*Show*
- Big Ideas Test Generator
- Game Closet at *BigIdeasMath.com*
- Vocabulary Puzzle Builder
- Resources by Chapter
 Puzzle Time
 Study Help

Answers

1. $\frac{3}{2}$; There are 3 butterflies for every 2 caterpillars.

2. $\frac{2}{1}$; There are 2 saxophones for every trumpet.

3. $\frac{3 \text{ stunts}}{1 \text{ movie}}$

4. $\frac{1166\frac{2}{3} \text{ stitches}}{1 \text{ minute}}$

Review of Common Errors

Exercises 1 and 2
- Remind students that reducing ratios is just like reducing fractions and they need to reduce until there are no more common factors in the numerator and denominator.

Exercises 3 and 4
- Students may not write the units on the rates. Compare rates to ratios, but with different units in the numerator and denominator. Illustrate the importance of the units with an examination of the speed of a car. In the United States speed is measured in miles per hour, but in most of Europe speed is measured in kilometers per hour. If you say that the speed of a car is 35, you do not know if it is miles or kilometers per hour.

Exercises 5 and 6
- Students may first take the distance part of the rate instead of finding the unit rate to solve the problem. Encourage students to find the unit rate first to help them solve the problem without errors.

Exercises 7–9
- Students may divide by the maximum number instead of the number of data values.

Exercises 10 and 11
- Students may try to identify the median without ordering the data first. Remind them that it is essential to order the data first and then find the median. This also makes finding the mode easier as well.

Exercises 12 and 13
- Students may only find the mean, median, and mode. Encourage them to put the numbers on a number line before deciding which measure of central tendency best represents the data.

5 Chapter Review

Review Key Vocabulary

ratio, *p. 192*
equivalent ratios, *p. 193*
rate, *p. 198*
unit rate, *p. 198*
unit cost, *p. 199*
mean, *p. 212*
outlier, *p. 213*
measure of central tendency, *p. 218*
median, *p. 218*
mode, *p. 218*
range, *p. 219*

Review Examples and Exercises

5.1 Ratios *(pp. 190–195)*

Write the ratio of sprinters to long distance runners on the track team.

Track Team	
Sprinters	26
Long distance	14

sprinters → $\dfrac{26}{14} = \dfrac{13}{7}$ Write in simplest form.
distance runners →

 The ratio of sprinters to long distance runners is $\dfrac{13}{7}$, 13 to 7, or 13 : 7.

Exercises

Write the ratio as a fraction in simplest form. Explain what the ratio means.

1. butterflies : caterpillars
2. saxophones : trumpets

5.2 Rates *(pp. 196–201)*

You receive 125 reward points for spending $25 in a store. How many reward points do you receive for every dollar spent?

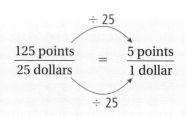

$\dfrac{125 \text{ points}}{25 \text{ dollars}} = \dfrac{5 \text{ points}}{1 \text{ dollar}}$

 You receive 5 points for every dollar spent.

Exercises

Write a unit rate for the situation.

3. 12 stunts in 4 movies
4. 3500 stitches in 3 minutes

5.3 Solving Rate Problems (pp. 202–207)

A horse can run at a speed of 55 feet per second. How far can it run in 5 seconds?

$$d = rt \quad \text{Write the formula for distance.}$$

$$= \frac{55 \text{ feet}}{1 \text{ second}} \times 5 \text{ seconds} \quad \text{Substitute the given values. The seconds divide out.}$$

$$= 275 \text{ feet} \quad \text{Multiply.}$$

∴ The horse can run 275 feet in 5 seconds.

Exercises

5. BIKING You bike 3 miles in 15 minutes. At this rate, how long does it take you to bike 10 miles?

6. MUSIC A song has 28 beats in 4 seconds. At this rate, how many beats are there in 30 seconds?

5.4 Mean (pp. 210–215)

The table shows the number of pictures of six students in a yearbook. What is the mean number of pictures per student?

$$\text{mean} = \frac{5 + 9 + 10 + 6 + 6 + 12}{6} \quad \begin{array}{l} \leftarrow \text{sum of the data} \\ \leftarrow \text{number of values} \end{array}$$

$$= \frac{48}{6}, \text{ or } 8 \quad \text{Simplify.}$$

Student	Pictures
1	5
2	9
3	10
4	6
5	6
6	12

∴ The mean number of pictures per person is 8.

Exercises

Find the mean of the data.

7. 4, 5, 7, 14, 17, 12, 18

8. 15, 5, 8, 12, 5, 9, 4, 10, 2, 11

9. WIND ENERGY The data show the amounts of wind energy that four states produced in one year. Find the mean of the data.

Wind Energy (megawatts)	
Texas	4356
California	2439
Minnesota	1299
Iowa	1273

Review Game
Calculating Mean, Median, or Mode

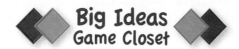
Big Ideas Game Closet

Materials
- 2 spinners labeled with the integers 0 through 9 (as shown below)
- 1 spinner labeled with *Mean, Median,* and *Mode* (as shown below)
- 2 pencils
- paper

Players: 2

Directions
1. One student spins to determine whether to calculate mean, median, or mode.
2. Each player spins his or her integer spinner 10 times, recording the results.
3. Each player calculates the mean, median, or mode, as determined by the spinner, of his or her numbers.

Who Wins?
The player with the greater mean, median, or mode wins a point for the round. Students can play several rounds or for a predetermined amount of time.

For the Student
Additional Practice
- Lesson Tutorials
- Study Help (textbook)
- Student Website
 - Multi-Language Glossary
 - Practice Assessments

Answers

5. 50 min
6. 210 beats
7. 11
8. 8.1
9. 2341.75 megawatts
10. median: 6; mode: 8; range: 4
11. median: 46; mode: none; range: 50
12. mean: 17.8; median: 14; mode: 12; *Sample answer:* The median is best because the mean is greater than most of the data and the mode is less than most of the data.
13. mean: $23.8\bar{3}$; median: 24; mode: 21; *Sample answer:* The mean or median is best, because they are very close and the mode is less than most of the data.

My Thoughts on the Chapter

What worked. . .

Teacher Tip
Not allowed to write in your teaching edition? Use sticky notes to record your thoughts.

What did not work. . .

What I would do differently. . .

5.5 Median, Mode, and Range (pp. 216–221)

Find the median and mode of the movie lengths in the table.

Order the data from least to greatest.

Movie Lengths (minutes)		
91	112	126
142	122	112
92	144	

Median: 91, 92, 112, 112, 122, 126, 142, 144

$$\frac{112 + 122}{2} = \frac{234}{2}, \text{ or } 117 \quad \text{Add the two middle values and divide by 2.}$$

Mode: 91, 92, 112, 112, 122, 126, 142, 144 ← The value 112 occurs most often.

∴ The median is 117 minutes and the mode is 112 minutes.

Exercises

Find the median, mode, and range of the data.

10. 8, 8, 6, 8, 4, 5, 6

11. 24, 74, 61, 29, 38, 27, 68, 54

5.6 Analyzing Data Sets (pp. 222–227)

Find the mean, median, and mode of the numbers of dancers at tryouts. Which measure best represents the data?

Number of Dancers at Tryouts			
102	92	99	96
92	105	94	200

Mean: $\dfrac{102 + 92 + 99 + 96 + 92 + 105 + 94 + 200}{8} = \dfrac{880}{8}$, or 110

Order the data from least to greatest.

Median: 92, 92, 94, 96, 99, 102, 105, 200

$$\frac{96 + 99}{2} = \frac{195}{2}, \text{ or } 97.5$$

Mode: 92, 92, 94, 96, 99, 102, 105, 200 ← The value 92 occurs most often.

∴ The median best represents the data. The mean is greater than most of the data and the mode is less than most of the data.

Exercises

Find the mean, median, and mode(s) of the set of data. Choose the measure that best represents the data. Explain your reasoning.

12. 36, 12, 14, 12, 15

13. 23, 25, 26, 21, 27, 21

5 Chapter Test

Write the ratio as a fraction in simplest form. Explain what the ratio means.

1. $\dfrac{\text{scooters}}{\text{bikes}}$

2. starfish : seashells

Write a unit rate for the situation.

3. $54.00 for 3 tickets

4. 210 miles in 3 hours

Find how far the object travels in the given amount of time.

5. 5 hours

Moves 50 miles in 1 hour.

6. 15 seconds

Climbs 88 feet in 5 seconds.

Find the mean, median, mode(s), and range of the data.

7. 2, 7, 7, 12, 4

8. 4, 5, 7, 5, 9, 9, 10, 6

Find the mean, median, and mode(s) of the data. Choose the measure that best represents the data. Explain your reasoning.

9. 5, 6, 4, 24, 18

10. 46, 27, 94, 56, 53, 65, 43

11. **RADIO** A radio DJ plays 34 songs in 3 hours. At this rate, how many songs will the DJ play in 5 hours?

12. **MP3 PLAYER** The data are the prices of MP3 players. What is the mean price?

 $89, $240, $120, $99, $150, $179, $75

13. **WATERFALLS** The table shows the four tallest waterfalls in the world.

 a. Describe how the median of the data will change by adding the fifth-tallest waterfall to the data.

 b. Without knowing the height of the fifth-tallest waterfall, describe the least amount that it will change the mean when added to the data.

Waterfall	Height (feet)
Angel Falls	3212
Tugela Falls	3110
Three Sisters Falls	3000
Olo'upena Falls	2953

Test Item References

Chapter Test Questions	Section to Review	Common Core State Standards
1, 2	5.1	6.RP.1
3, 4	5.2	6.RP.1, 6.RP.2, 6.RP.3b
5, 6, 11	5.3	6.RP.2, 6.RP.3b, 6.RP.3d
7–10, 12, 13	5.4	6.SP.3, 6.SP.5c
7–10	5.5	6.SP.3, 6.SP.5c
9, 10, 13	5.6	6.SP.4, 6.SP.5d

Test-Taking Strategies

Remind students to quickly look over the entire test before they start so that they can budget their time. Students should jot down the definitions of mean, median, and mode on the back of their test before they start. By doing this, they will not become confused when they are under pressure. Students need to use the **Stop** and **Think** strategy before answering questions.

Common Assessment Errors

- **Exercises 3 and 4** Students may write rates that are inverted and make the numerators 1 instead of the denominators. Remind them of familiar unit rates such as miles per gallon and ask them which quantity is 1.
- **Exercise 6** Students may write the distance as 88 feet per second multiplied by 15 seconds instead of $\frac{88}{5}$ feet per second multiplied by 15 seconds. Encourage them to find the unit rate before they find the distance.
- **Exercises 7–10** Students may have difficulty stating the mode(s). Remind them that a data set can have one mode (Exercise 7), more than one mode (Exercise 8) or no mode (Exercises 9 and 10). Point out that as with the median, the mode may be easier to find if the data is ordered.
- **Exercises 7–10** Students may incorrectly identify the range by finding the difference between the first and the last data values as they are given. Remind them that the range is the difference between the greatest and least value.
- **Exercises 9 and 10** Students may not graph the numbers on a number line before deciding which measure best represents the data. Even if they are correct, encourage them to graph the numbers to check their answers.

Reteaching and Enrichment Strategies

If students need help...	If students got it...
Resources by Chapter • Practice A and Practice B • Puzzle Time Record and Practice Journal Practice Differentiating the Lesson Lesson Tutorial Practice from the Test Generator Skills Review Handbook	Resources by Chapter • Enrichment and Extension • School-to-Work • Financial Literacy Game Closet at *BigIdeasMath.com* Start the next chapter

Answers

1. $\frac{1}{1}$; For each scooter there is 1 bike.
2. $\frac{2}{5}$; For every 2 starfish there are 5 shells.
3. $\frac{\$18.00}{1 \text{ ticket}}$
4. $\frac{70 \text{ miles}}{1 \text{ hour}}$
5. 250 mi
6. 264 ft
7. mean: 6.4; median: 7; mode: 7; range: 10
8. mean: 6.875; median: 6.5; modes: 5 and 9; range: 6
9. mean: 11.4; median: 6; mode: none; *Sample answer:* The median is best because there is no mode, the mean is greater than most of the data, and the median is close to 3 of 5 data values.
10. mean: about 54.86; median: 53; mode: none; The mean or median is best because there is no mode and the mean and median are close to each other.
11–13. See Additional Answers.

Assessment Book

Test-Taking Strategies
Available at *BigIdeasMath.com*

After Answering Easy Questions, Relax
Answer Easy Questions First
Estimate the Answer
Read All Choices before Answering
Read Question before Answering
Solve Directly or Eliminate Choices
Solve Problem before Looking at Choices
Use Intelligent Guessing
Work Backwards

About this Strategy

When taking a multiple choice test, be sure to read each question carefully and thoroughly. Sometimes it is easier to solve the problem and then look for the answer among the choices.

Answers

1. A
2. H
3. 23
4. A

Item Analysis

1. **A.** Correct answer
 B. The student finds the quotient of the whole numbers and the quotient of the fractions and combines these to get the answer.
 C. The student finds the product of the whole numbers and the product of the fractions and combines these to get the answer.
 D. The student rewrites the quotient as $\frac{76}{9} \div \frac{14}{3}$ but then changes this to a product without finding the reciprocal of the divisor.

2. **F.** The student rounds each addend up to the next whole number.
 G. The student chooses an approximate sum but not the one closest to the actual sum.
 H. Correct answer
 I. The student rounds each addend down to the next whole number.

3. **Gridded Response:** Correct answer: 23

 Common Errors:
 - The student finds the median of the six numbers shown, getting an answer of 24.5.
 - The student finds the mean of the six numbers shown, getting an answer of approximately 24.2.

4. **A.** Correct answer
 B. The student does not list the data in order and finds the median using 12 and 3, getting 7.5, which is less than the mean of 8.
 C. The student does not correctly determine that the mode is 10 and instead incorrectly determines a value that is less than the other measures.
 D. The student does not list the data in order and finds the range using 7 and 6, getting 1, which is less than the mean of 8.

5. **F.** The student lists the types of seashells in the order given.
 G. The student uses only the numbers given and ignores the symbols, ordering the numbers 1, 8, 9, and 11.
 H. The student thinks that 0.1 is less than 9% and reverses the order of the fighting conch and auger.
 I. Correct answer

Technology For the Teacher
Big Ideas Test Generator

5 Standardized Test Practice

1. What is the value of the expression below? (6.NS.1)

$$8\frac{4}{9} \div 4\frac{2}{3}$$

A. $1\frac{17}{21}$

B. $2\frac{2}{3}$

C. $32\frac{8}{27}$

D. $39\frac{11}{27}$

2. Which integer is closest to the value of the expression below? (6.NS.3)

$$4.18 + 6.225 + 5.7$$

F. 18

G. 17

H. 16

I. 15

3. The data set in the box below has no mode. One number is missing.

18, 24, 22, 30, 26, ____, 25

The mean of the set of data has the same value as the median of the set of data. What is the missing number? (6.SP.5c)

4. The number of hours each of 6 students spent reading last week is shown in the bar graph below.

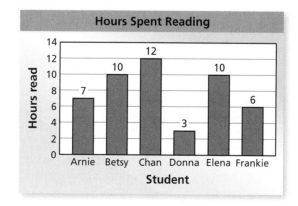

For the data in the bar graph, which measure is the *least*? (6.SP.5c)

A. mean

B. median

C. mode

D. range

5. You go to a beach and collect buckets of shells. Of the many shells you have collected, you notice the following.

- 9% of the seashells were auger shells.
- $\frac{1}{8}$ of the seashells were coquina shells.
- 11% of the seashells were rough scallop shells.
- 0.1 of the seashells were fighting conch shells.

Which list correctly shows the types of shells in order from least to greatest? *(6.NS.7b)*

F. auger, coquina, rough scallop, fighting conch

G. fighting conch, coquina, auger, rough scallop

H. fighting conch, auger, rough scallop, coquina

I. auger, fighting conch, rough scallop, coquina

6. What is the median of the data shown in the line plot below? *(6.SP.5c)*

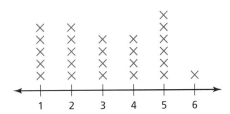

A. 3

B. 3.5

C. 4.2

D. 5

7. A service club wants to buy tickets to a baseball game. Tickets are available for the grandstand and for the bleachers.

Which expression represents the total cost, in dollars, for *g* grandstand tickets and *b* bleachers tickets? *(6.EE.6)*

F. $375(g + b)$

G. $40(g \times b)$

H. $25g + 15b$

I. $25g \times 15b$

234 Chapter 5 Ratios, Rates, and Data Analysis

Item Analysis (continued)

6. **A.** Correct answer

 B. The student identifies the middle of the line plot, but 3.5 lies between the 14th and 15th data points, whereas the median of 25 data points is the 13th data point.

 C. The student counts the number of Xs (25) and divides by the number of categories (6), and then rounds the quotient to the nearest tenth.

 D. The student finds the mode of the data.

7. **F.** The student multiplies the two ticket prices together and multiplies this product by the sum of the numbers of the two types of tickets.

 G. The student adds the two ticket prices together and multiplies this sum by the product of the numbers of the two types of tickets.

 H. Correct answer

 I. The student finds the correct expressions for the cost of each type of ticket but then multiplies these two expressions.

8. **Gridded Response:** Correct answer: 390

 Common Error: The student does not add a zero as a place holder, rewriting the quotient as 468 ÷ 12 instead of 4680 ÷ 12, getting an answer of 39.

9. **A.** The student uses 0.6 instead of 6 for a and 0.14 instead of 14 for b and finds $0.8 \times 0.6 + 0.02 \times 0.14$.

 B. The student uses 6 for the hundredths digit in $0.8a$ and 14 for the thousandths and ten-thousandths digits in $0.02b$ and finds $0.86 + 0.0214$.

 C. Correct answer

 D. The student finds the sum of the coefficients and the sum of the values of the variables and then finds the product of these two sums.

10. **F.** The student does not recognize that the Distributive Property is used to write the expression in the second line.

 G. Correct answer

 H. The student does not recognize that the Identity Property of Multiplication is used to find the final answer.

 I. The student does not recognize that the Commutative Property of Multiplication is used in the first line.

Answers

5. I
6. A
7. H

Answers

8. 390
9. C
10. G
11. Sample answer: 2, 3, 9, 10, 11; numbers sum to 35
12. B

Answer for Extra Example

1. **A.** The student rewrites the divisor as $\frac{17}{3}$ but does not find the reciprocal.
 B. The student only takes the reciprocal of the fraction part of the divisor and rewrites it as $5\frac{3}{2}$, which the student then rewrites as $\frac{13}{2}$.
 C. Correct answer
 D. The student only takes the reciprocal of the fraction part of the divisor and ignores the whole number part.

Item Analysis (continued)

11. **2 points** The student demonstrates a thorough understanding of analyzing sets of data with respect to the mean and median. The student correctly determines a data set that has a mean of 7 and a median of 9. One possible answer is the data set of 2, 3, 9, 10, and 11. The student clearly explains the steps used, and the method used is unambiguous. The student correctly shows that the data set has a mean of 7 and a median of 9.

 1 point The student demonstrates a partial understanding of analyzing sets of data with respect to the mean and median. The student shows some knowledge of how to determine the mean and/or median but does not successfully answer the question.

 0 points The student demonstrates insufficient understanding of analyzing sets of data with respect to the mean and median. The student makes many errors and shows little or no ability to work with the mean or the median.

12. **A.** The student thinks that 24.8% is approximately 25 and estimates the value of the expression by dividing 400 by 25.
 B. Correct answer
 C. The student estimates the value of the expression by dividing 400 by 0.25.
 D. The student thinks that 24.8% is approximately 25 and estimates the value of the expression by multiplying 25 by 400.

Extra Example for Standardized Test Practice

1. What number belongs in the box to make the equation true?

 $$3\frac{1}{2} \div 5\frac{2}{3} = \frac{7}{2} \times \Box$$

 A. $\frac{17}{3}$
 B. $\frac{13}{2}$
 C. $\frac{3}{17}$
 D. $\frac{3}{2}$

8. What is the value of the expression below? *(6.NS.3)*

$$46.8 \div 0.12$$

9. What is the value of the expression below when $a = 6$ and $b = 14$? *(6.EE.2c)*

$$0.8a + 0.02b$$

A. 0.4828

B. 0.8814

C. 5.08

D. 16.4

10. Which property was *not* used in the box below to simplify the expression? *(6.EE.3)*

$$0.3 \times 53 + 53 \times 0.7 = 53 \times 0.3 + 53 \times 0.7$$
$$= 53 \times (0.3 + 0.7)$$
$$= 53 \times 1$$
$$= 53$$

F. Distributive Property

G. Associative Property of Addition

H. Identity Property of Multiplication

I. Commutative Property of Multiplication

11. Determine a data set of 5 numbers that has the following measures:

- a mean of 7
- a median of 9

Explain how you determined your data set, and demonstrate that the mean of your data set is 7 and the median is 9. *(6.SP.5c)*

12. Which integer is closest to the value of the expression below? *(6.RP.3c)*

$$24.8\% \text{ of } 408$$

A. 16

B. 100

C. 1,600

D. 10,000

6 Circles and Area

- **6.1 Circles and Circumference**
- **6.2 Perimeters of Composite Figures**
- **6.3 Areas of Circles**
- **6.4 Areas of Composite Figures**

"Think of any number between 1 and 9."

"Okay, now add 4 to the number, multiply by 3, subtract 12, and divide by your original number."

"You end up with 3, don't you?"

"What do you get when you divide the circumference of a jack-o-lantern by its diameter?"

"Pumpkin pi, HE HE HE."

Connections to Previous Learning

- Describe and determine area, and recognize that a unit square is the standard unit for measuring area.
- Use appropriate measures and tools to estimate and solve real-world area problems.

- Use formulas to find the areas of parallelograms, triangles, and trapezoids.

- Understand pi and its estimates.
- Use values of pi to estimate and calculate the circumference and area of circles.
- Find perimeters and areas of composite two-dimensional figures, including semi-circles.

Math in History

Have you ever wondered why a circle is divided into 360 units (called degrees) instead of 100 units? There are three theories as to why this happened... each of which is related to the others. (1) About 360 days in a year, (2) Babylonians used a base 60 number system, and (3) 360 has many more divisors than 100.

★ Selecting 360 as the number of degrees in a circle was probably based on the fact that 360 is approximately the number of days in a year. Babylonian astronomers noticed that the stars in the sky, which circle the celestial pole every year, seem to advance in that circle by approximately one-360th of a circle, one degree, each day.

★ Another motivation for choosing the number 360 (which is mentioned in ancient writing found in India) is that 360 has many divisors, including every number from 1 to 10 except 7.

Divisors of 360: 1, 2, 3, 4, 5, 6, 8, 9, 10, 12, 15, 18, 20, 24, 30, 36, 40, 45, 60, 72, 90, 120, 180, 360

Pacing Guide for Chapter 6

Chapter Opener	1 Day
Section 1	
Activity	1 Day
Lesson	1 Day
Section 2	
Activity	2 Days
Lesson	1 Day
Study Help / Quiz	1 Day
Section 3	
Activity	1 Day
Lesson	1 Day
Section 4	
Activity	1 Day
Lesson	1 Day
Quiz / Chapter Review	1 Day
Chapter Test	1 Day
Standardized Test Practice	1 Day
Total Chapter 6	14 Days
Year-to-Date	101 Days

Check Your Resources

- Record and Practice Journal
- Resources by Chapter
- Skills Review Handbook
- Assessment Book
- Worked-Out Solutions

The Dynamic Planning Tool
Editable Teacher's Resources at
BigIdeasMath.com

Common Core State Standards

4.G.2 Classify two-dimensional figures based on the presence of absence of parallel or perpendicular lines, or the presence or absence of angles of a specified size. Recognize right triangles as a category, and identify right triangles.

6.EE.1 Write and evaluate numerical expressions involving whole-number exponents.

Additional Topics for Review

- Exponents

Try It Yourself

1. trapezoids
2. circles
3. trapezoid, triangle
4. triangles
5. rectangle, triangle
6. rectangle, triangles
7. 25
8. 144
9. 12
10. 196
11. 243
12. 188

Record and Practice Journal

1. trapezoid, triangle
2. rectangles
3. rectangle, trapezoids
4. trapezoid, rectangles, circles
5. trapezoids
6. 49
7. 121
8. 100
9. 700
10. 144
11. 405
12. 744
13. 566
14. 5000 meters

T-237

Math Background Notes

Vocabulary Review
- Evaluate
- Simplify
- Exponent

Classifying Figures
- Students should know how to classify figures.
- **Teaching Tip:** Encourage students to redraw each figure on a sheet of paper. By constructing it themselves, students are more likely to identify the polygons used in the figure.
- **Teaching Tip:** Tactile learners might benefit from doing these examples using polygon templates. Allow them to build the shapes they see using the polygon templates. This could help students to distinguish which polygons they are seeing.

Squaring Numbers and Using Order of Operations
- Students should know the order of operations.
- You may want to review the correct order of operations with students. Many students probably learned the pneumonic device *Please Excuse My Dear Aunt Sally*. Ask a volunteer to explain why this phrase is helpful.
- **Common Error:** Remind students that the exponent describes how many times the base acts a factor. The exponent itself will not appear as a factor.

Reteaching and Enrichment Strategies

If students need help...	If students got it...
Record and Practice Journal • Fair Game Review Skills Review Handbook Lesson Tutorials	Game Closet at *BigIdeasMath.com* Start the next section

What You Learned Before

Classifying Figures (4.G.2)

Identify the basic shapes in the figure.

Example 1

∴ Rectangle, right triangle

Example 2

∴ Semicircle, square, and triangle

Try It Yourself
Identify the basic shapes in the figure.

1.
2.
3.
4.
5.
6.

Squaring Numbers and Using Order of Operations (6.EE.1)

Example 3 Evaluate 4^2.

$$4^2 = 4 \cdot 4 = 16$$

4^2 means to multiply 4 by itself.

Example 4 Evaluate $3 \cdot 6^2$.

$$3 \cdot 6^2 = 3 \cdot (6 \cdot 6) = 3 \cdot 36 = 108$$

Use order of operations. Evaluate the exponent, and then multiply.

Try It Yourself
Evaluate the expression.

7. 5^2
8. 12^2
9. $3 \cdot 2^2$
10. $4 \cdot 7^2$
11. $3(1 + 8)^2$
12. $2(3 + 7)^2 - 3 \cdot 4$

6.1 Circles and Circumference

Essential Question How can you find the circumference of a circle?

COMMON CORE STATE STANDARDS
6.EE.2c

Archimedes was a Greek mathematician, physicist, engineer, and astronomer.

Archimedes discovered that in any circle the ratio of circumference to diameter is always the same. Archimedes called this ratio pi, or π (a letter from the Greek alphabet).

$$\pi = \frac{\text{Circumference}}{\text{Diameter}}$$

In Activities 1 and 2, you will use the same strategy Archimedes used to approximate π.

1 ACTIVITY: Approximating Pi

Work with a partner. Copy the table. Record your results in the table.

- Measure the perimeter of the large square in millimeters.
- Measure the diameter of the circle in millimeters.
- Measure the perimeter of the small square in millimeters.
- Calculate the ratios of the two perimeters to the diameter.
- The average of these two ratios is an approximation of π.

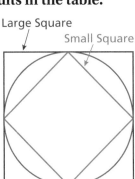

Sides of Polygon	Large Perimeter	Diameter of Circle	Small Perimeter	$\frac{\text{Large Perimeter}}{\text{Diameter}}$	$\frac{\text{Small Perimeter}}{\text{Diameter}}$	Average of Ratios
4						
6						
8						
10						

238 Chapter 6 Circles and Area

Laurie's Notes

Introduction

For the Teacher
- **Goal:** Students will investigate how to calculate pi using the same technique as Archimedes.
- **Big Idea:** Today's activity is not common in textbooks and is intended to give students the opportunity to understand that pi is a calculable number.
- The other big idea related to pi is that the ratio of the circumference to diameter is the same for all circles, and that ratio is pi. Activities using string and circular objects to develop this idea are common in many textbooks.

Motivate
- Share the history and some of Archimedes' inventions to interest students in the person whose work they will replicate today. (See the next page.)

Discuss
- ❓ "Does the diameter have to go through the center of the circle?" yes
- ❓ "What is the relationship between the radius and the diameter of a circle?" The diameter is twice the radius.

Activity Notes

Activity 1
- Discuss the first diagram involving the squares and a circle. A circle is inscribed in the large square—inscribed means the circle is entirely within the square and touches the square at exactly 4 points, the middle of each side of the square. The small square is inscribed in the circle—the small square is entirely within the circle and touches the circle at exactly 4 points, the vertices of the square.
- ❓ "What does perimeter mean?" distance around a figure "How do you calculate the perimeter of a square?" length of one side × 4
- ❓ "How will you measure the diameter of the circle?" measure the diagonal of the small square
- ❓ "What is another name for average?" mean "How do you calculate the average in this problem?" Sum the two ratios and divide by 2.
- Remind students to measure to the nearest millimeter.
- When students have finished the first row in the table, discuss the results and reflect on what the answers mean with reference to the diagram.
- ❓ "If the process were repeated for a different size square to start, would the perimeter of the larger square always be greater than the perimeter of the smaller square?" yes
- ❓ "How does the distance around the circle compare to the distance around the two squares?" (Note: The definition for circumference is presented in the next lesson.) The distance around the circle is a number between the perimeters of the two squares.

Common Core State Standards

6.EE.2c Evaluate expressions at specific values of their variables. Include expressions that arise from formulas used in real-world problems. Perform arithmetic operations, including those involving whole-number exponents, in the conventional order when there are no parentheses to specify a particular order (Order of Operations).

Previous Learning
Students should know how to find an average (mean).

Activity Materials
Textbook
• calculators • rulers

Start Thinking! and Warm Up

6.1 Record and Practice Journal

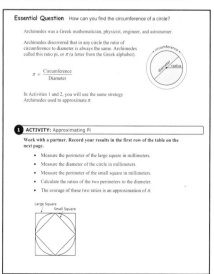

T-238

Differentiated Instruction

Kinesthetic

Have students bring in circular objects, such as cans or flying discs. Use string or a tape measure to find the diameter and circumference of each object. Have students divide the circumference by the diameter and record the results on the board. Using the classroom data, find the average to see how close the results are to π. Check to see if students make the connection that the ratio circumference : diameter is always the same.

6.1 Record and Practice Journal

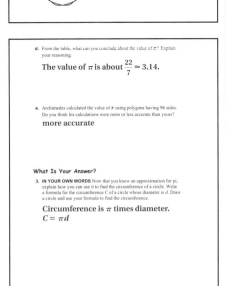

Laurie's Notes

Activity 2

- Students should be able to complete this activity on their own after the guidance given for Activity 1.
- **?** When students have finished ask, "Do you have any observations about the perimeters of the polygons in each problem?" *The two perimeters are getting closer together, and closer to the value of the circumference.*
- **?** The activity is called *Approximating Pi* and there is no reason why a student should know the value of pi, even though they may. Explain that the last column is the approximation of pi. Ask, "What do you observe about the values in the last column?" *As the number of sides of the polygons increase, the values get closer to the value of pi.*
- In discussing part (e), remind students that Archimedes did this investigation more than 2000 years ago without the aid of computer or calculator. Consider how long it might have taken him to construct the polygons with 96 sides!
- **Extension:** Read the short book referenced in the illustration. Your students will enjoy the story.

What Is Your Answer?

- **Big Idea:** Because the two perimeters converge to the circumference and each of the perimeters is divided by the diameter, it should seem reasonable to students that the circumference divided by the diameter approximates pi.

Closure

- Describe Archimedes' method for approximating pi.

History

- **Archimedes** is considered to be one of the greatest mathematicians of all time.
 - born in Syracuse, Italy; 287–212 B.C.
 - mathematician, physicist, engineer, and inventor
 - Archimedes was killed by a Roman soldier. Here are two accounts of his death.
 - He was working a math problem and refused to leave to meet the Roman general. So, the soldier killed him with his sword.
 - He was on his way to surrender and was carrying mathematical instruments. The soldier thought they were valuable and killed Archimedes.
- **His inventions include:**
 - Archimedes (water) screw for irrigating fields
 - Various instruments of war, such as catapults, cranes, and mirrors

The Dynamic Planning Tool
Editable Teacher's Resources at *BigIdeasMath.com*

A page from "Sir Cumference and the First Round Table" by Cindy Neuschwander.

2 ACTIVITY: Approximating Pi

Continue your approximation of pi. Complete the table using a hexagon (6 sides), an octagon (8 sides), and a decagon (10 sides).

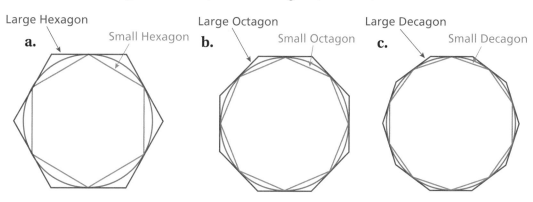

a. Large Hexagon / Small Hexagon
b. Large Octagon / Small Octagon
c. Large Decagon / Small Decagon

d. From the table, what can you conclude about the value of π? Explain your reasoning.

e. Archimedes calculated the value of π using polygons having 96 sides. Do you think his calculations were more or less accurate than yours?

What Is Your Answer?

3. **IN YOUR OWN WORDS** Now that you know an approximation for pi, explain how you can use it to find the circumference of a circle. Write a formula for the circumference C of a circle whose diameter is d. Draw a circle and use your formula to find the circumference.

Practice

Use what you learned about circles and circumference to complete Exercises 10–12 on page 243.

Section 6.1 Circles and Circumference 239

6.1 Lesson

A **circle** is the set of all points in a plane that are the same distance from a point called the **center**.

Key Vocabulary
circle, p. 240
center, p. 240
radius, p. 240
diameter, p. 240
circumference, p. 241
pi, p. 241
semicircle, p. 242

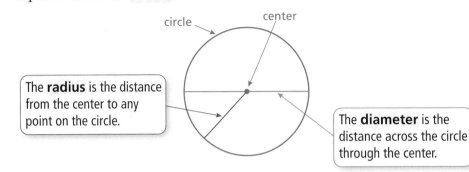

The **radius** is the distance from the center to any point on the circle.

The **diameter** is the distance across the circle through the center.

Key Idea

Radius and Diameter

Words The diameter d of a circle is twice the radius r. The radius r of a circle is one-half the diameter d.

Algebra Diameter: $d = 2r$ Radius: $r = \dfrac{d}{2}$

EXAMPLE 1 Finding a Radius and a Diameter

a. The diameter of a circle is 20 feet. Find the radius.

b. The radius of a circle is 7 meters. Find the diameter.

$r = \dfrac{d}{2}$ Radius of a circle

$= \dfrac{20}{2}$ Substitute 20 for d.

$= 10$ Divide.

∴ The radius is 10 feet.

$d = 2r$ Diameter of a circle

$= 2(7)$ Substitute 7 for r.

$= 14$ Multiply.

∴ The diameter is 14 meters.

On Your Own

Now You're Ready
Exercises 4–9

1. The diameter of a circle is 16 centimeters. Find the radius.
2. The radius of a circle is 9 yards. Find the diameter.

240 Chapter 6 Circles and Area Multi-Language Glossary at BigIdeasMath.com.

Laurie's Notes

Introduction

Connect
- **Yesterday:** Students investigated a technique for calculating the value of pi.
- **Today:** Students will use the formula for circumference to solve real-life problems.

Motivate
- **Whole Class Activity:** How observant are your students? Tell them that you are going to give them 1 minute to write a list of objects in the room that have a special characteristic. Give them time to get scrap paper and a pencil.
- Announce that they need to list objects that are circular or have a circle on them. Go!
- Items will vary but expected items include: clock and/or watch face, bottom of coffee cup, pencil's eraser, pupils of your eyes, Person X's glasses, metal feet on the chair, etc.

Lesson Notes

Key Idea
- Draw a circle on the board and label the center, a radius, and a diameter. Discuss each.
- Write the Key Idea.
- Discuss with students that if you know either the diameter or radius of a circle, you can find the other. There is a 2 : 1 relationship between **diameter** and **radius**.

Example 1
- Work through each example.
- Remind students to label answers with the appropriate units.
- **Extension:** If time permits, do an investigation to *discover* that the ratio of the circumference to diameter is the same for every circle, and the ratio is pi.

On Your Own
- It helps some students to draw a sketch of a circle and label the given information. These students need to see the relationship instead of only reading the given information.

Goal Today's lesson is estimating and calculating the **circumference** of **circles** using common estimates for **pi**.

Start Thinking! and Warm Up

Lesson 6.1 Warm Up For use before Lesson 6.1

Lesson 6.1 Start Thinking! For use before Lesson 6.1

Is it easier to measure the diameter or circumference of a tree trunk? Why?

Is it easier to measure the diameter or circumference of a quarter? Why?

In general, when is it easier to measure the diameter of a circular object? When is it easier to measure the circumference?

Extra Example 1
a. The diameter of a circle is 4 yards. Find the radius. 2 yd
b. The radius of a circle is 15 millimeters. Find the diameter. 30 mm

On Your Own
1. 8 cm
2. 18 yd

English Language Learners
Vocabulary
Use everyday objects, such as a clock, to discuss the meaning of *circle, center, radius, diameter,* and *circumference.* Have students add these terms to their notebooks along with a sketch of a circle with the parts labeled. Students should also include pi with its symbol, π, and its approximations, 3.14 and $\frac{22}{7}$, in their notebooks.

Extra Example 2
a. Find the circumference of a flying disc with a radius 8 centimeters. Use 3.14 for π. about 50.24 cm
b. Find the circumference of a clock with a diameter of 21 inches. Use $\frac{22}{7}$ for π. about 66 in.

On Your Own
3. about 12.56 cm
4. about 44 ft
5. about 28.26 in.

Laurie's Notes

Key Idea
- Read the information at the top of the page. Note the two approximations of pi and how each is used in the problems that follow. Note the *Study Tip*.
- Write the Key Idea.
- **Common Misconception:** Students may know that pi is a number; however, when they see it in a formula, they can become confused. Students may ask if π is a variable. Pi is a constant whose value is approximately 3.14 or $\frac{22}{7}$. In each formula, remind students that πd means π times the diameter and $2\pi r$ means 2 times π times the radius.
- **Big Idea:** The ratio of the circumference to the diameter is pi. To get the formula for circumference, multiply both sides of the equation by the diameter.

$$\frac{\text{circumference}}{\text{diameter}} = \text{pi}$$
$$\frac{C}{d} = \pi$$
$$C = \pi d$$

Example 2
- Work through each example.
- Point out to students when to use the different forms of the circumference formula given a radius or a diameter.
- Remind students that the symbol \approx means *approximately equal to*.
- ? "Why is the equal sign replaced by the approximately equal (\approx) sign?" 3.14 is an approximation for π.
- ? "Why is $2 \times 3.14 \times 5$ equal to $2 \times 5 \times 3.14$?" Commutative Property of Multiplication
- Note that in part (b), the diameter is a multiple of 7, so $\frac{22}{7}$ is used as the approximation for π.

On Your Own
- **Think-Pair-Share:** Students should read each question independently and then work with a partner to answer the questions. When they have answered the questions, the pair should compare their answers with another group and discuss any discrepancies.
- Check which estimate for π students used and why.

The distance around a circle is called the **circumference**. The ratio $\frac{\text{circumference}}{\text{diameter}}$ is the same for *every* circle and is represented by the Greek letter π, called **pi**. The value of π can be approximated as 3.14 or $\frac{22}{7}$.

Study Tip

When the radius or diameter is a multiple of 7, it is easier to use $\frac{22}{7}$ as the estimate of π.

Key Idea

Circumference of a Circle

Words The circumference C of a circle is equal to π times the diameter d or π times twice the radius r.

Algebra $C = \pi d$ or $C = 2\pi r$

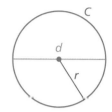

EXAMPLE 2 Finding Circumferences of Circles

a. **Find the circumference of the flying disc. Use 3.14 for π.**

$C = 2\pi r$ Write formula for circumference.

$\approx 2 \cdot 3.14 \cdot 5$ Substitute 3.14 for π and 5 for r.

$= 31.4$ Multiply.

∴ The circumference is about 31.4 inches.

b. **Find the circumference of the watch face. Use $\frac{22}{7}$ for π.**

$C = \pi d$ Write formula for circumference.

$\approx \frac{22}{7} \cdot 28$ Substitute $\frac{22}{7}$ for π and 28 for d.

$= 88$ Multiply.

∴ The circumference is about 88 millimeters.

On Your Own

Now You're Ready
Exercises 10–13

Find the circumference of the object. Use 3.14 or $\frac{22}{7}$ for π.

3.
2 cm

4.
14 ft

5.
9 in.

Section 6.1 Circles and Circumference **241**

EXAMPLE 3 **Standardized Test Practice**

The diameter of the new roll of caution tape decreases 3.25 inches after a construction worker uses some of the tape. Which is the best estimate of the circumference of the roll after the decrease?

Ⓐ 9 inches **Ⓑ** 16 inches **Ⓒ** 21 inches **Ⓓ** 30 inches

After the decrease, the diameter of the roll is 10 − 3.25 = 6.75 inches.

$C = \pi d$ Write formula for circumference.

$\approx 3.14 \cdot 6.75$ Substitute 3.14 for π and 6.75 for d.

$\approx 3 \cdot 7$ Round 3.14 down to 3. Round 6.75 up to 7.

$= 21$ Multiply.

 The correct answer is **Ⓒ**.

● **On Your Own**

6. **WHAT IF?** In Example 3, the diameter of the roll of tape decreases 5.75 inches. Estimate the circumference after the decrease.

EXAMPLE 4 **Finding the Perimeter of a Semicircular Region**

A semicircle is one-half of a circle. Find the perimeter of the semicircular region.

The straight side is 6 meters long. The distance around the curved part is half the circumference of a circle with a diameter of 6 meters.

$C = \dfrac{\pi d}{2}$ Divide the circumference by 2.

$\approx \dfrac{3.14 \cdot 6}{2}$ Substitute 3.14 for π and 6 for d.

$= 9.42$ Simplify.

 So, the perimeter is about 6 + 9.42 = 15.42 meters.

● **On Your Own**

Now You're Ready
Exercises 15 and 16

Find the perimeter of the semicircular region.

7. 8. 9.

Laurie's Notes

Example 3
- Ask a volunteer to read the given information.
- Note that this multiple choice question is asking about which *estimate* is best, so rounding the diameter and π is sufficient to select the correct answer.
- **Extension:** "If the diameter decreased by 3.25 inches, how much did the radius decrease?" $3.25 \div 2 = 1.625$ in.

Example 4
- Draw the diagram. It may be helpful to draw the other half of the semicircle as a dotted line.
- Another way to visualize the distance around the curved part is to overlap a circle with a diameter of 6 meters with a semicircle with a straight side of 6 meters.
- **Common Error:** Students may find half the circumference and forget to add the distance across the diameter.

Closure
- **Exit Ticket:** The peso has a diameter of 21 millimeters.
 a. What is the radius of a peso? 10.5 mm
 b. What is the circumference of a peso? about 66 mm

Extra Example 3
In Example 3, the diameter of the roll of tape decreases 1.5 inches. Estimate the circumference of the roll after the decrease. about 27 in.

 On Your Own
6. about 12 in.

Extra Example 4
Find the perimeter of a semicircular region with a radius of 5 yards. about 25.7 yd

 On Your Own
7. about 5.14 ft
8. about 18 cm
9. about 77.1 in.

The Dynamic Planning Tool
Editable Teacher's Resources at *BigIdeasMath.com*

Vocabulary and Concept Check

1. The radius is one-half the diameter.

2. the distance from the center to any point on the circle; This phrase describes the radius of a circle, whereas the other phrases describe the circumference of a circle.

3. *Sample answer:* A lawn game has two circular targets with 28-inch diameters. You lost one. You want to use a length of wire to make a replacement.
$C = \pi d \approx \frac{22}{7} \cdot 28 = 88$
You need a piece of wire 88 inches long.

Practice and Problem Solving

4. 2.5 cm 5. 14 mm

6. $1\frac{3}{4}$ in. 7. 12 cm

8. 4 in. 9. 1.6 ft

10. about 31.4 in.

11. about 44 in.

12. about 56.52 in.

Assignment Guide and Homework Check

Level	Day 1 Activity Assignment	Day 2 Lesson Assignment	Homework Check
Basic	10–12, 25–28	1–3, 5, 7, 9, 13–17	5, 7, 14, 16
Average	10–12, 25–28	1–3, 5, 7, 9, 13–15, 17, 20, 22	5, 7, 15, 20
Advanced	10–12, 25–28	1–3, 13, 16, 18, 20–24	16, 18, 20, 23

Common Errors

- **Exercises 4–9** Students may confuse what they are finding and double the diameter or halve the radius. Remind them that the radius is half the diameter. Encourage students to draw a line representing the radius or diameter for each problem so that they have a visual reference.
- **Exercises 10–12** Students may use the wrong formula for circumference when given a radius or diameter. Remind them of the different equations. If students are struggling with the two equations, tell them to use only one equation and to convert the dimension given to the one in the chosen formula.
- **Exercises 10–12** Students may use $\frac{22}{7}$ when it would be easier to use 3.14 and get frustrated. Remind them to use $\frac{22}{7}$ when the radius or diameter is a multiple of 7.

6.1 Record and Practice Journal

Technology For the Teacher
Answer Presentation Tool
Quiz*Show*

6.1 Exercises

Vocabulary and Concept Check

1. **VOCABULARY** What is the relationship between the radius and the diameter of a circle?

2. **WHICH ONE DOESN'T BELONG?** Which phrase does *not* belong with the other three? Explain your reasoning.

 | the distance around a circle | π times twice the radius |

 | π times the diameter | the distance from the center to any point on the circle |

3. **OPEN-ENDED** Choose a real-life circular object. Explain why you might need to know its circumference. Then find the circumference.

Practice and Problem Solving

Find the radius of the button.

4. 5 cm

5. 28 mm

6. $3\frac{1}{2}$ in.

Find the diameter of the object.

7. 6 cm

8. 2 in.

9. 0.8 ft

Find the circumference of the pizza. Use 3.14 or $\frac{22}{7}$ for π.

10. 10 in.

11. 7 in.

12. 18 in.

Section 6.1 Circles and Circumference

13. **SINKHOLE** A circular sinkhole has a radius of 12 meters. A week later, it has a diameter of 48 meters. How much greater is the circumference of the sinkhole compared to the previous week?

14. **REASONING** Consider the circles *A*, *B*, *C*, and *D*.

a. Copy and complete the table.
b. Which circle has the greatest circumference?
c. Which circle has the least circumference?

Circle	A	B	C	D
Radius		10 inches		50 inches
Diameter	8 feet		2 feet	

Find the perimeter of the window.

15.

16.

Find the circumferences of both circles.

17.

18.

19.

20. **WIRE** A wire is bent to form four semicircles. How long is the wire?

21. **CRITICAL THINKING** Because the ratio $\dfrac{\text{circumference}}{\text{diameter}}$ is the same for every circle, is the ratio $\dfrac{\text{circumference}}{\text{radius}}$ the same for every circle? Explain.

Common Errors

- **Exercise 13** Students may try to compare the circumference of the sink hole without realizing that first they are given a radius and then given the diameter. Make sure that students use the two different formulas to find the circumferences or that they change one to a radius or diameter before using the same circumference formula for both.
- **Exercise 14** Students may try to compare the radii or diameters without converting to the same units. Remind them that they need to have all the same units before comparing.
- **Exercises 15 and 16** Students may find the circumference and forget to divide it in half. Remind them that because it is not a whole circle, they must divide it in half.
- **Exercises 15 and 16** Students may forget to add the diameter onto the perimeter after they have found the circumference part. Remind them that the perimeter includes all the sides, not just the circular part.
- **Exercises 17–19** Students may use the incorrect radius or diameter for the larger or smaller circle. Remind them that they will need to figure out the radius or diameter of the other circle.

 Practice and Problem Solving

13. about 75.36 m

14. a.

Circle	A	B	C	D
Radius	4 feet	10 inches	1 foot	50 inches
Diameter	8 feet	20 inches	2 feet	100 inches

 b. D

 c. B

15. about 7.71 ft

16. about 102.8 cm

17. about 31.4 cm; about 62.8 cm

18. about 28.26 ft; about 44 ft

19. about 69.08 m; about 138.16 m

20. about 200.96 cm

21. yes; Because
$$\frac{\text{circumference}}{\text{radius}} = \frac{2\pi r}{r}$$
$$= \frac{2\pi \cancel{r}}{\cancel{r}}$$
$$= 2\pi,$$
the ratio is the same for every circle.

Differentiated Instruction

Auditory

For students that confuse radius and diameter, emphasize that a **di**ameter **di**vides a circle in half. In an exercise when the radius is needed in a formula and the diameter is given, emphasize that you **di**vide the **di**ameter to find the radius.

Practice and Problem Solving

22. See *Taking Math Deeper*.
23. a. small tire: about 127 rotations; large tire: about 38 rotations
 b. *Sample answer:* A bicycle with large wheels would allow you to travel farther with each rotation of the pedal.
24. a. about 254.34 mm; First find the length of the minute hand. Then find $\frac{3}{4}$ of the circumference of a circle whose radius is the length of the minute hand.
 b. See Additional Answers.

Fair Game Review

25. 22 ft 26. 20 m
27. 65 in. 28. D

Mini-Assessment

Find the circumference of the circle. Use 3.14 or $\frac{22}{7}$ for π.

1.
 about 75.36 ft

2.
 about 88 in.

3.
 about 37.68 m

4.
 about 12.56 cm

5. Find the perimeter of the window.
 about 20.56 ft

Taking Math Deeper

Exercise 22

Speed records for flights around the world are kept by Fédération Aeronautique Internationale. See *fai.org*. There are many different categories, depending on the type of plane and other factors.

 a. Find the circumference of the Tropic of Cancer.
$$C = 2\pi r$$
$$= 11{,}708\pi$$
$$\approx 36{,}763 \text{ km}$$

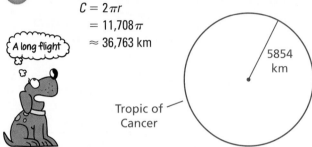

② Here is a route for a recent attempt to break the speed record for a flight around the world.

③ b. One recent record was set by Claude Delorme of France. He flew around the world at an average speed of 1231 kilometers per hour (westbound).

At this rate, it took Claude about 30 hours to fly around the world. If your students estimate a time that is less than this, you might point out that they should try for the next world record!

Project

Draw a picture of the path you would take to fly around the world in an attempt to set a new world record.

Reteaching and Enrichment Strategies

If students need help...	If students got it...
Resources by Chapter • Practice A and Practice B • Puzzle Time Record and Practice Journal Practice Differentiating the Lesson Lesson Tutorials Skills Review Handbook	Resources by Chapter • Enrichment and Extension Start the next section

22. **AROUND THE WORLD** "Lines" of latitude on Earth are actually circles. The Tropic of Cancer is the northernmost line of latitude at which the Sun appears directly overhead at noon. The Tropic of Cancer has a radius of 5854 kilometers.

 To qualify for an around-the-world speed record, a pilot must cover a distance no less than the circumference of the Tropic of Cancer, cross all meridians, and end on the same airfield where he started.

 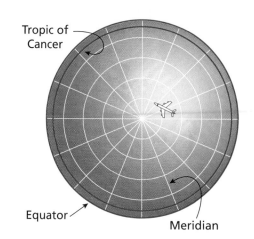

 a. What is the minimum distance that a pilot must fly to qualify for an around-the-world speed record?

 b. **CHOOSE TOOLS** Estimate the time it would take for a pilot to qualify for the speed record.

23. **BICYCLE** Bicycles in the late 1800s looked very different than they do today.

 a. How many rotations does each tire make after traveling 600 feet? Round your answers to the nearest whole number.

 b. Would you rather ride a bicycle made with two large wheels or two small wheels? Explain.

24. **Critical Thinking** The length of the minute hand is 150% of the length of the hour hand.

 a. What distance will the tip of the minute hand move in 45 minutes? Explain how you found your answer.

 b. In 1 hour, how much farther does the tip of the minute hand move than the tip of the hour hand? Explain how you found your answer.

 Fair Game Review What you learned in previous grades & lessons

Find the perimeter of the figure. *(Skills Review Handbook)*

25.

26.

27.

28. **MULTIPLE CHOICE** What is the median of the data set? *(Section 5.5)*

 12, 25, 16, 9, 5, 22, 27, 20

 Ⓐ 7 Ⓑ 16 Ⓒ 17 Ⓓ 18

6.2 Perimeters of Composite Figures

COMMON CORE STATE STANDARDS

6.EE.2c

Essential Question How can you find the perimeter of a composite figure?

1 ACTIVITY: Finding a Pattern

Work with a partner. Describe the pattern of the perimeters. Use your pattern to find the perimeter of the tenth figure in the sequence. (Each small square has a perimeter of 4.)

a.

b.

c.

2 ACTIVITY: Finding a Distance

Work with a partner.

a. Estimate the distance to the gold.
b. Estimate the distance to the silver.

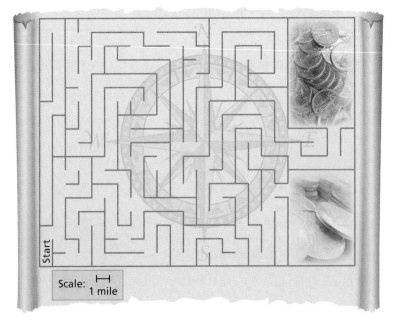

Scale: 1 mile

246 Chapter 6 Circles and Area

Laurie's Notes

Introduction

For the Teacher
- **Goal:** Students will develop an intuitive understanding of finding perimeters of composite figures by focusing on their component parts.
- Today's activities take a problem-solving approach to the topic of finding perimeters of composite figures.
- Explain to students that composite figures are composed of more than one geometric shape.

Motivate
- Draw the following "equations." Ask students how they would find the perimeter of the last figure knowing the dimensions of the figures on the left side of the equation.

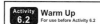

- These puzzles should help focus students' thinking on how they will find the perimeter of a composite figure, without defining composite figures!

Activity Notes

Activity 1
- Students should describe the pattern of the perimeter in words. For instance, the figures in part (a) have perimeters which are *increasing by 2* each time.
- **Common Error:** Remind students that perimeter is the distance *around* the figure and that interior segments are not included.
- **Representation:** For part (c), if students leave their answer in terms of π, they are more likely to see the pattern.
- **?** "How is the perimeter changing in part (a)? Can you explain why?"
 increasing by 2; Explanations will vary.
- **?** "How is the perimeter changing in part (b)? Can you explain why?"
 increasing by 4; Explanations will vary.
- **?** "How is the perimeter changing in part (c)? Can you explain why?"
 increasing by π; Explanations will vary.

Activity 2
- This is an adaptation of a perimeter problem. Students are finding the total distance traveled instead of the distance around a figure.
- Explain the scale for the maze. Have students use the edge of their paper to measure and approximate the perimeter of the maze. This will help students understand how to use the scale in this problem.
- **?** "Is the distance to the gold or silver greater than the perimeter of the maze?" yes for both the gold and silver

Common Core State Standards

6.EE.2c Evaluate expressions at specific values of their variables. Include expressions that arise from formulas used in real-world problems. Perform arithmetic operations, including those involving whole-number exponents, in the conventional order when there are no parentheses to specify a particular order (Order of Operations).

Previous Learning
Students should know how to find the perimeter and circumference for common shapes.

Activity Materials
Textbook
• square tiles

Start Thinking! and Warm Up

Activity 6.2 Start Thinking! For use before Activity 6.2

Activity 6.2 Warm Up For use before Activity 6.2

Find the perimeter or circumference of the figure described.

1. square with side length 4 cm
2. rectangle with length 5 ft and width 3.5 ft
3. rectangle with length 19 in. and width 7 in.
4. triangle with side lengths 7 m, 8 m, and 10 m
5. circle with radius 9 in.
6. circle with diameter 20 ft

6.2 Record and Practice Journal

Essential Question How can you find the perimeter of a composite figure?

1 ACTIVITY: Finding a Pattern

Work with a partner. Describe the pattern of the perimeters. Use your pattern to find the perimeter of the tenth figure in the sequence. (Each small square has a perimeter of 4.)

a. The perimeter of each figure is 2 greater than the last; 22.

b. The perimeter of each figure is 4 greater than the last; 40.

c.

The perimeter of each figure is π greater than the last; about 33.4.

T-246

English Language Learners
Visual
Have students find rectangular, trapezoidal, triangular, and circular regions in the school and on the school grounds. Then ask students to calculate the perimeter of these regions.

6.2 Record and Practice Journal

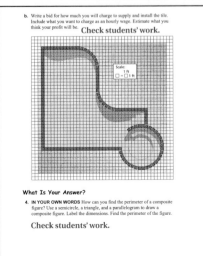

Laurie's Notes

Activity 3
- **Real World:** Share with your students that builders and contractors submit bids for work that they want to do. If more than one bid is received, the consumer selects the builder or contractor based upon a number of factors, one of which is the cost that is quoted.
- In this problem, assume that each pair of students is bidding on the job. They could even have a name for their two-person company. Their bid sheet (work done) should be neat and organized and easily understood by the pool's owner—you!
- Explain the term *$4 per linear foot* so that all students understand. In construction, when the width of material (in this case, the tile) is predetermined, the cost is given in terms of the length, not in terms of area (square foot).
- Students should count the number of tiles surrounding the pool and use the scale given on the diagram to determine how much tile is needed.
- Make sure students include a labor charge based on the information given. Students will set their own hourly wage. Is it realistic?
- Discuss general results with the whole class. Compare the quotes, separating out the material and labor.
- **Extension:** List the hourly wages and the number of hours of labor that each company charges. Use this data and find the mean, median, and range of the hourly wages and the total labor charge.

What Is Your Answer?
- **Neighbor Check:** Have students work independently and then have their neighbor check their work. Have students discuss any discrepancies.

Closure
- Find the perimeter of the arrow. 60 cm

Technology For the Teacher
Dynamic Classroom
The Dynamic Planning Tool
Editable Teacher's Resources at *BigIdeasMath.com*

T-247

3 ACTIVITY: Submitting a Bid

Work with a partner. You want to bid on a tiling contract. You will be supplying and installing the brown tile that borders the swimming pool.

- **Your cost for the tile is $4 per linear foot.**
- **It takes about 15 minutes to prepare, install, and clean each foot of tile.**

a. How many brown tiles are needed for the border?

b. Write a bid for how much you will charge to supply and install the tile. Include what you want to charge as an hourly wage. Estimate what you think your profit will be.

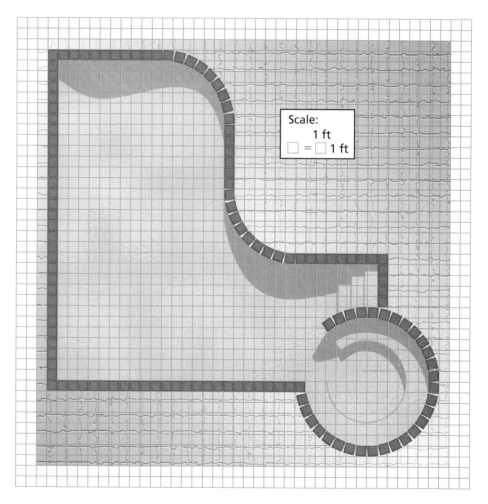

Scale:
1 ft
☐ = ☐ 1 ft

What Is Your Answer?

4. **IN YOUR OWN WORDS** How can you find the perimeter of a composite figure? Use a semicircle, a triangle, and a parallelogram to draw a composite figure. Label the dimensions. Find the perimeter of the figure.

Practice → Use what you learned about perimeters of composite figures to complete Exercises 3–5 on page 250.

Section 6.2 Perimeters of Composite Figures

6.2 Lesson

Key Vocabulary
composite figure, p. 248

A **composite figure** is made up of triangles, squares, rectangles, semicircles, and other two-dimensional figures. Here are two examples.

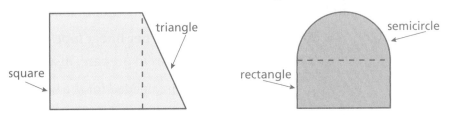

To find the perimeter of a composite figure, find the distance around the figure.

EXAMPLE 1 Finding a Perimeter Using Grid Paper

Each square on the grid paper is 1 square inch. Estimate the perimeter of the arrow.

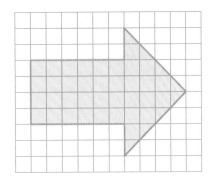

Count the number of grid square lengths around the arrow. There are 20.

Count the number of diagonal lengths around the arrow. There are 8.

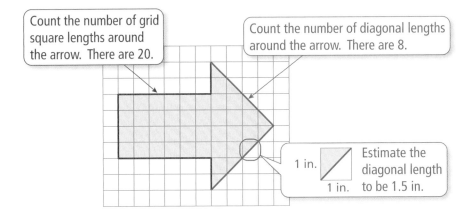

1 in. Estimate the diagonal length to be 1.5 in.
1 in.

Length of 20 grid square lengths: $20 \times 1 = 20$ inches

Length of 8 diagonal lengths: $8 \times 1.5 = 12$ inches

∴ The perimeter is about $20 + 12 = 32$ inches.

On Your Own

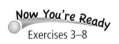
Exercises 3–8

1. Each square on the grid paper is 1 square foot. Estimate the perimeter of the red figure.

2. Measure the diagonal of a square whose area is exactly one square foot. Is the diagonal length closer to 1.5 feet or 1.4 feet? Explain.

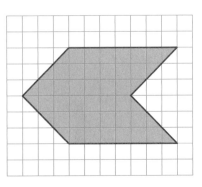

248 Chapter 6 Circles and Area ◀ Multi-Language Glossary at BigIdeasMath✓com.

Laurie's Notes

Introduction

Connect
- **Yesterday:** Students used a variety of problem-solving skills to find the perimeter of several composite figures.
- **Today:** Students will use formulas to find the perimeter of composite figures.

Motivate
- If you have a set of tangrams, use them to share a puzzle with students. Place the 7 pieces on the overhead and ask a volunteer to rearrange the pieces into a square. The solution is shown. Then rearrange the pieces to form the bird shown. Ask how the perimeter changes from the square to the bird.

- The goal is not to find the perimeter of the bird, only to recognize that as more segments are exposed (on the perimeter), the perimeter will increase.
- The figures that students work with today are composed of common geometric shapes. Students should be familiar with how to find the perimeter of each.
- Introduce the vocabulary word *composite figure*.

Lesson Notes

Example 1
- The arrow is a shape that students discussed at the beginning of yesterday's lesson.
- Students are asked to estimate the perimeter, because the slanted lengths are irrational numbers.
- **Common Error:** Students may think the diagonal of the square is 1 inch because the side lengths are 1 inch. Have them think about the *short-cut* of walking across the diagonal of a square versus walking the two side lengths.

On Your Own
- Question 1 is very similar to the first example.
- Ask a volunteer to share their thinking and solution to the problem.

Goal Today's lesson is finding the perimeters of **composite figures**.

Lesson Materials
Introduction
• tangrams

Start Thinking! and Warm Up

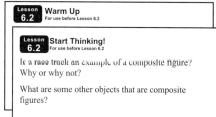

Extra Example 1
Each square on the grid paper is 1 square centimeter. Estimate the perimeter of the arrow.

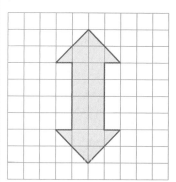

about 24 cm

On Your Own

1. about 32 ft

2. It is closer to 1.4 ft; The diagonal measures about 17 inches. By multiplying 17 inches by $\frac{1 \text{ ft}}{12 \text{ in.}}$, you find that 17 inches is about 1.4 feet.

T-248

Extra Example 2

The figure is made of a semicircle and a square. Find the perimeter.

5 m

5 m

about 22.85 m

Extra Example 3

Find the perimeter of the running track in Example 3 when the radius of the semicircle is 20 yards and the length of the rectangle is 50 yards. about 225.6 yd

On Your Own

3. about 41.12 m
4. about 291.3 seconds

Differentiated Instruction

Auditory
Review the two formulas for circumference, $C = \pi d$ and $C = 2\pi r$. On the board or overhead, derive the formulas for diameter, $d = \dfrac{C}{\pi}$, and radius, $r = \dfrac{C}{2\pi}$. Have students copy the four formulas in their notebooks to use as a quick reference when doing word problems. Remind students that the circumference of a semicircle is half of a full circle.

Laurie's Notes

Example 2
- Draw the diagram for the problem.
- ? "What is the diameter of the semicircle?" 10 ft
- Explain to students that although the third side of the triangle is shown (which is also the diameter of the circle), it is not part of the perimeter.
- ? "Could $\dfrac{22}{7}$ be used for π?" yes "Why do you think 3.14 was used?" The diameter is 10, which is not a multiple of 7.

Example 3
- Draw the diagram for the problem.
- ? "What are the dimensions of the rectangle?" 100 m by 64 m
- ? "What is the diameter of the semicircles?" 64 m
- Make sure students understand that they have to add both straight-aways of the running track. This is where the 100 + 100 comes from in the answer statement.
- Point out to students that in Examples 2 and 3 the answers are stated to be *about* a certain number. The answers are approximations because 3.14 is only an approximation for π.

On Your Own
- Students may think there is not enough information to solve Question 3. The quadrilateral is a square, so the diameter of the semicircles is 8 meters.
- Question 4 is challenging. The perimeter of the outfield is part of the circumference of a circle with a radius of 225 feet, not 300 feet.
- **Neighbor Check:** Have students work independently and then have their neighbor check their work. Have students discuss any discrepancies.
- When students have finished, ask volunteers to show their work at the board.

Closure
- Find the perimeter of the room shown. 64 ft

14 ft

4 ft

5.5 ft

12.5 ft

The Dynamic Planning Tool
Editable Teacher's Resources at *BigIdeasMath.com*

EXAMPLE 2 Finding a Perimeter

The figure is made up of a semicircle and a triangle. Find the perimeter.

The distance around the triangular part of the figure is 6 + 8 = 14 feet.

The distance around the semicircle is one-half the circumference of a circle with a diameter of 10 feet.

$$\frac{C}{2} = \frac{\pi d}{2}$$ Divide the circumference by 2.

$$\approx \frac{3.14 \cdot 10}{2}$$ Substitute 3.14 for π and 10 for d.

$$= 15.7$$ Simplify.

∴ The perimeter is about 14 + 15.7 = 29.7 feet.

EXAMPLE 3 Finding a Perimeter

The running track is made up of a rectangle and two semicircles. Find the perimeter.

The semicircular ends of the track form a circle with a radius of 32 meters. Find its circumference.

$$C = 2\pi r$$ Write formula for circumference.

$$\approx 2 \cdot 3.14 \cdot 32$$ Substitute 3.14 for π and 32 for r.

$$= 200.96$$ Multiply.

∴ The perimeter is about 100 + 100 + 200.96 = 400.96 meters.

On Your Own

3. The figure is made up of a square and two semicircles. Find the perimeter.

4. Running at 4 ft/sec, how long would it take a person to run around the baseball field?

Section 6.2 Perimeters of Composite Figures 249

6.2 Exercises

Vocabulary and Concept Check

1. **NUMBER SENSE** Is the perimeter of a composite figure *greater than*, *less than*, or *equal to* the sum of the perimeters of each figure separately? Explain.

2. **OPEN-ENDED** Draw a composite figure formed by a parallelogram and a trapezoid.

Practice and Problem Solving

Each square on the grid paper is 1 square inch. Estimate the perimeter of the figure.

① 3.

4.

5.

6.

7.

8.
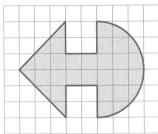

Find the perimeter of the figure.

② 9.

10.

11.

12. **ERROR ANALYSIS** Describe and correct the error in finding the perimeter of the figure.

 Perimeter = 4 + 3 + 4 + 5 + 4 + 5
= 25 in.

250 Chapter 6 Circles and Area

Assignment Guide and Homework Check

Level	Day 1 Activity Assignment	Day 2 Lesson Assignment	Homework Check
Basic	3–5, 20–24	1, 2, 6, 7, 9, 10, 12, 13, 15, 16	6, 9, 12, 13
Average	3–5, 20–24	1, 2, 6, 7, 9, 10, 12, 13, 15–17	6, 9, 12, 13
Advanced	3–5, 20–24	1, 2, 8–14 even, 17–19	8, 10, 12, 14

For Your Information
- **Exercise 1** Some students may include the choice *equal to*. This is true when composite figures do *not* share a common side, but do share a common vertex.

Common Errors
- **Exercises 3–8** Students may count the sides of the squares inside the figure. Remind them that they are finding the perimeter, so they only need to count the outside lengths.
- **Exercises 5 and 8** Students may have difficulty estimating the curved portions of the figure. Give students tracing paper and tell them to trace the line as straight instead of curved and compare it with the length of the side of a square to help them estimate the length.
- **Exercises 9–11, 13–15** Students may include the dotted lines in their calculation of the perimeter. Remind them that the dotted lines are for reference and sometimes give information to find another length. Only the outside lengths are counted.

Technology For the Teacher
Answer Presentation Tool
QuizShow

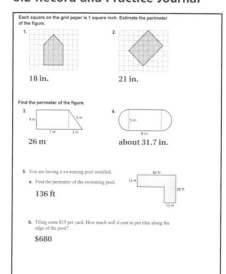
6.2 Record and Practice Journal

Vocabulary and Concept Check

1. less than and equal to; The perimeter is *less than* when figures making up a composite figure share a common side (dashed line).

 The perimeter is *equal to* when the figures making up a composite figure share a common vertex.

2. Sample answer:

Practice and Problem Solving

3. 19.5 in. 4. 20 in.
5. 25.5 in. 6. 28 in.
7. 19 in. 8. 30 in.
9. 56 m 10. 82 in.
11. 30 cm
12. The length of the rectangle was counted twice. Perimeter = 4 + 3 + 4 + 5 + 4 = 20 in.

Practice and Problem Solving

13. about 26.85 in.
14. about 50.26 in.
15. about 36.84 ft
16. a. 1875 ft b. $16,875
17. See *Taking Math Deeper*.
18. See Additional Answers.
19. Yes; *Sample answer:* By adding the triangle shown by the dashed line to the L-shaped figure, you *reduce* the perimeter.

Fair Game Review

20. 19.35 21. 279.68
22. 153.86 23. 205
24. D

Mini-Assessment
Find the perimeter of the figure.

1.

2.

3. Find the perimeter of the garden.

about 58 ft

T-251

Taking Math Deeper
Exercise 17

This problem asks students to add 15 feet to the perimeter of a flower garden. There are infinitely many ways to do this. This is a good problem to assign to pairs of students and then have students share the way they increased the perimeter.

 Draw a diagram. Label the lengths.

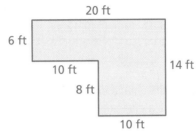

② a. Find the perimeter.

$P = 6 + 20 + 14 + 10 + 8 + 10 = 68$ ft

③ b. Add 15 feet to the perimeter.

Notice that the additional 10 feet is a "wash." It is both added and subtracted from the perimeter.

Project
Design a garden for your school. Prepare a poster that you can use to present your idea to the principal.

Reteaching and Enrichment Strategies

If students need help...	If students got it...
Resources by Chapter • Practice A and Practice B • Puzzle Time Record and Practice Journal Practice Differentiating the Lesson Lesson Tutorials Skills Review Handbook	Resources by Chapter • Enrichment and Extension Start the next section

Find the perimeter of the figure.

13.

14.

15.

16. **PASTURE** A section of land is to be fenced in for a horse pasture.

 a. Find the perimeter of the pasture.
 b. Fencing costs $27 per yard. How much will it cost to fence in the pasture?

17. **MODELING** A garden is built on a rooftop.

 a. Find the perimeter of the garden.
 b. You want to increase the perimeter by 15 feet. Draw a diagram of how you would do this. Is there more than one way? Explain.

18. In Example 3 on page 249, the track has six lanes. Explain why the starting points for the six runners are staggered. Draw a diagram as part of your explanation.

19. **Critical Thinking** Is it possible to add a figure to a composite figure without increasing its perimeter? Explain and draw a diagram to support your answer.

Fair Game Review What you learned in previous grades & lessons

Evaluate the expression. *(Section 3.2)*

20. $2.15(3)^2$
21. $4.37(8)^2$
22. $3.14(7)^2$
23. $8.2(5)^2$

24. **MULTIPLE CHOICE** Which expression represents "6 less than 5 times a number x?" *(Section 1.2)*

 Ⓐ $(6-5)x$ Ⓑ $6-5x$ Ⓒ $\dfrac{6}{5x}$ Ⓓ $5x-6$

6 Study Help

You can use a **word magnet** to organize formulas or phrases that are associated with a vocabulary word or term. Here is an example of a word magnet for circle.

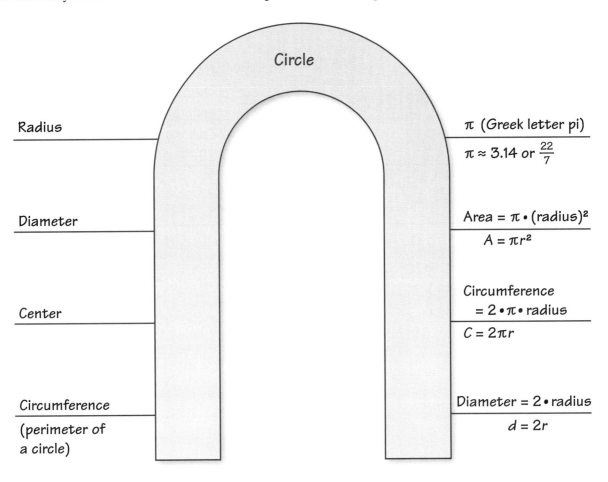

On Your Own

Make a word magnet to help you study these topics.

1. semicircle
2. composite figure
3. perimeter

After you complete this chapter, make word magnets for the following topics.

4. area of a circle
5. area of a composite figure

"I'm trying to make a word magnet for happiness, but I can only think of two words."

Sample Answers

1.

2.

3.

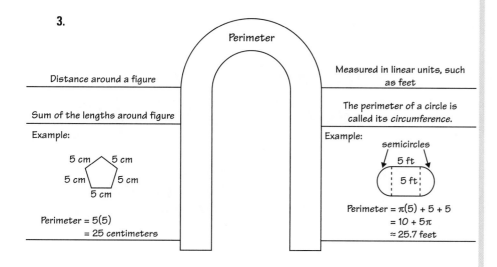

List of Organizers
Available at *BigIdeasMath.com*

Comparison Chart
Concept Circle
Definition (Idea) and Example Chart
Example and Non-Example Chart
Formula Triangle
Four Square
Information Frame
Information Wheel
Notetaking Organizer
Process Diagram
Summary Triangle
Word Magnet
Y Chart

About this Organizer

A **Word Magnet** can be used to organize information associated with a vocabulary word or term. As shown, students write the word or term inside the magnet. Students write associated information on the blank lines that "radiate" from the magnet. Associated information can include, but is not limited to: other vocabulary words or terms, definitions, formulas, procedures, examples, and visuals. This type of organizer serves as a good summary tool because any information related to a topic can be included.

Technology For the Teacher
Vocabulary Puzzle Builder

Answers

1. 18 cm
2. 22 in.
3. 18 in.
4. 21.42 in.
5. 21 in.
6. about 31.4 mm
7. about 4.71 ft
8. about 22 cm
9. 88 in.
10. about 49.12 ft
11. about 7.71 ft
12. about 22 mm
13. 60 ft
14. about 15.68 in.

Alternative Quiz Ideas

100% Quiz **Math Log**
Error Notebook Notebook Quiz
Group Quiz Partner Quiz
Homework Quiz Pass the Paper

Math Log
Ask students to keep a math log for the chapter. Have them include diagrams, definitions, and examples. Everything should be clearly labeled. It might be helpful if they put the information in a chart. Students can add to the log as new topics are introduced.

Assessment Book

Reteaching and Enrichment Strategies

If students need help...	If students got it...
Resources by Chapter • Study Help • Practice A and Practice B • Puzzle Time Lesson Tutorial *BigIdeasMath.com* Practice Quiz Practice from the Test Generator	Resources by Chapter • Enrichment and Extension • School-to-Work Game Closet at *BigIdeasMath.com* Start the next section

Technology For the Teacher

Answer Presentation Tool
Big Ideas Test Generator

6.1–6.2 Quiz

Check It Out
Progress Check
BigIdeasMath.com

1. The diameter of a circle is 36 centimeters. Find its radius. *(Section 6.1)*

2. The radius of a circle is 11 inches. Find its diameter. *(Section 6.1)*

Each square on the grid paper is 1 square inch. Estimate the perimeter of the figure. *(Section 6.2)*

3.
4.
5.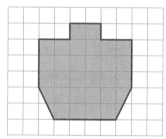

Find the circumference of the circle. Use 3.14 or $\frac{22}{7}$ for π. *(Section 6.1)*

6. 5 mm
7. 1.5 ft
8. 7 cm

Find the perimeter of the figure. *(Section 6.1 and Section 6.2)*

9.
10.
11.

12. **BUTTON** What is the circumference of a circular button with a diameter of 7 millimeters? *(Section 6.1)*

13. **GARDEN** Part of a yard is being fenced in to make a vegetable garden. How many feet of fencing are needed to surround the garden? *(Section 6.2)*

14. **BAKING** A baker is using two circular pans. The larger pan has a diameter of 12 inches. The smaller pan has a diameter of 7 inches. How much greater is the circumference of the larger pan than that of the smaller pan? *(Section 6.1)*

6.3 Areas of Circles

Essential Question How can you find the area of a circle?

COMMON CORE STATE STANDARDS
6.EE.2c

1 ACTIVITY: Estimating the Area of a Circle

Work with a partner. Each square in the grid is 1 unit by 1 unit.

a. Find the area of the large 10-by-10 square.

b. Copy and complete the table.

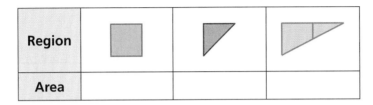

c. Use your results to approximate the area of the circle. Explain your reasoning.

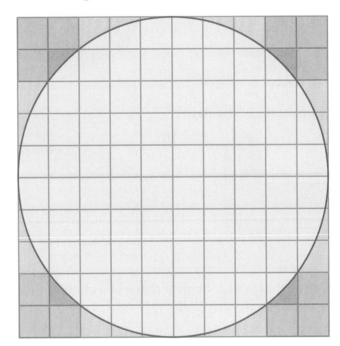

d. Fill in the blanks. Explain your reasoning.

Area of large square = ◻ • 5^2

Area of circle = ◻ • 5^2

e. What can you conclude?

254 Chapter 6 Circles and Area

Laurie's Notes

Introduction

For the Teacher

- **Goal:** Students will use a counting technique to find the area of a circle. This leads to the development of a formula for the area of a circle.
- Today's investigation uses a hands-on data collection approach for finding the area of a circle.
- **Preparation:** Trace and cut 2 circles from file folder weight paper. Cut one circle into 4 sectors (pie-shaped pieces) and one circle into 12 sectors. These will be used for Activity 2.

Motivate

- Sketch this diagram.
- ❓ "The lawn at my house is a square. A rotating sprinkler reaches each side but misses the four corners. What percent of my lawn do you think is getting watered?"
- Students should visually estimate at this stage. They should guess close to 75%. Record their guesses and return to them at the end of the class.

Activity Notes

Activity 1

- Students should observe that the circle is inscribed in a large square that has an area of 100 square units.
- As a class, discuss the area of each colored region. The orange square is a unit square. The green triangle is close to half of a 1 × 2 rectangle. The pink triangle is about half the unit square.
- ❓ "How does knowing the area of these pieces help you approximate the area of the circle?" Ask the question and let students begin their work.
- When groups have finished, have several record their work at the board. They should have Area of circle $= 100 - 4(3) - 8(1) - 4\left(\frac{1}{2}\right) = 78$ units2.
- ❓ "The area of the large square is 100 units2. Can you write 100 as the product of a number times 5^2?" $100 = 4(5^2)$
- ❓ "Let's do the same with the area of the circle. Can you write 78 as the product of a number times 5^2?" $78 = 3.12(5^2)$
- ❓ "Thinking about the circle, what part of the circle is 5 units long?" radius
- It is very likely that at this point students may guess that 3.12 is supposed to be π and may want to know what they did wrong because they got 3.12 instead of 3.14. It's important to say that they did nothing wrong. The areas of the two triangles are approximations because they aren't even triangles—there is a curved edge.
- The conclusion you are hoping for is that the area of this circle is πr^2.
- ❓ "Do you think this is a formula that works for all circles?" open-ended

Common Core State Standards

6.EE.2c Evaluate expressions at specific values of their variables. Include expressions that arise from formulas used in real-world problems. Perform arithmetic operations, including those involving whole-number exponents, in the conventional order when there are no parentheses to specify a particular order (Order of Operations).

Previous Learning

Students should know how to find areas of parallelograms.

Activity Materials
Textbook
• file folder

Start Thinking! and Warm Up

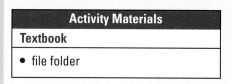

6.3 Record and Practice Journal

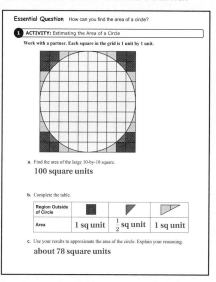

T-254

Differentiated Instruction

Visual

To help students visualize the relationship between area and radius, have them use a compass and grid paper to draw a circle with a radius of 5 units. Ask them to estimate the area of the circle by counting the squares. They should get an estimate that is slightly greater than $3r^2$. Then have students draw a square on the grid paper that has a vertex at the center of the circle and two perpendicular radii as sides. Students should see that one-quarter of the area is enclosed in the square. Because the area of the square is r^2, the area of the circle is less than $4r^2$. So, the area of the circle is between $3r^2$ and $4r^2$, or more precisely πr^2.

6.3 Record and Practice Journal

Laurie's Notes

Activity 2

- In this activity, a different approach is used to develop the area formula.
- To connect the two activities, tell students that you want to think about the circle in the large square.
- ❓ "If you cut the circle into 4 parts and rearranged them, do you think it would make a shape that was familiar?"
- Take out the 4 prepared pieces. Arrange the 4 pieces in a circle and then rearrange them in a fashion similar to the design in the book. The curvature of the circle is still very apparent, so students will not suggest a similarity to any figure they know.
- ❓ "What if the circle were cut into smaller pieces and then rearranged?"
- Take out the 12 prepared pieces. Arrange the 12 pieces in a circle and then rearrange them in a fashion similar to the design in the book. The curvature of the circle is much less apparent.
- **Big Idea:** The 24 pieces approximate a parallelogram. The curvature of the circle is not obvious. The area of a parallelogram is base × height.

 The base is half the circumference. (Pieces are alternated so that $\frac{1}{2}$ the circumference is on the top and $\frac{1}{2}$ is on the bottom.) The height is the radius of the original circle.

 $$\begin{aligned} \text{Area} &= \text{base} \times \text{height} \\ &= \frac{1}{2}C \cdot r \\ &= \frac{1}{2}(2\pi r)r \\ &= \pi r^2 \end{aligned}$$

What Is Your Answer?

- **Neighbor Check:** Have students work independently and then have their neighbor check their work. Have students discuss any discrepancies.

Closure

- What percent of the lawn is getting watered? 78%
- **Connection:** This is an example of a geometric probability question where the area of the circle is being compared to the area of the square. In Activity 1, students found the area of the square and the circle.

Technology For the Teacher

The Dynamic Planning Tool
Editable Teacher's Resources at *BigIdeasMath.com*

2 ACTIVITY: Approximating the Area of a Circle

Work with a partner.

a. Draw a circle. Label the radius as r.

b. Divide the circle into 24 equal sections.

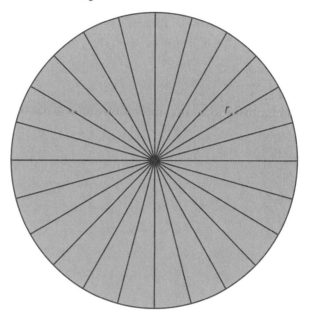

c. Cut the sections apart. Then arrange them to approximate a parallelogram.

d. What is the approximate height and base of the parallelogram?

e. Find the area of the parallelogram. What can you conclude?

What Is Your Answer?

3. **IN YOUR OWN WORDS** How can you find the area of a circle?

4. Write a formula for the area of a circle with radius r. Find an object that is circular. Use your formula to find the area.

Practice

Use what you learned about areas of circles to complete Exercises 3–5 on page 258.

6.3 Lesson

Check It Out
Lesson Tutorials
BigIdeasMath.com

Key Idea

Area of a Circle

Words The area A of a circle is the product of π and the square of the radius.

Numbers $A = \pi r^2$

EXAMPLE 1 Finding Areas of Circles

a. **Find the area of the circle. Use $\dfrac{22}{7}$ for π.**

Estimate $3 \times (7)^2 \approx 3 \times 50 = 150$

$$A = \pi r^2 \qquad \text{Write formula for area.}$$

$$\approx \dfrac{22}{7} \cdot (7)^2 \qquad \text{Substitute } \dfrac{22}{7} \text{ for } \pi \text{ and 7 for } r.$$

$$= \dfrac{22}{\cancel{7}} \cdot \overset{7}{\cancel{49}} \qquad \text{Evaluate } (7)^2. \text{ Divide out the common factor.}$$

$$= 154 \qquad \text{Multiply.}$$

∴ The area is about 154 square centimeters.

Reasonable? $154 \approx 150$ ✓

b. **Find the area of the circle. Use 3.14 for π.**

The radius is $26 \div 2 = 13$ inches.

Estimate $3 \times (13)^2 \approx 3 \times 170 = 510$

$$A = \pi r^2 \qquad \text{Write formula for area.}$$

$$\approx 3.14 \cdot (13)^2 \qquad \text{Substitute 3.14 for } \pi \text{ and 13 for } r.$$

$$= 3.14 \cdot 169 \qquad \text{Evaluate } (13)^2.$$

$$= 530.66 \qquad \text{Multiply.}$$

∴ The area is about 530.66 square inches.

Reasonable? $530.66 \approx 510$ ✓

On Your Own

Now You're Ready
Exercises 3–12

1. Find the area of a circle with a radius of 6 feet. Use 3.14 for π.

2. Find the area of a circle with a diameter of 28 meters. Use $\dfrac{22}{7}$ for π.

Laurie's Notes

Introduction

Connect
- **Yesterday:** Students gained an intuitive understanding about how to find the area of a circle.
- **Today:** Students will use the formula for area to solve real-life problems.

Motivate
- "At a pizza restaurant, you have a choice of ordering a 16-inch pizza or a 12-inch pizza for dinner. What measurement is being used to describe the pizza?" the diameter
- "Both pizzas are cut into the same number of pieces. Do both pizzas give you the same amount of pizza?" No, the size of the slices is different.
- "How would you figure out how many times more pizza you get with the 16-inch pizza than with the 12-inch pizza?" Listen for students to mention the idea of comparing the areas of the pizzas.
- Mention to students that after today's lesson, they will be able to answer this question. You may want to return to this question at the end of class.

Lesson Notes

Key Idea
- Discuss the formula, written in words and numbers.
- **Common Error:** Students square the product of π and r, instead of just the radius.

Example 1
- Work through each example. Remind students that pi can be rounded to 3 when estimating.
- **Common Error:** Make sure students are reading the figures correctly. A line segment from the center to the outside edge indicates a radius. A line segment across the circle through the center is a diameter.
- "What would a reasonable answer be?" Answers will vary, but should be similar to what is shown in the text.
- "In part(a), what estimate should be used for π? Why?" $\frac{22}{7}$; The radius is a multiple of 7.
- Remind students to label answers with the appropriate units.
- **Teaching Tip:** Many students view a diagram differently than a written problem: *Find the area of a circle with radius 7 centimeters.* The visual learner is aided by a simple diagram.
- **Common Error:** For part (b), students may forget to divide the diameter by 2 to find the radius before substituting in the area formula.
- If students are not using calculators, ask them to give a reasonable estimate for 3.14×169 before they multiply.

On Your Own
- **Neighbor Check:** Have students work independently and then have their neighbor check their work. Have students discuss any discrepancies.

Goal Today's lesson is finding the area of a circle and a semicircle.

Start Thinking! and Warm Up

Lesson 6.3 Warm Up For use before Lesson 6.3

Lesson 6.3 Start Thinking! For use before Lesson 6.3

Two approximations for π are $\frac{22}{7}$ and 3.14. In finding the area of a circle, when is it easier to use $\frac{22}{7}$? When is it easier to use 3.14?

Write a word problem involving the area of a circular object. Exchange problems with a classmate and solve your classmate's problem. Is it easier to use $\frac{22}{7}$ or 3.14 for π to solve your classmate's problem?

Extra Example 1
a. Find the area of a circle with a radius of 21 feet. Use $\frac{22}{7}$ for π.
 about 1386 ft^2
b. Find the area of a circle with a diameter of 16 meters. Use 3.14 for π.
 about 200.96 m^2

On Your Own
1. about 113.04 ft^2
2. about 616 m^2

T-256

Extra Example 2

The 1893 World's Fair in Chicago boasted the first ever Ferris wheel with a diameter of 250 feet. How far did a person travel in one revolution of the wheel? about 785 ft

 On Your Own

 3. the diameter of the tire

Extra Example 3

Find the area of a semicircle with a radius of 9 inches. about 127.17 in.²

 On Your Own

 4. about 25.12 m²

 5. about 9.8125 yd²

 6. about 189.97 cm²

English Language Learners

Vocabulary

Discuss with students the word *approximation*. Give and ask for examples of times they use approximations in their daily life. (e.g., "I'll meet you in ten minutes.") Discuss when using an approximation for an answer is as useful as an exact answer. Discuss the usefulness of approximations for π, 3.14 and $\frac{22}{7}$.

Laurie's Notes

Example 2
- This monster truck example will grab students' attention!

On Your Own
- Review this together as a class.

Example 3
- ❓ "How do you find the area of a semicircle?" Find $\frac{1}{2}$ the area of the circle.
- ❓ "What is the radius of the semicircle?" 15 ft
- Work through the problem.
- **Extension:** If time permits and students are using the area formula correctly, try the following problems.
 - ❓ "If you make a mistake and use the diameter instead of the radius in the area formula, is the area doubled? Explain." No; It is quadrupled.
 - ❓ "Find the area of the doughnut region." $8\pi \approx 25.12$ in.²

On Your Own
- Check to see that students recognize the difference in how the dimensions are labeled. The radius is marked with a line segment from the center to the outer edge. The diameter is marked with a line segment from one edge through the center to the opposite edge.

Closure
- Find the areas of a 16-inch pizza and a 12-inch pizza. How many times larger is the 16-inch pizza than the 12-inch pizza? about 1.8 times
- Students will find this easier to calculate if the area is kept in terms of π.
- This situation can also be thought of as "how many times more toppings are needed for the larger pizza?"

Technology For the Teacher

The Dynamic Planning Tool
Editable Teacher's Resources at *BigIdeasMath.com*

EXAMPLE 2 Standardized Test Practice

You want to find the distance the monster truck travels when the tires make one 360-degree rotation. Which best describes this distance?

Ⓐ the radius of the tire **Ⓑ** the diameter of the tire
Ⓒ the circumference of the tire **Ⓓ** the area of the tire

The distance the truck travels after one rotation is the same as the distance *around* the tire. So, the circumference of the tire best describes the distance in one rotation.

∴ The correct answer is **Ⓒ**.

On Your Own

3. You want to find the height of one of the tires. Which measurement would best describe the height?

EXAMPLE 3 Finding the Area of a Semicircle

Find the area of the semicircular orchestra pit.

The area of the orchestra pit is one-half the area of a circle with a diameter of 30 feet.

The radius of the circle is $30 \div 2 = 15$ feet.

$$\frac{A}{2} = \frac{\pi r^2}{2}$$ Divide the area by 2.

$$\approx \frac{3.14 \cdot (15)^2}{2}$$ Substitute 3.14 for π and 15 for r.

$$= \frac{3.14 \cdot 225}{2}$$ Evaluate $(15)^2$.

$$= 353.25$$ Simplify.

∴ The area of the orchestra pit is about 353.25 square feet.

On Your Own

Now You're Ready
Exercises 13–15

Find the area of the semicircle.

4.
 8 m

5.
 5 yd

6.
 11 cm

6.3 Exercises

Vocabulary and Concept Check

1. **VOCABULARY** Explain how to find the area of a circle given its diameter.

2. **DIFFERENT WORDS, SAME QUESTION** Which is different? Find "both" answers.

What is the area of a circle with a diameter of 1 m?	What is the area of a circle with a diameter of 100 cm?
What is the area of a circle with a radius of 100 cm?	What is the area of a circle with a radius of 500 mm?

Practice and Problem Solving

Find the area of the circle. Use 3.14 or $\frac{22}{7}$ for π.

3. 9 mm

4. 14 cm

5. 10 in.

6. 3 in.

7. 2 cm

8. 1.5 ft

9. Find the area of a circle with a diameter of 56 millimeters.

10. Find the area of a circle with a radius of 5 feet.

11. **PIZZA** The diameter of a pizza is 16 inches. What is the area?

12. **LIGHTHOUSE** The Hillsboro Inlet Lighthouse lights up how much more area than the Jupiter Inlet Lighthouse?

258 Chapter 6 Circles and Area

Assignment Guide and Homework Check

Level	Day 1 Activity Assignment	Day 2 Lesson Assignment	Homework Check
Basic	3–5, 22–25	1, 2, 6–13, 15	1, 6, 12, 13
Average	3–5, 22–25	1, 2, 7–17 odd, 12, 16	1, 7, 12, 13
Advanced	3–5, 22–25	1, 2, 12, 14, 16–21	12, 14, 16, 18

For Your Information
- **Exercise 16** Tell students that *radii* is the plural form of radius.
- **Exercise 16** Have students complete the table with their answers in terms of pi. The patterns in the table will be easier for the student to see.

Common Errors
- **Exercises 3–8, 13–15** Students may forget to divide the diameter by 2 to find the area. Remind them that they need the *radius* for the area formula of a circle.
- **Exercises 3–8, 13–15** Students may write the incorrect units for the area. Remind them to carefully check the units and to square the units as well.
- **Exercises 3–8** Students may refer to the formula for circumference and forget to square the radius when finding the area. Remind them that the area formula uses the radius squared.
- **Exercises 13–15** Students may forget to divide the area in half. Remind them that they are finding the area of half of a circle so the area is half of a whole circle.

6.3 Record and Practice Journal

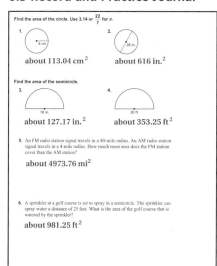

Technology For the Teacher
Answer Presentation Tool
QuizShow

Vocabulary and Concept Check

1. Divide the diameter by 2 to get the radius. Then use the formula $A = \pi r^2$ to find the area.
2. What is the area of a circle with a radius of 100 cm?; about 31,400 cm²; about 7850 cm²

Practice and Problem Solving

3. about 254.34 mm²
4. about 616 cm²
5. about 314 in.²
6. about 7.065 in.²
7. about 3.14 cm²
8. about 1.76625 ft²
9. about 2461.76 mm²
10. about 78.5 ft²
11. about 200.96 in.²
12. about 1444.4 mi²
13. about 628 cm²
14. about 226.08 in.²
15. about 1.57 ft²
16. a.

2	4π in.	4π in.²
4	8π in.	16π in.²
8	16π in.	64π in.²
16	32π in.	256π in.²

 b. The circumference doubles; The area becomes four times as great.

 c. The circumference triples; The area becomes 9 times as great.

 Practice and Problem Solving

17. See *Taking Math Deeper*.

18. greater than; The circle's diameter is one-half as long, so it equals the radius of the semicircle. A diagram shows that the area of the semicircle is greater.

19. about 9.8125 in.²; The two regions are identical, so find one-half the area of the circle.

20–21. See Additional Answers.

 Fair Game Review

22. 44 23. 53
24. 84.56 25. D

Mini-Assessment

Find the area of the circle. Use 3.14 or $\frac{22}{7}$ for π.

1.
about 3.14 ft²

2.
about 154 m²

3.
about 200.96 in.²

4.
about 314 cm²

5. Find the area of the rug.
about 28.26 ft²

Taking Math Deeper

Exercise 17
This problem is not difficult, once a student realizes that the dog's running area is three-quarters of a circle.

 Draw and label a diagram.

 Find the area.

$$\text{Area} = \frac{3}{4}(\text{Area of circle})$$
$$= \frac{3}{4} \cdot \pi \cdot 20^2$$
$$\approx 942 \text{ ft}^2$$

 Ask students to talk about the advantages and disadvantages of this type of outside exercise plan for a dog. What other ways can you allow your dog to be outside and get exercise?
- Build a dog run.
- Fence in a back yard.
- Use an invisible fence.
- Take the dog for a long walk each morning and night.

Project
Design a dog run for your back yard. Do some research to determine the cost of the run.

Reteaching and Enrichment Strategies

If students need help. . .	If students got it. . .
Resources by Chapter • Practice A and Practice B • Puzzle Time Record and Practice Journal Practice Differentiating the Lesson Lesson Tutorials Skills Review Handbook	Resources by Chapter • Enrichment and Extension • School-to-Work Start the next section

Find the area of the semicircle.

13.
 40 cm

14.
 24 in.

15.
 2 ft

16. **REPEATED REASONING** Consider five circles with radii of 1, 2, 4, 8, and 16 inches.

 a. Copy and complete the table. Write your answer in terms of π.
 b. Compare the areas and circumferences. What happens to the circumference of a circle when you double the radius? What happens to the area?
 c. What happens when you triple the radius?

Radius	Circumference	Area
1	2π in.	π sq. in.
2		
4		
8		
16		

20 ft

17. **DOG** A dog is leashed to the corner of a house. How much running area does the dog have? Explain how you found your answer.

18. **CRITICAL THINKING** Is the area of a semicircle with a diameter of x *greater than*, *less than*, or *equal to* the area of a circle with a diameter of $\frac{1}{2}x$? Explain.

Reasoning Find the area of the shaded region. Explain how you found your answer.

19.
 5 in.

20.
 9 m
 9 m

21.
 4 ft
 4 ft

Fair Game Review What you learned in previous grades & lessons

Evaluate the expression. *(Section 2.2 and Section 3.2)*

22. $\frac{1}{2}(7)(4) + 6(5)$

23. $\frac{1}{2} \cdot 8^2 + 3(7)$

24. $12(6) + 3.14 \cdot 2^2$

25. **MULTIPLE CHOICE** What is the product of $8\frac{1}{3}$ and $3\frac{2}{5}$? *(Section 2.4)*

 Ⓐ $2\frac{23}{51}$ Ⓑ $4\frac{1}{5}$ Ⓒ $24\frac{2}{15}$ Ⓓ $28\frac{1}{3}$

6.4 Areas of Composite Figures

COMMON CORE STATE STANDARDS
6.EE.2c
6.G.1

Essential Question How can you find the area of a composite figure?

1 ACTIVITY: Estimating Area

Work with a partner.

a. Choose a state. On grid paper, draw a larger outline of the state.

b. Use your drawing to estimate the area (in square miles) of the state.

c. Which state areas are easy to find? Which are difficult? Why?

Laurie's Notes

Introduction

For the Teacher
- **Goal:** Students will develop an intuitive understanding of the area of composite figures by focusing on their component parts.
- Select one or more of the activities based upon time available and student interest. You may also wish to spend two days on this lesson.

Motivate
- Share U.S. state trivia in question/answer format.
- Largest states: Alaska, Texas, California
- Smallest states: Rhode Island, Delaware, Connecticut
- Most densely populated states: New Jersey, Rhode Island, Massachusetts

Activity Notes

Activity 1
- ? Ask a variety of questions about the attributes of the states: state with most coastline; states with no coastline; state farthest north, south, east, and west.
- Questions are answered by using eyesight only. The goal is to get students thinking about the states and how each might be described.
- Distribute rulers, if requested.
- **Big Idea:** Each square in the grid is 2500 square miles. Each state is composed of whole squares and parts of whole squares. Students will find the area of the state by finding the area of composite figures.
- Your class size may dictate how you implement this activity. Here are a few strategies.
 - Students are assigned states within one particular region.
 - Students use $\frac{1}{2}$-inch or 1-inch grid paper to draw the larger outline.
 This is *not* a lesson on scale drawings. A larger copy of the state allows students to record their areas for each part on the drawing.
 - Students are given an enlarged photocopy of the state to use.
 - Assign the same state to two students. Have each student make a drawing and then compare their estimates for the non-square parts of the state.
- Discuss part (c). States with straighter borders are easier to draw and to estimate their areas.
- **Extension:** Drawings of adjacent states can be cut out and taped together on a bulletin board. Students can do outside research to gather information about their state: population, population density; or to rank the states by area and find the mean and median of the state areas.

Common Core State Standards

6.EE.2c Evaluate expressions at specific values of their variables. Include expressions that arise from formulas used in real-world problems. Perform arithmetic operations, including those involving whole-number exponents, in the conventional order when there are no parentheses to specify a particular order (Order of Operations).

6.G.1 Find the area of right triangles, other triangles, special quadrilaterals, and polygons by composing into rectangles or decomposing into triangles and other shapes; apply these techniques in the context of solving real-world and mathematical problems.

Previous Learning
Students should know area formulas.

Start Thinking! and Warm Up

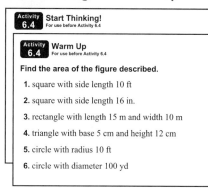

Activity 6.4 Start Thinking! For use before Activity 6.4

Activity 6.4 Warm Up For use before Activity 6.4

Find the area of the figure described.
1. square with side length 10 ft
2. square with side length 16 in.
3. rectangle with length 15 m and width 10 m
4. triangle with base 5 cm and height 12 cm
5. circle with radius 10 ft
6. circle with diameter 100 yd

6.4 Record and Practice Journal

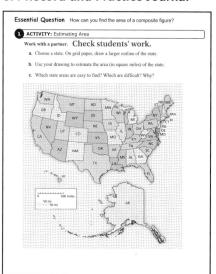

Essential Question How can you find the area of a composite figure?

1 ACTIVITY: Estimating Area
Work with a partner. Check students' work.
a. Choose a state. On grid paper, draw a larger outline of the state.
b. Use your drawing to estimate the area (in square miles) of the state.
c. Which state areas are easy to find? Which are difficult? Why?

T-260

Differentiated Instruction

Kinesthetic

Using grid paper, have students draw and cut out a square 4 units on each side. Next, have students draw and cut out a rectangle that is 3 units by 2 units. Have students place the two figures so that they are touching, but not overlapping. Ask them to find the combined area by finding the area of each piece. Have them draw other shapes that are touching on the grid paper, for example a square with a semicircle on top. Find the total area. Discuss how they could use this idea to find the area of other figures.

6.4 Record and Practice Journal

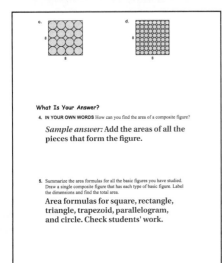

Laurie's Notes

Activity 2

- Students will work with actual-size pieces. The interlocking cuts are geometric shapes of which students should know how to find the area.
- Students will need to make accurate measurements to the nearest tenth of a centimeter.
- Scaled pieces are shown in the textbook.
- This activity reviews measurement, computation with decimals, and area formulas.
- Watch for computation errors.
- The completed puzzle is shown at the right.

Activity 3

- **Big Idea:** Each square has the same amount of green. Students can determine the radius of the circles. Then they can find the sum of the areas of circles in each diagram.
- You may need to suggest a table format to help guide student thinking.

Number of Circles	Radius of One Circle	Total Area
1	4	$1 \times \pi(4)^2 = 16\pi$ square units
4	2	$4 \times \pi(2)^2 = 16\pi$ square units
16	1	$16 \times \pi(1)^2 = 16\pi$ square units
64	$\frac{1}{2}$	$64 \times \pi\left(\frac{1}{2}\right)^2 = 16\pi$ square units

- Students find the results of this activity surprising.
- **Connection:** In the previous section, students found the area of a circle inscribed in a square. The circle was approximately 78% of the square. In this activity, compare the green area (16π square units) to the area of the square (64 square units): $\frac{16\pi}{64} \approx 0.785 = 78.5\%$.

What Is Your Answer?

- Question 5 is a nice summary problem that could become a small project.

Closure

- Does the area of a figure change when it is divided into smaller regions? Explain. No. As each region becomes smaller the number of regions increase, but the total area remains the same.

Technology For the Teacher

The Dynamic Planning Tool
Editable Teacher's Resources at *BigIdeasMath.com*

2 ACTIVITY: Estimating Areas

Work with a partner. The completed puzzle has an area of 150 square centimeters.

a. Estimate the area of each puzzle piece.

b. Check your work by adding the six areas. Why is this a check?

3 ACTIVITY: Filling a Square with Circles

Work with a partner. Which pattern fills more of the square with circles? Explain.

a.

b.

c.

d.

What Is Your Answer?

4. **IN YOUR OWN WORDS** How can you find the area of a composite figure?

5. Summarize the area formulas for all the basic figures you have studied. Draw a single composite figure that has each type of basic figure. Label the dimensions and find the total area.

Use what you learned about areas of composite figures to complete Exercises 3–5 on page 264.

Section 6.4 Areas of Composite Figures 261

6.4 Lesson

To find the area of a composite figure, split it up into figures with areas you know how to find. Then add the areas of those figures.

EXAMPLE 1 **Finding an Area Using Grid Paper**

Each square on the grid paper is 1 square meter. Find the area of the yellow figure.

Count the number of squares that lie entirely in the figure. There are 45.

Count the number of half-squares in the figure. There are 5.

The area of a half-square is $1 \div 2 = 0.5$ square meter.

Area of 45 squares: $45 \times 1 = 45$ square meters

Area of 5 half-squares: $5 \times 0.5 = 2.5$ square meters

∴ So, the area is $45 + 2.5 = 47.5$ square meters.

On Your Own

Now You're Ready
Exercises 3–8

Find the area of the shaded figure.

1.

2.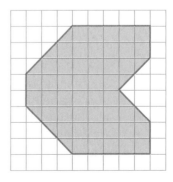

262 Chapter 6 Circles and Area

Laurie's Notes

Introduction

Connect
- **Yesterday:** Students gained an intuitive understanding about how to find the area of composite figures.
- **Today:** Students will divide composite figures into familiar geometric shapes and use known area formulas to find the total area.

Motivate
- Draw the following "equations." Ask students how they would find the area of the last figure knowing the areas of the figures on the left side of the equation.

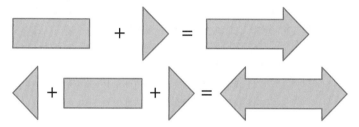

- These quick puzzles should help focus students' thinking on how they will find the area of a composite figure.
- Discuss the strategy of dividing a composite figure into smaller regions that are geometric figures with known area formulas.

Lesson Notes

Example 1
- Work through the example as shown and then try an alternate approach.
- **?** "What is the name of the yellow polygon?" heptagon
- **?** "Can you think of another way to find the area of the heptagon? Explain." Yes. Listen for a suggestion that divides the heptagon into a rectangle, a triangle, and a trapezoid.
- Note: The rectangle has dimensions 4×8, the triangle has a base of 2 and height of 2, and the trapezoid has bases of 3 and 6 and a height of 3. The total area is 47.5 square meters.
- **?** "Are there other strategies for finding the area of the yellow heptagon?" Yes, it could be divided into a rectangle and two trapezoids using two parallel horizontal lines (or two parallel vertical lines).
- Point out the solution is in square meters, because it was stated at the beginning that each square on the grid paper is 1 square meter.
- **Extension:** Ask students if the area of their classroom is more or less than 48 square meters. Hold two meter sticks perpendicular to each other as a visual clue.

On Your Own
- Encourage students to find more than one method to find the area of each.
- Have students share different strategies at the board.

Goal Today's lesson is finding the areas of composite figures.

Start Thinking! and Warm Up

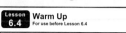

Lesson 6.4 Warm Up For use before Lesson 6.4

Lesson 6.4 Start Thinking! For use before Lesson 6.4

Draw a picture of a house with a roof and a chimney.

How can you use a ruler and some calculations to find the area covered by your drawing?

Use a ruler and the method you described to estimate the area of the house.

Extra Example 1
Each square on the grid paper is 1 square yard. Find the area of the figure.

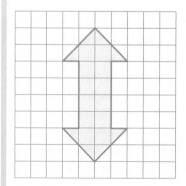

16 yd²

On Your Own
1. 37 square units
2. 51 square units

T-262

Extra Example 2

Find the area of the pool and the deck. The figure is made up of a right triangle, a square and a semicircle.

about 1061.33 ft²

Extra Example 3

Find the area of the figure made up of a triangle, a rectangle, and a trapezoid.

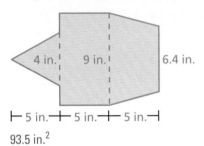

93.5 in.²

On Your Own

3. 90 m²
4. about 10.28 ft²

English Language Learners

Vocabulary
Formulas use letters to represent specific items or measures. For instance, in the formula for the area of a rectangle, $A = bh$, A represents area, b represents the base, and h represents the height. Review with your English learners the capital and lowercase letters used in formulas.

area (A)	base (b)
circumference (C)	diameter (d)
distance (d)	height (h)
length (ℓ)	perimeter (P)
radius (r)	rate (r)
time (t)	volume (V)
width (w)	

T-263

Laurie's Notes

Example 2
- After Example 1, it should be clear to students that this composite figure will be divided into a semicircle and a rectangle.
- ❓ "What are the dimensions of the rectangle?" 19 ft by 12 ft
- ❓ "What is the radius of the circle?" 6 ft
- In calculating the area of the semicircle, watch for student errors in multiplying and dividing decimals.
- **Extension:** Ask students if the area of a semicircle of radius 6 is the same as area of a circle of radius 3. The answer is no! 18π compared to 9π

Example 3
- Draw the composite figure without the interior dotted lines. Ask students to think about how the figure could be divided into regions with known area formulas. They may come up with more than three regions.
- Label the diagram with the dimensions given. I find it helpful to draw small arrows from the measurement to the segment it is associated with, especially for interior dimensions.
- Work through the problem, questioning students along the way about the process.
- ❓ "How do you find the area of a triangle? How do you multiply $\frac{1}{2}$ by 11.2? What units are used to label area?"
- **Common Misconception:** In a right triangle, the height is a side length. It is not an interior measurement.

On Your Own
- Have students share their strategies for these two problems. Question 4 might be done as 4 semicircles or 2 circles.

Closure
- **Exit Ticket:** Find the area. 100 units²

Technology For the Teacher

Dynamic Classroom

The Dynamic Planning Tool
Editable Teacher's Resources at *BigIdeasMath.com*

EXAMPLE 2 **Finding an Area**

Find the area of the portion of the basketball court shown.

The figure is made up of a rectangle and a semicircle. Find the area of each figure.

Area of rectangle

$A = \ell w$

$= (19)(12)$

$= 228$

Area of semicircle

$A = \dfrac{\pi r^2}{2}$

$\approx \dfrac{3.14 \cdot (6)^2}{2}$

$= 56.52$

The semicircle has a radius of $\dfrac{12}{2} = 6$ feet.

∴ So, the area is about $228 + 56.52 = 284.52$ square feet.

EXAMPLE 3 **Finding an Area**

Find the area of the figure.

The figure is made up of a triangle, a rectangle, and a parallelogram. Find the area of each figure.

Area of triangle

$A = \dfrac{1}{2}bh$

$= \dfrac{1}{2}(11.2)(4.5)$

$= 25.2$

Area of rectangle

$A = \ell w$

$= (8)(4.5)$

$= 36$

Area of parallelogram

$A = bh$

$= (8)(6.7)$

$= 53.6$

∴ So, the area is $25.2 + 36 + 53.6 = 114.8$ square centimeters.

● **On Your Own**

Now You're Ready
Exercises 9 and 10

Find the area of the figure.

3.

4.

Section 6.4 Areas of Composite Figures 263

6.4 Exercises

Vocabulary and Concept Check

1. **REASONING** Describe two different ways to find the area of the figure. Name the types of figures you used and the dimensions of each.

2. **REASONING** Draw a trapezoid. Suppose you can't remember the formula for the area of a trapezoid. Explain how you can think of the trapezoid as a composite figure to find its area.

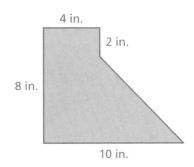

Practice and Problem Solving

Each square on the grid paper is 1 square inch. Find the area of the figure.

3.

4.

5.

6.

7.

8.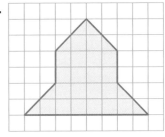

Find the area of the figure.

9.

10.

11. **OPEN-ENDED** Trace your hand and your foot on grid paper. Then estimate the area of each. Which one has the greater area?

264 Chapter 6 Circles and Area

Assignment Guide and Homework Check

Level	Day 1 Activity Assignment	Day 2 Lesson Assignment	Homework Check
Basic	3–5, 18–22	1, 2, 6–13	2, 6, 9, 12
Average	3–5, 18–22	1, 2, 8–13, 15, 16	8, 9, 12, 15
Advanced	3–5, 18–22	1, 2, 7, 8, 12–17	8, 12, 15, 16

Common Errors

- **Exercises 3–8** Students may forget to count all the squares inside the figure and just count the ones along the border because this is what they did for perimeter. Remind them that the area includes everything inside the figure.
- **Exercises 9 and 10** Students may forget to include one of the areas of the composite figures or may count one area more than once. Tell them to break apart the figure into several figures. Draw and label each figure and then find the area of each part. Finally add the areas of each part together for the area of the whole figure.
- **Exercises 12–14** Students may forget to subtract the unshaded area from the figure. Remind them that in this situation instead of adding on the area of a figure they must subtract a portion out. Give a real-life example to help students understand, like tiling a bathroom floor but taking out the part where the sink or toilet is.

6.4 Record and Practice Journal

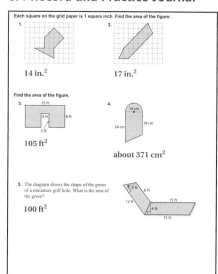

Technology For the Teacher
Answer Presentation Tool
QuizShow

Vocabulary and Concept Check

1. *Sample answer:* You could add the areas of an 8-inch × 4-inch rectangle and a triangle with a base of 6 inches and a height of 6 inches. Also you could add the area of a 2-inch × 4-inch rectangle to the area of a trapezoid with a height of 6 inches, and base lengths of 4 inches and 10 inches.

2. *Sample answer:* You can think of the trapezoid as a rectangle and two triangles.

Practice and Problem Solving

3. 28.5 in.2 4. 33 in.2

5. 25 in.2 6. 30 in.2

7. 25 in.2 8. 24 in.2

9. 132 cm^2

10. about 236.625 ft^2

11. *Answer will include but is not limited to:* Tracings of a hand and foot on grid paper, estimates of the areas, and a statement of which is greater.

T-264

Practice and Problem Solving

12. 89 m² **13.** 23.5 in.²

14. about 21.87 ft²

15. 24 m²

16. $P = $ about 94.2 ft
$A = $ about 628 ft²

17. See *Taking Math Deeper*.

Fair Game Review

18. $x - 12$ **19.** $y \div 6$

20. $b + 3$ **21.** $7w$

22. A

Mini-Assessment

Find the area of the figure.

1.

20 in.²

2.

about 14.28 yd²

3. Find the area of the red region.
about 3.44 in.²

Taking Math Deeper

Exercise 17

This problem is not conceptually difficult. However, it does have a lot of calculations and it is difficult to do without a calculator. Before starting the calculations, suggest that students think about which design is more efficient. Each design will fit on an 11×17 sheet of paper. By cutting and folding each pattern, a student can see how much overlap each design has. The design with the greatest overlap is the least efficient.

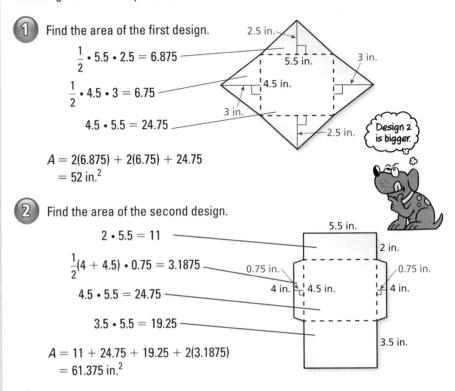

① Find the area of the first design.

$\frac{1}{2} \cdot 5.5 \cdot 2.5 = 6.875$

$\frac{1}{2} \cdot 4.5 \cdot 3 = 6.75$

$4.5 \cdot 5.5 = 24.75$

$A = 2(6.875) + 2(6.75) + 24.75$
$\quad = 52$ in.²

② Find the area of the second design.

$2 \cdot 5.5 = 11$

$\frac{1}{2}(4 + 4.5) \cdot 0.75 = 3.1875$

$4.5 \cdot 5.5 = 24.75$

$3.5 \cdot 5.5 = 19.25$

$A = 11 + 24.75 + 19.25 + 2(3.1875)$
$\quad = 61.375$ in.²

③ a. So, the second design has the greater area. Answer the question.

$500 \cdot 61.375 = 30{,}687.5$ in.² (500 envelopes using Design 2)

$\dfrac{30{,}687.5}{52} \approx 590.1$ (number of envelopes using Design 1)

b. You could make 90 more envelopes using Design 1.

Reteaching and Enrichment Strategies

If students need help...	If students got it...
Resources by Chapter • Practice A and Practice B • Puzzle Time Record and Practice Journal Practice Differentiating the Lesson Lesson Tutorials Skills Review Handbook	Resources by Chapter • Enrichment and Extension • School-to-Work Start the next section

Find the area of the figure.

12.

13.

14.

15. **AREA** The figure is made up of a square and a rectangle. Find the area of the shaded region.

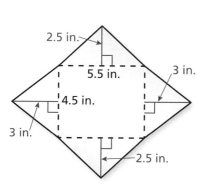

16. **FOUNTAIN** The fountain is made up of two semicircles and a quarter circle. Find the perimeter and area of the fountain.

17. **Critical Thinking** You are deciding on two different designs for envelopes.

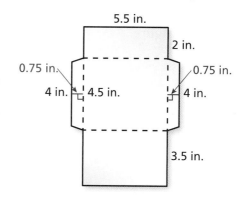

a. Which design has the greater area?

b. You make 500 envelopes using the design with the greater area. Using the same amount of paper, how many more envelopes can you make with the other design?

 Fair Game Review What you learned in previous grades & lessons

Write the phrase as an expression. *(Section 1.2)*

18. 12 less than a number x

19. a number y divided by 6

20. a number b increased by 3

21. the product of 7 and a number w

22. **MULTIPLE CHOICE** What is 0.02% of 50? *(Section 4.4)*

Ⓐ 0.01 Ⓑ 0.1 Ⓒ 1 Ⓓ 100

6.3–6.4 Quiz

Each square on the grid paper is 1 square centimeter. Find the area of the figure. *(Section 6.4)*

1.

2.

3.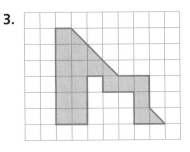

Find the area of the circle. Use 3.14 or $\frac{22}{7}$ for π. *(Section 6.3)*

4.

5. 6 cm

6.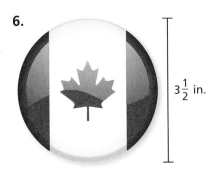

Find the area of the figure. *(Section 6.4)*

7.

8.

9.

10. **POTHOLDER** A knitted potholder is shaped like a circle. Its radius is 3.5 inches. What is its area? *(Section 6.3)*

11. **CARD** The heart-shaped card is made by placing a square and two semicircles together. What is the area of the card? *(Section 6.4)*

12. **DESK** A desk top is shaped like a semicircle with a diameter of 30 inches. What is the area of the desk top? *(Section 6.3)*

13. **RUG** The circular rug is placed on a square floor. The rug touches all four walls. How much of the floor space is *not* covered by the rug? *(Section 6.4)*

Alternative Assessment Options

Math Chat
Structured Interview
Student Reflective Focus Question
Writing Prompt

Math Chat

- Have students work in pairs. Assign Exercises 10–13 from the Quiz to each pair. Each student works through all four problems. After the students have worked through the problems, they take turns talking through the process that they used to get the answer. Students analyze and evaluate the mathematical thinking and strategies used.
- The teacher should walk around the classroom listening to the pairs and asking questions to ensure understanding.

Study Help Sample Answers

Remind students to complete Graphic Organizers for the rest of the chapter.

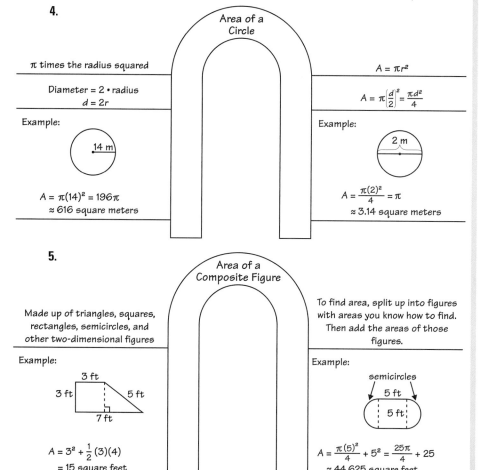

Reteaching and Enrichment Strategies

If students need help...	If students got it...
Resources by Chapter • Study Help • Practice A and Practice B • Puzzle Time Lesson Tutorial BigIdeasMath.com Practice Quiz Practice from the Test Generator	Resources by Chapter • Enrichment and Extension • School-to-Work Game Closet at BigIdeasMath.com Start the Chapter Review

Answers

1. 16 cm^2
2. 28 cm^2
3. 19 cm^2
4. about 314 in.^2
5. about 28.26 cm^2
6. about 9.625 in.^2
7. 198 ft^2
8. about 16.28 m^2
9. 2450 cm^2
10. about 38.465 in.^2
11. about 114.24 cm^2
12. about 353.25 in.^2
13. about 42 ft^2

Technology For the Teacher

Answer Presentation Tool

Assessment Book

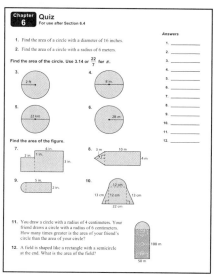

**For the Teacher
Additional Review Options**
- Quiz*Show*
- Big Ideas Test Generator
- Game Closet at *BigIdeasMath.com*
- Vocabulary Puzzle Builder
- Resources by Chapter
 Puzzle Time
 Study Help

Answers

1. 4 in.
2. 30 mm
3. 50 m
4. 1.5 yd
5. 40 ft
6. 10 m
7. 2 in.
8. 50 mm
9. about 18.84 ft
10. about 31.4 cm
11. about 132 in.

Review of Common Errors

Exercises 1–8
- Students may confuse what they are finding and double the diameter or halve the radius. Remind them that the radius is half the diameter.

Exercises 9–11
- Students may use the radius in the circumference formula that calls for diameter or the diameter in the circumference formula that calls for radius. Remind them of the different formulas.

Exercises 12–17
- Students may include the dashed lines in their calculation of the perimeter. Remind them that the dashed lines are for reference and sometimes give information to find another length. Only the outside lengths are counted.

Exercises 18–20
- Students may forget to divide the diameter by 2 to find the area. Remind them that they need the *radius* in the given formula for area of a circle.
- Students may use the formula for circumference or forget to square the radius when finding the area.

Exercises 21–23
- Students may forget to include the area of one or more parts of a composite figure or count a part more than once. Tell them to draw and label each part, find the area of each part, and add the areas of the parts to find the area of the composite figure.

6 Chapter Review

Review Key Vocabulary

circle, *p. 240*
center, *p. 240*
radius, *p. 240*
diameter, *p. 240*
circumference, *p. 241*
pi, *p. 241*
semicircle, *p. 242*
composite figure, *p. 248*

Review Examples and Exercises

6.1 Circles and Circumference *(pp. 238–245)*

Find the circumference of the circle. Use 3.14 for π.

The radius is 4 millimeters.

$C = 2\pi r$ Write formula for circumference.

$\approx 2 \cdot 3.14 \cdot 4$ Substitute 3.14 for π and 4 for r.

$= 25.12$ Multiply.

∴ The circumference is about 25.12 millimeters.

Exercises

Find the radius of the circle with the given diameter.

1. 8 inches
2. 60 millimeters
3. 100 meters
4. 3 yards

Find the diameter of the circle with the given radius.

5. 20 feet
6. 5 meters
7. 1 inch
8. 25 millimeters

Find the circumference of the circle. Use 3.14 or $\frac{22}{7}$ for π.

9.
10.
11.

Chapter Review 267

6.2 Perimeters of Composite Figures (pp. 246–251)

The figure is made up of a semicircle and a square. Find the perimeter. Use 3.14 for π.

The distance around the square part is $6 + 6 + 6 = 18$ meters. The distance around the curved part is one-half the circumference of a circle with $d = 6$ meters.

$$\frac{C}{2} = \frac{\pi d}{2} \qquad \text{Divide the circumference by 2.}$$

$$\approx \frac{3.14 \cdot 6}{2} \qquad \text{Substitute 3.14 for } \pi \text{ and 6 for } d.$$

$$= 9.42 \qquad \text{Simplify.}$$

∴ The perimeter of the figure is about $18 + 9.42 = 27.42$ meters.

Exercises

Find the perimeter of the figure.

12.

13.

14.

15.

16.

17.

6.3 Areas of Circles (pp. 254–259)

Find the area of the circle. Use 3.14 for π.

$$A = \pi r^2 \qquad \text{Write formula for area.}$$

$$\approx 3.14 \cdot (20)^2 \qquad \text{Substitute 3.14 for } \pi \text{ and 20 for } r.$$

$$= 1256 \qquad \text{Multiply.}$$

∴ The area of the circle is about 1256 square yards.

268 Chapter 6 Circles and Area

Review Game

Area, Perimeter, and Circumference

Big Ideas Game Closet

Materials per Pair:
- paper
- pencils (colored pencils optional)

Directions:
Each group of four divides into teams of two. Each team competes with the other team. The class is directed to draw and label a composite figure using selected shapes and a predetermined area. For example, the teacher could say, "Using squares, rectangles, and circles, draw and label a composite figure with an area of 50 square units." Each team then draws their figure and calculates the area. Papers are exchanged within the group and teams check each other's work. The team whose correctly-drawn figure has an area that is closest to the area specified receives 1 point. If both teams are equally close, both teams receive 1 point. Play for a set amount of time.

Who Wins?
When time is up, the team with the most points wins.

For the Student Additional Practice
- Lesson Tutorials
- Study Help (textbook)
- Student Website
 Multi-Language Glossary
 Practice Assessments

Answers

12. 24 in.
13. 96 ft
14. 62 cm
15. about 90.24 mm
16. about 28.56 in.
17. about 42.84 cm
18. about 50.24 in.2
19. about 379.94 cm^2
20. about 1386 mm^2
21. about 79.25 in.2
22. 29 in.2
23. about 31.625 ft^2

My Thoughts on the Chapter

What worked...

Teacher Tip
Not allowed to write in your teaching edition? Use sticky notes to record your thoughts.

What did not work...

What I would do differently...

Exercises

Find the area of the circle. Use 3.14 or $\frac{22}{7}$ for π.

18.

4 in.

19.

11 cm

20.

42 mm

6.4 Areas of Composite Figures (pp. 260–265)

Find the area of the figure.

The figure is made up of a rectangle, a triangle and a semicircle. Find the area of each figure.

Area of rectangle

$A = \ell w$

$= (26)(10)$

$= 260$

Area of triangle

$A = \frac{1}{2}bh$

$= \frac{1}{2}(10)(24)$

$= 120$

Area of semicircle

$A = \frac{\pi r^2}{2}$

$\approx \frac{3.14 \cdot (13)^2}{2}$

$= 265.33$

∴ So, the area of the figure is about $260 + 120 + 265.33 = 645.33$ square miles.

Exercises

Find the area of the figure.

21.

22.

23.

6 Chapter Test

Find the radius of the circle with the given diameter.

1. 10 inches
2. 5 yards

Find the diameter of the circle with the given radius.

3. 34 feet
4. 19 meters

Find the circumference and area of the circle. Use 3.14 or $\frac{22}{7}$ for π.

5. diameter of 4 feet
6. radius of 1 meter
7. diameter of 70 inches

8. Each square on the grid is 1 square inch. Estimate the perimeter of the figure. Then find the area.

Find the perimeter and area of the figure. Use 3.14 or $\frac{22}{7}$ for π.

9.
10.
11.

12. **MUSEUM** A museum plans to rope off the perimeter of the L-shaped exhibit. How much rope is needed?

13. **ANIMAL PEN** You unfold chicken wire to make a circular pen with a diameter of 2.9 meters. How many meters of chicken wire do you need?

14. **YIN AND YANG** In the Chinese symbol for yin and yang, the dashed curve shows two semicircles formed by the curve separating the yin (dark) and yang (light). Is the circumference of the entire yin and yang symbol *less than, greater than,* or *equal to* the perimeter of the yin?

Test Item References

Chapter Test Questions	Section to Review	Common Core State Standards
1–7, 13, 14	6.1	6.EE.2c
8–12, 14	6.2	6.EE.2c
5–7	6.3	6.EE.2c
8–11	6.4	6.EE.2c, 6.G.1

Test-Taking Strategies

Remind students to quickly look over the entire test before they start so that they can budget their time. There is a lot of vocabulary in this chapter, so students should have been making flash cards as they worked through the chapter. Words that get mixed up should be jotted on the back of the test before they start. (For instance, they should write down *radius* and *circumference*.) Students need to use the **Stop** and **Think** strategy before they answer the question.

Common Assessment Errors

- **Exercises 1–4** Students may confuse what they are finding and double the diameter or halve the radius. Remind them that the radius is half the diameter. Encourage students to draw a line segment representing each radius or diameter so that they have a visual for reference.
- **Exercises 5–7** Students may use the radius in a formula that calls for diameter, the diameter in a formula that calls for radius, the area formula when calculating circumference, or the circumference formula when calculating area. Again, remind students that the radius is half the diameter. Also, remind them of the circumference and area formulas.
- **Exercise 8** Students may count the sides of the squares inside the figure when finding the perimeter, or confuse area and perimeter. Remind students of how to find perimeter and area.
- **Exercises 9–11** When calculating the perimeter of the figure, students may count the dashed lines. Remind them that the dashed lines are for reference and that only the outside lengths are counted. When calculating the area of the figure, students may forget to include the area of one or more parts or count part(s) more than once. Tell them to draw and label each part, find the area of each part, and add the areas of the parts.

Reteaching and Enrichment Strategies

If students need help...	If students got it...
Resources by Chapter • Practice A and Practice B • Puzzle Time Record and Practice Journal Practice Differentiating the Lesson Lesson Tutorial Practice from the Test Generator Skills Review Handbook	Resources by Chapter • Enrichment and Extension • School-to-Work Game Closet at *BigIdeasMath.com* Start Standardized Test Practice

Answers

1. 5 in.
2. 2.5 yd
3. 68 ft
4. 38 m
5. $C \approx 12.56$ ft, $A \approx 12.56$ ft^2
6. $C \approx 6.28$ m, $A \approx 3.14$ m^2
7. $C \approx 220$ in., $A \approx 3850$ in.2
8. $P \approx 15$ in., $A = 9$ in.2
9. $P = 26$ m, $A = 44$ m^2
10. $P = 48$ in., $A = 96$ in.2
11. $P \approx 108$ m, $A \approx 623$ m^2
12. 200 ft
13. about 9.106 m
14. *equal to*

Assessment Book

T-270

Test-Taking Strategies
Available at *BigIdeasMath.com*

After Answering Easy Questions, Relax
Answer Easy Questions First
Estimate the Answer
Read All Choices before Answering
Read Question before Answering
Solve Directly or Eliminate Choices
Solve Problem before Looking at Choices
Use Intelligent Guessing
Work Backwards

About this Strategy
When taking a multiple choice test, be sure to read each question carefully and thoroughly. When taking a timed test, it is often best to skim the test and answer the easy questions first. Be careful that you record your answer in the correct position on the answer sheet.

Answers
1. A
2. F
3. $\dfrac{3}{4}$
4. D
5. H

Item Analysis

1. **A.** Correct answer
 B. The student thinks that 16 is equivalent to 0.16 and misinterprets the inverse relationship, multiplying instead of dividing.
 C. The student aligns the decimal point of the quotient with the decimal point of the dividend.
 D. The student misinterprets the inverse relationship, multiplying instead of dividing.

2. **F.** Correct answer
 G. The student rounds all the mixed numbers down to the next whole number.
 H. The student estimates $2\dfrac{9}{10}$ as 2 and then ignores the parentheses and divides before multiplying.
 I. The student ignores the parentheses and divides before multiplying.

3. **Gridded Response:** Correct answer: $\dfrac{3}{4}$
 Common Error: The student divides the greater number by the lesser number, getting an answer of $\dfrac{4}{3}$ or equivalent.

4. **A.** On the left side of the equation, the student writes an expression that represents the product of *n* and a number that is 7 less than 5.
 B. On the left side of the equation, the student writes an expression that represents the product of *n* and a number that is 5 less than 7.
 C. On the left side of the equation, the student misinterprets the use of "less than" and reverses the order of the subtraction.
 D. Correct answer

5. **F.** The student did not square the radius.
 G. The student multiplied the radius by 2 instead of squaring it.
 H. Correct answer
 I. The student squared the diameter instead of the radius.

Technology For the Teacher
Big Ideas Test Generator

6 Standardized Test Practice

1. What number belongs in the box to make the equation true? *(6.NS.3)*

 $$0.4 = 16 \times \boxed{}$$

 A. 0.025
 B. 0.064
 C. 2.5
 D. 6.4

2. Which integer is closest to the value of the expression below? *(6.NS.1)*

 $$35\tfrac{7}{8} \div \left(4\tfrac{1}{5} \times 2\tfrac{9}{10}\right)$$

 F. 3
 G. 4
 H. 18
 I. 27

3. What is the value of the expression below? *(6.NS.1)*

 $$4\tfrac{1}{8} \div 5\tfrac{1}{2}$$

4. Your mathematics teacher described an equation in words. Her description is in the box below.

 > "5 less than the product of 7 and an unknown number is equal to 42."

 Which equation matches your mathematics teacher's description? *(6.EE.6)*

 A. $(5 - 7)n = 42$
 B. $(7 - 5)n = 42$
 C. $5 - 7n = 42$
 D. $7n - 5 = 42$

5. What is the area of the circle below? $\left(\text{Use } \tfrac{22}{7} \text{ for } \pi.\right)$ *(6.EE.2c)*

 F. 44 cm²
 G. 88 cm²
 H. 616 cm²
 I. 2464 cm²

6. John was finding the area of the figure below.

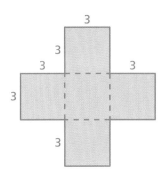

John's work is in the box below.

> area of horizontal rectangle
>
> A = 3 × (3 + 3 + 3)
>
> = 3 × 9
>
> = 27 square units
>
> area of vertical rectangle
>
> A = (3 + 3 + 3) × 3
>
> = 9 × 3
>
> = 27 square units
>
> total area of figure
>
> A = 27 + 27
>
> = 54 square units

What should John do to correct the error that he made? *(6.G.1)*

A. Add the area of the center square to the 54 square units.

B. Find the area of one square and multiply this number by 4.

C. Subtract the area of the center square from the 54 square units.

D. Subtract 54 from the area of a large square that is 9 units on each side.

7. You are baking cookies. For each batch of cookies, you use $2\frac{1}{4}$ cups of flour. How many cups of flour do you use for 6 batches of cookies? *(6.NS.1)*

F. $8\frac{1}{4}$ cups

G. $12\frac{1}{4}$ cups

H. $12\frac{1}{2}$ cups

I. $13\frac{1}{2}$ cups

Item Analysis (continued)

6. **A.** The student adds the area of the center square that was included twice, instead of subtracting it.

 B. The student does not include the area of the center square at all.

 C. Correct answer

 D. The student employs a plausible strategy but makes the same error as in John's work of counting the center square twice.

7. **F.** The student adds the number of batches to the number of cups of flour.

 G. The student multiplies only the whole numbers together.

 H. The student multiplies 6 by 2 to get 12 and 6 by $\frac{1}{4}$ to get $1\frac{1}{2}$, then multiplies 12 by 1 to get 12, and then combines the 12 and the $\frac{1}{2}$ to get $12\frac{1}{2}$.

 I. Correct answer

8. **Gridded Response:** Correct answer: 26 units

 Common Errors: The student finds only half the perimeter, getting 13 units.

9. **A.** The student divides instead of multiplying.

 B. The student subtracts instead of multiplying.

 C. The student adds instead of multiplying.

 D. Correct answer

10. **F.** Correct answer

 G. The student chooses the store offering the greatest percent off but ignores the fact that this store also had the greatest original price.

 H. The student chooses the store that had the least original price but ignores the fact that this store also offered the least percent off.

 I. The student chooses the store with the highest sale price.

Answers

6. C
7. I

Answers

8. 26
9. D
10. F
11. Part A 942 ft^2

 Part B 134.2 ft

Answers for Extra Examples

1. **A.** The student finds the area of only 1 circle.

 B. The student finds the area of only 1 circle and uses the diameter of the circle instead of the radius, finding $\pi \cdot 30^2$.

 C. Correct answer

 D. The student uses the diameters of the circles instead of the radii, finding $\pi \cdot 30^2$ for each of the circles.

2. **F.** Correct answer

 G. The student finds the range instead of the mode.

 H. The student finds the median instead of the mode.

 I. The student finds the mean instead of the mode.

Item Analysis (continued)

11. **4 points** The student demonstrates a thorough understanding of solving problems involving area and perimeter of composite figures. In Part A, the student correctly finds the area of the sprayed region to be approximately 942 square feet. In Part B, the student correctly finds the perimeter of the sprayed region to be approximately 134.2 feet. The student shows accurate, complete work for both parts and provides clear and complete explanations.

 3 points The student demonstrates an understanding of solving problems involving area and perimeter of composite figures, but the student's work and explanations demonstrate an essential but less than thorough understanding.

 2 points The student demonstrates a partial understanding of solving problems involving area and perimeter of composite figures. The student's work and explanations demonstrate a lack of essential understanding.

 1 point The student demonstrates very limited understanding of solving problems involving area and perimeter of composite figures. The student's response is incomplete and exhibits many flaws.

 0 points The student provided no response, a completely incorrect or incomprehensible response, or a response that demonstrates insufficient understanding of solving problems involving area and perimeter of composite figures.

Extra Examples for Standardized Test Practice

1. A North American hockey rink contains 5 face-off circles. Each of these circles has a radius of 15 feet. What is the total area, in square feet, of all the face-off circles? (Use 3.14 for π.)

 A. 706.5 ft^2

 B. 2826 ft^2

 C. 3532.5 ft^2

 D. 14,130 ft^2

2. What is the mode of the data set in the box below?

 > 52, 65, 37, 50, 59, 29, 64, 50, 80

 F. 50

 G. 51

 H. 52

 I. 54

8. What is the perimeter of the figure below? *(6.EE.2c)*

9. On a certain date in the year 2008, one U.S. dollar was equivalent to 47.775 Indian rupees. At this rate, ten U.S. dollars are equivalent to how many Indian rupees? *(6.RP.3d)*

 A. 4.7775

 B. 37.775

 C. 57.775

 D. 477.75

10. Four jewelry stores are selling an identical pair of earrings.

 - Store F: original price of $75; 20% off during sale
 - Store G: original price of $100; 35% off during sale
 - Store H: original price of $70; 10% off during sale
 - Store I: original price of $95; 30% off during sale

 Which store has the least sale price for the pair of earrings? *(6.NS.7b)*

 F. Store F

 G. Store G

 H. Store H

 I. Store I

11. A lawn sprinkler sprays water onto part of a circular region, as shown below. *(6.EE.2c)*

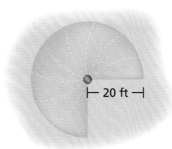

 Part A What is the area, in square feet, of the region that the sprinkler sprays with water? Show your work and explain your reasoning. (Use 3.14 for π.)

 Part B What is the perimeter, in feet, of the region that the sprinkler sprays with water? Show your work and explain your reasoning. (Use 3.14 for π.)

Standardized Test Practice 273

7 Equations

7.1 Writing Equations in One Variable
7.2 Solving Equations Using Addition or Subtraction
7.3 Solving Equations Using Multiplication or Division
7.4 Solving Two-Step Equations
7.5 Finding Dimensions of Plane Figures
7.6 Finding Dimensions of Prisms

"What is 7Q plus 3Q?" / Ten Q / "You're welcome."

"With the help of your twin brother, I think I have figured it out."

"You weigh 36 dog biscuits." / "Can someone call the society for cruelty to cats?"

Connections to Previous Learning

- Describe mathematical relationships using equations and visual representations.

- Use properties of equality to solve numerical and real-world situations.
- Find surface area and volume of prisms without formulas.
- Use a formula to find the areas of parallelograms, triangles, and trapezoids.

- Write, solve, and graph one-step and two-step linear equations.
- Solve problems using a formula.
- Use given information to find a missing dimension of a plane figure or prism.
- Find area or volume of a plane figure or prism.

Math in History

The beginnings of algebra can be traced back to ancient Babylonian mathematicians roughly 4000 years ago.

★ Babylonian algebra was concerned with the solution of linear, quadratic, and cubic equations. Ancient Babylon was located in the country that is now called Iraq.

★ The name algebra comes from an Arabian book called "Al-jabr wa'l muqabalah" or "The Compendious Book on Calculation by Completion and Balancing." It describes solving polynomials up to the second degree.

Pacing Guide for Chapter 7

Chapter Opener	1 Day
Section 1 Activity Lesson	 1 Day 1 Day
Section 2 Activity Lesson	 1 Day 2 Days
Section 3 Activity Lesson	 1 Day 1 Day
Section 4 Activity Lesson	 1 Day 2 Days
Study Help / Quiz	1 Day
Section 5 Activity Lesson	 1 Day 1 Day
Section 6 Activity Lesson Lesson b	 1 Day 1 Day 2 Days
Quiz / Chapter Review	1 Day
Chapter Test	1 Day
Standardized Test Practice	1 Day
Total Chapter 7	21 Days
Year-to-Date	122 Days

Check Your Resources

- Record and Practice Journal
- Resources by Chapter
- Skills Review Handbook
- Assessment Book
- Worked-Out Solutions

Technology For the Teacher

The Dynamic Planning Tool
Editable Teacher's Resources at
BigIdeasMath.com

Common Core State Standards

6.EE.2c Evaluate expressions at specific values of their variables. Include expressions that arise from formulas used in real-world problems. Perform arithmetic operations, including those involving whole-number exponents, in the conventional order when there are no parentheses to specify a particular order (Order of Operations).

6.EE.2a Write expressions that record operations with numbers and with letters standing for numbers.

Additional Topics for Review
- Order of Operations
- Finding the volume of a figure without a formula

Try It Yourself

1. 21
2. 1
3. 11
4. 46
5. $3w + 6$
6. $\dfrac{7}{p}$
7. $t - 2$
8. $5x$
9. $\dfrac{6}{r} + 5$
10. $3b - 4$

Record and Practice Journal

1. 30
2. 10
3. 17
4. 6
5. 6
6. 44
7. 16
8. 72
9. 85
10. $2k + 3$
11. $\dfrac{q}{2} + 8$
12. $p - 6$
13. $9x$
14. $\dfrac{5}{n}$
15. $1 + 3y$
16. $2c$
17. $\dfrac{p}{2} + 6$

Math Background Notes

Vocabulary Review
- Evaluate
- Substitute
- Expression
- Order of Operations

Evaluating Expressions

- Students should know how to evaluate expressions. Because this skill is fairly new to them, you may want to allot slightly more time for practice.
- Remind students that evaluate means to find a value.
- To evaluate an expression, students will need to substitute values into the expression.
- **Teaching Tip:** English Language Learners may find the vocabulary and writing involved in these opening exercises challenging. To ease the transition, help them make analogies. In the same way a sugar substitute replaces sugar, substituting numbers into an expression will replace the variables until you are able to evaluate.
- After they substitute, remind students that they must follow the correct order of operations as they simplify. You may wish to review order of operations prior to completing the examples.

Writing Expressions

- Students should know how to write expressions.
- Review key words with students. For example, words such as sum and product alert students to use addition and multiplication.
- **Common Error:** Many students will read Example 3, part (b) and incorrectly write the expression $12 - 4y$. Remind students that they cannot subtract 12 from a number unless they have a number to start with. Always write the number first and then add or subtract as indicated.

Reteaching and Enrichment Strategies

If students need help...	If students got it...
Record and Practice Journal • Fair Game Review Skills Review Handbook Lesson Tutorials	Game Closet at *BigIdeasMath.com* Start the next section

What You Learned Before

● Evaluating Expressions (6.EE.2c)

Example 1 Evaluate $7x + 3y$ when $x = 2$ and $y = 4$.

$7x + 3y = 7 \cdot 2 + 3 \cdot 4$ Substitute 2 for x and 4 for y.

$ = 14 + 12$ Using order of operations, multiply from left to right.

$ = 26$ Add 14 and 12.

Example 2 Evaluate $5x^2 - 2(y + 1) + 9$ when $x = 2$ and $y = 1$.

$5x^2 - 2(y + 1) + 9 = 5(2)^2 - 2(1 + 1) + 9$ Substitute 2 for x and 1 for y.

$ = 5(2)^2 - 2 \cdot 2 + 9$ Using order of operations, evaluate within the parentheses.

$ = 5 \cdot 4 - 2 \cdot 2 + 9$ Using order of operations, evaluate the exponent.

$ = 20 - 4 + 9$ Using order of operations, multiply from left to right.

$ = 25$ Subtract 4 from 20. Add the result to 9.

Try It Yourself
Evaluate the expression when $a = \dfrac{1}{2}$ and $b = 7$.

1. $6ab$ **2.** $16a - b$ **3.** $3b - 2a - 9$ **4.** $b^2 - 16a + 5$

● Writing Expressions (6.EE.2a)

Example 3 Write the phrase as an expression.

a. the sum of twice a number n and five

$2n + 5$

b. twelve less than four times a number y

$4y - 12$

Try It Yourself
Write the phrase as an expression.

5. six more than three times a number w

6. the quotient of seven and a number p

7. two less than a number t

8. the product of a number x and five

9. five more than six divided by a number r

10. four less than three times a number b

7.1 Writing Equations in One Variable

COMMON CORE STATE STANDARDS
6.EE.2b
6.EE.7

Essential Question
How does rewriting a word problem help you solve the word problem?

1 EXAMPLE: Rewriting a Word Problem

Read the problem several times. Rewrite the problem. Leave out information you do not need to solve the problem.

Given Problem (63 words)

Your minivan has a flat, rectangular space in the back. When you fold down the rear seats of the van and move them forward, the rectangular space in the van is increased by 2 feet, as shown in the diagram.

By how many square feet does the area of the rectangular space increase when the rear seats are folded down and moved forward?

Rewritten Problem (29 words)

When you fold down the back seats of a minivan, the added space is a 5-foot by 2-foot rectangle. What is the area of the added space?

Can you make the problem even simpler?

Rewritten Problem (16 words)

What is the area of a rectangle that is 5 feet long and 2 feet wide?

Added Area = 2 × 5 = 10 sq ft

Now the problem is easy to read. It asks you to find the area of the additional space after the back seats are folded down.

276 Chapter 7 Equations

Laurie's Notes

Introduction

For the Teacher
- **Goal:** Students will rewrite word problems using fewer words, focusing on the mathematics that must be done.
- Today's investigation models how to read through a word problem and identify the key information. Then the problem is rephrased to succinctly state the problem to be solved.

Motivate
- Because this investigation is about words in math problems, try a version of Word Scrabble as a warm-up. Start with a 4-letter word. Each player in succession changes just one letter to make a new word. The player may also rearrange the letters. Sample:
 MATH → MOTH → BOTH → BOAT → COAT → SCAT → MATS → MATH
- This sample comes back to the word it began with but that is not required.
- Try any of these words for starters: FOUR, MADE, BOOK, MALL, TALK.

Activity Notes

Example 1
- Students may become intimidated by problems that have a lot of words. In this example, they have the opportunity to see the stages needed to reduce the number of words to get at the real question.
- Explain that additional words are often added to a problem for interest, and/or to help clarify the problem. Once the problem is clear, it is rewritten using fewer words.
- Ask for volunteers to read each section.
- Pause at the end of each section to discuss what the words tell you and what is being asked. Discuss how the diagram aids understanding of the problem.
- ? "What is the answer to the question?" 10 ft^2

Common Core State Standards

6.EE.2b Identify parts of an expression using mathematical terms (sum, term, product, factor, quotient, coefficient); view one or more parts of an expression as a single entity.

6.EE.7 Solve real-world and mathematical problems by writing and solving equations of the form $x + p = q$ and $px = q$ for cases in which p, q and x are all nonnegative rational numbers.

Previous Learning
Students should be familiar with the vocabulary associated with the four operations.

Start Thinking! and Warm Up

Solve the problem.

1. A car travels at an average rate of 30 miles per hour for 2 hours. How far has the car traveled?

2. A picture frame is shaped like a rectangle with a length of 10 inches and a width of 8 inches. What is the area of the picture frame?

3. You are walking 1 mile home from school. When you are halfway home, you realize that you forgot a book in your locker and have to go back and get it. Once you finally get home, how far have you walked?

7.1 Record and Practice Journal

English Language Learners

Vocabulary

Have students create a table in their notebooks of common words used to indicate addition, subtraction, multiplication, and division. Sample:

Addition	Subtraction
added to	subtracted from
sum of	difference of
more than	less than
increased by	decreased by
total of	fewer than
and	take away

Multiplication	Division
multiplied by	divided by
times	quotient of
product of	
twice	
of	

7.1 Record and Practice Journal

2 ACTIVITY: Rewriting a Word Problem

Work with a partner. Rewrite each problem until it is easy to read. Then solve the problem.

a. (62 words)

A supermarket is having its grand opening on Saturday morning. Every fifth customer will receive a $20 coupon for a free turkey. Every seventh customer will receive a $25 coupon for 2 gallons of free ice cream. You are the manager of the store and you expect 400 customers. How many of each type of coupon should you plan to use?

80 $20 coupons and 57 $25 coupons

b. (72 words)

You and your friend are at a football game. The stadium is 4 miles from your home. You each brought 5 dollars to spend on refreshments. During the third quarter of the game, you say, "I read that the greatest distance that a baseball has been thrown is 445 feet 10 inches." Your friend says, "That's about one and a half times the length of the football field." Is your friend correct?

yes

c. (90 words)

You are visiting your cousin who lives in the city. To get back home, you take a taxi. The taxi charges $2.10 for the first mile and $0.90 for each additional mile. After riding 13 miles, you decide that the fare is going to be more than the $20 you have with you. So, you tell the driver to stop and let you out. Then you call a friend and ask your friend to come pick you up. After paying the driver, how much of your $20 is left?

$7.10

What Is Your Answer?

3. **IN YOUR OWN WORDS** How does rewriting a word problem help you solve the word problem? Make up a word problem that has more than 50 words. Then show how you can rewrite the problem using at most 25 words.

Leaves out information you do not need and is easier to read.

"Solving a math word problem is like making maple syrup."

"You need to boil down 40 gallons of sap from a sugar maple tree to get 1 gallon of syrup."

Laurie's Notes

Activity 2

- **Teaching Tip:** It would be helpful to provide a transparency and pen for each pair so that they can share their sample rewrites easily.
- Remind students to keep track of the number of words used in each successive rewrite. Encourage them to use a diagram.
- **FYI:** There is additional (extraneous) information in each problem. Some students may find it helpful to make a list of extraneous information.
- **Part (a):** The question pares down to multiples of 5 and 7, and how many times each divides into 400. The question could be extended to ask how many people will receive both coupons. multiples of 35
- **Part (b):** Students need to know that a football field is 100 yards or 300 feet.
- **Part (c):** Students need to be familiar with how taxis charge their fares—a base fare plus so much for each mile (or portion of a mile).
- Have students share their rewritten problems and their answers. Using transparencies makes it possible to share more than one version of each problem.

What Is Your Answer?

- **FYI:** Read the cartoon. Students may not be aware of how much sap it takes to make a gallon of maple syrup!

Closure

- How do adverbs and adjectives in an English sentence relate to today's activities? Adverbs and adjectives are descriptive words, but are not necessary to the meaning of the sentence, like extraneous information in a word problem.

Technology For the Teacher

Dynamic Classroom

The Dynamic Planning Tool
Editable Teacher's Resources at *BigIdeasMath.com*

2 ACTIVITY: Rewriting a Word Problem

Work with a partner. Rewrite each problem until it is easy to read. Then solve the problem.

a. (62 words)

A supermarket is having its grand opening on Saturday morning. Every fifth customer will receive a $20 coupon for a free turkey. Every seventh customer will receive a $25 coupon for 2 gallons of ice cream. You are the manager of the store and you expect to have 400 customers. How many of each type of coupon should you plan to use?

b. (72 words)

You and your friend are at a football game. The stadium is 4 miles from your home. You each brought 5 dollars to spend on refreshments. During the third quarter of the game, you say, "I read that the greatest distance that a baseball has been thrown is 445 feet, 10 inches." Your friend says, "That's about one and a half times the length of the football field." Is your friend correct?

c. (90 words)

You are visiting your cousin who lives in the city. To get back home, you take a taxi. The taxi charges $2.10 for the first mile and $0.90 for each additional mile. After riding 13 miles, you decide that the fare is going to be more than the $20 you have with you. So, you tell the driver to stop and let you out. Then you call a friend and ask your friend to come and pick you up. After paying the driver, how much of your $20 is left?

What Is Your Answer?

3. **IN YOUR OWN WORDS** How does rewriting a word problem help you solve the word problem? Make up a word problem that has more than 50 words. Then show how you can rewrite the problem using at most 25 words.

"Solving a math word problem is like making maple syrup."

"You need to boil down 40 gallons of sap from a sugar maple tree to get 1 gallon of syrup."

Practice

Use what you learned about writing equations to complete Exercises 4 and 5 on page 280.

Section 7.1 Writing Equations in One Variable

7.1 Lesson

Key Vocabulary
equation, p. 278

An **equation** is a mathematical sentence that uses an equal sign, =, to show that two expressions are equal.

Expressions	Equations
$4 + 8$	$4 + 8 = 12$
$x + 8$	$x + 8 = 12$

To write a word sentence as an equation, look for key words or phrases such as "is," "the same as," or "equals" to determine where to place the equal sign.

EXAMPLE 1 Writing Equations

Write the word sentence as an equation.

a. The sum of a number n and 7 is 15.

The sum of a number n and 7 is 15.

$n + 7 \qquad = 15$ "Sum of" means addition.

∴ An equation is $n + 7 = 15$.

b. A number y decreased by 4 is 3.

A number y decreased by 4 is 3.

$y - 4 \qquad = 3$ "Decreased by" means subtraction.

∴ An equation is $y - 4 = 3$.

c. 12 times a number p equals 48.

12 times a number p equals 48.

$12p \qquad = 48$ "Times" means multiplication.

∴ An equation is $12p = 48$.

On Your Own

Now You're Ready
Exercises 6–13

Write the word sentence as an equation.

1. 9 less than a number b equals 2.
2. The product of a number g and 5 is 30.
3. A number k increased by 10 is the same as 24.
4. The quotient of a number q and 4 is 12.

Laurie's Notes

Introduction

Connect
- **Yesterday:** Students practiced rewriting word problems using fewer words.
- **Today:** Students will translate words into mathematical equations.
- Previously, students learned to translate words into algebraic expressions. In this section, students take the process one step further and write an equation.

Motivate
- Ask four students to volunteer. Two of them will hold a sentence strip card with an English phrase and two will hold a sentence strip card with an English sentence. Don't tell them the difference. Simply say that two of them are similar in some way and the other two are similar in some way. They should sort themselves and explain their grouping.
 - **Examples:** WINTER IS HERE
 - THE LONG COLD WINTER (PHRASE)
 - ORANGE JUICE FOR HEALTHY LIVING (PHRASE)
 - I LIKE ORANGE JUICE
- **Discuss:** Discuss the difference between expressions (English phrases) and equations (English sentences). Note that variables are used in each, but equations and expressions can be written without variables. A phrase is a group of words that have meaning but are not a full sentence. A sentence is a group of words that must include a subject and a verb.

Lesson Notes

Example 1
- Note the use of color to connect the words and the operation. If possible, change chalk or marker color when writing the examples on the board.
- **Common Error:** Subtraction is not commutative so the order in which the terms are written does matter. *A number decreased by 4* is different from *4 decreased by a number*.
- **Representation:** The notation $12x$ is still somewhat new to students so be sure to ask, "What does $12x$ mean?" Multiply 12 by the value of x.
- **?** Ask a few questions about part (a) and part (c).
 - "Would $7 + n = 15$ be correct for part (a)? Explain." Yes; Addition is commutative.
 - "Would $p12 = 48$ be correct for part (c)? Explain." Yes; Multiplication is commutative. It is standard notation, however, to write the number (coefficient) before the variable.

On Your Own
- **Think-Pair-Share:** Students should read each question independently and then work with a partner to answer the questions. When they have answered the questions, the pair should compare their answers with another group and discuss any discrepancies.
- **Common Error:** In Question 4, division is not commutative. The quotient of a number q and 4, $q \div 4$, is not the same as the quotient of 4 and a number q, $4 \div q$.

Goal
Today's lesson is reading and translating a word problem into a mathematical **equation**.

Activity Materials
Introduction
- sentence strip cards

Start Thinking! and Warm Up

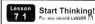

Describe some instances of when you have used two expressions that are equal to each other in your daily life.

Extra Example 1
Write the word sentence as an equation.
a. A number m added to 3 equals 15. $3 + m = 15$
b. The difference of a number w and 6 is the same as 12. $w - 6 = 12$
c. 22 is twice a number h. $22 = 2h$

On Your Own
1. $b - 9 = 2$
2. $5g = 30$
3. $k + 10 = 24$
4. $q \div 4 = 12$

T-278

Extra Example 2

Ten servers decorate 20 tables for a party. Each table has 4 red candles and 3 white candles. Let c be the total number of red and white candles. Write an equation that can be used to find c. $c = 20 \times (4 + 3)$

Extra Example 3

After three rounds, 15 students are eliminated from a geography bee. There are 55 students remaining. Write an equation you can use to find the number of students that started the geography bee. $s - 15 = 55$

 On Your Own

5. $x - 7 = 10$
6. $28 + x = 52$
7. $24,000 = (24x)(60)$

Differentiated Instruction

Auditory

On the board, write a sentence that translates to an algebraic expression. Discuss with students what each word in the sentence means and how it relates to an algebraic equation. Sentences similar to Example 1:
The quotient of 8 and a number p is 2.
The difference of a number s and 6 is 8.
Twice a number z is 12.
The total of a number y and 4 is 11.

Laurie's Notes

Example 2

- Students may need to read the problem more than once. Just as with the activity yesterday, there is additional information that is not important in solving the problem.
- Model rewriting the problem using fewer words.
- **?** When the problem is finished ask, "Would $25(4 + 6) = c$ also be correct?" yes
- Be mindful of the examples you do in class. You do not want to mislead students into thinking that all of the arithmetic occurs on the left side of the equation and the answer is on the right side of the equation. This is often referred to as the "rightward arrow." You do not want students to view the equation sign as an arrow. ($3 + 4 \rightarrow 7$ or $3 + x \rightarrow 7$)
- **? Extension:** Ask how many of each color of candle is needed. 100 white and 150 purple

Example 3

- Ask a volunteer to read through the problem. Note the *Reading* tip at the left.
- **Common Error:** Students may subtract 24 from 96.
- The words written are an example of a *verbal model*. The verbal model says in words how the problem can be solved. As I say to my students, "If you can't say in words how to solve the problem, don't try to write an equation. You need to understand the problem first."
- **Note:** The variable is defined to be *the number of students that started*, not simply *students*. Often students are too brief in describing what the variable represents.

On Your Own

- For each question the students should write the question in words (verbal model), define the variable, and then translate the words into an equation.
- For Question 7, students will need to determine the number of minutes in one day before writing the equation.
- Students could work in pairs to complete the questions.
- Share solutions at the board.

Closure

- **Exit Ticket:** Translate each sentence.
 The quotient of a number and 8 is 40. $n \div 8 = 40$
 The quotient of 40 and a number is 8. $40 \div n = 8$
 40 is the quotient of 8 and a number. $40 = 8 \div n$

The Dynamic Planning Tool
Editable Teacher's Resources at *BigIdeasMath.com*

EXAMPLE 2 Standardized Test Practice

Ten servers decorate 25 tables for a wedding. Each table is decorated as shown. Let *c* be the total number of white and purple candles. Which equation can be used to find *c*?

A $c = 25 + (4 \times 6)$ **B** $c = 25(4 + 6)$
C $c = 10(25 + 4 + 6)$ **D** $c = 10(4 + 6)$

Words The number is the number times the number
of candles of tables of candles on
 each table.

Variable Let *c* be the number of candles.

Equation *c* = 25 × (4 + 6)

∴ The correct answer is **B**.

EXAMPLE 3 Real-Life Application

After two rounds, 24 students are eliminated from a spelling bee. There are 96 students remaining. Write an equation you can use to find the number of students that started the spelling bee.

Words The number minus the number is the number
of students of students of students
that started eliminated remaining.

Reading
The word "eliminated" means subtraction.

Variable Let *s* be the number of students that started.

Equation *s* − 24 = 96

∴ An equation is $s - 24 = 96$.

On Your Own

5. You enter an elevator and go down 7 floors. You exit on the 10th floor. Write an equation you can use to find the floor where you entered the elevator.

6. Together you and a friend have $52. Your friend has $28. Write an equation you can use to find how much money you have.

7. A typical person takes about 24,000 breaths each day. Write an equation you can use to find the number of breaths a typical person takes each minute.

7.1 Exercises

Vocabulary and Concept Check

1. **VOCABULARY** What is the difference between an expression and an equation?
2. **DIFFERENT WORDS, SAME QUESTION** Which is different? Write "both" equations.

 | 4 less than a number *n* is 8. | A number *n* is 4 less than 8. |

 | A number *n* minus 4 equals 8. | 4 subtracted from a number *n* is 8. |

3. **OPEN-ENDED** Write a word sentence for the equation $28 - n = 5$.

Practice and Problem Solving

Rewrite the problem until it is easy to read. Then solve the problem.

4. You run at a steady rate of 7 minutes per mile in a cross-country race. You finish after 21 minutes in fourth place. How long is the race?

5. You drop a 4-ounce rock into a pond. It makes a circular ripple. The radius increases by 1 foot each second. You take a picture when the radius is 3 feet. What is the circumference of the ripple when you take the picture?

Write the word sentence as an equation.

① 6. The sum of a number *x* and 4 equals 12.
 7. A number *y* decreased by 9 is 8.
 8. 9 times a number *b* is 36.
 9. A number *w* divided by 5 equals 6.
 10. 54 equals 9 more than a number *t*.
 11. 5 is one-fourth of a number *c*.
 12. 11 is the quotient of a number *y* and 6.
 13. 9 less than a number *n* equals 27.

14. **ERROR ANALYSIS** Describe the error in writing the sentence as an equation.

 A number *n* is 5 more than 12.
 $n + 5 = 12$

15. **FUNDRAISING** Students and faculty raised $6042 for band uniforms. The faculty raised $1780. Write an equation you can use to find the amount *a* raised by the students.

16. **GOLF** You hit a golf ball 90 yards. It travels three-fourths of the distance to the hole. Write an equation you can use to find the distance *d* from the tee to the hole.

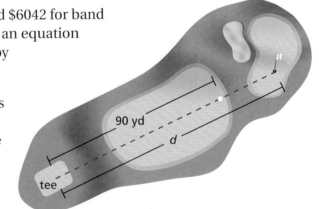

280 Chapter 7 Equations

Assignment Guide and Homework Check

Level	Day 1 Activity Assignment	Day 2 Lesson Assignment	Homework Check
Basic	4, 5, 23–27	1–3, 7–17 odd, 14, 16	7, 11, 16, 17
Average	4, 5, 23–27	1–3, 7–13 odd, 14, 17, 19, 20	7, 11, 17, 20
Advanced	4, 5, 23–27	1–3, 10–22 even, 21	12, 18, 20, 21

Common Errors

- **Exercises 6–13** Students may write the equations in the wrong order or forget parts of the equation. Remind them of the guidelines for writing expressions. Remind students that these are equations, so there will be an equal sign in their answer and numbers and/or variables on each side.
- **Exercise 20** Students may forget that the sales tax rate *r* is the decimal form of the percent of sales tax. Remind them that they are using the decimal form of the percent in part (a).

Technology For the Teacher
Answer Presentation Tool
QuizShow

7.1 Record and Practice Journal

Vocabulary and Concept Check

1. An equation has an equal sign and an expression does not.
2. A number *n* is 4 less than 8; $n = 8 - 4$; $n - 4 = 8$
3. *Sample answer:* A number *n* subtracted from 28 is 5.

Practice and Problem Solving

4. You run for 21 minutes at a rate of 7 minutes per mile. How far do you run? 3 mi
5. What is the circumference of a circular pond ripple with radius 3 feet? about 18.84 ft
6. $x + 4 = 12$
7. $y - 9 = 8$
8. $9b = 36$
9. $w \div 5 = 6$
10. $54 = t + 9$
11. $5 = \frac{1}{4}c$
12. $11 = y \div 6$
13. $n - 9 = 27$
14. The word *is* indicates an equal sign, so the equal sign is in the wrong position. $n = 12 + 5$
15. $6042 = 1780 + a$
16. $90 = \frac{3}{4}d$

T-280

Practice and Problem Solving

17. $16 = 3x$

18. $30 = 4x$

19. $326 = 12(14) + 6(5) + 16x$

20. a. Answer should include, but is not limited to: The equation should have the form total amount paid = total price + (total price × rate).

 b. yes; Multiplying r by 100 gives the percent for the sales tax.

21. See *Taking Math Deeper*.

22. $20 = 3s + \dfrac{\pi}{2}s$

Fair Game Review

23. 13 **24.** 3

25. 28 **26.** 5

27. C

Mini-Assessment

Write the word sentence as an equation.

1. The sum of a number x and 5 equals 20. $x + 5 = 20$

2. 64 equals 33 more than a number y. $64 = 33 + y$

3. A number c decreased by 12 is 4. $c - 12 = 4$

4. 10 is the quotient of a number s and 8. $\dfrac{s}{8} = 10$

5. You went grocery shopping with $40.65. You left the store with $5.10. Write an equation you can use to find the cost c of the groceries. *Sample answer:* $5.10 + c = 40.65$

T-281

Taking Math Deeper

Exercise 21

This problem is similar to real life because the given information is not presented in nice little bundles. Start by organizing the given information.

 Organize the given information.

Information about blocks

There are 16 blocks.
The time is 28 minutes.

In U.S. cities there are 12–16 blocks per mile.

Information about strawberries

 a. Yes, there is enough information to find the time it takes to walk each block. You walked 16 blocks in 28 minutes.

Rate: $\dfrac{16}{28} = \dfrac{4}{7}$ block per minute

 b. No, there is not enough information to find how many strawberries you ate while walking the last block. The rate at which you are eating the strawberries is not constant. You ate 12 in the first 8 blocks, but only 8 in the last 8 blocks. On the other hand, if a student estimates that eating 8 strawberries in 8 blocks means that the person probably ate 1 strawberry while walking the last block, that is still good thinking.

Reteaching and Enrichment Strategies

If students need help...	If students got it...
Resources by Chapter • Practice A and Practice B • Puzzle Time Record and Practice Journal Practice Differentiating the Lesson Lesson Tutorials Skills Review Handbook	Resources by Chapter • Enrichment and Extension Start the next section

GEOMETRY Write an equation that can be used to find the value of *x*.

17. Perimeter of triangle: 16 in.

18. Perimeter of square: 30 mm

19. **MUSIC** You sell instruments at a Caribbean music festival. You earn $326 by selling 12 sets of maracas, 6 sets of claves, and *x* djembe drums. Write an equation you can use to find the number of djembe drums you sold.

20. **SALES TAX** Find a sales receipt from a store in your county. It should show the total price and the total amount paid including sales tax.

 a. Write an equation you can use to find the sales tax rate *r* for your county.
 b. Can you use *r* to find the *percent* for the sales tax? Explain.

21. **STRAWBERRIES** You buy a basket of 24 strawberries. You eat them as you walk to the beach. It takes the same amount of time to walk each block. When you are halfway there, half of the berries are gone. After walking 3 more blocks, you still have 5 blocks to go. You reach the beach 28 minutes after you began. One-sixth of your strawberries are left.

 a. Is there enough information to find the time it takes to walk each block? Explain.
 b. Is there enough information to find how many strawberries you ate while walking the last block? Explain.

22. **Geometry** A semicircle is cut from a square. The perimeter of the shaded region is 20 inches. Write an equation you can use to find the side length *s*.

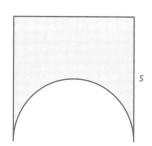

Fair Game Review What you learned in previous grades & lessons

Evaluate the expression when $a = 7$. *(Section 1.1)*

23. $6 + a$ 24. $a - 4$ 25. $4a$ 26. $\dfrac{35}{a}$

27. **MULTIPLE CHOICE** Which is the most convenient expression to use in estimating the quotient of 76.23 and 15.81? *(Section 3.1)*

 Ⓐ $76.23 \div 15.81$ Ⓑ $76 \div 16$ Ⓒ $80 \div 16$ Ⓓ $76.2 \div 15.8$

7.2 Solving Equations Using Addition or Subtraction

COMMON CORE STATE STANDARDS

6.EE.5
6.EE.7

Essential Question How can you use addition or subtraction to solve an equation?

If two sides of a scale weigh the same, the scale will balance.

If you add or subtract the same amount on each side of the scale, it will still balance.

1 EXAMPLE: Solving an Equation Using Subtraction

Use a scale to model and solve $n + 3 = 7$.

Think of a scale with both sides in balance. Whatever you add to or subtract from one side, you need to do the same on the other side to keep the scale in balance.

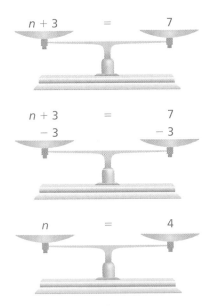

When $n + 3$ is on one side and 7 is on the other side, the scale is balanced.

Your goal is to get n by itself on one side. To do this, subtract 3 from each side.

Simplify each side. The remaining value is the solution.

∴ So, $n = 4$.

Check
You can check the solution by substituting 4 for n in the original equation.

$4 + 3 = 7$

282 Chapter 7 Equations

Laurie's Notes

Introduction

For the Teacher
- **Goal:** Students will use the concept of a pan balance (balance scale) to develop some intuition about what happens when adding or subtracting amounts to or from both sides of an equation.

Motivate
- Ask for 6 volunteers to come to the front of the room. Three students are Team A and the other 3 are Team B.
- After they are positioned say, "I changed my mind. I want [name a 7th student] to join Team A." Team B will immediately want someone else to join their team so it will be fair (balanced). A = 4 people, B = 3 people
- Change your mind again and have two students from Team A sit down. A = 2 people, B = 3 people Now Team A will say it is unfair (not balanced).
- Students have a sense about fairness and they want everyone to be treated the same. This idea can be used with equations. You want to treat both sides of the equation the same way.

Discuss
- Discuss the two scales in the context of being balanced. If any quantity is added to one side, the same amount must be added to the other. If any quantity is removed (subtracted) from one side, the same amount must be removed (subtracted) from the other side.

Activity Notes

Example 1
- **Representation:** If you are familiar with algebra tiles, you could use the tiles to model these problems. You could also draw the sketch shown or model with actual materials on a pan balance.
- **Note:** Physical objects on a pan balance make more sense to students than simply seeing a variable and numbers. The equation written above the balance scale in the book helps students think about two sides of an equation being balanced. The physical objects help students think what must be done mathematically (operations) to solve the equation.
- The concept of what it means to *solve an equation* may be very unfamiliar to students. They have solved many problems in math, and now they are being asked to solve an equation. Problems and equations are not the same, especially if students think of the equal sign as a rightward pointing arrow. The *problem* $4 + 3 \rightarrow ?$ and the *equation* $? + 3 = 7$ are different to students.
- Discuss the example shown and model it with physical objects. Note that the goal is stated: "Get *n* by itself on one side." You want the variable by itself. With the physical models you want the object representing the variable by itself on one side.

Common Core State Standards

6.EE.5 Understand solving an equation ... as a process of answering a question: which values from a specified set, if any, make the equation ... true? Use substitution to determine whether a given number in a specified set makes an equation ... true.

6.EE.7 Solve real-world and mathematical problems by writing and solving equations of the form $x + p = q$ and $px = q$ for cases in which p, q and x are all nonnegative rational numbers.

Previous Learning
Students should know how to evaluate expressions.

Activity Materials
Textbook
• pan balance (for modeling)

Start Thinking! and Warm Up

Activity 7.2 Start Thinking! For use before Activity 7.2

Activity 7.2 Warm Up For use before Activity 7.2

Evaluate the expression.
1. $3 + x$ when $x = 2$
2. $a + 16$ when $a = 4$
3. $10 + z$ when $z = 5$
4. $17 - c$ when $c = 9$
5. $8 - w$ when $w = 8$
6. $s - 20$ when $s = 34$

7.2 Record and Practice Journal

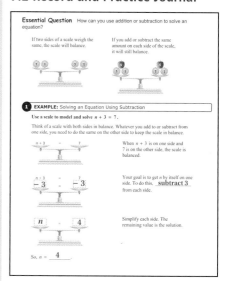

T-282

Differentiated Instruction

Kinesthetic

Have students play *The Solution Game*. On index cards, write equations that can be solved using mental math. Working in small groups, one student turns over a card. The other students suggest solutions. The solution should be checked by substitution into the equation.

7.2 Record and Practice Journal

Laurie's Notes

Activity 2

- Read the sample problem.
- ❓ "Can you ask a similar question for the equation $x + 3 = 7$?" Yes, you want students to understand that there is more than one way to phrase the question. What do you add to 3 to get 7?
- This is just mental math at this stage. Do not expect students to state inverse operations. Let students rely on their mental math.
- **Note:** Two of the problems have the variables on the right side of the equation. The question formed is still an equivalence: "8 is the sum of 3 and what number?"
- ❓ When students finish ask, "Could part (d) be phrased, what number do you subtract from 9 to get 21?" *No, subtracting a number from 9 and subtracting 9 from a number are not the same.*

Activity 3

- In the formal lesson, students will be expected to show the mathematical process. In other words, identify which number has been added to or subtracted from each side of the equation. At this stage, you might want students to simply note what is being done to each side of the equation, and then write the solution. For example, part (c) would be "add 3 to each side; $x = 9$."

What Is Your Answer?

- Have students work with a partner. To do a quick (but private) check of student understanding, have students give a thumbs up for *true* and a thumbs down for *false*. Discuss each question as needed.

Closure

- Select 3 people for Team A and 3 people for Team B and repeat the questions from the motivate activity.
 - "If I have 3 people join Team B, then …?" Students should say, "Have 3 people join Team A." 12 students are now standing.
 - "If I have 5 people from Team B sit down, then…?" Students should say, "Have 5 people from Team A sit down."

Technology For the Teacher

The Dynamic Planning Tool
Editable Teacher's Resources at *BigIdeasMath.com*

2 ACTIVITY: Solving Equations Using Mental Math

Work with a partner. Write a question that represents the equation. Use mental math to answer the question. Then check your solution.

	Equation	Question	Solution	Check
Sample:	a. $x + 3 = 7$	What number plus 3 equals 7?	$x = 4$	$4 + 3 = 7$ ✓
	b. $4 + m = 11$			
	c. $8 = a + 3$			
	d. $x - 9 = 21$			
	e. $13 = p - 4$			

3 ACTIVITY: Solving Equations Using Addition or Subtraction

Work with a partner. Solve the equation using Example 1 as a sample.

a.

b.

c.

d.

What Is Your Answer?

4. Decide whether the statement is *true* or *false*. If false, explain your reasoning.

 a. In an equation, any letter can be used as a variable.
 b. The goal in solving an equation is to get the variable by itself.
 c. In the solution, the variable must always be on the left side of the equal sign.
 d. If you add a number to one side, you should subtract it from the other side.
 e. If you add a number to one side, you should add it to the other side.

5. **IN YOUR OWN WORDS** How can you use addition or subtraction to solve an equation? Give two examples to show how your procedure works.

Practice

Use what you learned about solving equations to complete Exercises 12–17 on page 287.

7.2 Lesson

Key Vocabulary
solution, p. 284
inverse operations, p. 285

A **solution** of an equation is a value that makes the equation true.

Value of x	x + 3 = 7	Are both sides equal?
3	$3 + 3 \stackrel{?}{=} 7$ $6 \neq 7$ ✗	no
4	$4 + 3 \stackrel{?}{=} 7$ $7 = 7$ ✓	yes
5	$5 + 3 \stackrel{?}{=} 7$ $8 \neq 7$ ✗	no

Reading
The symbol \neq means "is not equal to."

So, the value $x = 4$ is a solution of the equation $x + 3 = 7$.

EXAMPLE 1 Checking Solutions

Tell whether the given value is a solution of the equation.

a. $p + 10 = 38; \ p = 18$

$18 + 10 \stackrel{?}{=} 38$ Substitute 18 for p.

$28 \neq 38$ ✗ Sides are *not* equal.

∴ So, $p = 18$ is not a solution.

b. $4y = 56; \ y = 14$

$4(14) \stackrel{?}{=} 56$ Substitute 14 for y.

$56 = 56$ ✓ Sides are equal.

∴ So, $y = 14$ is a solution.

On Your Own

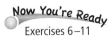
Exercises 6–11

Tell whether the given value is a solution of the equation.

1. $a + 6 = 17; \ a = 9$ **2.** $9 - g = 5; \ g = 3$

3. $35 = 7n; \ n = 5$ **4.** $\dfrac{q}{2} = 28; \ q = 14$

284 Chapter 7 Equations

Laurie's Notes

Introduction

Connect
- **Yesterday:** Students gained an intuitive understanding of how equations are solved using addition and subtraction.
- **Today:** Students will use Properties of Equality to solve equations.

For the Teacher
- Use file folder paper to make a quick model to help students think about what it means to solve an equation.
- Make a number strip with the numbers 3, 4, and 5. Make three equation strips with equations $x + 4 = 9$, $1 = x - 3$, and $5 + x = 0$. Cut a slit in the paper on either side of the variable so that the number strip can slide through, replacing the variable with a number.

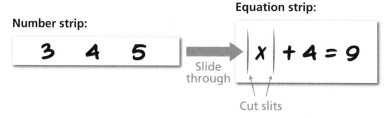

Motivate
- Tell the students you have several equations. They are to decide if the equation is *true* or *false*. They can vote with thumbs up/down, or on a personal white board.
- When the number slip is woven through the opening, it will read: $5 + 4 = 9$ (true), $4 + 4 = 9$ (false), and $3 + 4 = 9$ (false).
- Point out to students that equations can have only numbers, $5 + 4 = 9$, or numbers and variables, $x + 4 = 9$.
- Use the same number strip with the other equation strips.
- The goal is for students to recognize that one of the numbers make a true statement and the other numbers make a false statement.
- ? "If you use the equation $x - 3 = 4$, do you think any of the numbers (3, 4, or 5) would have made a true statement?" no

Lesson Notes

Example 1
- Work through each part.
- Be careful to use the correct language. You *substitute* for the value of the variable. You do not *plug it in*!
- ? "What does $4y$ mean, forty-something?" No, it means 4 times the value of y.

On Your Own
- **Neighbor Check:** Have students work independently and then have their neighbor check their work. Have students discuss any discrepancies.
- **Common Error:** For Question 4, students will quickly look at the numbers 2, 14, and 28 and say 14 is a solution without thinking about the operation. Caution students to work slowly and actually substitute the value of the variable.

Goal Today's lesson is using the **inverse operations** of addition and subtraction to find the **solution** of an equation.

Lesson Materials
Introduction
• file folder paper

Start Thinking! and Warm Up

Lesson 7.2 Warm Up
For use before Lesson 7.2

Lesson 7.2 Start Thinking!
For use before Lesson 7.2

Two operations are **inverse operations** if one operation "undoes" the other.

Of the four operations (addition, subtraction, multiplication, division), which are inverse operations? Why?

This idea can be applied to real-life situations. What are the inverses of the following actions?

1. Unbutton 2. Ascend
3. Deposit 4. Turn off

Describe another pair of real-life inverse actions.

Extra Example 1
Tell whether the given value is a solution of the equation.
a. $t - 5 = 17$; $t = 12$ no
b. $b \div 6 = 8$; $b = 48$ yes

On Your Own
1. no 2. no
3. yes 4. no

English Language Learners

Vocabulary

English learners may have difficulty understanding that addition *undoes* subtraction and that subtraction *undoes* addition. Emphasize that *undo* means to do the opposite of what was done before.

You get $x - 2$ by subtracting 2 from x.
You get $x - 2$ back to x by adding 2 to $x - 2$.
$x - 2 + 2 = x$

You get $x + 11$ by adding 11 to x.
You get $x + 11$ back to x by subtracting 11 from $x + 11$.
$x + 11 - 11 = x$

This will help students to better understand *inverse operations*.

Extra Example 2
a. Solve $p - 8 = 14$. $p = 22$
b. Solve $11 = y - 4$. $y = 15$

On Your Own
5. $k = 4$
6. $n = 14$
7. $r = 21$

Laurie's Notes

Key Ideas

- Discuss other inverse operations such as: putting your shoes on and taking them off; going to school and coming home from school; an airplane taking off and landing; turning a light switch on and off.
- Write the Key Ideas.
- Discuss the Words, Numbers, and Algebra.
- How you record what is being done mathematically to each side of the equation is very important. You should not be sloppy. Encourage students to develop good habits from the start.

Example 2

- Explain to students that although they can probably solve these problems using mental math, another equally important part of equation solving is the *process*. How can the problem be solved when the answer cannot be found using mental math?
- Write a few examples.
 (a) $x - 5 = 13$ (b) $x - 8.7 = 19.49$ (c) $x - 4\frac{3}{5} = 6\frac{2}{3}$
- You can use mental math for (a). For (b) and (c), you probably want to use paper and pencil. What is the process now?
- **Note:** Some textbooks use the horizontal format as well, but using just one format this year is fine.
- Discuss with students how to determine if their answer is correct by substituting the solution for the variable in the original equation.
- **Big Idea:** Whatever number is added to one side of the equation is also added to the other. Using an alternate color (chalk or overhead pen) to show this will help the many students who are visual learners. Notice in the examples that the *process* is annotated along the right side of the equation.

On Your Own

- Ask volunteers to share their work at the board.

T-285

You can use *inverse operations* to solve equations. **Inverse operations** "undo" each other. Addition and subtraction are inverse operations.

 Key Ideas

Addition Property of Equality

Words If you add the same number to each side of an equation, the two sides remain equal.

Numbers
$$8 = 8$$
$$+5 \quad +5$$
$$13 \quad 13$$

Algebra
$$x - 4 = 5$$
$$+4 \quad +4$$
$$x = 9$$

Subtraction Property of Equality

Words If you subtract the same number from each side of an equation, the two sides remain equal.

Numbers
$$8 = 8$$
$$-5 \quad -5$$
$$3 \quad 3$$

Algebra
$$x + 4 = 5$$
$$-4 \quad -4$$
$$x = 1$$

EXAMPLE 2 **Solving Equations Using Addition**

a. **Solve $x - 2 = 6$.**

$x - 2 = 6$	Write the equation.
$+2 \quad +2$	Add 2 to each side.
$x = 8$	Simplify.

Undo the subtraction.

Check
$$x - 2 = 6$$
$$8 - 2 \stackrel{?}{=} 6$$
$$6 = 6 \checkmark$$

∴ The solution is $x = 8$.

b. **Solve $18 = x - 7$.**

$18 = x - 7$	Write the equation.
$+7 \quad +7$	Add 7 to each side.
$25 = x$	Simplify.

Study Tip
You can check your solution by substituting it for the variable in the original equation.

Check
$$18 = x - 7$$
$$18 \stackrel{?}{=} 25 - 7$$
$$18 = 18 \checkmark$$

∴ The solution is $x = 25$.

On Your Own

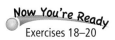
Exercises 18–20

Solve the equation. Check your solution.

5. $k - 3 = 1$ **6.** $n - 10 = 4$ **7.** $15 = r - 6$

Section 7.2 Solving Equations Using Addition or Subtraction **285**

EXAMPLE 3 Solving Equations Using Subtraction

a. Solve $x + 2 = 9$.

$$x + 2 = 9 \quad \text{Write the equation.}$$

Undo the addition. → $-2 \quad -2 \quad$ Subtract 2 from each side.

$$x = 7 \quad \text{Simplify.}$$

∴ The solution is $x = 7$.

Check
$x + 2 = 9$
$7 + 2 \stackrel{?}{=} 9$
$9 = 9$ ✓

b. Solve $26 = 11 + x$.

$$26 = 11 + x \quad \text{Write the equation.}$$
$$-11 \quad -11 \quad \text{Subtract 11 from each side.}$$
$$15 = x \quad \text{Simplify.}$$

∴ The solution is $x = 15$.

Check
$26 = 11 + x$
$26 \stackrel{?}{=} 11 + 15$
$26 = 26$ ✓

EXAMPLE 4 Real-Life Application

Wall height: 48 ft

You climb 23 feet on a rock climbing wall. Write and solve an equation to find the distance to the top.

Words The distance climbed plus the distance to the top is the height of the wall.

Variable Let d be the distance to the top.

Equation $23 \quad + \quad d \quad = \quad 48$

$$23 + d = 48 \quad \text{Write the equation.}$$
$$-23 \quad -23 \quad \text{Subtract 23 from each side.}$$
$$d = 25 \quad \text{Simplify.}$$

∴ The distance to the top is 25 feet.

On Your Own

Now You're Ready
Exercises 21–29

Solve the equation. Check your solution.

8. $s + 8 = 17$ **9.** $9 = y + 6$ **10.** $13 + m = 20$

11. You eat 8 blueberries from a package. There are 31 blueberries left. Write and solve an equation to find the number of blueberries in a full package.

Laurie's Notes

Example 3
? "How do you undo the operation of addition?" subtraction
- Work through the problems.
? "Why isn't 9 subtracted from each side of the equation in part (a)?" You could; however it wouldn't achieve the goal of getting the variable by itself on one side of the equation. You need to subtract 2 from each side of the equation.
- **Big Idea:** Whatever is subtracted from one side of the equation is also subtracted from the other. Using an alternate color (chalk or overhead pen) to show this will help visual learners. Notice that the *process* is annotated along the right side of the equation.

Example 4
- This example uses skills developed in the last section. Note how the words in the verbal model come directly from the written problem.
- Information to write the verbal model must be read from the photograph.
- Students will want to skip writing the equation and simply jump to $48 - 23$ to arrive at the answer. Remind them that the question asks for an equation to be written. The equation helps to explain why subtraction is the correct operation to use in solving.
- Students should check their answer in the original equation and within the context of the problem. Does the answer make sense?

On Your Own
- Ask volunteers to share their work at the board.

Closure
- **Exit Ticket:** Solve the equations.
 $17 = 6 + t$ $t = 11$
 $m + 6 = 17$ $m = 11$
? "Are these two problems the same? Explain." Yes; Both equations could be stated as *6 added to a number is 17*.

Extra Example 3
a. Solve $z + 5 = 14$. $z = 9$
b. Solve $18 = r + 6$. $r = 12$

Extra Example 4
You climb 12 feet on a rock climbing wall that is 25 feet high. Write and solve an equation to find the distance to the top. $12 + d = 25$; $d = 13$ ft

On Your Own
8. $s = 9$
9. $y = 3$
10. $m = 7$
11. $x - 8 = 31$; $x = 39$ blueberries

The Dynamic Planning Tool
Editable Teacher's Resources at *BigIdeasMath.com*

Vocabulary and Concept Check

1. Substitute your solution back into the original equation and see if you obtain a true statement.
2. addition
3. subtraction
4. subtraction
5. So that x is by itself. So that the two sides remain equal.

Practice and Problem Solving

6. yes
7. yes
8. no
9. no
10. no
11. yes
12. $n = 2$
13. $t = 9$
14. $c = 6$
15. What number plus 5 equals 12?; $a = 7$
16. What number plus 9 equals 18?; $v = 9$
17. 20 is what number minus 6?; $d = 26$
18. $y = 10$
19. $z = 16$
20. $r = 22$
21. $p = 3$
22. $k = 12$
23. $h = 34$
24. $f = 46$
25. $q = 11$
26. $j = 1\frac{1}{4}$
27. $x = \frac{7}{30}$
28. $m = 3.7$
29. $a = 11.8$
30. They should subtract 7 from both sides.
$$\begin{array}{r} x + 7 = 13 \\ -7 \quad -7 \\ \hline x = 6 \end{array}$$
31. They must apply the same operation to both sides.
$$\begin{array}{r} 34 = y - 12 \\ +12 \quad +12 \\ \hline 46 = y \end{array}$$

Assignment Guide and Homework Check

Level	Day 1 Activity Assignment	Day 2 Lesson Assignment	Homework Check
Basic	12–17, 55–59	1–5, 7, 9, 11, 19–25 odd, 30–34	7, 19, 21, 32
Average	12–17, 55–59	1–5, 19–31 odd, 32–34, 44–46	19, 27, 32, 44
Advanced	12–17, 55–59	1–5, 30, 34, 35, 40–50 even, 49, 52–54	34, 40, 50, 52

Common Errors

- **Exercises 6–11** Students may try to perform operations on both sides of the equation in order to get the same answer on both sides. Remind them that they are checking to see if the value of the variable makes the equation true, so they should focus on simplifying one side of the equation and then compare the answer to the other side.
- **Exercises 18–29** Instead of using the inverse operation, students may use the same operation in solving for x. Simplify the equation on the board or overhead to demonstrate that this will not work. Students may have ignored the side with the variable when they made this mistake. Remind them to check their answers in the original equation.
- **Exercises 18–29** Students may add or subtract the number on the side without the variable. For example, $-14 + 14 = k + 6 + 14$ instead of $-14 - 6 = k + 6 - 6$. Remind them that they are trying to get the variable by itself. They should start on the side with the variable and use the inverse of that operation.

7.2 Record and Practice Journal

Tell whether the given value is a solution of the equation.
1. $34 + x = 46; x = 12$ — yes
2. $y - 9 = 14; y = 22$ — no
3. $6d = 54; d = 9$ — yes
4. $\frac{n}{3} = 13; n = 39$ — yes

Solve the equation. Check your solution.
5. $7 + k = 11$; $k = 4$
6. $p - 24 = 13$; $p = 37$
7. $b - 16 = 7$; $b = 23$
8. $\frac{2}{5} + m = \frac{5}{6}$; $m = \frac{13}{30}$
9. In the heavyweight class of professional wrestling, the junior weight limit is 190 pounds. This is 15 pounds heavier than the light heavyweight limit. Write and solve an equation to find the weight limit of the light heavyweight class.
$w + 15 = 190$
$w = 175$

Technology For the Teacher
Answer Presentation Tool
QuizShow

T-287

7.2 Exercises

Check It Out
Help with Homework
BigIdeasMath.com

✓ Vocabulary and Concept Check

1. **WRITING** How can you check the solution of an equation?

Name the inverse operation you can use to solve the equation.

2. $x - 8 = 12$
3. $n + 3 = 13$
4. $b + 14 = 33$

5. **WRITING** When solving $x + 5 = 16$, why do you subtract 5 from the left side of the equation? Why do you subtract 5 from the right side of the equation?

Practice and Problem Solving

Tell whether the given value is a solution of the equation.

6. $x + 42 = 85$; $x = 43$
7. $8b = 48$; $b = 6$
8. $19 - g = 7$; $g = 15$
9. $\frac{m}{4} = 16$; $m = 4$
10. $w + 23 = 41$; $w = 28$
11. $s - 68 = 11$; $s = 79$

Use a scale to model and solve the equation.

12. $n + 7 = 9$
13. $t - 4 = 5$
14. $c + 2 = 8$

Write a question that represents the equation. Use mental math to answer the question. Then check your solution.

15. $a + 5 = 12$
16. $v + 9 = 18$
17. $20 = d - 6$

Solve the equation. Check your solution.

18. $y - 7 = 3$
19. $z - 3 = 13$
20. $8 = r - 14$
21. $p + 5 = 8$
22. $k + 6 = 18$
23. $64 = h + 30$
24. $f - 27 = 19$
25. $25 = q + 14$
26. $\frac{3}{4} = j - \frac{1}{2}$
27. $x + \frac{2}{3} = \frac{9}{10}$
28. $1.2 = m - 2.5$
29. $a + 5.5 = 17.3$

ERROR ANALYSIS Describe and correct the error in solving the equation.

30.

31.

32. PENGUINS An emperor penguin is 45 inches tall. It is 24 inches taller than a rockhopper penguin. Write and solve an equation to find the height of a rockhopper penguin.

33. ELEVATOR You get in an elevator and go down 8 floors. You exit on the 16th floor. Write and solve an equation to find what floor you got on the elevator.

34. AREA The area of Jamaica is 6460 square miles less than the area of Haiti. Write and solve an equation to find the area of Haiti.

35. REASONING The solution of the equation $x + 3 = 12$ is shown. Write the property used for each step.

$$x + 3 = 12$$
$$x + 3 - 3 = 12 - 3$$
$$x + 0 = 9$$
$$x = 9$$

Write the equation.

Write the word sentence as an equation. Then solve the equation.

36. 13 subtracted from a number w is 15.

37. A number k increased by 7 is 34.

38. 9 is the difference of a number n and 7.

39. 93 is the sum of a number g and 58.

Solve the equation. Check your solution.

40. $b + 7 + 12 = 30$

41. $y + 4 - 1 = 18$

42. $m + 18 + 23 = 71$

43. $v - 7 = 9 + 12$

44. $5 + 44 = 2 + r$

45. $22 + 15 = d - 17$

GEOMETRY Write and solve an addition equation to find x.

46. Perimeter = 48 ft

47. Perimeter = 132 in.

48. Perimeter = 93 ft

49. REASONING Explain why the equations $x + 4 = 13$ and $4 + x = 13$ have the same solution.

50. REASONING Explain why the equations $x - 13 = 4$ and $13 - x = 4$ do *not* have the same solution.

Common Errors

- **Exercises 32–34** Students may be struggling with determining what information is important in a word problem. Encourage them to simplify the word problem as they did in the previous section.
- **Exercises 36–39** Students may write the wrong equation and get an answer that is negative or have a variable that is negative. Remind them to read the word sentence carefully. Refer them to the previous section for help writing the equation.
- **Exercises 40–45** Students may lose one of the numbers when solving for the variable. For example,

$$\begin{array}{r} b + 7 + 12 = 30 \\ -7 \quad\quad -7 \\ \hline b = 23 \end{array}$$

Remind them to be very careful when solving for the variable. Encourage them to simplify and combine like terms before solving. Developing this skill will be valuable in future courses. Students should check their answer by substituting the value for the variable back into the original equation. This will help them to catch their mistake.

Practice and Problem Solving

32. $45 = h + 24$; 21 in.
33. $x - 8 = 16$; 24th floor
34. $h - 6460 = 4181$; 10,641 mi^2
35. Subtraction Property of Equality; Subtract.; Addition Property of Zero
36. $w - 13 = 15; w = 28$
37. $k + 7 = 34; k = 27$
38. $9 = n - 7; n = 16$
39. $93 = g + 58; g = 35$
40. $b = 11$ 41. $y = 15$
42. $m = 30$ 43. $v = 28$
44. $r = 47$ 45. $d = 54$
46. $x + 12 + 20 = 48$; 16 ft
47. $x + 34 + 34 + 16 = 132$; 48 in.
48. $x + 15 + 15 + 18 + 18 = 93$; 27 ft
49. Addition is commutative.
50. Subtraction is not commutative.

English Language Learners

Vocabulary
Make sure that students understand the definition of *solution*. They should understand that a solution needs to be checked in the original equation. Students should also understand the difference between solving an equation and simplifying an expression.

T-288

 Practice and Problem Solving

51. See *Taking Math Deeper*.

52. $a + 53 = 75$; $b + 44 = 75$;
 $c + 62 = 75$
 $a = 22$; $b = 31$; $c = 13$

53. **a.** $r = f + 1.25$; $g = 1.75 - 0.5$;
 $f = g + 1.5$
 $r = \$4$; $g = \$1.25$; $f = \$2.75$

 b. $5.25

54. decreases by 3

 Fair Game Review

55. 96 56. 208
57. 5 58. 24
59. B

Mini-Assessment

Solve the equation.

1. $y - 10 = 4$ $y = 14$
2. $25 = f + 13$ $f = 12$
3. $q - 67 = 3$ $q = 70$
4. $s + 24 = 35$ $s = 11$
5. You get in an elevator and go up 4 floors. You exit on the 12th floor. Write and solve an equation to find what floor you got on the elevator.
 $x + 4 = 12$; You got on the elevator on the 8th floor.

T-289

Taking Math Deeper

Exercise 51

It would be great if you could spend some extra time on this problem. Even in high school, many students are confused about the difference between simplifying an expression and solving an equation.

 Simplifying an expression.

Simplify the expression $2(x + 3) - 4$.
$$2(x + 3) - 4 = 2x + 6 - 4$$
$$= 2x + 2$$

Characteristics:
- Solution steps are "L" shaped.
- Write a sequence of equivalent expressions.
- Not solving for a value of *x*.
- There are *not two sides* of the equation.

 Solving an equation.

Solve the equation $x + 3 = 4$.
$$x + 3 = 4$$
$$-3 = -3$$
$$x = 1$$

Characteristics:
- Equation has two sides.
- Want two sides to be equivalent expressions.
- Trying to find values of *x* that make the equation true.
- Whatever you do to one side, you need to do the same thing to the other side.

 If I try to "solve" the equation,
$$2(x + 3) - 4 = 2x + 2$$
I will discover that the equation is true for *all* values of *x*. This happens because $2(x + 3) - 4$ and $2x + 2$ are equivalent expressions.

Reteaching and Enrichment Strategies

If students need help...	If students got it...
Resources by Chapter • Practice A and Practice B • Puzzle Time Record and Practice Journal Practice Differentiating the Lesson Lesson Tutorials Skills Review Handbook	Resources by Chapter • Enrichment and Extension Start the next section

51. SIMPLIFYING AND SOLVING Compare and contrast the two problems.

Simplify the expression $2(x + 3) - 4$.
$$2(x + 3) - 4 = 2x + 6 - 4$$
$$= 2x + 2$$

Solve the equation $x + 3 = 4$.
$$x + 3 = 4$$
$$-3 = -3$$
$$x = 1$$

a	37	16
19	25	b
34	c	28

52. PUZZLE In a *magic square*, the sum of the numbers in each row, column, and diagonal is the same. Write and solve equations to find the values of a, b, and c.

Bumper Cars: $1.75
Roller Coaster: $1.25 more than Ferris Wheel
Giant Slide: $0.50 less than Bumper Cars
Ferris Wheel: $1.50 more than Giant Slide

53. AMUSEMENT PARK You have $15.

 a. Write and solve three equations to find the cost r of the roller coaster, the cost g of the giant slide, and the cost f of the Ferris wheel.

 b. How much money do you have left if you ride each ride once?

54. **Critical Thinking** Consider the equation $x + y = 15$. The value of x increases by 3. What has to happen to the value of y so that $x + y = 15$ remains true?

Fair Game Review What you learned in previous grades & lessons

Find the product or quotient. *(Skills Review Handbook)*

55. 12×8 **56.** 13×16 **57.** $75 \div 15$ **58.** $72 \div 3$

59. MULTIPLE CHOICE The figure is made up of a trapezoid and a semicircle. What is the perimeter of the figure? Use 3.14 for π. *(Section 6.2)*

 Ⓐ 32 in.
 Ⓑ 39.41 in.
 Ⓒ 52.41 in.
 Ⓓ 59.82 in.

7.3 Solving Equations Using Multiplication or Division

Essential Question How can you use multiplication or division to solve an equation?

COMMON CORE STATE STANDARDS
6.EE.5
6.EE.7

1 ACTIVITY: Writing and Solving Multiplication Equations

Work with a partner. Solve for x. Check your answer.

a. Rectangle
Area = 24 square units

b. Parallelogram
Area = 20 square units

c. Triangle
Area = 28 square units

2 EXAMPLE: Using an Equation to Model a Story

The problem is represented by the equation.

Problem

Three people go out to lunch. They decide to share the $12 bill evenly. How much does each person pay?

Equation

$3x = 12$

- What does x represent?
- Solve for x.
- Answer the question.

3 people · Amount each person pays = $12

x is the amount each person pays.

$3x = 12$

$\dfrac{3x}{3} = \dfrac{12}{3}$

$x = 4$ ← Solve for x.

So, each person pays $4. ← Answer the question.

290 Chapter 7 Equations

Laurie's Notes

Introduction

For the Teacher
- **Goal:** Students will use familiar contexts including area formulas to develop an intuitive understanding about what happens when multiplying or dividing by the same amount on both sides of an equation.
- This activity is a nice review of area formulas from prior years.

Motivate
- ? "Have you ever thought about building a robot to perform a simple task? What skills do you think are necessary to build a robot?"
- Share information about a middle school competition called *FIRST* LEGO League (FLL), a robotics program for students ages 9 to 14.
- **Check This Out:** To find out more about this international program visit: *www.usfirst.org*
- Explain to students that several of today's word problems are about robots.

Activity Notes

Activity 1
- Introduce the activity. Explain that they are to solve for the missing dimension in each figure.
- ? "What is the name of each polygon?" rectangle, parallelogram, triangle
- ? Draw the parallelogram including the red line and right angle mark. You may want to make the red line dashed on the board. Then point to the right angle mark and ask, "What does this mean?" indicates a 90° angle
- The first two parts should be easy. For part (c), remind students that it is okay to use the problem solving strategy of *Guess, Check, and Revise*. In doing so, students are actually checking to see if a value is a solution to an equation. For the triangle, students are checking for a solution to $28 = \frac{1}{2} \cdot x \cdot 8$.

Example 2
- This example demonstrates how an equation can be used to model a real-life problem. There are three bulleted steps for the students to follow.
- Students may quickly say that the answer is $4 without bothering to write an equation. Remind them that by showing the process they can check their thinking.

Common Core State Standards

6.EE.5 Understand solving an equation ... as a process of answering a question: which values from a specified set, if any, make the equation ... true? Use substitution to determine whether a given number in a specified set makes an equation ... true.

6.EE.7 Solve real-world and mathematical problems by writing and solving equations of the form $x + p = q$ and $px = q$ for cases in which p, q and x are all nonnegative rational numbers.

Previous Learning
Students should know how to evaluate expressions and use area formulas.

Start Thinking! and Warm Up

7.3 Record and Practice Journal

T-290

English Language Learners
Graphic Organizer

When solving an equation, students must understand that they need to isolate the variable in the equation. To do this, they use inverse operations. Have students create a reference table with examples that show the operation needed to move the numbers to the other side of the equal sign.

Equation Type	Operation to Solve	Example
Subtraction	Addition	$b - 4 = 2$
Addition	Subtraction	$h + 3 = 7$
Division	Multiplication	$w \div 5 = 5$
Multiplication	Division	$m \cdot 3 = 6$

7.3 Record and Practice Journal

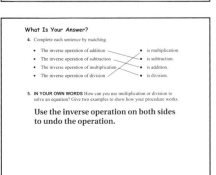

Laurie's Notes

Activity 3

- Explain to students that they are to use Example 2 as a model for completing this activity.
- Students need practice in identifying what the variable represents. Caution them to be complete in their description of the variable. For instance, in Example 2, x was equal to the amount each person paid. It would not be sufficient to say "x equals money" or "x equals pay."
- When students finish, ask for volunteers to share their solutions.
- ? "Is $3x = 11.91$ the same as $11.91 = 3x$?" yes

What Is Your Answer?

- Students should know that multiplication and division are inverse operations.
- The inverse operation must be done on both sides of the equation so that the two sides remain equal.

Closure

- **Writing prompt:** To solve $28 = 7x$, you . . . Divide both sides by 7.

 To solve $7 = \dfrac{x}{28}$, you . . . Multiply both sides by 28.

Technology For the Teacher

Dynamic Classroom

The Dynamic Planning Tool
Editable Teacher's Resources at *BigIdeasMath.com*

3 ACTIVITY: Using Equations to Model a Story

Work with a partner. Each problem is represented by the equation.

- What does x represent?
- Solve for x.
- Answer the question.

Problem	Equation
a. Three robots go out to lunch. They decide to share the $11.91 bill evenly. How much does each robot pay?	$3x = 11.91$
b. On Earth, objects weigh 6 times what they weigh on the moon. A robot weighs 96 pounds on Earth. What does it weigh on the moon?	$6x = 96$
c. At maximum speed, a robot runs 6 feet in 1 second. How many feet does the robot run in 1 minute?	$\dfrac{x}{60} = 6$
d. Four identical robots lie on the ground head-to-toe and measure 14 feet. How tall is each robot?	$4x = 14$

What Is Your Answer?

4. Complete each sentence by matching.

- The inverse operation of addition
- The inverse operation of subtraction
- The inverse operation of multiplication
- The inverse operation of division

- is multiplication.
- is subtraction.
- is addition.
- is division.

5. IN YOUR OWN WORDS How can you use multiplication or division to solve an equation? Give two examples to show how your procedure works.

 Use what you learned about solving equations to complete Exercises 15–18 on page 294.

7.3 Lesson

Key Idea

Multiplication Property of Equality

Words If you multiply each side of an equation by the same nonzero number, the two sides remain equal.

Numbers
$$\frac{8}{4} = 2$$
$$\frac{8}{4} \times 4 = 2 \times 4$$
$$8 = 8$$

Algebra
$$\frac{x}{4} = 2$$
$$\frac{x}{4} \cdot 4 = 2 \cdot 4$$
$$x = 8$$

Remember
Inverse operations "undo" each other. Multiplication and division are inverse operations.

EXAMPLE 1 — Solving Equations Using Multiplication

a. Solve $\frac{w}{4} = 12$.

$\frac{w}{4} = 12$ — Write the equation.

Undo the division. → $\frac{w}{4} \cdot 4 = 12 \cdot 4$ — Multiply each side by 4.

$w = 48$ — Simplify.

∴ The solution is $w = 48$.

Check
$\frac{w}{4} = 12$
$\frac{48}{4} \stackrel{?}{=} 12$
$12 = 12$ ✓

b. Solve $x \div 7 = 5$.

$x \div 7 = 5$ — Write the equation.

Undo the division. → $x \div 7 \cdot 7 = 5 \cdot 7$ — Multiply each side by 7.

$x = 35$ — Simplify.

∴ The solution is $x = 35$.

Check
$x \div 7 = 5$
$35 \div 7 \stackrel{?}{=} 5$
$5 = 5$ ✓

On Your Own

Now You're Ready
Exercises 7–10

Solve the equation. Check your solution.

1. $\frac{a}{8} = 6$
2. $z \div 2 = 10$
3. $9 = \frac{y}{15}$

4. Six friends share a box of pencils. Each person receives two pencils. Write and solve an equation to find the number of pencils in the box.

Laurie's Notes

Introduction

Connect
- **Yesterday:** Students gained an intuitive understanding about how equations are solved using multiplication and division.
- **Today:** Students will use Properties of Equality to solve equations.

Motivate
- **Race Time!** Have students time you as you cross the front of your room walking backwards, toe-to-heel, toe-to-heel. Invent some other form of race if you wish.
- **Collect Data:** Write the distance traveled and the amount of time on the board. Record data for several students. Revisit the data at the end of class.

Lesson Notes

Key Idea
- Discuss the *Words*, *Numbers*, and *Algebra*.
- **Representation:** Point out to students that division will commonly be represented in the fraction format. When writing and solving equations, do not write $x \div 4 = 2$.
- **Connection:** When multiplying, write the whole number over a denominator of 1. Divide out common factors. Then multiply the numerators and multiply the denominators.

Numbers: $\dfrac{8}{4} \times 4 = \dfrac{8}{\cancel{4}} \times \dfrac{\cancel{4}}{1} = 8$ Algebra: $\dfrac{x}{4} \cdot 4 = \dfrac{x}{\cancel{4}} \cdot \dfrac{\cancel{4}}{1} = x$

Example 1
- Explain to students that although they can probably solve these problems using mental math, another equally important part of equation solving is the *process*. How can the problem be solved when the answer cannot be found using mental math?
- Write a few examples.

 (a) $\dfrac{x}{5} = 20$ (b) $\dfrac{x}{3.2} = 5.6$ (c) $x \div 3\dfrac{1}{2} = 4\dfrac{2}{5}$

- You can use mental math for (a). For (b) and (c), you probably want to use paper and pencil. What is the process now?
- Discuss how to check that their answer is correct.
- **Big Idea:** Whatever is multiplied on one side of the equation is also multiplied on the other. Using an alternate color (chalk or overhead pen) to show this will help visual learners. Notice that the *process* is annotated along the right side of the equation.

On Your Own
- Ask volunteers to share their work at the board.

Goal Today's lesson is using the inverse operations of multiplication and division to find the solution of an equation.

Start Thinking! and Warm Up

Lesson 7.3 Warm Up
For use before Lesson 7.3

Lesson 7.3 Start Thinking!
For use before Lesson 7.3

Explain how to use the distance formula ($d = rt$) to find your average rate getting to school in the morning. Be sure to include units in your explanation.

Extra Example 1
a. Solve $\dfrac{t}{6} = 9$. $t = 54$
b. Solve $y \div 8 = 7$. $y = 56$

On Your Own
1. $a = 48$
2. $z = 20$
3. $y = 135$
4. $\dfrac{x}{6} = 2$; 12 pencils

T-292

Differentiated Instruction

Visual

Show that the inverse operations of multiplication and division produce the multiplicative identity, 1.

$$5b = 65$$

$$1\left\{\dfrac{\overset{1}{\cancel{5}}b}{\underset{1}{\cancel{5}}} = \dfrac{65}{5}\right.$$

$$b = 13$$

Extra Example 2

Solve $7c = 42$. $c = 6$

Extra Example 3

A driver travels 2500 miles at a rate of 550 miles per day. About how many days did it take to complete the trip?
about 5 days

On Your Own

5. $p = 6$
6. $q = 5$
7. $r = 9$
8. $10 = 50t$; 12 min

Laurie's Notes

Key Idea
- Write the Key Idea.
- Discuss the *Words*, *Numbers*, and *Algebra*.
- **Representation:** Remind students that $4x$ is how multiplication is represented.
- **Common Misconception:** Students sometimes think that when $\dfrac{4x}{4}$ is simplified to x, the 4s have just been "crossed out" because they are the same. Explain that the 4s represent common factors in the numerator and denominator and have been *divided out*, in the same way a common factor of 4 would be divided out when simplifying the fraction $\dfrac{12}{16}$.

Example 2
- **?** "How do you undo the operation of multiplication?" division
- Work through the problem.
- Discuss the last step in the solution. This can be confusing for students because of the 5 on each side of the equation. On the left, $5b$ means 5 times the value of b and on the right, 5 is the ones digit in a two-digit number. If the equation had been written $b5 = 65$, it would have been even more confusing. Remember that our comfort level with representing multiplication as $5b$ is not the same as students'. It's not uncommon to see a wrong answer here of $b = 6$. The 5s were "crossed out" on both sides! It might be helpful to write $\dfrac{5 \cdot b}{5} = \dfrac{5 \cdot 13}{5}$ as an intermediate step.

Example 3
- **FYI:** Marathon swimmer Martin Strel made history in April 2007, swimming the Amazon River from Atalaya, Peru to Belem, Brazil. The distance he swam (5268 km ≈ 3273 mi) is about the same as the distance across the Atlantic Ocean from Boston, MA to Spain.
- **?** "What is the distance formula?" distance = rate × time
- Use the distance formula to set up the equation.

On Your Own
- Note that variables can appear on either side of the equal sign.
- Ask volunteers to share their work at the board.

Closure
- Use the data collected at the beginning of class and calculate the rate, probably in feet per second. Set up the problem as an equation: distance = rate × time. Students should be able to solve for each trial.

The Dynamic Planning Tool
Editable Teacher's Resources at *BigIdeasMath.com*

 Key Idea

Division Property of Equality

Words If you divide each side of an equation by the same nonzero number, the two sides remain equal.

Numbers $8 \times 4 = 32$ **Algebra** $4x = 32$

$8 \times 4 \div 4 = 32 \div 4$ $\dfrac{4x}{4} = \dfrac{32}{4}$

$8 = 8$ $x = 8$

EXAMPLE 2 Solving an Equation Using Division

Solve $5b = 65$.

Undo the multiplication.

$5b = 65$ Write the equation.

$\dfrac{5b}{5} = \dfrac{65}{5}$ Divide each side by 5.

$b = 13$ Simplify.

Check
$5b = 65$
$5(13) \stackrel{?}{=} 65$
$65 = 65$ ✓

∴ The solution is $b = 13$.

EXAMPLE 3 Using the Formula for Distance

Martin Strel set a world record by swimming 5268 kilometers down the Amazon River at a rate of about 80 kilometers per day. About how many days did it take him to complete his journey?

$d = rt$ Write formula for distance.

$5268 = 80t$ Substitute 5268 for d and 80 for r.

$\dfrac{5268}{80} = \dfrac{80t}{80}$ Divide each side by 80.

$65.85 = t$ Simplify.

Study Tip

When solving a real-life problem, check that each side of your equation has the same units.

$5268 \text{ km} = \dfrac{80 \text{ km}}{\text{day}} \cdot t \text{ days}$

∴ It took him about 66 days to complete his journey.

 On Your Own

Now You're Ready
Exercises 11–22

Solve the equation. Check your solution.

5. $p \cdot 3 = 18$ **6.** $12q = 60$ **7.** $81 = 9r$

8. A subway train travels at a rate of 50 miles per hour. Write and solve an equation to find the number of minutes it takes the train to travel 10 miles.

7.3 Exercises

Vocabulary and Concept Check

1. **NUMBER SENSE** What number divided by 12 equals 1?

2. **WRITING** What property of equality would you use to solve $\frac{x}{6} = 7$? Explain how you would use the property.

Copy and complete the first step in the solution.

3. $4x = 24$

 $\dfrac{4x}{\square} = \dfrac{24}{\square}$

4. $\dfrac{x}{3} = 11$

 $\dfrac{x}{3} \cdot \square = 11 \cdot \square$

5. $8 = n \div 3$

 $8 \cdot \square = (n \div 3) \cdot \square$

6. **OPEN-ENDED** Write an equation that can be solved using the Division Property of Equality.

Practice and Problem Solving

Solve the equation. Check your solution.

7. $\dfrac{s}{10} = 7$

8. $6 = \dfrac{t}{5}$

9. $x \div 2 = 8$

10. $24 = \dfrac{r}{4}$

11. $3a = 12$

12. $5 \cdot z = 35$

13. $40 = 4y$

14. $42 = 7k$

15. $7x = 105$

16. $75 = 6 \cdot w$

17. $13 = d \div 6$

18. $9 = v \div 5$

19. $b \div 12 = 4.5$

20. $\dfrac{c}{15} = 8.8$

21. $12.5 \cdot n = 32$

22. $3.4m = 20.4$

23. **ERROR ANALYSIS** Describe and correct the error in solving the equation.

$$x \div 4 = 28$$
$$\dfrac{x \div 4}{4} = \dfrac{28}{4}$$
$$x = 7$$

24. **ANOTHER WAY** Show how you can solve the equation $3x = 9$ by multiplying each side by the reciprocal of 3.

25. **SNOWMOBILES** A snowmobile is traveling at a speed of 88 feet per second. Write and solve an equation to find the number of seconds s it takes for the snowmobile to travel 528 feet.

26. **MUSIC** The mean length of a song is 200 seconds. Write and solve an equation to find the total length s of 350 songs.

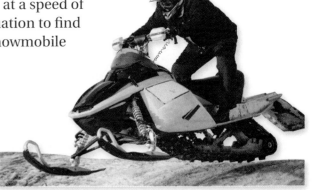

Assignment Guide and Homework Check

Level	Day 1 Activity Assignment	Day 2 Lesson Assignment	Homework Check
Basic	15–18, 34–38	1–6, 7–13 odd, 23–27 odd, 26	7, 11, 26, 27
Average	15–18, 34–38	1–6, 7–13 odd, 23, 27, 30, 31	7, 11, 27, 30
Advanced	15–18, 34–38	1–6, 19–23, 30–33	20, 22, 30, 32

Common Errors

- **Exercises 7–22** Students may do the same operation on both sides instead of the opposite operation. This is most common with problems like Exercise 9, where the equation is written in division format instead of fraction format. Remind them that to get the variable alone, they need to use the inverse operation.
- **Exercises 27–29** Students may use the formula for perimeter instead of area. Remind them of the two formulas and how to find both.

7.3 Record and Practice Journal

Solve the equation. Check your solution.

1. $7k = 77$
 $k = 11$
2. $\dfrac{p}{5} = 10$
 $p = 50$
3. $3 = \dfrac{m}{12}$
 $m = 36$

4. $4a = 36$
 $a = 9$
5. $5 \cdot x = 12$
 $x = 2.4$
6. $4.2 = \dfrac{c}{8}$
 $c = 33.6$

7. You earn $5 for every friendship bracelet you sell. Write and solve an equation to find the number of bracelets you have to sell to earn $85.
 $5a = 85$
 $a = 17$

8. You practice the piano for 30 minutes each day. Write and solve an equation to find the total time t you spend practicing the piano for a week.
 $\dfrac{t}{7} = 30$
 $t = 210$

Technology For the Teacher
Answer Presentation Tool
QuizShow

Vocabulary and Concept Check

1. 12
2. Multiplication Property of Equality; Multiply each side of the equation by 6.
3. $\dfrac{4x}{4} = \dfrac{24}{4}$
4. $\dfrac{x}{3} \cdot 3 = 11 \cdot 3$
5. $8 \cdot 3 = (n \div 3) \cdot 3$
6. *Sample answer:* $3x = 15$

Practice and Problem Solving

7. $s = 70$
8. $t = 30$
9. $x = 16$
10. $r = 96$
11. $a = 4$
12. $z = 7$
13. $y = 10$
14. $k = 6$
15. $x = 15$
16. $w = 12.5$
17. $d = 78$
18. $v = 45$
19. $b = 54$
20. $c = 132$
21. $n = 2.56$
22. $m = 6$
23. They should have multiplied by 4.
 $x \div 4 = 28$
 $(x \div 4) \cdot 4 = 28 \cdot 4$
 $x = 112$
24. $3x = 9$
 $3x \cdot \dfrac{1}{3} = 9 \cdot \dfrac{1}{3}$
 $x = 3$
25. $528 = 88s$; 6 sec
26. $\dfrac{s}{350} = 200$; 70,000 sec

T-294

Practice and Problem Solving

27. 9 units
28. 11 units
29. 8 units
30. 50 points
31. 20 cards
32. See *Taking Math Deeper*.
33. length: 20 in.; width: 5 in.

Fair Game Review

34. $x = 12$ 35. $t = 23$
36. $b = 61$ 37. $s = 16$
38. C

Mini-Assessment

Solve the equation.

1. $\dfrac{t}{5} = 7$ $t = 35$
2. $12d = 60$ $d = 5$
3. $16 = \dfrac{s}{2}$ $s = 32$
4. $35 = 5g$ $g = 7$
5. You pay $1.56 for 4 pounds of bananas. Write and solve an equation to find how much it costs for one pound of bananas. $4x = 1.56$; It costs $0.39 for one pound of bananas.

T-295

Taking Math Deeper

Exercise 32

The second part of this problem tends to be difficult for students. Before trying to answer part (b), ask students to decide the following.

- *The cones are smaller. So, can I fill more cones than cups? Or fewer cones than cups?*

 Organize the given information.

 Rate: 1440 cups in 24 hours

30 symbols represents 1440 cups

 Write and solve an equation for c.

$$30c = 1440$$
$$c = 48$$

 a. Each symbol represents 48 cups.

 How many cones can be filled in 24 hours?

x ounces $0.8x$ ounces

cups per 24 hours
cups per cone

$$\dfrac{1440}{0.8} = 1800 \text{ cones}$$

 b. You can fill 1800 cones in 24 hours.

Reteaching and Enrichment Strategies

If students need help...	If students got it...
Resources by Chapter • Practice A and Practice B • Puzzle Time Record and Practice Journal Practice Differentiating the Lesson Lesson Tutorials Skills Review Handbook	Resources by Chapter • Enrichment and Extension • School-to-Work Start the next section

Solve for *x*. Check your answer.

27. Rectangle

Area = 45 square units

28. Rectangle

Area = 176 square units

29. Parallelogram

Area = 104 square units

30. TEST SCORE On a test, you correctly answer six 5-point questions and eight 2-point questions. You earn 92% of the possible points on the test. How many points *p* is the test worth?

31. CARD GAME You use index cards to play a homemade game. The object is to be the first to get rid of all your cards. How many cards are in your friend's stack?

32. PROBLEM SOLVING A frozen juice machine fills 1440 cups in 24 hours.

 a. Write and solve an equation to find the number *c* of cups each symbol represents.

 b. To lower costs, the cups are replaced by paper cones that hold 20% less. Write and solve an equation to find the number *n* of paper cones that can be filled in 24 hours.

33. **Number Sense** The area of the picture is 100 square inches. Find the length $4x$ and width x of the picture.

Fair Game Review *What you learned in previous grades & lessons*

Solve the equation. Check your solution. *(Section 7.2)*

34. $x + 7 = 19$ **35.** $t - 12 = 11$ **36.** $51 = b - 10$ **37.** $22 = 6 + s$

38. MULTIPLE CHOICE What is the value of a^3 when $a = 4$? *(Skills Review Handbook)*

 (A) 12 **(B)** 43 **(C)** 64 **(D)** 81

7.4 Solving Two-Step Equations

Essential Question What is a "two-step" equation? How can you solve a two-step equation?

COMMON CORE STATE STANDARDS
6.EE.2b
6.EE.5
6.EE.7

Sir Isaac Newton's Third Law of Motion
For every action, there is an equal and opposite reaction.

A teddy bear
Sits in a chair.
Down pushes Teddy.

Chair says "I'm ready".
With a confident "Yup"
The chair pushes up.

Sir Isaac Newton
(1642–1727)

Because $5 - 5 = 0$, neither the bear nor the chair moves.

1 ACTIVITY: Identifying Inverse Operations

Work with a partner. Describe how you can "undo" the operation in blue.

a. **Sample:** $3x + 5 = 14$

Subtract 5 from each side. →
$$\begin{aligned} 3x + 5 &= 14 \\ -5 & -5 \\ \hline 3x &= 9 \end{aligned}$$

b. $2n - 6 = 4$ c. $2(m + 3) = 6$ d. $\dfrac{x - 2}{4} = 1$

2 ACTIVITY: Solving Two-Step Equations

Work with a partner. Solve each equation in Activity 1. Use substitution to check your answer.

a. $3x + 5 = 14$ b. $2n - 6 = 4$

c. $2(m + 3) = 6$ d. $(x - 2) \div 4 = 1$

Laurie's Notes

Introduction

For the Teacher
- **Goal:** Students will develop conceptual understanding of how two-step equations are solved.
- **Science:** Investigate and explain that when a force is applied to an object but it does not move, it is because another opposing force is being applied by something in the environment so that the forces are balanced.
- The introduction to this activity is not intended to be an application of Newton's Third Law of Motion, but a parallel example to the math.
- **Connection:** You want students to start thinking about inverse operations.

Motivate
- **FYI:** Share information about Sir Isaac Newton:
 - Lived December 25, 1642 to March 20, 1727, English mathematician and natural philosopher (physicist)
 - Considered by many the greatest scientist that has ever lived
 - Known for defining gravity and for his three Laws of Motion (one of which is referenced today)
 - Newton served as a Member of Parliament for exactly one year (1689–1690). During all of the lengthy proceedings, he spoke only once: he asked a nearby person to close an open window!
 - A classic example of Newton's Third Law of Motion is a space shuttle launch. The tremendous downward thrust of the rocket motors results in the upward motion of the shuttle.

Activity Notes

Activity 1
- This is a quick activity to review inverse operations.
- ? Ask for examples of inverse operations outside of math class like the space shuttle launching and then landing. *open-ended*
- ? Ask for examples of inverse operations related to equation solving. *addition and subtraction, multiplication and division*
- Part (c) may be a little tricky for students. Suggest to students that they should think about how the left side of the equation would be evaluated for a given value of m: add 3 and then multiply that sum by 2. To undo these steps, divide by 2 and subtract 3.
- Students could put the first step for each problem on the board.

Activity 2
- All students should have the first step correct. Now they can finish solving the equation with their partner.

Common Core State Standards

6.EE.2b Identify parts of an expression using mathematical terms (sum, term, product, factor, quotient, coefficient); view one or more parts of an expression as a single entity.

6.EE.5 Understand solving an equation . . . as a process of answering a question: which values from a specified set, if any, make the equation . . . true? Use substitution to determine whether a given number in a specified set makes an equation . . . true.

6.EE.7 Solve real-world and mathematical problems by writing and solving equations of the form $x + p = q$ and $px = q$ for cases in which p, q and x are all nonnegative rational numbers.

Previous Learning
Students should understand inverse operations and know how to solve one-step equations.

Start Thinking! and Warm Up

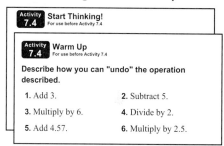

Activity 7.4 Start Thinking! For use before Activity 7.4

Activity 7.4 Warm Up For use before Activity 7.4

Describe how you can "undo" the operation described.
1. Add 3.
2. Subtract 5.
3. Multiply by 6.
4. Divide by 2.
5. Add 4.57.
6. Multiply by 2.5.

7.4 Record and Practice Journal

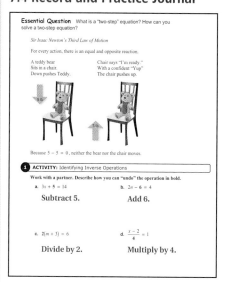

Differentiated Instruction

Auditory

Remind students that in order to solve an equation, the variable must be isolated on one side of the equation. This means that the operations on the same side as the variable are the ones that need to be undone.

7.4 Record and Practice Journal

Laurie's Notes

Activity 3

- Discuss the video challenge. Describe it in terms of making your way through a maze, and then having to reverse the movements to get back out of the maze.
- Ask if there are video gamers in the class who might provide some language that would help in describing the skills in this video game.
- Students will work with a partner to describe the forward motion of the challenges, perhaps with invented vocabulary.
- **Connection:** In reversing the steps, the order and direction of the skill is reversed. To solve an equation, reverse the order and use inverse operations.

What Is Your Answer?

- The flaps of the box in the cartoon are a nice example of inverse operations.

Closure

- At the beginning of class you came in, took your book out of your backpack, and opened up your book. What will you do now that it's the end of class? Close your book and put the book in your backpack.

The Dynamic Planning Tool
Editable Teacher's Resources at *BigIdeasMath.com*

3 ACTIVITY: Analyzing a Video Game

Work with a partner. For Level 1 in a video game, you have to accomplish a sequence of challenges. Then, you have to leave the level by undoing the challenges in reverse order.

a. Describe the challenges in order.

b. Describe the order of challenges to get out of the level.

c. This is Level 1. Make up challenges for Level 2. Draw the level and describe the reverse order to get back out of the level.

What Is Your Answer?

4. **IN YOUR OWN WORDS** What is a "two-step" equation? How can you solve a two-step equation? Give an example to show how your procedure works.

"Hey, it says 'Close this flap first,' but they closed it last!"

Practice — Use what you learned about solving two-step equations to complete Exercises 5–7 on page 301.

7.4 Lesson

Check It Out
Lesson Tutorials
BigIdeasMath **V**com

Key Vocabulary 🔊
two-step equation, p. 298
terms, p. 300
like terms, p. 300

 Key Idea

Solving Two-Step Equations

A **two-step equation** is an equation that contains two different operations. To solve a two-step equation, use inverse operations to isolate the variable.

EXAMPLE 1 Solving Two-Step Equations

a. Solve $2x - 5 = 13$.

	$2x - 5 = 13$	Write the equation.
Step 1: Undo the subtraction. →	$\underline{+5 +5}$	Add 5 to each side.
	$2x = 18$	Simplify.
Step 2: Undo the multiplication. →	$\dfrac{2x}{2} = \dfrac{18}{2}$	Divide each side by 2.
	$x = 9$	Simplify.

∴ The solution is $x = 9$.

Check
$2x - 5 = 13$
$2(9) - 5 \stackrel{?}{=} 13$
$18 - 5 \stackrel{?}{=} 13$
$13 = 13$ ✓

Common Error

Do not get confused when the variable is on the right side of the equation. The equation $4 = \dfrac{y}{8} + 1$ is solved the same way as the equation $\dfrac{y}{8} + 1 = 4$.

b. Solve $4 = \dfrac{y}{8} + 1$.

$4 = \dfrac{y}{8} + 1$		Write the equation.
$\underline{-1 -1}$		Subtract 1 from each side.
$3 = \dfrac{y}{8}$		Simplify.
$3 \cdot 8 = \dfrac{y}{8} \cdot 8$		Multiply each side by 8.
$24 = y$		Simplify.

∴ The solution is $y = 24$.

Check
$4 = \dfrac{y}{8} + 1$
$4 \stackrel{?}{=} \dfrac{24}{8} + 1$
$4 \stackrel{?}{=} 3 + 1$
$4 = 4$ ✓

● **On Your Own**

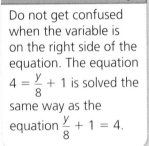
Now You're Ready
Exercises 5–16

Solve the equation. Check your solution.

1. $5c - 1 = 14$
2. $\dfrac{h}{4} + 9 = 20$
3. $3(x - 1) = 9$

Laurie's Notes

Introduction

Connect
- **Yesterday:** Students developed an intuitive understanding about how to solve two-step equations.
- **Today:** Students will solve equations by using inverse operations to isolate the variable.

Motivate
- **Preparation:** Prepare a sealed envelope containing a slip of paper with the number 7 written on it. Write the number 9 on the outside of the envelope.
- Hold the envelope against your forehead and say, "I'm thinking of a number that I doubled and from which I subtracted 5. I got an answer of 9."
- ❓ "What number did I start with? Can anyone explain how they know what number I was thinking of?" Listen for students to "undo" your process: add 5 to 9 and then divide by 2.
- Open the envelope and reveal the number 7 on your slip of paper.
- Write on the board what you said.
 Think of a number → double the number → subtract 5 → answer is 9.
- Also write how the students solved the puzzle.
 Start with 9 → add 5 → divide by 2 → answer is 7.
- Ask for observations now or wait until the lesson is introduced.

Lesson Notes

Key Idea
- Write the Key Idea on the board. As an example, you could use the problem from the beginning of class: $2x - 5 = 9$. In evaluating the expression on the left, you would multiply, then subtract. In solving the equation, use inverse operations to isolate the variable: add 5 and divide by 2.

Example 1
- Point out to students that the first operation is recorded below the operation that it is undoing. Adding 5 is underneath subtracting 5.
- Comment to students that equations are solved working down the paper. Students sometimes like to solve horizontally, failing to show what is being done to both sides of the equation. Insist that students develop good habits from the outset—work down the paper and keep the equal signs lined up!
- Discuss the *Common Error* box.
- ❓ Ask a student to read the equation in part (b). 4 equals y divided by 8 plus 1. Students should say y divided by 8 and not y over 8. You want to hear the operation.

On Your Own
- Circulate to check form and correctness. Look for solutions that demonstrate good models that can then be shared at the board.

Goal Today's lesson is using inverse operations to find the solution of a **two-step equation**.

Lesson Materials
Introduction
• envelope

Start Thinking! and Warm Up

 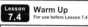

Lesson 7.4 Warm Up
For use before Lesson 7.4

Lesson 7.4 Start Thinking!
For use before Lesson 7.4
How are two-step equations related to the cost of purchasing prints from an online photo website?

Extra Example 1
a. Solve $3t + 4 = 10$. $t = 2$
b. Solve $\dfrac{a}{5} - 7 = 13$. $a = 100$

○ **On Your Own**
1. $c = 3$
2. $h = 44$
3. $x = 4$

English Language Learners
Vocabulary

English learners will benefit from understanding that a *term* is a number, a variable, or the product of a number and a variable. *Like terms* are terms that have identical variable parts.

3 and 16 are like terms because they contain no variable.

$4x$ and $7x$ are like terms because they have the same variable x.

$5a$ and $5b$ are *not* like terms because they have different variables.

Extra Example 2

You pay $112 for a game system and rent 3 video games for m dollars per game. Your total cost is $133. Using the equation $3m + 112 = 133$, find the cost of renting one video game. $7

Extra Example 3

The total cost of renting the tandem bike in Example 3 is $33. Write and solve an equation to find the number of extra hours you rented the bike. $24 + 4.5h = 33$, 2 extra hours

 On Your Own

4. $40 + 7.5h = 62.50$; 3 extra hours

Laurie's Notes

Words of Wisdom

- **Big Idea:** You want students to recognize how evaluating expressions and solving equations are related.
- For the expression $3x + 2$, when $x = 4$, think about how the expression is evaluated. First, multiply by 3 and then add 2 and get an answer of 14. Now, when solving the equation $3x + 2 = 14$, undo those steps in reverse order using inverse operations. First subtract 2 and then divide by 3.
- I often use the backtracking schematic to make this point. "I'm thinking of a number that I multiplied by 3, then added 2, and got an answer of 14. What number did I start with?"
- To solve this, backtrack by undoing the steps in reverse order.

Evaluating an expression

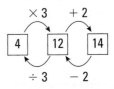
Solving an equation

Example 2

- Discuss why $12m + 80 = 188$ makes sense as the equation for this problem.
- What does m represent? cost in dollars for one month's rental
- Discuss the equation solving technique. Subtracting 80 is written underneath adding 80. It is not underneath $12m$. The equal signs are aligned.

Example 3

- Some information about the problem is to the left in the diagram. It is not written in the problem.
- The words written are an example of a *verbal model*. The verbal model says in words how the problem can be solved. The verbal model reads as a sentence from left to right.
- Note that the variable is defined to be *the number of extra hours*, not simply *extra hours*. Students are often too brief in describing what the variable represents.
- **Common Error:** Because the variable term, $4.5h$, comes after the constant term, 24, students sometimes think they should start by dividing by 4.5. Remind students that addition is commutative and this equation could be written as $4.5h + 24 = 42$.
- Discuss the reasonableness of the answer of 4 extra hours. If a student gets an answer of 81 (by multiplying by 4.5 instead of dividing), they can determine that the answer is unreasonable.

On Your Own

- **Neighbor Check:** Have students work independently and then have their neighbor check their work. Have students discuss any discrepancies.

EXAMPLE 2 Standardized Test Practice

You pay $80 for a game system. The monthly rental fee for games is m dollars. Your cost for the year is $188. Using the equation $12m + 80 = 188$, how much is your monthly fee?

(A) $8 **(B)** $9 **(C)** $12 **(D)** $22

$$12m + 80 = 188 \quad \text{Write the equation.}$$
$$\underline{-80 \quad -80} \quad \text{Subtract 80 from each side.}$$
$$12m = 108 \quad \text{Simplify.}$$
$$\frac{12m}{12} = \frac{108}{12} \quad \text{Divide each side by 12.}$$
$$m = 9 \quad \text{Simplify.}$$

∴ Your monthly fee is $9. The correct answer is **(B)**.

EXAMPLE 3 Real-Life Application

$24 for 3 hours
$4.50 each extra hour

You and your friend rent a tandem bike. Your total cost is $42. Write and solve an equation to find the number of extra hours you rented the bike.

Words The cost for three hours plus **the cost for each extra hour** times the number of extra hours is **the total cost.**

Variable Let h be the number of extra hours.

Equation $24 \; + \; 4.5 \; \cdot \; h \; = \; 42$

$$24 + 4.5h = 42 \quad \text{Write the equation.}$$
$$\underline{-24 \quad -24} \quad \text{Subtract 24 from each side.}$$
$$4.5h = 18 \quad \text{Simplify.}$$
$$\frac{4.5h}{4.5} = \frac{18}{4.5} \quad \text{Divide each side by 4.5.}$$
$$h = 4 \quad \text{Simplify.}$$

∴ You rented the bike for 4 extra hours.

On Your Own

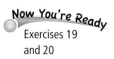
Exercises 19 and 20

4. You and your friend rent a kayak. It costs $40 for the first 4 hours and $7.50 for each extra hour. Your total cost is $62.50. Write and solve an equation to find the number of extra hours you rented the kayak.

Key Idea

Terms and Like Terms

In the equation $5x + 2x = 16 - 2$, $5x$, $2x$, 16, and 2 are called **terms**. $5x$ and $2x$ are called **like terms**. 16 and 2 are also like terms.

Like terms → $5x + 2x$ = $16 - 2$ ← Like terms

To solve, use the Distributive Property to *combine* like terms.

EXAMPLE 4 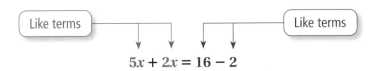 Solving Equations by Combining Like Terms

a. Solve the equation $3x + 6x = 45$.

$3x + 6x = 45$ Write the equation.

$(3 + 6)x = 45$ Use the Distributive Property to combine like terms.

$9x = 45$ Simplify.

$\dfrac{9x}{9} = \dfrac{45}{9}$ Divide each side by 9.

$x = 5$ Simplify.

∴ The solution is $x = 5$.

Check
$3x + 6x = 45$
$3(5) + 6(5) \stackrel{?}{=} 45$
$15 + 30 \stackrel{?}{=} 45$
$45 = 45$ ✓

b. Solve the equation $5a - 2a = 6$.

$5a - 2a = 6$ Write the equation.

$(5 - 2)a = 6$ Use the Distributive Property to combine like terms.

$3a = 6$ Simplify.

$\dfrac{3a}{3} = \dfrac{6}{3}$ Divide each side by 3.

$a = 2$ Simplify.

∴ The solution is $a = 2$.

Check
$5a - 2a = 6$
$5(2) - 2(2) \stackrel{?}{=} 6$
$10 - 4 \stackrel{?}{=} 6$
$6 = 6$ ✓

On Your Own

Exercises 21–29

Solve the equation. Check your solution.

5. $2x + 5x = 7$ **6.** $7c + 4c = 22$ **7.** $3w - 2w = 9$

Laurie's Notes

Key Idea
- Write the Key Idea on the board.
- **FYI:** **Like terms** are also referred to as *similar terms*. The formal definition for like terms is: two terms which have the same variables raised to the same power. Students will see this definition in future math courses.
- Explain how the Distributive Property is used to combine like terms. Usually you see the Distributive Property written with the common factor on the left of the parentheses. In this problem, the left side of the equation is given (sum of two terms) and it is rewritten.

$5x + 2x = 16 - 2$	Write the equation.
$(5 + 2)x = 16 - 2$	Use Distributive Property to combine like terms.
$7x = 14$	Simplify.
$\dfrac{7x}{7} = \dfrac{14}{7}$	Divide each side by 7.
$x = 2$	Simplify.

Example 4
- Adding like terms generally makes sense to students as the first step.
- In part (a), students will often say, "adding 3 of something to 6 of the same thing makes 9 of them." In part (b), "subtracting 2 dollars from 5 dollars leaves 3 dollars." The units are still the same.

On Your Own
- **Neighbor Check:** Have students work independently and then have their neighbor check their work. Have students discuss any discrepancies.

Closure
- **Writing Prompt:** To solve a two-step equation …

Extra Example 4
Solve the equation.
a. $7d + 4d = 55$ $d = 5$
b. $8r - 2r = 18$ $r = 3$

On Your Own
5. $x = 1$
6. $c = 2$
7. $w = 9$

The Dynamic Planning Tool
Editable Teacher's Resources at *BigIdeasMath.com*

Vocabulary and Concept Check

1. There are two different operations.
2. $3x$ and $4x$; Both have the variable x.
3. $11(x - 1) = 22$; The others all can be rewritten as $11x = 22$.
4. Use the Distributive Property to combine like terms.; $16x - 5x = (16 - 5)x = 11x$

Practice and Problem Solving

5. $z = 60$
6. $a = 63$
7. $c = 6$
8. $x = 125$
9. $b = 3$
10. $w = 3$
11. $t = 418$
12. $s = 17$
13. $t = 108$
14. $a = 8$
15. $s = 5.4$
16. $t = 6.7$
17. Subtraction should be the first step.
$$4 = \frac{y}{8} + 1$$
$$3 = \frac{y}{8}$$
$$24 = y$$
18. They should have subtracted 7 from both sides.
$$28y + 7 = 21$$
$$28y = 14$$
$$y = \frac{1}{2}$$
19. 16 lb

Assignment Guide and Homework Check

Level	Day 1 Activity Assignment	Day 2 Lesson Assignment	Homework Check
Basic	5–7, 49–53	1–4, 8–13, 17–25 odd, 30, 31, 41	8, 10, 21, 30, 41
Average	5–7, 49–53	1–4, 9, 11, 13, 17, 23–25, 31–37 odd, 43–45	9, 13, 24, 33, 44
Advanced	5–7, 49–53	1–4, 15–18, 22–28 even, 36–42 even, 46–48	16, 22, 38, 46

Common Errors

- **Exercises 5–16** Students may divide or multiply the coefficient first instead of adding or subtracting first. Tell them that while this is a valid method, they must remember to divide each part of the equation by the coefficient. Suggest that students begin away from the variable and work towards it. Also point out to them that they are going backwards through the order of operations to help them remember a system for solving for a variable.
- **Exercise 19** Students may get lost within the word problem and not realize that the equation is given to them. They may try to write the equation themselves and get an incorrect answer. Remind students to read word problems carefully and decide what information is necessary to solve the problem.

7.4 Record and Practice Journal

Solve the equation. Check your solution.

1. $4k + 2 = 6$ — $k = 1$
2. $\frac{p}{6} - 9 = 3$ — $p = 72$
3. $21 = \frac{r}{3} + 3$ — $r = 54$
4. $\frac{a}{5} - 8 = 1$ — $a = 45$
5. $7x + 9 = 51$ — $x = 6$
6. $37 = 6g - 11$ — $g = 8$
7. $2x + x = 33$ — $x = 11$
8. $49 = 10p - 3p$ — $p = 7$
9. $42.3 = 1.5d + 3.2d$ — $d = 9$

10. You need to save $5 more than 8 times your paycheck to buy a dirt bike. Your paycheck is $50. How much money do you need to save? — $405

11. You sell tickets for a fundraiser. You give half of the tickets to your sister to sell. You sell 25 of your half of the tickets and have 17 left. How many tickets did you start with? — 84 tickets

Technology For the Teacher

Answer Presentation Tool
QuizShow

7.4 Exercises

Vocabulary and Concept Check

1. **VOCABULARY** Why is the equation $5x - 12 = 23$ called a *two-step equation*?

2. **VOCABULARY** Identify the *like terms* in the equation $3x + 4x = 21$. Explain why they are like terms.

3. **WHICH ONE DOESN'T BELONG?** Which one does *not* belong with the other three? Explain your reasoning.

 | $16x - 5x = 22$ | $11x = 22$ | $11(x - 1) = 22$ | $(16 - 5)x = 22$ |

4. **WRITING** Describe a process you can use to combine the like terms in the equation $16x - 5x = 22$.

Practice and Problem Solving

Solve the equation. Check your solution.

5. $8 + \dfrac{z}{4} = 23$

6. $\dfrac{a}{3} - 9 = 12$

7. $4c - 7 = 17$

8. $6 + \dfrac{x}{5} = 31$

9. $4b - 12 = 0$

10. $12w - 8 = 28$

11. $\dfrac{t}{19} - 9 = 13$

12. $131 = 7s + 12$

13. $42 + \dfrac{t}{9} = 54$

14. $2.4a + 8 = 27.2$

15. $\dfrac{s}{3} - 0.6 = 1.2$

16. $5t - 17.2 = 16.3$

ERROR ANALYSIS Describe and correct the error in solving the equation.

17. ✗
$$4 = \dfrac{y}{8} + 1$$
$$32 = y + 1$$
$$31 = y$$

18. ✗
$$28y + 7 = 21$$
$$28y = 28$$
$$y = 1$$

19. **HIKING** You go on a hike with your uncle. Your backpack weighs 25 pounds. Your uncle is a math teacher and he tells you that your pack is 7 pounds less than twice as heavy as his pack. Use the equation $2p - 7 = 25$ to find the weight of your uncle's backpack.

Section 7.4 Solving Two-Step Equations 301

20. **TRAVEL** You drive from Chicago, IL to St. Louis, MO. On the return trip, you fly straight back to Chicago at a steady speed in 0.9 hour. The total distance is 525 miles. Write and solve an equation to find your speed from St. Louis to Chicago.

Solve the equation. Check your solution.

21. $c + 3c = 16$
22. $2x + 6x = 24$
23. $51 = 15y + 2y$
24. $6z - 5z = 20$
25. $18 = 8a - 5a$
26. $7t - t = 54$
27. $3.2x - 1.2x = 8$
28. $4.8 = 1.8n + 0.6n$
29. $15 = 3.5s - 2s$

30. **COMPUTERS** You help the owner of a computer store load monitors into a truck. You load 10 monitors and the owner loads 7 monitors. The total weight of the monitors is 765 pounds. Write and solve an equation to find the weight of each monitor.

31. **MODEL TRAIN** The model train track has 6 straight sections and 12 curved sections. The total length of the track is 351 centimeters. Each section is d centimeters long. Write and solve an equation to find the length of each section of the track.

Solve the equation. Check your solution.

32. $32y + 10 - 2 = 24$
33. $11 + \dfrac{g}{4} - 3 = 12$
34. $9.2 = 5.7 + \dfrac{h}{6} + 0.4$
35. $125 = 5(3 + x)$
36. $12(z - 7) = 60$
37. $\dfrac{z - 3}{10} = 10$
38. $7 = \dfrac{(5 + a)}{4}$
39. $6(11 + s) = 96$
40. $15 = \dfrac{22 + t}{3}$

Write and solve an equation to find x.

41. Perimeter = 18 inches

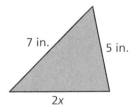

42. Perimeter = 35 feet

43. Perimeter = 132 yards

302 Chapter 7 Equations

Common Errors

- **Exercises 21–29** Students may have difficulty combining like terms. Remind them of the Distributive Property in Example 4 and show a few examples. Also, you could remind students that multiplying by the coefficient is like adding the variable that number of times. For example, $c + 3c = c + (c + c + c) = c + c + c + c = 4c$.
- **Exercises 32–40** Students may struggle with simplifying the equation and make mistakes. Remind them of the Associative, Commutative, and Distributive Properties to help them solve the equations. Also review order of operations.
- **Exercises 41–43** Students may need to be reminded of the formula for perimeter. They may also need to be reminded of the Associative and Commutative Properties for simplifying their equation.
- **Exercise 44** Students may recognize that they have two equations to solve, but may not realize that both equations are equal to 80 trading cards. A student may be able to see that the first equation is equal to 80, but will then think there is not enough information to solve the second equation. Encourage them to write the problem as two separate problems; one with their friend's equation and the other with their equation. Then ask if both equations give the same value for the variable.

Practice and Problem Solving

20. $525 = 300 + 0.9r$; 250 mi/h
21. $c = 4$
22. $x = 3$
23. $y = 3$
24. $z = 20$
25. $a = 6$
26. $t = 9$
27. $x = 4$
28. $n = 2$
29. $s = 10$
30. $10w + 7w = 765$; 45 lb
31. $6d + 12d = 351$; 19.5 cm
32. $y = \frac{1}{2}$
33. $g = 16$
34. $h = 18.6$
35. $x = 22$
36. $z = 12$
37. $z = 103$
38. $a = 23$
39. $s = 5$
40. $t = 23$
41. $2x + 5 + 7 = 18$; 3 in.
42. $x + x + 15 = 35$; 10 ft
43. $3x + 4x + 4x = 132$; 12 yd

Differentiated Instruction

Vocabulary

Write the equation $\frac{c}{5} + 6 = 21$ on the board. Ask students to name the steps needed to solve the problem. Repeat the steps out loud as you do them on the board. Then have students check the solution in the equation. Ask them if the equation might have been solved differently.

T-302

Practice and Problem Solving

44. yes; The solution to both $4x - 16 = 80$ and $3x + 8 = 80$ is $x = 24$.

45. 4.5 cups

46. a. 95

 b. greater than

 c. no; You would have to score a 91.25, which is not possible.

47. See *Taking Math Deeper*.

48. 11 pens; 15 pencils

Fair Game Review

49. $\frac{17}{20}$ **50.** $\frac{43}{50}$

51. $1\frac{7}{25}$ **52.** $\frac{3}{400}$

53. B

Mini-Assessment

Solve the equation.

1. $9 + 2y = 23$ $y = 7$
2. $4x - 4 = 20$ $x = 6$
3. $6 + \frac{s}{7} = 13$ $s = 49$
4. $\frac{f}{3} - 4 = 19$ $f = 69$
5. A dog weighs 30 pounds. Its weight is 6 pounds more than twice the weight w of a cat. Write and solve an equation to find the weight of the cat. $2w + 6 = 30$; The cat weighs 12 pounds.

T-303

Taking Math Deeper

Exercise 47

There are a couple things that students need to look out for in this problem. One is that the two given measurements have different units: centimeters and millimeters. The second thing is that there is a difference between the "number of pages in a book" and the "number of sheets of paper" in a book. The first is twice the second.

1 What other information is needed to find the thickness of a single page?
 a. I need to know how many pages the book has.

2 Choose a reasonable number of pages.
 b. A book that is about 1 inch thick has about 500 *numbered* pages. However, this means that it has only 250 sheets of paper.

3 Find the thickness of a sheet of paper.
 x = thickness of a sheet of paper
 3 cm = 30 mm

 30 mm − 2(3 mm) = thickness of 250 sheets of paper
 24 mm = thickness of 250 sheets of paper

 $250x = 24$
 $x = 0.096$ mm

Each page has 2 sides.

Each sheet of paper is about a tenth of a millimeter thick.

Common thickness for paper ranges from 0.08 millimeter to 0.2 millimeter. So, this answer is reasonable.

Project

Design a bookshelf that will hold 300 books that are the same size as the one in the problem.

Reteaching and Enrichment Strategies

If students need help...	If students got it...
Resources by Chapter • Practice A and Practice B • Puzzle Time Record and Practice Journal Practice Differentiating the Lesson Lesson Tutorials Skills Review Handbook	Resources by Chapter • Enrichment and Extension • School-to-Work Start the next section

44. TRADING CARDS You have 80 trading cards. Your friend says that you have 16 less than 4 times the number of cards that she has. You say that you have 8 more than 3 times as many cards as she has. Can you both be right? Explain.

45. RECIPE You want to make 3 batches of barbecue sauce, but you can't remember how much brown sugar you need. You know that 4 batches make about 17 cups of sauce. How much brown sugar do you need for 3 batches?

46. TESTS After four 100-point tests, you have 365 points.

 a. How many points do you need to score on your next 100-point test to have a mean score of 92 points?

 b. Would a mean score of 92 points after 5 tests be *greater than* or *less than* your mean score after 4 tests?

 c. Your score on each test is a whole number. Is it possible that your mean score does *not* change after the fifth test? Explain.

47. HARDCOVER BOOK Each page of the book has the same thickness t.

 a. What other piece of information do you need to find the thickness of one page?

 b. Choose a reasonable number for the missing piece of information.

 c. Use the number from part (b) to write and solve an equation to find the thickness t of one page. Does your answer seem reasonable?

48. A teacher has a box of pens and pencils. There are 8 more pencils than pens. After students take 1 pen and 5 pencils from the box, there are 26 pens and pencils left in the box. How many pens are in the box now? How many pencils?

Fair Game Review *What you learned in previous grades & lessons*

Write the percent as a fraction or mixed number in simplest form. *(Section 4.1)*

49. 85% **50.** 86% **51.** 128% **52.** 0.75%

53. MULTIPLE CHOICE Use a formula to find the area of the triangle. *(Section 1.5)*

 Ⓐ 36 m² Ⓑ 48 m² Ⓒ 72 m² Ⓓ 96 m²

7 Study Help

You can use an **example and non-example chart** to list examples and non-examples of a vocabulary word or term. Here is an example and non-example chart for equations.

Equations

Examples	Non-Examples
$x = 5$	5
$2a = 16$	$2a$
$3x + 4 = 19$	$3x + 4$
$5 = 8x + 3$	$8x + 3$
$12 - 7 = 5$	$12 - 7$
$\frac{3}{4}y = 6$	$\frac{3}{4}$

On Your Own

Make an example and non-example chart to help you study these topics.

1. solutions of equations
2. inverse operations
3. equations solved using addition or subtraction
4. equations solved using multiplication or division
5. two-step equations
6. like terms

After you complete this chapter, make example and non-example charts for the following topics.

7. plane figures
8. rectangular prisms
9. volume

"I need a good non-example of a cool animal for my *example and non-example chart*."

Sample Answers

1. Solutions of Equations

Examples	Non-Examples
$x - 1 = 5$ $x = 6$	$2x$ $x = 3$
$2a = 16$ $a = 8$	$x - 1 = 5$ $x = 4$
$3x + 4 = 19$ $x = 5$	$y + 9 = 12$ $y = 1$
$5 = 3x - 4$ $3 = x$	$2a = 16$ $a = 9$

2. Inverse Operations

Examples	Non-Examples
addition and subtraction	addition and multiplication
multiplication and division	addition and division
	subtraction and multiplication
	subtraction and division

3. Equations Solved Using Addition or Subtraction

Examples	Non-Examples
$y + 9 = 12$	$2a = 16$
$a - 11 = 3 - 2$	$0.4b = 4$
$10 = x - 19$	$\frac{3}{4}y = 6$
$5x - 3x - x + 1 = 6$	$\frac{y}{6} = 0$

4. Equations Solved Using Multiplication or Division

Examples	Non-Examples
$2a = 16$	$x - 1 = 5$
$\frac{y}{4} = 6$	$y + 9 = 12$
$0.4b = 4$	$5x - 3x - x + 1 = 6$
$\frac{6}{7} = \frac{x}{14}$	$3a - 2a = 10$

5. Two-Step Equations

Examples	Non-Examples
$3x + 4 = 19$	$2a = 16$
$2x - 3 = 5$	$x - 1 = 5$
$8x - 5x = 3$	$4 = \frac{b}{7}$
$1 = 0.1x - 9$	$y + 3 = 4$

6. Like Terms

Examples	Non-Examples
$2x$ and $5x$	$2x$ and $5y$
$6b$ and $6b$	$6b$ and 6
14 and 2	0 and x
$2x$ and $-2x$	$2x$ and -2

List of Organizers
Available at *BigIdeasMath.com*

Comparison Chart
Concept Circle
Definition (Idea) and Example Chart
Example and Non-Example Chart
Formula Triangle
Four Square
Information Frame
Information Wheel
Notetaking Organizer
Process Diagram
Summary Triangle
Word Magnet
Y Chart

About this Organizer
An **Example and Non-Example Chart** can be used to list examples and non-examples of a vocabulary word or term. Students write examples of the word or term in the left column and non-examples in the right column. This type of organizer serves as a good tool for assessing students' knowledge of pairs of topics that have subtle but important differences, such as equations and expressions. Blank example and non-example charts can be included on tests or quizzes for this purpose.

Technology For the Teacher
Vocabulary Puzzle Builder

Answers

1. $x - 3 = 5$
2. $a \div 7 = 14$
3. $3b = 9$
4. $n + 7 = 20$
5. $k = 10$
6. $s = 1\frac{1}{6}$
7. $w = 50$
8. $p = 12$
9. $x = 1$
10. $n = 20$
11. $y = 3$
12. $j = 0.6$
13. $\frac{2}{3}r = 30$
14. $360 = \ell - 30$; 390 m
15. $33 = 2c + 5$; 14 lb
16. $0.8m = 4(3) + 2(2)$; $20

Alternative Quiz Ideas

100% Quiz Math Log
Error Notebook Notebook Quiz
Group Quiz Partner Quiz
Homework Quiz Pass the Paper

Homework Quiz
A homework notebook provides an opportunity for teachers to check that students are doing their homework regularly. Students keep their homework in a notebook. They should be told to record the page number, problem number, and copy the problem exactly in their homework notebook. Each day the teacher walks around and visually checks that homework is completed. Periodically, without advance notice, the teacher tells the students to put everything away except their homework notebook.

Questions are from students' homework.

1. What is the answer to Exercise 14 on page 280?
2. What are the answers to Exercises 18–20 on page 287?
3. What are the answers to Exercises 15–18 on page 294?
4. What is the answer to Exercise 30 on page 302?

Assessment Book

Reteaching and Enrichment Strategies

If students need help...	If students got it...
Resources by Chapter • Study Help • Practice A and Practice B • Puzzle Time Lesson Tutorials BigIdeasMath.com Practice Quiz Practice from the Test Generator	Resources by Chapter • Enrichment and Extension • School-to-Work Game Closet at *BigIdeasMath.com* Start the next section

Technology For the Teacher
Answer Presentation Tool
Big Ideas Test Generator

7.1–7.4 Quiz

Write the word sentence as an equation. *(Section 7.1)*

1. A number x decreased by 3 is 5.
2. A number a divided by 7 equals 14.
3. The product of 3 and a number b is 9.
4. 7 more than a number n is 20.

Solve the equation. Check your solution. *(Section 7.2, Section 7.3, and Section 7.4)*

5. $4 + k = 14$
6. $\frac{5}{6} = s - \frac{1}{3}$
7. $10 = \frac{w}{5}$
8. $9p = 108$
9. $9x + 2 = 11$
10. $\frac{n}{2} - 4 = 6$
11. $\frac{y}{3} + 0.2 = 1.2$
12. $6j + 1.9 = 5.5$

13. **RIBBON** The length of the blue ribbon is two-thirds the length of the red ribbon. Write an equation you can use to find the length r of the red ribbon. *(Section 7.1)*

14. **BRIDGES** The main span of the Sunshine State Skyway is 360 meters long. The Skyway is 30 meters shorter than the main span of the Dames Point Bridge. Write and solve an equation to find the length ℓ of the main span of the Dames Point Bridge. *(Section 7.2)*

15. **CAT** A dog weighs 33 pounds. Its weight is 5 pounds more than twice the weight of a cat. Write and solve an equation to find the weight of the cat. *(Section 7.4)*

16. **SHOPPING** At a farmer's market, you buy 4 pounds of tomatoes and 2 pounds of sweet potatoes. You spend 80% of the money in your wallet. Write and solve an equation to find how much money is in your wallet before you pay. *(Section 7.3)*

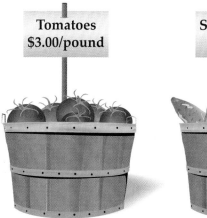

7.5 Finding Dimensions of Plane Figures

Essential Question How can you use area and perimeter formulas to find missing dimensions of plane figures?

COMMON CORE STATE STANDARDS
6.EE.7
6.G.1

1 ACTIVITY: Finding Missing Dimensions

Work with a partner. Match the equation with a figure. Then use the equation to solve for x.

a. Area: $24 = 4x$

b. Perimeter: $24 = 4x$

c. Circumference: $8\pi = 2\pi x$

d. Area: $20 = 5x$

e. Area: $16 = \frac{1}{2}x(5 + 3)$

f. Area: $10 = \frac{1}{2}(5x)$

g. Perimeter: $30 = 20 + 2x$

h. Circumference: $4\pi = \pi x$

Laurie's Notes

Introduction

For the Teacher
- **Goal:** Students will use a variety of strategies to solve for the missing dimension in a geometric figure.
- This activity is a nice review of area formulas from prior years. It also provides a practical context for practice in equation solving.

Motivate
- None of your students are old enough to drive, but they have seen many traffic signs. Traffic signs are geometric figures, similar to the shapes students will be working with today.
- ? "What traffic sign is an octagon (8-sided)?" STOP sign
- ? "What traffic sign is a circle?" railroad crossing
- **FYI:** The first traffic signs were installed in London in 1868. The first traffic fatality was in New York City in 1899. The first stop sign was installed in Detroit in 1915.

Activity Notes

Activity 1
- Begin with a few warm-up questions.
- ? "If you were looking for a perimeter formula for the orange triangle, what would the formula look like?" Listen for students explaining that because the triangle has three sides, the perimeter formula would have 3 numbers or variables being summed.
- ? "If you were looking for an area formula for the orange triangle, what would the formula look like?" Area of a triangle has a $\frac{1}{2}$. Then it would be multiplied by the base and height.
- Give students sufficient time to match the formulas and figures, and to solve for the variable x. In solving for x, it is very possible that *Guess, Check, and Revise* is used as a strategy and that's okay.
- Circulate around the classroom and listen to the strategies being used to match the formulas and to solve for x. This will provide valuable information about students' familiarity with the formulas, their number fluency, and their ability to solve equations by inspection and visual manipulation. Did students check their solutions? Good problem solvers do!
- ? "How did you know you matched the orange triangle correctly with $10 = \frac{1}{2}(5x)$? Of the four area formulas, only two had dimensions 5 and x. Only one had a $\frac{1}{2}$, so the correct match was (d).
- ? "What was your answer for x in the orange triangle? Explain how you found it." 4; Students might say "Half of 20 is 10, so I knew $5x$ had to be 20, so x was 4."
- Select different pairs to share their solution *and* strategy for solving for x.

Common Core State Standards
6.EE.7 Solve real-world and mathematical problems by writing and solving equations of the form $x + p = q$ and $px = q$ for cases in which p, q and x are all nonnegative rational numbers.
6.G.1 Find the area of right triangles, other triangles, special quadrilaterals, and polygons by composing into rectangles or decomposing into triangles and other shapes; apply these techniques in the context of solving real-world and mathematical problems.

Previous Learning
Students should know the formulas for area.

Start Thinking! and Warm Up

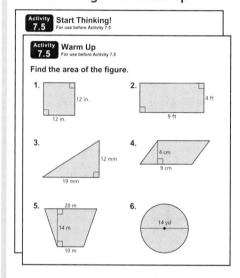

7.5 Record and Practice Journal

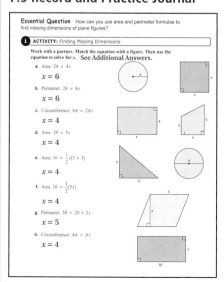

T-306

English Language Learners
Notebook Development
Have students create a formula page in their notebooks. Include the area and perimeter formulas used in this section. For each formula, include a description of what each variable represents and a sketch of the plane figure.

7.5 Record and Practice Journal

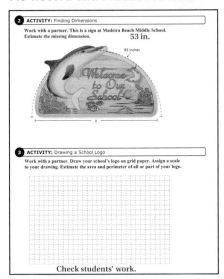

Laurie's Notes

Activity 2
- **?** "What shape does the school sign remind you of?" semicircle
- **?** "What does the 83 inches measure in the diagram?" It is half the circumference. Students might think the sign has actually been traced—it has not.
- Students are not expected to do any symbolic manipulation to solve for the diameter (the missing dimension). They should use *Guess, Check, and Revise*. Their thought process might be something like the following:

 $C = \pi d$. If the semicircle is 83 inches, the whole circumference is 166 inches; π times what would be about 166 inches?

- **Another Way:** An alternate approach might be:

 $C = 2\pi r$. Because $\frac{1}{2}$ the circumference is 83 inches, $\frac{1}{2}$ of $2\pi r$ is πr.

 π times what would be about 83 inches?

 In this case, the student would be solving for the radius and would need to double it to get the diameter x.

- **Whole Class Discussion:** This problem offers multiple strategies for solving and should be discussed by the whole class.

Activity 3
- Give examples of common logos that your students will know.
- Students have not studied scale drawings, so in this activity they will simply label the dimensions of their drawings.
- Hang all of the logos on the bulletin board. Students could vote on their favorite design. If no school sign exists, or a new sign is needed, submit a student design for consideration.

What Is Your Answer?
- Suggest to students that they label all of the dimensions, compute the area(s), and erase one of the dimensions. Exchange figures with another student and use algebra to find the missing dimension.

Closure
- **Exit Ticket:** What is the length of the street sign? 18 in.

 4 in.

Area = 72 in.²

Technology For the Teacher

The Dynamic Planning Tool
Editable Teacher's Resources at *BigIdeasMath.com*

② ACTIVITY: Finding Dimensions

Work with a partner. This is a sign at Madeira Beach Fundamental School. Estimate the missing dimension.

③ ACTIVITY: Drawing a School Logo

Work with a partner. Draw your school's logo on grid paper. Assign a scale to your drawing. Estimate the area and perimeter of all or part of your logo.

Challenger Middle School Westlake Middle School

What Is Your Answer?

4. **IN YOUR OWN WORDS** How can you use area and perimeter formulas to find missing dimensions of plane figures? Draw a plane figure and label its dimensions. Find the area or perimeter. Then erase one of the dimensions and show how you can use algebra to find it again.

Use what you learned about finding dimensions of plane figures to complete Exercises 3–5 on page 310.

Section 7.5 Finding Dimensions of Plane Figures

7.5 Lesson

Check It Out
Lesson Tutorials
BigIdeasMath.com

EXAMPLE 1 Finding Missing Dimensions

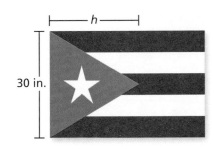

a. The blue triangle in the Puerto Rican flag has an area of 390 square inches. What is the height of the triangle?

$A = \dfrac{1}{2}bh$ Write formula for area of a triangle.

$390 = \dfrac{1}{2}(30)h$ Substitute 390 for A and 30 for b.

$390 = 15h$ Multiply $\dfrac{1}{2}$ and 30.

$\dfrac{390}{15} = \dfrac{15h}{15}$ Divide each side by 15.

$26 = h$ Simplify.

∴ The height of the triangle is 26 inches.

b. The circle in the Japanese flag has a circumference of 4.71 feet. What is the radius of the circle?

$C = \pi d$ Write formula for circumference.

$4.71 \approx 3.14d$ Substitute 4.71 for C and 3.14 for π.

$\dfrac{4.71}{3.14} = \dfrac{3.14d}{3.14}$ Divide each side by 3.14.

$1.5 = d$ Simplify.

∴ So, the radius is about $1.5 \div 2$, or 0.75 feet.

On Your Own

Now You're Ready
Exercises 3–11

Write and solve an equation to find the missing dimension of the figure. Check your solution.

1. Area = 45 cm²

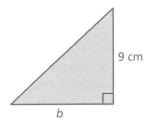

2. Circumference = 28.26 m

308 Chapter 7 Equations

Laurie's Notes

Introduction

Connect
- **Yesterday:** Students gained an intuitive understanding of how equations might be solved for the missing dimension in a geometric figure.
- **Today:** Students will use Properties of Equality to solve equations in order to solve for a missing dimension.

Motivate
- Share the symbolism about the Puerto Rican flag. The white star represents the Commonwealth of Puerto Rico. The three sides of the blue triangle stand for the three branches of the Republican government, legislative, executive, and judicial. The red stripes stand for the blood that nourishes the three branches. The white stripes represent individual liberty and the rights that keep their form of government in perfect balance.

Lesson Notes

Example 1
- Explain that students will work with the geometric formulas and Properties of Equality in order to solve for a variable.
- Ask questions to develop the problem.
 - "How do you solve $32 = 4x$? What is x?" Divide both sides by 4; $x = 8$
 - **Connection:** "If the area of a rectangle is 32 and the width is 4, how would you solve for the length?" You want students to see the connection between this question and the previous question.
 - "How do you find the area of the blue triangle on the Puerto Rican flag?" Listen for students to talk about the area formula for a triangle and the fact that the base is known.
- **Common Misconception:** When students see $\frac{1}{2}(30)h$, they sometimes think that they have to take one-half of 30 and one-half of h. I ask them how they would work with the expression $\frac{1}{2} \times 30 \times 4$ or $2 \times 30 \times 4$. Once one-half has been multiplied by 30, the product is then multiplied by h.
- In part (b), students may get nervous about the decimals. This is the same type of problem, just a different set of numbers.
- "Could the problem be solved using the formula $C = 2\pi r$?" Yes
- Remind students that they can check their solution by substituting it for the variable in the formula along with the other given dimensions.

On Your Own
- In Question 1, there are two common strategies for dealing with the fraction. Because multiplication is commutative, the right side can be rewritten as $\frac{1}{2}(9)b$, or $4.5b$. The second strategy is to begin by multiplying both sides of the original equation by 2.

Goal Today's lesson is determining a missing dimension of a plane figure, given its area and some of the dimensions.

Start Thinking! and Warm Up

Warm Up
For use before Lesson 7.5

Start Thinking!
For use before Lesson 7.5

One television screen has a height of 19 inches and an area of 665 square inches.

Another television screen has a height of 19 inches and a perimeter of 90 inches.

Are the television screens the same size?

Extra Example 1
a. A triangle has an area of 63 square feet and a height of 9 feet. What is the length of the base of the triangle? 14 ft
b. A circle has a circumference of 7.85 meters. What is the radius of the circle? about 1.25 m

On Your Own
1. $45 = \frac{1}{2}(9b)$; 10 cm
2. $28.26 = 3.14d$; about 9 m

T-308

Differentiated Instruction

Visual

When solving word problems, have students make sketches of the situation. By labeling their sketches, students will have a clearer picture of what they know and what they need to find to answer the question.

Extra Example 2

The top of a rectangular table has a length of 6 feet and a perimeter of 20 feet. What is the width? 4 ft

Extra Example 3

The triangular part of the dance studio in Example 3 covers 153 square feet. What is the area of the rectangular part? 765 ft²

On Your Own

3. $18 = 2\ell + 6$; 6 ft
4. 120 ft²

Laurie's Notes

For the Teacher

- **Big Idea:** The focus of this chapter has been solving equations in one variable. In this section, the formulas used involve two or three variables. It is important to help students recognize that if some of the variable(s) are known, the numbers can be substituted for the variable(s). Once this is done, the resulting equation has only one variable. The equation can then be solved using techniques from earlier in the chapter.

Example 2

? This example involves a formula with three variables.
- "What is known in this problem?" Students need to read from the diagram to see that the width is 3. The perimeter is 22.
- "In the expression $2\ell + 6$, what order are the operations performed?" multiplication, then addition
- "How do you undo these two steps?" Reverse the order of operations. Subtract 6 then divide by 2.
- Finish solving the problem.

Example 3

? Have students read the given information.
- "What do you know about the triangle?" The base is 12 feet.
- "What else do you need to know in order to find the area of the triangle?" the height
- "How can you find the height of the triangle?" Find the height of the rectangle.
- This is a multi-step problem. Work slowly.
- **Common Misconception:** Watch the step of multiplying $\frac{1}{2}(12)(16.5)$. Students will want to multiply each factor by $\frac{1}{2}$!

? "If the problem had been $\frac{1}{2}(16.5)(12)$ instead, how would you do the calculation?" It is good to check the idea of multiplication being commutative and associative. You want students to recognize that it is much easier to multiply $\frac{1}{2}$ and an even number (12) than a decimal (16.5). This is a helpful strategy.

On Your Own

- Ask volunteers to share their work at the board.

Closure

- In Example 1, if the area of the Puerto Rican flag is 1350 square inches, what is the length of the flag? 45 in.

The Dynamic Planning Tool
Editable Teacher's Resources at *BigIdeasMath.com*

EXAMPLE 2 Finding a Missing Dimension

The top of the table has a perimeter of 22 feet. What is the length?

$P = 2\ell + 2w$	Write formula for perimeter of a rectangle.
$22 = 2\ell + 2(3)$	Substitute 22 for P and 3 for w.
$22 = 2\ell + 6$	Multiply 2 and 3.
$\underline{-6 \quad\quad -6}$	Subtract 6 from each side.
$16 = 2\ell$	Simplify.
$\dfrac{16}{2} = \dfrac{2\ell}{2}$	Divide each side by 2.
$8 = \ell$	Simplify.

∴ The length of the table is 8 feet.

EXAMPLE 3 Real-Life Application

The rectangular part of a dance studio covers 495 square feet. What is the area of the triangular part?

Use the rectangle to solve for h.

$A = 30h$	Use the formula for area of a rectangle.
$495 = 30h$	Substitute 495 for A.
$\dfrac{495}{30} = \dfrac{30h}{30}$	Divide each side by 30.
$16.5 = h$	Simplify.

Use $h = 16.5$ to find the area of the triangle.

$A = \dfrac{1}{2}bh$	Write formula for area of a triangle.
$A = \dfrac{1}{2}(12)(16.5)$	Substitute 12 for b and 16.5 for h.
$A = 99$	Simplify.

∴ The area of the triangular part is 99 square feet.

On Your Own

3. **WHAT IF?** In Example 2, the tabletop has a perimeter of 18 feet. Write and solve an equation to find the length.

4. **WHAT IF?** In Example 3, the rectangular part covers 600 square feet. What is the area of the triangular part?

7.5 Exercises

Vocabulary and Concept Check

1. **WRITING** Explain what is wrong with the statement.

 "The length of the rectangle is 30 square centimeters."

2. **OPEN-ENDED** Give an example of a rectangle that has an area of 48 square inches. What is its length? its width?

Practice and Problem Solving

Write and solve an equation to find the missing dimension of the figure. Check your solution.

3. Area = 42 mm²

4. Circumference = 62.8 m

5. Area = 36 in.²

6. Circumference = 15.7 yd

7. Perimeter = 40 m

8. Perimeter = 24 ft

9. Area = 27 ft²

10. Perimeter = 18 in.

11. Area = 15 km²

12. **ERROR ANALYSIS** Describe and correct the error in finding the height h of the triangle.

13. **YIELD SIGN** The yield sign has a perimeter of 2.28 meters. What is the side length x?

Assignment Guide and Homework Check

Level	Day 1 Activity Assignment	Day 2 Lesson Assignment	Homework Check
Basic	3–5, 19–23	1, 2, 6–13	6, 7, 11, 13
Average	3–5, 19–23	1, 2, 6–12, 15, 16	6, 7, 11, 16
Advanced	3–5, 19–23	1, 2, 6–18 even, 17	6, 10, 14, 17

Common Errors

- **Exercises 3–11** Students may use an incorrect equation when solving for the unknown variable. For example, instead of using the equation for area, they may use the equation for perimeter. Tell students to read the question carefully. Encourage them to write down the equation they will use before substituting any values.
- **Exercises 3–11** Students may write the incorrect units or forget to include units in their answer. Remind them that they need to include units as part of their answer. This is a way to check if their answer is reasonable.
- **Exercise 15** Students may write an incorrect equation. Remind them about the different units that are given. Area is in square inches, so how can they write an equation that has square inches as the units?
(576 seeds ÷ 3 seeds per square inch).

7.5 Record and Practice Journal

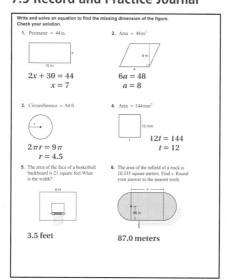

Technology For the Teacher
Answer Presentation Tool
QuizShow

 Vocabulary and Concept Check

1. Square centimeters represents an area, not a length.

2. *Sample answer:*

length = 8 in., width = 6 in.

 Practice and Problem Solving

3. $42 = 7x$; 6 mm

4. $62.8 = 3.14d$; about 20 m

5. $36 = \frac{1}{2}(8w)$; 9 in.

6. $15.7 = 3.14d$; about 5 yd

7. $40 = 2c + 24$; 8 m

8. $24 = 2y + 10$; 7 ft

9. $27 = 4a$; 6.75 ft

10. $18 = 4x$; 4.5 in.

11. $15 = \frac{1}{2}(6 + 4)s$; 3 km

12. They should have divided by 8 in the last step.
$$32 = \frac{1}{2}bh$$
$$32 = \frac{1}{2}(16)h$$
$$32 = 8h$$
$$4 = h$$

13. 0.76 m

14. about 4 in. 15. 24 in.

16. **a.** The circumference of the wheel is equal to the perimeter of the city block divided by the number of times your wheel rotates. Then you can solve for the diameter.

b. about 2.45 ft

T-310

Practice and Problem Solving

17. See *Taking Math Deeper*.
18. **a.** length: 6 ft, width: 4 ft
 b. *Sample answer:* Measure the length and width of the figure. Then find the perimeter of the figure. Divide the perimeter of the mural by the perimeter of the figure to find how many times larger the mural is. Multiply the length and width of the figure by this number to find the length and width of the mural.

Fair Game Review

19. $x = 16$ 20. $y = 153$
21. $h = 4.5$ 22. $w = 8$
23. B

Mini-Assessment

Write and solve an equation to find x.

1. Area = 12 in.² $12 = \frac{1}{2}(6)x$; $x = 4$ in.

2. Area = 32 in.²
 $32 = \frac{1}{2}(6 + 10)x$; $x = 4$ in.

3. The perimeter of the picture is 36 inches. What is the width w of the picture? $w = 8$ in.

Taking Math Deeper

Exercise 17

At first reading, this problem appears to be difficult. However, by taking the problem "one step at a time" it turns out to be a fairly straightforward problem. A big help in the problem is to draw a diagram and label the dimensions.

 Draw and label a diagram.

 Find the area of the rectangle.

Area of door	26.28 ft²	Given
− Area of semicircle	− 6.28 ft²	$\frac{1}{2}\pi r^2$
Area of rectangle	20.00 ft²	

 Solve for x and answer the question.

$4x = 20$
$x = 5$

The rectangle is 5 feet high.

So, the height of the door is $5 + 2 = 7$ feet.

Project

Draw a new door for the entryway to your home. The door cannot be rectangular nor can it have the same shape as the problem. Find the amount of wood it would take to make the new door.

Reteaching and Enrichment Strategies

If students need help...	If students got it...
Resources by Chapter • Practice A and Practice B • Puzzle Time Record and Practice Journal Practice Differentiating the Lesson Lesson Tutorials Skills Review Handbook	Resources by Chapter • Enrichment and Extension • School-to-Work • Financial Literacy Start the next section

14. **GLASS** You want to place your juice glass on a circular coaster. The circumference of the coaster is 12.56 inches. What is the diameter of the largest juice glass that you can fit on the coaster? Use 3.14 for π.

15. **GARDENING** You plant 576 flower seeds in a window box. You plant 3 seeds per square inch. The box is 8 inches wide. How long is it?

16. **BICYCLING** You ride your bicycle once around the city block and count the number of times the front wheel on your bike rotates.

 a. How can you find the diameter of the wheel?
 b. The wheel rotates 200 times. What is the diameter of the wheel? Use $\frac{22}{7}$ for π.

17. **DOOR** A door is made up of a semicircle and a rectangle. The area of the door is 26.28 square feet. What is the height h of the door? Use 3.14 for π.

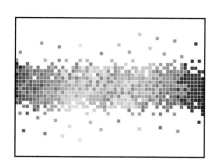

18. **MURAL** A mural has a perimeter of 20 feet and an area of 24 square feet.

 a. Find the length and width of the mural using trial and error. Organize your results in a table.
 b. **Reasoning** Explain how to find the length and width of the mural using the figure.

Fair Game Review What you learned in previous grades & lessons

Solve the equation. Check your solution. *(Section 7.3)*

19. $4x = 64$
20. $\frac{1}{3}y = 51$
21. $6(7)h = 189$
22. $\frac{1}{4}(7)w = 14$

23. **MULTIPLE CHOICE** What value is a solution of $7x = 42$? *(Section 7.3)*

 (A) $x = \frac{1}{6}$ (B) $x = 6$ (C) $x = 42$ (D) $x = 294$

Section 7.5 Finding Dimensions of Plane Figures

7.6 Finding Dimensions of Prisms

COMMON CORE STATE STANDARDS
6.G.2
6.G.4

Essential Question How can you use a volume formula to find missing dimensions of prisms?

1 ACTIVITY: Finding Missing Dimensions

Work with a partner. Solve the equation for x.

a. Volume: $264 = 24x$

b. Volume: $162 = 18x$

8 in. 3 in.

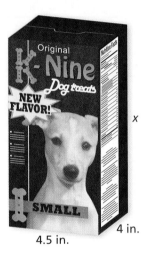
4.5 in. 4 in.

c. Volume: $45 = 9x$

d. Volume: $12 = 8x$

6 in. 1.5 in.

2 in. 4 in.

e. Volume: $288 = 48x$

12 in. 4 in.

312 Chapter 7 Equations

Laurie's Notes

Introduction

For the Teacher
- **Goal:** Students will develop an intuitive understanding of how to solve for a missing dimension in a prism.
- Today's investigation extends the skill work of the last section using the formula related to the volume of rectangular prisms.
- This activity requires that students recall how to find the volume of a prism.

Motivate
- **Time for the Midas Touch:** A standard 400 ounce (25 pounds) gold bullion bar measures $8\frac{1}{3}$ inches by $3\frac{1}{3}$ inches by $1\frac{7}{8}$ inches. This is about the size of a small tissue box. Bring in a tissue box and hold it.
- ❓ Share with students that if this were a bar of gold, it would weigh 400 ounces.
 - "How many pounds is this?" 25 lb
 - "If gold were selling at $800 per ounce, what would the approximate value of the gold bar be?" $320,000
 - "What shape is the gold bar? Rectangular prism, although students will likely say a prism without reference to the type.
 - "Could this gold bar (still holding the tissue box) be melted down and formed into another prism with different dimensions?" yes
 - "Would there still be 400 ounces of gold worth $320,000?" yes
- The last question is to get students thinking about the volume of prisms, the focus of today's lesson.

Discuss
- ❓ "What is a prism?" Listen for descriptions that suggest a box. It is unlikely that students will think about nonrectangular prisms, such as triangular prisms.
- ❓ "What features, or attributes does a prism have?" Listen for length, width, and height. Students may also say edges, rectangular faces, vertices, and so on.

Activity Notes

Activity 1
- Introduce the activity. Explain that they are to solve for the missing dimension in each figure.
- If students have forgotten the volume formula, the equation should be a good prompt. In part (a), it should be pretty obvious that the $24x$ comes from multiplying the length × width × height even though it is not written.
- Students are not expected to do symbolic manipulation to find the missing dimension. They should use *Guess, Check, and Revise*. In part (a), they should reason that 24×10 is 240, so the missing dimension is more than 10.
- In discussing the solutions, the formula should be discussed. This will provide a natural transition to the next activity.

Common Core State Standards

6.G.2 Find the volume of a right rectangular prism with fractional edge lengths by packing it with unit cubes of the appropriate unit fraction edge lengths, and show that the volume is the same as would be found by multiplying the edge lengths of the prism. Apply the formulas $V = \ell wh$ and $V = bh$ to find volumes of right rectangular prisms with fractional edge lengths in the context of solving real-world and mathematical problems.

6.G.4 Represent three-dimensional figures using nets made up of rectangles and triangles, and use the nets to find the surface area of these figures. Apply these techniques in the context of solving real-world and mathematical problems.

Previous Learning
Students should know how to find the volume of a prism.

Start Thinking! and Warm Up

Activity 7.6 Start Thinking! For use before Activity 7.6

Activity 7.6 Warm Up For use before Activity 7.6

Solve the equation. Check your solution.
1. $2x = 10$
2. $14x = 7$
3. $7x = 63$
4. $25x = 300$
5. $9x = 162$
6. $16x = 24$

7.6 Record and Practice Journal

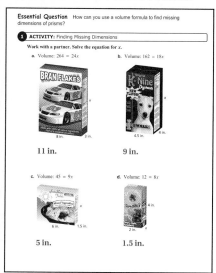

Essential Question How can you use a volume formula to find missing dimensions of prisms?

1 ACTIVITY: Finding Missing Dimensions

Work with a partner. Solve the equation for x.

a. Volume: $264 = 24x$
11 in., 8 in., 3 in.

b. Volume: $162 = 18x$
9 in., 4.5 in., 4 in.

c. Volume: $45 = 9x$
5 in., 6 in., 1.5 in.

d. Volume: $12 = 8x$
1.5 in., 2 in., 4 in.

T-312

English Language Learners
Forming Answers
Encourage English learners to form complete sentences in their responses. Students can use the question to help them form the answer.

Question: If you know the height and the width of a prism, what else do you need to know to find the volume?

Response: If you know the height and width of a prism, you need to know the length of the prism to find the volume.

7.6 Record and Practice Journal

Laurie's Notes

Activity 2

- **Estimation:** The buildings shown in this activity are essentially prisms, although each has certain architectural features so that the faces are not exact rectangles. This is an estimation problem.
- For each part, students are given the number of stories, the height of the building, and the total square footage of floor space.
- You will need to explain what square footage of floor space means. This could be done using the tissue box. Explain that the box might be the first floor of an office building. The area of the base would be the square footage of floor space on the first floor in the office building. (Mention the first floor of your school building.) Now, if the office building has a second floor, the square footage of floor space just doubled. (Put a second tissue box on top of the first.) This could be repeated for several floors.
- **Big Idea:** If you know the total square footage of floor space and divide it by the number of floors, you'll have the square footage of one floor, or the base of the prism. Multiplying the area of the base by the height of the building, you have the volume of the building.
- Students will need to spend some time thinking about this problem. If you have blueprint information about your own school, it might be helpful to use it as an example before students work on Activity 2.
- The key to this problem is finding the area of the base. The actual dimensions are not found, but the product of the dimensions (area) is found.
- By working through these two activities, students should start to see that the volume of the prism is the product of three numbers.

What Is Your Answer?

- Review the answers to the questions together as a class. Ask for volunteers to share their answers.

Closure

- Sketch the object.

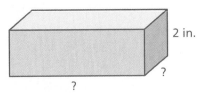

❓ "What are the missing dimensions, if the volume is 24 cubic inches?"
Look for two numbers whose product is 12.

The Dynamic Planning Tool
Editable Teacher's Resources at *BigIdeasMath.com*

2 ACTIVITY: Finding Dimensions

Work with a partner. Is there enough information given to estimate the volume of each building? If not, explain why. If there is enough information, estimate the volume of the building.

Republic Plaza
Denver, CO
56 Stories: 714 ft
1,340,000 ft² floor space

Seagram Building
New York, NY
38 Stories: 515 ft
820,000 ft² floor space

Espirito Santo Plaza
Miami, FL
37 Stories: 483 ft
750,000 ft² floor space

What Is Your Answer?

3. **IN YOUR OWN WORDS** How can you use a volume formula to find missing dimensions of prisms?

4. Design a skyscraper that will be 10% taller than one of the skyscrapers in Activity 2. Find its volume.

"Dear Sir. The question 'How much is half of 8?' is confusing."

"If you mean up and down, the answer is 3. If you mean sideways, the answer is 0."

Practice — Use what you learned about finding dimensions of prisms to complete Exercises 9–11 on page 316.

Section 7.6 Finding Dimensions of Prisms 313

7.6 Lesson

Check It Out
Lesson Tutorials
BigIdeasMath.com

EXAMPLE 1 Counting Cubes

How many cubes do you need to fill the box?

Key Vocabulary
rectangular prism, *p. 314*
volume, *p. 314*
cubic units, *p. 314*

The bottom layer of the box is 4 cubes long and 3 cubes wide. So, you need 4 × 3, or 12 cubes to cover the bottom layer.

3 cubes
4 cubes

To fill the box, you need two layers of 12 cubes.

12 cubes

∴ So, you need 2 × 12 = 24 cubes to fill the box.

12 cubes

On Your Own

Now You're Ready
Exercises 3–5

Find the number of cubes it takes to fill the box.

1.

2.

Remember
A three-dimensional figure has length, width, and height.

A **rectangular prism** is a three-dimensional figure that has six rectangular sides. The **volume** of a prism is a measure of the amount of space that it occupies. Volume is measured in **cubic units**.

🔑 Key Idea

Volume of a Rectangular Prism

Words The volume V of a rectangular prism is the product of its length ℓ, width w, and height h.

Algebra $V = \ell w h$

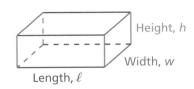
Height, h
Width, w
Length, ℓ

314 Chapter 7 Equations Multi-Language Glossary at BigIdeasMath.com.

Laurie's Notes

Introduction

Connect
- **Yesterday:** Students gained an intuitive understanding of how to solve for the missing dimension(s) of a prism.
- **Today:** Students will define volume of a rectangular prism and use the formula to solve different problems.

Motivate
- Place 24 cubes on a desk. Tell the students that you have 24 cubes and you want to build a prism so that none are left over. Ask a volunteer to build a prism.
- ? "What are the dimensions of the prism?" Possible prisms: $1 \times 1 \times 24$, $1 \times 2 \times 12$, $1 \times 3 \times 8$, $1 \times 4 \times 6$, $2 \times 2 \times 6$, and $2 \times 3 \times 4$ Record the dimensions on the board.
- ? "What is the volume of the prism?" Students should immediately say 24 cubic units because it was built with 24 cubes. Students will probably take the time to multiply the three dimensions.
- Use the dimensions as an example to talk about layering. For instance, assume the prism built has dimensions 4 by 3 by 2. The bottom layer has 12 cubes. The next layer doubles the number of cubes, so there are 24.
- **Model:** Use the cubes to model each of the dimensions you listed on the board.

Lesson Notes

Example 1
- Having cubes for students to manipulate is very helpful.
- **Big Idea:** Volume is a filling process. **Volume** is the amount of space inside a figure. Volume is measured in cubic units.
- **Misconception:** The largest face does not have to be the base. The prism can be rotated around and placed on a different face. In a prism, the words length, width, and height, are interchangeable. You do however often think of the height as a vertical distance perpendicular to the base. An alternate word for height is depth.

On Your Own
- **Neighbor Check:** Have students work independently and then have their neighbor check their work. Have students discuss any discrepancies.

Key Idea
- Write the Key Idea.
- Discuss the algebraic representation. When you write the formula, you generally write $V = \ell w h$. You do not write the multiplication symbol. Remind students that multiplication is still performed.

Goal Today's lesson is determining a missing dimension of a **rectangular prism**, given its **volume** and some of the dimensions.

Lesson Materials	
Introduction	Textbook
• 24 unit cubes	• 30 unit cubes

Start Thinking! and Warm Up

Tell whether *volume* applies to the situation.
- **A** Wrapping paper used to wrap a gift
- **B** Milk poured into a glass
- **C** Air needed to blow up a ball
- **D** Fabric for making a quilt
- **E** How much will fit in a suitcase

Describe another situation in which volume would apply.

Extra Example 1
How many cubes do you need to fill the box? **12 cubes**

On Your Own
1. 30 cubes
2. 27 cubes

T-314

Extra Example 2

Find the volume of a rectangular prism with a length of 3 inches, a width of 7 inches, and a height of 2 inches. 42 in.3

On Your Own

3. 144 ft^3
4. 704 yd^3

Extra Example 3

An MP3 player has a depth (length) of 0.6 centimeter, a width of 4 centimeters, and a volume of 21.6 cubic centimeters. Write and solve an equation to find the height of the MP3 player.
$21.6 = 0.6(4)h$, 9 cm

On Your Own

5. 6 in.
6. 12 cm

Differentiated Instruction

Kinesthetic

Divide the class into two groups. Give each group a piece of yarn or string 36 inches long. Have the first group use all of the yarn to form a rectangle with the largest possible area. Have the second group use all of the yarn to form a rectangle with the smallest possible area, with one side at least 1 inch long. Ask students:
What are the dimensions of your rectangle? 9 in. by 9 in.; 1 in. by 17 in.
If the rectangle was the base of a prism having a height of 5 inches, what would the volume of the prism be? 405 in.3; 85 in.3
What is the difference between the two volumes? 320 in.3

Laurie's Notes

Example 2
- Work through the example.
- **Extension:** Ask students to describe the difference between a foot, a square foot, and a cubic foot. They can use their hands to show the difference.

On Your Own
- **Think-Pair-Share:** Students should read each question independently and then work with a partner to answer the questions. When they have answered the questions, the pair should compare their answers with another group and discuss any discrepancies.

Example 3
- ❓ "What information is given for the computer tower?" length, width, volume
- ❓ "How could we solve for the height?" Students should connect this to the problems done in the previous section when they wrote an equation to solve for the missing dimension.
- Note how the formula is written: $\ell w h$ implies multiplication. The volume is the product of the three variables.
- Remind students to label their answers with the correct units.

On Your Own
- Ask volunteers to share their work at the board.

Closure
- **Exit Ticket:** A pizza box has dimensions of 12 inches by 12 inches by 1.5 inches. Find the volume. 216 in.3

The Dynamic Planning Tool
Editable Teacher's Resources at *BigIdeasMath.com*

EXAMPLE 2 **Finding the Volume of a Rectangular Prism**

Find the volume of the rectangular prism.

$V = \ell w h$ Write formula for volume.

$ = 7(4)(5)$ Substitute values.

$ = 140$ Multiply.

∴ The volume is 140 cubic meters.

On Your Own

Now You're Ready
Exercises 6–8

Find the volume of the rectangular prism.

3.

4.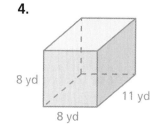

EXAMPLE 3 **Finding a Missing Dimension of a Rectangular Prism**

Write and solve an equation to find the height of the computer tower.

Volume = 1792 in.³

$V = \ell w h$ Write formula for volume.

$1792 = 7(16)h$ Substitute 1792 for V, 7 for ℓ, and 16 for w.

$1792 = 112h$ Simplify.

$\dfrac{1792}{112} = \dfrac{112h}{112}$ Divide each side by 112.

$16 = h$ Simplify.

∴ The height is 16 inches.

On Your Own

Now You're Ready
Exercises 9–13

Find the missing dimension of the rectangular prism.

5. Volume = 72 in.³

6. Volume = 1380 cm³

7.6 Exercises

Vocabulary and Concept Check

1. **VOCABULARY** What types of units are used to describe volume?
2. **VOCABULARY** What types of units are used to describe area?

Practice and Problem Solving

Find the number of cubes it takes to fill the box.

3.

4.

5.

Find the volume of the rectangular prism.

6.
 8 in., 6 in., 12 in.

7.
 9 ft, 5 ft, 5 ft

8.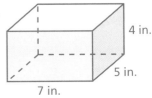
 4 in., 5 in., 7 in.

Write and solve an equation to find the missing dimension of the rectangular prism.

9. Volume = 1620 cm³

9 cm, 9 cm, h

10. Volume = 220.5 cm³

7 cm, 7 cm, w

11. Volume = 177,500 mm³
142 mm, 10 mm, h

12. Volume = 646 in.³

2 in., w, 19 in.

13. Volume = 936 in.³

3 in., 13 in., ℓ

14. **BUG TRAP** What is the volume of the trap used to collect bugs?

14 in., 12 in., 12 in.

316 Chapter 7 Equations

Assignment Guide and Homework Check

Level	Day 1 Activity Assignment	Day 2 Lesson Assignment	Homework Check
Basic	9–11, 21–24	1–8, 12–14	4, 6, 12, 14
Average	9–11, 21–24	1–3, 5, 7, 12–16	7, 12, 15, 16
Advanced	9–11, 21–24	1, 2, 12–20	12, 15, 16, 18

For Your Information
- **Exercise 17** A shadow box is an enclosed case that can hold pictures or other objects.

Common Errors
- **Exercises 3–5** Some students may not have good spatial reasoning skills and will struggle determining how many cubes can fill the box. Give them unit cubes that they can use to build the picture and count how many cubes will fit.
- **Exercises 6–8** Students may write the incorrect units. Remind them that volume has units cubed because there are three dimensions.
- **Exercises 9–13** Students may forget to simplify before solving for the variable and may lose or forget about one of the dimensions. For example,
$$\frac{1620}{9} = \frac{(9)(9)h}{9}$$
$$180 = h.$$
Remind them to simplify their equations before solving for a variable.

Vocabulary and Concept Check
1. cubic units
2. square units

Practice and Problem Solving
3. 32 cubes
4. 30 cubes
5. 27 cubes
6. 576 in.3
7. 225 ft^3
8. 140 in.3
9. $1620 = h \cdot 9 \cdot 9$; 20 cm
10. $220.5 = 7 \cdot 7 \cdot w$; 4.5 cm
11. $177,500 = 142 \cdot 10 \cdot h$; 125 mm
12. $646 = 2 \cdot 19 \cdot w$; 17 in.
13. $936 = 3 \cdot 13 \cdot \ell$; 24 in.
14. 2016 in.3

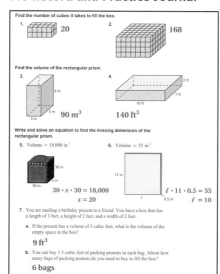

7.6 Record and Practice Journal

Technology For the Teacher
Answer Presentation Tool
QuizShow

Practice and Problem Solving

15. 216 cubes

16. 777.6 in.³

17. See *Taking Math Deeper*.

18. 1152 cm³

19. a. *Sample answer:* 324 in.³

 b. no; The container only holds 196 cubic inches.

20. 3080 in.³

Fair Game Review

21. yes **22.** no

23. no **24.** A

Mini-Assessment

Find the volume of the rectangular prism.

1. 32 ft³

2. 216 cm³

3. The volume of the cell phone is 75 cubic centimeters. Write and solve an equation to find the height of the cell phone. $75 = 10(5)h$; 1.5 cm

Taking Math Deeper

Exercise 17

There are many possible answers to this question. The most distinctive shell in the shadow box is the chambered nautilus shell. On the Internet, you will discover that nautilus shells range from 6 inches to 10 inches.

 Estimate the length of the nautilus shell. Use this to estimate the length and width of the shadow box.

 a. Draw a box with your dimensions.

 Find the volume of the box.

b. Volume = 18(11)(4)
 = 792 in.³

Answers will vary. However, answers less than 400 cubic inches are unrealistic.

Project

Choose at least 10 items and make a shadow box.

Reteaching and Enrichment Strategies

If students need help...	If students got it...
Resources by Chapter • Practice A and Practice B • Puzzle Time Record and Practice Journal Practice Differentiating the Lesson Lesson Tutorials Skills Review Handbook	Resources by Chapter • Enrichment and Extension • School-to-Work • Financial Literacy • Technology Connection Start the next section

15. **CUBE** How many 1-centimeter cubes are needed to create a cube with a side length of 6 centimeters?

16. **FISH TANK** The fish tank is 80% full. After adding stones to the bottom of the tank, the tank is 92% full. What is the volume of the stones?

17. **CHOOSE TOOLS** Use the Internet or some other resource to estimate the sizes of the seashells in the shadow box.

 a. Draw a box that will hold all of the shells.
 b. Find the volume of your box.

18. **GEOMETRY** The area of the shaded face is 96 square centimeters. What is the volume of the rectangular prism?

19. **FOOD STORAGE**

 a. Estimate the amount of macaroni and cheese left in the casserole.
 b. Will the macaroni and cheese fit in the storage container? Explain your reasoning.

20. **Critical Thinking** The outside dimensions of a rectangular cooler without its top are 26 inches by 14 inches by 16 inches. The cooler is 2 inches thick. What is the volume inside the cooler?

Fair Game Review What you learned in previous grades & lessons

Tell whether the given value is a solution of the equation. *(Section 7.2)*

21. $x + 17 = 24$; $x = 7$
22. $6x = 72$; $x = 13$
23. $x - 19 = 42$; $x = 21$

24. **MULTIPLE CHOICE** What is the area of the semicircular region? *(Section 6.3)*

 Ⓐ 127.17 ft²
 Ⓑ 254.34 ft²
 Ⓒ 506.68 ft²
 Ⓓ 1017.36 ft²

7.6b Volume and Surface Area

Check It Out
Lesson Tutorials
BigIdeasMath.com

To find the volume of a rectangular prism, multiply the number of cubes needed to fill the prism by the volume of one of the cubes.

EXAMPLE 1 Finding the Volume of a Rectangular Prism

Use cubes to find the volume of the rectangular prism.

Use cubes with an edge length of $\frac{1}{4}$ inch.

The bottom layer is 6 cubes long and 2 cubes wide. So, you need 6 × 2, or 12 cubes to cover the bottom layer.

Check
$V = Bh = \ell wh$
$= \left(1\frac{1}{2}\right)\left(\frac{1}{2}\right)\left(\frac{3}{4}\right)$
$= \left(\frac{3}{2}\right)\left(\frac{1}{2}\right)\left(\frac{3}{4}\right)$
$= \frac{9}{16}$ in.³ ✓

To fill the prism, you need three layers of 12 cubes. So, you need 3 × 12, or 36 cubes.

The volume of each cube is $\left(\frac{1}{4}\right)^3 = \frac{1}{64}$ cubic inch.

∴ So, the volume of the prism is $36 \times \frac{1}{64} = \frac{9}{16}$ cubic inch.

Practice

Use cubes with the given edge length to find the volume of the rectangular prism. Check your answer using the volume formula.

1. Edge length: $\frac{1}{5}$ cm

2. Edge length: $\frac{1}{6}$ ft

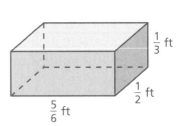

3. **SINK** A sink is shaped like a rectangular prism. How much water can the sink hold?

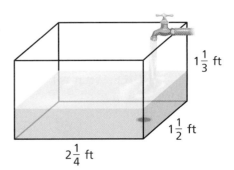

Laurie's Notes

Introduction

Connect
- **Yesterday:** Students used a formula for the volume of a rectangular prism.
- **Today:** Students will find volumes of prisms and use nets to find surface areas of prisms and pyramids.

Motivate
- Have available cardboard prisms and pyramids that have been folded.
- Holding the assembled solids, ask students to visualize and then describe the cardboard nets that result when the solids are "unfolded."
- Unfold one or more of the solids so that students can see the net of the prism or pyramid.
- Explain the connection between the cardboard net and the surface area of the prism or pyramid.

Lesson Notes

Example 1
- ❓ Arrange twelve $\frac{1}{4}$-inch cubes to make a rectangular prism that is 6 cubes by 2 cubes by 1 cube. "The edge length of each cube is $\frac{1}{4}$ inch. What are the dimensions of the prism?" $1\frac{1}{2}$ in. by $\frac{1}{2}$ in. by $\frac{1}{4}$ in.
- Use this first layer of cubes to explain that the cube edge length must be a factor of each dimension of the prism. In other words, you want to be able to build the prism with cubes that are the same size.
- Now build the three layers of the prism using the $\frac{1}{4}$-inch cubes.
- ❓ "In the example, how many $\frac{1}{4}$-inch cubes did it take to make the prism?" 36
- ❓ "What is the volume of each cube? Explain." $\frac{1}{64}$ cubic inch because each edge is $\frac{1}{4}$ inch and $\left(\frac{1}{4}\right)^3 = \frac{1}{64}$
- Discuss the Check box. In the example, the volume is found by multiplying the volume of each cube by the number of cubes. In the Check, the volume is found by multiplying the three dimensions.

Practice
- **Connection:** Students should be thinking about common factors. For example, in Exercise 3: What divides evenly into $2\frac{1}{4}$, $1\frac{1}{2}$, and $1\frac{1}{3}$? $\frac{1}{12}$

Goal
Today's lesson is finding volumes of prisms and **surface areas** of prisms and pyramids.

Lesson Materials

Introduction	Textbook
• cardboard prisms and pyramids	• $\frac{1}{4}$-inch cubes • polygon frames (optional) • file folders (optional)

Warm Up

Lesson 7.6b Warm Up
For use before Lesson 7.6b.

Find the area of the figure.

Extra Example 1
Use cubes with an edge length of $\frac{1}{3}$ inch to find the volume of the rectangular prism.

$4\frac{4}{9}$ in.³

Practice
1. $\frac{24}{125}$ cm³
2. $\frac{5}{36}$ ft³
3. $4\frac{1}{2}$ ft³

Record and Practice Journal Practice
See Additional Answers.

T-317A

Extra Example 2

Find the surface area of the rectangular prism.

150 cm²

 Practice

4. 258 m²
5. 286 ft²
6. 180 in.²

Laurie's Notes

Discuss
- Review faces of a solid. A face is a flat surface of a solid.
- Refer to the Motivate from the beginning of class. The nets are revealed when you unfold the prism. Each face of the prism has an area.
- Define surface area of a three-dimensional figure.

Key Idea
- **Teaching Tip:** Have several boxes available to disassemble.
- A rectangular prism has 6 faces which are informally referred to as the two sides, front, back, top, and bottom. Because the orientation of the prism can change, the labels are arbitrary.

Example 2
? Hold a cardboard box and ask, "How can you find the total amount of cardboard used to make this box?" Find the sum of the areas of the six faces.
- Work through the example as shown. Students may find it easier to consider the three pairs of opposite faces instead of finding the area of each of the six faces separately.
? "What units are used to label surface area?" square units

Practice
- **Common Error:** Students may forget to include the area of one or more of the faces when finding the surface area.
- Verify that decimal (Exercise 5) and fraction (Exercise 6) computations are done correctly.

The **surface area** S of a three-dimensional figure is the sum of the areas of its faces. Surface area is measured in **square units**. You can find the surface area of a three-dimensional figure by using a two-dimensional representation of the figure called a **net**.

Key Idea

Net of a Rectangular Prism

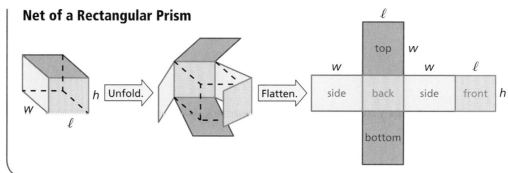

EXAMPLE 2 — Finding the Surface Area of a Rectangular Prism

Find the surface area of the rectangular prism.

Use a net to find the area of each face.

Top: $7 \cdot 4 = 28$
Bottom: $7 \cdot 4 = 28$
Front: $7 \cdot 3 = 21$
Back: $7 \cdot 3 = 21$
Side: $4 \cdot 3 = 12$
Side: $4 \cdot 3 = 12$

Find the sum of the areas of the faces.

Surface area = Area of top + Area of bottom + Area of front + Area of back + Area of a side + Area of a side

$S = 28 + 28 + 21 + 21 + 12 + 12$
$= 122$

∴ The surface area is 122 square inches.

Practice

Find the surface area of the rectangular prism.

4.

5.

6.

Volume and Surface Area **317B**

Key Idea

Net of a Triangular Prism

A **triangular prism** is a three-dimensional figure that has two triangular faces and three rectangular faces.

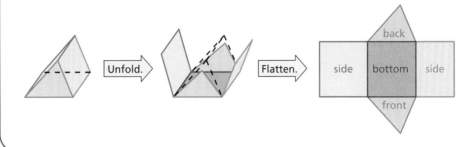

Remember

The area A of a triangle with base b and height h is $A = \frac{1}{2}bh$.

EXAMPLE 3 Finding the Surface Area of a Triangular Prism

Find the surface area of the triangular prism.

Use a net to find the area of each face.

Bottom: $12 \cdot 8 = 96$

Front: $\frac{1}{2} \cdot 12 \cdot 5 = 30$

Back: $\frac{1}{2} \cdot 12 \cdot 5 = 30$

Side: $13 \cdot 8 = 104$

Side: $8 \cdot 5 = 40$

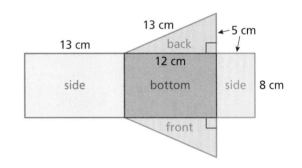

Find the sum of the areas of the faces.

Surface area = Area of bottom + Area of front + Area of back + Area of a side + Area of a side

$S = 96 + 30 + 30 + 104 + 40$

$= 300$

∴ The surface area is 300 square centimeters.

Practice

Find the surface area of the triangular prism.

7.

8.

9.

Laurie's Notes

Key Idea
- Have a net available as a visual, whether it is a cardboard box or a triangular prism made from polygon frames.
- Discuss with students that a triangular prism has two opposite faces that are triangles, and the rest of the faces are rectangles.

Example 3
- Note that the net is a visual reminder to find the area of each face. Color-coding the faces helps students keep track of their work.
- ❓ "How many faces does a triangular prism have?" 5
- ❓ "How do you find the area of a triangle?" $A = \dfrac{1}{2}bh$
- Encourage students to write a verbal model for the surface area of the prism and then substitute the areas of the faces as they are computed.

Practice
- Exercise 7 involves a right triangle. Some students may need to be reminded that one side is the base and one side is the height.
- Exercise 9 includes decimal dimensions. Students may find it helpful to think of the area formula for triangles as $A = (0.5)bh$.
- **Common Error:** Students may forget to include the area of one or more of the faces when finding the surface area.

Extra Example 3
Find the surface area of the triangular prism.

545 in.²

Practice
7. 60 yd²
8. 420 m²
9. 299.32 ft²

Extra Example 4

Find the surface area of the square pyramid.

256 mm²

Practice

10. 16 ft²
11. 75 cm²
12. 44.8 in.²

Mini-Assessment

1. Use cubes with an edge length of $\frac{1}{8}$ meter to find the volume of the rectangular prism. $\frac{21}{128}$ m³

Find the surface area of the solid.

2. 58 in.²

3. 576 ft²

4. 96 cm²

Laurie's Notes

Key Idea

- Have a net available as a visual, whether it is a cardboard pyramid or a square pyramid made from polygon frames. The cardboard version could be made from file folders.

Example 4

- Note that the net is a visual reminder to find the area of each face. Color-coding the faces helps students keep track of their work.
- ? "How many faces does a square pyramid have?" 5
- ? "Are any of the faces the same?" yes; the four triangles
- Encourage students to write a verbal model for the surface area of the pyramid and then substitute the areas of the faces as they are computed.
- Some students may wish to find the area of one triangular face and then multiply the answer by 4.

Practice

- **Common Error:** Students may forget to add the area of the base when finding the surface area of a pyramid.

Closure

- **Writing:** Hold a prism or pyramid from the Motivate section of today's lesson and ask students to write about how they would find the surface area of the prism or pyramid. Look for: finding the areas of the faces and adding those areas together.

The Dynamic Planning Tool
Editable Teacher's Resources at *BigIdeasMath.com*

Key Idea

Net of a Square Pyramid

A **square pyramid** is a three-dimensional figure that has one square face and four identical triangular faces.

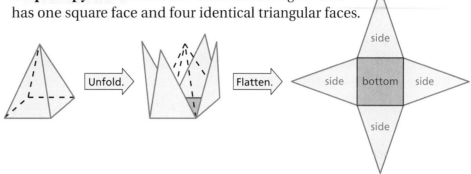

EXAMPLE 4 Finding the Surface Area of a Square Pyramid

Find the surface area of the square pyramid.

Use a net to find the area of each face.

Bottom: $7 \cdot 7 = 49$

Side: $\frac{1}{2} \cdot 7 \cdot 10 = 35$

Side: $\frac{1}{2} \cdot 7 \cdot 10 = 35$

Side: $\frac{1}{2} \cdot 7 \cdot 10 = 35$

Side: $\frac{1}{2} \cdot 7 \cdot 10 = 35$

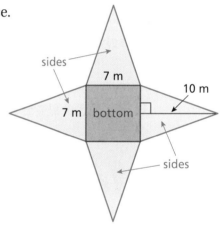

Find the sum of the areas of the faces.

Surface area = Area of bottom + Area of a side + Area of a side + Area of a side + Area of a side

$S = 49 + 35 + 35 + 35 + 35$

$= 189$

∴ The surface area is 189 square meters.

Practice

Find the surface area of the square pyramid.

10.

11.

12.

Volume and Surface Area **317D**

7.5–7.6 Quiz

1. Find the number of cubes it takes to fill the box. *(Section 7.6)*

2. Find the volume of the rectangular prism. *(Section 7.6)*

Write and solve an equation to find *x*. Check your solution. *(Section 7.5)*

3. Area = 15 in.²

4. Area = 112 yd²

5. Circumference = 157 in.

Write and solve an equation to find the missing dimension of the rectangular prism. *(Section 7.6)*

6. Volume = 1620 in.³

7. Volume = 154 in.³

8. Volume = 4250 in.³

9. **BATH MAT** A circular bath mat has a circumference of 75.36 inches. What is the diameter of the mat? *(Section 7.5)*

10. **TOY CHEST** A toy company sells two different toy chests. The toy chests have different dimensions, but the same volume. What is the width *w* of toy chest 2? *(Section 7.6)*

11. **CUBE** How many 1-inch cubes are needed to create a cube with a side length of 7 inches? *(Section 7.6)*

12. **FENCING** You put 50 feet of fencing around the rectangular garden. What is the length ℓ of the garden? *(Section 7.5)*

Alternative Assessment Options

Math Chat **Student Reflective Focus Question**
Structured Interview **Writing Prompt**

Math Chat
Ask students to use their own words to summarize what they know about solving one- and two-step equations. Be sure that they include examples. Select students at random to present their summary to the class.

Study Help Sample Answers
Remind students to complete Graphic Organizers for the rest of the chapter.

7. Plane Figures

Examples	Non-Examples
○	⊔
▭	▱
△	◯
⬭	✈

8. Rectangular Prisms

Examples	Non-Examples
▱	△
▱	△
▱	◯
▱	◯

9. Volume

Examples	Non-Examples
24 cubic inches	24 square inches
10 cubic meters	10 meters
39 cubic centimeters	$C = \pi \times d$
$V = \ell \times w \times h$	$A = \ell \times w$

Answers
1. 32 cubes
2. 336 ft³
3. $15 = 5x$; 3 in.
4. $112 = \frac{1}{2}(16x)$; 14 yd
5. $157 = (2)3.14x$; about 25 in.
6. $1620 = 12 \cdot 15 \cdot w$; 9 in.
7. $154 = 11 \cdot 2 \cdot \ell$; 7 in.
8. $4250 = 10 \cdot 25 \cdot h$; 17 in.
9. about 24 in.
10. 15 units
11. 343 cubes
12. 20 ft

Technology For the Teacher
Answer Presentation Tool

Reteaching and Enrichment Strategies

If students need help...	If students got it...
Resources by Chapter • Study Help • Practice A and Practice B • Puzzle Time Lesson Tutorials *BigIdeasMath.com* Practice Quiz Practice from the Test Generator	Resources by Chapter • Enrichment and Extension • School-to-Work Game Closet at *BigIdeasMath.com* Start the Chapter Review

Assessment Book

T-318

7.3 Solving Equations Using Multiplication or Division (pp. 290–295)

Solve $4c = 32$.

Undo the multiplication.

$4c = 32$ Write the equation.

$\dfrac{4c}{4} = \dfrac{32}{4}$ Divide each side by 4.

$c = 8$ Simplify.

Check
$4c = 32$
$4(8) \stackrel{?}{=} 32$
$32 = 32$ ✓

∴ The solution is $c = 8$.

Exercises

Solve the equation. Check your solution.

7. $7 \cdot q = 42$
8. $k \div 2 = 26$
9. $\dfrac{a}{14} = 3$

7.4 Solving Two-Step Equations (pp. 296–303)

Solve $\dfrac{x}{2} - 3 = 5$.

$\dfrac{x}{2} - 3 = 5$ Write the equation.

Undo the subtraction. $\underline{+3 \quad +3}$ Add 3 to each side.

$\dfrac{x}{2} = 8$ Simplify.

Undo the division. $\dfrac{x}{2} \cdot 2 = 8 \cdot 2$ Multiply each side by 2.

$x = 16$ Simplify.

Check
$\dfrac{x}{2} - 3 = 5$
$\dfrac{16}{2} - 3 \stackrel{?}{=} 5$
$8 - 3 \stackrel{?}{=} 5$
$5 = 5$ ✓

∴ The solution is $x = 16$.

Exercises

Solve the equation. Check your solution.

10. $6q + 1 = 13$
11. $\dfrac{d}{3} + 2 = 9$
12. $110 = 4p - 50$
13. $8 = \dfrac{z}{3} + 8$
14. $6.8c + 5 = 25.4$
15. $\dfrac{t}{12} - 0.5 = 2$

16. You buy three CDs and a $15 DVD. You spend $45. Write and solve an equation to find the price of each CD.

Review Game

Equations

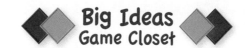

Materials per Group:
- a copy of the chapter test from the pupil's edition
- chalk or dry erase markers

Directions:
Divide the class into four teams. Team members gather in groups at the board. One member of each team works the first problem from the chapter test. Coaching is allowed from other team members. Check each student's work and award 2 points to the team if the work is correct. If the work needs to be corrected, award 1 point to the team after the student corrects it. Team members take turns working the problems and the process is repeated until the test is finished.

Who Wins?
The team with the most points wins.

**For the Student
Additional Practice**
- Lesson Tutorials
- Study Help (textbook)
- Student Website
 Multi-Language Glossary
 Practice Assessments

Answers

7. $q = 6$
8. $k = 52$
9. $a = 42$
10. $q = 2$
11. $d = 21$
12. $p = 40$
13. $z = 0$
14. $c = 3$
15. $t = 30$
16. $3x + 15 = 45$; \$10
17. $36 = 2x + 16$; 10 yd
18. $44 = \frac{1}{2}(11)x$; 8 cm
19. $62.8 = 2(3.14)x$; about 10 m
20. $312 = 13 \cdot 6 \cdot h$; 4 in.
21. $48 = 1 \cdot 8 \cdot w$; 6 in.

My Thoughts on the Chapter

What worked. . .

What did not work. . .

What I would do differently. . .

> **Teacher Tip**
> Not allowed to write in your teaching edition? Use sticky notes to record your thoughts.

7.5 Finding Dimensions of Plane Figures (pp. 306–311)

The area of the parallelogram is 12 square feet. What is its height?

$A = bh$ Write formula for area of a parallelogram.

$12 = 6h$ Substitute 12 for A and 6 for b.

$2 = h$ Divide each side by 6.

∴ The height is 2 feet.

Exercises

Write and solve an equation to find x. Check your solution.

17. Perimeter = 36 yd **18.** Area = 44 cm² **19.** Circumference = 62.8 m

7.6 Finding Dimensions of Prisms (pp. 312–317)

Write and solve an equation to find the length of the container.

$V = \ell wh$ Write formula for volume.

$840 = \ell(6)(14)$ Substitute 840 for V, 6 for w, and 14 for h.

$840 = 84\ell$ Simplify.

$10 = \ell$ Divide each side by 84.

∴ The length is 10 inches.

Exercises

Write and solve an equation to find the missing dimension of the rectangular prism.

20. Volume = 312 in.³ **21.** Volume = 48 in.³

Chapter Review **321**

7 Chapter Test

Write the word sentence as an equation.

1. 7 times a number s is 84.
2. 13 is one-third of a number m.

Solve the equation. Check your solution.

3. $15 = 7 + b$
4. $v - 6 = 16$
5. $5x = 70$
6. $\dfrac{m}{13} = 30$
7. $2b + 5 = 13$
8. $\dfrac{k}{3} - 7 = 9$

Write and solve an equation to find the missing dimension of the figure. Check your solution.

9. Perimeter = 42 mm

10. Area = 30 ft²

Write and solve an equation to find the missing dimension of the rectangular prism.

11. Volume = 960 in.³

12. Volume = 80 in.³

13. **SCHOOL DANCE** Each ticket to a school dance is $4. The total amount collected in ticket sales is $332. Write and solve an equation to find the number of students attending the dance.

14. **PLACE MAT** A circular place mat has a circumference of 56.52 centimeters. What is the diameter of the place mat?

15. **DICTIONARIES** You carry 7 boxes of dictionaries to the library and the librarian carries 8 boxes. Each box has the same weight. The total weight of all the boxes is 210 pounds. Write and solve an equation to find the weight of each box.

16. **GEOMETRY** The prism is made of identical cubes. The area of the shaded face is 72 square millimeters. What is the volume of the prism?

Test Item References

Chapter Test Questions	Section to Review	Common Core State Standards
1, 2	7.1	6.EE.2b, 6.EE.7
3, 4	7.2	6.EE.5, 6.EE.7
5, 6, 13	7.3	6.EE.5, 6.EE.7
7, 8, 15	7.4	6.EE.2b, 6.EE.5, 6.EE.7
9, 10, 14	7.5	6.EE.7, 6.G.1
11, 12, 16	7.6	6.G.2, 6.G.4

Test-Taking Strategies

Remind students to quickly look over the entire test before they start so that they can budget their time. When working with equations, they need to write all numbers and variables clearly. Some numbers and variables are easy to confuse. Have them pay attention to 6, *s*, 4, 9, 7, and *x*. Remind students to line up terms in each step and to not crowd their work. Have them use the **Stop** and **Think** strategy before they answer each question.

Common Assessment Errors

- **Exercises 1 and 2** Students may write parts of the equation in the wrong order or forget parts of the equation. Remind them that they should be writing *equations*, so they should have equal signs in their answers.
- **Exercises 3–6** Students may use the same operation that is in the equation, instead of the inverse operation, to solve for the variable. Remind them to check their solutions in the original equation.
- **Exercises 7 and 8** Students may divide or multiply by the coefficient first instead of adding or subtracting first. Tell them that while this is a valid method, they must remember to divide or multiply *each part* of the equation by the coefficient. Suggest that they begin away from the variable and work towards it.
- **Exercises 9–12** Students may use an incorrect formula when solving for the variable. For example, in Exercise 9, instead of using the formula for perimeter, they may use the formula for area. Tell them to read the question carefully and to write the correct formula before substituting values.

Reteaching and Enrichment Strategies

If students need help. . .	If students got it. . .
Resources by Chapter • Practice A and Practice B • Puzzle Time Record and Practice Journal Practice Differentiating the Lesson Lesson Tutorials Practice from the Test Generator Skills Review Handbook	Resources by Chapter • Enrichment and Extension • School-to-Work • Financial Literacy Game Closet at *BigIdeasMath.com* Start Standardized Test Practice

Answers

1. $7s = 84$
2. $13 = \frac{1}{3} \cdot m$
3. $b = 8$
4. $v = 22$
5. $x = 14$
6. $m = 390$
7. $b = 4$
8. $k = 48$
9. $42 = 2x + 18$; 12 mm
10. $30 = \frac{1}{2}(10x)$; 6 ft
11. $960 = 10 \cdot 6 \cdot h$; 16 in.
12. $80 = 5 \cdot 2 \cdot \ell$; 8 in.
13. $4x = 332$; 83 students
14. about 18 cm
15. $7x + 8x = 210$; 14 lbs
16. 432 mm^3

Assessment Book

Test-Taking Strategies
Available at *BigIdeasMath.com*

After Answering Easy Questions, Relax
Answer Easy Questions First
Estimate the Answer
Read All Choices before Answering
Read Question before Answering
Solve Directly or Eliminate Choices
Solve Problem before Looking at Choices
Use Intelligent Guessing
Work Backwards

About this Strategy

When taking a multiple choice test, be sure to read each question carefully and thoroughly. After skimming the test and answering the easy questions, stop for a few seconds, take a deep breath, and relax. Work through the remaining questions carefully, using your knowledge and test-taking strategies. Remember, you already completed many of the questions on the test!

Answers

1. D
2. F
3. 1.5 or $\frac{3}{2}$
4. D

Technology For the Teacher
Big Ideas Test Generator

Item Analysis

1. **A.** The student adds instead of subtracting and then divides instead of multiplying.
 B. The student performs the two operations in the wrong order.
 C. The student divides instead of multiplying.
 D. Correct answer

2. **F.** Correct answer
 G. The student divides before adding, dividing only $4g$ by 2 and finding $\frac{4 \times 6}{2} + 8$.
 H. The student thinks that the g is a digit, as opposed to $4g$ representing multiplication, and finds $\frac{46 + 8}{2}$.
 I. The student divides before adding, dividing only 8 by 2 and finding $4 \times 6 + \frac{8}{2}$.

3. **Gridded Response:** Correct answer: 1.5 or $\frac{3}{2}$
 Common Error: The student subtracts 4 from 6 and gets an answer of 2.

4. **A.** The student subtracts 12 from 26.
 B. The student takes half of 26.
 C. The student multiplies 12 by 2 and subtracts the product from 26.
 D. Correct answer

5. **F.** The student uses the digits of the decimal number as the denominator.
 G. Correct answer
 H. The student uses the digits of the decimal number as the numerator and the denominator.
 I. The student uses a numerator and denominator whose product is equal to the decimal number.

6. **A.** The student chooses an incorrect operation and does not convert the percent into a decimal number.
 B. The student chooses an incorrect operation and incorrectly converts the percent into a decimal number.
 C. The student incorrectly converts the percent into a decimal number.
 D. Correct answer

7 Standardized Test Practice

1. Which description is a correct way to solve the equation below? *(6.EE.7)*

 $$\frac{x}{6} + 15 = 57$$

 A. Add 15 to both sides and then divide both sides by 6.

 B. Multiply both sides by 6 and then subtract 15 from both sides.

 C. Subtract 15 from both sides and then divide both sides by 6.

 D. Subtract 15 from both sides and then multiply both sides by 6.

Test-Taking Strategy
After Answering Easy Questions, Relax

"After answering the easy questions, relax and try the harder ones. For this, $2x = 12$, so $x = 6$ hyenas."

2. What is the value of the expression below when $g = 6$? *(6.EE.2c)*

 $$\frac{4g + 8}{2}$$

 F. 16 **H.** 27

 G. 20 **I.** 28

3. What is the value of m that makes the equation below true? *(6.EE.7)*

 $$4m = 6$$

4. A rectangle and one of its dimensions are shown below.

 12 in. Not drawn to scale

 The perimeter of the rectangle is 26 inches. What is the value of x? *(6.EE.7)*

 A. 14 in. **C.** 2 in.

 B. 13 in. **D.** 1 in.

Standardized Test Practice **323**

5. Which fraction is equivalent to 0.35? *(6.NS.1)*

 F. $\dfrac{1}{35}$

 G. $\dfrac{7}{20}$

 H. $\dfrac{3}{5}$

 I. $\dfrac{5}{7}$

6. Helene was finding the percent of a number in the box below. *(6.RP.3c)*

 > What is 8% of 24?
 >
 > x = 24 ÷ 0.08
 >
 > x = 300

 What should Helene do to correct the error that she made?

 A. Divide 24 by 8.

 B. Divide 24 by 0.8.

 C. Multiply 24 by 0.8.

 D. Multiply 24 by 0.08.

7. Stock prices on the New York Stock Exchange used to be fractions with denominators of 2, 4, and 8 instead of decimals. What would be the total cost of 100 shares (units) of a stock with a price of 16\dfrac{5}{8}$ per share? *(6.NS.1)*

 F. $1600.58

 G. $1600.63

 H. $1658.00

 I. $1662.50

8. What is the value of *p* that makes the equation below true? *(6.EE.7)*

 $$25 = 5p - 15$$

9. The cost of a bowling party for 12 friends was $133. After using a coupon for $25 off the cost of the party, the 12 friends shared the remaining cost equally.

 Let *f* represent the amount of money, in dollars, that each friend paid. Which equation can be used to correctly determine the value of *f*? *(6.EE.7)*

 A. $12f + 25 = 133$

 B. $12f - 25 = 133$

 C. $\dfrac{f}{12} + 25 = 133$

 D. $\dfrac{f}{12} - 25 = 133$

Item Analysis (continued)

7. **F.** The student multiplies 100 by 16 only and then uses the numerator and denominator as the digits in the decimal part.

 G. The student multiplies 100 by 16 only and then uses the rounded decimal equivalent for $\frac{5}{8}$ as the decimal part.

 H. The student thinks that $16\frac{5}{8} = 16.58$ and then multiplies this number by 100.

 I. Correct answer

8. **Gridded Response:** Correct answer: 8

 Common Error: The student subtracts 15 from both sides and then divides both sides by 5, getting an answer of 2.

9. **A.** Correct answer

 B. The student misinterprets the inverse operation associated with the coupon.

 C. The student misinterprets the inverse operation associated with sharing.

 D. The student misinterprets the inverse operation associated with sharing, as well as the inverse operation associated with the coupon.

10. **F.** The student thinks that the fraction is equal to the percent.

 G. The student uses the numerator and denominator as digits of the percent.

 H. The student chooses an approximate answer that is not the exact answer choice.

 I. Correct answer

11. **A.** The student thinks that decimals are less than fractions, which are less than percents. The student misunderstands place value and thinks that 0.6 is less than 0.08. The student misunderstands denominators and thinks that $\frac{1}{7}$ is less than $\frac{1}{10}$.

 B. Correct answer

 C. The student orders the numbers based on 6, 7, 8, 9, 10, 12.

 D. The student thinks that $\frac{1}{7} = 0.07$ and that $9\% = 0.9$.

Answers

5. G
6. D
7. I
8. 8
9. A

Answers

10. I
11. B
12. H
13. *Sample answer:*
 length = 6, width = 6;
 length = 18, width = 2

Answers for Extra Examples

1. **A.** The student finds the product of 0.6 and 15 and then aligns the decimal point of the product with the decimal point of 0.6.

 B. The student ignores the decimal point in 0.6 and finds the quotient of 15 and 6.

 C. The student finds the product of 0.6 and 15.

 D. Correct answer

2. Correct answer: $19.14

 Common Error: The student multiplies the whole numbers together and the decimal parts together and finds $1 \times 15 + 2 \times 0.95$, which equals 16.90.

Item Analysis (continued)

12. **F.** The student chooses a line plot with a median of 7 but a mode of 9.

 G. The student chooses a line plot with a mode of 7 but a median of 6.

 H. Correct answer

 I. The student chooses a line plot with a maximum count of 7 and a middle row of 7 **X**s.

13. **2 points** The student demonstrates a thorough understanding of finding missing measurements in a right rectangular prism when given its volume and its height. The student correctly finds two pairs of measurements, in inches, whose product is 36, such as 6 inches \times 6 inches and 18 inches \times 2 inches.

 1 point The student demonstrates a partial understanding of finding missing measurements in a right rectangular prism when given its volume and its height. The student provides some correct work and explanation toward finding two correct pairs of measurements.

 0 points The student demonstrates insufficient understanding of finding missing measurements in a right rectangular prism when given its volume and its height. The student is unable to make any meaningful progress toward finding the missing length and width.

Extra Examples for Standardized Test Practice

1. What number belongs in the box to make the equation true?

 $0.6 \times \boxed{} = 15$

 A. 0.9 **C.** 9
 B. 2.5 **D.** 25

2. Ryan bought 1.2 pounds of chocolates for $15.95 per pound. How much did Ryan pay, in dollars?

10. What percent discount is the store offering?
 (6.RP.3c)

 F. $\frac{1}{3}$% H. 33%

 G. 13% I. $33\frac{1}{3}$%

11. Which list of numbers is in order from least to greatest? *(6.NS.7a)*

 A. 0.6, 0.08, $\frac{1}{7}$, $\frac{1}{10}$, 9%, 12%

 B. 0.08, 9%, $\frac{1}{10}$, 12%, $\frac{1}{7}$, 0.6

 C. 0.6, $\frac{1}{7}$, 0.08, 9%, $\frac{1}{10}$, 12%

 D. $\frac{1}{7}$, 0.08, $\frac{1}{10}$, 12%, 0.6, 9%

12. Which line plot represents a set of data that has a median of 7 and a mode of 7?
 (6.SP.5c)

 F.

 G.

 H.

 I.
 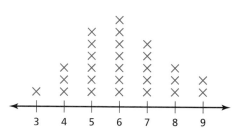

13. A right rectangular prism has a volume of 324 cubic inches and a height of 9 inches. Determine *two* possible pairs of measurements, in inches, for the length and width of the base of the prism. For *each* pair of measurements, show your work and explain your reasoning. *(6.G.2)*

Standardized Test Practice 325

8 Inequalities

8.1 Writing and Graphing Inequalities

8.2 Solving Inequalities Using Addition or Subtraction

8.3 Solving Inequalities Using Multiplication or Division

8.4 Solving Two-Step Inequalities

"If you reached into your water bowl and found more than $20..."

"And then reached into your cat food bowl and found more than $40..."

"What would you have?"

"Dear Precious Pet World: Your ad says 'Up to 75% off on selected items'."

"I select Yummy Tummy Bacon-Flavored Dog Biscuits."

Connections to Previous Learning

- Describe mathematical relationships using equations and visual representations.

- Use properties of equality to solve numerical and real-world situations.

- Write, solve, and graph one-step and two-step linear inequalities.

Math in History

Primitive cultures from all over the world tended to develop counting methods—often before the culture developed a written language. It was possible to count without naming many different numbers.

★ The Aboriginal Australians were able to count even though they had only a few names for numbers.

1 = Neecha
2 = Boolla
3 = Boolla Neecha (2 + 1)
4 = Boolla Boolla (2 + 2)
5 = Hand (not spoken)

With this system, a person could communicate the number 23 by saying "boolla neecha" and holding up both hands twice.

★ The ancient Incas used a knotted string called a *quipu* to record numbers. A number was represented by knots in the string, using a base 10 system. For example, the number 586 was represented as follows.

Pacing Guide for Chapter 8

Chapter Opener	1 Day
Section 1	
Activity	1 Day
Lesson	2 Days
Section 2	
Activity	1 Day
Lesson	1 Day
Study Help / Quiz	1 Day
Section 3	
Activity	1 Day
Lesson	1 Day
Section 4	
Activity	1 Day
Lesson	1 Day
Quiz / Chapter Review	1 Day
Chapter Test	1 Day
Standardized Test Practice	1 Day
Total Chapter 8	14 Days
Year-to-Date	136 Days

Check Your Resources

- Record and Practice Journal
- Resources by Chapter
- Skills Review Handbook
- Assessment Book
- Worked-Out Solutions

The Dynamic Planning Tool
Editable Teacher's Resources at
BigIdeasMath.com

Common Core State Standards

4.NF.1 Explain why a fraction a/b is equivalent to a fraction $(n \times a)/(n \times b)$ by using visual fraction models, with attention to how the number and size of the parts differ even though the two fractions themselves are the same size. Use this principle to recognize and generate equivalent fractions.

5.NBT.3b Compare two decimals to thousandths based on meanings of the digits in each place, using >, =, and < symbols to record the results of comparisons.

Additional Topics for Review
- Solving simple equations

Try It Yourself

1.

2.

3. > 4. = 5. <

6. *Sample answer:* $8.5, 9, \frac{19}{2}$

7. *Sample answer:* $-1, 0, 1$

8. *Sample answer:* $0.705, 0.749, 0.75$

Record and Practice Journal

1. $\frac{1}{2}, \frac{2}{3}, \frac{5}{6}, 1$ 2. $\frac{1}{4}, \frac{4}{7}, \frac{17}{28}, \frac{11}{14}$

3. $\frac{1}{6}, \frac{1}{3}, \frac{3}{5}, \frac{19}{30}, \frac{11}{15}$

4. $\frac{5}{12}, \frac{7}{12}, \frac{17}{24}, \frac{3}{4}, \frac{7}{8}$

5. $0, \frac{14}{25}, \frac{33}{50}, 1\frac{3}{10}, 1\frac{17}{20}$

6. $1, \frac{13}{12}, \frac{5}{4}, 1\frac{7}{9}, 1\frac{5}{6}$

7. $\frac{5}{32}, \frac{3}{16}, \frac{1}{4}, \frac{9}{32}$

8–20. See Additional Answers.

T-327

Math Background Notes

Vocabulary Review
- Common Denominator
- Mixed Number
- Improper Fraction

Ordering Fractions
- Students should be familiar with ordering fractions.
- You may choose to review converting mixed numbers into improper fractions and finding common denominators prior to covering Example 1.
- You may also want to discuss the scaling of the number line with students. Each tick mark should count the same amount and be equally spaced on the number line.
- **Common Error:** Some students may suggest converting each of the given numbers into decimals and ordering the decimals on a number line. While this is a valid approach, it is not optimal in this example. Because 3 and 6 appear in the denominator of fractions, this approach will produce two non-terminating decimals. Encourage students to find and write each number as a fraction using a common denominator.

Comparing Numbers
- Students should be familiar with comparing numbers.
- Remind students that it is easier to compare numbers if they are both in the same form. In Example 3, for instance, one number is a decimal and one is a fraction. Encourage students to convert one of the numbers so that either both are decimals or both are fractions.
- **Teaching Tip:** Many students benefit from imagery when using the greater than and less than signs to compare numbers. Remind them that the greater than and less than symbols act as hungry alligators. They always want to eat the bigger number. Encourage students to be sure that the mouth of the alligator always opens towards the greater number.

Reteaching and Enrichment Strategies

If students need help...	If students got it...
Record and Practice Journal • Fair Game Review Skills Review Handbook Lesson Tutorials	Game Closet at *BigIdeasMath.com* Start the next section

What You Learned Before

"Move farther back. We still have an inequality."

Ordering Fractions (4.NF.1)

Example 1 Use a number line to order $\frac{3}{2}, \frac{3}{4}, 0, 1\frac{2}{3},$ and $\frac{1}{6}$ from least to greatest.

Write each of the numbers using a common denominator.

Try It Yourself

Use a number line to order the numbers from least to greatest.

1. $1, \frac{3}{5}, \frac{3}{20}, \frac{3}{4}, \frac{7}{10}$

2. $\frac{1}{4}, \frac{13}{12}, \frac{5}{6}, \frac{5}{12}, \frac{1}{2}$

Comparing Numbers (5.NBT.3b)

Complete the number sentence with <, >, or =.

Example 2 10 ▢ 15

On a number line, 10 is closer to zero than 15.

∴ So, 10 < 15.

Example 3 0.875 ▢ $\frac{7}{8}$

$$0.875 = \frac{875}{1000} = \frac{875 \div 125}{1000 \div 125} = \frac{7}{8}$$

∴ So, $0.875 = \frac{7}{8}$.

Example 4 Find three numbers that make the number sentence $1\frac{2}{5} \leq$ ▢ true.

∴ Sample answer: $1\frac{3}{5}, \frac{5}{2}, 2$

Try It Yourself

Complete the number sentence with <, >, or =.

3. 2.01 ▢ 2.001

4. 4.5 ▢ $\frac{9}{2}$

5. π ▢ 3.2

Find three numbers that make the number sentence true.

6. $\frac{17}{2} \leq$ ▢

7. $1\frac{1}{2} >$ ▢

8. $0.75 \geq$ ▢

8.1 Writing and Graphing Inequalities

Essential Question How can you use a number line to represent solutions of an inequality?

COMMON CORE STATE STANDARDS
6.EE.5
6.EE.8

1 ACTIVITY: Understanding Inequality Statements

Work with a partner.

a. Consider the statement "Your friend is more than 3 minutes late."

 Circle each number that makes the statement true.

 −3 −2 −1 0 1 2 3 4 5 6

 • Write four other numbers that make the statement true.

b. Consider the statement "The temperature is at most 2 degrees."

 • Can the temperature be exactly 2 degrees? Explain.
 • Circle each number that makes the statement true.

 −5 −4 −3 −2 −1 0 1 2 3 4

 • Write four other numbers that make the statement true.

c. Consider the statement "You need at least 4 pieces of paper for your math homework."

 • Can you have exactly 4 pieces of paper? Explain.
 • Circle each number that makes the statement true.

 −3 −2 −1 0 1 2 3 4 5 6

 • Write four other numbers that make the statement true.

d. Consider the statement "After playing a video game for 20 minutes, you have fewer than 6 points."

 • Circle each number that makes the statement true.

 −2 −1 0 1 2 3 4 5 6 7

 • Write four other numbers that make the statement true.

328 Chapter 8 Inequalities

Laurie's Notes

Introduction

For the Teacher
- **Goal:** Students will explore the concept of inequality and develop an understanding of the difference between <, >, ≤, and ≥.
- Today's activities focus mostly on vocabulary. They may also raise questions about different sets of numbers, such as natural, whole, integers, rational, and real. Students do not have formal definitions for these sets of numbers, nor have they performed operations with them, but they clearly know numbers such as -5 and -2.4 exist.

Motivate
- As a quick introduction to the vocabulary in this section, ask a variety of questions. Each student will stand up if the answer is yes, otherwise they remain seated.
- **?** Sample questions to ask with appropriate inequalities provided for discussion later:
 - Is your birth month before May? $m < 5$ or $m \leq 4$
 - Is your height more than 5 feet? $h > 5$
 - Do you have at most 1 sibling? $x \leq 1$
 - Do you have at least 1 pet? $p \geq 1$
- In discussing each problem, students should observe that not all students who stood up had the same "answer," meaning that for the first example birth months of 1, 2, 3, and 4 all satisfied the statement. Students should also make note of the numbers which do not work. For instance, students with a birth month of 9 (September) did not stand.

Activity Notes

Activity 1
- In this activity, students must read a statement and decide what numbers make sense in the problem.
- Students should pay particular attention to the key words in red.
- Students may need additional examples of contexts where *at most* and *at least* are used. For instance, you may have *at most* 2 cookies. You must read *at least* 20 pages tonight. Ask students for more examples.
- **?** "For part (a), could your friend be 30 minutes late?" yes; "3 minutes late?" no; "$5\frac{1}{2}$ minutes late?" yes
- **?** "How many solutions are there for part (a)?" Students may or may not know the word infinite but should be able to describe the concept.
- **Note**: Students may only consider whole number solutions, but you should bring up rational numbers. Also, consider the boundary point such as $x = 3$ in part (a).
- **Big Idea**: When giving additional numbers that make the statement true, students must consider the context of the problem. For instance, in part (d), it may not make sense to have a negative number of points, and for part (c) you would use whole sheets of paper.

Common Core State Standards

6.EE.5 Understand solving an equation or inequality as a process of answering a question: which values from a specified set, if any, make the equation or inequality true? Use substitution to determine whether a given number in a specified set makes an equation or inequality true.

6.EE.8 Write an inequality of the form $x > c$ or $x < c$ to represent a constraint or condition in a real-world or mathematical problem. Recognize that inequalities of the form $x > c$ or $x < c$ have infinitely many solutions; represent solutions of such inequalities on number line diagrams.

Previous Learning
Students should know the symbols and vocabulary associated with inequalities.

Start Thinking! and Warm Up

8.1 Record and Practice Journal

T-328

English Language Learners
Vocabulary and Symbols
Students should review the vocabulary and symbols for inequalities. Have students add a table of symbols and what the symbols mean to their notebooks. Students should add to the table as new phrases are used in the chapter.

Symbol	Phrase
=	is equal to
≠	is not equal to
<	is less than
≤	is less than or equal to
>	is greater than
≥	is greater than or equal to

8.1 Record and Practice Journal

Laurie's Notes

Activity 2
- Write the less than symbol < on the board and ask the following questions.
 - "What does this symbol mean? How is the symbol read?" less than; is less than
 - "What inequality symbols are you familiar with?" They may respond with >, <, ≤, ≥; They may also know ≠.
- Write a few inequality statements on the board as a visual reference for students. Use simple statements such as 4 < 5 and 8 > 6.
- **Note**: This activity and Activity 1 try to discern between $x < 2$ and $x \leq 2$.
- "In part (a), would 1.5 make $x < 2$ true?" yes; "would $1\frac{3}{4}$?" yes
- "How many numbers are there that are less than 2?" Infinite if you are not restricted to only whole numbers.

Activity 3
- Discuss the difference in the symbols $x > 0$ and $x \geq 0$. Students should be able to describe the difference in words.
- Graphing the solution of an inequality is new. If decimals and fractions have been suggested throughout, the graph should not be a big stretch for them. The open and closed circle notation will be new, but generally the notation will make sense to students.
- **Big Idea**: The graph shows that the set of solutions is infinite. It is not possible to list all of the solutions, although the . . . notation suggests that the list continues in the same fashion.

What Is Your Answer?
- **Think-Pair-Share**: Students should read each question independently and then work with a partner to answer the questions. When they have answered the questions, the pair should compare their answers with another group and discuss any discrepancies.

Closure
- Use questions similar to those used to motivate the lesson. Ask "How would you write (and graph) the solutions of these inequalities?"
 - You need at least 8 hours of sleep. $h \geq 8$
 - You will have at most 2 hours of homework. $h \leq 2$

Technology For the Teacher

The Dynamic Planning Tool
Editable Teacher's Resources at *BigIdeasMath.com*

2 ACTIVITY: Understanding Inequality Symbols

Work with a partner.

a. Consider the statement "*x* is a number such that *x* < 2."
 - Can the number be exactly 2? Explain.
 - Circle each number that makes the statement true.
 $-5 \quad -4 \quad -3 \quad -2 \quad -1 \quad 0 \quad 1 \quad 2 \quad 3 \quad 4$
 - Write four other numbers that make the statement true.

b. Consider the statement "*x* is a number such that $x \geq 1$."
 - Can the number be exactly 1? Explain.
 - Circle each number that makes the statement true.
 $-5 \quad -4 \quad -3 \quad -2 \quad -1 \quad 0 \quad 1 \quad 2 \quad 3 \quad 4$
 - Write four other numbers that make the statement true.

3 ACTIVITY: How Close Can You Come to 0?

Work with a partner.

a. Which number line shows $x > 0$? Which number line shows $x \geq 0$? Explain your reasoning.

b. Write the least positive number you can think of that is still a solution of the inequality $x > 0$. Explain your reasoning.

What Is Your Answer?

4. **IN YOUR OWN WORDS** How can you use a number line to represent solutions of an inequality?

5. Write an inequality. Graph all solutions of your inequality on a number line.

Practice

Use what you learned about graphing inequalities to complete Exercises 17–20 on page 333.

Section 8.1 Writing and Graphing Inequalities 329

8.1 Lesson

Check It Out
Lesson Tutorials
BigIdeasMath com

Key Vocabulary
inequality, *p. 330*
solution of an inequality, *p. 331*
solution set, *p. 331*
graph of an inequality, *p. 332*

An **inequality** is a mathematical sentence that compares expressions. It contains the symbols $<$, $>$, \leq, or \geq. To write an inequality, look for the following phrases to determine where to place the inequality symbol.

Inequality Symbols				
Symbol	$<$	$>$	\leq	\geq
Key Phrases	• is less than • is fewer than	• is greater than • is more than	• is less than or equal to • is at most • is no more than	• is greater than or equal to • is at least • is no less than

EXAMPLE 1 Writing Inequalities

Write the word sentence as an inequality.

a. A number c is less than 4.

 A number c | is less than | 4.
 $\quad\quad c \quad\quad < \quad\quad 4$

 ∴ An inequality is $c < 4$.

b. A number k plus 5 is greater than or equal to 8.

 A number k plus 5 | is greater than or equal to | 8.
 $\quad\quad k + 5 \quad\quad \geq \quad\quad 8$

 ∴ An inequality is $k + 5 \geq 8$.

c. Four times a number q is at most 16.

 Four times a number q | is at most | 16.
 $\quad\quad 4q \quad\quad \leq \quad\quad 16$

 ∴ An inequality is $4q \leq 16$.

On Your Own

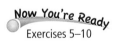
Now You're Ready
Exercises 5–10

Write the word sentence as an inequality.

1. A number n is greater than 1.

2. Twice a number p is fewer than 7.

3. A number w minus 3 is less than or equal to 10.

4. A number z divided by 2 is at least 6.

Laurie's Notes

Introduction

Connect
- **Yesterday:** Students used a variety of contexts to explore inequalities.
- **Today:** Students will translate words into mathematical inequalities and determine what numbers are solutions of an inequality.

Motivate
- Make eleven number cards using 3×5 index cards. Write the integers from -5 to 5 on the cards.
- Ask 11 students to stand at the front of the room. Each student holds one card and faces their classmates who are seated. Tell them you will describe a set of numbers and if their number is in the set described, they should take one step forward.
- "If your number is [sample] take one step forward on the word go."
- Samples:
 - greater than 2
 - less than or equal to 1
 - at least -1
 - at most 3
- This introduction helps to connect the vocabulary to the graphs that students will be drawing today.

Lesson Notes

Example 1
- Discuss the definition of inequality and the vocabulary associated with each symbol.
- If possible, change chalk or marker color as you do the problems. This will help students view the different components of writing an inequality.
- Work through each example. Note that the inequalities are *not* being solved.
- Part (c) will be the most challenging. Students may struggle with determining the correct inequality symbol for *at most*. Give them a context to relate to such as, "the wait time to enter is at most 15 minutes." This will help students translate the words using the correct inequality symbol, \leq.

On Your Own
- **Think-Pair-Share:** Students should read each question independently and then work with a partner to answer the questions. When they have answered the questions, the pair should compare their answers with another group and discuss any discrepancies.
- Ask volunteers to write their inequalities at the board.
- ? "In Question 4, the inequality is $\frac{z}{2} \geq 6$. Could $z = 4$? How do you know?"
 no; $4 \div 2$ is not greater than 6. "Could $z = 20$? How do you know?"
 yes; $20 \div 2$ is greater than 6.

Goal
Today's lesson is graphing and determining **solutions of an inequality**.

Lesson Materials
Introduction
• number cards (-5 to 5)

Start Thinking! and Warm Up

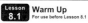

Lesson 8.1 Warm Up
For use before Lesson 8.1

Lesson 8.1 Start Thinking!
For use before Lesson 8.1

The speed limit in a school zone is 15 miles per hour.

How are speed limits related to inequalities?

Extra Example 1
Write the word sentence as an inequality.
a. A number t is no more than 7. $t \leq 7$
b. The difference of a number p and 3 is at least 10. $p - 3 \geq 10$
c. One-fourth of a number x is greater than 2. $\frac{1}{4}x > 2$

On Your Own
1. $n > 1$
2. $2p < 7$
3. $w - 3 \leq 10$
4. $\frac{z}{2} \geq 6$

T-330

Extra Example 2

Tell whether the given value is a solution of the inequality.

a. $d - 5 \leq 7$; $d = 9$ yes
b. $4y \geq 24$; $y = 6$ yes
c. $\frac{m}{5} < 6$; $m = 30$ no

On Your Own

5. no 6. yes
7. yes

Differentiated Instruction

Auditory

Stress to students the importance of reading a statement and translating it into an expression, equation, or inequality. The word "is" plays an important role in the meaning of the statement. For instance, *six less than a number* translates to $x - 6$. While *six **is** less than a number* translates to $6 < x$.

Laurie's Notes

Example 2

- **Vocabulary:** Explain the vocabulary words *solution of an inequality* and *solution set*.
- **Representation:** Discuss the notation used to show that 8 is not less than or equal to 7. ($8 \nleq 7$).
- Use the samples in the textbook or Example 1 to show the substitution for the value of the variable. If the inequality statement is true, the value of the variable is a solution, otherwise it is not a solution.
- Work through each part.
- **Language:** In each part, the *value of the variable is substituted* for the variable. Avoid using language such as, "plug in the number for *x*."
- Note that in part (c), 5 is greater than or equal to 5.

? "Would 15 be a solution of the inequality $\frac{z}{3} > 5$? Explain."
no; 5 is not greater than 5.

On Your Own

- **Neighbor Check:** Have students work independently and then have their neighbor check their work. Have students discuss any discrepancies.

A **solution of an inequality** is a value that makes the inequality true. An inequality can have more than one solution. The set of all solutions of an inequality is called the **solution set**.

Value of x	x + 3 ≤ 7	Is the inequality true?
3	$3 + 3 \stackrel{?}{\leq} 7$ $6 \leq 7$ ✓	yes
4	$4 + 3 \stackrel{?}{\leq} 7$ $7 \leq 7$ ✓	yes
5	$5 + 3 \stackrel{?}{\leq} 7$ $8 \not\leq 7$ ✗	no

Reading

The symbol ≰ means "is not less than or equal to."

EXAMPLE 2 Checking Solutions

Tell whether the given value is a solution of the inequality.

a. $x + 1 > 7$; $x = 8$

$x + 1 > 7$ Write the inequality.

$8 + 1 \stackrel{?}{>} 7$ Substitute 8 for x.

$9 > 7$ ✓ Add. 9 is greater than 7.

∴ So, 8 is a solution of the inequality.

b. $7y < 27$; $y = 4$

$7y < 27$ Write the inequality.

$7(4) \stackrel{?}{<} 27$ Substitute 4 for y.

$28 \not< 27$ ✗ Multiply. 28 is *not* less than 27.

∴ So, 4 is *not* a solution of the inequality.

c. $\frac{z}{3} \geq 5$; $z = 15$

$\frac{z}{3} \geq 5$ Write the inequality.

$\frac{15}{3} \stackrel{?}{\geq} 5$ Substitute 15 for z.

$5 \geq 5$ ✓ Divide. 5 is greater than or equal to 5.

∴ So, 15 is a solution of the inequality.

On Your Own

Now You're Ready
Exercises 11–16

Tell whether 3 is a solution of the inequality.

5. $b + 4 < 6$ **6.** $9 - n \geq 6$ **7.** $18 \div x \leq 10$

The **graph of an inequality** shows all of the solutions of the inequality on a number line. An open circle ○ is used when a number is *not* a solution. A closed circle ● is used when a number is a solution. An arrow to the left or right shows that the graph continues in that direction.

EXAMPLE 3 Graphing an Inequality

Graph $g > 2$.

EXAMPLE 4 Real-Life Application

The NASA Solar Probe can withstand temperatures up to and including 2600°F. Write and graph an inequality that represents the temperatures the probe can withstand.

Words	temperatures	up to and including	2600°F
Variable	Let t be the temperatures the probe can withstand.		
Inequality	t	\leq	2600

∴ An inequality is $t \leq 2600$.

On Your Own

Now You're Ready
Exercises 25–36

Graph the inequality on a number line.

8. $a < 4$ **9.** $f \leq 7$ **10.** $n > 0$ **11.** $p \geq -3$

Write and graph an inequality for the situation.

12. A cruise ship can carry at most 3500 passengers.

13. A board game is designed for ages 12 and up.

Laurie's Notes

Example 3
- **Vocabulary:** Explain the vocabulary word *graph of an inequality*.
- Discuss the difference between $x > 5$ and $x \geq 5$. Explain to students that they know how to describe the difference in words. Now you want to show them the differences when you graph the solution of each.
- Describe the difference between an open circle and a closed circle.
- Graph $x > 5$ and $x \geq 5$.
- Work through the example. Use the language of *test point* to describe how to determine which side of the boundary point (2) to shade.
- **Summarize:** Shading on the number line indicates a solution of the inequality, such as $g = 3$. No shading on the number line indicates that the number is not a solution, such as $g = 0$.

Example 4
- You may need to explain that *up to and including* means the same as *less than or equal to*.
- **Common Misconception:** Students may think that 0 must be included on the number line. In this example, the students use a scale on the number line that is not in increments of 1 and is not centered at 0.

On Your Own
- Students should work with a partner to graph the inequalities.
- **Teaching Tip:** If you have small individual white boards, do these questions together and have students display their graphs simultaneously.

Closure
- Repeat the inequalities used to motivate at the beginning of the lesson. Ask students to write the symbols and then graph the solution.

 - Greater than 2 $x > 2$
 - Less than or equal to 1 $x \leq 1$
 - At least -1 $x \geq -1$
 - At most 3 $x \leq 3$

The Dynamic Planning Tool
Editable Teacher's Resources at *BigIdeasMath.com*

Extra Example 3
Graph $h \leq 5$.

Extra Example 4
A classroom has enough room for up to 25 students. Write and graph an inequality that represents the number s of students in the classroom. $s \leq 25$

On Your Own

8.

9.

10.

11.

12. $p \leq 3500$;

13. $a \geq 12$;

Differentiated Instruction
Visual
Discuss with students the graph of the inequality for Exercise 40(a). Have them describe numbers that are solutions, like 139.99, and numbers that are not solutions, like 140.01. Ask students if the number -1 makes sense as a solution. Point out that in application problems, students need to determine whether the solution of an equation or inequality makes sense as an answer.

Vocabulary and Concept Check

1. Both phrases refer to numbers that are greater than a given number. The difference is that "greater than or equal to" includes the number itself, whereas "greater than" does not.
2. A number n is at least 3; $n \geq 3$; $n \leq 3$
3. The graph of $x \leq 6$ has a closed circle at 6. The graph of $x < 6$ has an open circle at 6.
4. Both graphs are the same, because both indicate that x is less than or equal to 5.

Practice and Problem Solving

5. $k < 10$
6. $a > 6$
7. $z < \dfrac{3}{4}$
8. $b \geq -3$
9. $1 + y \leq 13$
10. $\dfrac{x}{3} \leq 5$
11. yes
12. no
13. yes
14. no
15. no
16. yes
17. B
18. A
19. D
20. C
21. $x < 1$; A number x is less than 1.
22. $x \leq 1$; A number x is at most 1.
23. $x \geq -4$; A number x is at least -4.
24. $x > 0$; A number x is more than 0.

Assignment Guide and Homework Check

Level	Day 1 Activity Assignment	Day 2 Lesson Assignment	Homework Check
Basic	17–20, 47–51	1–4, 5–13 odd, 25–39 odd, 40	4, 5, 11, 25, 40
Average	17–20, 47–51	1–4, 5–15 odd, 29–41 odd, 40	4, 5, 11, 29, 40
Advanced	17–20, 47–51	1–4, 22–38 even, 41–46	22, 26, 42, 44

For Your Information
- **Exercise 46** Part (b) introduces the concept of compound inequalities.

Common Errors
- **Exercises 5–10** Students may struggle with knowing which symbol to use in the inequality. Encourage them to put the word sentence into a real-life context and to use the table in the lesson that explains what symbol matches each phrase.
- **Exercises 5–10** Students may write the subtraction inequalities in the wrong order. For example, a student will write $7 - x < 4$ instead of $x - 7 < 4$. Remind them that writing inequalities is similar to writing equations and expressions.
- **Exercises 11–16** Students may try to perform operations on both sides of the inequality (often not the same operation on both sides either) to get both sides to be the same answer. Remind them that they are checking to see if the value of the variable is true, so they are focusing on simplifying one side of the equation, then comparing the answer to the other side.

8.1 Record and Practice Journal

Technology For the Teacher
Answer Presentation Tool
QuizShow

8.1 Exercises

Vocabulary and Concept Check

1. **VOCABULARY** How are "greater than" and "greater than or equal to" similar? How are they different?

2. **DIFFERENT WORDS, SAME QUESTION** Which is different? Write "both" inequalities.

 | A number n is at most 3. | A number n is at least 3. |

 | A number n is less than or equal to 3. | A number n is no more than 3. |

3. **WRITING** Explain how the graph of $x \leq 6$ is different from the graph of $x < 6$.

4. **WRITING** Are the graphs of $x \leq 5$ and $5 \geq x$ the same or different? Explain.

Practice and Problem Solving

Write the word sentence as an inequality.

5. A number k is less than 10.

6. A number a is more than 6.

7. A number z is fewer than $\dfrac{3}{4}$.

8. A number b is at least -3.

9. One plus a number y is no more than 13.

10. A number x divided by 3 is at most 5.

Tell whether the given value is a solution of the inequality.

11. $x - 1 \leq 7$; $x = 6$

12. $y + 5 < 13$; $y = 17$

13. $3z > 6$; $z = 3$

14. $\dfrac{b}{2} \geq 6$; $b = 10$

15. $c + 2.5 < 4.3$; $c = 1.8$

16. $a \leq 0$; $a = -5$

Match each inequality with its graph.

17. $x \geq 2$ 18. $x < 2$ 19. $x > -2$ 20. $x \leq -2$

A.

B.

C.

D.

Section 8.1 Writing and Graphing Inequalities 333

Write an inequality and a word sentence that represent the graph.

21.
22.

23.
24.

Graph the inequality on a number line.

25. $a > 4$

26. $n \geq 8$

27. $3 \geq x$

28. $y < \dfrac{1}{2}$

29. $x < \dfrac{2}{9}$

30. $-3 \geq c$

31. $m > -5$

32. $b \geq 0$

33. $f < 1.5$

34. $t \geq -\dfrac{1}{2}$

35. $p > -1.6$

36. $z \leq -\dfrac{7}{3}$

ERROR ANALYSIS Describe and correct the error in graphing the inequality.

37.
38.

39. **FISHING** You are fishing and are allowed to catch at most 3 striped bass. Each striped bass must be no less than 18 inches in length.

a. Write and graph an inequality to represent the number of striped bass you are allowed to catch.

b. Write and graph an inequality to represent the length of each striped bass you are allowed to catch.

40. **LOW SODIUM** For a food to be labeled *low sodium*, there must be no more than 140 milligrams of sodium per serving.

 a. Write and graph an inequality to represent the amount of sodium in a low sodium serving.

 b. Write and graph an inequality to represent the amount of sodium in a serving that does not qualify as low sodium.

 c. Does the food represented by the label qualify as a low sodium food? Explain.

Nutrition Facts
Serving Size ½ cup (114g)
Servings Per Container 4

Amount Per Serving	
Calories 90	Calories from fat 30

	% Daily Value*
Total Fat 3g	5%
Saturated Fat 0g	0%
Cholesterol 0mg	0%
Sodium 300mg	13%
Total Carbohydrate 13g	4%
Dietary Fiber 3g	12%
Sugars 3g	
Protein 3g	

Vitamin A 80% • Vitamin C 60%
Calcium 4% • Iron 4%

Common Errors

- **Exercises 17–20** Students may mix up the open and closed circles when matching. Remind them that the open circle means the number is not included, and a closed circle means that the number is included.
- **Exercises 25–36** Students may shade the number line the wrong direction. This often happens when the variable is on the right of the inequality symbol instead of the left. Remind them to check numbers to the left and right of the given number to determine which direction to shade the number line.
- **Exercises 25–36** Students may draw a closed circle instead of an open circle and vice versa. Remind them of the definitions of open and closed circles.
- **Exercise 39** Students may mix up the inequalities given in the word problem. Remind them to read carefully and encourage them to write word sentences for the information given.

Practice and Problem Solving

25.
26.
27.
28.
29.
30.
31.
32.
33.
34.
35.
36.

37. $x \geq 1$ means that 1 is also a solution, so a closed circle should be used.

38. The graph shows $x < -1$, not $x > -1$.

39. **a.** $b \leq 3$;

 b. $\ell \geq 18$;

Practice and Problem Solving

40. a. $s \leq 140$;

b. $s > 140$;

c. no; There are 300 milligrams of sodium in one serving, which is greater than 140.

41. See *Taking Math Deeper*.

42. always; A number that is greater than 5 is always greater than or equal to 5.

43. sometimes; The only time this is not true is if $x = 5$.

44. $1.50x > 36$

45. $p \leq 375$

46. See Additional Answers.

Fair Game Review

47. $x = 9$ **48.** $x = 14$

49. $x = 28$ **50.** $x = 14.1$

51. D

Mini-Assessment

Write the word sentence as an inequality.

1. A number b is at most 5. $b \leq 5$
2. Twice a number s is less than 15. $2s < 15$

Tell whether the given value is a solution of the inequality.

3. $3x > 16$; $x = 6$ yes
4. $12 - n \geq 5$; $n = 8$ no
5. No more than 10 people will go to the movies. Write and graph an inequality to represent the number of movie goers. $x \leq 10$

Taking Math Deeper

Exercise 41

This problem gives students an opportunity to use an inequality to describe a real-life situation.

Item	Price (with tax)
T-shirt	$15
Book	$20
DVD	$13
Necklace	$16

① You have $33 and plan to buy a necklace for $16. One way to approach the problem is to let x represent the amount you have left after buying the necklace.

a.

$16 + x \leq 33$

Necklace Amount left Starting Amount

Students might also subtract the $16 from $33 and write the following inequality.

$x \leq 17$

② **b.** A T-shirt costs $15, so there *is* enough money left to buy a T-shirt.

③ **c.** A book costs $20, so there *is not* enough money left to buy a book.

Reteaching and Enrichment Strategies

If students need help...	If students got it...
Resources by Chapter • Practice A and Practice B • Puzzle Time Record and Practice Journal Practice Differentiating the Lesson Lesson Tutorials Skills Review Handbook	Resources by Chapter • Enrichment and Extension Start the next section

T-335

41. **SHOPPING** You have $33. You want to buy a necklace and one other item from the list.

 a. Write an inequality to represent the situation.
 b. Can the other item be a T-shirt? Explain.
 c. Can the other item be a book? Explain.

Item	Price (with tax)
T-shirt	$15
Book	$20
DVD	$13
Necklace	$16

Determine whether the statement is *sometimes*, *always*, or *never* true. Explain your reasoning.

42. A number that is a solution of the inequality $x > 5$ is also a solution of the inequality $x \geq 5$.

43. A number that is a solution of the inequality $5 \leq x$ is also a solution of the inequality $x > 5$.

44. **BUS RIDE** A bus ride costs $1.50. A 30-day bus pass costs $36. Write an inequality to represent the number of bus rides you would need to take for the bus pass to be a better deal.

45. **MOVIE THEATER** Fifty people are seated in a movie theater. The maximum capacity of the theater is 425 people. Write an inequality to represent the number of additional people who can still be seated.

46. **Critical Thinking** The map shows the elevations above sea level for an area of land.

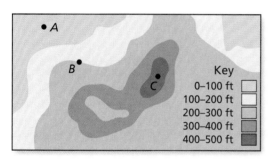

 a. Graph the possible elevations of *A*. Write the set of elevations as two inequalities.
 b. Graph the possible elevations of *C*. How can you write this set of elevations as a single inequality? Explain.
 c. What is the elevation of *B*? Explain.

Fair Game Review What you learned in previous grades & lessons

Solve the equation. Check your solution. *(Section 7.2)*

47. $x + 3 = 12$ 48. $x - 6 = 8$ 49. $16 + x = 44$ 50. $7.6 = x - 6.5$

51. **MULTIPLE CHOICE** A stack of boards is 24 inches high. Each board is $\frac{3}{8}$ of an inch thick. How many boards are in the stack? *(Section 2.5)*

 Ⓐ $\frac{1}{6}$ Ⓑ $\frac{1}{9}$ Ⓒ 9 Ⓓ 64

8.2 Solving Inequalities Using Addition or Subtraction

Essential Question How can you use an inequality to describe a real-life situation?

COMMON CORE STATE STANDARDS
6.EE.5
6.EE.8

1 ACTIVITY: Writing an Inequality

Work with a partner. In 3 years, your friend will still not be old enough to apply for a driver's license.

a. Which of the following represents your friend's situation? What does x represent? Explain your reasoning.

$x + 3 < 16$ $x + 3 \leq 16$ $x + 3 > 16$ $x + 3 \geq 16$

b. Graph the possible ages of your friend on a number line.

2 ACTIVITY: The Triangle Inequality

Work with a partner. Draw different triangles whose sides have lengths 10 cm, 6 cm, and x cm.

a. Which of the following describes how *small* x can be?

$6 + x < 10$ $6 + x \leq 10$

$6 + x > 10$ $6 + x \geq 10$

b. Which of the following describes how *large* x can be?

$x - 6 < 10$ $x - 6 \leq 10$ $x - 6 > 10$ $x - 6 \geq 10$

3 ACTIVITY: Writing an Inequality

Work with a partner. Baby manatees are about 4 feet long at birth. They grow to a maximum length of 13 feet.

a. Which of the following can represent a baby manatee's growth? What does x represent? Explain your reasoning.

$x + 4 < 13$ $x + 4 \leq 13$ $x - 4 > 13$ $x - 4 \geq 13$

b. Graph the solution on a number line.

336 Chapter 8 Inequalities

Laurie's Notes

Introduction

For the Teacher
- **Goal:** Students will use real-life contexts to explore inequalities that can be solved by addition or subtraction.
- **Teaching Tip:** Give students 15 minutes to complete the first three activities with their partner. Then discuss all three activities.

Motivate
- **FYI:** Share some historical notes about inequalities with students:
 - One of the earliest publications using the current day symbols < and > was in a posthumous publication by Thomas Harriot in 1631. Some claim Harriot invented these symbols while others suggest it was Harriot's editor who invented the symbols.
 - Forty years later, in 1670, John Wallis used ⩾ ("equal or greater") and ⩽ ("equal or less") in his publication.

Activity Notes

Activity 1
- **FYI:** In most states, the minimum age for a driver's license is 16, although some states grant a learner's permit at age 15. In South Dakota, the minimum age for a full driver's license is 14, and in Hawaii, Idaho, Montana, and New Mexico, the minimum age for a full license is 15.
- Students should be able to use their own age to think through the correct inequality for this problem.

Activity 2
- **FYI:** The Triangle Inequality in a geometry course would state that "The sum of any two legs of a triangle must exceed the length of the last side." You can model this by using 3 pencils of lengths (approximately) 5", 6", and 8". Selecting any two pencils, the sum of the lengths is longer than the third pencil. This would not be the case if the initial pencils were 3", 4", and 8". The sum of the two shortest pencils is shorter than the longest pencil.
- ? If students need a hint on this activity, ask the following questions to help them think about the range of numbers for x.
 - "Can x be 2 centimeters?" no
 - "Can x be 20 centimeters?" no
- **Note:** x must be longer than 4 cm and shorter than 16 centimeters.

Activity 3
- **FYI:** Adult manatees have been known to weigh over 3500 pounds.

Common Core State Standards

6.EE.5 Understand solving an equation or inequality as a process of answering a question: which values from a specified set, if any, make the equation or inequality true? Use substitution to determine whether a given number in a specified set makes an equation or inequality true.

6.EE.8 Write an inequality of the form $x > c$ or $x < c$ to represent a constraint or condition in a real-world or mathematical problem. Recognize that inequalities of the form $x > c$ or $x < c$ have infinitely many solutions; represent solutions of such inequalities on number line diagrams.

Previous Learning
Students should know how to solve equations using addition or subtraction.

Start Thinking! and Warm Up

Graph the inequality on a number line.
1. $a < 2$
2. $b > 1$
3. $c \leq 0$
4. $d \geq 2.5$
5. $f > -2$
6. $g \leq -5$

8.2 Record and Practice Journal

T-336

English Language Learners
Vocabulary

English learners may benefit by restating word problems so that the wording corresponds to the wording of inequality symbols. In Activity 3, the problem could be rewritten as:

Baby manatees are about 4 feet long at birth. They grow a certain number of feet to a length less than or equal to 13 feet.

This leads to writing the inequality $4 + x \leq 13$.

8.2 Record and Practice Journal

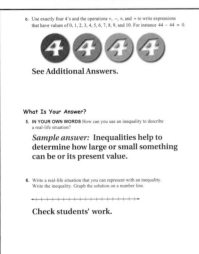

Laurie's Notes

Activity 4

- **Scaffolding:** If time is short, you could scaffold these puzzles or students could work some of them at home, such as part (c).
- **Part (a)**
 - Students are trying to decide what number is a solution of both inequalities in the colored rectangle. Once they have figured out the solution, they apply the classic alphabet code of A = 1, B = 2, C = 3, ... Z = 26. If solved correctly, the puzzle yields a common 4-letter word.
 - **Caution:** Students may misread the inequality when the variable is written on the right side of the inequality symbol. Remind students that the inequality $4 < x$ is read *4 is less than x* and is equivalent to saying *x is greater than 4*.
- **Part (b)**
 - If there are gaps in the square when arranging the pieces, the side lengths of the square are too long. If the pieces overlap, the side lengths of the square are too short.
- **Part (c)**
 - Remind students that the order of operations must be used, so they need to insert parentheses as needed.
 - If the arithmetic is done correctly, the answer is exactly the number desired, instead of greater or less than the number desired.

What Is Your Answer?

- **Neighbor Check:** Have students work independently and then have their neighbor check their work. Have students discuss any discrepancies.

Closure

- "How would you write and graph the solution of $x + 5 \leq 13$?" $x \leq 8$

Technology For the Teacher

The Dynamic Planning Tool
Editable Teacher's Resources at *BigIdeasMath.com*

4 ACTIVITY: Puzzles

Work with a partner.

a. Use the clues to find the word that is spelled by 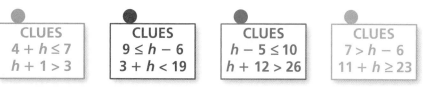. Assume A = 1, B = 2, and so on.

CLUES	CLUES	CLUES	CLUES
$4 + h \leq 7$	$9 \leq h - 6$	$h - 5 \leq 10$	$7 > h - 6$
$h + 1 > 3$	$3 + h < 19$	$h + 12 > 26$	$11 + h \geq 23$

b. Trace the pieces and cut them out. Rearrange them to make a square without overlapping the pieces.

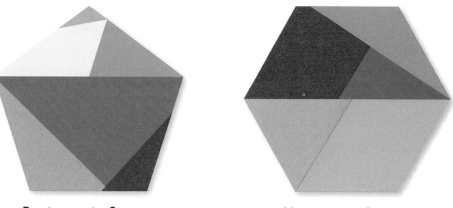

Pentagon to Square Hexagon to Square

c. Use exactly four 4's and the operations $+$, $-$, \times, and \div to write expressions that have values of 0, 1, 2, 3, 4, 5, 6, 7, 8, 9, and 10. For instance, $44 - 44 = 0$.

What Is Your Answer?

5. **IN YOUR OWN WORDS** How can you use an inequality to describe a real-life situation?

6. Write a real-life situation that you can represent with an inequality. Write the inequality. Graph the solution on a number line.

Practice — Use what you learned about solving inequalities using addition or subtraction to complete Exercises 5–7 on page 340.

8.2 Lesson

Key Ideas

Study Tip

You can solve inequalities the same way you solve equations. Use inverse operations to get the variable by itself.

Addition Property of Inequality

Words If you add the same number to each side of an inequality, the inequality remains true.

Numbers
$$\begin{array}{r} 3 < 5 \\ +2 \phantom{<} +2 \\ \hline 5 < 7 \end{array}$$

Algebra
$$\begin{array}{r} x - 4 > 5 \\ +4 +4 \\ \hline x > 9 \end{array}$$

Graph

Subtraction Property of Inequality

Words If you subtract the same number from each side of an inequality, the inequality remains true.

Numbers
$$\begin{array}{r} 3 < 5 \\ -2 \phantom{<} -2 \\ \hline 1 < 3 \end{array}$$

Algebra
$$\begin{array}{r} x + 4 > 5 \\ -4 -4 \\ \hline x > 1 \end{array}$$

Graph

These properties are also true for \leq and \geq.

EXAMPLE 1 Solving an Inequality Using Addition

Solve $x - 3 > 1$. Graph the solution.

$$\begin{array}{rl} x - 3 > 1 & \text{Write the inequality.} \\ +3 +3 & \text{Add 3 to each side.} \\ \hline x > 4 & \text{Simplify.} \end{array}$$

Undo the subtraction.

∴ The solution is $x > 4$.

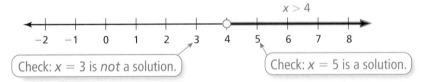

Study Tip

To check a solution, you check some numbers that are solutions and some that are not.

Check: $x = 3$ is *not* a solution. Check: $x = 5$ is a solution.

On Your Own

Solve the inequality. Graph the solution.

1. $x - 2 < 3$ 2. $x - 6 \geq 4$ 3. $10 \geq x - 1$

Laurie's Notes

Introduction

Connect
- **Yesterday:** Students used a variety of contexts to explore inequalities that could be solved using addition or subtraction.
- **Today:** Students will use the Properties of Inequality to solve inequalities.

Motivate
? Ask a series of questions that suggest the Properties of Inequality that will be used today. Samples:
 - "I am older than each of you. In 3 years, who will be older?" adding 3 years to each age did not change the fact that you are older than your students.
 - "Sarah is shorter than Matt. If they both grow 2 inches, who is shorter?" adding 2 inches to each height did not change the fact that Sarah is shorter than Matt.
 - "Dave has less money than Roberto. If they each pay $2, who has less money?" subtracting $2 from each amount does not change the fact that Dave has less money than Roberto.
- This introduction helps to create a context for the Properties of Inequality.

Lesson Notes

Key Ideas
- Work through the Words, Numbers, and Algebra.
- Explain that solving inequalities is very similar to solving equations (refer to the Study Tip).
- Remind students to line up their inequality symbols as they solve vertically.
- **Teaching Tip:** Students will want to solve the inequalities in their head and resist writing their work down. Remind students that you are trying to help them develop good problem solving techniques. The problems will become more involved later and students need to know how to record the process.

Example 1
- Work through the example.
? "How do you graph $x > 4$?" open circle at 4, shade to the right of 4
? "How do you know which side of the boundary point ($x = 4$) to shade?" test values on each side of 4 "Why is there an open circle?" 4 is not a solution.

On Your Own
- **Think-Pair-Share:** Students should read each question independently and then work with a partner to answer the questions. When they have answered the questions, the pair should compare their answers with another group and discuss any discrepancies.
- **Common Error:** In Question 3, because the variable is on the right, it is common for students to shade the wrong side of the inequality when graphing. Encourage students to rewrite the solution $11 \geq x$ as $x \leq 11$.

Goal Today's lesson is solving inequalities using addition or subtraction.

Start Thinking! and Warm Up

A classroom should have no more than 22 students.

A math classroom meets this requirement. After two students transferred into the class, it still meets the requirement.

Write an inequality that represents this situation. Solve it to find the possible numbers of students in the class before the two students transferred.

Extra Example 1
Solve $5 \leq x - 4$. Graph the solution.
$9 \leq x$ or $x \geq 9$

On Your Own

1. $x < 5$;

2. $x \geq 10$;

3. $11 \geq x$;

T-338

Extra Example 2

Solve $n + 7 < 8$. Graph the solution.
$n < 1$

● **On Your Own**

4. $x > 4$;

5. $y < 15$;

6. $7 \leq m$;

Extra Example 3

Your book report must be at least 500 words. So far, you have written 334 words. Write and solve an inequality to represent the number w of words you still need to write. $w + 334 \geq 500$; $w \geq 166$

● **On Your Own**

7. $24 + x \geq 75$; $x \geq 51$; You must spend at least $51 more to get free shipping.

Differentiated Instruction

Visual

To show students that the Addition and Subtraction Properties of Inequalities are true, graph two numbers, 2 and 5, on the number line. Write the ordered relationship, $2 < 5$, on the board. Now add 4 to each number. Both points move the same distance to the right, so their ordered positions remain the same, $6 < 9$. Now subtract 2 from each of the original numbers. Both points move 2 units to the left, so their ordered positions remain the same, $0 < 3$.

Laurie's Notes

Example 2

? Discuss the initial inequality with the variable written on the right side.
- "Is $15 \geq 6 + x$ equivalent to $15 \geq x + 6$? Explain."
 yes; Commutative Property
- "Is $15 \geq 6 + x$ equivalent to $6 + x \leq 15$? Explain."
 yes; If $a > b$, then $b < a$.
- Students could solve an equivalent inequality, or the inequality as it is stated in the textbook. If students solve the inequality as it is stated, note that the last step involves rewriting the solution so that the variable is on the left side of the inequality symbol.
- **Big Idea:** The solution $x \leq 9$ means that all numbers less than or equal to 9 are solutions. The graph conveys this idea with the left-pointing arrow, and the use of negative numbers.
- Discuss with students the scale used for the graph. Remind them that it is not necessary to use an increment of 1 for the scaling on the number line.

On Your Own

- **Neighbor Check:** Have students work independently and then have their neighbor check their work. Have students discuss any discrepancies.

Example 3

- **Note:** The last sentence in the problem identifies what you need to find. While reading the problems, students should be thinking about what the question is asking them.
- **Extension:** Ask students to graph the solution.

On Your Own

- **Think-Pair-Share:** Students should read each question independently and then work with a partner to answer the questions. When they have answered the questions, the pair should compare their answers with another group and discuss any discrepancies.
- Ask students to think of other examples where inequalities are used. An example is unlimited text messaging up to 500 messages.

● **Closure**

- Explain how solving $x - 8 > 14$ and $x - 8 = 14$ are the same and how they are different. Solve each. $x > 22$ and $x = 22$; *Sample answer:* For both, you add 8 to each side (Addition Property). The inequality has infinitely many solutions, but the equation has only one solution.

Technology For the Teacher

The Dynamic Planning Tool
Editable Teacher's Resources at *BigIdeasMath.com*

EXAMPLE 2 Solving an Inequality Using Subtraction

Solve $15 \geq 6 + x$. Graph the solution.

Undo the addition.

$$15 \geq 6 + x \quad \text{Write the inequality.}$$
$$\underline{-6 \quad -6} \quad \text{Subtract 6 from each side.}$$
$$9 \geq x \quad \text{Simplify.}$$

∴ The solution is $x \leq 9$.

Reading
The inequality $x \leq 9$ is the same as $9 \geq x$.

On Your Own

Now You're Ready
Exercises 5–16

Solve the inequality. Graph the solution.

4. $x + 3 > 7$ 5. $y + 2 < 17$ 6. $16 \leq m + 9$

EXAMPLE 3 Real-Life Application

A flea market advertises that it has more than 250 vending booths. Of these, 184 are currently filled. Write and solve an inequality to represent the number of vending booths still available.

Words	Booths filled	plus	the number of remaining booths	is greater than	the total number of booths.

Variable Let b be the number of remaining booths.

Inequality	184	+	b	>	250

$$184 + b > 250 \quad \text{Write the inequality.}$$
$$\underline{-184 \quad\quad -184} \quad \text{Subtract 184 from each side.}$$
$$b > 66 \quad \text{Simplify.}$$

∴ More than 66 vending booths are still available.

On Your Own

7. You have already spent $24 shopping online for clothes. Write and solve an inequality to represent the additional amount you must spend to get free shipping.

8.2 Exercises

Vocabulary and Concept Check

1. **OPEN-ENDED** Write an inequality that can be solved by subtracting 7 from each side.

2. **WRITING** Explain how to solve the inequality $x - 6 > 3$.

3. **WRITING** Describe the graph of the solution of $x + 3 \leq 4$.

4. **OPEN-ENDED** Write an inequality represented by the graph. Then use the Subtraction Property of Inequality to write another inequality represented by the graph.

Practice and Problem Solving

Solve the inequality. Graph the solution.

5. $x - 4 < 5$
6. $5 + h > 7$
7. $3 \geq y - 2$
8. $9 \leq c + 1$
9. $18 > 12 + x$
10. $37 + z \leq 54$
11. $y - 21 < 85$
12. $g - 17 \geq 17$
13. $7.2 < x + 4.2$
14. $12.7 \geq s - 5.3$
15. $\dfrac{3}{4} \leq \dfrac{1}{2} + n$
16. $\dfrac{1}{3} + b > \dfrac{3}{4}$

17. **ERROR ANALYSIS** Describe and correct the error in solving the inequality.

18. **AIR TRAVEL** Your carry-on bag can weigh at most 40 pounds. Write and solve an inequality to represent how much more weight you can add to the bag and still meet the requirement.

19. **SHOPPING** It costs x for a round-trip bus ticket to the mall. You have $24. Write and solve an inequality to represent the greatest amount of money you can spend for the bus fare and still have enough to buy the baseball cap.

Write the word sentence as an inequality. Then solve the inequality.

20. Five more than a number is less than 17.

21. Three less than a number is more than 15.

340 Chapter 8 Inequalities

Assignment Guide and Homework Check

Level	Day 1 Activity Assignment	Day 2 Lesson Assignment	Homework Check
Basic	5–7, 29–33	1–4, 9–21 odd, 18, 20	9, 11, 17, 18, 20
Average	5–7, 29–33	1–4, 9–17 odd, 18, 20, 21, 23, 25	9, 11, 17, 20, 23
Advanced	5–7, 29–33	1–4, 14–17, 22–28	14, 17, 22, 26

For Your Information
- **Exercise 26** In some cases, the weights of drivers and gasoline are included in a vehicle's weight.

Common Errors
- **Exercises 5–16** Students may use the same operation in solving for the variable instead of the inverse operation. Demonstrate to them that this will not work by simplifying the inequality on the board or overhead. Students most likely ignored the side with the variable when they made this mistake. Remind them to check their answers in the original inequality.
- **Exercises 5–16** Students may also switch the direction of the inequality when adding or subtracting. Remind them that the inequality does not change. Work through an example without variables to demonstrate that the inequality is still true when numbers are added or subtracted.

Technology For the Teacher
Answer Presentation Tool
QuizShow

8.2 Record and Practice Journal

Vocabulary and Concept Check

1. Sample answer: $x + 7 \geq 143$
2. Add 6 to both sides of the inequality.
3. By solving the inequality to obtain $x \leq 1$, the graph has a closed circle at 1 and an arrow pointing in the negative direction.
4. Sample answer: $x > 2$; $x - 1 > 1$

Practice and Problem Solving

5. $x < 9$;

6. $h > 2$;

7. $5 \geq y$;

8. $8 \leq c$;

9. $6 > x$;

10. $z \leq 17$;

11. $y < 106$;

12. $g \geq 34$;

13. $3 < x$;

14. $18 \geq s$;

15–16. See Additional Answers.

T-340

Practice and Problem Solving

17. To solve the inequality, 9 should be added to both sides, not subtracted.
$$28 \geq t - 9$$
$$\underline{+9 \quad +9}$$
$$37 \geq t$$

18. $22 + x \leq 40$; $x \leq 18$ lb

19. $x + 18.99 \leq 24$; $x \leq \$5.01$

20. $x + 5 < 17$; $x < 12$

21–26. See Additional Answers.

27. See *Taking Math Deeper*.

28. 25

Fair Game Review

29. $t = 48$
30. $54 = s$
31. $x = 9$
32. $6 = z$
33. A

Mini-Assessment

Solve the inequality. Graph the solution.

1. $x + 9 \leq 15$ $x \leq 6$

2. $5 > 2 + y$ $y < 3$

3. $z - 2.3 < 4.7$ $z < 7$

4. $6 \leq t + 4$ $t \geq 2$

5. You have 70 DVDs. The DVD cabinet you own holds 150 DVDs. Write and solve an inequality to represent how many DVDs you can buy and fit into the cabinet. $70 + x \leq 150$; $x \leq 80$

T-341

Taking Math Deeper

Exercise 27

This is an open-ended problem that allows students to use estimation, common sense, and real-life experiences to come up with a reasonable solution.

Triathlon	
Event	Your Time (minutes)
Swimming	18.2
Biking	45.4
Running	?

① Write an inequality.

a. $18.2 + 45.4 + x < 100$

 Swimming Biking Running Total

② $63.6 + x < 100$ Combine like terms.

 $x < 36.4$ Subtract 63.6 from each side.

③ b. There are many ways a student can estimate how long it takes to run a mile. However they do this, their estimate should be greater than 4 minutes, which is approximately the world record for running a mile.

Suppose a student estimates that it would take 8 minutes to run a mile. With this estimate, there is plenty of time to run 3.1 miles and still meet the goal of less than 100 minutes in the race.

Project

Research triathlons that are run in the United States. Report on the training, cost, and time commitments of the participants.

Reteaching and Enrichment Strategies

If students need help...	If students got it...
Resources by Chapter • Practice A and Practice B • Puzzle Time Record and Practice Journal Practice Differentiating the Lesson Lesson Tutorials Skills Review Handbook	Resources by Chapter • Enrichment and Extension Start the next section

Solve the inequality. Graph the solution.

22. $x + 9 - 3 \leq 14$ **23.** $44 > 7 + s + 26$ **24.** $6.1 - 0.3 \geq c + 1$

25. VIDEO GAME The high score for a video game is 36,480. Your current score is 34,280. Each dragonfly you catch is worth 1 point. You also get a 1000-point bonus for reaching 35,000 points. Write and solve an inequality to represent the number of dragonflies you must catch to earn a new high score.

26. PICKUP TRUCKS You can register a pickup truck as a passenger vehicle if the truck is not used for commercial purposes and the weight of the truck with its contents does not exceed 8500 pounds.

 a. Your pickup truck weighs 4200 pounds. Write an inequality to represent the number of pounds your truck can carry and still qualify as a passenger vehicle. Then solve the inequality.

 b. A cubic yard of sand weighs about 1600 pounds. How many cubic yards of sand can you haul in your truck and still qualify as a passenger vehicle? Explain your reasoning.

27. TRIATHLON You complete two events of a triathlon. Your goal is to finish with an overall time of less than 100 minutes.

 a. Write and solve an inequality to represent how many minutes you can take to finish the running event and still meet your goal.

 b. The running event is 3.1 miles long. Estimate how many minutes it would take you to run 3.1 miles. Would this time allow you to reach your goal? Explain your reasoning.

Triathlon	
Event	Your Time (minutes)
Swimming	18.2
Biking	45.4
Running	?

28. **Number Sense** The possible values of x are given by $x - 3 \geq 2$. What is the least possible value of $5x$?

Fair Game Review *What you learned in previous grades & lessons*

Solve the equation. Check your solution. *(Section 7.3)*

29. $\dfrac{t}{12} = 4$ **30.** $6 = \dfrac{s}{9}$ **31.** $8x = 72$ **32.** $9 = 1.5z$

33. MULTIPLE CHOICE How many times greater is the circumference of the large circle than the circumference of the small circle? *(Section 6.1)*

 Ⓐ 2 **Ⓑ** 3.14 **Ⓒ** 4 **Ⓓ** 6.28

8 Study Help

You can use a **Y chart** to compare two topics. List differences in the branches and similarities in the base of the Y. Here is an example of a Y chart that compares solving equations and solving inequalities.

Solving Equations
- The sign between two expressions is an equal sign, =.
- One number is the solution.

Solving Inequalities
- The sign between two expressions is an inequality symbol: <, >, ≤, or ≥.
- More than one number can be a solution.

- Use inverse operations to group numbers on one side.
- Use inverse operations to group variables on one side.
- Solve for the variable.

On Your Own

Make a Y chart to help you study and compare these topics.

1. writing equations and writing inequalities

2. graphing the solution of an equation and graphing the solution of an inequality

3. graphing inequalities that use > and graphing inequalities that use <

4. graphing inequalities that use > or < and graphing inequalities that use ≥ or ≤

5. solving inequalities using addition and solving inequalities using subtraction

"Hey Descartes, do you have any suggestions for the Y chart I am making?"

After you complete this chapter, make Y charts for the following topics.

6. solving inequalities using multiplication and solving inequalities using division

7. solving two-step equations and solving two-step inequalities

8. Pick two other topics that you studied earlier in this course and make a Y chart to compare them.

Sample Answers

1.

Writing Equations
- The sign between two expressions is an equal sign, =.
- One number is the solution.

Writing Inequalities
- The sign between two expressions is an inequality symbol: $<$, $>$, \leq or \geq.
- More than one number can be the solution.

- Write one expression on the left and one expression on the right.
- Look for key phrases to determine which operation(s) to use: $+$, $-$, \times, or \div.
- Look for key phrases to determine where to place the equal or inequality sign.

2.

Graphing the Solution of an Equation
- Each number that is a solution is represented by a closed circle, ●.
- One number is the solution.

Graphing the Solution of an Inequality
- The endpoint of the graph can be an open circle, ○, or a closed circle, ●.
- Use an arrow pointing to the left (for $<$ and \leq) or right (for $>$ and \geq).
- More than one number can be a solution.

- Solve for the variable.
- Graph the solution on a number line.

3.

Graphing Inequalities That Use >
- Use an arrow pointing to the right.

Graphing Inequalities That Use <
- Use an arrow pointing to the left.

- Use a number line to graph the solution.
- Use an open circle, ○.
- Use an arrow to show that the graph continues.
- Solution usually includes many numbers.

4–5. Available at *BigIdeasMath.com*.

List of Organizers
Available at *BigIdeasMath.com*

Comparison Chart
Concept Circle
Definition (Idea) and Example Chart
Example and Non-Example Chart
Formula Triangle
Four Square
Information Frame
Information Wheel
Notetaking Organizer
Process Diagram
Summary Triangle
Word Magnet
Y Chart

About this Organizer

A **Y Chart** can be used to compare two topics. Students list differences between the two topics in the branches of the Y and similarities in the base of the Y. As with an example and non-example chart, a Y chart serves as a good tool for assessing students' knowledge of a pair of topics that have subtle but important differences. You can include blank Y charts on tests or quizzes for this purpose.

Technology For the Teacher

Vocabulary Puzzle Builder

Answers

1. $x > 0$
2. $k < -2$
3. $\dfrac{y}{3} \leq 15$
4. $2c \geq 8$
5. yes
6. yes
7.
8.
9.
10. $x \leq 4$;

11. $g < 2$;

12. $p \geq 13$;

13–19. See Additional Answers.

Assessment Book

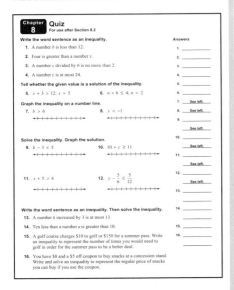

Alternative Quiz Ideas

100% Quiz Math Log
Error Notebook Notebook Quiz
Group Quiz Partner Quiz
Homework Quiz Pass the Paper

Group Quiz
Students work in groups. Give each group a large index card. Each group writes five questions that they feel evaluate the material they have been studying. On a separate piece of paper, students solve the problems. When they are finished, they exchange cards with another group. The new groups work through the questions on the card.

Reteaching and Enrichment Strategies

If students need help...	If students got it...
Resources by Chapter • Study Help • Practice A and Practice B • Puzzle Time Lesson Tutorials *BigIdeasMath.com* Practice Quiz Practice from the Test Generator	**Resources by Chapter** • Enrichment and Extension • School-to-Work Game Closet at *BigIdeasMath.com* Start the next section

Technology For the Teacher
Answer Presentation Tool
Big Ideas Test Generator

8.1–8.2 Quiz

Write the word sentence as an inequality. *(Section 8.1)*

1. A number x is greater than 0.
2. A number k is fewer than -2.
3. A number y divided by 3 is at most 15.
4. Twice a number c is at least 8.

Tell whether the given value is a solution of the inequality. *(Section 8.1)*

5. $2n > 16$; $n = 9$
6. $x - 1 \leq 9$; $x = 10$

Graph the inequality on a number line. *(Section 8.1)*

7. $n < 5$
8. $y > -4$
9. $m \leq \frac{3}{5}$

Solve the inequality. Graph the solution. *(Section 8.2)*

10. $x + 4 \leq 8$
11. $18 > 16 + g$
12. $7 \leq p - 6$
13. $x - 12 > 12$

Write the word sentence as an inequality. Then solve the inequality. *(Section 8.2)*

14. Two less than a number is more than 15.
15. Seven more than a number is less than or equal to 27.

16. **PARTY** At most 80 guests will attend a party. Write and graph an inequality to represent the number of party guests. *(Section 8.1)*

17. **DELIVERY** You were planning to spend $12 on a pizza. Write and solve an inequality to represent the additional amount you must spend to get free delivery. *(Section 8.2)*

18. **WATER PARK** Each visit to a water park costs $19.95. An annual pass to the park costs $89.95. Write an inequality to represent the number of times you would need to visit the park for the pass to be a better deal. *(Section 8.1)*

19. **COUPON** You have $40 and a $5 coupon. Write and solve an inequality to represent the regular price of a pair of jeans you can buy if you use the coupon. *(Section 8.2)*

8.3 Solving Inequalities Using Multiplication or Division

Essential Question How can you use multiplication or division to solve an inequality?

COMMON CORE STATE STANDARDS
6.EE.5
6.EE.8

1 ACTIVITY: Matching Inequalities

Work with a partner. Match the inequality with its graph.

a. $3x < 9$ b. $3x \leq 9$ c. $\dfrac{x}{2} \geq 1$

d. $6 < 2x$ e. $12 \leq 4x$ f. $\dfrac{x}{2} < 2$

A.
B.
C.
D.
E.
F.

2 ACTIVITY: Writing an Inequality

Work with a partner. One of your favorite stores is having a 75% off sale. You have $20. You want to buy a pair of jeans.

a. Which of the following represents your ability to buy the jeans with $20?

$0.25x < 20$ $0.25x \leq 20$

$0.25x > 20$ $0.25x \geq 20$

b. What does x represent? Graph the possible values of x on a number line.

c. Can you afford a pair of jeans that originally costs $100? Explain.

Laurie's Notes

Introduction

For the Teacher
- **Goal:** Students will develop an intuitive understanding of solving inequalities using multiplication or division.

Motivate
- Pac Man, one of the most famous video games of all time, was released in 1980. Ask how many students have played Pac Man or a similar video game. It is expected that virtually 100% of your students will answer yes.
- Historical facts to share: William Higinbotham created the first video game in 1958. His game, called *Tennis for Two,* was created and played on a Brookhaven National Laboratory oscilloscope. In 1962, Steve Russell invented *SpaceWar!,* the first game intended for computer use. Russell used a MIT PDP-1 mainframe computer to design his game.

Activity Notes

Activity 1
- Students should be able to use Guess, Check, and Revise to complete this matching activity. In the process, students should be able to confirm that you solve a multiplication inequality such as $3x < 9$ the same way you solve the equation $3x = 9$.
- Caution students to be careful with the inequalities that have the variable on the right side. Also remind students how division is represented $\left(\dfrac{x}{2} \text{ versus } x \div 2\right)$.
- ? "What are the inequalities graphed in (A) and (F)? (A) is $x \leq 3$ and (F) is $x < 3$.
- ? "What are the inequalities graphed in (B) and (E)?" (B) is $x \geq 3$ and (E) is $x > 3$.
- Given the similarity of all 6 solutions, it is important to have students discuss how they completed the matching activity.
- **Extension:** Ask students to name a value of x that would be a solution of (C) but would not be a solution of (F). $x = 3.5$ Name a value of x that would be a solution of (D) but would not be a solution of (E). $x = 2.5$ Name a value of x that is a solution of (A) and (B). $x = 3$

Activity 2
- Make sure students understand that x equals the original purchase price of the jeans.
- ? "Why is 0.25 used in this problem when the jeans are 75% off?" 25% is the amount you must pay when there is a 75% off sale; $25\% = 0.25$.
- **Extension:** Discuss the sales tax that would be added in your state.

Common Core State Standards

6.EE.5 Understand solving an equation or inequality as a process of answering a question: which values from a specified set, if any, make the equation or inequality true? Use substitution to determine whether a given number in a specified set makes an equation or inequality true.

6.EE.8 Write an inequality of the form $x > c$ or $x < c$ to represent a constraint or condition in a real-world or mathematical problem. Recognize that inequalities of the form $x > c$ or $x < c$ have infinitely many solutions; represent solutions of such inequalities on number line diagrams.

Previous Learning
Students should know how to solve equations using multiplication or division.

Start Thinking! and Warm Up

8.3 Record and Practice Journal

English Language Learners

Vocabulary

It is important that English learners understand the difference between *less than* and *less than or equal to,* as well as *greater than* and *greater than or equal to.* Starting with 1, give each student a unique number card. Tell students to stand up if their number is *less than* (say a number) and then *less than or equal to* that same number. Students should discuss the difference between the two. This can be repeated with *greater than* and *greater than or equal to.*

8.3 Record and Practice Journal

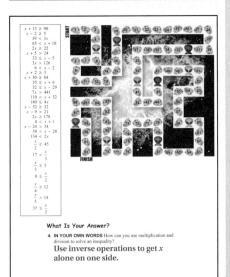

Laurie's Notes

Activity 3

- Give students time to read through the directions and ask questions.
- This activity provides fun practice with solving inequalities and with practicing mental math strategies!
- Students will be solving many inequalities and should become better at judging which inequalities to skip over as they scan for an appropriate inequality.
- **Big Idea:** The third bullet says that asteroids and aliens can only be destroyed in a straight line. In other words, you cannot turn a 90° angle to destroy additional asteroids or aliens.
- When students have finished, ask volunteers to explain their strategies for locating the next inequality to be solved.

What Is Your Answer?

- **Neighbor Check:** Have students work independently and then have their neighbor check their work. Have students discuss any discrepancies.

Closure

- **Exit Ticket:** Solve each inequality.

 $4x \geq 16$ $x \geq 4$ and $\dfrac{x}{4} \geq 16$ $x \geq 64$

Technology For the Teacher

The Dynamic Planning Tool
Editable Teacher's Resources at *BigIdeasMath.com*

3 ACTIVITY: Spaceman Game

$x + 13 \geq 90$
$x - 2 \geq 5$
$39 < 3x$
$65 < x + 10$
$2x \geq 22$
$x + 5 > 24$
$33 \leq x - 5$
$3x > 126$
$6 < x - 2$
$x + 2 \geq 3$
$x + 30 \geq 84$
$35 \leq x + 6$
$32 \leq x - 29$
$7x > 441$
$110 < x + 32$
$180 \leq 4x$
$x - 52 \geq 32$
$x - 9 > 21$
$2x \geq 178$
$4 < x + 1$
$x - 24 > 34$
$58 < x - 28$
$134 < 2x$
$\dfrac{x}{2} \geq 45$
$17 < \dfrac{x}{3}$
$\dfrac{x}{3} \geq 5$
$8 \leq \dfrac{x}{2}$
$\dfrac{x}{4} \geq 12$
$\dfrac{x}{5} > 14$
$37 \leq \dfrac{x}{2}$

Play with a partner.

- Player 1: Start the game by choosing an inequality from the list that has a solution of $x \geq 1$. This allows you to destroy Asteroids 1, 2, and 3.
- Player 2: Asteroid 3 is destroyed. Choose an inequality from the list that has a solution of $x > 3$ or $x \geq 4$. This allows you to destroy Asteroid 4, an alien, and Asteroid 6.
- You can only destroy asteroids and aliens in a straight line.
- Each asteroid is worth 1 point. Each alien is worth 2 points.
- Take turns until someone reaches the end of the maze. The player with the most points wins.
- If an incorrect inequality is chosen, then the player loses that turn.

What Is Your Answer?

4. **IN YOUR OWN WORDS** How can you use multiplication or division to solve an inequality?

Practice — Use what you learned about solving inequalities using multiplication or division to complete Exercises 8–11 on page 348.

Section 8.3 Solving Inequalities Using Multiplication or Division 345

8.3 Lesson

Key Ideas

Multiplication Property of Inequality

Words If you multiply each side of an inequality by the same *positive* number, the inequality remains true.

Numbers $8 > 6$

$8 \times 2 > 6 \times 2$

$16 > 12$

Algebra $\dfrac{x}{4} < 2$

$\dfrac{x}{4} \cdot 4 < 2 \cdot 4$

$x < 8$

Division Property of Inequality

Words If you divide each side of an inequality by the same *positive* number, the inequality remains true.

Numbers $8 > 6$

$8 \div 2 > 6 \div 2$

$4 > 3$

Algebra $4x < 8$

$\dfrac{4x}{4} < \dfrac{8}{4}$

$x < 2$

These properties are also true for ≤ and ≥.

Remember
Multiplication and division are inverse operations.

EXAMPLE 1 Solving an Inequality Using Multiplication

Solve $\dfrac{x}{5} \leq 2$. Graph the solution.

$\dfrac{x}{5} \leq 2$ Write the inequality.

$\dfrac{x}{5} \cdot 5 \leq 2 \cdot 5$ Multiply each side by 5. *(Undo the division.)*

$x \leq 10$ Simplify.

∴ The solution is $x \leq 10$.

Check: $x = 5$ is a solution. Check: $x = 15$ is *not* a solution.

On Your Own

Solve the inequality. Graph the solution.

Now You're Ready
Exercises 6–9

1. $p \div 3 > 2$
2. $\dfrac{q}{6} \leq 5$
3. $1 < \dfrac{s}{7}$

346 Chapter 8 Inequalities

Laurie's Notes

Introduction

Connect
- **Yesterday:** Students developed an intuitive understanding about how multiplication and division inequalities are solved.
- **Today:** Students will use the Properties of Inequality to solve inequalities.

Motivate
- Present the following problem: If you were the owner of a theme park that included many different rides, would you sell individual tickets for each ride or would you sell a pass that allowed an unlimited number of rides?
- Allow students to discuss this problem. It is related to Example 3. The hope is that students will recognize that there is a point at which the individual tickets start to cost more for the consumer than if they had purchased a pass. The other factor to consider is the consumer who only wishes to take a few rides. They may not want to purchase a pass.

Lesson Notes

Key Ideas
- Work through the Words, Numbers, and Algebra.
- Explain that solving inequalities is similar to solving equations.
- Today's inequalities involve multiplication and division which are inverse operations.
- **Teaching Tip:** Students will want to solve the inequalities in their head and resist writing their work down. Remind students that you are trying to help them develop good problem solving techniques. The problems will become more involved later and students need to know how to record the process.
- The inequalities solved in this lesson do *not* involve multiplying or dividing by a negative quantity. This type of inequality is taught in a future course.

Example 1
- Work through the example.
- **?** "How do you graph $x \leq 10$?" closed circle at 10, shade to the left of 10
- **?** "How do you know which side of $x = 10$ to shade?" test values "Why is there a closed circle?" 10 is a solution

On Your Own
- **Think-Pair-Share:** Students should read each question independently and then work with a partner to answer the questions. When they have answered the questions, the pair should compare their answers with another group and discuss any discrepancies.
- Note that division in Question 1 is represented differently than in the next two questions.
- **Common Error:** In Question 3, because the variable is on the right, it is common for students to shade the wrong side of the inequality when graphing.

Goal Today's lesson is solving inequalities using multiplication or division.

Start Thinking! and Warm Up

Lesson 8.3 Warm Up For use before Lesson 8.3

Lesson 8.3 Start Thinking! For use before Lesson 8.3

On some cell phone plans, text messages cost $0.20 per message.

Write a word problem involving the cost of text messages that can be solved with an inequality using division.

Exchange problems with a classmate and solve your classmate's problem.

Extra Example 1
Solve $4 \geq r \div 2$. Graph the solution.
$8 \geq r$ or $r \leq 8$

On Your Own

1. $p > 6$;

2. $q \leq 30$;

3. $7 < s$;

Extra Example 2

Solve $8h < 56$. Graph the solution. $h < 7$

Extra Example 3

A one-way bus ride costs $1.50. A 30-day bus pass costs $30.

a. Write and solve an inequality to find the least number of one-way rides you must take for the 30-day pass to be a better deal. $1.5r > 30$; $r > 20$

b. You ride the bus an average of 24 times each month. Is the pass a better deal? Explain. Yes. The cost of 24 one-way rides is more than $30. So, the pass is a better deal.

On Your Own

4. $k \leq 3$;

5. $j > 4$;

6. $25 \leq m$;

7. $2.5x \leq 15$; $x \leq 6$

Differentiated Instruction

Visual

Students may question why they are asked to graph the solution of an inequality, but not asked to graph the solution of an equation. The graph of an equation is one point on the number line. The graph of an inequality provides more information, because of the infinite number of possible solutions.

Laurie's Notes

Example 2

? "How do you undo a multiplication inequality?" divide
? "Why is the circle open?" The solution does not include 8.

Example 3

? Discuss how multiple-day bus passes and single ride tickets work.
 • "How many rides can you take with a 30-day bus pass?" unlimited
 • "If you took 42 rides, what would the cost be per ride? Explain your thinking." $1.00; Divide the cost of $42 by the number of rides.
 • "If you took 42 rides and paid each time, what would the total cost be? Explain your thinking." $73.50; Multiply 42 rides by $1.75, the cost of a single ride.

? Discuss how to find the least number of rides you would need to take in order to make the cost of the 30-day pass a better deal.
 • "How many rides do you need to take before the cost of the individual tickets is at least as expensive as the 30-day pass?" Set up the inequality and solve.

• **Estimate:** Encourage students to estimate. $42 \div 1.75$ is approximately $42 \div 2 = 21$.

On Your Own

• **Think-Pair-Share:** Students should read each question independently and then work with a partner to answer the questions. When they have answered the questions, the pair should compare their answers with another group and discuss any discrepancies.

• Note that Question 5 is written as $5 \cdot j$ instead of $5j$. The inequality is still solved by dividing both sides by 5.

• Question 6 is written with the variable on the right side. Students should rewrite the solution as $m \geq 25$.

Closure

• **Writing Prompt:** To solve inequalities, … Make sure that students include all 4 operations in their description.

Technology For the Teacher

The Dynamic Planning Tool
Editable Teacher's Resources at *BigIdeasMath.com*

EXAMPLE 2 Solving an Inequality Using Division

Solve $4n > 32$. Graph the solution.

Undo the multiplication.

$4n > 32$ Write the inequality.

$\dfrac{4n}{4} > \dfrac{32}{4}$ Divide each side by 4.

$n > 8$ Simplify.

∴ The solution is $n > 8$.

EXAMPLE 3 Real-Life Application

A one-way bus ride costs $1.75. A 30-day bus pass costs $42.

a. Write and solve an inequality to find the least number of one-way rides you must take for the 30-day pass to be a better deal.

b. You ride the bus an average of 20 times each month. Is the pass a better deal? Explain.

a. **Words** The price of a one-way ride times the number of one-way rides is more than $42.

Variable Let r be the number of one-way rides.

Inequality 1.75 • r > 42

$1.75r > 42$ Write the inequality.

$\dfrac{1.75r}{1.75} > \dfrac{42}{1.75}$ Divide each side by 1.75.

$r > 24$ Simplify.

∴ You need to take more than 24 one-way rides for the pass to be a better deal.

b. No. The cost of 20 one-way rides is less than $42. So, the pass is not a better deal.

On Your Own

Solve the inequality. Graph the solution.

4. $11k \leq 33$ **5.** $5 \cdot j > 20$ **6.** $50 \leq 2m$

7. The sign shows the toll for driving on Alligator Alley. Write and solve an inequality to represent the number of times someone can drive on Alligator Alley with $15.

Passenger Cars Toll $2.50

Section 8.3 Solving Inequalities Using Multiplication or Division

8.3 Exercises

Vocabulary and Concept Check

1. **REASONING** How is the graph of the solution of $2x \geq 10$ different from the graph of the solution of $2x = 10$?

Name the property you should use to solve the inequality.

2. $3x \leq 27$
3. $7x > 49$
4. $\dfrac{x}{2} < 36$

5. **OPEN-ENDED** Write two inequalities that have the same solution set: one that can be solved using division and one that can be solved using multiplication.

Practice and Problem Solving

Solve the inequality. Graph the solution.

 6. $\dfrac{m}{8} < 4$ 7. $\dfrac{n}{6} > 2$ 8. $\dfrac{t}{3} \geq 15$ 9. $\dfrac{c}{11} \geq 9$

 10. $12x < 96$ 11. $5x \geq 25$ 12. $8w \leq 72$ 13. $7p \leq 42$

14. $\dfrac{b}{4} > 13$ 15. $6x < 90$ 16. $3s \geq 36$ 17. $\dfrac{v}{9} \leq 45$

18. $4t > 72$ 19. $\dfrac{q}{5} \leq 35$ 20. $12m < 132$ 21. $\dfrac{x}{9} \geq 18$

22. **ERROR ANALYSIS** Describe and correct the error in solving the inequality.

23. **GEOMETRY** The length of a rectangle is 8 feet and its area is less than 168 square feet. Write and solve an inequality to represent the width of the rectangle.

24. **PLAYGROUND** Students at a playground are divided into five equal groups with at least six students in each group. Write and solve an inequality to represent the number of students at the playground.

Write the word sentence as an inequality. Then solve the inequality.

25. Eight times a number n is less than 72.

26. A number t divided by 32 is at most 4.25.

27. 225 is no less than 12 times a number w.

Assignment Guide and Homework Check

Level	Day 1 Activity Assignment	Day 2 Lesson Assignment	Homework Check
Basic	8–11, 39–42	1–5, 7, 13–21 odd, 22–26	7, 13, 22, 24, 26
Average	8–11, 39–42	1–5, 13–21 odd, 22, 26, 27, 29–31	13, 17, 22, 26, 29
Advanced	8–11, 39–42	1–5, 20–22, 28, 29, 31, 32, 34–38	20, 28, 32, 34

For Your Information
- **Exercises 35–38** Encourage students to use numerical values for the variables to test if the statement is true.

Common Errors
- **Exercises 6–21** Students may do the same operation on both sides instead of the opposite operation. Remind them that to get the variable alone, they need to use the opposite operation.
- **Exercises 6–21** Students may also switch the direction of the inequality when multiplying or dividing.
- **Exercises 28 and 29** Students may not understand how to graph these two inequalities together. Encourage them to draw the graph for each inequality separately, one above the other, and then lay them on top of each other and remove the parts that do not overlap.

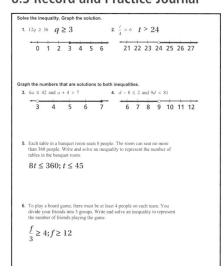

8.3 Record and Practice Journal

Technology For the Teacher
Answer Presentation Tool
QuizShow

Vocabulary and Concept Check

1. The solution of $2x \geq 10$ includes the solution of $2x = 10$, $x = 5$, and all other x values that are greater than 5.
2. Div. Prop. of Ineq.
3. Div. Prop. of Ineq.
4. Mult. Prop. of Ineq.
5. Sample answer: $\dfrac{x}{2} \geq 4$, $2x \geq 16$

Practice and Problem Solving

6. $m < 32$;

7. $n > 12$;

8. $t \geq 45$;

9. $c \geq 99$;

10. $x < 8$;

11. $x \geq 5$;

12. $w \leq 9$;

13. $p \leq 6$;

14. $b > 52$;

15. $x < 15$;

16. $s \geq 12$;

17. See Additional Answers.

T-348

Practice and Problem Solving

18. $t > 18$;

19. $q \leq 175$;

20. $m < 11$;

21–29. See *Additional Answers*.

30. See *Taking Math Deeper*.

31–38. See *Additional Answers*.

Fair Game Review

39. $x = 5$

40. $w = 96$

41. $x = 12$

42. C

Mini-Assessment

Solve the inequality. Graph the solution.

1. $\dfrac{r}{3} \geq 2$ $r \geq 6$

2. $1 > \dfrac{w}{7}$ $w < 7$

3. $5g > 25$ $g > 5$

4. $60 \geq 10s$ $s \leq 6$

5. You use a square section of your yard for a sandbox. You have at most 24 feet of wood to build the sides of the sandbox. Write and solve an inequality to represent the possible lengths of each side of the sandbox. $4x \leq 24$; $x \leq 6$

T-349

Taking Math Deeper

Exercise 30

This problem gives students an opportunity to estimate time. A good way to begin is to organize the given information.

 Organize the given information.

| 12 people per ride | 15,000 people per day |

 How many rides are needed for 15,000 people?

$$\dfrac{15,000}{12} = 1250 \text{ rides}$$

Let x be the number of rides per day.

a. $x \geq 1250$

 The park is open for 12 hours per day. To have 1250 rides in a day, the thrill ride would have to operate

$$\dfrac{1250}{12} \approx 104.2 \text{ times per hour.}$$

b. This means that the thrill ride would have to operate once every 35 seconds. Even without being at the ride and seeing it, it seems unreasonable to unload 12 people, load another 12 people, and conduct the ride in only 35 seconds.

Project

There are several drop towers in amusement parks around the country. Research the height of the different towers. Draw a bar graph that compares the heights.

Reteaching and Enrichment Strategies

If students need help...	If students got it...
Resources by Chapter • Practice A and Practice B • Puzzle Time Record and Practice Journal Practice Differentiating the Lesson Lesson Tutorials Skills Review Handbook	Resources by Chapter • Enrichment and Extension • School-to-Work Start the next section

Graph the numbers that are solutions to both inequalities.

28. $x + 7 > 9$ and $8x \leq 64$

29. $x - 3 \leq 8$ and $6x < 72$

30. THRILL RIDE A thrill ride at an amusement park holds a maximum of 12 people per ride.

 a. Write and solve an inequality to find the least number of rides needed for 15,000 people.

 b. Do you think it is possible for 15,000 people to ride the thrill ride in one day? Explain.

31. FOOTBALL A winning football team more than doubled the offensive yards gained by its opponent. The opponent gained 272 offensive yards. The winning team had 80 offensive plays. Write and solve an inequality to find the possible number of yards per play for the winning team.

Park Hours
10:00 A.M.–10:00 P.M.

32. LOGIC Explain how you know that $7x < 7x$ has no solution.

33. OPEN-ENDED Give an example of a real-life situation in which you can list all the solutions of an inequality. Give an example of a real-life situation in which you cannot list all the solutions of an inequality.

34. FUNDRAISER You are selling items from a catalog for a school fundraiser. Write and solve two inequalities to find the range of sales that will earn you between $40 and $50.

 Let $a > b$ and $x > y$. Tell whether the statement is *always* true. Explain your reasoning.

35. $a + x > b + y$ **36.** $a - x > b - y$ **37.** $ax > by$ **38.** $\dfrac{a}{x} > \dfrac{y}{b}$

Fair Game Review What you learned in previous grades & lessons

Solve the equation. Check your solution. *(Section 7.4)*

39. $3x - 8 = 7$ **40.** $\dfrac{w}{4} - 3 = 21$ **41.** $4x + 5 = 53$

42. MULTIPLE CHOICE On a normal day, 12 airplanes arrive at an airport every 15 minutes. Which rate does not represent this situation? *(Section 5.2)*

 Ⓐ $\dfrac{12 \text{ airplanes}}{15 \text{ minutes}}$ Ⓑ $\dfrac{4 \text{ airplanes}}{5 \text{ minutes}}$ Ⓒ $\dfrac{6 \text{ airplanes}}{9 \text{ minutes}}$ Ⓓ $\dfrac{48 \text{ airplanes}}{1 \text{ hour}}$

8.4 Solving Two-Step Inequalities

COMMON CORE STATE STANDARDS
6.EE.5
6.EE.8

Essential Question How can you use inequalities to classify different species of animals?

1 ACTIVITY: Classifying Dinosaurs

Work with a partner. Let L represent the length of an adult dinosaur.

a. Copy the diagram. Put each species of dinosaur shown on the next page into the correct region in the diagram.

b. Are all the dinosaur species that are in the "$L \leq 35$ feet" category also in the "$L \leq 50$ feet" category? Explain your reasoning.

c. Are all the dinosaur species that are in the "$L \leq 35$ feet" category also in the "$L \leq 20$ feet" category? Explain your reasoning.

d. Draw a different diagram that classifies the six dinosaur species by weight.

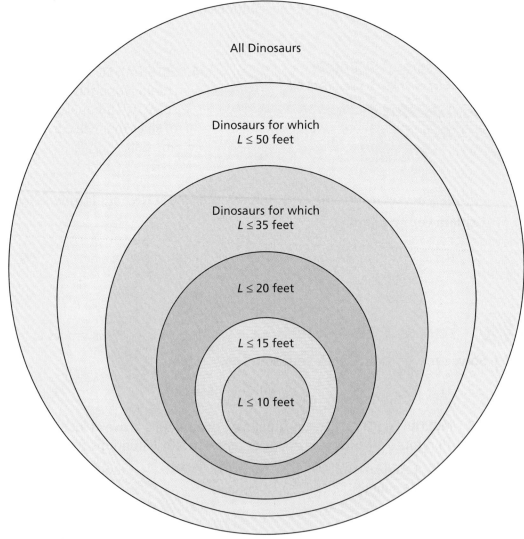

350 Chapter 8 Inequalities

Laurie's Notes

Introduction

For the Teacher
- **Goal:** Students will use a Venn diagram to organize dinosaurs by their lengths.
- **FYI:** A good reference for information about dinosaurs is *dinosaurs.about.com*.

Motivate
- It is important that students understand how Venn diagrams are used to organize data and to show logical relationships between data, or groups of things. Venn diagrams help to classify data.
- Place a large loop of yarn on the floor. Say, "This represents the students who are 6th graders." Ask about 10 students to participate. They should all stand inside the loop. You, on the other hand, are outside the loop.
- Place another loop inside the first. State that this represents the students who like vegetables. Students step inside the loops. If they like vegetables, they will be inside the inner loop; otherwise they'll be between the two loops of yarn.
- ? Place a third loop of yarn inside the second. State that this represents the students who like green beans. Students step inside the loops.
 - "Where are the students who only like broccoli?"
 - "Where are the students who like green beans?"
 - "Where are the students who don't like vegetables?"
- This should help students to understand that belonging to the smallest region in a nested Venn diagram makes you a member of every larger set.

Activity Notes

Activity 1
- There are two parts to this activity—classifying the dinosaurs by length and then classifying the dinosaurs by weight.
- Give students time to read the Venn diagram and ask questions before they begin.
- ? If students don't ask clarifying questions, you might ask the following.
 - "If you belong to the pink set, do you necessarily belong to the green set?" no "the blue set?" yes
 - "If you belong to the blue set, do you necessarily belong to the pink set?" no "Give an example." A dinosaur that is 40 feet long belongs to the blue set, but would not belong to the pink set.
- **Big Idea:** The Venn diagram shows the logical relationship between sets. In this context, if a dinosaur is less than 20 feet long, then it is also less than 40 feet long. The converse is not true: a dinosaur that is less than 40 feet long may be greater than 20 feet long.

Common Core State Standards

6.EE.5 Understand solving an equation or inequality as a process of answering a question: which values from a specified set, if any, make the equation or inequality true? Use substitution to determine whether a given number in a specified set makes an equation or inequality true.

6.EE.8 Write an inequality of the form $x > c$ or $x < c$ to represent a constraint or condition in a real-world or mathematical problem. Recognize that inequalities of the form $x > c$ or $x < c$ have infinitely many solutions; represent solutions of such inequalities on number line diagrams.

Previous Learning
Students should know how to solve two-step equations.

Start Thinking! and Warm Up

Activity 8.4 Start Thinking! For use before Activity 8.4

Activity 8.4 Warm Up For use before Activity 8.4

Tell whether the given value is a solution of the inequality.

1. $x + 1 > 8; x = 6$
2. $y - 8 \leq 9; y = 8$
3. $3r \geq 9; r = 3$
4. $\dfrac{s}{6} < 10; s = 66$
5. $5 \geq z - 1; z = -1$
6. $4 \geq p + 2; p = -4$

8.4 Record and Practice Journal

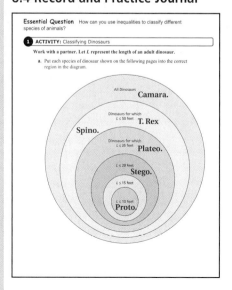

T-350

Differentiated Instruction

Kinesthetic

Teachers should have students use algebra tiles to work through the steps for solving these inequalities. Make sure that students have an inequality sign in the model instead of an equal sign.

8.4 Record and Practice Journal

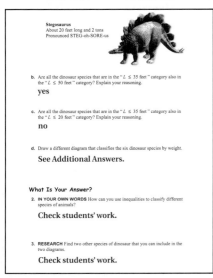

b. Are all the dinosaur species that are in the "$L \leq 35$ feet" category also in the "$L \leq 50$ feet" category? Explain your reasoning.

yes

c. Are all the dinosaur species that are in the "$L \leq 35$ feet" category also in the "$L \leq 20$ feet" category? Explain your reasoning.

no

d. Draw a different diagram that classifies the six dinosaur species by weight.

See Additional Answers.

What Is Your Answer?

2. IN YOUR OWN WORDS How can you use inequalities to classify different species of animals?

Check students' work.

3. RESEARCH Find two other species of dinosaur that you can include in the two diagrams.

Check students' work.

Laurie's Notes

Activity 1 Continued

- Discuss answers to part (b) and part (c).
- To classify the dinosaurs by weight, students must first decide what the categories should be.
- ? "How many pounds are in 1 ton?" 2000
- ? "What would be a good strategy for deciding what categories to use for your new Venn diagram?" By listing the weights in order, or somehow clustering the weights, students will have a better sense as to how the categories could be defined.
- Share the results of different groups as there will be more than one correct solution.

What Is Your Answer?

- **Think-Pair-Share:** Students should read each question independently and then work with a partner to answer the questions. When they have answered the questions, the pair should compare their answers with another group and discuss any discrepancies.

Closure

- "What other attributes about a dinosaur could be classified using a Venn diagram?" *Sample answer:* Height, type of food it eats, what prehistoric era the dinosaur lived in

The Dynamic Planning Tool
Editable Teacher's Resources at *BigIdeasMath.com*

Tyrannosaurus Rex
About 40 feet long and 7 tons
Pronounced tih-RAN-oh-SORE-us REX

Spinosaurus
About 50 feet long and 7 tons
Pronounced SPINE-oh-SORE-us

Camarasaurus
About 60 feet long and 20 tons
Pronounced cam-AH-rah-SORE-us

Plateosaurus
About 25 feet long and 2 tons
Pronounced PLATT-ee-oh-SORE-us

Protoceratops
About 6 feet long and 400 pounds
Pronounced PRO-toe-SER-ah-tops

Stegosaurus
About 20 feet long and 2 tons
Pronounced STEG-oh-SORE-us

What Is Your Answer?

2. **IN YOUR OWN WORDS** How can you use inequalities to classify different species of animals?

3. **RESEARCH** Find two other species of dinosaur that you can include in the two diagrams.

8.4 Lesson

You can solve two-step inequalities the same way you solve two-step equations.

EXAMPLE 1 Solving Two-Step Inequalities

a. Solve $3x + 2 > 17$. Graph the solution.

	$3x + 2 >$	17	Write the inequality.
Step 1: Undo the addition. →	-2	-2	Subtract 2 from each side.
	$3x >$	15	Simplify.
Step 2: Undo the multiplication. →	$\dfrac{3x}{3} >$	$\dfrac{15}{3}$	Divide each side by 3.
	$x >$	5	Simplify.

∴ The solution is $x > 5$.

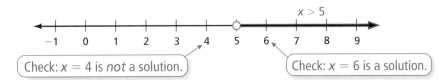

Check: $x = 4$ is *not* a solution. Check: $x = 6$ is a solution.

b. Solve $\dfrac{y}{6} - 9 \le 1$. Graph the solution.

$\dfrac{y}{6} - 9 \le$	1	Write the inequality.
$+9$	$+9$	Add 9 to each side.
$\dfrac{y}{6} \le$	10	Simplify.
$\dfrac{y}{6} \cdot 6 \le$	$10 \cdot 6$	Multiply each side by 6.
$y \le$	60	Simplify.

∴ The solution is $y \le 60$.

On Your Own

Solve the inequality. Graph the solution.

1. $4b - 1 < 7$ **2.** $8 + 5c \ge 28$ **3.** $\dfrac{n}{2} + 11 > 12$

Laurie's Notes

Introduction

Connect
- **Yesterday:** Students reviewed inequalities by using inequalities and Venn diagrams to classify dinosaurs.
- **Today:** Students will use the Properties of Inequality to solve two-step inequalities.

Motivate
- Hold a sealed envelope against your forehead. The envelope has a slip of paper inside with the inequality $x \leq 7$ written on it. Say, "I'm thinking of a lot of numbers that I doubled and then I subtracted 5. All of my answers were 9 or less." Show that $x \leq 9$ is written on the outside of the envelope.
- ? "With what numbers did I start? Explain how you know." Listen for students to "undo" your process in reverse: add 5 to numbers 9 or less and the result is 14 or less; then divide by 2 and the numbers are 7 or less.
- Open the envelope and reveal the inequality $x \leq 7$ on your slip of paper.
- Record on the board what you said and how the students solved it:
 (you) Think of numbers → double the numbers → subtract 5 → answer is 9 or less.
 (students) Start with 9 or less → add 5 → divide numbers by 2 → answer is 7 or less.
- ? "What inequality did you just solve?" $2x - 5 \leq 9$

Lesson Notes

Example 1
- **Big Idea:** Discuss the idea that you can solve two-step *inequalities* the same way you solve two-step *equations*.
- **Teaching Tip:** Switch colors of chalk or overhead pens to help focus on the operations performed on each side of the inequality.
- **Common Error:** Some students may divide by 3 before subtracting 2. Remind students that in solving the inequality, you undo the operations in the reverse order of the order of operations.
- ? For part (b), "What is the first step in solving this inequality?"
 Add 9 to each side of the inequality.
- **Representation:** When writing the second step, multiplication may be represented using the multiplication dot or parentheses.

On Your Own
- Have volunteers write their solutions on the board.
- **Common Misconception:** In Question 2, students may begin by subtracting $5c$ from each side of the inequality. While this can be done, it does not help in solving the inequality. Some students see the $5c$ being added to 8, but not the 8 being added to the $5c$. Remind students that $8 + 5c$ is equivalent to $5c + 8$ (Commutative Property of Addition).

Goal Today's lesson is solving two-step inequalities.

Start Thinking! and Warm Up

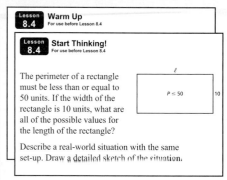

Extra Example 1
Solve the inequality. Graph the solution.

a. $2d - 7 < 5$ $d < 6$

b. $18 \geq \dfrac{t}{4} + 10$ $32 \geq t$ or $t \leq 32$

On Your Own

1. $b < 2$;

2. $c \geq 4$;

3. $n > 2$;

T-352

Extra Example 2

Members of the school spirit club are selling pompoms for $3 each. The club has raised $99 towards its goal of $300.

a. Write an inequality to represent the number of additional pompoms the club must sell to reach or exceed its goal. $3p + 99 \geq 300$

b. How many additional pompoms does the club need to sell to reach or exceed its goal? $p \geq 67$

On Your Own

4. $3x + 55 \leq 85$; $x \leq 10$ pitches

English Language Learners

Vocabulary

Give English learners the opportunity to use the precise language to solve an inequality. Write the inequality $2x + 4 > 8$ on the board. Have one student come to the board. For each step of the solution, call on another student to give the instructions for solving. The instructions should be given in complete sentences. The instructions for the inequality are:
(1) Subtract 4 from each side.
(2) Simplify.
(3) Divide each side by 2.
(4) Simplify.

Laurie's Notes

Example 2

- Discuss the thermometer and what information it conveys.
- Discuss why you need to know how to find the number of additional pizzas that must be sold in order to reach or surpass the goal.
- ❓ "How much will be raised if 100 pizzas are sold? Explain your thinking." $800, 100 times the $8 per pizza; "How much will you have total?" $800 + $1200 = $2000
- Write the inequality.
- ❓ "If the club wanted to only reach their goal, what *problem* would you solve?" $1200 + 8p = 3600$
- ❓ "Could the inequality also be written as $8p + 1200 \geq 3600$? Explain." yes; Commutative Property
- Solve the inequality.
- Remember to switch colors of chalk or overhead pens as you solve the inequality.
- ❓ "How would the inequality $p \leq 300$ be interpreted for this same problem?" The club sells at most 300 pizzas.

On Your Own

- Students may want to work with a partner to write and solve this inequality.
- **FYI:** In Little League baseball, players are restricted during the regular season to the number of pitches they may deliver.

League Age	Pitches per day
17–18	105 pitches per day
13–16	95 pitches per day
11–12	85 pitches per day
9–10	75 pitches per day
7–8	50 pitches per day

Closure

- **Exit Ticket:** Solve and graph the inequality $3x - 4 \geq 14$ $x \geq 6$

Technology For the Teacher

The Dynamic Planning Tool
Editable Teacher's Resources at *BigIdeasMath.com*

EXAMPLE 2 **Real-Life Application**

Members of a club are selling pizzas for $8 each.

a. Write an inequality to represent the number of additional pizzas the club must sell to reach or exceed its goal.

b. How many additional pizzas does the club need to sell to reach or exceed its goal?

a. From the chart, you know that the club has already raised $1200 of its $3600 goal. Because the club wants to *reach or exceed* $3600, use the symbol \geq.

Words | Amount plus the cost times the number is greater the already of each of additional than or goal. raised pizza pizzas sold equal to

Variable Let p be the number of additional pizzas sold.

Inequality 1200 + 8 · p \geq 3600

∴ An inequality is $1200 + 8p \geq 3600$.

b. Solve the inequality to find the number of additional pizzas the club must sell to reach or exceed its goal.

$$1200 + 8p \geq 3600 \quad \text{Write the inequality.}$$
$$\underline{-1200} \quad \underline{-1200} \quad \text{Subtract 1200 from each side.}$$
$$8p \geq 2400 \quad \text{Simplify.}$$
$$\frac{8p}{8} \geq \frac{2400}{8} \quad \text{Divide each side by 8.}$$
$$p \geq 300 \quad \text{Simplify.}$$

∴ The club must sell at least 300 additional pizzas.

● **On Your Own**

4. A baseball player throws 55 pitches and plans to pitch three more innings. The coach will not allow the player to throw more than 85 pitches in a game. Write and solve an inequality to find the average number of pitches the player can throw in each of the next three innings.

8.4 Exercises

Vocabulary and Concept Check

1. **REASONING** What is the first operation you would undo to solve the inequality $4x - 9 \leq 15$? Explain your reasoning.

2. **REASONING** Describe the steps you could use to solve the inequality $\frac{t}{7} + 4 > 6$.

3. **WHICH ONE DOESN'T BELONG?** Which one does *not* belong with the other three? Explain your reasoning.

$\frac{s}{4} - 7 \leq 3$ \quad $\frac{s}{5} \leq 8$ \quad $s - 7 \leq 12$ \quad $s \leq 40$

Practice and Problem Solving

Solve the inequality. Graph the solution.

4. $\frac{z}{3} + 2 \leq 5$

5. $5t - 3 > 7$

6. $8 \leq 8 + 4x$

7. $5 > \frac{s}{3} - 5$

8. $3n - 14 \geq 1$

9. $20 + 4a < 40$

10. $15 \leq \frac{b}{4} + 9$

11. $2y - 8 \leq 12$

12. $27 < 8w - 5$

13. $0.2 + 2.6c \leq 8$

14. $3.3 > \frac{d}{5} - 1.8$

15. $4k + 10.2 > 22.6$

ERROR ANALYSIS Describe and correct the error in solving the inequality.

16.

17.

18. **UNICYCLE** The first jump in a unicycle high jump contest is shown. The bar is raised 2 centimeters after each jump. Solve the inequality $2n + 10 \geq 26$ to find the number of additional jumps needed to meet or exceed the goal of clearing a height of 26 centimeters.

10 cm

354 Chapter 8 Inequalities

Assignment Guide and Homework Check

Level	Day 1 Activity Assignment	Day 2 Lesson Assignment	Homework Check
Basic	26–30	1–3, 5–17 odd, 18	5, 7, 17, 18
Average	26–30	1–3, 7–21 odd, 18, 22	7, 17, 18, 19
Advanced	26–30	1–3, 14–17, 20–25	14, 16, 20, 24

Common Errors

- **Exercises 4–15** Students may divide or multiply the coefficient first instead of adding or subtracting. Tell them that while this is a valid method, they must remember to divide each part of the equation by the coefficient. Suggest that they begin away from the variable and work toward it. Also point out to them that they are going backward through the order of operations.
- **Exercises 19–21** Students may try to find the perimeter instead of the area. Remind them that they are finding the possible areas.

Technology For the Teacher
Answer Presentation Tool
QuizShow

Vocabulary and Concept Check

1. Add 9 to each side to get $4x$ by itself.
2. Subtract 4 from each side and then multiply each side by 7.
3. $s - 7 \leq 12$ does not belong because the solution is $s \leq 19$ and the solutions of the other three are all $s \leq 40$.

Practice and Problem Solving

4. $z \leq 9$;
5. $t > 2$;
6. $x \geq 0$;
7. $s < 30$;
8. $n \geq 5$;
9. $a < 5$;
10. $b \geq 24$;
11. $y \leq 10$;
12. $w > 4$;
13. $c \leq 3$;
14. $d < 25.5$;
15. See Additional Answers.

T-354

Practice and Problem Solving

16. The x and the 1 were swapped without flipping the sign.
 $8 > 2x + 6$
 $2 > 2x$
 $1 > x$
 $x < 1$

17–21. See Additional Answers.

22. See *Taking Math Deeper*.

23–25. See Additional Answers.

Fair Game Review

26. 3 in. 27. 2 ft
28. 12 cm 29. 6.5 m
30. D

Mini-Assessment

Solve the inequality. Graph the solution.

1. $2x + 1 > 5$ $x > 2$

2. $4 \geq \dfrac{y}{3} + 2$ $y \leq 6$

3. $4z - 3 \geq 13$ $z \geq 4$

4. $\dfrac{t}{2} - 3 < 2$ $t < 10$

5. You want to sell at least $225 worth of cookies for a school fundraiser. So far, you have sold $66 worth of cookies. Each box of cookies costs $3. Write and solve an inequality to find how many more boxes you need to sell. $3x + 66 \geq 225; x \geq 53$

Taking Math Deeper

Exercise 22

Inequality problems can throw students off, simply because of the "inequality." A good way to approach the problem is to imagine that it is an "equality" problem. After the "equation" is written, decide which way the inequality symbol should point.

 Write an equation.

 a. Write an inequality and solve it. Recall the whale needs to eat at least 140 pounds of fish each day.

 $75 + 15x \geq 140$ Write inequality.
 $15x \geq 65$ Subtract 75 from each side.
 $x \geq 4\dfrac{1}{3}$ Divide each side by 15.

The whale needs to eat at least $4\dfrac{1}{3}$ more buckets of fish.

 b. If you only have a choice of whole buckets, then the whale should be given 5 more buckets of food. With 4 buckets, the whale would only get $75 + 60 = 135$ pounds of fish.

Project

Select a marine park. Research the amount of fish needed to feed all of the animals at the park. Determine the cost of the fish, not including storage and personnel needed to feed the animals. Using the entrance fee to the park, how many visitors are needed every day to meet the cost? What are some ways that park officials might help offset the daily costs?

Reteaching and Enrichment Strategies

If students need help...	If students got it...
Resources by Chapter • Practice A and Practice B • Puzzle Time Record and Practice Journal Practice Differentiating the Lesson Lesson Tutorials Skills Review Handbook	Resources by Chapter • Enrichment and Extension • School-to-Work Start the next section

T-355

AREA The area of the figure is at least 400 square meters. Write and solve a two-step inequality to represent the possible values of *x*. Use 3.14 for π.

19.
20.
21.

22. **KILLER WHALES** A killer whale has eaten 75 pounds of fish today. It needs to eat at least 140 pounds of fish each day.

 a. A bucket holds 15 pounds of fish. Write and solve an inequality to represent how many more buckets of fish the whale needs to eat.

 b. Should the whale be given *four* or *five* more buckets of fish? Explain.

23. **RECYCLING** The sixth grade goal is to collect at least 250 pledges in the Recycling Pledge Card Contest. Your class has 22 students. The other classes have already completed their collections with a total of 180 pledges. Write and solve an inequality to find how many more pledges each student needs for your class to reach its goal.

24. **FRIENDSHIP BRACELET** The bracelet is formed from knots. Each bracelet has two large knots on the ends and six smaller knots per diagonal row. What does the solution of the inequality $6n + 2 \leq 200$ represent?

25. Write and solve an inequality to help you decide when paying per visit to the zoo is a better deal than an individual membership. When is buying a friend membership a better deal?

Zoo Admission	
1. Pay per visit	$7 per visit
2. Individual membership	$35 per year
3. Friend membership	$65 per year
Member *and* 1 friend free each visit	

Fair Game Review What you learned in previous grades & lessons

Find the radius of a circle with the given diameter. *(Section 6.1)*

26. 6 inches 27. 4 feet 28. 24 centimeters 29. 13 meters

30. **MULTIPLE CHOICE** Write the ordered pair that represents the point. *(Skills Review Handbook)*

 Ⓐ (4, 0) Ⓑ (3, 3)
 Ⓒ (3, 4) Ⓓ (4, 3)

8.3–8.4 Quiz

Solve the inequality. Graph the solution. *(Section 8.3)*

1. $\dfrac{a}{3} < 5$
2. $121 \geq 11s$
3. $12p > 24$

Write the word sentence as an inequality. Then solve the inequality. *(Section 8.3)*

4. Three times a number x is more than 18.

5. A number d divided by 24 is at most 1.25.

6. 84 is no less than 7 times a number k.

Solve the inequality. Graph the solution. *(Section 8.4)*

7. $2n + 14 < 15$
8. $\dfrac{n}{2} + 3 > 5$
9. $5 \leq 6k - 1$

10. **BAND** A band director wants to divide members of the band into 4 equal groups with no more than 9 students in each group. Write and solve an inequality to represent the total number of band members. *(Section 8.3)*

11. **CHARITY WALK** You want to raise at least $300 for a charity walk. So far, you have raised $135. You ask donors to give $15. Write and solve an inequality to find how many more $15 donations you need. *(Section 8.4)*

12. **GARDEN** You want to use a square section of your yard for a garden. You have at most 52 feet of fencing to surround the garden. Write and solve an inequality to represent the possible lengths of each side of the garden. *(Section 8.3)*

13. **BOTTLED WATER** You need at least 24 bottles of water for a camping trip. You have 3 bottles left over from a previous camping trip. *(Section 8.4)*

 a. A store sells water in packages of 8 bottles each. Write and solve an inequality to represent the number of packages you need to buy for the trip.

 b. Would it be better to buy *two* or *three* packages of bottled water for the trip? Explain.

Alternative Assessment Options

Math Chat
Structured Interview
Student Reflective Focus Question
Writing Prompt

Student Reflective Focus Question
Ask students to use their own words to summarize the similarities and differences between solving equations and solving inequalities. Be sure that they include examples. Select students at random to present their summary to the class.

Study Help Sample Answers
Remind students to complete Graphic Organizers for the rest of the chapter.

6.

7.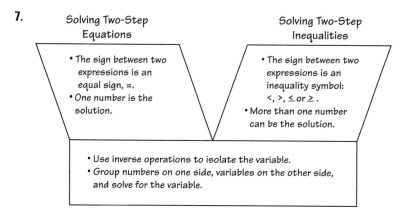

8. Available at *BigIdeasMath.com*

Reteaching and Enrichment Strategies

If students need help...	If students got it...
Resources by Chapter • Study Help • Practice A and Practice B • Puzzle Time Lesson Tutorials *BigIdeasMath.com* Practice Quiz Practice from the Test Generator	Resources by Chapter • Enrichment and Extension • School-to-Work Game Closet at *BigIdeasMath.com* Start the Chapter Review

Answers

1. $a < 15$;
2. $s \leq 11$;
3. $p > 2$;
4. $3x > 18$;
 $x > 6$
5. $\dfrac{d}{24} \leq 1.25$;
 $d \leq 30$
6. $84 \geq 7k$;
 $12 \geq k$
7. $n < \dfrac{1}{2}$;
8. $n > 4$;
9. $k \geq 1$;
10–13. See Additional Answers.

Technology For the Teacher
Answer Presentation Tool

Assessment Book

T-356

For the Teacher
Additional Review Options
- *QuizShow*
- Big Ideas Test Generator
- Game Closet at *BigIdeasMath.com*
- Vocabulary Puzzle Builder
- Resources by Chapter
 Puzzle Time
 Study Help

Answers

1. $m < 5$
2. $h \geq 12$
3. $k + 7 \leq 19$
4. $3b > 6$
5. yes
6. no
7. no
8. yes
9.
10. (number line: closed circle at 2, arrow left)
11. (number line: closed circle at -2, arrow right)
12. $x > 2$;
13. $k \leq 7$;
14. $y \geq 1$;
15. $x > 13$;
16. $n \geq 8$;

T-357

Review of Common Errors

Exercises 1–4
- Students may have difficulty choosing the correct inequality symbol when writing the word sentence as an inequality. Encourage them to put the word sentence into a real-life context.

Exercises 5–8
- Students may perform operations on both sides of the inequality (often not the same operation) to try to get both sides to be the same. Remind them that they are checking to see if the given value is a solution, so they should be substituting for the variable, simplifying, and determining whether the inequality is true.

Exercises 9–11
- Students may draw the arrow in the wrong direction or use the wrong circle when graphing an inequality. Remind them to test numbers in the inequality to determine the correct direction of the arrow as well as whether to use an open or a closed circle.

Exercises 12–26
- Students may use the same operation that is in the inequality, instead of the inverse operation, to find the solution. Remind them to use inverse operations, just as they did when solving equations. Also, remind them to check numbers that are solutions, as well as numbers that are *not* solutions, in the original inequality.

Exercises 15, 16, 24, 30, and 33
- Students may draw the arrow in the wrong direction when graphing a solution that has the variable on the right side of the inequality symbol. Suggest that if it is easier to see the correct direction of the arrow, they can rewrite their solutions with the variable on the left. Point out that if they do this, they must remember to reverse the direction of the inequality symbol.

Exercises 27–35
- Students may divide or multiply by the coefficient first instead of adding or subtracting first. Tell them that while this is a valid method, they must remember to divide or multiply *each part* of the inequality by the coefficient. Suggest that they begin away from the variable and work towards it.

8 Chapter Review

Review Key Vocabulary

inequality, *p. 330*
solution of an inequality, *p. 331*
solution set, *p. 331*
graph of an inequality, *p. 332*

Review Examples and Exercises

8.1 Writing and Graphing Inequalities (pp. 328–335)

Write the word sentence as an inequality.

a. A number x is more than 9.

A number x is more than 9.

$x > 9$

∴ An inequality is $x > 9$.

b. A number r divided by 2 is at most 4.

A number r divided by 2 is at most 4.

$\dfrac{r}{2} \leq 4$

∴ An inequality is $\dfrac{r}{2} \leq 4$.

Exercises

Write the word sentence as an inequality.

1. A number m is less than 5.
2. A number h is at least 12.
3. A number k plus 7 is no more than 19.
4. Three times a number b is greater than 6.

Tell whether the given value is a solution of the inequality.

5. $x + 3 \leq 4$; $x = 1$
6. $n - 8 > 3$; $n = 10$
7. $6k < 36$; $k = 6$
8. $\dfrac{t}{3} \geq 5$; $t = 18$

Graph the inequality on a number line.

9. $x < 0$
10. $a \geq 3$
11. $n \leq -1$

8.2 Solving Inequalities Using Addition or Subtraction (pp. 336–341)

Solve $1 \leq x - 4$. Graph the solution.

$$1 \leq x - 4 \quad \text{Write the inequality.}$$
$$\underline{+4 \quad +4} \quad \text{Add 4 to each side.}$$
$$5 \leq x \quad \text{Simplify.}$$

Undo the subtraction.

The inequality $5 \leq x$ is the same as $x \geq 5$.

∴ The solution is $x \geq 5$.

Exercises

Solve the inequality. Graph the solution.

12. $x + 1 > 3$
13. $k - 7 \leq 0$
14. $y + 8 \geq 9$
15. $24 < 11 + x$
16. $4 \leq n - 4$
17. $x - 20 > 24$
18. $b + 12 \leq 26$
19. $s - 1.5 < 2.5$
20. $\frac{1}{4} + m \leq \frac{1}{2}$

8.3 Solving Inequalities Using Multiplication or Division (pp. 344–349)

Solve $7n < 42$. Graph the solution.

$$7n < 42 \quad \text{Write the inequality.}$$
$$\frac{7n}{7} < \frac{42}{7} \quad \text{Divide each side by 7.}$$
$$n < 6 \quad \text{Simplify.}$$

Undo the multiplication.

∴ The solution is $n < 6$.

Exercises

Solve the inequality. Graph the solution.

21. $x \div 2 < 4$
22. $9n \geq 63$
23. $\frac{x}{5} > 10$
24. $9 \geq 3b$
25. $10p > 40$
26. $\frac{k}{4} \geq 15$

Review Game
Writing and Solving Inequalities

Materials
- color coded spinner with operation symbols
- 10 cards numbered 0 through 9
- 2 pencils
- paper

Players: 2

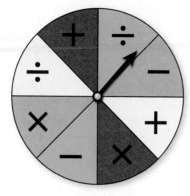

Directions
One player spins the spinner and draws two cards. The spinner is used to determine the operation and inequality symbol to be used. The inequality symbol is determined by color, as follows.

 Blue: <
 Yellow: >
 Red: ≤
 Green: ≥

The other player writes an inequality using the operation, the inequality symbol, the numbers on the cards, and *x*. For example, if a player spins red + and draws a 2 and a 3, the other player could form the inequality $2 + x \leq 3$.

Each player solves the inequality, exchanges their work with the other player, and checks the other player's work. Players receive 1 point for the round if their solution is correct. Each round can result in a total of 0, 1, or 2 points received between the two players. Players take turns spinning/drawing and writing the inequalities. Students can play several rounds or for a predetermined amount of time.

Who Wins?
The player with the most points wins.

For the Student
Additional Practice
- Lesson Tutorials
- Study Help (textbook)
- Student Website
 Multi-Language Glossary
 Practice Assessments

Answers

17. $x > 44$;

18. $b \leq 14$;

19. $s < 4$;

20. $m \leq \dfrac{1}{4}$;

21. $x < 8$;

22. $n \geq 7$;

23. $x > 50$;

24. $b \leq 3$;

25. $p > 4$;

26. $k \geq 60$;

27. $x < 1$;

28. $n \leq 24$;

29–35. See Additional Answers.

T-358

My Thoughts on the Chapter

What worked. . .

What did not work. . .

What I would do differently. . .

Teacher Tip

Not allowed to write in your teaching edition? Use sticky notes to record your thoughts.

8.4 Solving Two-Step Inequalities (pp. 350–355)

a. Solve $\frac{y}{3} + 1 > 6$. Graph the solution.

$\frac{y}{3} + 1 > 6$	Write the inequality.
Undo the addition. → $\quad -1 \quad -1$	Subtract 1 from each side.
$\frac{y}{3} > 5$	Simplify.
Undo the division. → $\frac{y}{3} \cdot 3 > 5 \cdot 3$	Multiply each side by 3.
$y > 15$	Simplify.

∴ The solution is $y > 15$.

b. Solve $10 \geq 3w - 5$. Graph the solution.

$10 \geq 3w - 5$	Write the inequality.
Undo the subtraction. → $\quad +5 \quad\quad +5$	Add 5 to each side.
$15 \geq 3w$	Simplify.
Undo the multiplication. → $\frac{15}{3} \geq \frac{3w}{3}$	Divide each side by 3.
$5 \geq w$	Simplify.

∴ The solution is $w \leq 5$.

Exercises

Solve the inequality. Graph the solution.

27. $4x + 1 < 5$
28. $\frac{n}{2} - 3 \leq 9$
29. $6w - 4 \geq 8$

30. $18 > \frac{t}{5} + 11$
31. $3c - 4 \geq 14$
32. $\frac{p}{4} + 1 > 1$

33. $25 \geq 7s + 4$
34. $\frac{x}{9} - 0.7 < 1.3$
35. $\frac{d}{6} - 1.4 > 0.1$

8 Chapter Test

Write the word sentence as an inequality.

1. A number x is at most 5.
2. A number m is more than -8.

Tell whether the given value is a solution of the inequality.

3. $5n > 20$; $n = 3$
4. $k - 8 \leq 6$; $k = 14$

Graph the inequality on a number line.

5. $x \geq 5$
6. $m \leq -2$

Solve the inequality. Graph the solution.

7. $x - 3 < 7$
8. $12 \geq n + 6$
9. $\dfrac{b}{3} \leq 12$
10. $\dfrac{y}{2} + 1 \geq 3$
11. $72 > 12p$
12. $9j - 3 \leq 15$

13. **HURRICANE** A hurricane has wind speeds that are greater than or equal to 74 miles per hour. Write an inequality to represent the possible wind speeds during a hurricane.

14. **T-SHIRTS** A soccer team will sell T-shirts for a fundraiser. The company that makes the T-shirts charges $8 per shirt plus a $20 shipping fee per order. The team plans to spend no more than $400.

 a. Write and solve an inequality to represent the number of T-shirts the team could order.
 b. What is the maximum number of T-shirts the team can order? Explain.

15. **LETTERS** The U.S. Postal Service delivers letters that weigh up to 3.5 ounces. Write an inequality to represent the weight of a letter you can mail.

LIBRARY FINES
$0.10 for each day late
$20.00 to replace lost book

16. **LIBRARY** The sign shows the fines for a book that is d days late. Write and solve an inequality to represent when paying to replace a book is less expensive than paying the fine.

17. **NUMBER SENSE** The possible values of n are given by $n - 4 \leq 10$. What is the greatest possible value of $\dfrac{n}{2}$?

Test Item References

Chapter Test Questions	Section to Review	Common Core State Standards
1–6, 13, 15	8.1	6.EE.5, 6.EE.8
7, 8, 17	8.2	6.EE.5, 6.EE.8
9, 11, 16, 17	8.3	6.EE.5, 6.EE.8
10, 12, 14	8.4	6.EE.5, 6.EE.8

Test-Taking Strategies

Remind students to quickly look over the entire test before they start so that they can budget their time. When writing word phrases as inequalities, students can get confused by subtle differences in wording, such as "is no more than" and "is no less than." Encourage students to think very carefully about which inequality symbol is implied by the wording. Students need to use the **Stop** and **Think** strategy before answering questions.

Common Assessment Errors

- **Exercises 1 and 2** Students may have difficulty choosing the correct inequality symbol when writing the word sentence as an inequality. Encourage them to put the word sentence into a real-life context.
- **Exercises 5 and 6** Students may draw the arrow in the wrong direction or use the wrong circle when graphing an inequality. Remind them to test numbers in the inequality to determine the correct direction of the arrow as well as whether to use an open or a closed circle.
- **Exercises 7–9, 11** Students may use the same operation that is in the inequality, instead of the inverse operation, to find the solution. Remind them to use inverse operations, just as they did when solving equations. Also, remind them to check numbers that are solutions, as well as numbers that are *not* solutions, in the original inequality.
- **Exercises 10 and 12** Students might first divide or multiply by the coefficient instead of adding or subtracting. Tell them that while this is a valid method, they must remember to divide or multiply *each part* of the inequality by the coefficient. Suggest that they begin away from the variable and work towards it.

Reteaching and Enrichment Strategies

If students need help...	If students got it...
Resources by Chapter • Practice A and Practice B • Puzzle Time Record and Practice Journal Practice Differentiating the Lesson Lesson Tutorials Practice from the Test Generator Skills Review Handbook	Resources by Chapter • Enrichment and Extension • School-to-Work Game Closet at *BigIdeasMath.com* Start Standardized Test Practice

Answers

1. $x \leq 5$
2. $m > -8$
3. no
4. yes
5. [number line with closed circle at 5, arrow left; marks -2, 0, 2, 4, 6, 8, 10]
6. [number line with closed circle at -2, arrow left; marks -8, -6, -4, -2, 0, 2, 4]
7. $x < 10$; [number line with open circle at 10, arrow left; marks -4, 0, 4, 8, 12, 16, 20]
8. $n \leq 6$; [number line with closed circle at 6, arrow left; marks -2, 0, 2, 4, 6, 8, 10]
9. $b \leq 36$; [number line with closed circle at 36, arrow left; marks 0, 8, 16, 24, 32, 40, 48]
10. $y \geq 4$; [number line with closed circle at 4, arrow right; marks -2, 0, 2, 4, 6, 8, 10]
11. $p < 6$; [number line with open circle at 6, arrow left; marks -4, 0, 4, 8, 12, 16, 20]
12. $j \leq 2$; [number line with closed circle at 2, arrow left; marks -6, -4, -2, 0, 2, 4, 6]
13–17. See Additional Answers.

Assessment Book

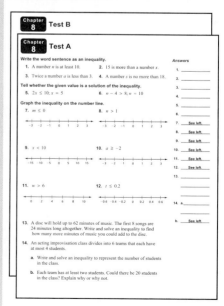

T-360

Test-Taking Strategies
Available at *BigIdeasMath.com*
After Answering Easy Questions, Relax
Answer Easy Questions First
Estimate the Answer
Read All Choices before Answering
Read Question before Answering
Solve Directly or Eliminate Choices
Solve Problem before Looking at Choices
Use Intelligent Guessing
Work Backwards

About this Strategy

One way to answer the question is to work backwards. Try putting the responses into the question, one at a time, and see if you get a correct solution.

Answers

1. A
2. H
3. 5
4. C
5. G

Item Analysis

1. A. Correct answer
 B. The student multiplies $2 \times 3 \times 5$ to get 30, but instead of writing 0.0030, the student incorrectly drops the 0 in 30 and gets 0.0003.
 C. Because there are 7 total decimal places in the factors, the student thinks that there should be 7 decimal places in the product.
 D. Because there are 7 total decimal places in the factors, the student thinks that there should be 7 zeros in the product.

2. F. The student finds only 10% of 25.
 G. The student divides 25 by 5, or the student makes a place value error while dividing 25 by 0.5.
 H. Correct answer
 I. The student divides 25 by 0.5.

3. **Gridded Response:** Correct answer: 5

 Common Error: The student misunderstands inequality signs and thinks that $42 < 42$, getting an answer of 6.

4. A. The student writes an expression that represents the difference of 18 and 5 instead of the sum. The student misinterprets "less than" and writes the expression on the right side of the equation in the wrong order.
 B. The student misinterprets "less than" and writes the expression on the right side of the equation in the wrong order.
 C. Correct answer
 D. The student writes an expression that represents the quotient of y and 3 instead of the product.

5. F. The student uses the incorrect type of circle on the number line.
 G. Correct answer
 H. The student adds 7 instead of subtracting, uses the incorrect type of circle, and shades in the wrong direction on the number line.
 I. The student adds 7 instead of subtracting and shades in the wrong direction on the number line.

6. A. Correct answer
 B. The student incorrectly multiplies by 6 where division is the correct inverse operation.
 C. The student incorrectly subtracts 9 where addition is the correct inverse operation.
 D. The student makes an error in logic with regard to the direction of the inequality.

Technology For the Teacher
Big Ideas Test Generator

8 Standardized Test Practice

1. Which decimal is closest to the value of the expression below? *(6.NS.3)*

 $$0.21 \times 0.31 \times 0.049$$

 A. 0.003 **C.** 0.0000003

 B. 0.0003 **D.** 0.00000003

2. What is 50% of 25? *(6.RP.3c)*

 F. 2.5 **H.** 12.5

 G. 5 **I.** 50

3. What is the greatest whole number value of x that will make the inequality below true? *(6.EE.5)*

 $$7x < 42$$

4. Which equation represents the word sentence below? *(6.EE.7)*

 The sum of 18 and 5 is equal to 9 less than the product of 3 and a number y.

 A. $18 - 5 = 9 - 3y$ **C.** $18 + 5 = 3y - 9$

 B. $18 + 5 = 9 - 3y$ **D.** $18 + 5 = \dfrac{y}{3} - 9$

5. Which number line is a graph of the solution to the inequality below? *(6.EE.8)*

 $$x + 7 \geq 12$$

 F.

 G.

 H. (number line with open circle at 19)

 I. (number line with closed circle at 19)

6. Your friend was solving the inequality in the box below.

$15 \leq 6n - 9$
$24 \leq 6n$
$n \leq 4$

What should your friend do to correct the error that she made? *(6.EE.8)*

 A. Rewrite the answer as $n \geq 4$.
 B. Multiply both sides of the inequality by 6.
 C. Subtract 9 from both sides of the inequality.
 D. Rewrite the given inequality as $6n - 9 \leq 15$.

7. An inequality is graphed on the number line below.

What is the least whole number value that is a solution to the inequality? *(6.EE.8)*

8. A recipe for pizza dough requires $4\frac{1}{2}$ cups of flour per pizza. How many cups of flour are required to make 5 pizzas using this recipe? *(6.NS.1)*

 F. $9\frac{1}{2}$ cups
 G. $20\frac{1}{2}$ cups
 H. $20\frac{2}{5}$ cups
 I. $22\frac{1}{2}$ cups

9. A hockey puck and its diameter are shown below.

What is the circumference of the hockey puck? (Use 3.14 for π.) *(6.EE.2c)*

 A. 9.42 in.
 B. 7.07 in.
 C. 6.14 in.
 D. 4.71 in.

Item Analysis (continued)

7. **Gridded Response:** Correct answer: 9

 Common Error: The student misinterprets the meaning of an open circle and gets an answer of 8.

8. **F.** The student adds $4\frac{1}{2}$ and 5 instead of multiplying.

 G. The student multiplies 5 by only the 4 and not by the $\frac{1}{2}$.

 H. The student multiplies 5 by 4 to get 20, but then incorrectly multiplies 5 by $\frac{1}{2}$ to get $\frac{2}{5}$.

 I. Correct answer

9. **A.** Correct answer

 B. The student finds the area, rounded to the nearest hundredth.

 C. The student finds the sum of 3 and 3.14.

 D. The student multiplies 3.14 by the radius instead of the diameter.

10. **F.** Correct answer

 G. The student chooses a benchmark percent that visually approximates the number of squares shaded red.

 H. The student thinks that the quantity of 48 squares is the same as 48%.

 I. The student chooses a benchmark percent that approximates the number of squares shaded red.

Answers

6. A
7. 9
8. I
9. A

Answers

10. F

11. *Part A* $750 + 150w \geq 4000$ or an equivalent inequality

 Part B 22 weeks

 Part C $\frac{3}{4}(20x) = 150$

 Part D $x = \$10$ per hour

Answers for Extra Examples

1. **A.** Correct answer

 B. The student does not correctly compare the unit cost in answer choice B, $0.325, with the unit cost in answer choice A, $0.30.

 C. The student does not correctly compare the unit cost in answer choice C, $0.375, with the unit cost in answer choice A, $0.30.

 D. The student does not correctly compare the unit cost in answer choice D, $0.33, with the unit cost in answer choice A, $0.30.

2. **F.** The student does not recognize the equation as an example of the Commutative Property of Addition.

 G. Correct answer

 H. The student does not recognize the equation as an example of the Commutative Property of Addition.

 I. The student does not recognize the equation as an example of the Commutative Property of Addition.

Item Analysis (continued)

11. **4 points** The student demonstrates a thorough understanding of writing and solving equations and inequalities. In Part A, the student writes the inequality, $750 + 150w \geq 4000$, or an equivalent inequality. In Part B, the student correctly solves and interprets the inequality to get an answer of 22 weeks.

 In Part C, the student writes the equation, $\frac{3}{4}(20x) = 150$, or an equivalent equation. In Part D, the student correctly solves and interprets the equation to get an answer of $10 per hour. The student provides accurate work and clear, complete explanations.

 3 points The student demonstrates an understanding of writing and solving equations and inequalities, but the student's work and explanations demonstrate an essential but less than thorough understanding.

 2 points The student demonstrates a partial understanding of writing and solving equations and inequalities. The student's work and explanations demonstrate a lack of essential understanding.

 1 point The student demonstrates a limited understanding of writing and solving equations and inequalities. The student's response is incomplete and exhibits many flaws.

 0 points The student provided no response, a completely incorrect or incomprehensible response, or a response that demonstrates insufficient understanding of writing and solving equations and inequalities.

Extra Examples for Standardized Test Practice

1. You are comparing the costs of buying bottles of water at the supermarket. Which of the following has the least cost per liter?

 A. six 1-liter bottles for $1.80

 B. one 2-liter bottle for $0.65

 C. eight $\frac{1}{2}$-liter bottles for $1.50

 D. twelve $\frac{1}{2}$-liter bottles for $1.98

2. Which property is demonstrated by the equation below?

 $$723 + (884 + 277) = 723 + (277 + 884)$$

 F. Associative Property of Addition

 G. Commutative Property of Addition

 H. Distributive Property

 I. Identity Property of Addition

10. In the mural below, the squares that are painted red are marked with the letter R.

What percent of the mural is painted red? (6.RP.3c)

F. 24%

H. 48%

G. 25%

I. 50%

11. Jeff wants to save enough money to buy a used car. He has already saved $750, and he plans to continue to save $150 per week. He plans to spend at least $4000 on a used car. (6.EE.7, 6.EE.8)

Part A Write an inequality to represent the number of weeks it will take Jeff to be able to afford a used car. Explain your reasoning.

Part B How many weeks will it take Jeff to be able to afford a used car? Show your work and explain your reasoning.

Jeff plans to save $\frac{3}{4}$ of what he earns at his job. He can work 20 hours per week.

Part C Write an equation to represent the amount per hour that Jeff must earn to save $150 per week. Explain your reasoning.

Part D What is the amount per hour that Jeff must earn? Show your work and explain your reasoning.

9 Tables, Graphs, and Functions

9.1 Mapping Diagrams
9.2 Functions as Words and Equations
9.3 Input-Output Tables
9.4 Graphs
9.5 Analyzing Graphs

"Here's a math anagram."

"Here's another one."

"The price of dog biscuits is up again this month."

"But I have a really good feeling about November."

Connections to Previous Learning

- Recognize and write algebraic expression for functions with two operations.

- Graph and describe continuous data, such as a quantity that changes over time.

- Construct and analyze tables, graphs, and equations to describe linear functions and other simple relations.
- Write, solve, and graph one-step and two-step linear equations.

Math in History

All ancient cultures had a concept of how long a day is. However, dividing a day into smaller units such as hours and minutes varied in different cultures.

★ The Babylonians seem to have been the first to divide the day into 24 units (hours) and then subdivide each hour into 60 minutes. They even went further and divided each minute into 60 seconds. This was not surprising because the Babylonian number system was written in base 60.

★ Up until 1000 A.D., the time keeping system in India was based purely on the number 60. There were 60 hours in a day, 60 minutes in an hour, and 60 seconds in a minute. Around 1000 A.D., Ali Ahmad Nasawi suggested using the ancient Babylonian system. It didn't take long before this system was accepted as the world standard.

Pacing Guide for Chapter 9

Chapter Opener	1 Day
Section 1 Activity	1 Day
Lesson	1 Day
Section 2 Activity	1 Day
Lesson	1 Day
Section 3 Activity	1 Day
Lesson	1 Day
Study Help / Quiz	1 Day
Section 4 Activity	1 Day
Lesson	1 Day
Section 5 Activity	1 Day
Lesson	1 Day
Quiz / Chapter Review	1 Day
Chapter Test	1 Day
Standardized Test Practice	1 Day
Total Chapter 9	**15 Days**
Year-to-Date	**151 Days**

Check Your Resources

- Record and Practice Journal
- Resources by Chapter
- Skills Review Handbook
- Assessment Book
- Worked-Out Solutions

Technology For the Teacher

The Dynamic Planning Tool
Editable Teacher's Resources at
BigIdeasMath.com

Common Core State Standards

4.OA.5 Generate a number or shape pattern that follows a given rule. Identify apparent features of the pattern that were not explicit in the rule itself.

5.G.1 Use a pair of perpendicular number lines, called axes, to define a coordinate system Understand that the first number indicates how far to travel from the origin in the direction of one axis, and the second number indicates how far to travel in the direction of the second axis,

Additional Topics for Review

- Writing word sentences as equations
- Solving simple equations
- Evaluating expressions

Try It Yourself

1. The In value times 2 equals the Out value.
2. The In value plus 0.5 equals the Out value.
3. The In value minus 13 equals the Out value.
4. (4, 1)
5. (3, 2)
6. (4, 0)
7. (0, 3)

Record and Practice Journal

1. The In value plus three equals the Out value.
2. The In value times three equals the Out value.
3. The In value minus 10 equals the Out value.
4. The In value divided by two equals the Out value.
5. See Additional Answers.
6. (3, 2)
7. (0, 4)
8. (1, 5)
9. (2, 0)
10. (5, 3)
11–15.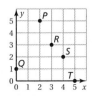

Math Background Notes

Vocabulary Review
- Ordered Pair

Identifying Patterns
- Students should be familiar with recognizing and describing patterns.
- The skill of identifying patterns using the In and Out Table is a precursor to working with functions. Try to foster the upcoming skill by using words such as "relationship" and "rule" when describing how the In value relates to the Out value.
- **Common Error:** Some students may search for a pattern among the In values or among the Out values. This could cause confusion in a table such as Exercise 1 where the In values do not have a common relationship. Remind students that they are searching for the relationship between In and Out values. Encourage them to relate the rows of the table, not the columns.

Plotting Points
- Students should know how to plot points in Quadrant I.
- If students struggle to complete Example 2, you may want to review moving in a coordinate plane before moving forward.
- **Common Error:** Students frequently confuse the x- and y-coordinates when writing ordered pairs. Remind them to always move over and then up. It may help to point out that the O in over appears before the U in up in the alphabet. Similarly, because x comes before y in the alphabet, the x-coordinate should come before the y-coordinate in the ordered pair. Encourage your students to write their points in alphabetical order!
- **Teaching Tip:** Consider turning the floor of your classroom into an interactive, life-sized Quadrant I. Use masking tape to form axes and tick marks on the floor. Then, ask students to start at (0, 0) and move to different points within the quadrant. Stress the importance of moving over, then up. This activity is especially beneficial to kinesthetic learners.

Reteaching and Enrichment Strategies

If students need help. . .	If students got it. . .
Record and Practice Journal • Fair Game Review Skills Review Handbook Lesson Tutorials	Game Closet at *BigIdeasMath.com* Start the next section

What You Learned Before

Identifying Patterns (4.OA.5)

Example 1 Using the numbers from the In and Out Table, find and state the rule in words.

In	Out
30	0
40	10
50	20
60	30

Each Out value is 30 less than the In value.

∴ The In value minus 30 equals the Out value.

Try It Yourself

Using the numbers from the In and Out Table, find and state the rule in words.

1.

In	Out
5	10
7	14
10	20
40	80

2.

In	Out
0.5	1
1.5	2
3	3.5
9.5	10

3.

In	Out
13	0
15	2
30	17
45	32

Plotting Points (5.G.1)

Example 2 Write an ordered pair corresponding to Point S.

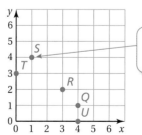

Move 1 unit right.
Move 4 units up.

∴ The ordered pair (1, 4) corresponds to Point S.

Try It Yourself

Use the graph in Example 2 to write an ordered pair corresponding to the point.

4. Point Q **5.** Point R **6.** Point U **7.** Point T

"Here's an interesting survey about favorite dog toys."

9.1 Mapping Diagrams

Essential Question What is a mapping diagram? How can it be used to represent a function?

COMMON CORE STATE STANDARDS
6.EE.9

1 ACTIVITY: Constructing Mapping Diagrams

Work with a partner. Copy and complete the mapping diagram.

a. Area A

b. Perimeter P

c. Circumference C

d. Volume V

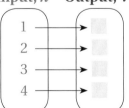

366 Chapter 9 Tables, Graphs, and Functions

Laurie's Notes

Introduction

For the Teacher
- **Goal:** Students will develop an understanding of mapping diagrams.
- This chapter has very little new material. It is intended to be a review of the material learned throughout the year.

Motivate
- **?** "What is a vending machine and how does it work?" Give students time to explain how vending machines operate. Have them discuss the idea of making a selection (input) and getting the desired item (output).
- Share some history of vending machines.
 - A Greek mathematician invented a machine in 215 B.C. to vend holy water in Egyptian temples.
 - During the early 1880s, the first commercial coin-operated vending machines were introduced in London, England and dispensed postcards.
 - Vending machines soon offered other things, including stamps. In Philadelphia, a completely coin-operated restaurant called Horn & Hardart was opened in 1902 and served customers until 1962.

Activity Notes

Activity 1
- **?** "What is meant by the area and perimeter of a rectangle? How do you compute each?" Area is the amount of surface covered. Perimeter is the distance around the rectangle.
- For each part, students will use a familiar formula. If they are given a particular dimension, they will be able to solve for a certain measurement, such as area or perimeter.
- Students do not need to label the output with units as none are given for the input.
- **Big Idea:** The mapping diagram is one way of showing the result of evaluating an expression or formula for a set of numbers. The results are recorded in two ovals with an arrow connecting the input with the output. From the diagrams shown, students will know the formulas to use. A mapping diagram is similar to making a table of values.
- For each problem, have students state the formula used.
 a. $A = 2x$ **b.** $P = 2x + 4$ **c.** $C = 2\pi r$ **d.** $V = 9h$
- **?** "Can the input of the mapping diagram be extended to other numbers, say 5, 6, and 7? Explain." yes; The dimensions in each figure could be 5, 6, or 7.
- **?** "Can the input of the mapping diagram be extended to other numbers, say 0, −1, and −2? Explain." no; For each figure, the missing dimension must be a positive number.

Common Core State Standards

6.EE.9 Use variables to represent two quantities in a real-world problem that change in relationship to one another; write an equation to express one quantity, thought of as the dependent variable, in terms of the other quantity, thought of as the independent variable. Analyze the relationship between the dependent and independent variables using graphs and tables, and relate these to the equation.

Previous Learning

Students should have an understanding of fractions, decimals, and percent operations.

Start Thinking! and Warm Up

9.1 Record and Practice Journal

Differentiated Instruction

Auditory

Help students understand the concept of a function by showing them how it is used in their everyday life. For example, the number of plates to set on the dinner table is a function of the number of people expected to eat. A person earning an hourly wage has an income that is a function of the number of hours worked. Discuss with students other instances of functions in life.

9.1 Record and Practice Journal

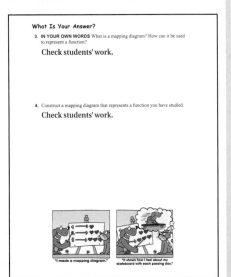

Laurie's Notes

Activity 2

- Students should be able to read a mapping diagram from the previous activity. In this activity, there are two things to do. First, determine what the pattern is for the given data. Then, use the pattern to finish the output column.
- Students may be able to recognize a pattern but are not able to describe a context or operation that generates the data. Have students look at the Table of Contents for ideas.
- **Note:** There is more than one correct pattern that can describe the data.
- **Example:** The pattern for part (a) could be described as
 - multiplying a decimal by 100 ($0.1 \times 100 = 10$).
 - finding a percent of 100 (10% of 100 = 10).
 - finding the area of a rectangle with length 100 and width d ($100 \times 0.1 = 10$).
- **Example:** The pattern for part (c) could be described as
 - multiplying a fraction by 3 $\left(\dfrac{1}{2} \times 3 = \dfrac{3}{2}\right)$.
 - finding the area of a rectangle with width 3 and length n $\left(3 \times \dfrac{1}{2} = \dfrac{3}{2}\right)$.
 - the total length of 3 boards if each board has length n feet $\left(3 \times \dfrac{1}{2} = \dfrac{3}{2} \text{ ft}\right)$.
- Challenge students to think of more than two contexts for each problem.
- ? "In any of the examples, did any input have more than one output?" no

What Is Your Answer?

- **Neighbor Check:** Have students work independently and then have their neighbor check their work. Have students discuss any discrepancies.

Closure

- Describe the pattern in the mapping diagram. Complete the mapping diagram. As each input increases by 1, the output increases by 2. The last 2 numbers are 8 and 10.

Technology For the Teacher

The Dynamic Planning Tool
Editable Teacher's Resources at *BigIdeasMath.com*

2 ACTIVITY: Interpreting Mapping Diagrams

Work with a partner. Describe the pattern in the mapping diagram. Copy and complete the diagram. Find two earlier lessons where you used a similar function.

a. Input, d Output, P

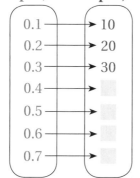

b. Input, t Output, M

c. Input, n Output, S

d. Input, x Output, A

What Is Your Answer?

3. **IN YOUR OWN WORDS** What is a mapping diagram? How can it be used to represent a function?

4. Construct a mapping diagram that represents a function you have studied.

"I made a mapping diagram."

"It shows how I feel about my skateboard with each passing day."

Practice

Use what you learned about mapping diagrams to complete Exercises 3–5 on page 370.

Section 9.1 Mapping Diagrams 367

9.1 Lesson

Key Vocabulary
input, *p. 368*
output, *p. 368*
function, *p. 368*
mapping diagram, *p. 368*

Ordered pairs can be used to show **inputs** and **outputs**.

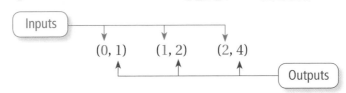

Key Idea

Functions and Mapping Diagrams

A **function** is a relationship that pairs each input with exactly one output. A function can be represented by ordered pairs or a **mapping diagram**.

Ordered Pairs
(0, 1)
(1, 2)
(2, 4)

Mapping Diagram

Input	Output
0	1
1	2
2	4

EXAMPLE 1 Listing Ordered Pairs

List the ordered pairs shown in the mapping diagram.

a.

b.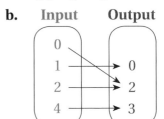

a. The ordered pairs are (1, 3), (2, 6), (3, 9), and (4, 12).

b. The ordered pairs are (0, 2), (1, 0), (2, 2), and (4, 3).

On Your Own

Now You're Ready
Exercises 6–8

List the ordered pairs shown in the mapping diagram.

1.

2.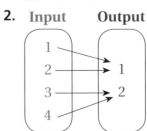

Laurie's Notes

Introduction

Connect
- **Yesterday:** Students explored mapping diagrams.
- **Today:** Students will use ordered pairs and mapping diagrams to display functions.

Motivate
- Time to play *Guess My Rule*.

Problem 1	Problem 2
(Santa Fe, New Mexico)	(Mexico, North America)
(Boise, Idaho)	(Japan, Asia)
(Frankfort, ?)	(Brazil, ?)
(Columbus, ?)	(Germany, ?)

? "If I gave another input for either problem, could you give the output?" Depends upon how well they know their state capitals or continents.

? "Could you make a mapping diagram from either problem?" yes

Lesson Notes

Key Idea
- Use either of the problems in the Motivate or the example in the text to explain the vocabulary.
- Stress that an ordered pair has an *order* associated with it, and the order matters! In the first problem, the ordered pairs are (state capital, state) and in the second problem, the ordered pairs are (country, continent).
- Spend time discussing the idea of a function. Each input has exactly one output. For example, each state capital is in one state. Each country is in one continent.
- Take time to show examples of functions represented as ordered pairs and as a mapping diagram. Discuss how the mapping diagram can be made from the ordered pairs and vice versa.

Example 1
? "Can you use a mapping diagram to list the ordered pairs?" yes
- Note the use of color to differentiate between the input and output.
- **Common Question:** Why are the outputs not listed more than once when they are used more than once as an output, as in Example 1(b)? The outputs are a set of numbers. It is not customary to list elements in a set more than once.

On Your Own
- Ask volunteers to put their ordered pairs on the board.

Goal
Today's lesson is representing a **function** as ordered pairs or a **mapping diagram**.

Start Thinking! and Warm Up

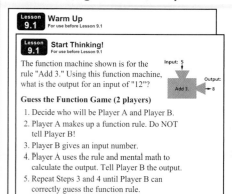

English Language Learners

Vocabulary
Students will find it helpful to relate the words *input* and *output* with the prepositions *in* and *out*.

Extra Example 1

List the ordered pairs shown in the mapping diagram.

a.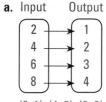

(2, 1), (4, 2), (6, 3), (8, 4)

b.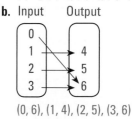

(0, 6), (1, 4), (2, 5), (3, 6)

On Your Own

1. (0, 12), (2, 10), (4, 8), (6, 6)
2. (1, 1), (2, 1), (3, 2), (4, 2)

Extra Example 2

Draw a mapping diagram of (1, 5), (2, 10), (3, 15), and (4, 20).

Extra Example 3

Describe the pattern of inputs and outputs in the mapping diagram from Extra Example 2. *As each input increases by 1, the output increases by 5.*

On Your Own

3.
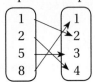

4. As each input increases by 2, the output decreases by 3.

5. See Additional Answers.

Extra Example 4

Use the table to draw a mapping diagram.

Square	
Side Length (m)	Area (m²)
3	9
4	16
5	25
6	36

Laurie's Notes

Example 2
- Now students are given the ordered pairs and are asked to draw the mapping diagram.
- Remind students to draw the arrows to show how the input pairs with the output.
- Note that the inputs and outputs are written in order. Even though the numbers are not paired in this manner, it is customary to list them in order. The arrows indicate how they are paired.

Example 3
- In the previous activity, students were asked to describe the pattern in the mapping diagram. Students may not be able to write an algebraic rule describing how to generate the output given an input. They should be able to describe in words what the pattern is.
- **Common Error:** Students often describe only what is happening to the output. In fact, the input is also changing. Encourage students to describe the change in output in terms of what is happening to the input. In the example, the description indicates that as the input increases by 1, the output increases by 15.
- The curved arrows are a convenient way to record what is changing as you move from one input to the next and from one output to the next.

Example 4
- Note that the data is coming from a table and is being listed in a mapping diagram.
- ? "Were there any bands that played the same number of songs?" no
- ? "Were there any bands that played the same amount of time?" yes
- ? "Is each input (number of songs) paired with one output (time on stage)?" yes

On Your Own
- Encourage students to copy the mapping diagram and use the curved arrows to show how the inputs and outputs are changing.
- For Question 5, have students work independently and then have their neighbor check their work. Have students discuss any discrepancies.

Closure
- Describe the pattern. Make a mapping diagram. List the inputs. List the outputs.
 - (40, Ronald Reagan) (41, George H.W. Bush) (42, William Clinton)
 - (43, George W. Bush) (44, Barack Obama)

The Dynamic Planning Tool
Editable Teacher's Resources at *BigIdeasMath.com*

EXAMPLE 2 **Drawing a Mapping Diagram**

Draw a mapping diagram of (0, 4), (2, 6), (3, 5), and (5, 9).

List the inputs and outputs in order from least to greatest.

 Inputs: 0, 2, 3, 5

 Outputs: 4, 5, 6, 9

Draw arrows from the inputs to their outputs.

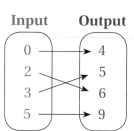

EXAMPLE 3 **Describing a Mapping Diagram**

Describe the pattern of inputs and outputs in the mapping diagram.

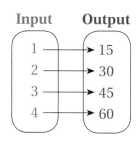

Look at the relationship between the inputs and outputs.

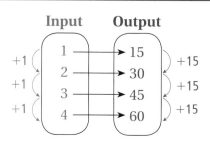

∴ As each input increases by 1, the output increases by 15.

On Your Own

Now You're Ready
Exercises 9–17

3. Draw a mapping diagram of (1, 2), (2, 4), (5, 3), and (8, 1).

4. Describe the pattern of inputs and outputs in the mapping diagram shown.

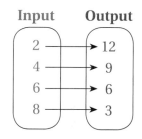

EXAMPLE 4 **Real-Life Application**

Number of Songs Played	Time Onstage (minutes)
8	45
10	60
7	45
14	90

The table shows the number of songs played by four bands at a festival and the amount of time each band played. Use the table to draw a mapping diagram.

Let the number of songs played be the inputs and the times onstage be the outputs.

 Inputs: 7, 8, 10, 14

 Outputs: 45, 60, 90

On Your Own

5. **WHAT IF?** In Example 4, a fifth band plays 12 songs and is onstage for 70 minutes. Draw a mapping diagram for the five bands.

9.1 Exercises

Vocabulary and Concept Check

1. **VOCABULARY** In an ordered pair, which number represents the input? the output?

2. **OPEN-ENDED** Draw a mapping diagram where the number of inputs is greater than the number of outputs.

Practice and Problem Solving

Describe the pattern in the mapping diagram. Copy and complete the diagram.

3.

4.

5.

List the ordered pairs shown in the mapping diagram.

6.

7.

8.

Draw a mapping diagram of the set of ordered pairs.

9. (1, 3), (5, 7), (8, 10), (14, 16)

10. (0, 10), (4, 6), (6, 4), (7, 3)

11. (0, 11), (1, 8), (4, 15), (6, 19)

12. (1, 0), (2, 0), (3, 0), (4, 2), (5, 2)

13. **ERROR ANALYSIS** Describe and correct the error in drawing a mapping diagram of the set of ordered pairs.

 (5, 4), (6, 4), (7, 4), (8, 4)

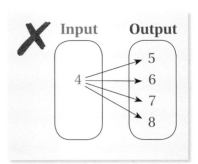

14. **AREA** The table shows the radius and approximate area for four circles. Use the table to draw a mapping diagram.

Radius (ft)	Area (ft²)
2	12.56
3	28.26
7	153.86
9	254.34

370 Chapter 9 Tables, Graphs, and Functions

Assignment Guide and Homework Check

Level	Day 1 Activity Assignment	Day 2 Lesson Assignment	Homework Check
Basic	3–5, 21–23	1, 2, 7–17 odd, 14	2, 7, 9, 14, 15
Average	3–5, 21–23	1, 2, 7–17 odd, 18	2, 7, 9, 15, 18
Advanced	3–5, 21–23	1, 2, 6–12 even, 13, 16, 19, 20	2, 6, 10, 16, 19

Common Errors

- **Exercises 6–8** Students may mix up the ordered pairs and write the output first and then the input. Encourage them to use the arrow as a guide. The arrow points from the first number in the ordered pair, the input, to the second number in the ordered pair, the output.
- **Exercises 9–12** Students may forget to write and order the inputs and outputs before making the mapping diagram. This often leads to mixing inputs and outputs. Remind them that the inputs and outputs should be in order from least to greatest (and sometimes greatest to least). The arrows will not always go directly across to the output.
- **Exercises 15–17** Students may only analyze the outputs. Remind them that the inputs are changing as well and they need to be able to describe how the inputs are changing for later in the chapter and in future courses.

Technology For the Teacher
Answer Presentation Tool
Quiz*Show*

9.1 Record and Practice Journal
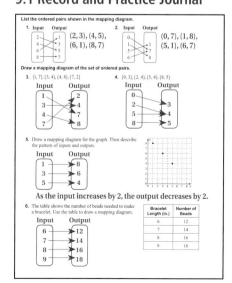

Vocabulary and Concept Check

1. the first number; the second number
2. *Sample answer:*
 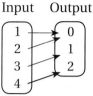

Practice and Problem Solving

3. As each input increases by 1, the output increases by 4. 16; 20; 24
4. As each input increases by 1, the output increases by 6. 20; 26; 32
5. As each input increases by 1, the output increases by 7. 21; 28; 35
6. (0, 4), (3, 5), (6, 6), (9, 7)
7. (1, 8), (3, 4), (5, 6), (7, 2)
8. (6, 1), (7, 1), (8, 2), (9, 2)
9. Input Output

10. Input Output

11. Input Output
 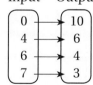

12–14. See Additional Answers.

T-370

Practice and Problem Solving

15–17. See Additional Answers.

18. a–b. See *Taking Math Deeper*.

 c. Answers will vary.

 d. See *Taking Math Deeper*.

19. See Additional Answers.

20. 1025; *Sample answer:* As each input increases by 1, the output increases by 5. From 0 to 200 is an input increase of 200. So, the output can be found by 25 + 200(5).

Fair Game Review

21. $x + 7 = 15$; $x = 8$

22. $3n = 24$; $n = 8$

23. C

Mini-Assessment

1. List the ordered pairs shown in the mapping diagram. (1, 2), (2, 4), (3, 6), (4, 8)

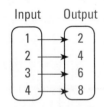

2. The table shows the ages of four children and their weights. Use the table to draw a mapping diagram.

Age (years)	Weight (pounds)
1	25
2	30
3	34
4	39

Taking Math Deeper

Exercise 18

This problem looks forward to the concept of a linear function. Students can see this in two ways.
(1) As the input increases by a constant amount, the output also increases by a constant amount.
(2) The graph of the inputs and outputs lie on a line.

 a. Complete the mapping diagram.

 b. List the ordered pairs. Then plot the ordered pairs in a coordinate plane.

 d. An amateur recreational diver should not go below 30 meters. The safe limit for a trained expert diver should be around 60 meters. Scuba diving can be dangerous. No one should attempt it without proper training.

Project

Use your school's library or the Internet to research scuba diving. Write a report that shows areas that are good for scuba diving. Include the cost of the sport in your report.

Reteaching and Enrichment Strategies

If students need help...	If students got it...
Resources by Chapter • Practice A and Practice B • Puzzle Time Record and Practice Journal Practice Differentiating the Lesson Lesson Tutorials Skills Review Handbook	Resources by Chapter • Enrichment and Extension Start the next section

Draw a mapping diagram for the graph. Then describe the pattern of inputs and outputs.

3️⃣ **15.** **16.** **17.**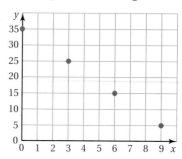

18. SCUBA DIVING The normal pressure at sea level is one atmosphere of pressure (1 ATM). As you dive below sea level, the pressure increases by 1 ATM for each 10 meters of depth.

a. Complete the mapping diagram.

b. List the ordered pairs. Then plot the ordered pairs in a coordinate plane.

c. Compare the mapping diagram and graph. Which do you prefer? Why?

d. **RESEARCH** What are common depths for people who are just learning to scuba dive? What are common depths for experienced scuba divers?

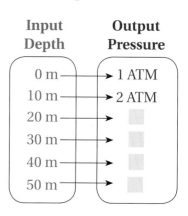

19. MOVIES A store sells previously viewed movies. The table shows the cost of buying 1, 2, 3, or 4 movies.

a. Use the table to draw a mapping diagram.

b. Describe the pattern. How does the cost per movie change as you buy more movies?

Movies	Cost
1	$10
2	$18
3	$24
4	$28

20. The table shows the outputs for several inputs. What do you think the output would be for an input of 200? Explain.

Input, x	0	1	2	3	4
Output, y	25	30	35	40	45

Fair Game Review *What you learned in previous grades & lessons*

Write the word sentence as an equation. Then solve. *(Section 7.2 and Section 7.3)*

21. The sum of a number x and 7 is 15.

22. 3 times a number n is 24.

23. MULTIPLE CHOICE Which inequality represents the word sentence? *(Section 8.1)*

"The sum of a number x and 7 is at least 25."

Ⓐ $x + 7 > 25$ Ⓑ $x + 7 \leq 25$ Ⓒ $x + 7 \geq 25$ Ⓓ $x + 7 < 25$

9.2 Functions as Words and Equations

Essential Question How can you describe a function with words? How can you describe a function with an equation?

COMMON CORE STATE STANDARDS
6.EE.9

1 ACTIVITY: Describing a Function

Work with a partner. Two mapping diagrams related to the rectangle are shown. Describe each function in words. Then write an equation for each function.

a. Area A

b. Perimeter, P

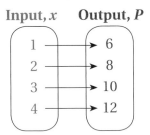

2 ACTIVITY: Describing a Function

Work with a partner. Copy and complete the mapping diagram for the area of the figure. Then write an equation that describes the function.

a.

b.

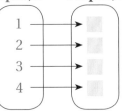

372 Chapter 9 Tables, Graphs, and Functions

Laurie's Notes

Introduction

For the Teacher
- **Goal:** Students will develop an understanding of what a mapping diagram is and how it can be extended.
- Students will work with functions again. In this lesson, students will describe functions using words and equations.

Motivate
- Make a card for each shape and formula in the table.

(circle with r)	Area $= \pi r^2$	Circumference $= 2\pi r$
(square with s)	Area $= s^2$	Perimeter $= 4s$

- Continue the list, include rectangle, triangle, parallelogram, and trapezoid.
- Give a card to each student. Students should walk around and match up with the two other cards in their set.
- Have each group share their set (aloud) with the rest of the class. If possible, tape the sets on the board for students to view as they are working through today's activity.

Activity Notes

Activity 1
- Model this problem by holding two pencils in a vertical position. Move the pencils toward one another and then away. This simulates the changing length of a rectangle that has a fixed width of 2.
- ? "Suppose the width of a rectangle is always 2 and the length is changing. Does the area change?" yes "Does the perimeter change?" yes
- Students should now work with a partner to complete the first activity.
- **Teaching Tip:** Use the word *rule* as well as *equation*.
- ? "What rule can describe the input and output?" double the input
- ? "How do you write double the input, or double x?" $2x$
- Writing a rule for the perimeter is likely to be more challenging. Suggest that they think about the formula. If the width is 2, the perimeter must be greater than 4.
- Writing the function in words does not mean writing a paragraph. It is sufficient to write "Area equals 2 times the length."

Activity 2
- ? "What information is needed in order to find the area of a triangle?" length of base and height of triangle "to find the area of a trapezoid?" length of both bases and height of trapezoid
- The suggestion I give students is to put x in the left input box. What would it be paired with in the right output box? That is the equation!

Common Core State Standards

6.EE.9 Use variables to represent two quantities in a real-world problem that change in relationship to one another; write an equation to express one quantity, thought of as the dependent variable, in terms of the other quantity, thought of as the independent variable. Analyze the relationship between the dependent and independent variables using graphs and tables, and relate these to the equation.

Previous Learning
Students should have an understanding of area and perimeter.

Activity Materials
Introduction
• index cards

Start Thinking! and Warm Up

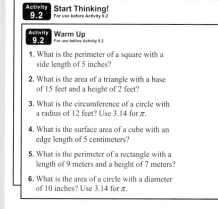

Activity 9.2 Start Thinking! For use before Activity 9.2

Activity 9.2 Warm Up For use before Activity 9.2

1. What is the perimeter of a square with a side length of 5 inches?
2. What is the area of a triangle with a base of 15 feet and a height of 2 feet?
3. What is the circumference of a circle with a radius of 12 feet? Use 3.14 for π.
4. What is the surface area of a cube with an edge length of 5 centimeters?
5. What is the perimeter of a rectangle with a length of 9 meters and a height of 7 meters?
6. What is the area of a circle with a diameter of 10 inches? Use 3.14 for π.

9.2 Record and Practice Journal

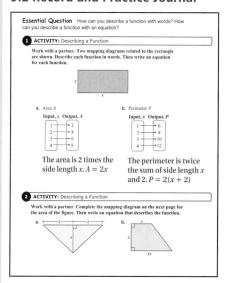

T-372

Differentiated Instruction

Visual

Students may be able to describe how the sequence of output numbers is changing (*Start with 2 and add 3.*), but may find it difficult to write a function rule for changing an input value to an output value. If students determine the output increases or decreases by a constant value as the input increases, the function will have an ax term. Have students create function tables for equations, such as $y = x + 3$, $y = 4x - 1$, and $y = 0.5x$, to see this pattern.

9.2 Record and Practice Journal

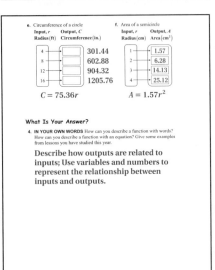

Laurie's Notes

Activity 3

- The polygon is now replaced with words. The words describe a function, just as the labeled polygon suggested a function.
- Note that the *names* and *units* associated with each input and output are listed. This should be helpful to students.
- ❓ "Can the input for part (b), phone bill, be something other than hours?" yes; can be number of calls made or number of minutes; How would the equation change?

What Is Your Answer?

- Here is a reference for students who need to look back in the book in order to complete this problem.

Section 4.4	Finding a percent of a number
Section 3.2 and 3.3	Multiplying decimals
Section 6.1	Circumference of a circle
Section 6.3	Area of a circle

Closure

- Complete the mapping diagram. Then write an equation that describes the function.

 Admission cost $12 per person

 Input, n Output, C
 # of people Cost ($)

 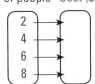

 24, 48, 72, 96; $C = 12n$

Technology For the Teacher

The Dynamic Planning Tool
Editable Teacher's Resources at *BigIdeasMath.com*

T-373

3 ACTIVITY: Describing a Function

Work with a partner. Copy and complete the mapping diagram. Then write an equation that describes the function.

a. Sales tax of 6%

Input, x — Cost ($) Output, T — Sales Tax ($)
- 10 →
- 20 →
- 30 →
- 40 →

b. Phone bill of $5 per hour

Input, x — Hours Output, B — Bill ($)
- 0.5 →
- 1.0 →
- 1.5 →
- 2.0 →

c. Perimeter of a square

Input, s — Side (in.) Output, P — Perimeter (in.)
- 4 →
- 8 →
- 12 →
- 16 →

d. Surface area of a cube

Input, s — Side (m) Output, S — Surface Area (m^2)
- 1 →
- 2 →
- 3 →
- 4 →

e. Circumference of a circle

Input, r — Radius (ft) Output, C — Circumference (in.)
- 4 →
- 8 →
- 12 →
- 16 →

f. Area of a semicircle

Input, r — Radius (cm) Output, A — Area (cm^2)
- 1 →
- 2 →
- 3 →
- 4 →

What Is Your Answer?

4. IN YOUR OWN WORDS How can you describe a function with words? How can you describe a function with an equation? Give some examples from lessons you have studied this year.

Use what you learned about writing functions as words and equations to complete Exercises 3–5 on page 376.

Section 9.2 Functions as Words and Equations 373

9.2 Lesson

Key Vocabulary
function rule, p. 374

Key Idea

Function Rule

A **function rule** describes the relationship between inputs and outputs. A function can be written as an equation in two variables.

$$y = 3x$$
(Output) (Input)

EXAMPLE 1 — Writing an Equation in Two Variables

Write an equation for "The output is five less than the input."

Words The output is five less than the input.

Equation $y\ =\ x - 5$

∴ An equation is $y = x - 5$.

On Your Own

Now You're Ready
Exercises 6–9

Write an equation that describes the function.

1. The output is eight more than the input.
2. The output is four times the input.
3. The output is the area of the rectangle.

EXAMPLE 2 — Evaluating a Function

What is the value of $y = 2x + 5$ when $x = 3$?

$y = 2x + 5$ Write the equation.
$ = 2(3) + 5$ Substitute 3 for x.
$ = 11$ Simplify.

∴ When $x = 3$, $y = 11$.

On Your Own

Now You're Ready
Exercises 10–15

Find the value of y when $x = 5$.

4. $y = 4x - 1$
5. $y = 10x$
6. $y = 3 + 7x$

374 Chapter 9 Tables, Graphs, and Functions

Laurie's Notes

Introduction

Connect
- **Yesterday:** Students used mapping diagrams to generate ordered pairs that were described in words and equations.
- **Today:** Students will use equations in two variables to write functions.

Motivate
- The daily admission at a theme park is $73 for ages 10 and up.
- ❓ "What would it cost for a family of 4, ages 10 and up, to visit for one day?" $4 \times \$73 = \292
- ❓ "What would an equation be that calculates the cost for n people (all older than 10 years) to visit the theme park?" $\text{cost} = \$73n$

Lesson Notes

Key Idea
- The term **function rule** is used to introduce the idea of writing an equation to represent the relationship between the input and output, the two variables in the equation.
- It is common to write equations using the variables x and y, where x is the input and y is the output.
- **Discuss:** In the motivate section, the equation was written as $\text{cost} = 73n$. The equation could have been written as $c = 73n$. Notice the use of two variables. While any variable names could have been chosen, n and c were chosen to represent number of people and cost of admission.

Example 1
- If it helps, read and write the function rule as: The output, y, is five less than the input, x.
- Note the use of color to help students connect the words and symbols.

On Your Own
- **Neighbor Check:** Have students work independently and then have their neighbor check their work. Have students discuss any discrepancies.

Example 2
- ❓ "For the equation $y = 2x + 5$, describe in words what operations are performed on the input x." The value of x is multiplied by 2 and then 5 is added.
- **Connection:** The rule in words for this function is "double the input and add 5," which students can make a mapping diagram for.

On Your Own
- **Connection:** Students already know how to evaluate algebraic expressions. Students would have been asked to evaluate $4x - 1$ when $x = 5$. Now the same problem is written as an equation with the answer called the output.

Goal Today's lesson is writing equations that represent functions.

Start Thinking! and Warm Up

> **Lesson 9.2 Warm Up** For use before Lesson 9.2
>
> **Lesson 9.2 Start Thinking!** For use before Lesson 9.2
>
> Using two variables, write an equation for a function that describes a real-life situation. Explain the situation and what each of the variables represents.
> Which variable represents input values?
> Which variable represents output values?
> Then make a mapping diagram that includes four different input values.

Extra Example 1
Write an equation for "The output is one-fourth of the input." $y = \frac{1}{4}x$

On Your Own

1. $y = x + 8$
2. $y = 4x$
3. $A = 5x$

Extra Example 2
What is the value of $y = 5x - 2$ when $x = 4$? 18

On Your Own

4. 19 5. 50
6. 38

Extra Example 3

Tell whether $(x, y) = (2, 5)$ is a solution of the equation.

a. $y = 4x - 3$ solution
b. $y = 2x + 2$ *not* a solution

Extra Example 4

A restaurant can serve 150 dinners an hour.

a. Write an equation that relates the dinners and time. $y = 150x$
b. How many dinners can be served in 6 hours? 900

On Your Own

7. yes; By substituting 2 for x and 5 for y, the equation is $5 = 2 + 3$, which is true.

8. a. $r = 1500h$
 b. 18,000 people

English Language Learners

Vocabulary

Have students add the key vocabulary words *input*, *output*, *function*, *mapping diagram*, and *function rule* to their notebooks. Definitions, examples, and pictures should accompany the words.

Laurie's Notes

Example 3
- Work through each example. Substitution of the ordered pair (3, 4) means both x and y are replaced with numerical values.

Example 4
- This example connects to the opening Motivate problem.
- **FYI:** Cedar Point is often called the roller coaster capital of the world because it has 17 roller coasters.
- Ask a student to read the problem and information in the caption. Ask what information will help in this problem. For instance, does the location of Cedar Point or the name of the ride add necessary information to the problem?
- ❓ "What key information is stated?" 1000 riders can ride in one hour. The park is open from 10 A.M. to 10 P.M.
- ❓ "If the ride accommodates 1000 riders per hour, how many can ride in 2 hours?" 2000 "in 5 hours?" 5000 "in h hours?" $1000h$
- Write the function in words. Note that the variables are defined.
- **Extension:** When the ride opened in 2005, there were 842,796 riders. Did 12,000 people ride each day in 2005? Explain. no; $12,000 \times 365 = 4,380,000$

On Your Own
- **Neighbor Check:** Have students work independently and then have their neighbor check their work. Have students discuss any discrepancies.
- **Question 8 part (b):** Students need to refer to the park hours in Example 4.

Closure
- Write an equation for the daily income from visitors age 10 and older at the theme park. $y = 73x$

Technology For the Teacher

The Dynamic Planning Tool
Editable Teacher's Resources at *BigIdeasMath.com*

EXAMPLE 3 Checking Solutions

Tell whether $(x, y) = (3, 4)$ is a solution of the equation.

Study Tip
A solution of an equation with two variables is an ordered pair that makes the equation true.

a. $y = x - 2$

$y = x - 2$ Write equation.
$(4) \stackrel{?}{=} 3 - 2$ Substitute.
$4 \neq 1$ ✗ Simplify.

∴ (3, 4) is *not* a solution.

b. $y = 5x - 11$

$y = 5x - 11$ Write equation.
$(4) \stackrel{?}{=} 5(3) - 11$ Substitute.
$4 = 4$ ✓ Simplify.

∴ (3, 4) is a solution.

EXAMPLE 4 Real-Life Application

Cedar Point Park Hours
10:00 A.M.–10:00 P.M.

The maXair ride can accommodate 1000 riders in an hour.

a. Write an equation that relates riders and time.
b. How many people can ride the maXair each day?

a. **Words** Number of riders equals the number of riders per hour times the number of hours.

 Variables Let r be the number of riders.
 Let h be the number of hours.

 Equation r = 1000 • h

∴ An equation is $r = 1000h$.

b. The park is open for 12 hours. Substitute 12 for h in the equation from part (a).

$r = 1000h$ Write the equation.
$ = 1000(12)$ Substitute 12 for h.
$ = 12{,}000$ Multiply.

∴ Each day, 12,000 people can ride the maXair.

On Your Own

Now You're Ready
Exercises 16–21

7. Is (2, 5) a solution of $y = x + 3$? Explain.

8. The Millennium Force roller coaster at Cedar Point can accommodate 1500 riders in an hour.

 a. Write an equation that relates riders and time.
 b. How many people can ride the Millennium Force each day?

9.2 Exercises

Vocabulary and Concept Check

1. **VOCABULARY** Identify the input variable and the output variable for the function rule $y = 2x + 5$.

2. **REASONING** Explain why (2, 6) is a solution of $y = x + 4$, but (6, 2) is *not* a solution.

Practice and Problem Solving

Write an equation that describes the function.

3. Input → Output
0 → 0
1 → 4
2 → 8
3 → 12

4.

5.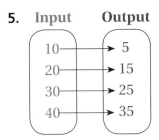

Write an equation that describes the function.

6. The output is three less than the input.
7. The output is six times the input.
8. The output is half of the input.
9. The output is eleven more than the input.

Find the value of y for the given value of x.

10. $y = x + 5$; $x = 3$
11. $y = 7x$; $x = 6$
12. $y = 2x - 1$; $x = 9$
13. $y = 3x + 2$; $x = 0.5$
14. $y = 4x + 7$; $x = \dfrac{5}{2}$
15. $y = \dfrac{x}{2} + 9$; $x = 8$

Tell whether the ordered pair is a solution of the equation.

16. $y = x - 6$; (6, 0)
17. $y = x + 7$; (2, 8)
18. $y = 11x$; (2, 22)
19. $y = \dfrac{x}{5}$; (5, 25)
20. $y = 9x + 8$; (1, 18)
21. $y = \dfrac{x}{3} - 6$; (18, 0)

22. **ERROR ANALYSIS** Describe and correct the error in checking whether (13, 3) is a solution of $y = 3x + 4$.

$y = 3x + 4$
$13 = 3(3) + 4$
$13 = 13$ ✓
(13, 3) is a solution.

23. **SPEED** A Scrub-Jay flies at a rate of 18 feet per second.

 a. Write an equation that relates the distance d traveled in s seconds.

 b. How many feet does a Scrub-Jay fly in 30 seconds?

376 Chapter 9 Tables, Graphs, and Functions

Assignment Guide and Homework Check

Level	Day 1 Activity Assignment	Day 2 Lesson Assignment	Homework Check
Basic	3–5, 31–33	1, 2, 6–9, 11–23 odd, 22	2, 6, 11, 17, 22
Average	3–5, 31–33	1, 2, 7–27 odd, 22, 28	7, 11, 17, 22, 25
Advanced	3–5, 31–33	1, 2, 10–30 even, 29	10, 16, 22, 24, 28

Common Errors

- **Exercises 6–9** Students may mix up the *x* and *y* variables when writing the equation. Remind them that *x* represents the input and *y* represents the output. Students may also need to be reminded about how to write expressions from word phrases.
- **Exercises 10–15** Students may forget the order of operations for some of the exercises and try to add or subtract before multiplying or dividing after substituting the value of *x*. Remind them of the order of operations and how to evaluate an expression.
- **Exercises 16–21** Students may substitute the ordered pair for the wrong variable. Ask them to identify which coordinate is the *x*-coordinate and which is the *y*-coordinate. Encourage them to write the values of *x* and *y* before substituting into the equation. For example, the ordered pair (8, 12) means *x* = 8 and *y* = 12.

9.2 Record and Practice Journal

Technology For the Teacher
Answer Presentation Tool
*Quiz*Show

 Vocabulary and Concept Check

1. input variable: *x*; output variable: *y*
2. (2, 6) is a solution because $6 = 2 + 4$. (6, 2) is *not* a solution because $2 \neq 6 + 4$.

 Practice and Problem Solving

3. $y = 4x$
4. $y = x + 7$
5. $y = x - 5$
6. $y = x - 3$
7. $y = 6x$
8. $y = \frac{1}{2}x$
9. $y = x + 11$
10. 8
11. 42
12. 17
13. 3.5
14. 17
15. 13
16. yes
17. no
18. yes
19. no
20. no
21. yes
22. The numbers were substituted for the wrong variables.
$y = 3x + 4$
$3 \stackrel{?}{=} 3(13) + 4$
$3 \neq 43$
(13, 3) is *not* a solution.
23. a. $d = 18s$
 b. 540 ft
24. 4
25. 5
26. 36

T-376

 Practice and Problem Solving

27. See *Taking Math Deeper*.

28. **a.** $G = 35 + 10h$
 b. $S = 25h$
 c. Snake Tours; For 2 hours, Gator Tours cost $55 and Snake Tours cost $50.

29. no; Many rectangles have the same perimeter but different areas.

30. 44 square units; 45 square units; *Sample answer:* The "green area" of an even numbered square is equal to 4 more than the product of 2 and the square number. The "green area" of an odd numbered square is equal to 3 more than the product of 2 and the square number.

 Fair Game Review

31. 8; 9; 10 32. 3; 5; 10
33. C

Mini-Assessment

Find the value of *y* for the given value of *x*.

1. $y = x - 7$; $x = 10$ $y = 3$
2. $y = 4x - 3$; $x = 2$ $y = 5$
3. $y = 6x$; $x = 4$ $y = 24$
4. $y = \dfrac{x}{5} - 8$; $x = 50$ $y = 2$
5. **You are selling magazines to raise money for your school. Each subscription you sell earns $8 for your school.**
 a. Write an equation you can use to find the total amount *d* in dollars that you can raise for your school after selling *s* subscriptions.
 $d = 8s$
 b. How much money will you raise if you sell 20 subscriptions? $160

Taking Math Deeper

Exercise 27

This is a common type of problem in business. Write an equation that represents the profit for a business.

 a. Write an equation for the profit.

$$\text{Profit} = (\text{Income}) - (\text{Expenses})$$
$$P = 3.50b - 84$$

Number of bracelets Expenses

Notice that when the number *b* of bracelets is small, the profit is negative. A negative profit is a *loss*.

② Set income equal to expenses and solve for *b*.

$$\text{Income} = \text{Expenses}$$
$$3.50b = 84$$
$$b = 24 \text{ bracelets}$$

b. You will break even when you sell 24 bracelets.

 Extension: Make a table that shows the profit for different numbers of bracelets sold.

b	24	25	26	27
$P = 3.50b - 84$	$0.00	$3.50	$7.00	$10.50

Project

Draw a design for a bracelet that you would like to make. Estimate the cost and the selling price of 1 bracelet.

Reteaching and Enrichment Strategies

If students need help...	If students got it...
Resources by Chapter • Practice A and Practice B • Puzzle Time Record and Practice Journal Practice Differentiating the Lesson Lesson Tutorials Skills Review Handbook	Resources by Chapter • Enrichment and Extension Start the next section

Find the value of x for the given value of y.

24. $y = 5x - 7$; $y = 13$

25. $y = 7x + 2$; $y = 37$

26. $y = \dfrac{x}{4} - 7$; $y = 2$

27. BRACELETS You decide to make and sell bracelets. The cost of your materials is $84. You charge $3.50 for each bracelet.

 a. Write an equation you can use to find the profit P for selling b bracelets.

 b. You will *break even* if the cost of your materials equals your income. How many bracelets must you sell to break even?

28. AIRBOAT TOURS You want to take a two-hour airboat tour.

 a. Write an equation that represents the cost G of a tour at Gator Tours.

 b. Write an equation that represents the cost S of a tour at Snake Tours.

 c. Which is a better deal? Explain.

29. CRITICAL THINKING Can you write a function for the area of any rectangle given the perimeter of the rectangle? Explain.

30. Puzzle The blocks that form the diagonals of each square are shaded. Each block is one square unit. Find the "green area" of Square 20. Find the "green area" of Square 21. Explain your reasoning.

Square 1 Square 2 Square 3 Square 4 Square 5

Fair Game Review *What you learned in previous grades & lessons*

Copy and complete the table. *(Section 1.1)*

31.

x	1	2	3
x + 7			

32.

x	6	8	13
x − 3			

33. MULTIPLE CHOICE You want to volunteer for at most 20 hours each month. So far, you have volunteered for 7 hours this month. Which inequality represents the number of hours you can volunteer for the rest of this month? *(Section 8.2)*

 Ⓐ $h \geq 13$ Ⓑ $h \geq 27$ Ⓒ $h \leq 13$ Ⓓ $h < 27$

9.3 Input-Output Tables

Essential Question How can you use a table to describe a function?

COMMON CORE STATE STANDARDS
6.EE.9

1 ACTIVITY: Using a Function Table

Work with a partner.

a. Copy and complete the table for the perimeter of the rectangle.

Input, x	1	2	3	4	5
Output, P					

b. Write an equation that describes the function.

c. Use your equation to find the value of x for which the perimeter is 50.

2 ACTIVITY: Using a Function Table

Work with a partner. Use the strategy shown in Activity 1 to make a table that shows the pattern for the area. Write an equation that describes the function. Then use your equation to find which figure has an area of 81.

a.

Figure 1 Figure 2 Figure 3 Figure 4

b.

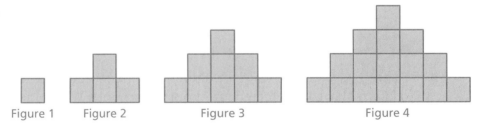

Figure 1 Figure 2 Figure 3 Figure 4

Laurie's Notes

Introduction

For the Teacher
- **Goal:** Students will develop an understanding of how an input-output table displays ordered pairs that describe a function.
- Students will work with functions again. In this lesson, students will describe the function using input-output tables.

Motivate
- **Story Time:** Tell students that you recently went to a store and brought an item to the cash register that costs $19.95. When you reached in your pocket you realized you only had a $20 bill and you had to return the item to the shelf. Why? This will determine if students are aware of sales taxes.
- One of the activities today involves sales tax. A sales tax can be assessed at different levels: state, county, and city (municipality).

Activity Notes

Activity 1
- **?** "How could the values in the input-output table be used to write ordered pairs?" Students should describe the ordered pairs using language such as (length, perimeter).
- **?** "Any observations about the input-output table?" Students might mention that the output is always even. Output is increasing by 2.
- **Big Idea:** Writing the function in an input-output table allows you to see many ordered pairs that satisfy the function.
- **Extension:** Ask students to describe how the perimeter changes as the input changes. To help students develop the terminology, say, "As the input increases by 1 unit, the perimeter . . ."
- **?** How do you write the equation that describes this function? $P = 2x + 6$

Activity 2
- If available, provide square tiles for students to manipulate.
- The input for each table is the figure number.
- Students often recognize the pattern, but writing the equation is more difficult.
- The first step is writing an x (or any variable) in the input table. Now ask, "for the x-th figure, what would the area be?"
- With practice, students will understand that to generate even numbers you multiply by 2. To generate odd numbers, you multiply by 2 and add or subtract 1 depending upon where the numbers begin.
- **FYI:** Part (b) has outputs known as square numbers. These are the only numbers that can make a square with the given number of square tiles. 20 tiles can be rearranged into a rectangle (1×20, 2×10, 4×5), but not a square. 16 tiles can be rearranged into a square that is 4×4.

Common Core State Standards

6.EE.9 Use variables to represent two quantities in a real-world problem that change in relationship to one another; write an equation to express one quantity, thought of as the dependent variable, in terms of the other quantity, thought of as the independent variable. Analyze the relationship between the dependent and independent variables using graphs and tables, and relate these to the equation.

Previous Learning

Students should know how to translate expressions to equations, and to solve two-step equations.

Activity Materials
Textbook
• square tiles

Start Thinking! and Warm Up

9.3 Record and Practice Journal

T-378

English Language Learners

Classroom
This chapter provides English learners a chance to share with the rest of the class and the opportunity to build their confidence. Many examples and exercises use tables and graphs giving English learners a rest from interpreting sentences.

9.3 Record and Practice Journal

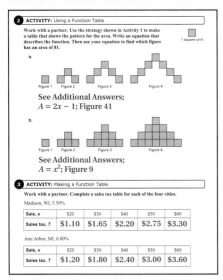

Laurie's Notes

Activity 3

- **Note:** These are the sales tax rates from 2008.
- This activity reviews finding a percent of a number and multiplication of decimals.
- ❓ "Can someone explain what a sales tax is? How is it calculated?" Sales tax is an amount of money that is added on to the purchase price of an item. The sales tax is a percent of the purchase price.
- ❓ "How do you find a percent of a number?" You write the percent as a fraction or decimal, then multiply.
- **Common Error:** Students write 7% as 0.7, not 0.07.
- Note that each percent is written with two decimal places. This will help students recognize the pattern: 7.00% = 0.07; 7.50% = 0.0750. Each of these percents begins with 0.07.
- Remind students that sales tax is an amount of money. The answers need to be labeled correctly. An answer of 1.2 is written as $1.20.
- **Extension:** Discuss the difference of 0.50% on a $20 item, meaning how much more was the sales tax in Ann Arbor than Madison? $0.10 How much more was the sales tax in Norman than Edison? $0.10
- **Extension:** You can ask students to find the sales tax rate in their area, and then compute the sales tax for the inputs in the table.

What Is Your Answer?

- **Neighbor Check:** Have students work independently and then have their neighbor check their work. Have students discuss any discrepancies.

Closure

- Make an input-output table for leaving a 15% tip.

Cost of Meal(s), x	$20	$40	$50	$60	$100
15% Tip, y	$3	$6	$7.50	$9	$15

Technology For the Teacher

Dynamic Classroom

The Dynamic Planning Tool
Editable Teacher's Resources at *BigIdeasMath.com*

③ ACTIVITY: Making a Function Table

Work with a partner. Copy and complete a sales tax table for each of the four cities.

Madison, WI, 5.50%

Sale, x	$20	$30	$40	$50	$60
Sales Tax, T					

Ann Arbor, MI, 6.00%

Sale, x	$20	$30	$40	$50	$60
Sales Tax, T					

Edison, NJ, 7.00%

Sale, x	$20	$30	$40	$50	$60
Sales Tax, T					

Norman, OK, 7.50%

Sale, x	$20	$30	$40	$50	$60
Sales Tax, T					

What Is Your Answer?

4. IN YOUR OWN WORDS How can you use a table to describe a function? Describe an example of a function table in real life.

Amount of Sale	Tax
.10 – .16	.01
.17 – .33	.02
.34 – .50	.03
.51 – .66	.04
.67 – .83	.05
.84 – 1.09	.06

"Dear Sir: Yesterday, I bought a piece of 9-cent candy six times and paid NO tax. Today, I bought six pieces at once and you charged me $0.04 tax. What's going on?"

Practice — Use what you learned about input-output tables to complete Exercises 3 and 4 on page 382.

9.3 Lesson

Key Vocabulary
input-output table, p. 380

Key Idea

Input-Output Tables

A function can be represented by an **input-output table**. The table below is for the function $y = x + 2$.

Input, x	Output, y
1	3
2	4
3	5
4	6

← $y = x + 2$
← $3 = 1 + 2$
← $4 = 2 + 2$
← $5 = 3 + 2$
← $6 = 4 + 2$

EXAMPLE 1 Completing Input-Output Tables

Write an equation for the function. Then copy and complete the table.

a. The output is 1 less than the input.

Input, x	2	3	4	5
Output, y				

b. The output is twice the input.

Input, x	0	3	6	9
Output, y				

a. An equation is $y = x - 1$.

Input, x	2	3	4	5
Output, y	1	2	3	4

↑ $y = x - 1$

b. An equation is $y = 2x$.

Input, x	0	3	6	9
Output, y	0	6	12	18

↑ $y = 2x$

On Your Own

Now You're Ready
Exercises 5 and 6

Write an equation for the function. Then copy and complete the table.

1. The output is 5 more than the input.

Input, x	1	3	5	7
Output, y				

2. The output is the product of 7 and the input.

Input, x	0	2	4	6
Output, y				

Laurie's Notes

Introduction

Connect
- **Yesterday:** Students performed an activity that resulted in collecting and recording data in an input-output table.
- **Today:** Students will use input-output tables to represent functions and connect the table to equations.

Motivate
- **Matching Activity:** Match the function rule with the ordered pair.
 - A. $y = 2x + 4$ — 4) (2, 6)
 - B. $y = x + 4$ — 3) (2, 8)...
 - C. $y = 2x$ — 1) (2, 4)
 - D. $y = 2x - 3$ — 2) (2, 1)

Lesson Notes

Key Idea
- Write the Key Idea.
- Students should recognize that *Output, y* and *Output, x + 2* are the same thing. Over time, they should become more comfortable using *y* instead of the function rule.
- **Connection:** In previous sections, students wrote functions as ordered pairs and mapping diagrams. In an ordered pair, you write (x, y) where x is the input and y is the output. The input was paired with the output in the mapping diagram.

Example 1
- Students must read and translate words into an algebraic equation. They need to substitute the input for x and calculate the output.
- Discuss with students the values selected for the input. Notice that instead of beginning with 0 or 1, the first example begins with an input of 2. Ask, "Could the input be 10?" yes "Could the input be 50.5?" yes
- **Big Idea:** There is no context for the two functions in this example. So, unless otherwise restricted, the input could be from any set of numbers the students are familiar with. If there were a context, such as "number of tickets purchased for a show," the input would need to be a whole number. You cannot purchase $3\frac{1}{2}$ tickets.
- ? "In Example 1(a), what is the output for an input of 10?" 9 "for 50.5?" 49.5
- ? "What pattern do you observe in the output?" increasing by 1
- ? "In Example 1(b), what pattern do you observe in the input?" increasing by 3 "the output?" increasing by 6

On Your Own
- **Neighbor Check:** Have students work independently and then have their neighbor check their work. Have students discuss any discrepancies.
- **Common Error:** Students interchange x and y. Instead of writing $y = x + 5$, they may write $x = y + 5$.

Goal
Today's lesson is representing a function as an **input-output table**.

Start Thinking! and Warm Up

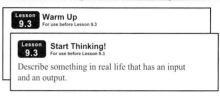

Extra Example 1
Write an equation for the function. Then copy and complete the table.

a. The output is 4 more than the input. $y = x + 4$

Input, x	2	4	6	8
Output, y	6	8	10	12

b. The output is the difference of the input and 3. $y = x - 3$

Input, x	3	6	9	12
Output, y	0	3	6	9

On Your Own
1. $y = x + 5$; 6; 8; 10; 12
2. $y = 7x$; 0; 14; 28; 42

Extra Example 2

Which function rule in Example 2 is shown by the table? D

Input, x	Output, y
0	0
4	40
8	80
12	120

Extra Example 3

Each output in the table is 3 more than half the input. Find the missing input. 10

Input, x	Output, y
2	4
4	5
6	6
8	7
?	8

● **On Your Own**

3. $y = x + 4$; 21

4. $y = \dfrac{x}{2}$; 8

Differentiated Instruction

Visual

Explain to students that representing a function table as a list of ordered pairs is for convenience. Once the function is represented by ordered pairs, it can be graphed in a coordinate plane for a visual representation of the function. This is an excellent way to show students the connection between algebra and geometry.

Laurie's Notes

Example 2

- Discuss a strategy for determining a rule for the table. Students may look *down* the input-output table instead of across. Remind students that they want to connect an input with its output.
- If you know the input is 1 and the output is 5, what has to happen? There are many things that can happen: add 4, multiply by 5, double and add 3, multiply by 6 and subtract 1. To help narrow the list, remind students to look at another ordered pair, (2, 10). Of the possibilities listed, are there any others that work for this ordered pair? Yes, multiply by 5, which is the rule.
- One strategy for a multiple choice question of this type is to try each rule. This is not always efficient, but it would lead to a solution.

Example 3

- Write the input-output table on the board.
- ❓ "Each output in the table is five more than twice the input. How can we find the missing input?" Students will want to guess instead of trying to write an equation. Remind students that the known output might be 153 versus only 53. We want to develop a strategy that works for all numbers.
- This is good practice for students to translate from words to symbols. Note the use of colorcoding. Remind students of the importance of defining their variables.
- ❓ "If you know the equation for this input-output table, how can you use it to find the missing input?" It may not be natural at this stage for students to think about substituting 53 for *y* and solving for *x*.
- Discuss with students what knowing the equation for this function means. It is the rule that tells us how to generate ordered pairs that have the relationship—"each output is five more than twice the input."
- This example is a good review of solving a two-step equation.
- **Extension:** If the output is 153, what is the input? 74

On Your Own

- Ask volunteers to do these problems at the board. Have two groups do each problem to see if strategies are different.

Closure

- **Exit Ticket:** How is a mapping diagram similar to an input-output table? *Sample answer:* They are both visual representations.

The Dynamic Planning Tool
Editable Teacher's Resources at *BigIdeasMath.com*

EXAMPLE 2 Standardized Test Practice

Which function rule is shown by the table?

A $y = 5x$ **B** $y = \dfrac{x}{5}$

C $y = x + 4$ **D** $y = 10x$

Input, x	Output, y
1	5
2	10
4	20
8	40

Look at the relationship between the inputs and outputs. Each output y is 5 times the input x. So, the function rule is $y = 5x$.

∴ The correct answer is **A**.

EXAMPLE 3 Finding a Missing Input

Input, x	Output, y
1	7
5	15
10	25
20	45
?	53

Each output in the table is 5 more than twice the input. Find the missing input.

Step 1: Write an equation for the function shown by the table.

Words Output is five more than twice the input.

Variables Let y be the output value and x be the input value.

Equation $y\ =\ 5 +\ 2 \cdot\ x$

An equation is $y = 5 + 2x$.

Step 2: Substitute 53 for y. Then solve for x.

$y = 5 + 2x$ Write the equation.

$53 = 5 + 2x$ Substitute 53 for y.

$48 = 2x$ Subtract 5 from each side.

$24 = x$ Divide each side by 2.

Check

$2x + 5 = 53$

$2(24) + 5 \stackrel{?}{=} 53$

$48 + 5 \stackrel{?}{=} 53$

$53 = 53$ ✓

∴ The missing input is 24.

On Your Own

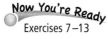
Exercises 7–13

Use the first three input values to write an equation for the function shown by the table. Then find the missing input.

3.
Input, x	Output, y
1	5
3	7
7	11
?	25

4.
Input, x	Output, y
2	1
4	2
6	3
?	4

Section 9.3 Input-Output Tables

9.3 Exercises

Vocabulary and Concept Check

1. **VOCABULARY** Explain how you can use an input-output table to represent a function.

2. **DIFFERENT WORDS, SAME QUESTION** Which is different? Find "both" answers.

What output is 4 more than twice the input 3?	What output is twice the sum of the input 3 and 4?
What output is the sum of 2 times the input 3 and 4?	What output is 4 increased by twice the input 3?

Practice and Problem Solving

Copy and complete the input-output table for the function.

3. $y = x + 5$

Input, x	1	2	3	4
Output, y				

4. $y = 4x$

Input, x	0	2	4	6
Output, y				

Write an equation for the function. Then copy and complete the table.

5. The output is 3 more than the input.

Input, x	0	1	2	3
Output, y				

6. The output is 5 times the input.

Input, x	1	3	5	7
Output, y				

Write an equation for the function shown by the table.

7.
Input, x	1	2	3	4
Output, y	9	10	11	12

8.
Input, x	2	4	6	8
Output, y	4	8	12	16

9.
Input, x	0	3	6	9
Output, y	0	1	2	3

10.
Input, x	3	5	7	9
Output, y	1	3	5	7

11. **ERROR ANALYSIS** Describe and correct the error in writing an equation for the function shown by the table.

Input, x	0	4	8	12
Output, y	0	1	2	3

$y = 4x$

Assignment Guide and Homework Check

Level	Day 1 Activity Assignment	Day 2 Lesson Assignment	Homework Check
Basic	3, 4, 17–21	1, 2, 5–11, 13, 14	2, 6, 8, 13
Average	3, 4, 17–21	1, 2, 5–11, 13, 14	2, 6, 8, 13
Advanced	3, 4, 17–21	1, 2, 6, 8, 10, 11–16	6, 8, 12, 14

For Your Information
- **Exercise 15** The input-output table for Greenwich Mean Time and home time may be different depending on daylight savings time.

Common Errors
- **Exercises 5 and 6** Students may switch the x and y variables. For example, they may write $x = y + 3$ instead of $y = x + 3$. Remind them of the methods for writing equations that they learned in the previous section.
- **Exercises 7–10** Students may focus on the changes in the input and the output, instead of finding the relationship between the input and output. For example, a student may notice that the input and output both change by 1 each time, instead of the output values being 8 more than the input values. Remind them to compare the inputs and outputs, not how each changes individually.
- **Exercises 12 and 13** Students may use the word sentence incorrectly when trying to find the missing input values. Encourage them to write the equation from the word sentence, substitute the given value for y, and then solve the equation for x.

Vocabulary and Concept Check
1. Choose the inputs that represent the situation or show the pattern of the function. Pair each input in the table with its resulting output.
2. What output is twice the sum of the input 3 and 4?; $2(3 + 4) = 14$; $2(3) + 4 = 10$

Practice and Problem Solving
3. 6; 7; 8; 9
4. 0; 8; 16; 24
5. $y = x + 3$; 3; 4; 5; 6
6. $y = 5x$; 5; 15; 25; 35
7. $y = x + 8$
8. $y = 2x$
9. $y = \dfrac{x}{3}$
10. $y = x - 2$
11. Each output in the table is one-fourth of the input, but the equation would make each output four times each input; $y = \dfrac{x}{4}$

9.3 Record and Practice Journal

Technology For the Teacher
Answer Presentation Tool
QuizShow

T-382

Practice and Problem Solving

12. 10; 15
13. 0; 16; 26
14. See *Taking Math Deeper*.
15. See Additional Answers.
16. *Sample answer:* $2y = 4x + 6$

Fair Game Review

17–20.

21. B

Mini-Assessment

Write an equation for the function. Then copy and complete the table.

1. The output is 4 more than the input.

Input, x	1	2	3	4
Output, y	5	6	7	8

$y = x + 4$

2. The output is 2 times the input.

Input, x	1	2	3	4
Output, y	2	4	6	8

$y = 2x$

3. You walk at a steady pace of 1 mile every 20 minutes. Copy and complete the input-output table. Then write a function rule in which x is the input and y is the output.

Distance, x	1	2	3	4
Time, y	20	40	60	80

$y = 20x$

Taking Math Deeper

Exercise 14

The basic concept for this problem is straightforward. Each time I travel a mile, I have one less mile to travel. It's a little like the old proverb *"The longest journey starts with a single step."*

 a. Copy and complete the input-output table.

Distance from Key West, x	0	30	47	82	100
Distance to Key Largo, y	100	70	53	18	0

 b. Write a function.

(distance to go) = 100 − (distance traveled)
$y = 100 - x$

c. The function rule does not give the distance to Florida City. Because the mile marker at Florida City is 127, you can use the following function rule to give the distance to Florida City.

$y = 127 - x$

NOTE: If you haven't been to Key West, you might not know that chickens roam the streets of the city. That is the reason the rooster is pictured with the problem.

Project

Write an input-output table for the distance from your town to a major city. Write a function rule in which x represents the input and y represents the output.

Reteaching and Enrichment Strategies

If students need help...	If students got it...
Resources by Chapter • Practice A and Practice B • Puzzle Time Record and Practice Journal Practice Differentiating the Lesson Lesson Tutorials Skills Review Handbook	Resources by Chapter • Enrichment and Extension • School-to-Work Start the next section

In Exercises 12 and 13, copy and complete the table.

12. For each output, multiply the input by 4, then subtract 5.

Input, x	2	3	4	7		
Output, y	3	7	11	23	35	55

13. For each output, divide the input by 2, then add 4.

Input, x			2	4	10		
Output, y	4	5	6	9	12	17	

14. **GEOGRAPHY** You travel along US Highway 1 from mile marker 0 in Key West to mile marker 100 in Key Largo.

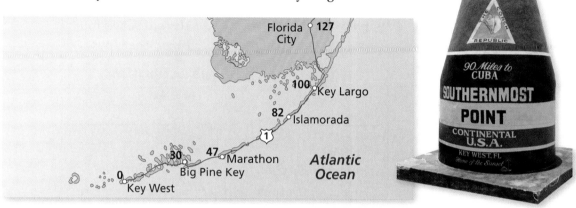

a. Copy and complete the input-output table.

Distance from Key West, x	0	30	47	82	100
Distance to Key Largo, y					

b. Write a function rule in which x is the input and y is the output.

c. Can you use your function rule to find the distance to Florida City? If not, write a function rule that you can use.

15. **TIME** Make an input-output table with the Greenwich Mean Time (GMT) hourly times as inputs, and times where you live as outputs. Write a function rule for the data.

16. **Critical Thinking** Write an equation with the same outputs as $y = 2x + 3$ for $x = 0, 1, 2, 3,$ and 4.

Fair Game Review What you learned in previous grades & lessons

Plot the ordered pairs in the same coordinate plane. *(Skills Review Handbook)*

17. (1, 2) 18. (0, 7) 19. (2, 3) 20. (6, 5)

21. **MULTIPLE CHOICE** Which is the solution of the inequality $6x \leq 24$? *(Section 8.3)*

 Ⓐ $x < 4$ Ⓑ $x \leq 4$ Ⓒ $x < 144$ Ⓓ $x \leq 144$

9 Study Help

You can use a **four square** to organize information about a topic. Each of the four squares can be a category, such as *definition, vocabulary, example, non-example, words, algebra, table, numbers, visual, graph,* or *equation*. Here is an example of a four square for a function.

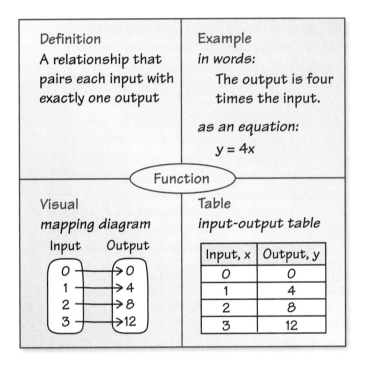

On Your Own

Make a four square to help you study these topics.

1. function
2. mapping diagram
3. input-output table

After you complete this chapter, make four squares for the following topics.

4. graph of a function
5. linear function
6. Choose two other topics that you studied earlier in this course. Make a four square for each topic.

"I'm taking a survey for my four square. How many fleas do you have?"

Sample Answers

1.

2.

3.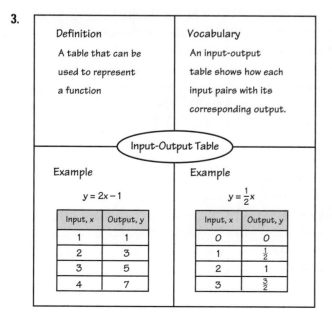

List of Organizers
Available at *BigIdeasMath.com*

Comparison Chart
Concept Circle
Definition (Idea) and Example Chart
Example and Non-Example Chart
Formula Triangle
Four Square
Information Frame
Information Wheel
Notetaking Organizer
Process Diagram
Summary Triangle
Word Magnet
Y Chart

About this Organizer

A **Four Square** can be used to organize information about a topic. Students write the topic in the "bubble" in the middle of the four square. Then students write concepts related to the topic in the four squares surrounding the bubble. Any concept related to the topic can be used. Encourage students to include concepts that will help them learn the topic. Students can place their four squares on note cards to use as a quick study reference.

Technology For the Teacher
Vocabulary Puzzle Builder

T-384

Answers

1. (10, 0), (20, 0), (30, 10), (40, 5)
2. (0, 8), (1, 9), (2, 10), (3, 11)
3. 40
4. 16
5. 7
6. no
7. yes
8. yes
9. $y = 9x$; 9; 18; 27; 36
10. $y = \dfrac{x}{8}$
11. $x = 8$; $y = 60$; $y = 6x$
12.
13. a. $d = 45 - 0.9s$
 b. 50 songs

Assessment Book

T-385

Alternative Quiz Ideas

100% Quiz
Error Notebook
Group Quiz
Homework Quiz

Math Log
Notebook Quiz
Partner Quiz
Pass the Paper

100% Quiz
This is a quiz where students are given the answers and then they have to explain and justify each answer.

Reteaching and Enrichment Strategies

If students need help...	If students got it...
Resources by Chapter • Study Help • Practice A and Practice B • Puzzle Time Lesson Tutorials BigIdeasMath.com Practice Quiz Practice from the Test Generator	Resources by Chapter • Enrichment and Extension • School-to-Work Game Closet at BigIdeasMath.com Start the next section

Technology For the Teacher

Answer Presentation Tool
Big Ideas Test Generator

9.1–9.3 Quiz

List the ordered pairs shown in the mapping diagram. *(Section 9.1)*

1.

2.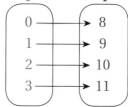

Find the value of y for the given value of x. *(Section 9.2)*

3. $y = 10x$; $x = 4$

4. $y = 2x - 6$; $x = 11$

5. $y = 4x + 5$; $x = \dfrac{1}{2}$

Tell whether the ordered pair is a solution of the equation. *(Section 9.2)*

6. $y = x - 1$; $(5, 6)$

7. $y = 5x + 3$; $(3, 18)$

8. $y = \dfrac{x}{7}$; $(42, 6)$

9. Write an equation for the function "The output is the product of 9 and the input." Then copy and complete the table. *(Section 9.3)*

Input, x	1	2	3	4
Output, y				

10. Write an equation for the function shown by the table. *(Section 9.3)*

Input, x	0	8	16	24
Output, y	0	1	2	3

11. **RACE** You run a 10-kilometer race at a steady pace of 1 kilometer every 6 minutes. Copy and complete the input-output table. Then write a function rule in which x is the input and y is the output. *(Section 9.3)*

Distance, x	1	2	6		10
Time, y	6	12	36	48	

12. **PUPPIES** The table shows the ages of four puppies and their weights. Use the table to draw a mapping diagram. *(Section 9.1)*

Age (weeks)	Weight (oz)
3	11
4	85
6	85
10	480

13. **GIFT CARD** You have a $45 gift card for an online music store. Each song costs $0.90. *(Section 9.2)*

 a. Write an equation you can use to find the number of dollars d remaining on the card after you buy s songs.

 b. What is the greatest number of songs you can buy with the gift card?

9.4 Graphs

COMMON CORE STATE STANDARDS
6.EE.9

Essential Question How can you use a graph to describe a function?

1 ACTIVITY: Interpreting a Graph

Work with a partner. Use the graph to test the truth of each statement. If the statement is true, write an equation that shows how to get one measurement from the other measurement.

a. "You can find the horsepower of a race car engine if you know its volume in cubic inches."

b. "You can find the volume of a race car engine in cubic centimeters if you know its volume in cubic inches."

2 ACTIVITY: Interpreting a Graph

Work with a partner. The table shows the average speeds of the winners of the Daytona 500. Graph the data. Does the graph allow you to predict future winning speeds? Explain why or why not.

Year	2000	2001	2002	2003	2004	2005	2006	2007	2008
Speed (mi/h)	156	162	143	134	156	135	143	149	153

386 Chapter 9 Tables, Graphs, and Functions

Laurie's Notes

Introduction

For the Teacher
- **Goal:** Students will develop an understanding of how a graph displaying ordered pairs can describe a function.
- **Big Idea:** When the ordered pairs are plotted, each input will be paired with exactly one output. A quick way to determine if this has happened is to use what is called the *Vertical Line Test*. As a vertical line passes over the graph, it should intersect the plotted points only once. If the vertical line intersects two or more ordered pairs, it is *not* a function.

Motivate
- Ask students to guess the mileage a race car gets during a long race at Indianapolis or Daytona. less than 2 mi/gal
- Today's activity includes two problems related to race cars.
- **FYI:** While race cars at Indy or Daytona need to fill up frequently, a hybrid gets about 500 miles on one tank!

Activity Notes

Activity 1
- The two graphs are different. The first has scattered points with no apparent pattern. The second appears to have a pattern.
- The challenge will be in writing the equation for part (b). Suggest to students that they should make an input-output table. Students may recognize and say, "every 100 in.3 adds another 1640 cm^3," but not know how to write an equation for the pattern.
- **?** "What if only 1 cubic inch was added; how many cubic centimeters would be added?" one-hundredth of 1640 or 16.4
- This may be enough to help students recognize the equation: $y = 16.4x$.
- Discuss student answers. Focus on how they determined if the statement was true. Discuss the equation written for part (b).

Activity 2
- The challenging part of this activity is deciding how to scale each axis. You might want to step students through this process. The years will be the input on the horizontal axis. The speed will be the output on the vertical axis.
- **?** "What is the range of the speeds you need to graph?" 134 to 162
- **Big Idea:** This graph does not have a predictable pattern. Unlike part (a) in the previous activity, each input (year) has a unique output (average speed of winners). Even though the speeds repeat (143 miles per hour was the average speed in two different years), the years do not. This is a function, but not one that can be written as an equation.
- **Extension:** Ask how the average speed is determined. 500 miles divided by racing time in hours

Common Core State Standards
6.EE.9 Use variables to represent two quantities in a real-world problem that change in relationship to one another; write an equation to express one quantity, thought of as the dependent variable, in terms of the other quantity, thought of as the independent variable. Analyze the relationship between the dependent and independent variables using graphs and tables, and relate these to the equation.

Previous Learning
Students should know how to plot ordered pairs.

Activity Materials
Textbook
• metal washers • string • stop watch

Start Thinking! and Warm Up

9.4 Record and Practice Journal

T-386

Differentiated Instruction

Kinesthetic

When plotting an ordered pair in the coordinate plane, have students start with their finger on (0, 0). Ask them how many units they are moving to the right. Then ask how many units they are moving up. Have students plot the point and write the ordered pair next to the point.

9.4 Record and Practice Journal

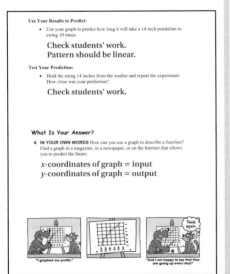

Laurie's Notes

Activity 3

- Tell students they will be conducting an experiment where they will collect data and graph it. Because pairs of students are collecting their own data, there will be variation in results.
- **FYI:** The relationship between pendulum length and period (time) is quadratic (or square root), *not* linear. However, for short periods of time, a linear approximation is good. There is a very rich history in mathematics on the motion of a pendulum. Its study corresponded to the Scientific Revolution in Europe (1500s) in which people began to understand that events in nature can be predicted.
- **FYI:** Note that the period (swing back and forth) of a pendulum does *not* depend on the arc length. The swing time will be the same whether the student holds the string several inches from the washer or a few inches from the washer. It is this law that makes a pendulum clock work.
- **Teaching Tip:** I usually have students let one swing go by before timing. They will be timing for 10 swings, where "a swing" is the movement back and forth. One student manages the pendulum while their partner does the timing and recording.
- Be sure students label the axes with words.
- Have several groups read the description of their graph.
- Students should be able to make a reasonable prediction from their graph for a pendulum of length 14 inches.
- ? "Is this the graph of a function? Explain." Yes, each input (length of pendulum) is paired with a unique output (time of 10 swings).

What Is Your Answer?

- **Cartoon:** This is a classic error that students make!

Closure

- Describe the graph. Use the graph to predict the cost of 8 tickets. $48

Technology For the Teacher

Dynamic Classroom

The Dynamic Planning Tool
Editable Teacher's Resources at *BigIdeasMath.com*

T-387

3 ACTIVITY: Conducting an Experiment

Work with a partner.

Collect Materials:
- Metal washer
- String (at least 15 in. long)
- Stopwatch

Perform the Experiment:
- Tie one end of the string securely around the washer.
- Hold the string 6 inches from the washer. Swing the washer and measure the time it takes to swing back and forth 10 times.
- Record your result in a table.
- Repeat the experiment when holding the string at lengths of 7 in., 8 in., 9 in., 10 in., 11 in., and 12 in.

Analyze the Results:
- Make a graph of your data.
- Describe the graph.

Use Your Results to Predict:
- Use your graph to predict how long it will take a 14-inch pendulum to swing 10 times.

Test Your Prediction:
- Hold the string 14 inches from the washer and repeat the experiment. How close was your prediction?

What Is Your Answer?

4. **IN YOUR OWN WORDS** How can you use a graph to describe a function? Find a graph in a magazine, in a newspaper, or on the Internet that allows you to predict the future.

"I graphed our profits."

"And I am happy to say that they are going up every day!"

Use what you learned about graphs to complete Exercises 4–7 on page 390.

9.4 Lesson

Key Vocabulary
graph, p. 388

Graph of a Function

A function can be represented by a **graph**. The graph below is for the function $y = x + 2$.

Input, x	Output, y	Ordered Pair, (x, y)
1	3	(1, 3)
2	4	(2, 4)
3	5	(3, 5)

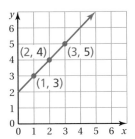

When you draw a line through the points, you graph *all* of the solutions of $y = x + 2$.

EXAMPLE 1 Graphing a Function

Graph $y = 2x + 1$.

Make an input-output table. Use the values 1, 2, and 3 for *x*.

x	y = 2x + 1	y	(x, y)
1	y = 2(1) + 1	3	(1, 3)
2	y = 2(2) + 1	5	(2, 5)
3	y = 2(3) + 1	7	(3, 7)

Plot the ordered pairs. Draw a line through the points.

On Your Own

Graph the function.

1. $y = x + 1$ 2. $y = 3x$ 3. $y = 3x + 2$

388 Chapter 9 Tables, Graphs, and Functions

Laurie's Notes

Introduction

Connect
- **Yesterday:** Students performed an activity that resulted in collecting and graphing data.
- **Today:** Students will use a graph to represent functions and use the graph to make predictions.

Motivate
- ❓ "How many miles do each of you travel to get to school each day?" Answers will vary.
- Record the least and farthest distances.
- Use distances known to your students. Talk about gas consumption and the production of carbon dioxide. This is the context for Example 2.

Lesson Notes

Key Idea
- Write the Key Idea.
- **Connection:** In previous sections, students wrote functions as mapping diagrams, words and equations, and tables. In today's lesson, the ordered pairs from a table are plotted. The input is x and is found on the horizontal axis. The output is y and is found on the vertical axis. The graph shows the relationship between inputs and outputs.
- Discuss with students why connecting the points with a line is the graph of all of the solutions of the equation $y = x + 2$. For instance, if $x = 1.5$, $y = 3.5$. Find this point on the graph.
- ❓ "Is the ordered pair (3, 1) a solution of $y = x + 2$?" no "Is (3, 1) on the graph?" no

Example 1
- Discuss the steps necessary to graph a function. Evaluate the function (equation) for the values (inputs) given. Record the information in a table. Plot the ordered pairs and draw a line through the points.
- Work through the example with students.
- **Note:** Graphing in the coordinate plane (all four quadrants) is taught in a future course. It is unlikely that students will question why the input and output are always positive, although they may.

On Your Own
- **Extension:** Ask students to compare their graphs for Questions 2 and 3. How are the graphs alike? If the scales are the same for both graphs, students may observe that the graph in Question 3 is shifted up 2 units from the graph in Question 2.

Goal Today's lesson is using a **graph** to represent a function.

Start Thinking! and Warm Up

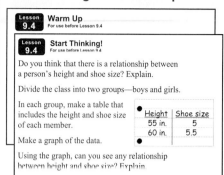

Extra Example 1

Graph $y = x + 3$.

On Your Own

1.

2.

3.

T-388

Extra Example 2

Use the function $d = 50t$ to find the number d of miles traveled in t hours. Graph the function.

On Your Own

4.

English Language Learners

Auditory
Put the alphabet to use for your English learners. Because X comes before Y in the alphabet, students can alphabetize pairs of words and associate them with the x-coordinate and the y-coordinate. So, *input* comes before *output* and *horizontal* comes before *vertical*.

Laurie's Notes

Example 2

- This example relates to the topic of carbon dioxide produced by gasoline, which was discussed in the Motivate.
- The equation can be written as $y = 20x$. It is common to use variable names to remind us of what the variable represents. In this problem, the ordered pairs are: (gallons of gasoline, pounds of carbon dioxide), or (g, p).
- **Common Misconception:** The axes have to be scaled in the same units.
- ? "Is it possible for $g = 1.5$?" yes "What is the output for $g = 1.5$?" 30 "Is the ordered pair (1.5, 30) on the graph?" yes
- **Extension:** Is it possible to estimate g when $p = 50$? yes; $g = 2.5$ by reading the graph

On Your Own

- Have one group graph their solution on a transparency to share with the class.

Summary

- Discuss the *Summary* box with students. They have represented functions using words, equations, input-output tables, mapping diagrams, and graphs.
- **Connection:** In future mathematics courses, students will study connections between the various forms of functions.
- **FYI:** It is not always possible to represent the function in a different form. For instance, the data for the Daytona 500 winning speeds was represented in a table and graph, but could not be represented as an equation.

Closure

- Divide the class into groups of 4. Say, "an output is 1 less than twice the input." Each group should represent the function in four different ways.
Sample answer:
Words: An output is 1 less than twice the input.
Equation: $y = 2x - 1$
Mapping Diagram: **Graph:**

The Dynamic Planning Tool
Editable Teacher's Resources at *BigIdeasMath.com*

EXAMPLE 2 Graphing a Function

Use the function $p = 20g$ to find the number of pounds p of carbon dioxide produced by burning g gallons of gasoline. Graph the function.

Make an input-output table.

Input, g	$p = 20g$	Output, p	Ordered Pair, (g, p)
1	$p = 20(1)$	20	$(1, 20)$
2	$p = 20(2)$	40	$(2, 40)$
3	$p = 20(3)$	60	$(3, 60)$

Plot the ordered pairs. Draw a line through the points.

Reading

Because you cannot have a negative number of gallons, use only positive values of g.

On Your Own

Now You're Ready
Exercise 17

4. Use the function $d = 35t$ to find the distance d (in miles) and time t (in hours) for a boat. Graph the function.

Summary

Representing a Function

There are several ways to represent a function.

Words Each output is 2 more than the input.

Equation $y = x + 2$

Input-Output Table

Input, x	Output, y
1	3
2	4
3	5
4	6

Mapping Diagram

Input → Output
1 → 3
2 → 4
3 → 5
4 → 6

Graph

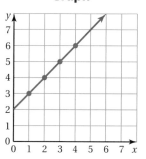

Section 9.4 Graphs

9.4 Exercises

Vocabulary and Concept Check

1. **VOCABULARY** Describe steps you can use to draw a graph of a function.
2. **OPEN-ENDED** Describe four ways to represent the function $y = 4x + 5$.
3. **WRITING** You are given the graph of a function. Explain how you could find the function rule that describes the graph.

Practice and Problem Solving

Graph the data.

4.
Input, x	0	2	4	6
Output, y	0	4	8	12

5.
Input, x	0	1	2	3
Output, y	1	2	3	4

6.
Input, x	Output, y
1	0
3	2
5	4
7	6

7.
Input, x	Output, y
2	0.5
4	1
6	1.5
8	2

Graph the function.

① 8. $y = x + 4$ 9. $y = 2x$ 10. $y = 5x + 3$ 11. $y = \dfrac{x}{4}$

12. $y = 2x + 5$ 13. $y = x - 1$ 14. $y = \dfrac{3}{2}x + 1$ 15. $y = 1 + 0.5x$

16. **ERROR ANALYSIS** Describe and correct the error in graphing the line from the input-output table.

Input, x	1	2	3	4
Output, y	1	3	5	7

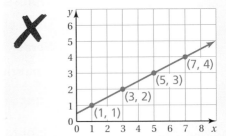

② 17. **DOLPHIN** Use the function $p = 30d$ to find the number of pounds p of fish that a dolphin eats in d days. Graph the function.

Assignment Guide and Homework Check

Level	Day 1 Activity Assignment	Day 2 Lesson Assignment	Homework Check
Basic	4–7, 25–29	1–3, 9–15 odd, 16–20	9, 16, 17, 18
Average	4–7, 25–29	1–3, 9–15 odd, 16, 18–22	9, 16, 18, 22
Advanced	4–7, 25–29	1–3, 8–20 even, 22–24	8, 16, 18, 22

Common Errors

- **Exercises 8–15** Students may mix up their axes or label one or both axes inconsistently. Remind them that the input (x) values are horizontal and the output (y) values are vertical. Encourage students to label each axis in specific increments that easily show the points plotted and will give a line through the points.
- **Exercise 17** Students may get the input and output confused because they are not using x and y as variables. Ask them which variable is the input and which is the output. Then ask which variable goes on each axis. This should help them graph the equation.
- **Exercises 18–20** Students may just guess which equation goes with each graph without testing any values. Encourage them to find a point or two on the graph and substitute the values into the equations to find which one matches.

9.4 Record and Practice Journal

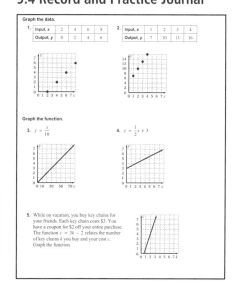

Technology For the Teacher
Answer Presentation Tool
QuizShow

Vocabulary and Concept Check

1. Make an input-output table. Plot the ordered pairs. Draw a line through the points.
2. *Sample answer:* graph, equation, input-output table, mapping diagram
3. Find points on the graph. Make a mapping diagram or input-output table to show the pattern. Use the pattern to write a function rule.

Practice and Problem Solving

4.

5.

6.

7.

8.

9–17. See Additional Answers.

T-390

 Practice and Problem Solving

18. B **19.** C

20. A

21. See *Taking Math Deeper*.

22. a.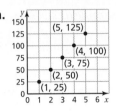

b. $y = 25x$

23. 17

24. See Additional Answers.

 Fair Game Review

25. 6 **26.** 9

27. 19 **28.** 112

29. C

Mini-Assessment

1. Graph $y = x + 2$.

2. Graph the function $a = 10d$ to find the amount a of money you raise in d days.

Taking Math Deeper

Exercise 21

This problem gives students an opportunity to practice reading the questions carefully. Notice that parts (a) and (b) ask for the discount d, but part (c) asks for the sale price.

 a. Write a function for the discount.

(Discount) = (Percent)(Regular Price)
$$d = 0.4p$$

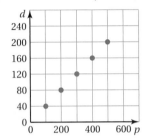

b. Graph the function for $p = 100, 200, 300, 400$, and 500.

 c. Find the sale price of a bookshelf.
$$d = 0.4p$$
$$= 0.4(85)$$
$$= \$34$$
Sale price = $85 - 34$
$$= \$51$$
The sale price is $51.

Save $34

Reteaching and Enrichment Strategies

If students need help...	If students got it...
Resources by Chapter • Practice A and Practice B • Puzzle Time Record and Practice Journal Practice Differentiating the Lesson Lesson Tutorials Skills Review Handbook	Resources by Chapter • Enrichment and Extension • School-to-Work Start the next section

Match each graph with a function.

18.

19.

20.

A. $y = \dfrac{x}{3}$

B. $y = x + 1$

C. $y = 2x - 1$

21. **SALE** A furniture store is having a sale where everything is 40% off.

 a. Write a function that you can use to find the amount of discount d on an item with a regular price p.

 b. Graph the function using the inputs 100, 200, 300, 400, and 500 for p.

 c. You buy a bookshelf that has a regular price of $85. What is the sale price of the bookshelf?

22. **WIND TURBINE** The table shows the number of rotations a wind turbine makes in x minutes.

Minutes, x	1	2	3	4	5
Rotations, y	25	50	75	100	125

 a. Graph the data.

 b. Find a function to describe the data.

23. **REASONING** The graph of a function is a straight line that goes through the points (3, 2), (5, 8), and (8, y). What is the value of y?

24. **Critical Thinking** Make an input-output table where the input is the side length of a square and the output is the *perimeter*. Make a second input-output table where the input is the side length of a square and the output is the *area*. Graph both functions in the same coordinate plane. Compare the functions and graphs.

Fair Game Review What you learned in previous grades & lessons

Find the value of y for the given value of x. *(Section 9.2)*

25. $y = x + 2$; $x = 4$

26. $y = 4x - 3$; $x = 3$

27. $y = 3x + 1$; $x = 6$

28. $y = 16x$; $x = 7$

29. **MULTIPLE CHOICE** The rectangular rug has a perimeter of 224 inches. What is the width of the rug? *(Section 7.5)*

 Ⓐ 3.5 in. Ⓑ 36 in.

 Ⓒ 48 in. Ⓓ 96 in.

9.5 Analyzing Graphs

Essential Question How can you analyze a function from its graph?

COMMON CORE STATE STANDARDS
6.EE.9

1 ACTIVITY: Analyzing Graphs

Work with a partner. Copy and complete the table for the given situation. Then make a graph of the data. Write an equation for the function. Describe the characteristics of the graph.

a. Find the area of a square with side length s.

Side, s	1	2	3	4
Area, A				

Side, s

b. Find the amount earned for working h hours at $3 per hour.

Hour, h	1	2	3	4
Amount, A				

Hour, h

c. You start with $20 in a savings account. Find the amount left in the account when you withdraw $2 each day d.

Day, d	1	2	3	4
Amount, A				

Day, d

d. You start with $10 in a savings account. Find the amount in the account when you deposit $2 each day d.

Day, d	1	2	3	4
Amount, A				

Day, d

Laurie's Notes

Introduction

For the Teacher
- **Goal:** Students will gain additional practice generating and plotting ordered pairs, and begin to understand that not all graphs are linear.

Activity Notes

Activity 1
- Explain to students that they will be generating ordered pairs for a function from a word description of the function. They will then graph the data.
- **?** "Do the scales on the axes need to be equal?" no
- Students will need to decide on a scale for the vertical axis.
- **Common Error:** Students scale the axis by using the output values instead of making equal intervals. For example, in the first problem, students make equally-spaced tick marks up the vertical axis and label the tick marks 1, 4, 9, and 16. Doing this will make the graph linear, which it is not!
- Note that the axes are labeled with the input and output variables. Remind students that it is necessary to label the axes.
- You may need to explain to students what it means to "describe the characteristics of a graph." You could sketch some graphs on the board unrelated to the graphs in this activity. You might suggest that students look for graphs that are lines versus graphs that are not lines. Is the graph steep? Is it increasing or decreasing? Does the graph touch either axis?
- **?** "What characteristics did you notice about the graph in part (a)?" Students should mention that the data do *not* form a line. They should mention that the graph is getting steep quickly.
- **FYI:** The first graph is known as a parabola, only half of which has been drawn. The equation is a quadratic or second-degree equation.
- **Extension:** Ask what the side length would be for an area of 64. 8
- In describing the second graph, students should use language such as: every time the hours increase by 1, the amount earned increases by 3. This will help students in their preparation for future courses and their study of slope.
- **?** "Do the first two graphs go through (0, 0)?" yes "Does it make sense that (0, 0) would be on each graph?" yes; No side length means no area. No time spent working means no money earned.
- The graphs in parts (c) or (d) do *not* have (0, 0) as a solution. Students should observe that in part (c) the graph is descending (going down), while in part (d) the graph is ascending (going up).
- **Extension:** For each of these graphs, ask prediction questions. *Example:* "For the graph in part (c), when will your account balance be zero?" after 10 days "For the graph in part (d), when will your account balance be $30?" after 10 days

Common Core State Standards

6.EE.9 Use variables to represent two quantities in a real-world problem that change in relationship to one another; write an equation to express one quantity, thought of as the dependent variable, in terms of the other quantity, thought of as the independent variable. Analyze the relationship between the dependent and independent variables using graphs and tables, and relate these to the equation.

Previous Learning
Students should know how to plot ordered pairs.

Activity Materials
Textbook
• wood ramp
• books
• stop watch
• toy car

Start Thinking! and Warm Up

9.5 Record and Practice Journal

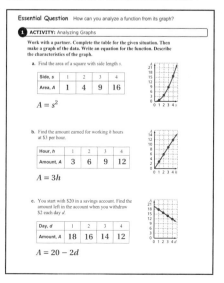

T-392

Differentiated Instruction

Kinesthetic

Have students gather real-life data using a meter stick and a table tennis ball. Drop the ball 10 times from different initial heights and measure the initial bounce with the meter stick. Tell students to record the data in a table with the initial height as the input and the initial bounce as the output, and then have them graph the data.

9.5 Record and Practice Journal

Laurie's Notes

Activity 2

- **Whole Class Activity:** Depending upon space and materials, you may need to do this as a whole class activity.
- **Management Tip:** Have the materials ready to go in advance.
- Ask students to help move desks so that an area is in view of all students.
- One student should record the data at the board. All students should record the data at the conclusion of the 4th trial.
- After students have made their prediction for 5 books, complete the 5th trial.
- **Teaching Tip:** An alternate approach is to use an 8-foot table that has the legs folded under. My students have used the 8-foot table and two different cars racing at once with two different timers. Two sets of data are collected and two graphs plotted. One car is always faster than the other, and the steepness of the graphs show this.
- **FYI:** The relationship between the number of books and time is a square root, *not* linear. However, for short periods of time, a linear approximation is good.

What Is Your Answer?

- **Think-Pair-Share:** Students should read each question independently and then work with a partner to answer the questions. When they have answered the questions, the pair should compare their answers with another group and discuss any discrepancies.

Closure

- Graph $y = x + 2$ and $y = x - 2$ for inputs of 2, 3, 4, and 5. Describe the characteristics of each graph.

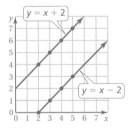

Both graphs represent a linear function. Both graphs have the same steepness. The graph of $y = x + 2$ is four units above the graph of $y = x - 2$.

Technology For the Teacher

The Dynamic Planning Tool
Editable Teacher's Resources at *BigIdeasMath.com*

2 ACTIVITY: Conducting an Experiment

Work with a partner.

Collect Materials:
- A board at least 8 feet long
- Five books of the same thickness
- Toy car
- Stopwatch

Perform the Experiment:
- Place one book underneath one end of the board.
- Put the car at the top of the ramp. Measure the time (in seconds) it takes the car to roll down the ramp.
- Record your result in a table.
- Repeat the experiment with two, three, and four books.

Analyze the Results:
- Make a graph of your data.
- Does the graph have the characteristics of any of the graphs in Activity 1? Explain.

Use Your Results to Predict:
- Use your graph to predict how long it will take the car to roll down the ramp when five books are placed under the board.

Test Your Prediction:
- Repeat the experiment with five books. How close was your prediction?

What Is Your Answer?

3. **IN YOUR OWN WORDS** How can you analyze a function from its graph? Give a real-life example of how a graph can help you make a decision.

Practice — Use what you learned about analyzing graphs to complete Exercises 3 and 4 on page 396.

9.5 Lesson

Check It Out
Lesson Tutorials
BigIdeasMath **V**com

A function whose graph is a straight line is a **linear function**.

EXAMPLE 1 Identifying Linear Functions

Key Vocabulary
linear function, p. 394

Does the graph represent a linear function? Explain.

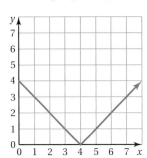

∴ The graph is not a straight line. So, the graph does *not* represent a linear function.

∴ The graph is a straight line. So, the graph does represent a linear function.

EXAMPLE 2 Identifying a Linear Function

Input, x	Output, y
0	0
1	1
4	2
9	3

Does the input-output table represent a linear function? Explain.

The ordered pairs in the table are (0, 0), (1, 1), (4, 2), and (9, 3). Plot the ordered pairs and draw a graph through the points.

∴ The graph is not a straight line. So, the function is *not* linear.

On Your Own

Exercises 5–14

Does the graph or table represent a linear function? Explain.

1.

2.

3.
Input, x	0	3	6	8
Output, y	1	4	7	9

4.
Input, x	0	1	2	3
Output, y	1	2	5	10

394 Chapter 9 Tables, Graphs, and Functions 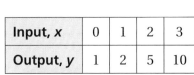 Multi-Language Glossary at BigIdeasMath✓com.

Laurie's Notes

Introduction

Connect
- **Yesterday:** Students gained an understanding that not all graphs are lines.
- **Today:** Students will graph linear and nonlinear data.

Motivate
- Ask 5 students to stand at the front wall facing their classmates.
- Hand students a card with an input number from the set {1, 2, 3, 4, 5}. Students should be in numerical order. Tell the students they are to step forward n steps, where $n =$ the input number plus 1. So, the student with card 1 will take 2 steps, student with card 2 will take 3 steps, and so on.
- If students all take approximately the same size step, the students should still be in a line but at a diagonal, or slant, from the wall. The students have just graphed the linear function $y = x + 1$.
- Ask for observations. Students should observe they are in a line.

Lesson Notes

Example 1
- **Vocabulary:** Write the description of a linear function.
- The intent of this lesson is not to study a variety of functions. These functions will be presented in a future class. The names of the different functions shown do *not* have to be used. The focus is on noting which graphs are linear and which are not.
- **FYI:** The graph in part (a) is known as an absolute value graph. The equation is $y = |x - 4|$. A contextual example of how it might be used would be the following: The distance from a highway exit is a function of the amount of time traveling on a highway. As you drive at a constant speed, you continue to get closer to the exit, and once you pass the exit, you start to get farther away.

Example 2
- Students are often surprised by this graph. Because x is increasing much faster than y, students think the graph should be decreasing. The graph does *not* decrease. It just does *not* increase very rapidly!
- **FYI:** This graph is the square root graph. The equation is $y = \sqrt{x}$.

On Your Own
- Students will be able to answer Questions 1 and 2 by visual inspection. Students will need to graph Questions 3 and 4.
- ❓ "Describe the characteristics of the graph in Question 1." linear, decreasing
- Remind students to connect the plotted points for Questions 3 and 4.
- ❓ "The graph of Question 3 is linear. Can you think of an equation that represents this data?" $y = x + 1$
- ❓ **Extension:** "The graph of Question 4 is *not* linear, but it is a function. Can you think of an equation that represents this data?" $y = x^2 + 1$

Goal
Today's lesson is graphing linear and non-linear data.

Start Thinking! and Warm Up

Lesson 9.5 Start Thinking! For use before Lesson 9.5

When you hear the term *linear*, what do you think of?

Explain what you think that the graph of a *linear function* looks like. Make a graph to illustrate your answer.

Extra Example 1
Does the graph represent a linear function? Explain.

a.
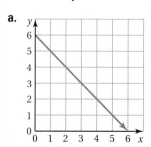

The graph is a line and does represent a linear function.

b.

The graph is *not* a line and does *not* represent a linear function.

Extra Example 2
Does the input-output table represent a linear function? Explain.

Input, x	1	2	3	4
Output, y	8	4	2	1

The graph of the function is *not* a line. So, the function is *not* linear.

On Your Own
1–4. See Additional Answers.

T-394

Extra Example 3

Is the function relating the diagram number x to the number of dots y linear?

The function is *not* linear.

Extra Example 4

Your sister earns $10 per hour. Your brother earns $12 per hour. The functions $m = 10h$ and $m = 12h$ show the relationship between the number of hours h they work and the money m they earn. Which graph is steeper? Explain.

The graph $m = 12h$ is steeper. Your brother's hourly rate is greater than your sister's hourly rate.

On Your Own

5–6. See Additional Answers.

English Language Learners

Vocabulary

Give extra credit to students who can give acceptable definitions of words in this chapter. For instance, a definition of linear *function* could be "A linear function is a line when you graph it."

T-395

Laurie's Notes

Example 3

- This problem works through the process of collecting data, representing the data in an input-output table, writing the ordered pairs, and plotting the ordered pairs.
- The ordered pairs are connected in order to determine whether the function is linear. The graph curves and therefore it is *not* linear.
- **FYI:** This is an example of *discrete* data versus *continuous* data. From the context of the problem, you know that there is no ordered pair associated with an input of 1.5. The diagram number is a counting number only.
- **Extension:** The list of outputs is a set of numbers known as the triangular numbers. You can have students research triangular numbers. This set of numbers shows up in several places in future mathematics.

Example 4

- This example relates to the concept of slope, a topic which students will study in future courses.
- To help students, it may be necessary to make a table of values for each equation. This also shows how much faster the output is growing for $m = 10h$ than for $m = 7h$.
- ? "Approximate the amount each person makes working 6 hours." This can be approximated from the graph or calculated using the equations.
- **Extension:** Ask approximately how long each person must work to make $40. 4 hours; 6 hours

On Your Own

- **Think-Pair-Share:** Students should read each question independently and then work with a partner to answer the questions. When they have answered the questions, the pair should compare their answers with another group and discuss any discrepancies.
- The function in Question 5 is generating odd numbers, a pattern students have seen before.
- The function in Question 6 relates to the distance traveled on a highway with speed limits set at 65 mi/h and 55 mi/h.

Closure

- **Writing prompt:** A linear function is . . . (Include an example of an equation that would be a linear function.) *Sample answer:* $y = 2x$

Technology For the Teacher

The Dynamic Planning Tool
Editable Teacher's Resources at *BigIdeasMath.com*

EXAMPLE 3 Identifying a Linear Function

Is the function relating the diagram number x to the number of dots y linear?

Diagram 1 Diagram 2 Diagram 3 Diagram 4

Make an input-output table. Then plot the ordered pairs and draw the graph.

Diagram, x	Dots, y	(x, y)
1	1	(1, 1)
2	3	(2, 3)
3	6	(3, 6)
4	10	(4, 10)

The graph is not a straight line. So, the function is *not* linear.

EXAMPLE 4 Comparing Linear Functions

Your sister earns \$10 per hour. Your brother earns \$7 per hour.

The functions $m = 10h$ and $m = 7h$ show the relationship between the numbers of hours h they work and the money m they earn. Which graph is steeper? Explain.

The graph of $m = 10h$ is steeper. The reason it is steeper is that your sister's hourly rate is greater than your brother's hourly rate.

On Your Own

Now You're Ready
Exercises 15–19

5. Make an input-output table for the pattern. Is the function relating the diagram number x to the number of dots y linear? Explain.

 Diagram 1 Diagram 2 Diagram 3 Diagram 4

6. The functions $d = 65t$ and $d = 55t$ show the relationship between the distances d (in miles) traveled and the times t (in hours) for two cars. Graph the functions. Which graph is steeper? Explain.

9.5 Exercises

Vocabulary and Concept Check

1. **VOCABULARY** Why are some functions called *linear functions*?
2. **WRITING** How can you decide whether or not an input-output table represents a linear function?

Practice and Problem Solving

Copy and complete the table. Then make a graph of the data. Write an equation for the function.

3. Find the diameter of a circle with radius r.

Radius, r	1	2	3	4
Diameter, d				

4. Find the cost of renting roller blades for h hours at $6 per hour.

Hours, h	1	2	3	4
Cost, c				

Does the graph represent a linear function? Explain.

5.

6.

7.

8.

9.

10.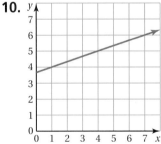

Does the input-output table represent a linear function? Explain.

11.
Input, x	1	2	3	4
Output, y	1	3	5	7

12.
Input, x	0	2	4	6
Output, y	10	9	8	7

13.
Input, x	1	4	7	10
Output, y	5	2	2	5

14.
Input, x	3	4	5	6
Output, y	5	8	9	8

Assignment Guide and Homework Check

Level	Day 1 Activity Assignment	Day 2 Lesson Assignment	Homework Check
Basic	3, 4, 23–27	1, 2, 5–19 odd, 20	2, 7, 11, 15, 20
Average	3, 4, 23–27	1, 2, 5–21 odd, 20	2, 7, 11, 15, 20
Advanced	3, 4, 23–27	1, 2, 6–22 even, 21	6, 12, 16, 20

Common Errors

- **Exercises 5–10** Students may say that a graph is linear if parts of the graph are lines. Remind them that for a graph to be linear the graph must not curve at all or change direction.
- **Exercises 11–14** Students may incorrectly guess that a table is linear or nonlinear without graphing. Encourage them to plot the points given in the table to make sure that they know the equation is linear.
- **Exercises 15–17** Students may mix up which equation is represented by each graph. Encourage them to label or colorcode the graphs of both equations.

9.5 Record and Practice Journal

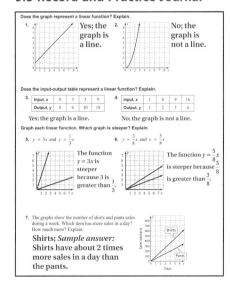

Technology For the Teacher
Answer Presentation Tool
*Quiz*Show

Vocabulary and Concept Check

1. A function is called a linear function if its graph is a line.
2. Plot the ordered pairs from the input-output table. If the points are all on one line, then the input-output table represents a linear function.

Practice and Problem Solving

3. 2; 4; 6; 8

 $d = 2r$

4. 6; 12; 18; 24

 $c = 6h$

5. yes; The graph is a line.
6. no; The graph is *not* a line.
7. no; The graph is *not* a line.
8. no; The graph is *not* a line.
9. no; The graph is *not* a line.
10. yes; The graph is a line.
11. yes; The graph is a line.
12. yes; The graph is a line.
13. no; The graph is *not* a line.
14. no; The graph is *not* a line.

T-396

15–19. See Additional Answers.

20. See *Taking Math Deeper*.

21–22. See Additional Answers.

 Fair Game Review

23. $\dfrac{11}{25}$ **24.** 0.27

25. 0.802 **26.** $66\dfrac{2}{3}\%$

27. C

Mini-Assessment

1. Does the graph represent a linear function? Explain.

The graph is a line. So, the graph does represent a linear function.

2. Your sister earns $10 per hour. Your cousin earns $8 per hour. The functions $m = 10h$ and $m = 8h$ show the relationship between the number of hours h they work and the money m they earn. Which graph is steeper? Explain.

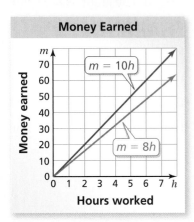

$m = 10h$; Your sister's hourly rate is greater than your cousin's hourly rate.

Taking Math Deeper

Exercise 20

This is a classic type of problem that students will study in a future course. There are two linear functions that represent two rates. One runner starts later than the other, but eventually catches up. The point at which the runner catches up corresponds to the point of intersection of the two graphs. This type of problem is called a *System of Linear Equations*.

① **a.** Write a function for your distance.
 Distance = (Rate)(Time)
 $d = 6t$

② **b.** Write a function for your friend's distance.
 Distance = (Head start) + (Rate)(Time)
 $d = 10 + 5t$

③ **c.** Graph both functions.

d.

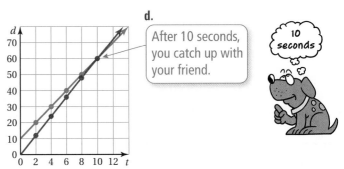

After 10 seconds, you catch up with your friend.

Reteaching and Enrichment Strategies

If students need help...	If students got it...
Resources by Chapter • Practice A and Practice B • Puzzle Time Record and Practice Journal Practice Differentiating the Lesson Lesson Tutorials Skills Review Handbook	Resources by Chapter • Enrichment and Extension • School-to-Work • Financial Literacy Start the next section

Graph each linear function. Which graph is steeper? Explain.

15. $y = 5x$ and $y = \frac{1}{5}x$

16. $y = \frac{4}{5}x$ and $y = \frac{3}{5}x$

17. $y = x$ and $y = 2x + 1$

Make an input-output table for the pattern. Is the function relating the figure number x to the area y linear? Explain.

18.

Figure 1 Figure 2 Figure 3 Figure 4 1 square unit

19.

Figure 1 Figure 2 Figure 3 Figure 4 1 square unit

20. MODELING In a 100-meter race, you give your friend a head start. Your speeds are shown.

- 10 m
- Lane 1: 5 m per second — Friend
- Lane 2: 6 m per second — You

a. Write a function that represents your distance d after t seconds.

b. Write a function that represents your friend's distance d after t seconds.

c. Graph your distance and your friend's distance in the same coordinate plane.

d. What does the intersection of the two graphs represent?

21. AIRPORT USE The graph shows the numbers of flights that arrive at and depart from two regional airports. Which airport has more flights in a day? How many more? Explain.

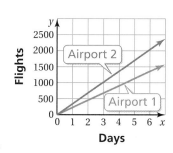

22. Reasoning Use the graph.

a. Copy and complete the table. Then find the differences.

x	0	1	2	3	4	5
y	1	1.5	2	2.5		

Differences: 0.5 0.5 ? ? ?

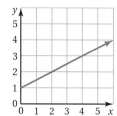

b. Graph other lines and find the "differences." Describe a property suggested by your results.

Fair Game Review *What you learned in previous grades & lessons*

Tell which number is greater. *(Section 4.3)*

23. 40%, $\frac{11}{25}$

24. 0.27, 2.8%

25. $\frac{4}{5}$, 0.802

26. $\frac{33}{50}$, $66\frac{2}{3}\%$

27. MULTIPLE CHOICE For which inequality is $x = 7$ a solution? *(Section 8.1)*

Ⓐ $x < 7$ Ⓑ $x + 4 \geq 12$ Ⓒ $21 \leq 3x$ Ⓓ $4x - 5 > 23$

9.4–9.5 Quiz

Graph the function. *(Section 9.4)*

1. $y = x + 5$
2. $y = 2x + 3$
3. $y = \dfrac{x}{2}$

Does the graph represent a linear function? Explain. *(Section 9.5)*

4.
5.
6.

Does the input-output table represent a linear function? Explain. *(Section 9.5)*

7.
Input, x	0	1	2	3
Output, y	2	3	6	11

8.
Input, x	2	4	6	8
Output, y	7	9	11	13

9. **PATTERN** Make an input-output table for the pattern. Is the function relating the figure number x to the number of squares y linear? Explain. *(Section 9.5)*

 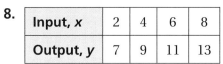

 Figure 1 Figure 2 Figure 3 Figure 4

10. **BABY** The function $m = 32d$ gives the number of ounces m of milk that a baby drinks in d days. Graph the function. *(Section 9.4)*

11. **SALES TAX** The sales tax rate is 6%. *(Section 9.4)*

 a. Write a function you can use to find the amount of sales tax t on the price p of each item at a gift shop.

 b. Graph the function using the inputs 0, 10, 20, 30, and 40 for p.

 c. You buy a souvenir that costs $4.00. How much do you pay for the souvenir after sales tax is added?

12. **HELICOPTER** The graph shows the numbers of gallons of gasoline that two helicopters use after 5 hours of flying. Which helicopter uses more fuel in an hour? Explain. *(Section 9.5)*

Alternative Assessment Options

Math Chat
Structured Interview
Student Reflective Focus Question
Writing Prompt

Math Chat
- Have individual students work problems from the quiz on the board. The student explains the process used and justifies each step. Students in the class ask questions of the student presenting.
- The teacher probes the thought process of the student presenting, but does not teach or ask leading questions.

Study Help Sample Answers
Remind students to complete Graphic Organizers for the rest of the chapter.

4.

Words	Vocabulary
A way to represent a function visually by plotting ordered pairs (input, output) = (x, y)	When you connect the points, you graph all of the solutions of the function.

Graph of a Function

Example	Example
$y = x + 1$	$y = \frac{1}{2}x$
points: (0,1), (1,2), (2,3), (3,4)	points: $(1, \frac{1}{2})$, (2,1), $(3, \frac{3}{2})$, (4,2)

5–6. Available at *BigIdeasMath.com*

Reteaching and Enrichment Strategies

If students need help...	If students got it...
Resources by Chapter • Study Help • Practice A and Practice B • Puzzle Time Lesson Tutorials *BigIdeasMath.com* Practice Quiz Practice from the Test Generator	Resources by Chapter • Enrichment and Extension • School-to-Work Game Closet at *BigIdeasMath.com* Start the Chapter Review

Technology For the Teacher
Answer Presentation Tool

Answers

1–3. See Additional Answers.

4. yes; The graph is a line.

5. no; The graph is not a line.

6. no; The graph is not a line.

7. no; The graph is not a line.

8. yes; The graph is a line.

9.

Figure, x	Squares, y
1	1
2	4
3	7
4	10

yes; The graph is a line.

10.

11. a. $t = 0.06p$

b.

c. $4.24

12. helicopter 1; The graph is steeper.

Assessment Book

T-398

For the Teacher
Additional Review Options
- Quiz*Show*
- Big Ideas Test Generator
- Game Closet at *BigIdeasMath.com*
- Vocabulary Puzzle Builder
- Resources by Chapter
 Puzzle Time
 Study Help

Review of Common Errors

Exercises 1–3
- Students may mix up the ordered pairs and write the output first and then the input.

Exercises 4 and 5
- Students may forget to write and order the inputs and outputs before making the mapping diagram. This often leads to mixing inputs and outputs.

Exercises 6–14
- Students may substitute for the wrong variable.

Exercises 15–18
- Students may focus on the changes in the input or the changes in the output instead of finding the relationship between the input and output.

Exercises 19–21
- Students may mix up their axes or label one or both axes with inconsistent increments.

Exercises 22–24
- Students may think that a graph is linear if parts of the graph are straight.
- Students may guess whether a table represents a linear function without graphing.

Answers

1. (0, 1), (3, 4), (6, 7), (9, 10)
2. (1, 6), (2, 8), (3, 7), (4, 9), (5, 9)
3. (0, 1), (2, 1), (4, 1), (6, 3), (8, 3)
4. Input Output

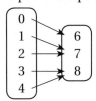

Input	Output
1	30
2	60
3	90
4	120

5. Input Output

Input	Output
0	
1	6
2	7
3	8
4	

9 Chapter Review

Review Key Vocabulary

input, *p. 368*
output, *p. 368*
function, *p. 368*
mapping diagram, *p. 368*
function rule, *p. 374*
input-output table, *p. 380*
graph, *p. 388*
linear function, *p. 394*

Review Examples and Exercises

9.1 Mapping Diagrams *(pp. 366–371)*

List the ordered pairs shown in the mapping diagram.

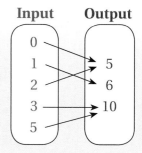

The ordered pairs are (0, 5), (1, 6), (2, 5), (3, 10), and (5, 10).

Exercises

List the ordered pairs shown in the mapping diagram.

1.
2.
3.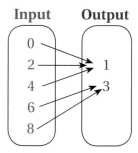

Draw a mapping diagram of the set of ordered pairs.

4. (1, 30), (2, 60), (3, 90), (4, 120)
5. (0, 6), (1, 7), (2, 7), (3, 8), (4, 8)

9.2 Functions as Words and Equations *(pp. 372–377)*

What is the value of $y = 3x + 4$ when $x = 8$?

$y = 3x + 4$ Write the equation.

$ = 3(8) + 4$ Substitute 8 for x.

$ = 28$ Simplify.

When $x = 8$, $y = 28$.

Exercises

Find the value of y for the given value of x.

6. $y = 9x$; $x = 5$
7. $y = x - 2$; $x = 9$
8. $y = 2x - 3$; $x = 4$
9. $y = 5x + 4$; $x = 2$
10. $y = 9x + 2$; $x = \frac{2}{3}$
11. $y = \frac{x}{3} + 5$; $x = 6$

Tell whether $(x, y) = (2, 5)$ is a solution of the equation.

12. $y = x + 4$
13. $y = 4x - 4$
14. $y = 2x + 1$

Input-Output Tables (pp. 378–383)

Write an equation for the function. Then copy and complete the table.

The output is 6 more than the input.

Input, x	1	3	5	7
Output, y				

∴ An equation is $y = x + 6$.

$y = x + 6$ →

Input, x	1	3	5	7
Output, y	7	9	11	13

Exercises

Write an equation for the function. Then copy and complete the table.

15. The output is 8 times the input.

Input, x	1	3	5	7
Output, y				

16. The output is 2 less than the input.

Input, x	2	4	6	8
Output, y				

Write an equation for the function shown by the table.

17.

Input, x	1	2	3	4
Output, y	7	14	21	28

18.

Input, x	5	7	9	11
Output, y	2	4	6	8

Review Game
Equations and Graphs

Big Ideas Game Closet

Materials per Group:
- graph paper
- pencil
- marker

Directions:
Divide the class into groups of four. Each group thinks of a word consisting of 3 to 5 letters. Each group 'draws' their word on grid paper and writes linear functions for the parts of each letter.

Things for students to remember:
- They should only use line segments that rise from left to right to form the letters.
- They should realize that some letters will be challenging to draw this way. They can use upper and lower case letters.
- They should draw axes so that the coordinates of the endpoints are greater than or equal to 0.
- They should record the endpoints (ordered pairs) of each line segment.
- They should record each linear function that they use.
- They can use colored markers to make their word look unique.

Who Wins?
Winner will be decided by a class vote. Judging criteria will be accuracy, difficulty, and creativity.

Example:
These linear functions can be used to draw the "b" shown below.

$y = 2x$
$y = x$
$y = \frac{1}{2}x + 3$

Segment endpoints:
(5, 10)
(0, 0)
(2, 4)
(6, 6)

Graph:

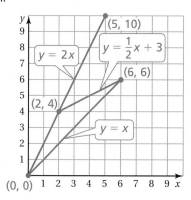

For the Student
Additional Practice
- Lesson Tutorials
- Study Help (textbook)
- Student Website
 Multi-Language Glossary
 Practice Assessments

Answers

6. 45 7. 7
8. 5 9. 14
10. 8 11. 7
12. no 13. no
14. yes
15. $y = 8x$; 8; 24; 40; 56
16. $y = x - 2$; 0; 2; 4; 6
17. $y = 7x$ 18. $y = x - 3$
19.
20.
21.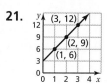
22. yes; The graph is a line.
23. yes; The graph is a line.
24. no; The graph is not a line.

T-400

My Thoughts on the Chapter

What worked. . .

What did not work. . .

What I would do differently. . .

Teacher Tip
Not allowed to write in your teaching edition? Use sticky notes to record your thoughts.

9.4 Graphs (pp. 386–391)

Graph $y = 2x + 2$.

Make an input-output table. Use the values 1, 2, 3, and 4 for x.

x	y = 2x + 2	y	(x, y)
1	y = 2(1) + 2	4	(1, 4)
2	y = 2(2) + 2	6	(2, 6)
3	y = 2(3) + 2	8	(3, 8)
4	y = 2(4) + 2	10	(4, 10)

Plot the ordered pairs. Draw a line through the points.

Exercises

Graph the function.

19. $y = x + 3$
20. $y = 5x$
21. $y = 3x + 3$

9.5 Analyzing Graphs (pp. 392–397)

Does the graph represent a linear function? Explain.

 The graph is not a straight line. So, the graph does *not* represent a linear function.

Exercises

Does the graph or table represent a linear function? Explain.

22.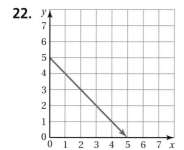

23.
Input, x	Output, y
0	0
1	2
2	4
3	6

24.

9 Chapter Test

Draw a mapping diagram of the set of ordered pairs.

1. (1, 0), (2, 1), (3, 0), (4, 1)
2. (3, 10), (6, 8), (9, 9), (12, 5)

3. Write an equation for the function "The output is 6 less than the input." Then copy and complete the table.

Input, x	10	11	12	13
Output, y				

4. Tell whether (3, 2) is a solution of $y = 2x - 4$.

5. Write an equation for the function shown by the table.

Input, x	3	4	5	6
Output, y	9	10	11	12

Graph the function.

6. $y = x + 8$
7. $y = 3x + 1$
8. $y = x - 4$

Does the graph or table represent a linear function? Explain.

9.

10.
Input, x	Output, y
6	4
7	1
8	0
9	1

11.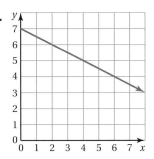

12. **GRASSHOPPER** The table shows the lengths of four grasshoppers and their jump heights. Use the table to draw a mapping diagram.

Body Length, (mm)	5	6	7	8
Jump Height, (mm)	50	60	70	80

13. **WATER SKI** The table shows the number of meters a water skier travels in x minutes.

Minutes, x	1	2	3	4	5
Meters, y	600	1200	1800	2400	3000

 a. Graph the data.
 b. Find a function to describe the data.
 c. At this rate, how many kilometers would the water skier travel in 1 hour?

14. **PATTERN** Make an input-output table for the pattern. Is the function relating the figure number x to the number of dots y linear? Explain.

Figure 1 Figure 2 Figure 3 Figure 4

Test Item References

Chapter Test Questions	Section to Review	Common Core State Standards
1, 2, 12	9.1	6.EE.9
4	9.2	6.EE.9
3, 5	9.3	6.EE.9
6–8, 13	9.4	6.EE.9
9–11, 14	9.5	6.EE.9

Test-Taking Strategies

Remind students to quickly look over the entire test before they start so that they can budget their time. Have them use the **Stop** and **Think** strategy before they answer each question.

Common Assessment Errors

- **Exercises 1 and 2** Students may mix inputs and outputs in their mapping diagrams. Remind them that for each ordered pair, the first number is always the input and the second number is always the output.
- **Exercises 3 and 5** Students may write x as the output and y as the input in their equations. For example, they may write $x = y - 6$ instead of $y = x - 6$. Remind them how to write a function as an equation in two variables.
- **Exercise 4** Students may substitute the value of x for y and the value of y for x. Remind them that for each ordered pair, the first number is always the value of x and the second number is always the value of y.
- **Exercises 6–8** When graphing a function, students may mix up their axes, or scale one or both axes with inconsistent increment widths. Remind them that the x-axis is horizontal and the y-axis is vertical. Encourage them to scale their axes consistently so that a line can be drawn through the ordered pairs that they plot.
- **Exercise 9** Students may think that a graph represents a linear function if part(s) of the graph appear to be straight. Point out that the graph of a linear function is not curved and does not change direction.
- **Exercise 10** Students may guess that a table represents a linear function. Encourage them to plot the given points so they can tell for sure.

Reteaching and Enrichment Strategies

If students need help...	If students got it...
Resources by Chapter • Practice A and Practice B • Puzzle Time Record and Practice Journal Practice Differentiating the Lesson Lesson Tutorials Practice from the Test Generator Skills Review Handbook	Resources by Chapter • Enrichment and Extension • School-to-Work • Financial Literacy Game Closet at *BigIdeasMath.com* Start Standardized Test Practice

Answers

1. Input Output

2. Input Output
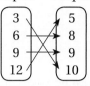

3. $y = x - 6$; 4; 5; 6; 7
4. yes
5. $y = x + 6$
6–8. See Additional Answers.
9. no; The graph is not a line.
10. no; The graph is not a line.
11. yes; The graph is a line.
12. Input Output

13–14. See Additional Answers.

Assessment Book

Test-Taking Strategies
Available at *BigIdeasMath.com*

After Answering Easy Questions, Relax
Answer Easy Questions First
Estimate the Answer
Read All Choices before Answering
Read Question before Answering
Solve Directly or Eliminate Choices
Solve Problem before Looking at Choices
Use Intelligent Guessing
Work Backwards

About this Strategy

When taking a multiple choice test, be sure to read each question carefully and thoroughly. One way to answer the question is to work backwards. Try putting the responses in to the question, one at a time, and see if you get a correct solution.

Answers

1. B
2. G
3. 76
4. C

Item Analysis

1. **A.** The student does not realize that the cost for 2 hours is not twice the cost for 1 hour and that answer choice B is correct.

 B. Correct answer

 C. The student does not realize that the cost for 2 hours is not twice the cost for 1 hour and that answer choice B is correct.

 D. The student does not realize that the cost for 2 hours is not twice the cost for 1 hour and that answer choice B is correct.

2. **F.** The student confuses the perimeter of a parallelogram with the area of a rectangle and finds a width of a rectangle that would give an area of 30.

 G. Correct answer

 H. The student finds three other lengths that would give a perimeter of 30, but ignores the fact that opposite sides of parallelograms are equal in length.

 I. The student does not divide 30 by 2 but instead subtracts 6 from 30, getting side lengths that give a perimeter of 60.

3. **Gridded Response:** Correct answer: 76

 Common Errors: The student chooses an input that is halfway between 50 and 106, getting an answer of 78. The student takes half of 30 and then subtracts 8 from the result, getting an answer of 7.

4. **A.** The student uses the incorrect type of circle and shades in the wrong direction.

 B. The student uses the incorrect type of circle.

 C. Correct answer

 D. The student shades in the wrong direction, possibly due to confusion resulting from the variable being on the right side of the inequality.

5. **F.** The student moves the decimal point in the wrong direction.

 G. The student counts the number of zeroes to the right of the decimal point in 0.005, rather than the number of decimal places.

 H. The student moves the decimal point in the wrong direction.

 I. Correct answer

Technology For the Teacher
Big Ideas Test Generator

9 Standardized Test Practice

1. An outdoor adventure company offers horseback riding. The total cost C for h hours of horseback riding is represented by the equation below.

 $$C = 12h + 26$$

 Which statement is *false*? *(6.EE.5)*

 A. The cost for 1 hour of horseback riding is $38.

 B. The cost for 2 hours of horseback riding is $76.

 C. The cost for 5 hours of horseback riding is $86.

 D. The cost for 8 hours of horseback riding is $122.

Test-Taking Strategy
Work Backwards

"You went x days without clawing the sofa where 5x + 3 = 18. How long was that?
Ⓐ 1 Ⓑ 2 Ⓒ 3 Ⓓ 4"

"Work backwards by trying 1, 2, 3, and 4. You will see that 5·3 + 3 = 18. So, C is correct."

2. A parallelogram has a perimeter of 30 inches. The length of one of its sides is 6 inches. What are the lengths, in inches, of its three other sides? *(6.EE.2c)*

 F. 5, 5, and 6

 G. 6, 9, and 9

 H. 7, 8, and 9

 I. 6, 24, and 24

3. Each output in the input-output table below is equal to 8 less than $\frac{1}{2}$ the input.

Input, x	20	34	50	?	106
Output, y	2	9	17	30	45

 What is the value of the missing input? *(6.EE.9)*

4. Which inequality is represented by the number line below? *(6.EE.8)*

 A. $x > 7$

 B. $x < 7$

 C. $7 \geq x$

 D. $7 \leq x$

5. José was dividing a whole number by a decimal in the box below.

$$0.005\overline{)705} \rightarrow 5\overline{)705}^{141}$$

What should José do to correct the error that he made? *(6.NS.3)*

F. Move the decimal point in the quotient three places to the left to get 0.141.

G. Move the decimal point in the quotient two places to the right to get 14,100.

H. Move the decimal point in the dividend three places to the left and find $5\overline{)0.705}$.

I. Move the decimal point in the dividend three places to the right and find $5\overline{)705,000}$.

6. Which value of *a* makes the equation below true? *(6.EE.5)*

$$24 = \frac{a}{3} - 9$$

A. 99

B. 45

C. 11

D. 5

7. A mapping diagram is shown.

What number belongs in the box below so that the equation will correctly describe the function represented by the mapping diagram? *(6.EE.9)*

$$y = \boxed{}x + 5$$

8. There are four sixth-grade classrooms in your school. You counted the number of boys and the number of girls in each class. Your results are shown in the table below.

Class	Number of Boys	Number of Girls
A	14	12
B	18	15
C	10	15
D	10	8

Based on the table, which class has the greatest ratio of boys to girls? *(6.RP.1)*

F. Class A

G. Class B

H. Class C

I. Class D

Item Analysis (continued)

6. **A.** Correct answer
 B. The student subtracts 9 from both sides instead of adding.
 C. The student divides both sides by 3 instead of multiplying.
 D. The student subtracts 9 from both sides instead of adding. The student then divides both sides by 3 instead of multiplying.

7. **Gridded Response:** Correct answer: 7

 Common Error: The student does not divide the differences between the outputs by 3, getting an answer of 21.

8. **F.** Of the three classes where boys outnumber girls, the student chooses the smallest ratio.
 G. The student chooses the class with the greatest number of boys.
 H. The student chooses the class with the largest ratio of girls to boys.
 I. Correct answer

9. **A.** The student multiplies the two lengths instead of dividing.
 B. The student subtracts the two lengths instead of dividing.
 C. The student divides 6 by 3, and then divides 32 by the result.
 D. Correct answer

10. **2 points** The student demonstrates a thorough understanding of creating, analyzing, and applying mapping diagrams. The student correctly maps the inputs of 5, 6, 7, and 8, to 14, 17, 20, and 23, respectively. The student correctly notices a pattern, such as the output increases by 3 whenever the input increases by 1. The student correctly determines another input and output such as (9, 26).

 1 point The student demonstrates a partial understanding of creating, analyzing, and applying mapping diagrams. The student is able to map the inputs and outputs correctly but makes errors in analyzing and applying the mapping diagram.

 0 points The student demonstrates insufficient understanding of creating, analyzing, and applying mapping diagrams. The student is unable to map the inputs and outputs correctly.

Answers

5. I
6. A
7. 7
8. I

Answers

9. D

10. *Part A*

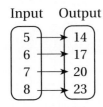

Part B The output increases by 3 when the input increases by 1.

Part C Sample answer: (9, 26); 8 + 1 = 9 and 23 + 3 = 26

11. H

Answer for Extra Example

1. **A.** The student does not multiply the radius by two to find the circumference but then correctly divides by 6, or the student correctly finds the circumference but then divides by 12 to find how far the tip of the hour hand will travel in 1 hour.

 B. Correct answer

 C. The student does not multiply the radius by two to find the circumference and then uses the incorrect circumference to find how far the tip of the hour hand will travel in 12 hours.

 D. The student finds how far the tip of the hour hand will travel in 12 hours.

Item Analysis (continued)

11. **F.** The student chooses an input-output table that does not show a linear relationship.

 G. The student chooses an input-output table that shows a quadratic, not a linear, relationship.

 H. Correct answer

 I. The student chooses an input-output table that shows an exponential, not a linear, relationship.

Extra Example for Standardized Test Practice

1. The hour hand of a clock has a length of 84 millimeters. How far, in millimeters, will the tip of the hour hand travel in 2 hours? $\left(\text{Use } \frac{22}{7} \text{ for } \pi.\right)$

 A. 44 mm **C.** 264 mm

 B. 88 mm **D.** 528 mm

T-405

9. The crazy ant and the green anole are both found in Florida. A biologist measured the lengths of a crazy ant and a green anole that he had in his laboratory. His measurements are shown below.

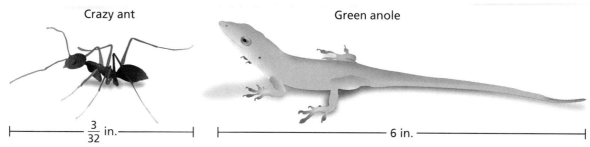

The length of the green anole is how many times greater than the length of the crazy ant? *(6.NS.1)*

A. $\dfrac{9}{16}$

B. $5\dfrac{29}{32}$

C. 16

D. 64

10. *Part A* Draw a mapping diagram for (5, 14), (6, 17), (7, 20), and (8, 23).

 Part B Describe any patterns that you see in the mapping diagram.

 Part C Determine another input and output that follows the patterns you describe in Part B. Explain your reasoning. *(6.EE.9)*

11. Which input-output table represents a linear function? *(6.EE.9)*

F.

Input, x	1	2	3	4
Output, y	1000	200	30	4

G.

Input, x	1	2	3	4
Output, y	1	4	9	16

H.

Input, x	2	4	6	8
Output, y	7	13	19	25

I.

Input, x	3	4	5	6
Output, y	8	4	2	1

Additional Topics

Topic 1 The Number Line
Topic 2 Comparing Numbers
Topic 3 Absolute Value
Topic 4 The Coordinate Plane

Connections to Previous Learning

- Compare two fractions with different numerators and different denominators.
- Compare two decimals to the hundredths place by reasoning about their size.

- Use a pair of perpendicular number lines, called axes, to define a coordinate system.
- Represent real-world and mathematical problems by graphing points in the first quadrant of the coordinate plane.

- Find and position integers and other rational numbers on a horizontal or vertical number line diagram.
- Understand that positive and negative numbers are used together to describe quantities having opposite directions or values.
- Recognize opposite signs of numbers as indicating locations on opposite sides of 0 on the number line.
- Interpret statements of inequality as statements about the relative position of two numbers on a number line diagram.
- Write, interpret, and explain statements of order for rational numbers in real-world contexts.
- Understand the absolute value of a rational number as its distance from 0 on the number line.
- Understand signs of numbers in ordered pairs as indicating locations in quadrants of the coordinate plane.
- Find and position pairs of integers and other rational numbers on a coordinate plane.
- Solve real-world and mathematical problems by graphing points in all four quadrants of the coordinate plane.
- Draw polygons in the coordinate plane given coordinates for the vertices.

Pacing Guide for Additional Topics

Opener	1 Day
Topic 1	1 Day
Topic 2	1 Day
Topic 3	1 Day
Topic 4	1 Day
Total Additional Topics	5 Days
Year-to-Date	156 Days

Math in History

The coordinate plane provides the link between algebra and geometry.

★ Its origins can be traced to the Greek philosopher and cartographer, Dicaearchus of Messana (350–285 B.C.). He was the first to use geographical coordinates in maps.

★ The father of analytic geometry, Rene Descartes (1596–1650 A.D.), is often given credit for discovering the coordinate plane, or as it is sometimes called, the Cartesian plane.

The Dynamic Planning Tool
Editable Teacher's Resources at
BigIdeasMath.com

T-406

Common Core State Standards

2.MD.6 Represent whole numbers as lengths from 0 on a number line diagram with equally spaced points corresponding to the numbers 0, 1, 2, . . . , and represent whole-number sums and differences within 100 on a number line diagram.

5.G.1 Use a pair of perpendicular number lines, called axes, to define a coordinate system Understand that the first number indicates how far to travel from the origin in the direction of one axis, and the second number indicates how far to travel in the direction of the second axis

4.MD.3 Apply the area and perimeter formulas for rectangles in real world and mathematical problems.

Additional Topics for Review

- Comparing whole numbers
- Comparing fractions and decimals
- Finding area

Try It Yourself

1. 1, 2, 4, 6, 7
2. 3, 4, 6, 7, 9
3. 5, 10, 20, 25, 35, 40
4. Point K
5. Point N
6. 48 in.

Record and Practice Journal

1–9. See Additional Answers.

10. Point C		**11.** Point D	
12. Point A		**13.** Point E	
14. 48 ft		**15.** 31 cm	
16. 26 m		**17.** 56 in.	

T-407

Math Background Notes

Vocabulary Review
- Whole Numbers
- Number Line
- Coordinate Plane
- Ordered Pair
- Perimeter

Ordering Whole Numbers
- Students should know how to order whole numbers and work with the number line.
- You may want to discuss the scaling of the number line with students. Each tick mark should count the same amount and be equally spaced on the number line.
- **Common Error:** Students may not use a number line and end up with the numbers out of order. Encourage students to graph the numbers on a number line, as shown in Example 1, to help them correctly order the numbers from least to greatest.

Plotting Points
- Students should know how to plot points in Quadrant I.
- **Common Error:** Students frequently confuse the x- and y-coordinates in an ordered pair. Remind them that x comes before y in the alphabet, so the x-coordinate comes before the y-coordinate in an ordered pair.
- If you have square tiles in the classroom, you can use masking tape to create a life-sized Quadrant I. Ask students to start at (0, 0) and move to different points. This activity is especially beneficial to kinesthetic learners.

Finding Perimeter
- Students should know how to find the perimeter of a figure.
- **Common Error:** Students may find the area of the figure. Remind them that the perimeter is the distance around a figure.

Reteaching and Enrichment Strategies

If students need help. . .	If students got it. . .
Record and Practice Journal • Fair Game Review Skills Review Handbook Lesson Tutorials	Game Closet at *BigIdeasMath.com* Start the next section

What You Learned Before

"The Dog License Department wrote and said that it has run out of positive license numbers. So, they assigned me negative 3."

● Ordering Whole Numbers (2.MD.6)

Example 1 Use a number line to order 5, 2, 0, 8, and 3 from least to greatest.

∴ The numbers in order from least to greatest are 0, 2, 3, 5, and 8.

Try It Yourself

Use a number line to order the numbers from least to greatest.

1. 6, 2, 4, 1, 7
2. 9, 3, 7, 6, 4
3. 25, 10, 35, 40, 20, 5

● Plotting Points (5.G.1)

Example 2 Which point is located at (4, 7)?

Start at the origin. Move 4 units right and 7 units up.

∴ Point J is located at (4, 7).

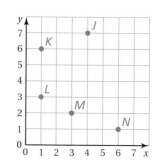

Try It Yourself

Use the graph to answer the question.

4. Which point is located at (1, 6)?
5. Which point is located at (6, 1)?

● Finding Perimeter (4.MD.3)

Example 3 Find the perimeter P of the rectangle.

$P = 2 + 5 + 2 + 5$

$ = 14$

∴ The perimeter of the rectangle is 14 centimeters.

Try It Yourself

6. The side length of a square is 12 inches. Find the perimeter of the square.

Topic 1 The Number Line

COMMON CORE STATE STANDARDS
6.NS.5
6.NS.6a
6.NS.6c

Positive numbers are greater than 0. They can be written with or without a positive sign (+).

+1 5 +20 10,000

Negative numbers are less than 0. They are written with a negative sign (−).

−1 −5 −20 −10,000

Two numbers that are the same distance from 0 on a number line, but on opposite sides of 0, are called **opposites**.

The Meaning of a Word

Opposite

When you sit across from your friend at the lunch table, you sit **opposite** your friend.

Key Idea

Integers

Words **Integers** are the set of whole numbers and their opposites.

Graph

Zero is neither negative nor positive. Zero is its own opposite.

EXAMPLE 1 Writing Positive and Negative Integers

Write a positive or negative integer that represents the situation.

a. A contestant gains 250 points on a game show.

"Gains" indicates a number greater than 0. So, use a positive integer.

∴ +250, or 250

b. Gasoline freezes at 40 degrees below zero.

"Below zero" indicates a number less than 0. So, use a negative integer.

∴ −40

Practice

Write a positive or negative integer that represents the situation.

1. A hiker climbs 900 feet up a mountain.
2. You have a debt of $24.
3. A student loses 5 points for being late to class.
4. A savings account earns $10.

408 Additional Topics

Laurie's Notes

Introduction

Connect
- **Previously:** Students wrote and graphed whole numbers.
- **Today:** Students will write and graph integers.

Motivate
- Ask for 7 volunteers to hold onto a piece of rope. They should be equally spaced. Ask the person in the middle to hold a card with the number 0 written on it.
- Say to the students:
 - "If you are 2 people away from 0, raise your hand."
 - "If you are 3 people away from 0, raise your hand."
- Now hand the card with the number 1 written on it to the person on the right of the person holding the 0 card.
- Say to the students:
 - "If you are the number -1, raise your hand."
 - "If you are the number 3, raise your hand."
 - "Keep your hand in the air if you are the least of the two numbers."
 - Repeat the instructions to compare 2 and 3, and to compare -2 and -3.
- This gives students a strong visual image of how a number line looks and where positive and negative numbers are located.

Lesson Notes

Key Idea
- Define integers. Make it clear to students that you can separate integers into 3 sets: positive integers, negative integers, and 0. Zero is neither positive nor negative.
- Remind students that whole numbers are 0, 1, 2, 3, . . .
- Draw a number line, using two colors to differentiate positive and negative integers.
- Point out the symmetry of a number line. If the number line were drawn on a piece of paper and folded at 0, opposites would fold on top of one another.
- ? "Does every number have an opposite?" yes; 0 is its own opposite.

Example 1
- Discuss each part of the example. The context for each part should be familiar to students.
- Point out that the + sign is optional in writing a positive number. It generally is not written.

Practice
- **Extension:** Ask students to state the opposite of each situation. Example: A hiker descends 900 feet down a mountain.

Common Core State Standards

6.NS.5 Understand that positive and negative numbers are used together to describe quantities having opposite directions or values; use positive and negative numbers to represent quantities in real-world contexts, explaining the meaning of 0 in each situation.

6.NS.6a Recognize opposite signs of numbers as indicating locations on opposite sides of 0 on the number line; recognize that the opposite of the opposite of a number is the number itself, and that 0 is its own opposite.

6.NS.6c Find and position integers and other rational numbers on a horizontal or vertical number line diagram; find and position pairs of integers and other rational numbers on a coordinate plane.

Goal
Today's lesson is reading and graphing **integers**.

Warm Up

Extra Example 1

Write a positive or negative integer that represents the situation.

a. A balloon floats 5 feet above the ground. 5
b. An anchor is 15 feet underwater. -15

Practice
1. 900
2. -24
3. -5
4. 10

Record and Practice Journal Practice
See Additional Answers.

T-408

Extra Example 2

Graph each integer and its opposite.

a. 4

b. −10

Extra Example 3

You start a game with 10 points. You gain 4 points in the first round and lose 7 points in the second round. After the third round, your total score is 10 points. Write an integer that represents how many points you gained or lost in the third round. 3

Practice

5.
6.
7.
8.

9. See Additional Answers.
10. 150

Mini-Assessment

Write a positive or negative integer that represents the situation.

1. A fish is 12 meters underwater. −12
2. You deposit $25 into your savings account. 25

Graph the integer and its opposite.

3. 7

4. −15

5. While playing a game, you move 6 steps forward and then 10 steps backward. Write each number as an integer. 6; −10

T-409

Laurie's Notes

Example 2

? "What is the opposite of 5?" −5 "What is the opposite of −13?" 13

- **Common Error:** When asked to graph the numbers 2 and −2, students may draw a number line and label −2, 0, and 2 as shown, and write nothing else.

Make it clear to students that they should decide on a scale and then put a closed circle on the number line to graph a number.

? "What number is halfway between a number and its opposite?" 0

Example 3

- Draw a vertical number line on the board as a visual aid for students as you work through this problem.
- **FYI:** Students may ask if underground floors are really numbered using negative numbers. It is most common to use a letter such as L or P to denote a lower level or a parking level.

Practice

- **Common Error:** Students may incorrectly label the number line when graphing an integer and its opposite.

Closure

- **Exit Ticket:** Graph each integer on the same number line and write a context for each: −6 and 4.

 −6 A student missed 6 questions on a quiz.
 4 There are 4 new students in the class.

Technology For the Teacher

The Dynamic Planning Tool
Editable Teacher's Resources at *BigIdeasMath.com*

EXAMPLE 2 Graphing Integers

Graph each integer and its opposite.

Reading
You can think of the negative sign (−) as referring to the opposite of a number. For example, you can read −2 as "the opposite of 2."

a. 3

−3 is the same distance from 0 as 3.

b. −2

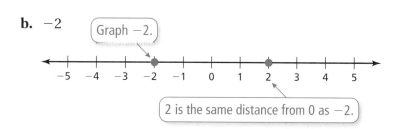

2 is the same distance from 0 as −2.

EXAMPLE 3 Real-Life Application

You deliver flowers to an office building. You enter at ground level and go down two floors to make the first delivery. Then you go up seven floors to make the second delivery. Write an integer that represents the return to ground level.

Use a vertical number line to model your movement, as shown.

The second delivery is on the fifth floor. You must go down five floors to return to ground level.

∴ The integer representing "down five floors" is −5.

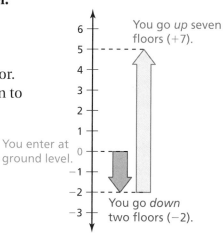

Practice

Graph the integer and its opposite.

5. 6 6. −4 7. −12 8. 1

9. **REASONING** Choose any positive integer.

 a. Find the opposite of the integer. b. Find the opposite of the integer in part (a).

 c. What can you conclude about the opposite of the opposite of the integer? Is this true for all integers? Use a number line to justify your answer.

10. **SUBMARINE** A submarine descends 100 meters from the ocean's surface. The submarine ascends 40 meters and then descends 90 meters. Write an integer that represents the return to the surface.

Topic 2 Comparing Numbers

COMMON CORE STATE STANDARDS
6.NS.6c, 6.NS.7a
6.NS.7b

On a number line, numbers to the left are less than numbers to the right. Numbers to the right are greater than numbers to the left.

EXAMPLE 1 Comparing Integers

a. Compare 2 and −6.

2 is to the right of −6. So, 2 > −6.

b. Compare −5 and −3.

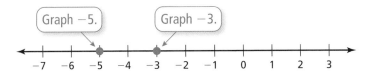

−5 is to the left of −3. So, −5 < −3.

Remember
The symbol < means "is less than" and the symbol > means "is greater than."

EXAMPLE 2 Comparing Decimals

Compare −3.08 and −3.8.

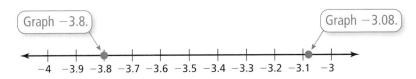

−3.08 is to the right of −3.8. So, −3.08 > −3.8.

Practice

Copy and complete the statement using < or >.

1. 0 ▢ −4
2. −5 ▢ 5
3. −8 ▢ −7
4. −0.5 ▢ 0.3
5. −2 ▢ −0.2
6. −3.24 ▢ −3.42

7. **TEMPERATURE** The freezing temperature of nitrogen is −210°C and the freezing temperature of oxygen is −223°C. Which temperature is warmer?

410 Additional Topics

Laurie's Notes

Introduction

Connect
- **Yesterday:** Students wrote and graphed integers.
- **Today:** Students will use a number line to compare rational numbers.

Motivate
- Place a piece of masking tape on the floor to represent a number line. Identify the location of 0 by having a student stand at zero.
- Have two more students stand at two other integers. Have the students change their locations several times to represent the following scenarios: two positive integers, one positive integer and one negative integer, and both negative integers.

Discuss
- **Big Idea:** The big idea for students to remember is that the farther to the right a number is on the number line, the greater it is. Likewise, the farther to the left a number is on the number line, the less it is.
- Define a rational number as a number that can be written as $\frac{a}{b}$ where a and b are integers and $b \neq 0$. Draw the Venn diagram shown and have students give examples of rational numbers.

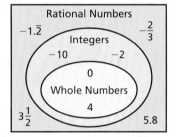

Lesson Notes

Example 1
- Draw a number line and graph -6 and 2.
- Discuss with students that 0 does not have to be in the middle of the number line. Sometimes number lines do not show 0.
- **Extension:** Ask the students how far -6 and 2 are from 0. -6 is 6 units from 0, 2 is 2 units from 0.

Example 2
- Use correct language in reading the decimals: negative 3 and 8 hundredths, and negative 3 and 8 tenths.
- **Teaching Tip:** To help students compare decimals with different place values, write trailing zeros as shown:

It is not obvious to all students that -3.1 is equivalent to -3.10, or in words, negative 3 and 1 tenth is equivalent to negative 3 and 10 hundredths. The trailing zeros help students recognize where -3.08 is in relation to -3.1 (or -3.10).

Practice
- **Common Error:** Students may incorrectly graph decimals when comparing.

Common Core State Standards
6.NS.6c Find and position integers and other rational numbers on a horizontal or vertical number line diagram; find and position pairs of integers and other rational numbers on a coordinate plane.
6.NS.7a Interpret statements of inequality as statements about the relative position of two numbers on a number line diagram.
6.NS.7b Write, interpret, and explain statements of order for rational numbers in real-world contexts.

Goal Today's lesson is comparing rational numbers on a number line.

Warm Up

Extra Example 1
Compare -2 and -10. $-2 > -10$

Extra Example 2
Compare 0.3 and -1.2. $0.3 > -1.2$

Practice
1. $0 > -4$
2. $-5 < 5$
3. $-8 < -7$
4. $-0.5 < 0.3$
5. $-2 < -0.2$
6. $-3.24 > -3.42$
7. $-210°C$

Record and Practice Journal Practice
See Additional Answers.

Extra Example 3

a. Compare $-\frac{5}{8}$ and $-\frac{3}{8}$.

$-\frac{5}{8} < -\frac{3}{8}$

b. Compare $-2\frac{2}{3}$ and $-3\frac{1}{3}$.

$-2\frac{2}{3} > -3\frac{1}{3}$

Extra Example 4

Which is the least, $-1\frac{3}{4}$, $-1\frac{9}{16}$, or $-1\frac{5}{8}$?

$-1\frac{3}{4}$

Practice

8. $\frac{2}{3} > -2$
9. $-\frac{4}{7} < -\frac{1}{7}$
10. $-1\frac{2}{3} > -1\frac{5}{6}$
11. $-3, -1, 0, 2, 3$
12. $-7.2, -4.1, 2.1, 3.5, 6.8$
13. $-3\frac{1}{4}, -\frac{3}{2}, -\frac{5}{4}, \frac{1}{2}, 1\frac{3}{4}$

Mini-Assessment

Copy and complete the statement.

1. $-4 \quad\square\quad -3$ $-4 < -3$
2. $3.1 \quad\square\quad -1.7$ $3.1 > -1.7$
3. $-\frac{16}{5} \quad\square\quad -\frac{12}{5}$ $-\frac{16}{5} < -\frac{12}{5}$

Order the numbers from least to greatest.

4. $-5, 0, 5, -3, -9$ $-9, -5, -3, 0, 5$
5. $2.3, -4.6, 0.1, -0.06, -1.4$
 $-4.6, -1.4, -0.06, 0.1, 2.3$

T-411

Laurie's Notes

Example 3

- Work through each part of the example.
- In part (a), note that the number line is incremented in fourths. Remind students that for negative numbers, you work from right to left: $0, -\frac{1}{4}, -\frac{1}{2}, -\frac{3}{4}, -1$.
- In part (b), note that the number line is incremented in sixths. Remind students that for negative numbers, you work from right to left: $-4, -4\frac{1}{6}, -4\frac{2}{6}$, and so on. Because $-4\frac{5}{6}$ is farther to the left on the number line than $-4\frac{1}{6}$, it is less.

? "Which number is farther from 0, $-4\frac{5}{6}$, or $-4\frac{1}{6}$?" $-4\frac{5}{6}$

Example 4

- **FYI:** The coldest recorded temperature in Arkansas was $-29°F$ on February 13, 1905.
- Students should recognize that -7.8 is closer to -8 than -7.
- **Extension:** Ask students to approximate the difference in temperatures between Jonesboro and Texarkana.
- **Extension:** Have students research the coldest and warmest recorded temperatures for their town or city.

Practice

- Remind students that they can explain their thinking by graphing each number on a number line.
- **Common Error:** Students may incorrectly locate fractions with different denominators on the number line.

Closure

- Explain why $2 > -42$. 2 is to the right of -42 on a number line.

Technology For the Teacher

Dynamic Classroom

The Dynamic Planning Tool
Editable Teacher's Resources at *BigIdeasMath.com*

EXAMPLE 3 Comparing Fractions and Mixed Numbers

a. Compare $-\frac{1}{2}$ and $-\frac{3}{4}$.

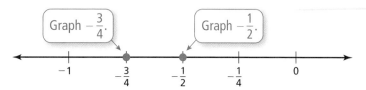

$-\frac{1}{2}$ is to the right of $-\frac{3}{4}$. So, $-\frac{1}{2} > -\frac{3}{4}$.

b. Compare $-4\frac{5}{6}$ and $-4\frac{1}{6}$.

$-4\frac{5}{6}$ is to the left of $-4\frac{1}{6}$. So, $-4\frac{5}{6} < -4\frac{1}{6}$.

EXAMPLE 4 Real-Life Application

- Rogers: −6°C
- Jonesboro: −7.8°C
- Little Rock: 0°C
- Hot Springs: −2.5°C
- Texarkana: 3°C

The diagram shows a day's low temperatures for several cities in Arkansas. Which city had the coldest low temperature?

Graph each temperature on a number line.

−7.8 is farthest to the left on the number line. So, Jonesboro had the coldest low temperature.

Practice

Copy and complete the statement using < or >.

8. $\frac{2}{3}$ ▭ -2

9. $-\frac{4}{7}$ ▭ $-\frac{1}{7}$

10. $-1\frac{2}{3}$ ▭ $-1\frac{5}{6}$

Order the numbers from least to greatest.

11. $0, -1, 2, 3, -3$

12. $-7.2, 2.1, 6.8, -4.1, 3.5$

13. $1\frac{3}{4}, \frac{1}{2}, -3\frac{1}{4}, -\frac{5}{4}, -\frac{3}{2}$

Comparing Numbers

Topic 3 Absolute Value

COMMON CORE STATE STANDARDS
6.NS.7c
6.NS.7d

Key Idea

Absolute Value

Words The **absolute value** of a number is the distance between the number and 0 on a number line. The absolute value of a number a is written as $|a|$.

Numbers $|-2| = 2$ $|2| = 2$

EXAMPLE 1 Finding Absolute Value

a. Find the absolute value of 3.

∴ So, $|3| = 3$.

b. Find the absolute value of $-2\frac{1}{2}$.

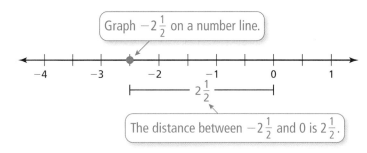

∴ So, $\left|-2\frac{1}{2}\right| = 2\frac{1}{2}$.

Practice

Find the absolute value.

1. $|5|$
2. $|-6|$
3. $|0|$
4. $|1.9|$
5. $|-12.6|$
6. $\left|\frac{1}{5}\right|$
7. $\left|-\frac{9}{2}\right|$
8. $\left|-6\frac{1}{3}\right|$

Laurie's Notes

Introduction

Connect
- **Yesterday:** Students compared rational numbers.
- **Today:** Students will find absolute values of rational numbers.

Motivate
- Have two students (A and B) stand at the front of the room holding a piece of string between them. You hold a piece of paper with the number 0 written on it. Share that the distance between the two students is 10 units.
- **?** Position yourself at various points along the line so that A and B have various values, such as A = 3, 5, or 8 and B = $-7, -5,$ or -2, respectively.
 - "If A is _____ [3, 5, 8], what number does B represent?" $-7, -5, -2$
 - "Who is closer to me (meaning 0)?" A; Neither; B
 - "How far away from me is each person?" Answers will vary depending on the numbers chosen.
- **?** Without the string, ask students "What number or numbers are 4 units from 0?" 4 and -4
- Discuss positive and negative directions from 0 on a number line. For instance, it is possible to be 5 units from 0 in either direction.

Lesson Notes

Key Idea
- Write the Key Idea on the board.
- $|a|$ is read, "the absolute value of a."
- Stress that the **absolute value** of a number is the *distance* from 0. Distance is a positive number or 0. Although directed distances are addressed in science, the concept of distance here is that it is a positive number or 0.

Example 1
- Work through each example.
- **Common Misconception:** Students may say, "Absolute values are always positive." This statement can cause problems when students find $|0|$, which is 0.
- **?** After part (b) ask, "What is the absolute value of -2.4?" 2.4

Words of Wisdom
- Point out that when you write the notation for the absolute value, it means *take the absolute value of the number inside the symbols*.

Practice
- **Common Error:** Students may incorrectly find that the absolute value of a negative number is negative.

Common Core State Standards

6.NS.7c Understand the absolute value of a rational number as its distance from 0 on the number line; interpret absolute value as magnitude for a positive or negative quantity in a real-world situation.

6.NS.7d Distinguish comparisons of absolute value from statements about order.

Goal Today's lesson is finding the **absolute value** of a number.

Lesson Materials
Introduction
• string

Warm Up

> **Topic 3 Warm Up**
> For use before Topic 3
>
> Copy and complete the statement using < or >.
>
> 1. 0 _?_ -4 2. -7 _?_ 7
> 3. -0.6 _?_ -0.4 4. 2.9 _?_ -6.2
> 5. $-\frac{1}{6}$ _?_ $\frac{1}{2}$ 6. $-1\frac{1}{4}$ _?_ $-1\frac{2}{3}$

Extra Example 1
a. Find the absolute value of -10. 10
b. Find the absolute value of 8.3. 8.3

Practice

1. 5
2. 6
3. 0
4. 1.9
5. 12.6
6. $\frac{1}{5}$
7. $\frac{9}{2}$
8. $6\frac{1}{3}$

Record and Practice Journal Practice
See Additional Answers.

Extra Example 2

Compare $-\frac{1}{2}$ and $\left|-1\frac{1}{4}\right|$.

$-\frac{1}{2} < \left|-1\frac{1}{4}\right|$

Extra Example 3

The table shows the elevations of a bird and a fish. Is the bird or the fish closer to sea level? fish

Animal	Elevation (m)
Bird	10.6
Fish	−9.8

Practice

9. $|-4| = 4$
10. $-2 < |2.4|$
11. $|-9.72| > 9.7$
12. $\frac{3}{4} < \left|-\frac{3}{2}\right|$
13. Scientist B
14. B; You owe $25, so debt = $25.
15. A; You owe more than $25, so debt > $25.
16. C; You owe less than $25, so debt < $25.

Mini-Assessment

Find the absolute value.

1. $\left|-4\frac{2}{5}\right|$ $4\frac{2}{5}$
2. $|3.1|$ 3.1

Copy and complete the statement using <, >, or =.

3. $-2 \;\square\; |-2.5|$ $-2 < |-2.5|$
4. $\frac{5}{6} \;\square\; \left|-\frac{4}{3}\right|$ $\frac{5}{6} < \left|-\frac{4}{3}\right|$

5. The temperature in town A is 8°C and the temperature in town B is −8.9°C. In which town is the temperature closer to the freezing point of water, 0°C? town A

T-413

Laurie's Notes

Example 2
- Explain that one way to compare numbers is to graph the numbers on a number line. The farther to the right a number is, the greater it is. The farther to the left a number is, the less it is.
- ❓ "What is $|-5|$?" 5
- **Common Error:** Students may graph $|-5|$ at −5. Explain that you find the absolute value of the number first and then graph it.
- Note that it would also be correct to conclude that $|-5| > 2$.

Example 3
- Discuss the number line. Make sure students observe that, for instance, −22 is to the left of −20. It is not uncommon for students to graph −22 to the right of −20 because 22 is to the right of 20.
- Work through each part of the problem.

Practice
- For Exercises 14–16, explain that a debt is an amount that you owe. So, if your account balance is less than −$100, you owe more than $100.
- **Common Error:** Students may forget to find the absolute value of a number when comparing.

Closure
- **Exit Ticket:** Write these numbers in order from least to greatest.

$$|-12|,\; -8,\; -10,\; |6|,\; |-4|$$

$-10,\; -8,\; |-4|,\; |6|,\; |-12|$

The Dynamic Planning Tool
Editable Teacher's Resources at *BigIdeasMath.com*

EXAMPLE 2 Comparing Values

Compare 2 and $|-5|$.

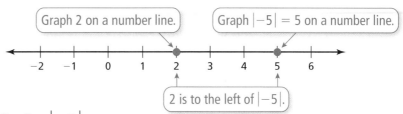

So, $2 < |-5|$.

EXAMPLE 3 Real-Life Application

The table shows the elevations of several animals.

Animal	Elevation (ft)
Shark	−4
Sea lion	5
Seagull	56
Shrimp	−65
Turtle	−22

a. Which animal is the deepest? Explain.

Graph each elevation.

The lowest elevation represents the animal that is the deepest. The integer farthest to the left on the number line is −65.

So, the shrimp is the deepest.

b. Is the shark or the sea lion closer to sea level?

Because sea level is at 0 feet, use absolute values.

Shark: $|-4| = 4$ **Sea lion:** $|5| = 5$

The shark is 4 feet from sea level and the sea lion is 5 feet from sea level.

Because 4 is less than 5, the shark is closer to sea level than the sea lion.

Practice

Copy and complete the statement using <, >, or =.

9. $|-4|$ ____ 4 10. -2 ____ $|2.4|$ 11. $|-9.72|$ ____ 9.7 12. $\dfrac{3}{4}$ ____ $\left|-\dfrac{3}{2}\right|$

13. **CAVES** Three scientists explore a cave. Which scientist is farthest underground?

 Scientist A: −48 ft Scientist B: −62 ft Scientist C: −53 ft

MATCHING Match the account balance with the debt that it represents. Explain your reasoning.

14. Account balance = −$25
15. Account balance < −$25
16. Account balance > −$25

 A. Debt > $25 B. Debt = $25 C. Debt < $25

Absolute Value

Topic 4 The Coordinate Plane

COMMON CORE STATE STANDARDS
6.G.3
6.NS.6b
6.NS.6c

Key Idea

The Coordinate Plane

A **coordinate plane** is formed by the intersection of a horizontal number line and a vertical number line. The number lines intersect at the **origin** and separate the coordinate plane into four regions called **quadrants**.

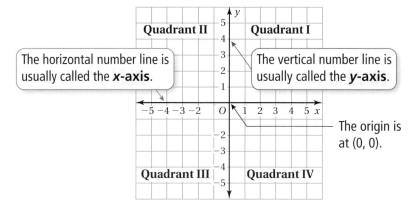

The horizontal number line is usually called the **x-axis**.

The vertical number line is usually called the **y-axis**.

The origin is at (0, 0).

An **ordered pair** is a pair of numbers that is used to locate a point in a coordinate plane.

ordered pair

The **x-coordinate** corresponds to a number on the x-axis.

$(4, -2)$

The **y-coordinate** corresponds to a number on the y-axis.

EXAMPLE 1 Plotting Ordered Pairs

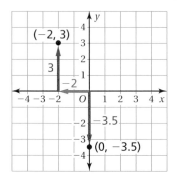

Plot (a) (−2, 3) and (b) (0, −3.5) in a coordinate plane. Describe the location of each point.

a. Start at the origin. Move 2 units left and 3 units up. Then plot the point.

∴ The point is in Quadrant II.

b. Start at the origin. Move 3.5 units down. Then plot the point.

∴ The point is on the y-axis.

Practice

Plot the ordered pair in a coordinate plane. Describe the location of the point.

1. $J(3, -1)$
2. $K(-5, 0)$
3. $L(-2.5, -1)$
4. $M\left(-1\frac{1}{2}, \frac{1}{2}\right)$

5. **REASONING** Both coordinates of a point are negative. In which quadrant is the point located? Describe the signs of the coordinates in each of the other three quadrants.

414 Additional Topics

Laurie's Notes

Introduction

Connect
- **Yesterday:** Students found the absolute value of a number.
- **Today:** Students will learn the vocabulary associated with the coordinate plane and how to plot points in the coordinate plane.

Motivate
- Create a coordinate grid on your classroom floor. Use two strips of masking tape to create an *x*- and *y*-axis. Use a marker to label the axes, leaving 1 foot between each label. Write each of the following points on a piece of paper: (4, 3), (−4, 3), (−4, −3), (4, −3).
- Ask for four volunteers. Give each student an ordered pair. Have them start at the origin and explain how they get to each point as they move.
- ? "What figure do the students form?" a rectangle

Lesson Notes

Key Idea
- It is important to have a model of the coordinate *grid* versus only a model of scaled axes. The grid is essential in helping students understand that a point is plotted by moving in two directions (horizontal and vertical).
- Identify the key vocabulary: coordinate plane, origin, quadrants, *x*-axis, *y*-axis, ordered pair, *x*-coordinate, and *y*-coordinate.
- **Connections:** Hold your arms out horizontally and relate them to the horizon to help students remember which way is horizontal. Origin means "where something starts." Relate the quadrants to a *four square* court.
- Stress that ordered pairs are (*x, y*). The order of the coordinates is in alphabetical order, *x* and then *y*.

Example 1
? Ask questions about how to position the point in part (a):
- "What is the *x*-coordinate?" −2 "Which direction do you move from (0, 0)?" Left 2 units
- "What is the *y*-coordinate?" 3 "Which direction do you move from (0, 0)?" Up 3 units
- **Common Error:** Students often plot points on the axes incorrectly. Make sure the plotted point in part (b) is (0, −3.5)
- Explain to students that ordered pairs are in one of the four quadrants, on an axis, or at the origin. The ordered pair in part (b) is on an axis, not in a quadrant.

Practice
- **Common Error:** Students may move the wrong direction, or mix up the *x*- and *y*-coordinates and plot the wrong point.

Common Core State Standards

6.G.3 Draw polygons in the coordinate plane.
6.NS.6b Understand . . . ordered pairs . . . in quadrants of the coordinate plane
6.NS.6c Find . . . pairs of integers and other rational numbers on a coordinate plane.

Goal Today's lesson is plotting **ordered pairs** in a **coordinate plane**.

Warm Up

Topic 4 Warm Up
For use before Topic 4

Graph the integer on the vertical number line.
1. 0
2. −2
3. 4
4. −8

Extra Example 1

Plot (2, −1.5) in a coordinate plane. Describe the location of the point.

The point is in Quadrant IV.

Practice

1–4.

1. Quadrant IV
2. *x*-axis
3. Quadrant III
4. Quadrant II
5. See Additional Answers.

Record and Practice Journal Practice

See Additional Answers.

T-414

Extra Example 2

An interior designer maps out a room design using a coordinate plane in which each unit represents 1 yard. The corners of the room are located at $(-3, 2)$, $(5, 2)$, $(5, -3)$, and $(-3, -3)$. What are the dimensions of the room? 8 yards long and 5 yards wide

 Practice

6–7. See Additional Answers.

8. 12 square units

9. 22 units

10. 1 meter

11. a. *x*-coordinates have different signs, *y*-coordinates are the same; *x*-coordinates are the same, *y*-coordinates have different signs

 b. reflection in both axes; The signs of the coordinates are different.

Mini-Assessment

The points $A(-2, 3)$, $B(4, 3)$, $C(-2, -4)$, and $D(4, -4)$ represent the vertices of a garden.

1. Plot the ordered pairs in a coordinate plane.

2. What shape does the garden form? a rectangle

3. Describe the location of each point. *A:* Quadrant II, *B:* Quadrant I, *C:* Quadrant III, *D:* Quadrant IV

Laurie's Notes

Example 2
- Relate this example to the Motivate section from the beginning of class.
- **Connection:** Point out that the absolute values for the length come from the *x*-coordinates of the points. The absolute values for the width come from the *y*-coordinates of the points. The absolute values represent the distance from 0 for that coordinate, i.e. the distance from the axis.
- Note that you can also use the points $(-4, 1)$ and $(2, 1)$ to find the length and the points $(-4, 5)$ and $(-4, 1)$ to find the width.
- Explain why you add the absolute values to find the length (points are in different quadrants), but subtract to find the width (points are in the same quadrant).
- Explain to students that even though two of the ordered pairs are in Quadrant II where *x*-coordinates are negative, the dimensions of the rectangle are positive.

Practice
- For Exercises 6 and 7, have students name the polygon.
- **Common Error:** Students may use incorrect dimensions to answer Exercises 8 and 9.

Closure
- Graph a rectangle with coordinates in Quadrants III and IV. Name the coordinates.

The Dynamic Planning Tool
Editable Teacher's Resources at *BigIdeasMath.com*

EXAMPLE 2 Real-Life Application

Reading
An *archaeologist* studies ancient ruins and objects to learn about people and cultures.

An *archaeologist* divides an area using a coordinate plane in which each unit represents 1 meter. The corners of a secret chamber are found at $(-4, 5)$, $(2, 5)$, $(2, 1)$, and $(-4, 1)$. What are the dimensions of the secret chamber?

Draw the secret chamber in a coordinate plane.

The length of the chamber is the distance between $(-4, 5)$ and $(2, 5)$. The width of the chamber is the distance between $(2, 5)$ and $(2, 1)$.

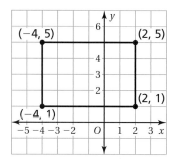

Use absolute values to find the distances between the points.

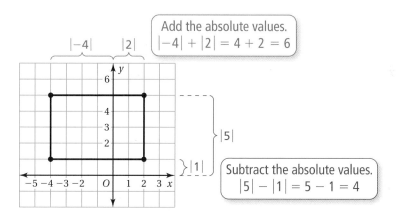

∴ The secret chamber is 6 meters long and 4 meters wide.

Practice

The points represent vertices of a polygon. Graph the polygon in a coordinate plane.

6. $J(3, 4)$, $K(3, -2)$, $L(-1, -2)$

7. $P(-5, 3)$, $Q(-1, 3)$, $R(-1, -4)$, $S(-5, -4)$

8. **GEOMETRY** Find the area of the figure in Exercise 6.

9. **GEOMETRY** Find the perimeter of the figure in Exercise 7.

10. **ARCHAEOLOGY** In Example 2, a gold coin is found at $(-1, 4)$, a silver coin is found at $(-4, 2)$, and pottery is found at $(-4, 4)$. How much closer is the pottery to the silver coin than to the gold coin?

11. **REASONING** Point A is a reflection of point B in the y-axis. Point C is a reflection of point B in the x-axis.

 a. What do you notice about the signs of the coordinates of the reflection in the y-axis? x-axis?

 b. Describe reflections that relate point A and point C. What do you notice about the signs of the coordinates?

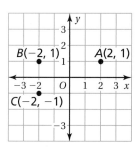

Appendix A
My Big Ideas Projects

A.1 **Literature Project**
Alice's Adventures in Wonderland

A.2 **History Project**
Mathematics in Ancient Egypt

About the Appendix

- The interdisciplinary projects can be used anytime throughout the year.
- The projects offer students an opportunity to build on prior knowledge, to take mathematics to a deeper level, and to develop organizational skills.
- Students will use the Essential Questions to help them form "need to knows" to focus their research.

Essential Question

- **Literature Project**
 How does mathematical problem solving influence a story plot?
- **History Project**
 How is mathematical knowledge that was originally discovered by the ancient Egyptians used today?

Additional Resources

BigIdeasMath.com

Additional Resources

BigIdeasMath.com

Essential Question

- **Art Project**
 How have circles influenced ancient and modern art?
- **Science Project**
 How does floating ice affect you and the world in which you live?

My Big Ideas Projects

A.3 Art Project
 Circle Art

A.4 Science Project
 Why Does Ice Float?

A.1 Literature Project

Alice's Adventures in Wonderland

1 Getting Started

Lewis Carroll was a mathematician who also wrote literature books for children.

Essential Question How does mathematical problem solving influence a story plot?

Read *Alice's Adventures in Wonderland* by Lewis Carroll. In each chapter, rewrite one of the episodes so that it contains some of the math that you have studied this year.

Sample: You could rewrite Alice's shrinking adventure as follows.

Alice's normal height was 4′7″ or 55 inches. After drinking the bottle marked "Drink Me," Alice could feel herself shrinking more and more. After several minutes, Alice was only two-elevenths of her original height. Would Alice have to bend over to walk through a door that is only 1 foot high?

Project Notes

Introduction

For the Teacher

- **Goal:** Students will rewrite episodes from *Alice's Adventures in Wonderland* using some of the mathematics they studied this year. Samples from each chapter are included below.
- **Management Tip:** You may want to have students work together in groups. Depending on the number of groups you have, you can assign one or more chapters to each group.

Essential Question

- How does mathematical problem solving influence a story plot?

Things to Think About

- **Chapter 2: The Pool of Tears** When Alice was caught in the hallway, she had grown to be over 9 feet tall. She began "shedding gallons of tears, until there was a large pool all round her, about four inches deep and reaching half down the hall." She held onto a fan and began shrinking. How tall was Alice when she fell into the pool if she was "up to her chin in salt water?" (Her head is about 12.5% of her height.) approximately 4.57 inches
- **Chapter 3: A Caucus-Race and a Long Tale** The race course for the Caucus-Race was circular. Alice and her friends ran for about 30 minutes. If they ran approximately 1000 meters per hour and made 50 laps around the course, how long would it take them to run once across the center of the circle? approximately 11.5 seconds
- **Chapter 4: The Rabbit Sends in a Little Bill** Alice was stuck in the Rabbit's house when the Rabbit and his friends decided to send Bill down the chimney. They got two ladders, tied them together, and put them at the corner of the house. Even so, they did not reach half high enough. They had enough rope to enable Bill to get up to the roof and down the chimney. The ladders were each 2.4 feet tall. What other information would you need to find the length of the rope? Answers will vary.
- **Chapter 5: Advice from a Caterpillar** Alice began her day about 4 feet 7 inches tall. She began growing and shrinking and growing and shrinking, so much so that she was quite confused. As she happened upon the Caterpillar, she was approximately 3 inches tall. After eating from one side of the mushroom, she was as short as she could be, with her chin upon her feet. Eating from the other side of the mushroom resulted in her being taller than the tallest tree in the woods. She nibbled her way back to her original height and was quite happy until she happened upon a house that was four feet tall. She then began nibbling the mushroom until she was 9 inches tall. Overall, how much did Alice's height change after she began nibbling the mushroom and why did she decide that 9 inches was a good height? 6 inches; She did not want to frighten the owner of the house.

References

Go to *BigIdeasMath.com* to access links related to this project.

Cross-Curricular Instruction

Meet with a reading or language arts teacher and review curriculum maps to identify whether students have read *Alice's Adventures in Wonderland*. If the book has been read, you may want to discuss the work students have completed and then review the book with them. If the book has not been read, perhaps you can both work simultaneously and share notes. Or, you may want to explore activities that the reading or language arts teacher has done in the past to support student learning in this particular area.

Project Notes

- **Chapter 6: Pig and Pepper** The Cheshire Cat used reasoning to prove to Alice that he was mad. Alice used reasoning to decide that visiting the March Hare was the better choice. She also used reasoning to determine when she had arrived at the March Hare's house. What was Alice's reasoning? Do you agree with her? She thought that visiting the March Hare would be more interesting than visiting the Hatter because she had seen hatters before; She thought it must be the right house, because the chimneys were shaped like ears and the roof was thatched with fur; *Answers will vary.*

- **Chapter 7: A Mad Tea-Party** During the tea-party, all the attendees moved one place on. The result of the translation was that the Hatter was the only one who had a clean cup. What translation would have been necessary for all of the guests to have a clean cup? four places on

- **Chapter 8: The Queen's Croquet Ground** The Queen's croquet court was quite different as, among other things, the soldiers were used as the arches and they kept moving about. Use the Internet or some other reference to find what a croquet court looks like. Using a coordinate plane with the center of the court as the origin, draw and label the location of each arch (wicket) and stake as well as the ordered pairs that would mark the corners of the court. See the diagram at the left.

- **Chapter 9: The Mock Turtle's Story** When the Mock Turtle went to school, his lessons were quite unusual, as was the length of the school day. The first day was 10 hours long, the next was 9 hours long, and so on to the end. How many hours were the lessons on the 5th day? How many hours were the lessons all together? What do you think happened on the 12th day? 6 hours; 55 hours; *Answers will vary.*

- **Chapter 10: The Lobster Quadrille** The Lobster Quadrille was a very curious dance. Find the definition of the word *quadrille*. Name and describe as many quadrilaterals as possible. Explain how a quadrille is related to a quadrilateral. *Sample answers:* a square dance for 4 couples (see illustration); *parallelogram:* a quadrilateral with 2 pairs of parallel sides; *rectangle:* a parallelogram with 4 right angles; *rhombus:* a parallelogram with 4 sides of equal length; *square:* a parallelogram with 4 right angles and 4 sides of equal length; *trapezoid:* a quadrilateral with one pair of parallel sides; A quadrille is a square dance and a square is a type of quadrilateral. Also, a quadrille requires 4 couples and a quadrilateral has 4 sides.

- **Chapter 11: Who Stole the Tarts?** During court, Alice began to grow. If she was a foot tall as court began and she grew at a rate of 2 inches per minute, how long did it take for her to be back to her original height of 4 feet 7 inches? 21.5 minutes

Closure

- **Rubric** An editable rubric for this project is available at *BigIdeasMath.com*.
- Students may present their reports to the class or school as a television report or public information broadcast.

The illustrations on these two pages are from the 1907 edition of *Alice's Adventures in Wonderland*. They were drawn by Arthur Rackham (1867–1939), a book illustrator from England.

2 Things to Remember

- You can download each chapter of the book at *BigIdeasMath.com*.
- Add your own illustrations to your project.
- Try to include as many different math concepts as possible. Your goal is to include at least one concept from each of the chapters you studied this year.
- Organize your math stories in a folder, and think of a title for your report.

A.2 History Project

Mathematics in Ancient Egypt

1 Getting Started

The ancient Egyptian civilization in North Africa began around 3150 B.C. with the union of Upper and Lower Egypt under the first pharaoh. The rule of the pharaohs ended in 31 B.C. when the Romans conquered Egypt and made it a province.

The ancient Egyptians' achievements included a system of mathematics, glass technology, medicine, literature, irrigation and agricultural techniques, and art. Their mastery of surveying, quarrying, and construction techniques allowed them to build pyramids, temples, and obelisks.

Essential Question How is mathematical knowledge that was originally discovered by the ancient Egyptians used today?

Sample: To multiply two whole numbers, the ancient Egyptians repeatedly multiplied by 2. The "doubling lists" used to multiply 53 by 85 are shown. The lists stop when the last number in the first list is greater than the first number in the second list.

- 1 53
- 2 106
- 4 212
- 8 424
- 16 848
- 32 1696
- 64 3392

Because $85 = 1 + 4 + 16 + 64$,
$85 \cdot 53 = 53 + 212 + 848 + 3392 = 4505$.

Ancient Papyrus

The Blue Sphinx

Ancient Egyptian Numbers

Project Notes

Introduction

For the Teacher
- **Goal:** Students will discover how ancient Egyptians wrote and used mathematics.
- **Management Tip:** Students can work in groups to create a report about the mathematics used in ancient Egypt.

Essential Question
- How is mathematical knowledge that was originally discovered by the ancient Egyptians used today?

Things to Think About

? How did the ancient Egyptians multiply two whole numbers?
- Egyptians only knew how to multiply and divide by 2. For example, in order to find 27 • 5, they did the following:

 (27 • 2) + (27 • 2) + (27 • 1) = 54 + 54 + 27 = 135.

- For more complicated numbers, they used powers of 2 and doubling. For example, the doubling pattern below shows how they found 45 • 67.

1	2	4	8	16	32	64	128
67	134	268	536	1072	2144	4288	8576

 Because 45 = 1 + 4 + 8 + 32, 45 • 67 = 67 + 268 + 536 + 2144 = 3015.

? How did the ancient Egyptians divide two whole numbers?
- The process the Egyptians used to divide was similar to the process they used to multiply. Suppose they wanted to divide 308 by 28. They thought about this problem as "28 times some number equals 308."

1	2	4	8	16	32
28	56	112	224	448	896

 Because 308 = 28 + 56 + 224 and 1 + 2 + 8 = 11, 308 ÷ 28 = 11.

? How did the ancient Egyptians write whole numbers?
- The numbers we use today are far different from the ones used by the Egyptians. The Egyptians used hieroglyphics to represent numbers. They used a vertical line (a staff) to represent 1, an upside-down U (a whale bone) to represent 10, a coil (a coil of rope) to represent 100, and a flower (a lotus flower) to represent 1000.
- To write the number 234, draw, from right to left, two coils of rope, three whale bones, and four staffs.

 234 = ||||| ∩∩∩ ☙☙

References
Go to *BigIdeasMath.com* to access links related to this project.

Cross-Curricular Instruction

Meet with a history teacher and review curriculum maps to identify whether students have covered ancient Egypt. If the topic has been covered, you may want to discuss the work students have completed and review prior knowledge with them. If the history teacher has not discussed these concepts, perhaps you can both work simultaneously on these concepts and share notes. Or, you may want to explore activities that the history teacher has done in the past to support student learning in this particular area.

Project Notes

- To write the number 1302, draw, from right to left, one lotus flower, three coils of rope, and two staffs.

 $1302 =$ (hieroglyphs: two staffs, three coils of rope, one lotus flower)

? How did the ancient Egyptians write fractions?

- Egyptians understood the concept of fractions but did not write them as we do today. They wrote most fractions with a numerator of 1. To do this, they wrote the number the same as the whole number, but placed an oval symbol above the number to show that it was a fraction.

 $\frac{1}{10} =$ (oval above hieroglyph)

- There were three exceptions to this rule:

 $\frac{1}{2} =$, $\frac{2}{3} =$, and $\frac{3}{4} =$.

? How did the ancient Egyptians use ratios to measure the height of a pyramid?

- Thales, an ancient philosopher, discovered a way to measure the heights of very tall objects, such as the pyramids, using shadows. He noticed that the ratio of his height to the length of his shadow was equal to the ratio of a pyramid's height to the length of its shadow at the same time of day. So, to find the height of a pyramid, he used the proportion

 $$\frac{\text{Height of Thales}}{\text{Length of Thales' Shadow}} = \frac{\text{Height of Pyramid}}{\text{Length of Pyramid's Shadow}}.$$

? How did the ancient Egyptians use mathematics?

- The ancient Egyptians did not have electric clocks or wrist watches, so they developed a way to measure time based on shadow clocks. They built obelisks and used the moving shadows created by the obelisks to help tell time.
- Instead of using yardsticks and measuring tape to measure lengths, the ancient Egyptians used parts of their body. The length of a person's forearm was called a royal cubit. A handbreadth (the length of a palm) was considered to be one seventh of a royal cubit. A fingerbreadth (the length of a finger) was even smaller; it was one fourth of a handbreadth. They even had a measurement for the volume of grains called a hekat.
- Because rulers did not exist, ancient Egyptians used rods with markings on them to measure the depth of the Nile River. They also invented a calendar to help them predict when the Nile would flood.

Closure

- **Rubric** An editable rubric for this project is available at *BigIdeasMath.com*.
- You may hold a class debate where students compare, defend, and discuss their findings with another student or group of students

2 Things to Include

- Explain how the ancient Egyptians multiplied two whole numbers. Give an example.
- Explain how the ancient Egyptians divided two whole numbers. Give an example.
- How did the ancient Egyptians write whole numbers? Give some examples.
- How did the ancient Egyptians write fractions? Give some examples.
- How did the ancient Egyptians use ratios to measure the height of a pyramid? Give an example.
- Write your name using ancient Egyptian hieroglyphics.
- Describe how ancient Egyptians used mathematics. How does this compare with the ways in which mathematics is used today?

Ancient Egyptian Alphabet

3 Things to Remember

- Add your own illustrations to your project.
- Try to include as many different math concepts as possible. Your goal is to include at least one concept from each of the chapters you studied this year.
- Organize your report in a folder, and think of a title for your report.

What does this spell?

A.3 Art Project

Circle Art

1 Getting Started

Circles have been used in art for thousands of years.

Essential Question How have circles influenced ancient and modern art?

Find examples of art in which circles were used. Describe how the artist might have used properties of circles to make each piece of art.

Sample: Here is a technique for making a pattern that uses circles in a stained glass window.

1. Use a compass to draw a circle.
2. Without changing the radius, draw a second circle whose center lies on the first circle.
3. Without changing the radius, draw a third circle whose center is one of the points of intersection.
4. Without changing the radius, continue to draw new circles at the points of intersection.
5. Color the design to make a pattern.

Mexican Tile

Ancient Roman Mosaic

Project Notes

Introduction
For the Teacher
- **Goal:** Students will discover how circles are used in art and architecture.
- **Management Tip:** Students may work in groups to find mosaic tile patterns. They can brainstorm as a group to come up with a way to measure the angles contained in the tile patterns.

Essential Question
- How have circles influenced ancient and modern art?

Things to Think About
? Where have artists used circles?
- Artists have used the circle as a basis for paintings, jewelry, wrought iron designs, architecture, and religious beliefs.
- Stonehenge on the Salisbury Plain in southwestern England is made of large upright stones set in a circular pattern. It is estimated that Stonehenge was built around 2000 B.C.

? How significant is the history of one of the most used circles, the ring?
- In ancient Greece, the *signet ring* was used to make an impression indicating authenticity.
- In Rome, citizens could wear rings of iron but slaves were forbidden to wear rings.
- Some churches use rings to signify the office of the wearer.
- The Egyptians wore rings of jasper or bloodstone for success in battle.
- Some cultures believe a magic ring will provide the wearer with power over others.
- The circle is a symbol of eternity because it has no beginning or end.
- The circle is also a symbol of mobility because of the wheel.
- The wheel is associated with chance and fortune.

References
Go to *BigIdeasMath.com* to access links related to this project.

Cross-Curricular Instruction

Meet with an art teacher and review curriculum maps to identify whether students have covered circles in art. If the topic has been covered, you may want to discuss the work students have completed and review prior knowledge with them. If the art teacher has not discussed these concepts, perhaps you can both work simultaneously on these concepts and share notes. Or, you may want to explore activities that the art teacher has done in the past to support student learning in this particular area.

Project Notes

? How did ancient artists draw circles?
- To draw a circle, ancient artists could have used two sticks connected by a length of rope or string. One stick would be placed at the center of the desired circle. The rope, which represented the radius, would be stretched out and the other stick would be used to draw the circle.

? What are some examples of how circles are used in mosaic tile patterns?

? What is an example of how circles are used in modern art or architecture?
- The Rotunda at the United States Capital is a circular room 96 feet in diameter and 180 feet in height.

Closure

- **Rubric** An editable rubric for this project is available at *BigIdeasMath.com*.
- Students may present their reports to a parent panel or community members.

2 Things to Include

- Describe how ancient artists drew circles.
- Describe the symbolism of circles in ancient art.
- Find examples of how circles are used to create mosaic tile patterns.
- Measure the angles that are formed by the patterns in the circle art you find. For instance, you might describe the angles formed by the netting in the Native American dreamcatcher.
- Find examples of the use of circles in modern art.
- Use circles to create your own art. Describe how you used mathematics and the properties of circles to make your art.

Native American Dreamcatcher

3 Things to Remember

- Add your own illustrations to your project.
- Try to include as many different math concepts as possible. Your goal is to include at least one concept from each of the chapters you studied this year.
- Organize your report in a folder, and think of a title for your report.

Section A.3 Art Project A7

A.4 Science Project

Why Does Ice Float?

1 Getting Started

Among the planets in our solar system, Earth is clearly the "water planet." Water occurs on Earth's surface as a liquid, solid, and gas. Ocean waters cover about 70% of Earth's surface. Fresh water in lakes and rivers covers less than 1%. Thick sheets of ice permanently cover Earth's polar regions, and glaciers occur in its higher mountains. Water in the form of clouds covers about half of Earth's surface at any time.

Essential Question How does floating ice affect you and the world in which you live?

There are three parts to this question. First, you have to understand why things float in water. Make lists of things that float in water and things that do not float in water. How can you describe, scientifically, the difference between these two types of objects?

Second, you have to discover what is special about water. Almost all other elements and compounds have the property that their solid forms sink in their liquid forms.

Third, you have to research how water in solid form influences the environment and living organisms. You may want to explore ocean currents, weather, and various cycles including the water cycle.

Earth's North Polar Ice Cap

Project Notes

Introduction
For the Teacher
- **Goal:** Students will discover why objects sink or float in water.
- **Management Tip:** You can have students discover other items that float in water. Allow students to bring items to school for this experiment. You can use a sink or bucket to collect the water.

Essential Question
- How does floating ice affect you and the world in which you live?

Things to Think About

? Why do some objects float in water and others do not?
- Objects float in water because they are less dense than water. Some objects may float in salt water and not in tap water, because salt makes water more dense.

Do these things float in water?	
Yes	No
bread	egg
apple	chocolate chip
vegetable oil	rock
pepper	penny
aluminum can	spaghetti
flour	key

? Do all compounds in a solid state float in their liquid state?
- With gold, nitrogen, solid ethanol, and most other compounds, the solid form is smaller than the liquid form. So the solid form is more dense and will sink when placed in the liquid form.
- Water is the opposite. It expands when it goes from liquid to solid, so ice is less dense and floats in water.

? Why does ice float in water?
- Even though ice is a solid state, ice (density = 0.92 g/mL) is less dense than water (density = 1.0 g/mL). Objects less dense than 1.0 g/mL will float. Objects more dense than 1.0 g/mL sink. Ice is less dense than water because water expands as it freezes.

? Does ice float in other liquids?
- Ice will float in vegetable oil because ice is less dense than vegetable oil.

Materials
- Bucket or sink
- Objects to place in water

References
Go to *BigIdeasMath.com* to access links related to this project.

Cross-Curricular Instruction

Meet with a science teacher and review curriculum maps to identify whether students have covered density. If the topic has already been covered, you may want to discuss the work students have completed and review prior knowledge with them. If the science teacher has not discussed these concepts, perhaps you can both work simultaneously on these concepts and share notes. Or, you may want to explore activities that the science teacher has done in the past to support student learning in this particular area.

Project Notes

? Are the majority of icebergs below the surface?
- The phrase "just the tip of the iceberg" came from the fact that most of an iceberg is below water. Research indicates that 80%-90% of an iceberg is below water.
- Very dense icebergs are deep blue. Less dense icebergs are white because they contain many trapped air bubbles. Icebergs that are greenish-black have calved off the bottom of the glacier. Dark brown stripes in an iceberg are an indication that the iceberg came from two glaciers that joined together.

? What would happen if ice did not float?
- If icebergs did not float, the oceans and waterways would freeze from the bottom up. The result would be a loss of biological life throughout the world.
- If icebergs were completely under water the sea level would rise and there would be great flooding, particularly in the low coastal regions. This would be devastating to the plants and animals in those areas.

? How does floating ice keep the oceans warm?
- Floating ice acts as a blanket in the winter, keeping the warm ocean from releasing all of its heat into the colder atmosphere.

Closure

- **Rubric** An editable rubric for this project is available at *BigIdeasMath.com*.
- Students may present their reports to the class or compare their report with other students' reports.

2 Things to Include

- Does vegetable cooking oil float in water? Does ice float in vegetable oil?
- Compare the density of water and the density of ice. How do these densities relate to the fact that ice floats in water?
- What percent of an iceberg is above the water? What percent is below?
- What environmental and biological systems would not work if ice sunk in the oceans instead of floating?
- Explain how the floating of ice helps keep the oceans warm and therefore helps aquatic life exist in Earth's polar oceans.

3 Things to Remember

- Add your own illustrations to your project.
- Try to include as many different math concepts as possible. Your goal is to include at least one concept from each of the chapters you studied this year.
- Organize your report in a folder, and think of a title for your report.

Floating Iceberg

Selected Answers

Section 1.1 Evaluating Algebraic Expressions
(pages 6 and 7)

1.

Algebraic Expression	Numbers	Variables	Operations
$x - 8$	8	x	Subtraction
$3w + 9$	3, 9	w	Multiplication and addition
$6y - 12$	6, 12	y	Multiplication and subtraction

3. decrease; When you subtract greater and greater values from 20, you will have less and less left.

5. $120 **7.** $8
9. 10 **11.** 9
13. 17 **15.** 2

17. 9 **19.** 24 **21.** $15; $105

23.

x	3	6	9
$x \cdot 8$	24	48	72

25. 23 **27.** 6 **29.** 22

31. 46 **33.** 24

35. What shape could have an area of 128 square feet? What shape could have an area of s^2 square feet?

37. (3, 2)

39. (5, 1)

41. C

 Hint

Section 1.2 Writing Expressions
(pages 12 and 13)

1. x take away 12; $x - 12$; $x + 12$
3. $8 - 5$
5. $28 \div 7$
7. $18 - 3$
9. $x - 13$
11. $18 \div a$
13. $7 + w$ or $w + 7$
15. $y + 4$ or $4 + y$
17. $2 \cdot z$ or $z \cdot 2$

19. The expression is not written in the correct order; $\dfrac{8}{y}$

21. a. $x \div 5$

b. *Sample answer:* If the total cost is $30, then the cost per person is $x \div 5 = 30 \div 5 = \$6$. The result is reasonable.

23–25. Sample answers are given.

23. The sum of n and 6; 6 more than a number n
25. A number b less than 15; 15 take away a number b

27. $\dfrac{y}{4} - 3$; 2
29. $8x + 6$; 46

A10 Selected Answers

31. a.

Game	1	2	3	4	5
Cost	$5	$8	$11	$14	$17

b.

c. $2 + 3g$

d. $26

33. It might help to see the pattern if you make a table of the data in the bar graph.

35. $\dfrac{x}{4}$ **37.** 59

39. 140

Section 1.3 Properties of Addition and Multiplication
(pages 18 and 19)

1. *Sample answer:* $\dfrac{1}{5} + \dfrac{3}{5} = \dfrac{3}{5} + \dfrac{1}{5}$
$\dfrac{4}{5} = \dfrac{4}{5}$

3. *Sample answer:* $(5 \cdot x) \cdot 1 = 5 \cdot (x \cdot 1)$
$= 5x$

5. Comm. Prop. of Mult. **7.** Assoc. Prop. of Mult. **9.** Add. Prop. of Zero

11. The grouping of the numbers did not change. The statement illustrates the Commutative Property of Addition because the order of the addends changed.

13. $(14 + y) + 3 = (y + 14) + 3$ Comm. Prop. of Add.
$= y + (14 + 3)$ Assoc. Prop. of Add.
$= y + 17$ Add 14 and 3.

15. $7(9w) = (7 \cdot 9)w$ Assoc. Prop. of Mult.
$= 63w$ Multiply 7 and 9.

17. $(0 + a) + 8 = a + 8$ Add. Prop. of Zero

19. $(18.6 \cdot d) \cdot 1 = 18.6 \cdot (d \cdot 1)$ Assoc. Prop. of Mult.
$= 18.6d$ Mult. Prop. of One

21. $(2.4 + 4n) + 9 = (4n + 2.4) + 9$ Comm. Prop. of Add.
$= 4n + (2.4 + 9)$ Assoc. Prop. of Add.
$= 4n + 11.4$ Add 2.4 and 9.

23. $z \cdot 0 \cdot 12 = (z \cdot 0) \cdot 12$ Assoc. Prop. of Mult.
$= 0 \cdot 12$ Mult. Prop. of Zero
$= 0$ Mult. Prop. of Zero

25. a. x represents the cost of a box of cookies. **b.** $120x$

27. $7 + (x + 5) = x + 12$ **29.** $(7 \cdot 2) \cdot y$ **31.** $(17 + 6) + 2x$

33. $w \cdot 16$ **35.** 98 **37.** 90

39. 37 is already prime. **41.** 3×7^2 **43.** B

Section 1.4
The Distributive Property
(pages 26 and 27)

1. *Sample answer:* You must distribute or give the number outside the parentheses to the numbers inside the parentheses.

3. $4 + (x \cdot 4)$ does not belong because it doesn't represent the Distributive Property.

5. 684 7. 440 9. 216 11. 196 13. $10b - 60$

15. $56 + 7y$ 17. $9n + 9$ 19. $18w + 90$ 21. $70 + 7x$ 23. $78 + 6z$

25. $29 + 8x$ 27. $5x + 52$

29. a. $30(8 + x) = 240 + 30x$
 b. *Sample answer:* $2; It is less than the regular price to the exhibit.
 c. *Sample answer:* $300; yes

31. $10(103 - x) = 1030 - 10x$ 33. $13(7 - x) = 91 - 13x$

35. $x = 8$ 37. $x = 3$

39. $2(3 + x)$ 41. $7(1 + 2x + 3)$

43. The expression for the profit will contain an expression for the large candles and an expression for the small candles.

45. 14 47. 120

49. no; $\frac{2}{3}$ 51. no; $\frac{19}{31}$

53. C

Section 1.5
Using Formulas to Solve Problems
(pages 32 and 33)

1. *Sample answer:* You substitute value(s) for the variable(s) to find the value of the formula.

3. 48 in.2 5. 108 in.2 7. 30 ft^2

9. a. 234 ft^2 b.

 c. 26 ft; The base of the parking space is related to the length of the car.

11. a. 192 in.3 b. almost 13 bowls

13. $4x - 9$ 15. $32x - 40 - x^2$

17. 24 karats; If you let $k = 24$, then $P = 100$.

19. Area of black = 252 in.2
 Area of yellow = 244 in.2
 Area of each blue stripe = 328 in.2

21. $\frac{1}{2}$ 23. 1

Section 2.1 — Fractions and Estimation
(pages 48 and 49)

1. rounding; The product will be easier to compute.

3. rounding; The product will be easier to compute.

5.

How to Round	Estimate
Round 77 to the nearest hundred.	$100 \div 4 = 25$
Round 77 to the nearest ten.	$80 \div 4 = 20$
Round 77 to the nearest compatible number.	$76 \div 4 = 19$

7. 0

9. 0

11. $\frac{1}{2}$

13. $\frac{1}{3}$

15. $\frac{1}{2}$

17. $\frac{1}{2}$

19. 27

21. 6

23. $\frac{5}{12}$ is closer to $\frac{1}{2}$ than to 0. $\frac{5}{12} \times \frac{9}{10} \approx \frac{1}{2} \times 1 = \frac{1}{2}$

25. 8

27. 203

29. 7

31. 20

33. Which operation should you use?

35. 27 in.2; underestimate

37. 18

39. 5

41. *Sample answer:* low estimate: 234 in.2; high estimate: 320 in.2
To find a low estimate, round the dimensions down.
To find a high estimate, round the dimensions up.

43. 36

45. $4\frac{1}{2}$

Section 2.2 — Multiplying Fractions and Whole Numbers
(pages 54 and 55)

1. Multiply the numerator of the fraction by the whole number. Then write the product over the denominator.

3. $\frac{1}{3} \times 24$; because $\frac{1}{3} > \frac{1}{4}$

5. $\frac{5}{8}$

7. $1\frac{7}{9}$

9. 15

11. $17\frac{1}{2}$

13. $2\frac{1}{10}$

15. 26

17. 9

19. $13\frac{1}{2}$

21. 9 should be multiplied by 3, not 7. $9 \times \frac{3}{7} = \frac{9 \times 3}{7} = \frac{27}{7}$, or $3\frac{6}{7}$

23. $2\frac{2}{3}$ cups

25. 6

27. 20

29. Multiply $25 \times \frac{2}{5}$ first by the Comm. Prop. of Mult.; 60

31. Multiply $\frac{3}{7} \times 14$ first by the Comm. Prop. of Mult.; 78

Section 2.2 Multiplying Fractions and Whole Numbers
(continued) (pages 54 and 55)

33. $1\frac{1}{2}$ **35.** $2\frac{2}{3}$ **37.** $22\frac{1}{2}$ **39.** $4\frac{1}{6}$

41. yes; If you have more money than your friend, then $\frac{1}{3}$ of your money could be greater than $\frac{1}{2}$ of your friend's money.

43. $1\frac{7}{8}$ **45.** $\frac{32}{175}$ **47.** D

Section 2.3 Multiplying Fractions
(pages 60 and 61)

1. Multiply numerators and multiply denominators, then simplify the fraction.

3. 4 **5.** $\frac{2}{21}$ **7.** $\frac{1}{10}$ **9.** $\frac{8}{15}$ **11.** $\frac{1}{24}$

13. $4\frac{1}{6}$ **15.** $\frac{2}{5}$ **17.** $\frac{9}{49}$ **19.** $\frac{13}{21}$

21. You did not multiply the denominators. $\frac{2}{5} \times \frac{3}{10} = \frac{2 \times 3}{5 \times 10} = \frac{\overset{1}{2} \times 3}{5 \times \underset{5}{10}} = \frac{3}{25}$

23. $\frac{1}{4}$ **25.** $\frac{2}{21}$ **27.** $\frac{3}{10}$ **29.** $\frac{7}{10}$ **31.** $\frac{21}{40}$

33. $\frac{9}{80}$ **35.** $\frac{7}{45}$ **37.** $\frac{27}{125}$ **39.** $\frac{25}{196}$

41. $\left(\frac{5}{8} \times \frac{22}{15}\right) > \frac{5}{8}$; Because $\frac{22}{15} > 1$, the product will be greater than $\frac{5}{8}$.

43. *Sample answer:* $\frac{1}{3}$ **45. a.** $\frac{3}{50}$ **b.** 45 people

47. $\frac{35}{8}$ **49.** $\frac{23}{6}$

Section 2.4 Multiplying Mixed Numbers
(pages 66 and 67)

1. a fraction with a numerator that is greater than or equal to the denominator

3. *Sample answer:* $3\frac{1}{2} \times 3\frac{1}{7} = 11$

5. 2 **7.** $\frac{3}{4}$ **9.** 2 **11.** 2 **13.** 2

15. $1\frac{1}{2}$ **17.** $1\frac{3}{14}$ **19.** $36\frac{2}{3}$ **21.** $6\frac{4}{9}$ **23.** $11\frac{3}{8}$

A14 Selected Answers

25. You must first rewrite the mixed numbers as improper fractions and then multiply.

$$2\frac{1}{2} \times 7\frac{4}{5} = \frac{5}{2} \times \frac{39}{5}$$
$$= \frac{\overset{1}{\cancel{5}} \times 39}{2 \times \underset{1}{\cancel{5}}}$$
$$= \frac{39}{2}, \text{ or } 19\frac{1}{2}$$

27. a. 7 ft^2 **b.** $10\frac{1}{3} \text{ ft}^2$

29. $13\frac{3}{4}$ **31.** $8\frac{13}{24}$ **33.** 155 **35.** $26\frac{2}{5}$

37. No; Positive mixed numbers are greater than 1 and the product of two numbers greater than 1 is always greater than 1.

39. a. $5\frac{1}{10}$ hours **b.** 11:39 A.M.

41. $\frac{1}{6}$ **43.** $\frac{7}{36}$ **45.** $9x - 27$ **47.** $12x - 96$

Section 2.5 Dividing Fractions
(pages 75–77)

1. Sample answer: $\frac{2}{5}, \frac{5}{2}$ **3.** B **5.** A

7. $\frac{1}{8}$ **9.** $\frac{5}{2}$ **11.** $\frac{1}{2}$ **13.** 16 **15.** $\frac{1}{14}$

17. $\frac{1}{3}$ **19.** 3 **21.** $\frac{2}{27}$ **23.** $\frac{27}{28}$ **25.** $20\frac{1}{4}$

27. You need to invert the second fraction before you multiply.

$$\frac{4}{7} \div \frac{13}{28} = \frac{4}{7} \times \frac{28}{13} = \frac{4 \times \overset{4}{\cancel{28}}}{\underset{1}{\cancel{7}} \times 13} = \frac{16}{13}, \text{ or } 1\frac{3}{13}$$

29. Round $\frac{2}{5}$ to $\frac{1}{2}$ and $\frac{8}{9}$ to 1. $\frac{1}{2} \div 1 = \frac{1}{2}$, which is not close to the incorrect answer of $\frac{20}{9}$.

31. $5\frac{5}{8}$ times **33.** $\frac{5}{16}$ **35.** $2\frac{1}{2}$ **37.** yes **39.** yes **41.** $\frac{1}{3}$

43. >; When you divide a number by a fraction less than 1, the quotient is greater than the number.

45. >; When you divide a number by a fraction less than 1, the quotient is greater than the number.

47. $\frac{1}{216}$ **49.** $1\frac{1}{6}$ **51.** 2 **53.** $\frac{3}{26}$ **55.** $\frac{2}{3}$

57. when the fraction has, or can be simplified to have, a 1 in the numerator; The reciprocal will have, or can be simplified to have, a 1 in the denominator, so it is a whole number.

59. It might help to make a table to find how many bowls and plates you can glaze so there is no glaze left over.

61. 3 **63.** 10 **65.** C

Section 2.6 — Dividing Mixed Numbers
(pages 82 and 83)

1. $\dfrac{3}{22}$
3. sometimes; The reciprocal of $\dfrac{2}{2}$ is $\dfrac{2}{2}$, which is improper.
5. 3
7. $9\dfrac{3}{4}$
9. $3\dfrac{18}{19}$
11. $\dfrac{9}{10}$
13. $12\dfrac{1}{2}$
15. $1\dfrac{1}{5}$
17. $\dfrac{2}{7}$
19. $1\dfrac{5}{18}$
21. The mixed number $1\dfrac{2}{3}$ was not written as an improper fraction before inverting.
$$3\dfrac{1}{2} \div 1\dfrac{2}{3} = \dfrac{7}{2} \div \dfrac{5}{3}$$
$$= \dfrac{7}{2} \times \dfrac{3}{5}$$
$$= \dfrac{7 \times 3}{2 \times 5}$$
$$= \dfrac{21}{10}, \text{ or } 2\dfrac{1}{10}$$
23. 14 hamburgers
25. $\dfrac{8}{15}$
27. $\dfrac{22}{27}$
29. $1\dfrac{1}{3}$
31. $5\dfrac{1}{6}$
33. $\dfrac{7}{54}$
35. $12\dfrac{1}{2}$
37. $\dfrac{22}{35}$
39. **a.** 6 ramps; *Sample answer:* The estimate is reasonable because $12\dfrac{1}{2}$ was rounded down.
 b. 6 ramps; $1\dfrac{1}{4}$ feet left over
41. 0.43
43. 3.8
45. C

Section 2.7 — Writing Decimals as Fractions
(pages 88 and 89)

1. *Sample answer:* $0.3 = \dfrac{3}{10}$
3. $6\dfrac{7}{10}$
5. $7\dfrac{9}{10}$
7. $\dfrac{9}{10}$
9. $\dfrac{2}{5}$
11. $\dfrac{16}{25}$
13. $\dfrac{3}{8}$
15. $\dfrac{11}{100}$
17. $\dfrac{87}{200}$
19. $2\dfrac{3}{4}$
21. $8\dfrac{113}{1000}$
23. Because 3 is in the thousandths' place, the denominator should be 1000. $0.073 = \dfrac{73}{1000}$
25. $2\dfrac{3}{10}$ hours
27. $\dfrac{36}{5}$
29. $\dfrac{341}{50}$
31. $\dfrac{19}{4}$
33. $\dfrac{138}{25}$
35. $0.05; \dfrac{1}{20}$
37. $0.08; \dfrac{2}{25}$
39. $0.32; \dfrac{8}{25}$
41. $0.55; \dfrac{11}{20}$
43. It may help to organize the information in a table.
45. 5
47. 17
49. $\dfrac{5}{7}$
51. $\dfrac{2}{7}$
53. B

Section 2.8 — Writing Fractions as Decimals
(pages 94 and 95)

1. When the remainder is 0 or the decimal begins repeating

3. terminating **5.** repeating **7.** $0.\overline{1}$ **9.** $0.5\overline{3}$

11. 0.7 **13.** 0.62 **15.** $0.9\overline{4}$ **17.** 0.84

19. 0.15 **21.** 0.375 **23.** $0.6\overline{3}$ **25.** 4.56

27. The digit 5 does not repeat; $\frac{8}{15} = 0.5\overline{3}$.

29. $0.7\overline{3}$ **31.** > **33.** $\frac{5}{8}$; 0.625 **35.** $\frac{9}{14}$, $0.6\overline{428571}$

37. *Sample answer:* $\frac{3}{5}$; 0.6

39. a. $\frac{3}{9} = 0.333\ldots;\ \frac{4}{9} = 0.444\ldots;\ \frac{5}{9} = 0.555\ldots;\ \frac{6}{9} = 0.666\ldots;$

$\frac{7}{9} = 0.777\ldots;\ \frac{8}{9} = 0.888\ldots;\ \frac{9}{9} = 0.999\ldots$

b. 1; $\frac{9}{9} = 1$

41. $0.00\overline{345}$ **43.** $0.00\overline{125125}$ **45.** 7 **47.** 12

Section 3.1 — Decimals and Estimation
(pages 110 and 111)

1. *Sample answer:* You use an estimate to approximate the price of an item after a 25% discount.

3. $3.9 \div 1.1,\ 7\frac{3}{5} \div 1\frac{7}{8},\ 11.3 \div 2.9$

5. Compatible numbers, because rounding does not give numbers that are easy to use.

7. Rounding, because it gives numbers that are easy to use.

9. 21 **11.** 25 **13.** 30 **15.** 90

17. 27 **19.** 9 **21.** 72 **23.** 2

25–39. Sample answers are given.

25. 600 **27.** 6 **29.** 450 **31.** 6

33. 320 lb; too much **35.** 35 m² **37.** 96 ft² **39.** 36

41. The calories burned in the table are for five minutes, not one minute.

43. *Sample answer:* Your cousins, from out of town, leave at 9:00 A.M. to visit you. To be sure you are home when they arrive, you underestimate their travel time.

45. 192 **47.** 195 **49.** 8.92 **51.** 10.004

Section 3.2 Multiplying Decimals and Whole Numbers
(pages 116 and 117)

1. 4; The decimal factor has 4 decimal places.
3. 7.6 is how much more than 3?; 4.6; 22.8
5. 5.6 **7.** 40.3 **9.** 54.9 **11.** 22.29 **13.** 47.5 **15.** 2.48
17. 31.5 **19.** 18.27 **21.** 29.45 **23.** 98.256 **25.** 0.091 **27.** 0.085
29. 0.076 **31.** 0.0558
33. The decimal is in the wrong place. $0.32 \times 5 = 1.60$
35. 8.75 ft **37.** 28.3 **39.** 9670 **41.** $5x + 1.55$
43. $7b + 0.364$ **45.** $12w + 0.0264$ **47.** 28.4 **49.** 282.24
51. $7.12 \times 8.22 \times 100 = 7.12 \times 822 = 5852.64$
53. *Answers should include, but are not limited to:*
 a. menu with main items, desserts, beverages, and prices
 b. guest check for 5 people showing items, prices, and subtotal
 c. tax and total with tax are shown
 d. amount rounded to nearest dollar, 20% tip, and total amount paid are shown
55. 18,272 **57.** 47,107

Section 3.3 Multiplying Decimals
(pages 122 and 123)

1. Place the decimal point so that there are two decimal places. $1.2 \times 2.4 = 2.88$
3. 6.2832 **5.** 3 **7.** 5 **9.** 0.024 **11.** 0.000072 **13.** 0.03
15. 0.000012 **17.** 109.74 **19.** 3.886 **21.** 7.677824 **23.** 51.358363 **25.** $3.24
27. $741.79 **29.** 0.576 **31.** 0.0222 **33.** 0.2664 **35.** 6.2
37. Each number is 0.1 times the previous number; 0.0015, 0.00015, 0.000015
39. Each number is 1.5 times the previous number; 25.3125, 37.96875, 56.953125
41. $0.45 \times 0.75 = 0.3375$ **43.** $5.525 \times 6.64 = 36.686$
45. a. 190.06 miles **b.** 91.29 miles **47.** Which framing is thicker?
49. 5 **51.** 7

A18 Selected Answers

Section 3.4
Dividing Decimals by Whole Numbers
(pages 130 and 131)

 1. 4.3; 12.9; 3
 3. 18.6 ÷ 4 = 4.65
 5. 88.27 ÷ 7 = 12.61
 7. 43.254 ÷ 9 = 4.806
 9. 4.2
 11. 0.5
 13. 4.3
 15. 6.2
 17. 5.58
 19. 0.15
 21. 2.165
 23. 8.757
 25. They brought down 2 zeros instead of 1.

$$\begin{array}{r} 3.112 \\ 9\overline{)28.008} \\ \underline{27} \\ 10 \\ \underline{9} \\ 10 \\ \underline{9} \\ 18 \\ \underline{18} \\ 0 \end{array}$$

 27. $0.12
 29. 7.945
 31. 25.2
 33. 2.35
 35. the 12-pack; The price per unit is $0.74 for the 4-pack, $0.72 for the 12-pack, and $0.73 for the 24-pack. So, the 12-pack is the best buy.
 37. Swimming a quarter second faster means to subtract 0.25 second from each swimmer's time.
 39. 6
 41. 7
 43. A

Section 3.5
Dividing Decimals
(pages 136 and 137)

 1. 10
 3. 21)1766
 5. 156 ÷ 47
 7. 5590 ÷ 647
 9. 12
 11. 12
 13. 12
 15. 36.5
 17. 18.7
 19. 52.1
 21. 3.525
 23. 7.2
 25. The decimal point in the dividend should be moved to the right instead of the left;
 0.32)146.4 ⟶ 32)14,640
 27. 4.8 ft
 29. 35.84
 31. 3.11
 33. about 6.04
 35. about 0.78
 37. =
 39. <
 41. 5357 bees
 43. When dividing, make sure your units cancel.
 45. 25%
 47. 21%
 49. B

Section 4.1 Percents and Fractions
(pages 152 and 153)

1. $\frac{2}{12}$, because it is equal to $0.1\overline{6}$, and the others are equal to 0.16.

3. yes; You can rewrite $1\frac{1}{4}$ as the improper fraction $\frac{125}{100}$, which is equal to 125%.

5. $\frac{9}{10}$ 7. $\frac{31}{40}$ 9. $\frac{79}{100}$ 11. $1\frac{22}{25}$ 13. $2\frac{6}{25}$ 15. $\frac{1}{250}$

17. 10%; method of Example 2 19. 55%; method of Example 2
21. 54%; method of Example 2 23. 37.5%; method of Example 3
25. 85%; method of Example 2 27. 82.5%; method of Example 3

29. The decimal point should not have been added to the percent expression.
$$\frac{14}{25} = \frac{14 \times 4}{25 \times 4} = \frac{56}{100} = 56\%$$

31. 40% 33. $\frac{13}{25} = 52\%$ 35. 294% 37. 392%

39. The answer depends on the goals for each year.

41. $8\frac{1}{3}\%$ or $8.\overline{3}\%$; *Sample answer:* Multiply the numerator and the denominator by $\frac{25}{3}$. Then write the result as a mixed number.

43. 0.225 45. 0.9375

Section 4.2 Percents and Decimals
(pages 158 and 159)

1. B 3. C

5. *Sample answer:* 0.11, 0.13, 0.19

7. 0.78 9. 0.185 11. 0.33 13. 0.4763 15. 1.66 17. 0.0006
19. 74% 21. 89% 23. 99% 25. 48.7% 27. 368% 29. 3.71%

31. The decimal point was moved in the wrong direction. $0.86 = 0.86 = 86\%$

33. 34% 35. $\frac{9}{25} = 0.36$ 37. $\frac{203}{1250} = 0.1624$ 39. 40%

41. a. $16.\overline{6}\%$ or $16\frac{2}{3}\%$ b. $\frac{5}{6}$ 43. $\frac{31}{100}$ 45. $4\frac{8}{25}$

47. $(7 + y) + 14 = (y + 7) + 14$ Comm. Prop. of Add.
 $\qquad\qquad\qquad = y + (7 + 14)$ Assoc. of Add.
 $\qquad\qquad\qquad = y + 21$ Add 7 and 14.

49. $6 \cdot c \cdot 5 = 6 \cdot 5 \cdot c$ Comm. Prop. of Mult.
 $\qquad\quad = 30c$ Multiply 6 and 5.

A20 Selected Answers

Section 4.3 — Comparing and Ordering Fractions, Decimals, and Percents (pages 164 and 165)

1.

Fraction	Decimal	Percent
$\frac{18}{25}$	0.72	72%
$\frac{17}{20}$	0.85	85%
$\frac{13}{50}$	0.26	26%
$\frac{31}{50}$	0.62	62%
$\frac{9}{20}$	0.45	45%

3. 0.04; 0.04 = 4%, but 40%, $\frac{2}{5}$, and 0.4 are all equal to 40%.

5. 20% 7. $\frac{13}{25}$ 9. 76%

11. 0.12 13. 140% 15. 80%

17.

19.

21.

23. Japan, Brazil, United States, China

25. 21%, $0.2\overline{1}$, $\frac{11}{50}$, $\frac{2}{9}$

27. D 29. C

31. Write the numbers as percents or decimals to make the ordering easier.

33. 4.8 35. 6.66 37. C

Section 4.4 — Finding the Percent of a Number (pages 172 and 173)

1. What is 20 multiplied by 30?; 600; 6

3. 12 5. 12.6 7. 7.2 9. 13

11. 21 13. 20.25 15. 24 17. 14

19. The percent was not written as a decimal or fraction before multiplying: 40% × 75 = 0.40 × 75 = 30.

21. 35.2 in. 23. 84 25. 94.5 27. 2.25

29. 4.2 31. = 33. > 35. 48 min

37. a. 432 in.2 b. 37.5%

39. *Sample answer:* Because 30% of *n* is equal to 2 times 15% of *n* and 45% of *n* is equal to 3 times 15% of *n*, you can write 30% of *n* = 2 × 12 = 24 and 45% of *n* = 3 × 12 = 36.

41. a. 243 points b. 97.2%

43. $1\frac{1}{2}$ 45. 10

Section 4.5 — Percents and Estimation
(pages 178 and 179)

1–3. Sample answers are given.

1. 25% **3.** 40%

5. 64% of 37 does not belong because it is about 24, and the others are about 8.

7–29. Sample answers are given.

7. 15 **9.** 4 **11.** 207 **13.** 12

15. 45 **17.** 1 **19.** about $2 **21. a.** 150 **b.** 280

23. 80 **25.** 30 **27.** 180 **29.** 70

31. sometimes true; It depends on how much the percent is rounded down compared to how much the number is rounded up.

33. *Sample answer:* Northeast: 100 children
Midwest: 50 children
South: 80 children
West: 160 children

35. Your friend's estimate is closer, because $\frac{1}{3} = 0.3\overline{3}$, which is closer to 33% (0.33) than 0.3.

37. $\frac{5}{6}$ **39.** $\frac{2}{7}$

Section 5.1 — Ratios
(pages 194 and 195)

1. 3 to 7, 3 : 7

3. *Sample answer:* Smith: vowels to consonants $= \frac{1}{4}$.

5. $\frac{6}{4}$, 6 : 4, 6 to 4 or $\frac{3}{2}$, 3 : 2, 3 to 2; There are 3 basketballs for every 2 soccer balls.

7. $\frac{3}{7}$, 3 : 7, 3 to 7; There are 3 shirts for every 7 pants.

9. $\frac{3}{4}$ **11.** $\frac{7}{5}$

13–19. Sample answers are given.

13. $\frac{1}{3}, \frac{2}{6}$ **15.** $\frac{1}{3}, \frac{22}{66}$ **17.** $\frac{18}{20}, \frac{27}{30}$ **19.** $\frac{3}{2}, \frac{6}{4}$

21. The numerator and denominator should be multiplied by the same number; $\frac{2}{3} = \frac{2 \times 2}{3 \times 2} = \frac{4}{6}$

23. 6 black pieces; The ratio of black to red is 3 : 5, so the ratio of black to all is 3 : 8. An equivalent ratio is 6 : 16.

25. It may be helpful to organize your results in a table.

27. 5 girls **29.** 3.29 **31.** 1.478

Section 5.2 Rates
(pages 200 and 201)

1. *Sample answer:* You walk at a rate of $\frac{2 \text{ blocks}}{1 \text{ min}}$, so you walk 12 blocks in 6 minutes.

3. $\frac{105 \text{ words}}{35 \text{ min}}$ 5. $\frac{36 \text{ students}}{16 \text{ computers}}$ 7. $\frac{3 \text{ baskets}}{45 \text{ min}}$ 9. $\frac{\$48}{6 \text{ tickets}}$

11. $\frac{\$7}{1 \text{ week}}$ 13. $\frac{45 \text{ mi}}{1 \text{ h}}$ 15. $\frac{140 \text{ kilobytes}}{1 \text{ sec}}$ 17. $\frac{72 \text{ mi}}{1 \text{ gal}}$

19. 6000 strikes per minute

21. The 15-pack of energy bars is the better buy because it has a lesser unit rate ($1.28 per bar) than the 10-pack ($1.35 per bar).

23. equivalent 25. not equivalent

27. Use Guess, Check, and Revise to find how many more items Homeroom B needs to collect.

29. 66

31 and 33. Sample answers are given.

31. $\frac{2}{6}, \frac{3}{9}$ 33. $\frac{4}{10}, \frac{8}{20}$ 35. B

Section 5.3 Solving Rate Problems
(pages 206 and 207)

1. Distance equals speed times time.

3. 105 mi 5. 276 ft 7. 5 in./sec

9. 58 ft/min 11. 12 m 13. 7240 ft

15. $6 17. 50 city blocks 19. $18.\overline{6}$ in./min

21. Time is given in two different units. Be sure to convert one of them.

23. 86 25. 141

Section 5.4 Mean
(pages 214 and 215)

1. No; Dividing the sum of the data by the number of data values to find the mean does not necessarily result in one of the data values.

3. Add the data values then divide by the number of data values.

5. The "average" could be 0 if you consider the most common value, or $\frac{8}{7}$ if you consider the mean.

7. 3 brothers and sisters 9. 16 visits 11. 3.45 minutes

13. *Sample answer:* 20, 21, 21, 21, 21, 22; 20, 20.5, 20.5, 21.5, 21.5, 22

Section 5.4 — Mean (continued)
(pages 214 and 215)

15. 3.9 inches; No, neither team has a height that is much shorter or taller than the other heights. So, you can say that the Tigers are taller than the Dolphins on average.

17. 11 pounds; Use the mean of each collection to work backwards to find the total weight of the collection. Divide the sum of the weights of the two collections by the total number of backpacks (20) to find the mean.

19. 30

21. 15.5

Section 5.5 — Median, Mode, and Range
(pages 220 and 221)

1. *Sample answer:* 1, 2, 3, 4, 5, 6

3. The greatest and least data values; The range is the difference of these values.

5. 25.4; No, the mode would be better because most of the data values are the mode of 30.

7. median: 7; mode: 3; range: 8

9. median: 92.5; mode: 94; range: 28

11. median: 17; mode: 12; range: 31

13. The data was not ordered from least to greatest; The median is 55.
49, 50, 51, 55, 58, 59, 63; The range is 63 − 49 = 14.

15. singing

17. *Sample answer:* An outlier increases the range of a data set because there is a wider spread between the greatest and least value.

19. mean: 14.9; median: 17.5; modes: 5, 19; range: 28

21. mean: 7.61; median: 7.42; no mode; range: 4.94

23. median: 11 years; mode: 12 years; range: 2 years

25. Find the mean, median, and mode with the original values and then again with the new value.

27. *Sample answer:* 18, 21, 28, 30, 35, 36

29. Outlier: 35; With outlier: 62.125; Without outlier: 66

Section 5.6

Analyzing Data Sets
(pages 226 and 227)

1. The mean is likely to be most affected unless the middle values are far apart, because then the median can be affected more by removing the outlier. The mode is unaffected unless the mode consists of outliers.

3. *Sample answer:* 190, 190, 190, 190, 192, 192, 195, 195

5.
Number	Tally	Total
6	I	1
7	II	2
8	III	3
9	IIII II	7
10	II	2

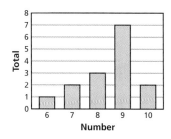

7. mean: 50; median: 40; mode: 95
 Sample answer: The mean is probably best, because the mode is the greatest value and the median is too far from the greater values.

9. mean: 110; median: 114.5; mode: 144
 Sample answer: Either the mean or median is best, because they are both at the middle of the data, while the mode is the greatest value.

11. *With Outlier* *Without Outlier*
 mean: 83 mean: 69.14
 median: 69.5 median: 67
 mode: 72 mode: 72

 The outlier makes the mean greater than most of the data, increases the median slightly, and does not change the mode.

13. *With Outlier* *Without Outlier*
 mean: 102 mean: 84
 median: 85 median: 85
 mode: 85 mode: 85

 The outlier makes the mean greater than all of the other values, but does not affect the median or mode.

15. *Answer should include, but is not limited to:* Prices of 10 cereals with 1 outlier; unit prices calculated.
 a. Calculation of mean, median, and mode and explanation of which is best.
 b. Outlier identified; mean, median, and mode calculated; measure most affected described.

17. a. No; The price is the mode, but it is the lowest price. Most cameras cost more.
 b. By advertising the lowest price, they are likely to draw more customers to the store.
 c. Knowing all the measures can help you to know whether the store has many models in your price range.

19. The mean, median, and mode increase by 2, but the range does not change.

21. 5.2 23. 55.2

Lesson 5.6b Data Displays
(pages 227A–227D)

1.
 Most of the fouls are 0, 1, 2, or 3.

3.

5.

7.

9. a. $\dfrac{3}{4}$ b. 80; The middle half of the heights vary by no more than 80 feet.

Section 6.1 Circles and Circumference
(pages 243–245)

1. The radius is one-half the diameter.

3. *Sample answer:* A lawn game has two circular targets with 28-inch diameters. You lost one. You want to use a length of wire to make a replacement.
 $C = \pi d \approx \dfrac{22}{7} \cdot 28 = 88$
 You need a piece of wire 88 inches long.

5. 14 mm 7. 12 cm 9. 1.6 ft
11. about 44 in. 13. about 75.36 m 15. about 7.71 ft
17. about 31.4 cm; about 62.8 cm 19. about 69.08 m; about 138.16 m

21. yes; Because
 $\dfrac{\text{circumference}}{\text{radius}} = \dfrac{2\pi r}{r}$
 $= \dfrac{2\pi \cancel{r}}{\cancel{r}}$
 $= 2\pi,$
 the ratio is the same for every circle.

23. a. small tire: about 127 rotations; large tire: about 38 rotations
 b. *Sample answer:* A bicycle with large wheels would allow you to travel farther with each rotation of the pedal.

25. 22 ft 27. 65 in.

Section 6.2

Perimeters of Composite Figures
(pages 250 and 251)

1. less than and equal to; The perimeter is *less than* when figures making up a composite figure share a common side (dashed line).

 The perimeter is *equal to* when the figures making up a composite figure share a common vertex.

3. 19.5 in. 5. 25.5 in. 7. 19 in. 9. 56 m
11. 30 cm 13. about 26.85 in. 15. about 36.84 ft

17. Remember to subtract the original garden side that you now cover up with the new portion of the flower garden when trying to add 15 feet to the perimeter.

19. Yes; *Sample answer:* By adding the triangle shown by the dashed line to the L-shaped figure, you *reduce* the perimeter.

21. 279.68 23. 205

Section 6.3

Areas of Circles
(pages 258 and 259)

1. Divide the diameter by 2 to get the radius. Then use the formula $A = \pi r^2$ to find the area.

3. about 254.34 mm^2 5. about 314 in.2 7. about 3.14 cm^2
9. about 2461.76 mm^2 11. about 200.96 in.2 13. about 628 cm^2
15. about 1.57 ft^2

17. What fraction of the circle is the dog's running area?

19. about 9.8125 in.2; The two regions are identical, so find one-half the area of the circle.

21. about 4.56 ft^2; Find the area of the shaded regions by subtracting the areas of both unshaded regions from the area of the quarter-circle containing them. The area of each unshaded region can be found by subtracting the area of the smaller shaded region from the semicircle. The area of the smaller shaded region can be found by drawing a square about the region.

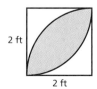

Subtract the area of a quarter-circle from the area of the square to find an unshaded area. Then subtract both unshaded areas from the square's area to find the shaded region's area.

23. 53 25. D

Section 6.4 — Areas of Composite Figures
(pages 264 and 265)

1. *Sample answer:* You could add the areas of an 8-inch × 4-inch rectangle and a triangle with a base of 6 inches and a height of 6 inches. Also you could add the area of a 2-inch × 4-inch rectangle to the area of a trapezoid with a height of 6 inches, and base lengths of 4 inches and 10 inches.

3. 28.5 in.^2 5. 25 in.^2 7. 25 in.^2 9. 132 cm^2

11. *Answer will include but is not limited to:* Tracings of a hand and foot on grid paper, estimates of the areas, and a statement of which is greater.

13. 23.5 in.^2 15. 24 m^2

17. Each envelope can be broken up into 5 smaller figures to find the area.

19. $y \div 6$ 21. $7w$

Section 7.1 — Writing Equations in One Variable
(pages 280 and 281)

1. An equation has an equal sign and an expression does not.

3. *Sample answer:* A number n subtracted from 28 is 5.

5. What is the circumference of a circular pond ripple with radius 3 feet? about 18.84 ft

7. $y - 9 = 8$ 9. $w \div 5 = 6$ 11. $5 = \frac{1}{4}c$ 13. $n - 9 = 27$

15. $6042 = 1780 + a$ 17. $16 = 3x$ 19. $326 = 12(14) + 6(5) + 16x$

21. It might be helpful to organize the given information visually.

23. 13 25. 28 27. C

Section 7.2 — Solving Equations Using Addition or Subtraction
(pages 287–289)

1. Substitute your solution back into the original equation and see if you obtain a true statement.

3. subtraction 5. So x is by itself. So the two sides remain equal.

7. yes 9. no 11. yes 13. $t = 9$

15. What number plus 5 equals 12?; $a = 7$ 17. 20 is what number minus 6?; $d = 26$

19. $z = 16$ 21. $p = 3$ 23. $h = 34$ 25. $q = 11$ 27. $x = \frac{7}{30}$ 29. $a = 11.8$

31. They must apply the same operations to both sides.

$$34 = y - 12$$
$$\underline{+\ 12 \quad +\ 12}$$
$$46 = y$$

33. $x - 8 = 16$; 24th floor

35. Subtraction Property of Equality; Subtract.; Addition Property of Zero

37. $k + 7 = 34$; $k = 27$ **39.** $93 = g + 58$; $g = 35$

41. $y = 15$ **43.** $v = 28$ **45.** $d = 54$

47. $x + 34 + 34 + 16 = 132$; 48 in.

49. Addition is commutative.

51. Begin by writing the characteristics of each problem.

53. a. $r = f + 1.25$; $g = 1.75 - 0.5$; $f = g + 1.5$
 $r = \$4$; $g = \$1.25$; $f = \$2.75$

 b. $5.25

55. 96 **57.** 5 **59.** B

Section 7.3 — Solving Equations Using Multiplication or Division (pages 294 and 295)

1. 12 **3.** $\dfrac{4x}{4} = \dfrac{24}{4}$ **5.** $8 \cdot 3 = (n \div 3) \cdot 3$

7. $s = 70$ **9.** $x = 16$ **11.** $a = 4$ **13.** $y = 10$

15. $x = 15$ **17.** $d = 78$ **19.** $b = 54$ **21.** $n = 2.56$

23. They should have multiplied by 4.
$$x \div 4 = 28$$
$$(x \div 4) \cdot 4 = 28 \cdot 4$$
$$x = 112$$

25. $528 = 88s$; 6 sec

27. 9 units **29.** 8 units **31.** 20 cards

33. length: 20 in.; width: 5 in. **35.** $t = 23$ **37.** $s = 16$

Section 7.4 — Solving Two-Step Equations (pages 301–303)

1. There are two different operations.

3. $11(x - 1) = 22$; The others all can be rewritten as $11x = 22$.

5. $z = 60$ **7.** $c = 6$ **9.** $b = 3$ **11.** $t = 418$ **13.** $t = 108$ **15.** $s = 5.4$

Section 7.4

Solving Two-Step Equations
(continued) (pages 301–303)

17. Subtraction should be the first step.

$4 = \dfrac{y}{8} + 1$

$3 = \dfrac{y}{8}$

$24 = y$

19. 16 lb

21. $c = 4$

23. $y = 3$

25. $a = 6$

27. $x = 4$

29. $s = 10$

31. $6d + 12d = 351$; 19.5 cm

33. $g = 16$

35. $x = 22$

37. $z = 103$

39. $s = 5$

41. $2x + 5 + 7 = 18$; 3 in.

43. $3x + 4x + 4x = 132$; 12 yd

45. 4.5 cups

47. The measurements are given in two different units.

49. $\dfrac{17}{20}$

51. $1\dfrac{7}{25}$

53. B

Section 7.5

Finding Dimensions of Plane Figures
(pages 310 and 311)

1. Square centimeters represents an area, not a length.

3. $42 = 7x$; 6 mm

5. $36 = \dfrac{1}{2}(8w)$; 9 in.

7. $40 = 2c + 24$; 8 m

9. $27 = 4a$; 6.75 ft

11. $15 = \dfrac{1}{2}(6 + 4)s$; 3 km

13. 0.76 m

15. 24 in.

17. Draw a diagram and label all of the dimensions you know.

19. $x = 16$

21. $h = 4.5$

23. B

Section 7.6

Finding Dimensions of Prisms
(page 316 and 317)

1. cubic units

3. 32 cubes

5. 27 cubes

7. 225 ft³

9. $1620 = h \cdot 9 \cdot 9$; 20 cm

11. $177{,}500 = 142 \cdot 10 \cdot h$; 125 mm

13. $936 = 3 \cdot 13 \cdot \ell$; 24 in.

15. 216 cubes

17. Try to find the length of the largest shell and use that to estimate the length and width of the shadow box.

19. **a.** *Sample answer:* 324 in.³

 b. no; The container only holds 196 cubic inches.

21. yes

23. no

Lesson 7.6b — Volume and Surface Area
(pages 317A–317D)

1. $\frac{24}{125}$ cm^3 3. $4\frac{1}{2}$ ft^3 5. 286 ft^2 7. 60 yd^2 9. 299.32 ft^2 11. 75 cm^2

Section 8.1 — Writing and Graphing Inequalities
(page 333–335)

1. Both phrases refer to numbers that are greater than a given number. The difference is that "greater than or equal to" includes the number itself, whereas "greater than" does not.

3. The graph of $x \leq 6$ has a closed circle at 6. The graph of $x < 6$ has an open circle at 6.

5. $k < 10$ 7. $z < \frac{3}{4}$ 9. $1 + y \leq 13$

11. yes 13. yes 15. no 17. B 19. D

21. $x < 1$; A number x is less than 1. 23. $x \geq -4$; A number x is at least -4.

25.

27.

29. (number line from $-\frac{4}{9}$ to $\frac{8}{9}$)

31. (number line from -6 to 6)

33. (number line from -3 to 3)

35. (number line from -1.8 to 1.4)

37. $x \geq 1$ means that 1 is also a solution, so a closed circle should be used.

39. **a.** $b \leq 3$; (number line 1 to 5)

 b. $\ell \geq 18$; (number line 17 to 21)

41. The cost of the necklace and another item should be less than or equal to $33.

43. sometimes; The only time this is not true is if $x = 5$.

45. $p \leq 375$ 47. $x = 9$ 49. $x = 28$ 51. D

Section 8.2 — Solving Inequalities Using Addition or Subtraction
(pages 340 and 341)

1. *Sample answer:* $x + 7 \geq 143$

3. By solving the inequality to obtain $x \leq 1$, the graph has a closed circle at 1 and an arrow pointing in the negative direction.

5. $x < 9$;

7. $5 \geq y$;

9. $6 > x$;

Section 8.2 Solving Inequalities Using Addition or Subtraction (continued) (pages 340 and 341)

11. $y < 106$;

13. $3 < x$;

15. $\frac{1}{4} \leq n$;

17. To solve the inequality, 9 should be added to both sides, not subtracted.

$$28 \geq t - 9$$
$$\underline{+9 \quad +9}$$
$$37 \geq t$$

19. $x + 18.99 \leq 24$; $x \leq \$5.01$

21. $x - 3 > 15$; $x > 18$

23. $11 > s$;

25. $34{,}280 + d + 1000 > 36{,}480$; $d > 1200$ dragonflies

27. The estimate for running a mile should be greater than 4 minutes, because the world record is under 4 minutes.

29. $t = 48$ **31.** $x = 9$ **33.** A

Section 8.3 Solving Inequalities Using Multiplication or Division (pages 348 and 349)

1. The solution to $2x \geq 10$ includes the solution to $2x = 10$, $x = 5$, and all other x values that are greater than 5.

3. Div. Prop. of Ineq.

5. Sample answer: $\frac{x}{2} \geq 4$, $2x \geq 16$

7. $n > 12$;

9. $c \geq 99$;

11. $x \geq 5$;

13. $p \leq 6$;

15. $x < 15$;

17. $v \leq 405$;

19. $q \leq 175$;

21. $x \geq 162$;

23. $8x < 168$; $x < 21$ ft

25. $8n < 72$; $n < 9$

27. $225 \geq 12w$; $18.75 \geq w$

29.

31. $80x > 2 \cdot 272$; $x > 6.8$ yards per play

33. *Sample answer:* The number of gallons of milk you can buy with $20.; The length of a park that has an area of at least 500 square feet.

35. yes; Because $a > b$ and $x > y$.

37. yes; Because $a > b$ and $x > y$.

39. $x = 5$

41. $x = 12$

Section 8.4 — Solving Two-Step Inequalities
(pages 354 and 355)

1. Add 9 to each side to get $4x$ by itself.

3. $s - 7 \leq 12$ does not belong because the solution is $s \leq 19$ and the solutions of the other three are all $s \leq 40$.

5. $t > 2$;

7. $s < 30$;

9. $a < 5$;

11. $y \leq 10$;

13. $c \leq 3$;

15. $k > 3.1$;

17. They should have subtracted first.
$$1 + \frac{a}{6} > 2$$
$$\frac{a}{6} > 1$$
$$a > 6$$

19. $10x + 240 \geq 400$; $x \geq 16$ m

21. $10x + 157 \geq 400$; $x \geq 24.3$ m

23. $22p + 180 \geq 250$; $p \geq 3.18$; Each student needs at least 4 more pledges.

25. $7x > 35$; $x > 5$ visits a year; An individual membership is better if you go 6 or more times a year. When you visit more than 9 times a year.

27. 2 ft

29. 6.5 m

Section 9.1 — Mapping Diagrams
(pages 370 and 371)

1. the first number; the second number

3. As each input increases by 1, the output increases by 4.

 Input Output
 1 → 4
 2 → 8
 3 → 12
 4 → 16
 5 → 20
 6 → 24

5. As each input increases by 1, the output increases by 7.

 Input Output
 1 → 0
 2 → 7
 3 → 14
 4 → 21
 5 → 28
 6 → 35

7. (1, 8), (3, 4), (5, 6), (7, 2)

9. Input Output
 1 → 3
 5 → 7
 8 → 10
 14 → 16

11.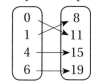

Section 9.1 — Mapping Diagrams (continued)
(pages 370 and 371)

13. The first number of each ordered pair should be an input and the second number should be the output that corresponds to the input.

15.

As each input increases by 2, the output increases by 2.

17.

As each input increases by 3, the output decreases by 10.

19. a. Input Output

1 → 10
2 → 18
3 → 24
4 → 28

b. The pattern is that for each input increase of 1, the output increases by $2 less than the previous increase. For each additional movie you buy, your cost per movie decreases by $1.

21. $x + 7 = 15$; $x = 8$

23. C

Section 9.2 — Functions as Words and Equations
(pages 376 and 377)

1. input variable: x; output variable: y

3. $y = 4x$

5. $y = x - 5$

7. $y = 6x$

9. $y = x + 11$

11. 42

13. 3.5

15. 13

17. no

19. no

21. yes

23. a. $d = 18s$ **b.** 540 ft

25. 5

27. The profit is equal to the revenue minus the expenses.

29. no; Many rectangles have the same perimeter but different areas.

31.

x	1	2	3
x + 7	8	9	10

33. C

A34 Selected Answers

Section 9.3 — Input-Output Tables
(pages 382 and 383)

1. Choose the inputs that represent the situation or show the pattern of the function. Pair each input in the table with its resulting output.

3.
Input, x	1	2	3	4
Output, y	6	7	8	9

5. $y = x + 3$

Input, x	0	1	2	3
Output, y	3	4	5	6

7. $y = x + 8$

9. $y = \dfrac{x}{3}$

11. Each output in the table is one-fourth of the input, but the equation would make each output four times each input; $y = \dfrac{x}{4}$

13.
Input, x	0	2	4	10	16	26
Output, y	4	5	6	9	12	17

15. *Sample answer:*

GMT, x	6:00	7:00	8:00	9:00	10:00
Eastern Standard Time, y	1:00	2:00	3:00	4:00	5:00

$y = x - 5$

17 and 19.

21. B

Section 9.4 — Graphs
(pages 390 and 391)

1. Make an input-output table. Plot the ordered pairs. Draw a line through the points.

3. Find points on the graph. Make a mapping diagram or input-output table to show the pattern. Use the pattern to write a function rule.

5.

7.

9.

11.

13.

15.

17.

19. C

21. Part (c) asks for the sale price, not the discount.

23. 17

25. 6

27. 19

29. C

Section 9.5 — Analyzing Graphs
(pages 396 and 397)

1. A function is called a linear function if its graph is a line.

3.
Radius, r	1	2	3	4
Diameter, d	2	4	6	8

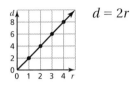

5. yes; The graph is a line. 7. no; The graph is *not* a line. 9. no; The graph is *not* a line.

11. yes; The graph is a line. 13. no; The graph is *not* a line.

15.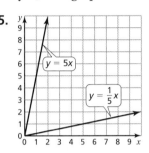

$y = 5x$; 5 is greater than $\frac{1}{5}$.

17.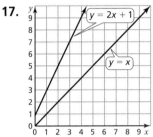

$y = 2x + 1$; 2 is greater than 1.

19.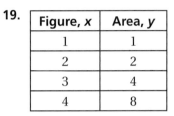

Figure, x	Area, y
1	1
2	2
3	4
4	8

no; The graph is *not* a line.

21. Airport 2; about 100 flights; *Sample answer:* From the graph, Airport 2 has about 350 flights each day and Airport 1 has about 250 flights each day.

23. $\frac{11}{25}$ 25. 0.802 27. C

Topic 1 — The Number Line
(pages 408 and 409)

1. 900 3. −5

5. (number line with points at −6 and 4) 7. (number line with points at −12 and 9)

9. a. *Sample answer:* Choosing 8, the opposite is −8.
 b. *Sample answer:* 8
 c. The opposite of the opposite of an integer is the integer.; Yes; *Sample answer:*

Case 1:

Case 2: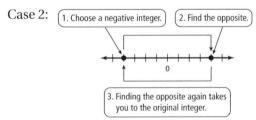

Case 3: Choose 0. The opposite of 0 is 0, so the opposite of the opposite of 0 is 0.

Topic 2 — Comparing Numbers
(pages 410 and 411)

1. $0 > -4$ **3.** $-8 < -7$ **5.** $-2 < -0.2$ **7.** $-210°C$

9. $-\frac{4}{7} < -\frac{1}{7}$ **11.** $-3, -1, 0, 2, 3$ **13.** $-3\frac{1}{4}, -\frac{3}{2}, -\frac{5}{4}, \frac{1}{2}, 1\frac{3}{4}$

Topic 3 — Absolute Value
(pages 412 and 413)

1. 5 **3.** 0 **5.** 12.6 **7.** $\frac{9}{2}$ **9.** $|-4| = 4$

11. $|-9.72| > 9.7$ **13.** Scientist B **15.** A; You owe more than $25, so debt > $25.

Topic 4 — The Coordinate Plane
(pages 414 and 415)

1–3.

1. Quadrant IV

3. Quadrant III

5. Quadrant III; In Quadrant I, both coordinates are positive. In Quadrant II, the x-coordinate is negative and the y-coordinate is positive. In Quadrant IV, the x-coordinate is positive and the y-coordinate is negative.

7.

9. 22 units

11. a. x-coordinates have different signs, y-coordinates are the same; x-coordinates are the same, y-coordinates have different signs

b. reflection in both axes; The signs of the coordinates are different.

Key Vocabulary Index

Mathematical terms are best understood when you see them used and defined *in context*. This index lists where you will find key vocabulary. A full glossary is available in your Record and Practice Journal and at *BigIdeasMath.com*.

absolute value, 412
algebraic expression, 4
box-and-whisker plot, 227C
center, 240
circle, 240
circumference, 241
compatible numbers, 47
composite figure, 248
coordinate plane, 414
cubic units, 314
diameter, 240
equation, 278
equivalent expressions, 16
equivalent ratios, 193
evaluate, 4
formula, 30
function, 368
function rule, 374
graph, 388
graph of an inequality, 332
histogram, 227B
inequality, 330
input, 368
input-output table, 380
integers, 408
interquartile range, 227D
inverse operations, 285
like terms, 300
line plot, 227A
linear function, 394
mapping diagram, 368
mean, 212
measure of central tendency, 218
median, 218
mode, 218
negative numbers, 408
net, 317B
numerical expression, 4

opposites, 408
ordered pair, 414
origin, 414
outlier, 213
output, 368
overestimate, 47
percent, 150
pi, 241
positive numbers, 408
quadrant, 414
quartiles, 227C
radius, 240
range, 219
rate, 198
ratio, 190, 192
reciprocals, 72
rectangular prism, 314
repeating decimal, 92
semicircle, 242
solution, 284
solution of an inequality, 331
solution set, 331
solve a formula, 30
square pyramid, 317D
square units, 317B
surface area, 317B
terminating decimal, 84
terms, 300
triangular prism, 317C
two-step equation, 298
underestimate, 47
unit cost, 199
unit rate, 198
volume, 314
x-axis, 414
x-coordinate, 414
y-axis, 414
y-coordinate, 414

Student Index

This student-friendly index will help you find vocabulary, key ideas, and concepts. It is easily accessible and designed to be a reference for you whether you are looking for a definition, real-life application, or help with avoiding common errors.

Absolute value, 412–413
 defined, 412
 practice problems, 412, 413
 real-life application, 413
 symbol for, 412
Addition
 Associative Property of, 15–16
 Commutative Property of, 14–17
 real-life application, 17
 of decimals, 106
 equations
 solving, 282–289
 inequalities
 solving, 336–341
 properties of, 14–19
Addition Property
 of Equality, 285
 of Inequality, 338
 of Zero, 17
Algebra
 algebraic expressions
 evaluating, 2–7
 writing, 8–13
 equations
 writing, 276–281
Algebraic expression(s), 2–13
 defined, 4
 error analysis, 6, 12
 evaluating, 2–7
 modeling, 2
 practice problems, 6
 vocabulary, 6
 writing, 8–13
Area
 of a circle, 254–259
 formula, 256
 practice problems, 258
 of composite figures, 29, 260–265
 practice problems, 264
 estimation of, 49, 110–111
 composite figures, 260–261
 formulas for, *See* Formulas
 of a parallelogram, 28–29
 practice problems, 310

 practice problems, 310
 of a semicircle, 257
 of a square, 28–29
 practice problems, 310
 of a trapezoid, 28, 31
 of a triangle, 28–29, 317C
 practice problems, 310
Associative Property, 15–19
 of Addition, 15–19
 practice problems, 18
 error analysis, 18
 of Multiplication, 15–19
Average, *See* Mean

Box-and-whisker plot, 227C–227D
 analyzing, 227D
 defined, 227C
 interquartile range, 227D
 making, 227C
 median of, 227C
 practice problems, 227C–227D
 quartiles, 227C

Center of a circle, defined, 240
Circle(s), 238–245
 area of, 254–259
 estimating, 254–255
 formula, 256
 center of, 240
 circumference of, 238, 241
 defined, 240
 diameter of, 238
 radius of, 238
 semicircle, 242
 area of, 257, 259
Circumference, 238–245
 defined, 241
 formula for, 241, 242
 solving, 306–311
 practice problems, 243, 310
 research, 245
Common errors
 expressions, 10, 17
 two-step equations, 298
Common factor, *See* Factor(s), common

Commutative Property
 of Addition, 14–19
 real-life application, 17
 of Multiplication, 14–19
Comparing, *See also* Ordering
 decimals, 160–165, 410
 fractions, 160–165, 411
 integers, 410
 real-life application, 411
 linear functions, 395
 means, 212
 percents, 160–165
 unit costs, 199
Compatible number(s), 176–177
 defined, 47
 estimating, 110
 decimals, 108–109
 using, 47–48
Composite figure(s)
 area of, 260–265
 practice problems, 264
 defined, 248
 perimeter of, 246–251
 practice problems, 250
Connections to math strands, *Throughout. For example, see:*
 algebra, 6, 7, 27, 55, 60, 67, 83, 122
 geometry, 18, 19, 27, 49, 110, 173, 195, 281, 288, 317, 348
Converting
 decimals
 to fractions, 84–89
 to mixed numbers, 87
 to percents, 155–156
 fractions
 to decimals, 86, 90–95
 to percents, 149–151
 mixed numbers, to decimals, 87
 percents
 to decimals, 154–156
 to fractions, 148–151
Coordinate plane(s)
 ordered pair, 414
 origin, 414
 plotting ordered pairs in, 414–415
 practice problems, 414–415

A40 Student Index

quadrants, 414
real-life application, 415
x-axis, 414
x-coordinate, 414
y-axis, 414
y-coordinate, 414
Cubic units, defined, 314

D

Data
 analyzing, 222–227
 interpreting, mean, median, mode, 212–215, 216–221, 224–227
 organizing, 222
Decimal(s)
 comparing, 160–165, 410
 dividing
 by a decimal, 132–137
 error analysis, 130, 136
 modeling, 126, 132
 practice problems, 110, 130, 136
 real-life applications, 129, 135
 by a whole number, 126–131
 estimating, 106–111
 practice problems, 110
 products, 107, 108
 quotients, 108
 real-life application, 109
 research, 111
 sums, 106
 using compatible numbers, 108
 as fractions, 84–89
 error analysis, 88
 practice problems, 88
 real-life application, 93
 fractions as, 90–95
 real-life applications, 162–163
 as mixed numbers, 87
 practice problems, 88
 multiplying, 118–123
 error analysis, 116, 122
 by powers of ten, 112, 118–119
 practice problems, 110, 116, 122
 real-life application, 121
 using powers of ten, 112, 118–119
 by a whole number, 112–117, 118

ordering, 127, 160–165
 as percents
 error analysis, 158
 real-life application, 157
 percents and, 154–159
 modeling, 154–156
 percents as, practice problems, 158
 repeating, 90–92, 94–95
 defined, 92
 research, 159
 terminating, 84–86, 92, 95
 writing as fractions, 84, 86
 writing as mixed numbers, 87
 writing as percents, 155–156
 writing fractions as, 90–91
 writing percents as, 154–156
Definition and example chart, 208
Diameter
 of a circle, 238
 defined, 240
 formula, 240
 practice problems, 243
Distance formula, 204, 293
Distributive Property, 22–27
 practice problems, 26
 real-life application, 25
 using, 64
 vocabulary, 26
Dividend, 128, 129
Division
 of decimals
 by a decimal, 132–137
 error analysis, 130, 136
 modeling, 126, 132–133
 practice problems, 110, 130
 real-life application, 135
 by a whole number, 126–131
 dividend, 128, 129
 divisor, 128
 equations, solving, 290–295
 estimating
 practice problems, 110
 quotients, 45
 of fractions, 48, 70–77
 modeling, 70–71
 with fractions, 70–77
 inequalities, solving, 344–349
 of mixed numbers, 78–83
 practice problems, 82
 quotients
 error analysis, 82
 estimating decimal product, 108
 vocabulary, 8, 10

Division Property
 of Equality, 293
 of Inequality, 346
Divisor, 128

E

Equation(s)
 addition, solving, 282–289
 defined, 278
 dimensions, finding, 306–311
 division, solving, 290–295
 functions as, 372–377
 modeling, 282, 290–291
 multiplication
 solving, 290–295
 practice problems, 280, 316
 solving
 addition, 282–289
 dimensions, 306–311
 division, 290–295
 error analysis, 287, 294
 multiplication, 290–295
 practice problems, 287, 294
 real-life application, 286
 subtraction, 282–289
 two-step, 296–303
 volume of a prism, 312–317
 subtraction, solving, 282–289
 volume
 practice problems, 316
 solving, 312–317
 writing, 276–281
 error analysis, 280
Equivalent expressions, *See also* Expression(s)
 defined, 16
Equivalent ratios, *See also* Ratio(s)
 defined, 193
Error analysis, *Throughout. For example, see:*
 algebraic expressions, 6, 12, 26
 Associative Property of Addition, 18
 Commutative Property of Addition, 18
 decimals
 dividing, 130, 136
 as fractions, 88
 multiplying, 116, 122
 as percents, 158
 Distributive Property, 26
 division
 of decimals, 130, 136
 of fractions, 75

Student Index **A41**

equations
 solving, 287, 294
fractions
 dividing, 75, 94
 estimating, 48
 multiplying, 54, 60, 66
 percents and, 152
functions, 376
 graphing, 390
 input-output tables, 382
 mapping diagram, 370
inequalities, 334
 solving, 340
input-output tables, 382
median and range, 220
multiplying
 decimals and whole numbers, 116
 fractions, 60, 66
 fractions and whole numbers, 54
 whole numbers and fractions, 54
percents
 fractions and, 152
 of a number, 172
perimeters, 250
plane figures, dimensions of, 310
quotients, 82
ratios, equivalent, 195
two-step equations, 301
two-step inequalities, 354
Estimation
 area
 of a circle, 254–255
 of a composite figure, 260–261
 of area, 110–111
 with decimals, 106–111
 with fractions, 44–49
 modeling, 44–45
 with mixed numbers, 46
 overestimate, 47
 with percents, 174–179
 practice problems, 178
 real-life application, 177
 research, 153
 of products, 46, 114
 of quotients, 45
 research, 111
 by rounding, 46, 48, 108
 underestimate, 47
 using compatible numbers, 47–48, 108–109

Evaluating
 algebraic expressions, defined, 4
 expressions, 4–5, 121
Example and non-example chart, 304
Exponents, *See* Powers
Expression(s)
 equivalent
 defined, 16
 error analysis, 26
 evaluating, 121
 numerical, 10
 operations and, 5
 real-life application, 11
 variables and, 4
 writing, 8–13

F

Factor(s), common
 defined, 52
 multiplying fractions with, 58
Formulas, 28–33
 area, 28–31
 of a circle, 256
 of a parallelogram, 28–29
 of a rectangle, 30
 solving, 28–32, 306–311
 of a square, 28
 of a trapezoid, 28, 31
 of a triangle, 28–29, 37
 circumference of a circle, 241
 practice problems, 243
 solving, 306–308, 310–311
 defined, 30
 diameter of a circle, 240
 distance, 204, 293
 perimeter, 127
 solving, 306–311
 radius of a circle, 240
 speed, 204
 volume
 of a prism, 312–317
 of a rectangular prism, 32
Four square, 384
Fraction(s)
 comparing, 160–165, 411
 as decimals, 90–95
 real-life application, 93
 decimals and, 84–89
 decimals as, 94
 error analysis, 88
 practice problems, 88
 real-life applications, 162–163

 dividing, 70–77
 error analysis, 75, 94
 modeling, 70–71
 practice problems, 75
 estimating, 44–49
 differences, 44
 practice problems, 48
 products, 45, 46
 quotients, 45
 sums, 44
 modeling, 44–45
 division, 70–71
 multiplying, 56–61
 with common factors, 58
 error analysis, 54, 60, 66
 by a mixed number, 62, 64
 practice problems, 54
 real-life application, 59
 by a whole number, 50–55
 ordering, 160–165
 percents and, 148–153
 error analysis, 152
 modeling, 148–150
 practice problems, 152, 158, 164
 real-life application, 151
 writing as decimals, 90–93
 writing decimals as, 84–89
 writing percents as, 148–151
Frequency table, 227B
Function(s)
 defined, 368
 describing
 as equation, 372–377
 error analysis, 376
 practice problems, 376
 real-life application, 375
 in words, 372–373
 graphing, 388–391
 error analysis, 390
 practice problems, 390
 reading, 389
 input-output tables, 378–383
 defined, 380
 error analysis, 382
 practice problems, 382
 linear, 394–397
 defined, 394
 practice problems, 396
 with mapping diagrams, 366–371
 defined, 368
 error analysis, 370
 practice problems, 370
 real-life application, 369
 research, 371

Function rule, defined, 374
Function tables, *See* Input-output table(s)

Geometry
 area, *See* Area
 circles, *See* Circle(s)
 perimeter, *See* Perimeter
 rectangular prism, volume of, 19, 32, 317
 triangle, area of, 28–29, 32, 37
Graphic organizer(s)
 definition and example chart, 208
 example and non-example chart, 304
 four square, 384
 information frame, 20
 notetaking organizer, 68
 process diagram, 124
 summary triangle, 166
 word magnet, 252
 Y chart, 342
Graphing
 error analysis, 390
 inequalities, 328–335
 error analysis, 334
 practice problems, 333, 340, 348
 practice problems, 390
 reading, 389
Graphs, 386–397
 analyzing, 392–397
 bar, 222–223
 of functions, 388–397
 error analysis, 390
 practice problems, 390
 reading, 389
 of inequalities, 328–335, *See also* Inequalities
 defined, 332
 interpreting, 386

Height,
 of a rectangular prism, 315
Histogram, 227B
 defined, 227B
 frequency table, 227B
 practice problems, 227B

Inequalities
 defined, 330
 graphing, 328–335, 344
 defined, 332
 error analysis, 334
 practice problems, 333, 340, 348
 real-life applications, 332, 339, 347
 solution of, 331
 defined, 331
 solution sets, defined, 331
 solving
 addition, 336–341
 division, 344–349
 error analysis, 340
 multiplication, 344–349
 real-life application, 347
 subtraction, 336–341
 symbols, 329–330, 410
 triangle, 336
 two-step, *See* Two-step inequalities
 writing, 328–335
 practice problems, 333
Information frame, 20
Input, 368
Input-output table(s), 378–383
 defined, 380
 error analysis, 382
 practice problems, 382, 396
Integer(s), 408, 409
 comparing, 410
 graphing, 409
 practice problems, 409
 real-life application, 409
Interquartile range, defined, 227D
Inverse operations, 285, 291, 292, 296, 346
 defined, 285

Like terms
 defined, 300
 to solve equations, 300
Line plot
 defined, 227A
 practice problems, 227A
Linear function(s), 394–397
 defined, 394
 practice problems, 396

Mapping diagram(s), 368
Mean, 210–215
 defined, 212
 error analysis, 220
 outlier, 213, 224–225
 practice problems, 214
Meaning of a word
 associate, 15
 commute, 14
 distribute, 22
 invert, 72
 opposite, 408
 percent, 148
 terminate, 84
Measures of central tendency, 224
 defined, 218
Median, 216–221
 of a box-and-whisker plot, 227C
 defined, 218
 error analysis, 220
Mental math, *Throughout. For example, see:*
 Addition
 Associative Property of, 15
 Commutative Property of, 15
 Distributive Property, 22–24
 estimating, percents, 175
 finding products, 117
 multiplying decimals and whole numbers, 115, 117
 percents, 169, 171
 estimating, 175
 using, 22–24
Mixed number(s)
 as decimals, 93
 decimals as, 87–88
 dividing, 78–83
 by a fraction, 80
 practice problems, 82
 real-life application, 81
 estimating with, 46
 multiplying, 62–67
 with fractions, 62, 64
 real-life application, 65
Mode, 216–221
 defined, 218
Modeling
 algebraic expressions, 2
 decimals
 division of, 126, 132–133
 equations, 282–283, 290–291
 fractions, 44–45
 division of, 70–71

estimation of, 44–45
multiplication of, 59
percent of, 148–149
whole numbers
division of, 126
Multiplication
Associative Property of, 15–17
Commutative Property of, 14–16
of decimals, 107–108, 118–123
error analysis, 122
by powers of ten, 112, 118–119
practice problems, 110, 122
real-life application, 121
using powers of ten, 112, 118–119
and whole numbers, 112–117
equations, solving, 290–295
of fractions, 45, 48, 56–61
error analysis, 54, 60, 66
and mixed numbers, 64–65
practice problems, 54
real-life applications, 53, 59
and whole numbers, 50–55
fractions and whole numbers, 50–55
inequalities, solving, 344–349
of mixed numbers, 62–67, 65
and fractions, 64–65
products
estimating, 46, 114
finding, 117
properties of, 14–19
using Distributive Property, 22–27
vocabulary, 8, 10
of whole numbers
and fractions, 50–55
Multiplication Property
of Equality, 292
of Inequality, 346
of One, 17
of Zero, 17

N

Negative numbers, 408–409
defined, 408
Net
defined, 317B
of a rectangular prism, 317B
of a square pyramid, 317D
of a triangular prism, 317C
Notetaking organizer, 68
Number(s)
compatible, 47, 176–177

defined, 47
using, 47–48, 108–109
finding percent of, 168–173
negative, 408–409
opposites, 408–409
ordering, 160–165
real-life applications, 162–163
positive, 408–409
Number line(s), 408–409
integers on, 408–409
opposites, 408
real-life application, 409
Number sense, *Throughout. For example, see:*
Commutative Property of Addition, 18
composite figure, 250
decimal places, 116
decimals, 122
estimating, 136
dividends, 130
division, of fractions, 48
divisors, 130
equations, 294
fractions, 77
as decimals, 94
dividing, 48, 82
multiplying, 54, 60
mean, median, and mode, 220, 226
order of operations, 6
outlier, 214
percents and fractions, 164
products, 116
quotients, 110, 130
writing expressions, 13
Numerical expression(s)
defined, 4
evaluating, 53

O

Open-ended, *Throughout. For example, see:*
algebraic expressions, 18, 26
area
estimating, 264
Associative Property of Addition, 26
circumference, 243
composite figures, 250
decimals
as fractions, 88
fractions as, 95

as mixed numbers, 88
percents and, 158
dimensions of a plane figure, 310
Distributive Property, 26
fractions
as decimals, 95
multiplying, 60, 61
reciprocal, 75
fractions and percents, 152
functions, 370
describing, 390
inequalities, 340, 349
mean, 215
mode, 226
multiplication,
of fractions, 61
of mixed numbers, 66
ratios, 194
writing equations, 280
Opposites, *See also* Number line(s)
defined, 408
Order of operations, 14, 15, 74, 81
expressions, 53, 130
Ordered pair(s), 368
defined, 414
plotting in a coordinate plane, 414–415
practice problems, 414–415
practice problems, 414–415
real-life application, 415
Ordering, 160–165
decimals, 160–165
fractions, 160–165
numbers, 160–165
percents, 160–165
real-life applications, 162–163
Origin, defined, 414
Outlier, defined, 213
Output, 368
Overestimate, defined, 47

P

Parallelogram(s)
area of, 28–29
practice problems, 310
Patterns
of perimeters, 246
Percent(s)
comparing, 160–165
converting to fractions, 176
decimals and, 154–159
modeling, 154–156

decimals as
 error analysis, 158
 practice problems, 158
 real-life application, 157
 defined, 148, 150
 estimating, 174–179
 practice problems, 178
 real-life application, 177
 research, 153
 fractions and, 148–153
 error analysis, 152
 modeling, 148–150
 practice problems, 152, 164
 real-life application, 151
 of a number
 error analysis, 172
 practice problems, 172
 ordering, 160–165
 research, 159
Perimeter, 242, 244
 of composite figures, 246–251
 practice problems, 250
 error analysis, 250
 formula, 127
 solving, 306–311
 practice problems, 310–311
Pi, 238–239
 defined, 241
 estimating, 241
Plane figure(s)
 error analysis, 310
 finding dimensions of, 306–311
Positive numbers, 408–409
 defined, 408
Powers
 of ten, 112
 multiplying decimals, 112, 118, 119
Practice problems, *Throughout. For example, see:*
 algebraic expressions, 6
 area, 310
 of a composite figure, 264
 Associative Property of Addition, 18
 circle
 area of, 258
 circumference of, 243, 310
 diameter of, 243
 radius of, 243
 coordinate plane, 414–415
 decimals
 dividing, 110, 130, 136
 estimating, 110
 as fractions, 88, 94

 multiplying, 110, 116, 122
 as percents, 158
 Distributive Property, 26
 division
 of decimals, 130
 of mixed numbers, 82
 equations, 280, 316
 solving, 287, 294
 estimating
 percents, 178
 perimeter of a composite figure, 250
 formulas
 distance, 206
 speed, 206
 fractions
 decimals as, 94
 dividing, 48, 75
 multiplying, 48, 54, 60
 and whole numbers, 54
 functions
 describing, 376
 graphing, 390
 input-output tables, 382
 linear, 396
 mapping diagrams, 370
 inequalities, 333
 graphing, 340, 348
 integers, graphing, 409
 linear functions, 396
 mean, 214
 median and mode, 226
 median and range, 220
 mixed numbers
 dividing, 82
 multiplying, 66
 multiplication
 of decimals, 122
 of fractions, 54, 60
 of mixed numbers, 66
 of whole numbers, 54
 percents
 as decimals, 158
 estimating, 178
 fractions and, 164
 of a number, 172
 perimeter, 310
 rates, 200, 206
 ratios, 194
 reciprocals, 75
 solving, equations, 287, 294
 two-step equations, 301
 two-step inequalities, 354
 volume of rectangular prisms, 316

 whole numbers
 multiplying, 54
Prism(s)
 defined, 314
 finding dimensions of, 312–317
 rectangular, 312–317
 net of, 317B
 research, 317
 triangular, 317C
 net of, 317C
 volume of, 32
Process diagram, 124
Products, *See* Multiplication
Properties
 Addition Property
 of Equality, 285
 of Inequality, 338
 of Zero, 17
 Associative Property
 of Addition, 15–16
 of Multiplication, 15–17
 Commutative Property
 of Addition, 14–17
 of Multiplication, 14–16
 Distributive Property, 22–27
 Division Property
 of Equality, 293
 of Inequality, 346
 Multiplication Property
 of Equality, 292
 of Inequality, 346
 of One, 17
 of Zero, 17
 Subtraction Property
 of Equality, 285
 of Inequality, 338
 Pyramid, square, 317D
 net of, 317D

Q

Quadrants, defined, 414
Quartile(s), defined, 227C
Quotients, *See* Division

R

Radius, 238
 of a circle, formula, 240
 defined, 240
 practice problems, 243
Range, 216–221
 defined, 219
 error analysis, 220

Student Index **A45**

Rate(s), 196–201, 202–207
 defined, 198
 practice problems, 200, 206
 research, 203
 unit, 197–201, 205
 cost, 199–200
 defined, 198
 reading, 198
Ratio(s), 190–195
 comparing, 193
 defined, 190, 192
 equivalent, 191
 defined, 193
 error analysis, 195
 writing, 191–193
 practice problems, 194
 reading, 193
 simplifying, 192
Reading
 algebraic expressions, 2–3
 graphing functions, 389
 inequalities, 339
 mean, 212
 negative sign, 409
 range of a panther, 47
 ratio(s), 193
 repeating decimals, 92
 subtraction, 279
 symbols
 approximately equal to, 46
 inequality, 329–331
 not equal to, 284
 unit rates, 198
Real-life applications, *Throughout.*
 For example, see:
 algebraic equations, 279
 area of a triangle, 308, 309
 Commutative Property of
 Addition, 17
 comparing fractions, decimals,
 and percents, 162, 163, 171
 coordinate plane, 415
 decimals, 93
 dividing, 129, 135
 estimating with, 109
 multiplying, 121
 percents as, 157
 Distributive Property, 25
 division of mixed numbers, 81
 equations, solving, 286
 estimating, 177
 expressions, 5
 fractions
 multiplying, 59
 writing as decimals, 93

 functions
 describing, 375
 mapping diagrams, 369
 inequalities, 332, 339
 solving, 347
 mixed numbers,
 dividing, 81
 multiplying, 65
 multiplying
 decimals, 121
 fractions, 59
 mixed numbers, 65
 number line, 409
 percents, 151
 as decimals, 157
 estimating, 177
 two-step equations, 299
 two-step inequalities, 353
 writing expressions, 11
 writing fractions as decimals, 93
Reasoning, *Throughout. For*
 example, see:
 area, 264
 area of a trapezoid, 264
 circles, 259
 division
 of fractions, 76
 equations, solving, 288
 expressions, simplifying, 27
 formulas
 evaluating, 32
 of a parallelogram, 33
 fractions, 55
 dividing, 76
 multiplying, 60
 function rule, 376
 functions, graphing, 391
 inductive, 51, 57, 85, 90
 inequalities, 348
 mean, 214
 mean and median, 220
 multiplying
 decimals, 123
 decimals and whole numbers,
 112
 mixed numbers, 67
 percents, 172–173
 quotients, 75
 range, 220
 two-step inequalities, 354
Reciprocals, 72–77
 defined, 72
 practice problems, 75
Rectangle(s)
 area of, 30

Rectangular prisms, 312–317
 defined, 314
 net of, 317B
Repeating decimals, 90–92, 94–95
 defined, 92
Right angle(s)
 symbol for, 31
Rounding
 mixed numbers, 46
 percents, 175, 178
 practice problems, 110
 products, 46

S

Semicircle
 area of, 257, 259
 defined, 242
Simplifying
 expressions, 16–17, 24
Solution
 of an equation
 defined, 284
 of an inequality
 defined, 331
Solution set, defined, 331
Solving
 equations, *See* Equation(s)
 formulas, 28, 30–33
 defined, 30
 inequalities, *See* Inequalities
Speed formula, 204
Square(s)
 area of, 28
Square pyramids, 317D
 net of, 317D
Square units, 317B
Standardized test practice
 algebraic equations, 278
 algebraic expressions, 11
 circumference, 242
 circumference of a circle, 256
 comparing fractions, decimals,
 and percents, 170
 Distributive Property, 25
 expressions, 25
 fractions, multiplying, 59
 functions, 381
 input-output tables, 381
 mean, 212, 225
 multiplying
 decimals and whole numbers,
 115
 fractions, 53, 59
 fractions and whole numbers,
 53

percents as decimals, 157
range, 219
two-step equations, 299
Study tips
 algebraic expressions, 4
 box-and-whisker plots, 227C
 compatible numbers, 108
 decimal points, 156
 decimals
 dividing, 134, 135
 multiplying, 115
 ordering, 162
 equations
 solution of, 375
 solving, 285, 293
 estimating pi, 241
 inequalities
 solutions, 338
 solving, 338
 measures of central tendency, 218
 mode, 218
 multiplying
 decimals, 115
 fractions, 58
 by zero, 72
 percents
 ordering, 162
 reciprocals, 72
Subtraction
 equations
 solving, 282–289
 inequalities
 solving, 336–341
 reading, 279
Subtraction Property
 of Equality, 285
 of Inequality, 338
Summary triangle, 166
Surface Area
 defined, 317B
 net
 of a rectangular prism, 317B
 of a square pyramid, 317D
 of a triangular prism, 317C
 of a rectangular prism, 317B
 practice problems, 317B
 of a square pyramid, 317D
 practice problems, 317D
 of a triangular prism, 317C
 practice problems, 317C
Symbol(s)
 approximately equal, 46
 of inequality, 329–330, 331, 339, 410

not equal, 284
 for right angle, 31

Terminating decimals, 84–86, 92, 95
 defined, 84
 as fractions, 84, 86
Terms, *See also* Like terms,
 defined, 300
Trapezoid(s)
 area of, 28, 31
Triangle(s)
 area of, 28–29, 37
 practice problems, 310
 real-life application, 309
Triangle inequality, 336
Triangular prisms, 317C
 net of, 317C
Two-step equation(s)
 defined, 298
 error analysis, 301
 practice problems, 301
 real-life application, 299
 solving, 296–303
Two-step inequalities
 error analysis, 354
 practice problems, 354
 real-life application, 353
 solving, 350–355

Underestimate, defined, 47
Unit cost, 199–200
 defined, 199
Unit rate, *See also* Rate(s),
 defined, 198

Variable(s), 276–281
Volume
 of a rectangular prism, 32, 312–317
 defined, 314
 practice problems, 316

Whole number(s)
 dividing
 by a decimal, 126–131

 by a fraction, 52, 73
 modeling, 126
 multiplying
 by a decimal, 112–117
 error analysis, 54
 by a fraction, 50–55
 practice problems, 54
Writing, *Throughout. For example, see:*
 algebraic expressions, 10
 decimals, as fractions, 84
 Distributive Property, 26
 equations, 287
 expressions, 32
 formulas, 32
 fractions, decimals as, 84
 inequalities
 graphing, 333, 340
 solving, 349
 like terms, 301
 linear functions, 396
 mean and median, 220
 multiplication
 of fractions and whole numbers, 54
 of mixed numbers, 66
 plane figures, 310
 properties, of equality, 294
 rates, 200, 206
 ratios, 190–193
 unit rates, 200
Word magnet, 252

***x*-axis,** defined, 414
***x*-coordinate,** defined, 414

Y chart, 342
***y*-axis,** defined, 414
***y*-coordinate,** defined, 414

Zero
 Addition Property of, 17
 Multiplication Property of, 17, 72
 as a place holder, 115, 129, 135, 156

Additional Answers

Chapter 1

Section 1.1
Vocabulary and Concept Check

1.

Algebraic Expression	Numbers	Variables	Operations
$x - 8$	8	x	Subtraction
$3w + 9$	3, 9	w	Multiplication and addition
$6y - 12$	6, 12	y	Multiplication and subtraction

Fair Game Review

37–40.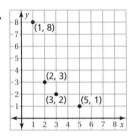

Section 1.2
Record and Practice Journal

4. *Sample answer:*

+	−
added to, plus, increased by	subtracted from, minus, less than

×	÷
multiplied by, product of, times	divided by, quotient of

Practice and Problem Solving

31. a.

Game	1	2	3	4	5
Cost	$5	$8	$11	$14	$17

b.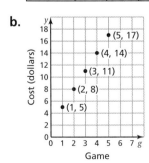

c. $2 + 3g$ d. $26

32. a. $5a - 8$

b. $2f + 25$

c. Florida has 67 counties. Georgia has 159 counties.

Section 1.3
Record and Practice Journal

5. *Sample answer:* You can use the Commutative Property of Addition to reorder the numbers: $11 + 9 + 7 + 13 + 12 + 8$. Then you can use the Associative Property of Addition to group the numbers and add:
$(11 + 9) + (7 + 13) + (12 + 8)$
$= 20 + 20 + 20 = 60$. Because each group of numbers adds to 20, it is easier to add the sum in your head.

Practice and Problem Solving

17. $(0 + a) + 8 = a + 8$ Add. Prop. of Zero

18. $9 \cdot c \cdot 4 = 9 \cdot 4 \cdot c$ Comm. Prop. of Mult.
$= (9 \cdot 4) \cdot c$ Assoc. Prop. of Mult.
$= 36c$ Multiply 9 and 4.

19. $(18.6 \cdot d) \cdot 1 = 18.6 \cdot (d \cdot 1)$ Assoc. Prop. of Mult.
$= 18.6 d$ Mult. Prop. of One

20. $(3k + 4.2) + 8.6 =$ Assoc. Prop. of Add.
$3k + (4.2 + 8.6)$
$= 3k + 12.8$ Add 4.2 and 8.6.

21. $(2.4 + 4n) + 9 =$ Comm. Prop. of Add.
$(4n + 2.4) + 9$
$= 4n + (2.4 + 9)$ Assoc. Prop. of Add.
$= 4n + 11.4$ Add 2.4 and 9.

22. $(3s) \cdot 8 = (s \cdot 3) \cdot 8$ Comm. Prop. of Mult.
$= s \cdot (3 \cdot 8)$ Assoc. Prop. of Mult.
$= s \cdot 24$ Multiply 3 and 8.
$= 24s$ Comm. Prop. of Mult.

23. $z \cdot 0 \cdot 12 = (z \cdot 0) \cdot 12$ Assoc. Prop. of Mult.
$= 0 \cdot 12$ Mult. Prop. of Zero
$= 0$ Mult. Prop. of Zero

24. $x + 16$

25. a. x represents the cost of a box of cookies.

b. $120x$

27. $7 + (x + 5) = x + 12$

28. $8 \cdot (y \cdot 9) = 72y$

29. $(7 \cdot 2) \cdot y$ 30. $13.2 \cdot (1 \cdot x)$

31. $(17 + 6) + 2x$ **32.** $2 + c$

33. $w \cdot 16$

34. a. $x + 37$

 b. In the expression $37(14) + 10x$, 37 represents the total number of hats you sold, 14 represents the price per hat you charged, 10 represents the price per hat your friend charged, and x represents the number of hats sold by your friend. Because $14 > $10, your friend was selling the hats at a discounted price.

 c. $x \le 51$

1.1–1.3 Quiz

13. $6 \cdot (10 \cdot k) = (6 \cdot 10) \cdot k$ Assoc. Prop. of Mult.

 $= 60k$ Multiply 6 and 10.

14. $68

15. a. $33 + 0.9m$

 b. $303

16. $2x + 50$

Section 1.4

Practice and Problem Solving

44. $7(x + 3) + 8 \cdot x + 3 \cdot x + 8 - 9 = 2(9x + 10)$

Section 1.5

Record and Practice Journal

3. a. *Sample answer:*

$A = 4 + 6 + 4 + 3(2)$
$= 20$ square units

 b. *Sample answer:*

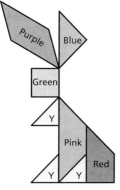

$A = 4 + 8 + 4 + 8 + 3(2) + 6$
$= 36$ square units

 c. *Sample answer:*

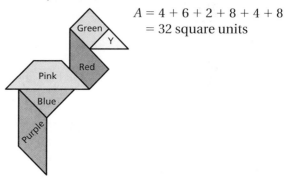
$A = 4 + 6 + 2 + 8 + 4 + 8$
$= 32$ square units

5. *Sample answer:*

Note that triangle 1 and triangle 2 have equal area. Therefore the area of the rectangle equals the area of the parallelogram which is $A = bh$.

6. *Sample answer:*

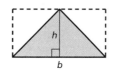

Note that the area of the triangle is half of the area of the rectangle. So $A = bh \div 2$.

Chapter 1 Review

20. $5.3 + (w + 1.2) = 5.3 + (1.2 + w)$ Comm. Prop. of Add.

 $= (5.3 + 1.2) + w$ Assoc. Prop. of Add.

 $= 6.5 + w$ Add 5.3 and 1.2.

21. $(0 + t) + 9 = t + 9$ Add. Prop. of Zero

Chapter 1 Test

11. $3(15w) = (3 \cdot 15)w$ Assoc. Prop. of Mult.

 $= 45w$ Multiply 3 and 15.

12. $40 + (x + 12) = 40 + (12 + x)$ Comm. Prop. of Add.

 $= (40 + 12) + x$ Assoc. Prop. of Add.

 $= 52 + x$ Add 40 and 12.

Chapter 2

Section 2.1
Record and Practice Journal

1. a. $\frac{5}{8}$;
 b. $\frac{3}{4}$; $\frac{5}{12}$; $\frac{4}{6}$;
 c. $\frac{3}{5}$; $\frac{4}{5}$; $\frac{7}{10}$;
 d. $\frac{1}{2}$; $\frac{3}{0}$; $\frac{3}{4}$;
 e. $\frac{7}{8}$; $\frac{1}{8}$; $\frac{1}{4}$;

Section 2.2
Practice and Problem Solving

41. yes; If you have more money than your friend, then $\frac{1}{3}$ of your money could be greater than $\frac{1}{2}$ of your friend's money.

Section 2.3
Practice and Problem Solving

42. $\frac{5}{6} = \left(\frac{5}{6} \times \frac{7}{7}\right)$; Because $\frac{7}{7} = 1$, by the Multiplication Property of One, the two expressions are equal.

Section 2.4
Practice and Problem Solving

25. You must first rewrite the mixed numbers as improper fractions and then multiply.
$$2\frac{1}{2} \times 7\frac{4}{5} = \frac{5}{2} \times \frac{39}{5}$$
$$= \frac{\cancel{5}^1 \times 39}{2 \times \cancel{5}_1}$$
$$= \frac{39}{2}, \text{ or } 19\frac{1}{2}$$

34. $5\frac{4}{9}$

35. $26\frac{2}{5}$

36. 462 in.²

37. No; Positive mixed numbers are greater than 1 and the product of two numbers greater than 1 is always greater than 1.

Chapter 4

Section 4.2
Practice and Problem Solving

38. a. car: 0.2, school bus: 0.48, bicycle: 0.08
 b. car: $\frac{1}{5}$, school bus: $\frac{12}{25}$, bicycle: $\frac{2}{25}$
 c. 24%
 d. *Answer should include, but is not limited to:* A bar graph showing either the *number* of students or the *portion* of students in the class that get to school in various ways.

Section 4.3
On Your Own

4. Washington, Michigan, Ohio, Illinois, New York, Texas, and California

Practice and Problem Solving

19. 0.15% 0.015 = 1.5% $\frac{3}{20}$ = 15%
 0% 5% 10% 15%

20. 271% = 2.71
 $2\frac{2}{5}$ = 2.4
 26.8% = 0.268 2.26 2.62
 0 1 2 3

21. 43.7% = 0.437
 $\frac{21}{50}$ = 0.42 $\frac{87}{200}$ = 0.435 0.44
 0.42 0.43 0.44

Chapter 5

Section 5.1
Practice and Problem Solving

26. a. Player 1: $\frac{5}{8}$ Player 2: $\frac{2}{5}$
 Player 3: $\frac{5}{8}$ Player 4: $\frac{3}{5}$
 b. No; Player 4 had the most hits and had a batting average of 0.600. Players 1 and 3, however, had a higher batting average of 0.625.

27. 5 girls

Section 5.4
Practice and Problem Solving
17. 11 pounds; Use the mean of each collection to work backwards to find the total weight of the collection. Divide the sum of the weights of the two collections by the total number of backpacks (20) to find the mean.

Section 5.5
Practice and Problem Solving
22. mean: $6\frac{3}{8}$;
 median: $6\frac{3}{16}$;
 mode: $6\frac{5}{8}$;
 range: $7\frac{1}{2}$

23. median: 11 years;
 mode: 12 years;
 range: 2 years

24. a. *Sample answer*: 54, 60, 45, 59, 55, 65, 68
 mean: 58 minutes;
 median: 59 minutes;
 mode: none;
 range: 23 minutes

 b. *Sample answer*: Success is strongly related to the median, which shows the typical effort; It is not related to the mode, because there is none; It is related to the range, because it is fairly small and shows consistency.

26. 32; There are 6 values, so the median is the mean of the middle two values. Because 42 is the middle value of the five given values, it is one of the two middle values of all six values. Because the mean of 29 and 42 is *not* the known median 37, the missing value must be the other middle value. To find the missing middle value, work backwards using the median 37, and the other middle value 32.

27. *Sample answer*: 18, 21, 28, 30, 35, 36

Record and Practice Journal Practice
5. median: 3;
 mode: 3 and 4;
 range: 4;
 The data set contains 26 values and consists of the dots used to make each letter. Order the data set to find the median, mode, and range.

Section 5.6
Record and Practice Journal
2. a.
| Interval | Tally | Total |
|---|---|---|
| 10.0–11.9 | | 0 |
| 12.0–13.9 | III | 3 |
| 14.0–15.9 | I | 1 |
| 16.0–17.9 | II | 2 |
| 18.0–19.9 | IIII I | 6 |
| 20.0–21.9 | IIII | 4 |
| 22.0–23.9 | II | 2 |
| 24.0–25.9 | II | 2 |

Practice and Problem Solving
5.
Number	Tally	Total
6	I	1
7	II	2
8	III	3
9	IIII II	7
10	II	2

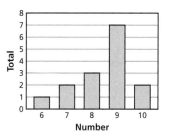

7. mean: 50
 median: 40
 mode: 95
 Sample answer: The mean is probably best, because the mode is the greatest value and the median is too far from the greater values.

8. mean: 12
 median: 8
 mode: 2
 The median is the best measure, because the mean is greater than most of the data and the mode is the least value.

9. mean: 110
 median: 114.5
 mode: 144
 Sample answer: Either the mean or median is best, because they are both at the middle of the data, while the mode is the greatest value.

10. With Outlier Without Outlier
 mean: 47 mean: 54.25
 median: 52 median: 54.5
 mode: none mode: none

 The outlier reduces the median slightly, but reduces the mean more. There is no mode with or without the outlier.

11. With Outlier Without Outlier
 mean: 83 mean: 69.14
 median: 69.5 median: 67
 mode: 72 mode: 72

 The outlier makes the mean greater than most of the data, increases the median slightly, and does not change the mode.

12. With Outlier Without Outlier
 mean: 15.8 mean: $8.\overline{6}$
 median: 7.5 median: 7
 mode: 5 mode: 5

 The outlier makes the mean greater than most of the data, increases the median slightly, and does not change the mode.

13. With Outlier Without Outlier
 mean: 102 mean: 84
 median: 85 median: 85
 mode: 85 mode: 85

 The outlier makes the mean greater than all of the other values, but does not affect the median or mode.

14. a. mean: 94°F; median: 91°F; mode: 91°F
 Sample answer: Both the median and mode are the best measures for the data because both are very close to most of the data.

 b. mean: 77°F;
 median: 77°F;
 modes: 77°F and 78°F
 Sample answer: Both the mean and median are the best measures for the data, because there are two modes.

15. *Answer should include, but is not limited to:*
 Prices of 10 cereals with 1 outlier; unit prices calculated.

 a. Calculation of mean, median, and mode and explanation of which is best.

 b. Outlier identified; mean, median, and mode calculated; measure most affected described.

Lesson 5.6b

Practice

1.

 Most of the fouls are 0, 1, 2, or 3.

2.

 Most of the camper registrations are 21, 22, 23, 24, or 25.

Mini-Assessment

2.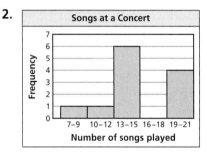

Record and Practice Journal Practice

1.

 Most of the games won are 7, 8, and 9 games.

2.

 The distribution of the number of students is pretty even.

3.

4.

5.

6.

7. a. $\frac{1}{4}$ **b.** $\frac{3}{4}$

 c. 3; The middle half of the summer camp enrollment varies by no more than 3 children.

5.4–5.6 Quiz

10. mean: 64;
median: 56.5;
mode: none;
Sample answer: The median is best because there is no mode and the mean is greater than most of the data.

11. 5 emails

12. mean: 18;
mode: 20;
range: 7

13. With the outlier, the median best represents the data, because the mean is greater than most of the data and the mode is less than most of the data.
Without the outlier, the mean or median is best, because they are very close and the mode is less than most of the values.

Chapter 5 Test

11. $56\frac{2}{3}$ songs

12. $136

13. a. The median would change from 3055 to 3000

 b. decrease 23.35 ft

Chapter 6

Section 6.1
Practice and Problem Solving

24. b. about 320.28 mm; Subtract $\frac{1}{12}$ the circumference of a circle whose radius is the length of the hour hand from the circumference of a circle whose radius is the length of the minute hand.

Section 6.2
Practice and Problem Solving

18. The starting points are staggered so that each runner can run the same distance and use the same finish line. This is necessary because the circumference is different for each lane. The diagram shows this because the diameter is greater in the outer lanes.

Section 6.3
Practice and Problem Solving

20. about 17.415 m²; Subtract the area of a circle with a 9-meter diameter from the area of the square.

21. about 4.56 ft²; Find the area of the shaded regions by subtracting the areas of both unshaded regions from the area of the quarter-circle containing them. The area of each unshaded region can be found by subtracting the area of the smaller shaded region from the semicircle. The area of the smaller shaded region can be found by drawing a square about the region.

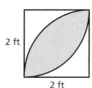

Subtract the area of a quarter-circle from the area of the square to find an unshaded area. Then subtract both unshaded areas from the square's area to find the shaded region's area.

Chapter 7

Section 7.5
Record and Practice Journal
1. a. b.
 c. d.
 e. f.
 g. h.

Lesson 7.6b
Record and Practice Journal Practice
1. $1\frac{3}{4}$ in.³
2. $\frac{20}{27}$ cm³
3. 19 yd²
4. 2732 in.²
5. 270 m²
6. 103.2 in.²
7. 72 cm²
8. 144 ft²

Chapter 8

Record and Practice Journal Fair Game Review
8. <
9. >
10. <
11. =
12. >
13. <
14. Sample answer: $\frac{1}{2}$, 0.36, and 0.4
15. Sample answer: $\frac{2}{9}, \frac{1}{9}$, and 0.2
16. Sample answer: $2\frac{4}{5}$, 3, and 2.6
17. Sample answer: $\frac{3}{10}$, 0.4, and $\frac{1}{2}$

18. Sample answer: $\frac{2}{5}$, 0.3, and 0
19. Sample answer: $5\frac{9}{10}$, 6, and 10
20. $\frac{18}{25} > 0.7$

Section 8.1
Practice and Problem Solving
46. a.
 $e \geq 0, e \leq 100$

 b. ;
 $400 \leq e \leq 500$; The elevation is between 400 and 500. This compound inequality maintains all characteristics of the separate inequalities, $e \geq 400$ and $e \leq 500$.

 c. 200 ft; B falls on the border of two ranges of elevation:
 $100 \leq e \leq 200$
 $200 \leq e \leq 300$
 The only elevation that is a solution of both inequalities is 200.

Section 8.2
Record and Practice Journal
4. b.
 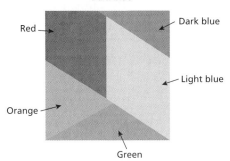

c. Sample answers:
 $0 = 44 - 44$
 $1 = 44 \div 44$
 $2 = (4 \div 4) + (4 \div 4)$
 $3 = ((4 \times 4) - 4) \div 4$
 $4 = ((4 - 4) \times 4) + 4$
 $5 = ((4 \times 4) + 4) \div 4$
 $6 = ((4 + 4) \div 4) + 4$
 $7 = (44 \div 4) - 4$
 $8 = (4 \times 4) - 4 - 4$
 $9 = 4 + 4 + (4 \div 4)$
 $10 = (44 - 4) \div 4$

Practice and Problem Solving

15. $\frac{1}{4} \le n$;

16. $b > \frac{5}{12}$;

21. $x - 3 > 15; x > 18$

22. $x \le 8$;

23. $11 > s$;

24. $4.8 \ge c$;

25. $34{,}280 + d + 1000 > 36{,}480; d > 1200$ dragonflies

26. a. $4200 + x \le 8500; x \le 4300$ lb

 b. at most 2.6875 cubic yards; You can add at most 4300 pounds to your truck and each cubic yard of sand weighs 1600 pounds. So, you can haul at most $4300 \div 1600 = 2.6875$ cubic yards.

8.1–8.2 Quiz

13. $x > 24$;

14. $x - 2 > 15$;
 $x > 17$

15. $7 + x \le 27$;
 $x \le 20$

16. $x \le 80$;

17. $12 + x \ge 30; x \ge \$18$

18. $19.95x > 89.95$ 19. $x - 5 \le 40; x \le \$45$

Section 8.3
Practice and Problem Solving

17. $v \le 405$;

21. $x \ge 162$;

22. The right side should have been multiplied by 6.
 $\frac{x}{6} \le 30$
 $\frac{x}{6} \cdot 6 \le 30 \cdot 6$
 $x \le 180$

23. $8x < 168; x < 21$ ft

24. $\frac{x}{5} \ge 6; x \ge 30$ students

25. $8n < 72; n < 9$ 26. $t \div 32 \le 4.25; t \le 136$

27. $225 \ge 12w; 18.75 \ge w$

28.

29.

31. $80x > 2 \cdot 272$
 $x > 6.8$ yards per play

32. A number is never less than itself.

33. *Sample answer:* The number of gallons of milk you can buy with $20.; The length of a park that has an area of at least 500 square feet.

34. $\frac{5x}{50} \le 50$ $\frac{5x}{50} \ge 40$
 $x \le \$500$ $x \ge \$400$
 Selling between $400 and $500 will earn you between $40 and $50.

35. yes; Because $a > b$ and $x > y$.

36. no; *Sample answer:* $a = 9, b = 8, x = 3, y = 2$;
 $a - x = 6, b - y = 6$; so, $a - x = b - y$.

37. yes; Because $a > b$ and $x > y$

38. no; *Sample answer:* $a = 6, b = 5, x = 10, y = 4$
 $\frac{a}{x} \overset{?}{>} \frac{y}{b}$
 $\frac{6}{10} \overset{?}{>} \frac{4}{5}$
 $0.6 \not> 0.8$

Section 8.4
Record and Practice Journal
1. d.

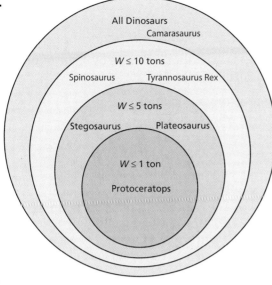

Practice and Problem Solving

15. $k > 3.1$;

17. They should have subtracted first.
$$1 + \frac{a}{6} > 2$$
$$\frac{a}{6} > 1$$
$$a > 6$$

18. $n \geq 8$ additional jumps

19. $10x + 240 \geq 400; x \geq 16$ m

20. $12x + 180 \geq 400; x \geq 18\frac{1}{3}$ m

21. $10x + 157 \geq 400; x \geq 24.3$ m

23. $22p + 180 \geq 250; p \geq 3.18$; Each student needs at least 4 more pledges.

24. The number of diagonal rows that form no more than 200 knots.

25. $7x > 35; x > 5$ visits a year; An individual membership is better if you go 6 or more times a year. When you visit more than 9 times a year.

8.3–8.4 Quiz

10. $\frac{x}{4} \leq 9; x \leq 36$ band members

11. $15x + 135 \geq 300; x \geq 11$; You need at least 11 more $15 donations.

12. $4s \leq 52; s \leq 13$ ft

13. a. $8p + 3 \geq 24; p \geq 2.625$

b. 3 packages; Because 2 packages will not give you enough bottles of water for the trip.

Chapter 8 Review

29. $w \geq 2$;

30. $t < 35$;

31. $c \geq 6$;

32. $p > 0$;

33. $s \leq 3$;

34. $x < 18$;

35. $d > 9$;

Chapter 8 Test

13. $w \geq 74$ miles per hour

14. a. $8x + 20 \leq 400; x \leq 47.5$

b. 47 T-shirts; Because the maximum they can buy is 47.5 shirts, but you cannot buy half of a shirt, so round down to the nearest whole number.

15. $x \leq 3.5$ ounces

16. $0.1d > 20; d > 200$ days

17. 7

Chapter 9

Record and Practice Journal Fair Game Review

5.

In	Out
0.65	0
0.70	0.05
0.75	0.10
1.00	0.35

The In value minus 0.65 equals the Out value.

Section 9.1
On Your Own
5. Input Output

Practice and Problem Solving
12. Input Output

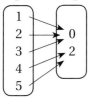

13. The first number of each ordered pair should be an input and the second number should be the output that corresponds to the input.

 Input Output

14. Input Output

15. Input Output

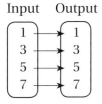

 As each input increases by 2, the output increases by 2.

16. Input Output

 As each input increases by 2, the output is 8.

17. Input Output

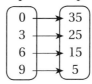

 As each input increases by 3, the output decreases by 10.

19. a. Input Output

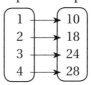

 b. The pattern is that for each input increase of 1, the output increases by $2 less than the previous increase. For each additional movie you buy, your cost per movie decreases by $1.

Section 9.3
Record and Practice Journal

2. a.
| Input, x | 1 | 2 | 3 | 4 |
|---|---|---|---|---|
| Output, A | 1 | 3 | 5 | 7 |

b.
Input, x	1	2	3	4
Output, A	1	4	9	16

Practice and Problem Solving
15. Sample answer:

GMT, x	6:00	7:00	8:00	9:00	10:00
Eastern Standard Time, y	1:00	2:00	3:00	4:00	5:00

$y = x - 5$

Section 9.4
Practice and Problem Solving

9.

10.

A58 Additional Answers

11.

12.

13.

14.

15.

16. The order of the *x*- and *y*-coordinates is reversed in each coordinate pair.

17.

24. *Sample answer:*

Input, s	1	2	3	4	5
Output, P	4	8	12	16	20

Input, s	1	2	3	4	5
Output, A	1	4	9	16	25

Sample answer: The perimeter function appears to form a line, the area function appears to form a curve. When the side length is less than 4, the perimeter function is greater. When the side length is greater than 4, the area function is greater. When the side length is 4, the two functions are equal.

Section 9.5

On Your Own

1. yes; The graph is a line.

2. no; The graph is *not* a line.

3. yes; The graph is a line.

4. no; The graph is *not* a line.

5.
Diagram, x	Dots, y
1	1
2	3
3	5
4	7

yes; The graph is a line.

6.

The graph of $d = 65t$ is steeper. It is steeper because 65 miles per hour is greater than 55 miles per hour.

Practice and Problem Solving

15.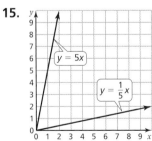

$y = 5x$; 5 is greater than $\frac{1}{5}$.

16.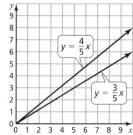

$y = \frac{4}{5}x$; $\frac{4}{5}$ is greater than $\frac{3}{5}$.

17.

$y = 2x + 1$; 2 is greater than 1.

18.
Figure, x	Area, y
1	3
2	5
3	7
4	9

yes; The graph is a line.

19.
Figure, x	Area, y
1	1
2	2
3	4
4	8

no; The graph is *not* a line.

21. Airport 2; about 100 flights; *Sample answer:* From the graph, Airport 2 has about 350 flights each day and Airport 1 has about 250 flights each day.

22. a. 3; 3.5; differences: 0.5, 0.5, 0.5

b. *Sample answer:*

x	0	1	2	3	4
y	0	2	4	6	8

2 2 2 2

x	1	2	3	4	5
y	0	4	8	12	16

4 4 4 4

Property: In a linear function, the differences between outputs for consecutive whole number inputs are the same.

9.4–9.5 Quiz

1.

2.

3.

Chapter 9 Test

6.

7.

8.

13. a.

b. $y = 600x$

c. 36 km

14.
Figure, x	Dots, y
1	1
2	5
3	13
4	25

no; The graph is not a line.

Additional Topics

Record and Practice Journal Fair Game Review

1. 0, 1, 3, 5, 6

2. 2, 4, 6, 7, 8

3. 5, 6, 12, 13, 14 4. 32, 35, 37, 38, 40
5. 5, 10, 15, 25, 30 6. 19, 30, 47, 55, 62
7. 80, 82, 86, 87, 88, 92 8. 87, 90, 94, 98, 102, 105
9. 20, 43, 85, 115, 180

Topic 1
Practice
9. a. *Sample answer:* Choosing 8, the opposite is −8.
 b. *Sample answer:* 8
 c. The opposite of the opposite of an integer is the integer.; Yes; *Sample answer:*

 Case 1:

 Case 2:
 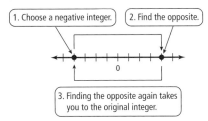

 Case 3: Choose 0. The opposite of 0 is 0, so the opposite of the opposite of 0 is 0.

Record and Practice Journal Practice
1. −42 2. 37,500 3. −56 4. 5
5.
6.
7.
8.
9. −2 10. 0 11. 7 12. −7
13. 50; −20; Zero represents when you have no money left.
14. −318; Zero represents sea level.
15. −4

Topic 2
Record and Practice Journal Practice
1. > 2. > 3. < 4. < 5. > 6. <
7. > 8. < 9. < 10. > 11. > 12. <
13. −4, −2, 0, 2, 3 14. −6.3, −3.9, 3.4, 4.2, 7.7
15. $-3\frac{7}{8}, -\frac{7}{9}, -\frac{3}{5}, \frac{1}{2}, 1\frac{4}{5}$ 16. $-3, -1\frac{2}{3}, 0, \frac{1}{4}, 0.6$
17. $-44\frac{3}{5} < -38$, and the pot is deeper than the arrowhead.
18. a. Florida, Louisiana, Arkansas, Tennessee, California; California
 b. California, Louisiana, Florida, Arkansas, Tennessee; California

Topic 3
Record and Practice Journal Practice
1. 3 2. 23 3. 11 4. 9
5. 0.5 6. 2.3 7. $\frac{1}{4}$ 8. $3\frac{2}{3}$
9. < 10. = 11. >
12. > 13. < 14. <
15. Account B 16. Account A
17. Account A 18. Account B
19. a. Fish D; The lowest elevation represents the fish that is the deepest.
 b. Fish A; Fish A is closest to sea level because the absolute value of its distance from 0 is less than the absolute value of the distances of the other fish.

Topic 4
Practice
5. Quadrant III; In Quadrant I, both coordinates are positive. In Quadrant II, the *x*-coordinate is negative and the *y*-coordinate is positive. In Quadrant IV, the *x*-coordinate is positive and the *y*-coordinate is negative.

6. 7.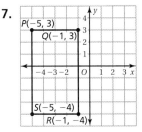

Record and Practice Journal Practice

1–6.

1. Quadrant I
2. x-axis
3. Quadrant III
4. y-axis
5. Quadrant II
6. Quadrant IV

7. a. b. 12 units c. 9 square units

8. C 9. A 10. B 11. 6 mi 12. 14 mi
13. 4 mi 14. 10 mi 15. 4 mi 16. 10 mi

A62 Additional Answers

Photo Credits

iv Big Ideas Learning, LLC; **vi** *top* ©iStockphoto.com/Angel Rodriguez, ©iStockphoto.com/Ann Marie Kurtz; *bottom* Evok20; **vii** *top* ©iStockphoto.com/Angel Rodriguez, ©iStockphoto.com/Ann Marie Kurtz; *bottom* Apollofoto; **viii** *top* ©iStockphoto.com/Stefan Klein; *bottom* ©iStockphoto.com/daaronj; **ix** *top* ©iStockphoto.com/Michael Flippo, ©iStockphoto.com/Ann Marie Kurtz; *bottom* Apollofoto; **x** *top* ©iStockphoto.com/Alistair Cotton; *bottom* ©iStockphoto.com/ronen; **xi** *top* ©iStockphoto.com/Martin Strmko, ©iStockphoto.com/Andreas Gradin, ©iStockphoto.com/Nico Smit; *bottom* Heather Prosch-Jensen; **xii** *top* ©iStockphoto.com/Jonathan Larsen; *bottom* Peter Close; **xiii** *top* ©iStockphoto.com/Ron Sumners, ©iStockphoto.com/Ann Marie Kurtz; *bottom* ©iStockphoto.com/ranplett; **xiv** *top* ©iStockphoto.com/Lisa Thornberg, ©iStockphoto.com/Ann Marie Kurtz; *bottom* Jane Norton; **xv** *top* Varina and Jay Patel, ©iStockphoto.com/Ann Marie Kurtz; *bottom* ©iStockphoto.com/Thomas Perkins; **xvii** Gelpi; **xlvi, xlvii** Big Ideas Learning, LLC; **l** ©iStockphoto.com/Ekaterina Monakhova; **1** ©iStockphoto.com/Angel Rodriguez, ©iStockphoto.com/Ann Marie Kurtz; **33** ©iStockphoto.com/Brandon Alms; **42** ©iStockphoto.com/Angel Rodriguez, ©iStockphoto.com/Ann Marie Kurtz; **T-49** ©iStockphoto.com/susaro; **69** ©iStockphoto.com/Andrey Shadrin; **T-77** ©iStockphoto.com/MegapixelMedia; **77** ©iStockphoto.com/MegapixelMedia; **81** ©iStockphoto.com/Karin Lau; **91** *top* ©Rafael Laguillo/123RF; *bottom* ©Romanchuck|Dreamstime.com; **95** ©iStockphoto.com/Ron Masessa; **96** ©iStockphoto.com/Russell Du parcq; **104** ©iStockphoto.com/Stefan Klein; **109** ©iStockphoto.com/Olivier Tuffé; **117** ©iStockphoto.com/tcp; **125** ©iStockphoto.com/Phil Berry; **131** ©iStockphoto.com/Joseph Abejon; **137** ©iStockphoto.com/proxyminder; **142** ©iStockphoto.com/Graham Heywood; **146** ©iStockphoto.com/Michael Flippo, ©iStockphoto.com/Ann Marie Kurtz; **151** ©iStockphoto.com/Aldo Murillo; **165** ©iStockphoto.com/Eric Isselée; **188** ©iStockphoto.com/Alistair Cotton; **195** ©iStockphoto.com/Vicki Reid; **199** ©iStockphoto.com/ene, ©iStockphoto.com/Jacob Wackerhausen; **200** ©iStockphoto.com/Gord Horne; **201** ©iStockphoto.com/Birgitte Magnus; **202** ©iStockphoto.com/Apatrimonio; **204** ©iStockphoto.com/John Pitcher; **205** ©iStockphoto.com/Bart Wolczyk; **207** ©iStockphoto.com/Katherine Moffitt; **209** ©iStockphoto.com/Ermin Gutenberger; **213** ©iStockphoto.com/Eric Isselée; **219** ©iStockphoto.com/Dave Rodriguez; **221** ©iStockphoto.com/Sue Mack; **227** ©iStockphoto.com/Murat Koc; **227C** ©iStockphoto.com/rusm; **236** ©iStockphoto.com/Martin Strmko, ©iStockphoto.com/Andreas Gradin, ©iStockphoto.com/Nico Smit; **238** ©iStockphoto.com/Hulton Archive/Getty Images; **239** Sir Cumference and the First Round Table Text copyright © 1997 by Cindy Neuschwander Illustrations copyright © 1997 by Wayne Geehan Used with permission by Charlesbridge Publishing, Inc. All rights reserved.; **243** *top row left and center* ©iStockphoto.com/Zentilia; *top row right* ©iStockphoto.com/Dave Hopkins; *center row left* ©iStockphoto.com/Boris Yankov; *center row center* ©iStockphoto.com/Angel Rodriguez; *center row right* ©iStockphoto.com/iLexx; *bottom row* ©iStockphoto.com/alexander mychko; **245** *center* ©iStockphoto.com/Hulton Archive/Getty Images; *bottom* ©iStockphoto.com/Marc Dietrich; **246** ©iStockphoto.com/Jerry Koch, ©iStockphoto.com/Chad Anderson; **251** ©iStockphoto.com/Scott Slattery; **253** *bottom row right* Kalamazoo (Michigan) Public Library; *center row left* ©iStockphoto.com/DivaNir4a; *center row center* ©iStockphoto.com/saw; *center row right* ©iStockphoto.com/Creativeye99; **257** ©iStockphoto.com/Brian Sullivan; **258** *top row left* ©iStockphoto.com/zentilia; *top row center and right* ©iStockphoto.com/Andrey Pustovoy; *bottom row left* ©iStockphoto.com/Marie-france Bélanger; *bottom row center* ©iStockphoto.com/Dave Hopkins; *bottom row right* ©iStockphoto.com/7nuit; **264** Big Ideas Learning, LLC; **266** *center row left* ©iStockphoto.com/Mr_Vector; *center row center* ©iStockphoto.com/Sándor F. Szabó; *center row right* ©iStockphoto.com/Sándor F. Szabó; **274** ©iStockphoto.com/Jonathan Larsen; **280** ©iStockphoto.com/Vladimir Vladimirov; **281** *top right* ©iStockphoto.com/Steve Goodwin, ©iStockphoto.com/Joop Snijder; ©iStockphoto.com/Michael MacFadden; *center left* ©iStockphoto.com/Kenneth C. Zirkel; **286** ©iStockphoto.com/Mel Stoutsenberger; **288** ©iStockphoto.com/Jan Will, ©iStockphoto.com/Jeremy Wee; **289** ©iStockphoto.com/kreicher; **291** ©iStockphoto.com/Leo Blanchette; **294** ©iStockphoto.com/Sergei Butorin; **295** ©iStockphoto.com/Eric Isselée; **296** © Godfey Kneller–Isaac Newton–1689/Wikipedia Commons; **301** ©iStockphoto.com/Don Nichols; **T-303** ©iStockphoto.com/Matthew Cole; **303** *top left* ©iStockphoto.com/Daniel Loiselle; *center right* ©iStockphoto.com/Matthew Cole; **305** ©iStockphoto.com/David Verinder; **307** Madeira Beach Fundamental School; **311** *top right* ©iStockphoto.com/Alexandru Magurean; *center left* ©iStockphoto.com/simon2579; **312** ©iStockphoto.com/SUSARO; **313** *top left* stevecadman; *top center* David Shankbone; *top right* ©iStockphoto.com/ROBERTO ADRIAN; **T-317** ©iStockphoto.com/Eduardo Jose Bernardino; **317** *center left* ©iStockphoto.com/Eduardo Jose Bernardino; *center bottom* ©iStockphoto.com/Jill Chen, ©iStockphoto.com/Long Ha; **317B** Niki Crucillo; **317D** Peter G; **326** ©iStockphoto.com/Ron Sumners, ©iStockphoto.com/Ann Marie Kurtz; **328** *first* ©iStockphoto.com/Anneke Schram; *second* ©iStockphoto.com/nicholas belton; *third* ©iStockphoto.com/Chris Schmidt; *fourth* ©iStockphoto.com/Robert Dant; **332** NASA; **334** ©iStockphoto.com/George Peters; **335** ©iStockphoto.com/Anthony Ladd; **336** ©iStockphoto.com/bratan007; **341** *top* ©iStockphoto.com/o-che; *center* ©iStockphoto.com/sunygraphics; **343** ©iStockphoto.com/Algimantas Balezentis; **344** ©iStockphoto.com/Lisa Fletcher; **345** NASA; **348** ©iStockphoto.com/rami ben ami; **349** Karin Lau; **351** *top left and right, center left and right, bottom left* Andreas Meyer; *bottom right* Ozja; **353** *top left* ©iStockphoto.com/Oleg Prikhodko; *bottom right* ©iStockphoto.com/Rob Friedman; **354** ©Keddie. BigStockPhoto.com; **355** ©Lars Christenen/123RF; **356** ©iStockphoto.com/P_Wei; **360** ©iStockphoto.com/Jani Bryson; **364** ©iStockphoto.com/Lisa Thornberg, ©iStockphoto.com/Ann Marie Kurtz; **371** ©iStockphoto.com/Kevin Panizza; **375** Cedar Point; **376** ©iStockphoto.com/Julie Vader; **377** *top right* ©iStockphoto.com/Sergey Lemeshencko; *center left* ©iStockphoto.com/Robert Rushton; **383** ©iStockphoto.com/Liz McClanahan, ©iStockphoto.com/Duncan Noakes; **385** ©iStockphoto.com/Mehmet Salih Guler; **386** ©iStockphoto.com/Jacom Stephens; **389** ©iStockphoto.com/DivaNir4a; **390** ©iStockphoto.com/Manuel Angel Diaz Blanco; **391** ©iStockphoto.com/José Luis Gutiérrez; **393** Michael Flippo; **398** ©iStockphoto.com/landbysea; **402** ©iStockphoto.com/Tomasz Zachariasz; **406** Varina and Jay Patel, ©iStockphoto.com/Ann Marie Kurtz; **409** ©iStockphoto.com/Egor Mopanko; **413** Shane W Thompson; **A0** ©iStockphoto.com/Björn Kindler; *center left and bottom* ©iStockphoto.com/Grafissimo; **A1** *top right* ©iStockphoto.com/Andrey Krasnov; *bottom left* ©iStockphoto.com/Dan Van Oss; *bottom right* ©iStockphoto.com/Victor Paez; **A4** *top right* ©iStockphoto.com/Grafissimo, *center right* ©iStockphoto.com/Ragnarocks; *bottom left* ©iStockphoto.com/Linda Steward; *bottom right* ©iStockphoto.com/Rackermann; **A5** *top right* ©iStockphoto.com/Rackermann; *bottom left* ©iStockphoto.com/Grafissimo; *bottom right* ©iStockphoto.com/Rackermann; **A6** *top right* ©iStockphoto.com/Keith Webber Jr.; *bottom left* ©iStockphoto.com/Eliza Snow; *bottom right* ©iStockphoto.com/David Seymour; **A7** *top right* ©iStockphoto.com/Andrey Krasnov; *bottom left* ©iStockphoto.com/Victor Paez; **A8** *top right* ©iStockphoto.com/Dan Van Oss; *bottom left* ©iStockphoto.com/Tomasz Tulik; *bottom right* © Geo Swan/Wikepedia Commons; **A9** *top right* ©iStockphoto.com/Tim Stapleton; *bottom left* ©iStockphoto.com/Liz Leyden; *bottom right* ©iStockphoto.com/Pauline Mills

Cartoon illustrations Tyler Stout
Cover image Lechner & Benson Design

K

- **Counting and Cardinality** — Count to 100 by Ones and Tens; Compare Numbers
- **Operations and Algebraic Thinking** — Understand and Model Addition and Subtraction
- **Number and Operations in Base Ten** — Work with Numbers 11–19 to Gain Foundations for Place Value
- **Measurement and Data** — Describe and Compare Measurable Attributes; Classify Objects into Categories
- **Geometry** — Identify and Describe Shapes

1

- **Operations and Algebraic Thinking** — Represent and Solve Addition and Subtraction Problems
- **Number and Operations in Base Ten** — Understand Place Value for Two-Digit Numbers; Use Place Value and Properties to Add and Subtract
- **Measurement and Data** — Measure Lengths Indirectly; Write and Tell Time; Represent and Interpret Data
- **Geometry** — Draw Shapes; Partition Circles and Rectangles into Two and Four Equal Shares

2

- **Operations and Algebraic Thinking** — Solving One- and Two-Step Problems Involving Addition and Subtraction; Build a Foundation for Multiplication
- **Number and Operations in Base Ten** — Understand Place Value for Three-Digit Numbers; Use Place Value and Properties to Add and Subtract
- **Measurement and Data** — Measure and Estimate Lengths in Standard Units; Work with Time and Money
- **Geometry** — Draw and Identify Shapes; Partition Circles and Rectangles into Two, Three, and Four Equal Shares

3

Operations and Algebraic Thinking	– Represent and Solve Problems Involving Multiplication and Division; Solve Two-Step Problems Involving Four Operations
Number and Operations in Base Ten	– Round Whole Numbers; Add, Subtract, and Multiply Multi-Digit Whole Numbers
Number and Operations – Fractions	– Understand Fractions as Numbers
Measurement and Data	– Solve Time, Liquid Volume, and Mass Problems; Understand Perimeter and Area
Geometry	– Reason with Shapes and Their Attributes

4

Operations and Algebraic Thinking	– Use the Four Operations with Whole Numbers to Solve Problems; Understand Factors and Multiples
Number and Operations in Base Ten	– Generalize Place Value Understanding; Perform Multi-Digit Arithmetic
Number and Operations – Fractions	– Build Fractions from Unit Fractions; Understand Decimal Notation for Fractions
Measurement and Data	– Convert Measurements; Understand and Measure Angles
Geometry	– Draw and Identify Lines and Angles; Classify Shapes

5

Operations and Algebraic Thinking	– Write and Interpret Numerical Expressions
Number and Operations in Base Ten	– Perform Operations with Multi-Digit Numbers and Decimals to Hundredths
Number and Operations – Fractions	– Add, Subtract, Multiply, and Divide Fractions
Measurement and Data	– Convert Measurements within a Measurement System, Understand Volume
Geometry	– Graph Points in the First Quadrant of the Coordinate Plane; Classify Two-Dimensional Figures